Modeling Volcanic Processes

The Physics and Mathematics of Volcanism

Understanding the physical behavior of volcanoes is key to mitigating the hazards active volcanoes pose to the ever-increasing populations living nearby. The processes involved in volcanic eruptions, from magma generation at depth to eruption and emplacement of deposits at the surface, are driven by a series of interlinked physical phenomena. To fully understand volcanic behavior, volcanologists must employ a range of physics subdisciplines, including thermodynamics, fluid dynamics, solid mechanics, ballistics, and wave theory.

This book provides the first advanced-level, one-stop resource that examines the physics of volcanic behavior and reviews the state of the art in modeling volcanic processes. Each chapter begins by explaining simple modeling formulations, to provide students with the necessary physics and mathematics, and progresses to present cutting-edge research illustrated by case studies. Individual chapters cover subsurface magmatic processes through to eruption in various environments, and conclude with the application of modeling to understanding the other volcanic planets of our Solar System. The book is also supported by a range of online materials, including additional exercises and solutions to those in the book, extra images, movies and data from the field, and links to other relevant resources.

Providing an accessible and practical text for graduate students of physical volcanology, this book is also an important resource for researchers and professionals working in the fields of volcanology, geophysics, geochemistry, petrology, and natural hazards.

SARAH A. FAGENTS is a Researcher in the Hawaii Institute of Geophysics and Planetology at the University of Hawaii. She specializes in volcanic fluid dynamics, combining numerical modeling with field and remote sensing studies on Earth and other volcanic planets of the Solar System. In addition to frequent visits to the volcanoes of Hawaii, her research includes studying lahar emplacement in New Zealand, investigating explosive lava–water interactions in Iceland and on Mars, and deciphering the signatures of cryovolcanism on the icy moons of the outer Solar System. She has shared her enthusiasm for planetary volcanism through TV documentaries and public lectures.

TRACY K. P. GREGG is an Associate Professor of Geology and a member of the Volcanology Research Group at the University at Buffalo. She has studied volcanoes around the world, including Iceland, Hawaii, and the bottom of the ocean – the latter using the HOV Alvin. Her area of expertise is the study of lava flow emplacement around the Solar System, particularly on Mars, but she enjoys studying lava flows of any composition, in any tectonic setting, on any planet. She is the co-editor of two other volcanology books that focus on volcanic behaviors in the range of environments found in our Solar System.

ROSALY M. C. LOPES is a Senior Research Scientist at NASA's Jet Propulsion Laboratory and Deputy Manager for Planetary Science. An expert on volcanism and cryovolcanism in the Solar System, she is the author of five other books, including *The Volcano Adventure Guide* (Cambridge University Press). She is a Fellow of the American Association for the Advancement of Science and has received numerous other honors, including the Carl Sagan Medal from the American Astronomical Society. She is a frequent contributor to TV documentaries and lectures widely on volcanoes and space exploration.

"This elegant text fills an important gap in volcanologists' libraries. It is a quantitative and accessible introduction to the physical modelling of volcanic and magmatic processes that illuminates the science behind the glamour and razzmatazz of volcanic eruptions. The coverage is broad: from magma chambers to eruptive processes; and from eruptions under water, air, and ice to planetary volcanism. Chapters are advanced primers to each topic, and can as easily be read alone, or as a part of the whole. The text will be accessible to readers from advanced undergraduate level, and up; and the inclusion of problem sets and up-to-date references means that this is destined to become the working textbook for advanced courses in volcanic processes. The editors have done a magnificent job, matched by the high quality production that we expect from Cambridge University Press."

– Professor David Pyle,
University of Oxford

"For much too long, quantitative volcanology has had to be taught through collections of individual articles from the literature. Finally, here is, in one place, an excellent summary of the quantitative aspects of physical volcanology, well integrated with field observations. Madame Pele will be proud and delighted with the contributions in this volume and its editors!"

– Professor Susan Kieffer,
University of Illinois.

"All those entering this challenging and exciting field will find this an essential reference. The comprehensive scope and the depth of the presentations make the book invaluable to senior researchers as well."

– Dr Stephen M. Baloga,
Proxemy Research, Maryland

Modeling Volcanic Processes

The Physics and Mathematics of Volcanism

Edited by

Sarah A. Fagents
University of Hawaii

Tracy K. P. Gregg
University at Buffalo

Rosaly M. C. Lopes
Jet Propulsion Laboratory

CAMBRIDGE
UNIVERSITY PRESS

CAMBRIDGE
UNIVERSITY PRESS

University Printing House, Cambridge CB2 8BS, United Kingdom

One Liberty Plaza, 20th Floor, New York, NY 10006, USA

477 Williamstown Road, Port Melbourne, VIC 3207, Australia

314-321, 3rd Floor, Plot 3, Splendor Forum, Jasola District Centre, New Delhi - 110025, India

79 Anson Road, #06-04/06, Singapore 079906

Cambridge University Press is part of the University of Cambridge.

It furthers the University's mission by disseminating knowledge in the pursuit of
education, learning and research at the highest international levels of excellence.

www.cambridge.org
Information on this title: www.cambridge.org/9781108812658

© Cambridge University Press 2013

First published 2013
Reprinted 2015
First paperback edition 2021

A catalogue record for this publication is available from the British Library

Library of Congress Cataloging in Publication data
    Modeling volcanic processes : the physics and mathematics of volcanism/
    edited by Sarah A. Fagents, Tracy K. P. Gregg, Rosaly M. C. Lopes.
        pages   cm
    Includes bibliographical references and index.
    ISBN 978-0-521-89543-9
    1. Volcanism – Mathematical models.   I. Fagents, Sarah A., editor of
    compilation.   II. Gregg, Tracy K. P., editor of compilation.   III. Lopes,
    Rosaly M. C., 1957 – editor of compilation.
    QE522.M637   2013
    551.2101´51–dc23   2012018835

ISBN 978-0-521-89543-9 Hardback
ISBN 978-1-108-81265-8 Paperback

Additional resources for this publication at www.cambridge.org/fagents

To Ron Greeley (1939–2011)

Just as we were putting the finishing touches on this book, we learnt of the unexpected passing of Professor Ronald Greeley, Regents' Professor in the School of Earth and Space Exploration at Arizona State University. Ron was a friend and mentor to all three of us editors and a pioneer in the study of planetary geology and volcanology. He was involved in lunar and planetary exploration since 1967 and contributed significantly to our understanding of the geology of the Moon, Mars, Europa, and several other Solar System bodies. He participated in nearly every major space mission flown in the Solar System since the Apollo Moon landing. Above all, Ron was a superb teacher and mentor who touched many lives and enabled many careers in planetary science. His reach was extensive and it is not surprising that all three of us learnt much from him, although we came from different scientific backgrounds and even different countries.

Ron hired Sarah as a post-doc in 1996, shortly after she had moved to the United States from the United Kingdom. He provided unparalleled mentorship as she bridged the chasm between wide-eyed graduate student and fully fledged independent research scientist. Ron challenged all his students and post-docs to really stretch themselves. His in-at-the-deep-end approach introduced Sarah to whole new research areas, from lava–ice interactions in Iceland and on Mars, to the icy satellites of Jupiter. His generosity with research, fieldwork, and travel opportunities allowed her to engage with research scientists around the globe, and to gain those vital first footholds en route to becoming established in the scientific community. She attributes her subsequent research successes to Ron's exquisite, though sometimes painful, tutelage in the fine art of proposal-writing. Perhaps the fondest memories she has of Ron are of being in the field with him, from scrambling up impossibly loose cinder cones in the high desert of Arizona, to some eventful wilderness camping in the Pinacate Volcanic Field in Sonora, Mexico. Here his playful side would come to the fore, and we would all learn as much as we had fun.

Tracy met Ron at the insistence of Dr. Peter Schultz, who was Tracy's undergraduate research advisor at Brown University. Pete asserted that Ron's experience and knowledge in planetary volcanism was a perfect match for Tracy's graduate school interests. Ron welcomed Tracy into his research group at Arizona State University when she arrived to begin her graduate program in 1990. Although Tracy ultimately received her PhD under the advisement of Dr. Jon Fink, Ron coached her through the writing and publication of her first peer-reviewed journal article, and she still uses the lessons he taught her about scientific writing today: Ron's words and phrases literally flow through her fingers onto the pages of this very book. After Tracy became a professor at the University at Buffalo, she found herself turning to Ron frequently for professional and diplomatic advice in navigating the waters of proposals, papers, and grants. His experience was invaluable and will be sorely missed.

Rosaly first met Ron while she was starting her graduate studies at University College London. Her advisor, John Guest, was a close friend of Ron's and the two of them did field work together on several volcanoes around the world. (Sadly, John passed away while this book was in press.) Rosaly joined Ron and John in Hawaii in 1981 to help them with field work on Kilauea. Coming from a physics and astronomy background, she had much to learn about geology field work and benefited not only from

Ron's expertise but also his patience and true gift for teaching. She had been warned about Ron and John's love for playing pranks on unsuspecting students, such as the story of the dreaded, poisonous Hawaiian ahu, a snake that liked to camouflage itself in the ropy texture of pahoehoe lava, its bite deadly. Ron and John's pranks were clever and easy for us young students to fall for, but the lesson they imparted was that not everything professors say should be believed. In their jovial way, they taught us to think critically before accepting any of their teachings as true. Rosaly went on to a career at NASA's Jet Propulsion Laboratory, where she interacted with Ron often, working with him on the Galileo mission to Jupiter and on proposals for a mission to Io and a camera for the Europa Jupiter System Mission.

We are grateful to Ron for the way he touched our lives and know that his legacy will live on.

Contents

Color plates are to be found between pages 246 and 247.

Contributors

Costanza Bonadonna, Section des Sciences de la Terre et de l'Environnement, Université de Genève, 13 rue des Maraîchers, CH-1205 Genève, Switzerland.

Ralf Büttner, Physikalisch Vulkanologisches Labor, Universität Würzburg, Pleicherwall 1, Würzburg, D-97070, Germany.

Bernard Chouet, United States Geological Survey, Volcano Science Center, 345 Middlefield Road, MS 910, Menlo Park, CA 94025.

Antonio Costa, Osservatorio Vesuviano, INGV, Napoli, Italy, and Environmental Systems Science Centre, University of Reading, Reading RG6 6AL, UK.

Amanda B. Clarke, School of Earth and Space Exploration, Arizona State University, Box 871404, Tempe, AZ 85287.

Josef Dufek, School of Earth and Atmospheric Science, Georgia Institute of Technology, 311 Ferst Drive, Atlanta, GA 30332.

Sarah A. Fagents, Hawaii Institute of Geophysics and Planetology, University of Hawaii at Manoa, 1680 East-West Road, Honolulu, HI 96822.

David Fee, Wilson Infrasound Observatories, Geophysical Institute, University of Alaska Fairbanks, 903 Koyukuk Drive, Fairbanks, AK 99775–7320.

Milton A. Garcés, Infrasound Laboratory, Hawaii Institute of Geophysics and Planetology, University of Hawaii, 73–4460 Queen Kaahumanu Hwy., #119, Kailua-Kona, HI 96740.

Helge M. Gonnermann, Department of Earth Science, Rice University, MS 126, 6100 Main Street, Houston, Texas, 77005.

Tracy K. P. Gregg, Department of Geology, 411 Cooke Hall, University at Buffalo, Buffalo, NY 14260.

Andrew J. L. Harris, Laboratoire Magmas et Volcans, Université Blaise Pascal, 5 rue Kessler, 63038 Clermont-Ferrand, France.

James W. Head, Department of Geological Sciences, Brown University, 324 Brook St., Providence, RI 02912.

Bruce F. Houghton, Department of Geology and Geophysics, University of Hawaii at Manoa, 1680 East-West Road, Honolulu, HI 96822.

Christian Huber, School of Earth and Atmospheric Science, Georgia Institute of Technology, 311 Ferst Drive, Atlanta, GA 30332.

Mike R. James, Lancaster Environment Centre, Lancaster University, Lancaster LA1 4YQ, UK.

Leif Karlstrom, Department of Earth and Planetary Science, 307 McCone Hall, University of California, Berkeley, Berkeley, CA 94720.

Karim Kelfoun, Laboratoire Magmas et Volcans, Université Blaise Pascal – CNRS – IRD, 5 rue Kessler, 63038 Clermont-Ferrand, France.

Steve J. Lane, Lancaster Environment Centre, Lancaster University, Lancaster LA1 4YQ, UK.

Rosaly M. C. Lopes, Earth and Space Sciences Division, Mail Stop 183–601, Jet Propulsion Laboratory, Pasadena, CA 91109.

Michael Manga, Department of Earth and Planetary Science, UC Berkeley, 307 McCone Hall, Berkeley, CA 94720.

Jon J. Major, US Geological Survey, 1300 SE Cardinal Court, Bldg 10, Suite 100, Vancouver, WA 98683.

Vernon Manville, School of Earth and Environment, University of Leeds, Leeds LS2 9JT, UK.

Robin Matoza, Institute of Geophysics and Planetary Physics, Scripps Institution of Oceanography UC San Diego, La Jolla, CA 92093-0225.

Karl L. Mitchell, Jet Propulsion Laboratory, California Institute of Technology, Mail Stop 183–601, 4800 Oak Grove Dr., Pasadena, CA 91109–8099.

Jeremy C. Philips, School of Earth Sciences, University of Bristol, Wills Memorial Building, Queens Road, Bristol BS8 1RJ, UK.

Olivier Roche, Laboratoire Magmas et Volcans, Université Blaise Pascal, CNRS, IRD, 5 rue Kessler, 63038 Clermont-Ferrand, France.

John L. Smellie *, British Antarctic Survey, High Cross, Madingley Road, Cambridge CB3

0ET, UK. *Now at: Department of Geology, University of Leicester, University Road, Leicester LE1 7RH, UK.

Benoit Taisne, Institut de Physique du Globe de Paris, 1 rue Jussieu, 75238 Paris Cedex 05, France.

Steve Tait, Institute de Physique du Globe de Paris, 1 rue Jussieu, 75238 Paris Cedex 05, France.

Lionel Wilson, Lancaster Environment Centre, Lancaster University, Lancaster LA1 4YQ, UK.

Ken Wohletz, Geophysics Group, EES-17, MS F665, Los Alamos National Laboratory, Los Alamos, NM 87545.

Andrew W. Woods, BP Institute, University of Cambridge, Bullard Laboratories, Madingley Rise, Madingley Road, Cambridge CB3 0EZ, UK.

Bernd Zimanowski, Physikalisch Vulkanologisches Labor, Universität Würzburg, Pleicherwall 1, Würzburg, D-97070, Germany.

Chapter 1

Introduction

Sarah A. Fagents, Tracy K. P. Gregg, and Rosaly M. C. Lopes

Scope of this book

The processes involved in volcanic eruptions, from magma generation at depth to eruption and emplacement of deposits at the surface, comprise a suite of interlinked physical phenomena. In seeking to understand volcanic behavior, volcanologists call on a diversity of physics subdisciplines, including fluid dynamics, thermodynamics, solid mechanics, ballistics, and acoustics, to name just a few. Understanding the physical behavior of volcanoes is critical to assessing the hazards posed to the ever-increasing populations living in close proximity to active volcanoes, and thus to mitigating the risks posed by those hazards.

The motivation for producing this book arises in part from the editors' experiences as educators, as well as our interests in keeping current with developments in volcanologic subdisciplines outside our own. Modeling volcanic processes, and the resulting improvements in our understanding of how volcanoes work, has advanced in leaps and bounds over the past decade or two. This is a result of both a maturation of our field- and laboratory-based understanding of volcanic processes, as well as

vast improvements in computational capabilities. Synergies with the similarly rapidly evolving knowledge of geochemical processes also greatly enhance the physical understanding of volcanoes. However, in developing courses for advanced undergraduates and graduate students, we were struck by the lack of an up-to-date single-source book from which we could draw for instructional purposes. While several existing notable books cover the physics of specific aspects of volcanic behavior in detail (e.g., Sparks et al., 1997; Gilbert and Sparks, 1998; Freundt and Rosi, 2001), an advanced, quantitative text covering a wide range of volcanic phenomena is currently lacking in the instructor's armory. Furthermore, with the rapid developments in the field, the existing books have become outdated. We view this book as the natural next step for students pursuing volcanology beyond introductory level, for which there are several excellent texts (e.g., Cas and Wright, 1988; Francis and Oppenheimer, 2004; Schmincke, 2004; Parfitt and Wilson, 2008), as well as an update on recent developments in the field.

In parallel with the educational aims of this book, we sought to develop a resource that the

Modeling Volcanic Processes: The Physics and Mathematics of Volcanism, eds. Sarah A. Fagents, Tracy K. P. Gregg, and Rosaly M. C. Lopes. Published by CAMBRIDGE UNIVERSITY PRESS. © Cambridge University Press 2013.

research or professional volcanologist could draw on in seeking up-to-date information on modeling approaches to volcanic phenomena or a point of access to a new volcanologic subdiscipline. While the book therefore has elements aimed at students, it does not have the traditional textbook style and is rich in references, both historic and contemporary.

The increasing populations living near active volcanoes imply rising volcanic risk worldwide. Therefore, the role of the well-prepared volcanologist, and the early training thereof, is becoming increasingly important. Thus this book is conceived to bridge the gap in the literature by offering quantitative treatments of volcanic processes, from the subsurface through deposition and post-eruptive modification, at a level designed for students, their professors, and professional volcanologists alike to come efficiently up to speed on the state of the art in understanding the physics and mathematics of volcanism.

Content

This book treats a wide variety of volcanic processes, from the dynamics of magma accumulation deep beneath the Earth's surface through ascent, eruption, and deposition of volcanic products. In compiling this volume, we sought out experts in each subdiscipline of quantitative physical volcanology in order to ensure that each subject area was treated comprehensively. Thus, each of the chapters that follow is devoted to a specific volcanic process and authored by specialists in that subject matter. Each chapter is designed to take the reader from the basic physics through to the state of the art in understanding the volcanic process through modeling, illustrated with field examples. Most chapters include a set of exercises at the end, designed to consolidate the reader's understanding of the subject matter.

The order of the book chapters logically follows the sequence of volcanic processes observed in nature: we begin with what happens at great depths beneath the Earth's surface and track the magma through ascent,

eruption, and deposition at the surface, followed by post-deposition modification. We thus cover the thermal and dynamic processes taking place in accumulations of magma at depth in the crust (Chapter 2), before considering the ways in which magma ascends to the surface (Chapters 3 and 4). Processes taking place within the conduit, such as the evolution of magma rheology and the behavior of magmatic (or external) volatiles, determine the subaerial manifestation of the eruption; whether as an effusion of lava (Chapter 5), or explosive manifestation such as the unsteady, transient processes of strombolian (Chapter 6) and vulcanian explosions (Chapter 7), or the sustained explosive columns of sub-plinian or plinian eruptions, and lava fountains of hawaiian activity (Chapter 8). Chapter 9 considers the fallout of tephra ejected in sustained explosive eruptions, while Chapter 10 considers the emplacement of material as pyroclastic density currents. The interaction of magma with its environment can produce a diverse range of eruption styles. For example surface or near-surface water interacting with erupting magma has the potential to produce violently explosive activity with distinctive eruptive products (Chapter 11), while deep-sea eruptions produce a variety of effusive and weakly explosive deposits that dominate Earth's volcanic record on a volumetric basis (Chapter 12). The interaction of magma with ice is considered in Chapter 13, while Chapter 14 addresses the syn- or post-eruptive phenomena of lahars. Volcano monitoring techniques (i.e., seismology and infrasound) exploit the physics of volcanic processes to elucidate a broad spectrum of volcanic behavior (Chapters 15 and 16). Looking outward from planet Earth, we also consider how models of terrestrial volcanism can be adapted to help us understand the volcanism of other planets (Chapter 17).

While we followed this logical order, there are many commonalities among chapters. For example, the intricacies of and feedbacks among volatile exsolution, magma rheologic behavior, crystal growth, and multiphase flow are critical to eruptive behavior, and so these themes are addressed in several chapters. In addition, the dynamics of granular flow are common to quantitative treatments of both pyroclastic density currents and lahars, even though the fluid phase differs. Moreover, the physics of volcano seismology and volcano infrasound share common roots. Chapters dealing primarily with fluid flow discuss the governing conservation equations (mass, momentum, and energy) in their various formulations (such as the Navier–Stokes equations or St. Venant equations). The use of certain dimensionless numbers to describe key properties or behaviors of physical systems also appears in most chapters. Many chapters distinguish between one-, two-, and three-dimensional modeling approaches, as well as criteria for treating steady vs. unsteady flow. Where appropriate, therefore, extensive use has been made of cross-referencing among the chapters to assist the reader in consolidating their understanding of the processes in question. Extensive treatment of computational methods of solution of governing equations is left to works referenced in each chapter, or to any of the existing texts on numerical methods.

Robust scientific investigations always raise more questions than they answer, and clearly identifying and articulating those nascent questions is essential to improving our understanding of volcanic processes and the hazards they represent. As authors for this book reveal, physical volcanology is a complex and multi-faceted field of study that requires an integrated, multi-disciplinary approach to better understand how volcanoes work. In reading each chapter in this book, the reader is encouraged to consider the big question: what is the next step? Whereas each subdiscipline has specific questions that remain to be answered, there are significant commonalities among the issues raised. For example, although computational power has vastly increased in recent years, thereby allowing very sophisticated numerical models to be developed and run to high degrees of precision, there remains a great need for detailed field and laboratory data to aid in the development of theoretical treatments, or to serve as constraints or validation for the models. The variety and complex, multiphase nature of volcanic fluids mean that a thorough,

quantitative understanding remains elusive. Another common issue is that users need to consider the purposes of modeling a particular process, whether it is to expose the fundamental physical behavior of the phenomenon in question, to provide rapid hazard assessment during volcanic crises (such as aviation safety during explosive eruptions), or a longer-term understanding of the behavior of a particular volcano. All of these approaches are key to advancing our understanding of volcanic processes, but no single model fits all needs. The authors have addressed many of these issues in the final sections of each chapter, identifying the outstanding questions in each subdiscpline and making recommendations for the most fruitful paths of future research.

A particular challenge with this book was to incorporate sufficient detail and referencing to cover the range of phenomena in question, from basic formulations through to the state of the art, with sufficient explanation, illustration, and applications of the methods used, all within the limitations of the printed volume. Realizing that this was an impossible task, we developed a website to accompany the book (www.cambridge.org/fagents). There the reader will find a variety of resources and additional materials relevant to each chapter, including images and video of volcanic phenomena and laboratory simulations, links to additional sites of interest, additional exercises, answers to exercises, data tables, and extensive additional reading bibliographies. We hope that this makes the treatment of modeling volcanic processes far more comprehensive than could be achieved in a single printed volume.

Acknowledgments

A great many people contributed their time and expertise to this project. As well as the authors, to whom we are deeply indebted for their generosity in sharing their expertise, each chapter was reviewed by at least two expert reviewers; thus a small army of volcanologists was recruited to help craft the book contents into publishable form. We are deeply grateful to the following for their time and wisdom in enhancing the value of this book: Steve Baloga, Jackie Caplan-Auerbach, Bill Chadwick, Chuck Connor, Tim Davies, Pierfrancesco Dellino, Arnau Folch, Bruce Houghton, Gert Lube, Emaneule Marchetti, Larry Mastin, Brian McArdell, Meghan Morrissey, Harry Pinkerton, Maurizio Ripepe, Elise Rumpf, Shan da Silva, Adam Soule, Hugh Tuffen, David A. Williams, Lionel Wilson, James White, and two anonymous reviewers.

Finally, we acknowledge the enthusiastic support and infinite patience of Susan Francis and Laura Clark at Cambridge University Press, who maintained a steady pressure while we endeavored to make this the book that we envisioned.

References

Cas, R. A. F. and Wright, J. V. (1988). *Volcanic Successions.* London: Unwin Hyman.

Francis, P. W. and Oppenheimer, C. (2004). *Volcanoes.* Oxford University Press.

Freundt, A. and Rosi, M. (2001). *From Magma to Tephra: Modeling Physical Processes of Explosive Eruptions.* Amsterdam: Elsevier.

Gilbert, J. S. and Sparks, R. S. J. (1998). *The Physics of Explosive Eruptions.* Geological Society of London Special Publication 145, London.

Parfitt, E. A. and Wilson, L. (2008). *Fundamentals of Physical Volcanology.* Wiley-Blackwell.

Schmincke, H.-U. (2004). *Volcanism.* Berlin: Springer-Verlag.

Sparks, R. S. J., Bursik, M. I., Carey, S. N. *et al.* (1997). *Volcanic Plumes.* Chichester: Wiley.

Chapter 2

Magma chamber dynamics and thermodynamics

Josef Dufek, Christian Huber, and Leif Karlstrom

Overview

Magma chambers are continuous bodies of magma in the crust where magma accumulates and differentiates. Both geophysical and geochemical techniques have illuminated many aspects of magma chambers since they were first proposed. In this chapter, we review these observations in the context of heat and mass transfer theory. This chapter reviews heat transfer calculations from magma chamber to the surrounding crust, and also considers the coupled stress fields that are generated and modified by the presence of magmatic systems. The fluid dynamics of magma chambers has received considerable attention over the last several decades and here we review the ramifications of convection in magma chambers. Multiphase flow (melt + crystals + bubbles) plays a particularly important role in the evolution of magma chambers. The large density differences between melt and discrete phases such as bubbles and crystals, and the resulting flow fields generated by buoyancy are shown to be an efficient mechanism to generate mixing in chamber systems. Finally we discuss integrative approaches between geophysics and geochemistry and future directions of research.

2.1 | Introduction

The compositional diversity of melts that reach the surface of the Earth, and diversity in eruptive style, are largely determined through processing of these melts as they ascend and sometimes stall in the crust. Most eruptive products and intrusive suites have been modified substantially from their progenitor mantle magmas, either through preferential removal of crystal phases during fractionation, assimilation of crustal melts, or a combination of these processes (Daly, 1914; Anderson, 1976; Wyllie, 1977; Hildreth and Moorbath, 1988; DePaolo et al., 1992; Feeley et al., 2002). Much of the evolution of these magmas likely occurs where they spend the most time: where magma has either permanently or temporarily stalled in magma chambers. This accumulation is fundamental to the genesis of large eruptions, as the background flux from the mantle cannot explain the voluminous outbursts of magma at the surface of the Earth. Magma chamber dynamics largely control the compositional evolution of these magmas, and ultimately a better understanding of magma chambers may provide clues to the triggering of eruptions. In this chapter we

Modeling Volcanic Processes: The Physics and Mathematics of Volcanism, eds. Sarah A. Fagents, Tracy K. P. Gregg, and Rosaly M. C. Lopes. Published by Cambridge University Press. © Cambridge University Press 2013.

consider the conditions for the formation and evolution of magma chambers from thermal, solid stress, and fluid mechanics perspectives.

Magma chambers are commonly envisaged as high melt-fraction bodies (> 40% melt). However, thermal and dynamic arguments typically favor these intrusive bodies existing at lower melt fractions for considerable proportions of their histories (Marsh, 1981). Large, high melt-fraction chambers have existed at least transiently in the crust, as evidenced by voluminous ignimbrite sheets and corresponding caldera collapse structures (Spera and Crisp, 1981; Druitt and Francavigilia, 1992). However, buffering by latent heat release and diminishing thermal gradients during cooling of a magma intrusion implies longer periods of time at low melt fraction. This chapter adopts the definition of a magma chamber as a spatially connected body of magma in a suprasolidus state, and uses this term generically regardless of whether the chamber is composed mostly of melt or a crystal-rich mush.

Speculation on chamber processes dates to the recognition of some intrusive suites as being of fluidal, magmatic origin (Hutton, 1788). Consideration of the dynamics of convection in magma chambers dates back to at least the late nineteenth century (Grout, 1918). Modern, rock-record-based investigations of chamber dynamics are largely based on two parallel sources of data: the examination of exhumed plutonic rocks and the examination of eruptive products (Bachmann et al., 2007a). Both analyses have been useful in delineating aspects of chamber processes, although they provide very different types of data. The plutonic realm provides a time-integrated view of magmatic processes, but reflects different parts of the temporal record with unequal fidelity. Eruptive products tend to provide closer to an instantaneous snapshot of the compositional field of a chamber at the time of the eruption. However, some time-integrative information still exists in the form of phenocrysts that crystallized and were advected with the magma (Wallace and Bergantz, 2005). Interpretation of the compositional field from eruptive products can be somewhat confounded by complex dynamics

in the eruption and emplacement process. The very fact that the magmas are eruptible, at relatively low crystallinity, may suggest that this state is not necessarily representative of a large portion of the magmatic history (Marsh, 1981).

While many difficulties exist in interpreting both the plutonic and volcanic record to infer dynamics, geochemical tracers continue to be among the best sources of information about the dynamics of these systems (Wyllie, 1977; Marsh, 1989; Bachmann et al., 2007a). Importantly, diffusion-based and radioactive decay-based chronometers indicate that phenocrysts in magmatic systems can grow over a range of timescales and can exhibit prolonged histories at suprasolidus temperatures in the crust, sometimes in excess of hundreds of ka (Costa et al., 2003; Vazquez, 2004; Simon and Reid, 2005, 2008). The decay of uranium-series nuclides has also shown that different crystal populations can have different residence times in magma chambers, can indicate mixing of complex populations from in situ crystallization and assimilation from older wall rock (Cooper and Reid, 2003), and can indicate much shorter bursts of volatile exsolution and gas transport (Berlo et al., 2006).

However, these measurements of magmatic residence times do not give precise information about the spatial and temporal variability in melt fraction or the dynamics of these systems. Real-time geophysical observations, though still sparse, are beginning to provide important snapshots of the crust and short-duration magmatic system evolution. These investigations are primarily conducted by observing how seismic waves interact with the crust and by observing the deformation of the ground due to the changes in magmatic systems (geodesy). Gravity and changes in the conductivity have also proven to be useful tools to describe magmatic systems in certain localities (Ajakaiye, 1970; Manzella et al., 2000; Gudmundsson and Hognadottir, 2007).

Seismic tomographic inversion has identified low-velocity regions beneath volcanic edifices that may correspond to a combination of regions of partial melt and elevated temperatures. Nevertheless, the elastic properties of rock, as shown by laboratory measurements, depend on several other variables, such as temperature, pressure, rock type, and the presence of hydrothermal fluids. The non-unique interpretations for spatial variation of seismic velocities make it generally difficult to attribute those effects solely to the presence of melt, and a good a priori knowledge of the geology of the region is required to make accurate estimates of melt fraction (Lees, 2007). Additionally, the problem of resolution due to both ray coverage and the large-scale filtering of fine structures by seismic wavelengths generally on the order of a few hundred meters to kilometers introduces blurring of the geometry and size of the magma body.

The measurement of ground deformation in volcanic areas provides another observational indication of the presence of magma chambers. Geodetic studies fit time-series of ground deformation to models, which are then used to back out processes, stresses, and volumes of intruded magma. Data are collected primarily by GPS units deployed on the edifice (Segall and Davis, 1997), or by Interferometric Synthetic Aperture Radar (InSAR) (Poland et al., 2006). Although sensitive to rheological variations, including those induced by different melt fractions, ground deformation is primarily used to indicate changes in volume of magma chambers due to either inflation or deflation of a chamber as a result of magmatic flux. Despite difficulties in gathering and interpreting these data, a growing database of geodetic observations and increasing sophistication of inversion methods provide unique insight into active magma chamber dynamics.

Diverse observations, from geochemically inferred residence times to remote imaging methods, show that subsurface magmatic systems are active on a variety of time and length scales. However, our lack of direct observation of these systems has limited our ability to understand how they evolve dynamically in time. Therefore, mathematical, experimental, and computational models are widely used to integrate and explain some of the disparate observations and provide a framework that can help guide future observations of magma chambers. Modeling also helps us to understand these

bodies in the broader context of crustal evolution and volcano–pluton connections. Section 2.2 discusses the thermal evolution of the crust in magmatically active regions and its relationship to the evolution of chambers and geochemical observations of chamber residence times. Section 2.3 examines the relationship between magma chambers and crustal stress fields, and how this relates to geodetic observations. Section 2.4 describes magma motion inside magma chambers, and the multiphase interactions between melt and discrete phases (such as crystals and bubbles). Finally, Section 2.5 is devoted to a discussion of future directions in the physical examination of magma chambers.

2.2 | Heat transfer and magmatic intrusions

The magmas that accumulate in crustal magma chambers are either generated by mantle melting with subsequent crustal transport, or are a result of crustal-level melting due to anomalous thermal conditions. Although many crustal melts can be generated in proximity to intrusions that have heated surrounding rocks (Bergantz, 1995), or result from over-thickening of the crust in orogenic zones (Patiño Douce and Harris, 1998), the advection of mass and heat from the mantle ultimately drives crustal-level magmatism.

After accumulation, the long-term viability of magma chambers in the crust is critically dependent on the surrounding thermal environment, and the thermodynamic as well as rheological properties of the melt and country rock (Patiño Douce and Harris, 1998; Newman et al., 2006; Karlstrom et al., 2009). Relatively cold country rock will drive large thermal gradients between the magma body and its margins, limit the time a body of magma remains melt-dominated, and also inhibit the partial melting and assimilation of wall rock material. Likewise, hydrothermal circulation transports thermal energy away from the immediate vicinity of an intrusion, leading to enhanced solidification in the shallow crust. In contrast, all else being

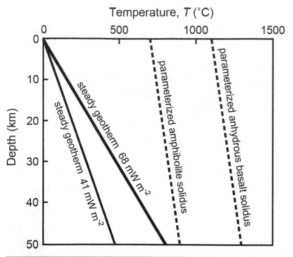

Figure 2.1 Steady-state geotherms and the solidus of amphibolites and basalts. Two geotherms are plotted for surface heat flows of 68 mW m^{-2} and 41 mW m^{-2}. Amphibolite solidus is parameterized from the experiments of Rushmer (1991), Wolf and Wyllie (1994), Patiño Douce and Beard (1995), Rapp (1995), and Rapp and Watson (1995), and anhydrous basalt solidus is parameterized from anhydrous experiments (Green, 1973).

or melting events (Rudnick *et al.*, 1998; Blundy and Wood, 2003). Because of this, it is commonly assumed that radiogenic heat-producing elements can be described by an exponential declining concentration relationship in the crust (Chapman, 1986; Chapman and Furlong, 1992). Although an exponential decline in radioactive nuclide concentration with depth is almost certainly not correct in detail, this profile does result in the linear decay of heat flow in the crust that is observed in many locations (Turcotte and Schubert, 1982).

Using the approach of Chapman and Furlong (1992) two steady-state geotherms have been plotted in Figure 2.1, along with an idealized linearization of the solidus of amphibolites constructed from a number of petrologic experiments (Rushmer, 1991; Wolf and Wyllie, 1994; Patiño Douce and Beard, 1995; Rapp and Watson, 1995) and an anhydrous solidus curve for basaltic melt produced from the melting of the primitive upper mantle (Green, 1973). The geotherms plotted assume 0.94 µW m^{-3} heat source production due to radioactive nuclides at the surface, and an exponential decrease in magnitude with a characteristic length scale of 15 km. Two geotherms are plotted, one with a surface heat flux of 68 mW m^{-2}, which is approximately equal to average heat flow in the Cascade arc (Touloukian *et al.*, 1981), and one with a surface flux of 41 mW m^{-2}, approximately equal to average heat flow from continental crust (Rudnick *et al.*, 1998).

To gain a conceptual understanding of the residence times of magma bodies emplaced at different depths in the crust we consider a series of two-dimensional "square" bodies of basaltic magma, all of which have a length of 2 km. We first consider only conductive transfer of heat from the magmatic system to the surrounding country rock, but Section 2.4 revisits internal convection inside the chamber. However, we note that the broad scale heat budget, especially in the country rock, is relatively insensitive to the convective chamber motion (although the compositional heterogeneity and internal melt fraction will be very sensitive to convection). The thermal co-evolution of the magma body and surrounding rock provides

equal, intrusions deeper in the crust have the potential for longer residence times and greater potential for assimilation of the crust, given that the solidus for many mineral assemblages is close to steady-state geotherms (Figure 2.1). Enthalpy excursions due to magmas emplaced from the mantle can more readily melt the lower crust than upper crust, although prolonged heating by successive intrusions at all depths has important implications for the stability of magma chambers, as well as the dynamic evolution of the crust (Dufek and Bergantz, 2005a; Karlstrom *et al.*, 2009).

The background thermal condition of the crust in the absence of magmatic intrusions is largely determined by conductive transfer of heat from depth and heating due to the decay of radiogenic nuclides, modulated by shallow hydrothermal circulation. Most radiogenic nuclides (e.g., uranium, thorium, and potassium) are much more compatible in silicate melts relative to solid phases and hence tend to accumulate in the more silicic, upper crust that has grown by either multiple fractionation the boundary condition for the heat budget, and in this sense is a fully coupled problem. In contrast, a specified, constant-temperature boundary condition at the margin of a magma chamber would generate unphysical amounts of crustal melt.

The thermal interface between the magma and wall rock can be described as (Carrigan, 1988):

$$k \frac{\partial T}{\partial z}\bigg|_b = h(T_b - T_R). \qquad (2.1)$$

All notation is summarized in Section 2.7. Here T_b is the temperature at the boundary, T_R is the temperature of the thermally unperturbed wall rock some distance from the chamber, k is the thermal conductivity of the magma (W K^{-1} m^{-1}), and h a heat transfer coefficient (W K^{-1}). This boundary condition relation can be made dimensionless by introducing $z^* = z/\delta$, where δ is the thickness of the boundary layer inside the chamber where convective motion ceases, and by setting $T^* = T/T_R$. This then results in the equation,

$$\frac{\partial T^*}{\partial z^*}\bigg|_b = \text{Bi}(T^* - 1). \qquad (2.2)$$

The dimensionless term, the Biot number (Bi), gives the ratio of thermal resistance of a magma chamber to the thermal resistance of the surrounding rock:

$$\text{Bi} = \frac{h\delta}{k}. \qquad (2.3)$$

A summary of dimensionless numbers used commonly in magma dynamics is given in Online Supplement 2A (see end of chapter). Most estimates of magma chamber conditions indicate Bi $\ll$ 1 (Carrigan, 1988). Under these conditions heat transfer is limited by the thermal resistance of the wall rock, even in the presence of hydrothermal circulation. Described as the "thermos-bottle effect", the amount of heat lost into the wall rock is limited by the more insulating wall rock. Accordingly, we focus first on conduction of heat into the wall rock and the limitations this places on the thermal evolution of the magma chamber.

The generalized equation for enthalpy evolution takes the form

$$\frac{\partial H_T}{\partial t} + \frac{\partial}{\partial x_i}(\mathbf{u}_i H_T) = \frac{\partial}{\partial x_i} k\left(\frac{\partial T}{\partial x_i}\right), \qquad (2.4)$$

where the enthalpy (H_T) is composed of a sensible (heat that results in a temperature change) and latent heat component:

$$H_T = \rho \overbrace{\int_{Tref}^{T} c_p \, dT}^{\text{Sensible Heat}} + \overbrace{\rho f L}^{\text{Latent Heat}}. \qquad (2.5)$$

Here ρ is the magma density (kg m^{-3}), c_p is the magma specific heat capacity (J kg^{-1} K^{-1}), L is the latent heat (J kg^{-1}), f is the melt fraction (volumetric), and $\mathbf{u}_i$ is the velocity vector. For the case of constant heat capacity and conductivity, normalizing density by a reference density (ρ_0), characteristic length scale (δ), velocity (u_0), and temperature (T_0), the heat conservation can be recast in dimensionless form to give:

$$\rho^*\left(\frac{\partial T^*}{\partial t^*} + \mathbf{u}_i^* \frac{\partial T^*}{\partial x_i^*}\right) = \frac{1}{\text{Pe}} \frac{\partial^2 T^*}{\partial x_i^{*2}} - \frac{R^*}{\text{Ste}}, \qquad (2.6)$$

where R^* is the dimensionless rate of production of melt, given by:

$$R^* = \rho^* \frac{\partial f}{\partial t^*}. \qquad (2.7)$$

Two dimensionless parameters are introduced in this formulation, the Stefan number and the Peclet number. The Stefan number (Ste) is a ratio of the sensible heat to the latent heat contributions in the flow, given as

$$\text{Ste} = \frac{c_p T_0}{L}. \qquad (2.8)$$

The Peclet number (Pe) is defined as the ratio of advective to diffusive heat transport:

$$\text{Pe} = \frac{c_p \rho_0 \delta}{k} \mathbf{u}_0. \qquad (2.9)$$

In the case of heat transport in the solid crust, the Peclet number is small and we can neglect

Table 2.1 Summary of magma intrusion models, modified from Dufek and Bergantz (2005a).

Study	Model type[1]	Intr. style[2]	Total intr.[3] (km)	T_{init} (°C)	Rock type[4]	T_{melt}, T_{solid} (°C)	$E_\%$
Younker and Vogel, 1976	1D, cond., no bottom heat loss	single intrusion	2.0	500	basalt, biotite–granite	L:1200, S:1100 L:1100, S:800	32
Wells, 1980	1D, cond., over-accretion	multiple intrusion	40.0	200	tonalite	L:1050, S:800	8
Huppert and Sparks, 1988	1D, param. convection, no bottom heat loss	single intrusion	0.5	500	basalt, granodiorite	L:1200, S:1091 L:1000, S:850	44
Bergantz, 1989	1D, cond., no bottom heat loss	single intrusion	16.6	700	basalt, pelite	L:1250, S: 980 L:1200, S: 725	38
Bittner and Schmeling, 1995	2D, convection	single intrusion	5.0	756	basalt, granite	L:1100, S: 950 L:1050, S:760	NA
Barboza and Bergantz, 1996	2D, convection	fixed T bottom boundary	NA	600	pelite	L:1200, S:750	NA
Raia and Spera, 1997	2D, convection	fixed T bottom boundary	NA	1195	$(CaAl_2Si_2O_8-CaMgSi_2O_6)$	L:1547, S:1277	NA
Pedersen et al., 1988	1D, cond., over-accretion	multiple intrusion	10.0	650	basalt, granodiorite	L:1250, S:1100 L:1000, S:710	5
Petford and Gallagher, 2001	1D, cond., over-accretion	multiple intrusion	1.0	650	basalt, amphibolite	L:1250, S:1050 L:1075, S:1010	4
Annen and Sparks, 2002	1D, cond., over-accretion	multiple intrusion	8.0	variable, based on depth (600)	basalt, amphibolite	L:1300, S:620 L:1075, S:1010	8
Dufek and Bergantz, 2005a	2D, cond. and convection, stochastic	multiple intrusion	variable (5.0)	variable, based on depth (640)	basalt, amphibolite	pressure dependent L:1240, S:640 L: 1100, S:850	0.5–10.4 (7.1)

[1] Dimension, heat conduction or convection.
[2] Intruding magma, physical configuration of intrusion (sill, etc.) or specified temperature boundary condition.
[3] Integrated melt volume/basal area, or for 1D models integrated melt height.
[4] Intruded magma listed first, then country-rock.
[5] Parentheses indicate specific example.

the advective term in Eq. (2.6). Furthermore, in the absence of solidification or melting (e.g., phase changes) the latent heat term in Eq. (2.6) equals zero.

The melt fraction, f, needs to be specified as a function of temperature to solve Eq. (2.6). The form of the f–T relationship depends on the mineral phases that either melt or crystallize as the melting or solidification reaction proceeds. Ultimately, information to develop these critical relationships comes from experiments. Two main approaches have been used to construct these closure relationships: simplified parameterizations of experiments for specific melting/ solidification relationships, or the development of self-consistent thermodynamic models that incorporate experimental results. This latter method provides more accurate interpolation and extrapolation of existing experiments for a more general composition and specified thermodynamic variables; the widely used MELTS thermodynamic software is an example of this approach (Ghiorso and Sack, 1995). The use of melt-fraction relationships was reviewed in Bergantz (1990) and many melting scenarios and lithologies are considered in thermal modeling of magma intrusions (see Table 2.1).

Here, we make the simplest possible assumption – a linear relationship between melt fraction and the liquidus and solidus temperatures. This assumption can also be cast as a modified thermal diffusivity when melt is present,

$$\kappa_{eff} = \frac{k}{\rho c_p + \rho L (T_{liq} - T_{sol})^{-1}}, \qquad (2.10)$$

where T_{liq} is the liquidus and T_{sol} is the solidus temperature. In this formulation, one-dimensional analytical solutions can be found for simple initial conditions. When the melt fraction has a nonlinear relationship with temperature, representing more realistic melting reactions, the solution to the thermal problem usually requires iterative numerical solution procedures (Voller, 1985; Prakash, 1990; Voller and Swaminathan, 1991; Voller and Cross, 1993). The linear relationship, however, demonstrates several key points. For illustrative purposes we assume that the intruding magma is anhydrous basalt, the surrounding crust is also basaltic, and we use

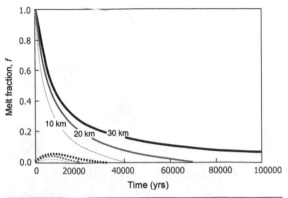

Figure 2.2 Individual en masse 2 × 2 km intrusions introduced into a crust with a surface heat flow of 68 mW m^{-2} as described in Figure 2.1, at depths of 10, 20, and 30 km. The solid lines, labeled with the intrusion depth, show the average melt fraction (by volume) relative to the initial intruded volume. These lines represent the cooling and fractionating melt volumes through time. The dotted lines represent the equivalent country rock melt volumes relative to the initial intrusion volume, for the same depths.

thermal conditions for a 68 mW m^{-2} surface heat flux (Fig. 2.1). The individual 2-km-square intrusions are emplaced instantaneously at 10, 20, and 30 km depth.

Figure 2.2 shows that deep intrusions persist longer with greater melt fraction, and produce greater fractions of crustal melt. However, when the crust is composed of the same material as the intruding magma, as in this example, very little crustal melting can be expected. In all scenarios considered in Figure 2.2 the crustal melt component never exceeds ~5% of the intruded volume at any point in time. More silicic and hydrous crustal compositions with lower solidi are more prone to crustal melting, although large amounts of enthalpy typically have to be supplied to explain pervasive melting scenarios.

In most magmatically active environments, evidence exists for multiple intrusive episodes overlapping in space and time (Hopson and Mattinson, 1994). Three scales of dike–crust interaction are shown in Figure 2.3, illustrating the integrated expression of multiple intrusion events at the Black Canyon of the Gunnison in Colorado, USA and in the Grand Canyon, Arizona, USA. In such settings, sequential intrusions can generate a thermal anomaly, and allow bodies

Figure 2.3 Three scales of melt–crust interaction. (a) Dike network of 1.43 Ga Vernal Mesa pegmatite cross-cutting 1.73 Ga gneisses on the 670 m high Painted Wall, Black Canyon of the Gunnison, Colorado. Dikes were emplaced at about 10 km depth in tensional opening during dextral/oblique shear on the Black Canyon shear zone (Jessup et al., 2006). (b) Orthogonal dike network of ~1.69 Ga granite and pegmatite cross-cutting 1.73 Ga granodiorite within Tuna Canyon, tributary to Upper Granite Gorge of Grand Canyon. Dike networks in Grand Canyon reflect late-syntectonic granite dike swarms emplaced at about 20 km depth (Dumond et al., 2007) during late stages of accretionary assembly of southwestern North American lithosphere (note geologist for scale) (Ilg et al., 1996); photograph courtesy of Laurie Crossey. (c) Four generations of ~1.69 Ga granite dikes in Spaghetti Canyon of Lower Granite Gorge of Grand Canyon, note pen for scale. Earlier dikes underwent ductile deformation before later dikes were emplaced, suggesting an interplay of diking and ductile flow at middle crustal depths (Karlstrom and Williams, 2006); photograph courtesy of Laurie Crossey. See color plates section.

of magma to reside at higher melt fraction for longer periods of time than in en masse scenarios (Petford and Gallagher, 2001). The flux of magma in such settings becomes an important parameter.

2.2.1 Multiple intrusions and crustal melting efficiency

One of the more interesting questions regarding the growth of continental crust is to what extent do the contents of magma chambers represent fractionated melts extracted from the mantle or melts of the existing crust. Melting of the crust, and subsequent segregation and advection of this melt, can change the density structure of the crust, but does not, in itself, cause growth of the crust. Injection of mantle melts, and fractionation to form low-density melts and then rocks, however, can contribute mass to the crust over time (Grunder et al., 2006). As shown in Figure 2.2, during single intrusion scenarios when the intruding magma has the same composition as the country rock, little crustal melting can be expected. However, real melting scenarios can involve different lithologies and multiple intrusive events.

Regions of strong isotopic contrast, such as old continental crust, can sometimes provide a clear indication of crustal melting. However, mafic portions of the crust, especially the lower crust, can have little isotopic contrast relative to mantle melts, and major- and trace-element consideration of the magmas can lead to ambiguous interpretations of whether a magma is primarily of mantle or crustal origin (Hart *et al.*, 2002). Also, cryptic assimilation, or assimilation of crystal phases without completely melting them, can modify whole-rock isotopic values and generate mixed signatures even when the full crustal melting reaction has not proceeded to high melt fraction (Wolff *et al.*, 2002; Beard *et al.*, 2005). A complementary tool to geochemical investigation of fractionation and melting involves thermal calculations (e.g., Fig. 2.2). Many thermal models have been considered for different melting scenarios, including analytical solutions applied to simplified scenarios, parametric and numerical convection simulations, and numerical conduction simulations (Younker and Vogel, 1976; Wells, 1980; Huppert and Sparks, 1988; Bergantz, 1989; Bittner and Schmeling, 1995; Barboza and Bergantz, 1996; Raia and Spera, 1997; Pedersen *et al.*, 1998; Petford and Gallagher, 2001; Annen and Sparks, 2002; Dufek and Bergantz, 2005a). One way that these models can be compared is in their relative efficiency in melting the crust. An enthalpy balance of a volume of basalt (V_B) intruded at its liquidus (T_{liqB}) and cooling to a temperature T gives the volume of crust melted:

$$V_{Ceff} = \frac{\rho_B V_B \left(c_{pB}(T_{liqB} - T) + L_B \right)}{\rho_C c_{pC}(T_{solC} - T_R) + \rho_C \left(c_{pC}(T - T_{solC}) + L_C f \right)}. \quad (2.11)$$

Here the B and C subscripts refer to basalt and crust, respectively, and T_R is the initial rock temperature. This volume can be considered the perfectly efficient end-member, as all the energy from the intrusion goes into heating only the portion of crust that becomes molten.

In other words, energy is not "wasted" by heating parts of the crust that never become molten. Conductive heat transfer is less efficient than this end-member, however, as energy will diffuse over large areas, and eventually radiate from the surface of the Earth. Therefore, only a small fraction of the region that experiences temperature increases will reach the crustal solidus. Using published results for a range of melt compositions, intrusion volumes, thermodynamic variables, and melt fraction relationships, we can compare the amount of crustal melt predicted by these models (V_{Cmod}) with the completely efficient end-member:

$$E_\% = 100 \frac{V_{Cmod}}{V_{Ceff}}. \quad (2.12)$$

A 100% efficient scenario would convert all the enthalpy from an intrusion into melting the crust. A compilation of several models is shown in Table 2.1.

An almost universal conclusion from thermal models, especially those considering multidimensional thermal diffusion, is that injection of mantle melts in the crust, even under optimal conditions, is a relatively inefficient method of melting the crust. Although there are obvious differences depending on the intruding magma and country-rock compositions considered, most are only around 10–40% efficient. This means that even in relatively fertile, hydrous lithologies, large volumes of intruded mantle basalts are required to produce voluminous crustal melts.

Making the space required to accommodate intrusions has long been recognized as a constraint on magma system growth, especially where extensive fractionation is required, and is sometimes termed the "room problem". Although crustal melting has been invoked as a mechanism to alleviate the room problem and explain the absence of observed fractionate residua, it can be as inefficient as fractionation and sometimes more inefficient than fractionation at producing silicic melts. (Fractionation of a basalt to high-silica rhyolite requires ~90% removal of crystals; Grout, 1926; Winter, 2001.)

The crustal lithology, thickness, melt flux, and evolving thermal anomaly all play a role in the synoptic view of crustal melting and fractionated mantle melts. When considering the random intrusion of mafic melts in arc crustal conditions, Dufek and Bergantz (2005a) found that, in general, thin, immature arc crust undergoes relatively little melting, and that a random sampling of melt from the crustal column is likely to contain < 10% crustal melt by volume. Thick and mature crust, after several million years of intrusion for typical arc fluxes, results in greater crustal melting: with a crust of ~50 km and fertile amphibolite lower crustal lithology, they found a random sampling could result in roughly equal probabilities of intersecting crustal melts or fractionated mantle melts.

Due to the relative inefficiency of fractionation and crustal melting to produce silicic melts (the former requires large volumes of basalt as a direct mass source and the latter requires large volumes to account for the required enthalpy source), the accommodation mechanism and tectonic regime of the crust are also likely important factors in the production of evolved melts in the crust. The thermal and mass requirements for magma chamber development and preservation are inextricably linked to the tectonic and chamber-induced stress regimes that permit accommodation.

2.3 | Crustal stresses and magma chambers

Stresses around magma chambers provide the mechanical means by which transport of magma to and from the chamber occurs, and determine the nature of deformation and failure of surrounding country rocks. Sufficiently high stresses above lithostatic confinement (deviatoric stresses) will induce fracture and chamber rupture (Tait *et al.*, 1989), whereas isotropic magmatic stresses such as uniform loading may variously cause or suppress chamber rupture (Vigneresse and Tikoff, 1999). Chamber stresses are also responsible for the dynamic evolution of magma chambers in the broader context of

the entire volcanic system, and are strongly coupled to thermal, chemical and rheological processes within and around the chamber.

Magma chamber stresses are generated internally by the pressurization or buoyancy of magma relative to lithostatic stresses, through a variety of processes. Chamber over-pressurization, ΔP, can occur through injection of magma into the chamber (Parfitt *et al.*, 1993; Woods, 1995), thermal expansion of magma in the chamber or melting of country rocks (Bonafede, 1990), and volume change upon crystallization of mineral phases (Fowler and Spera, 2008) following the thermodynamic relationship

$$\Delta P = \frac{1}{\beta}\frac{\Delta V}{V} + \frac{\alpha}{\beta}\Delta T, \qquad (2.13)$$

where ΔV and ΔT are changes in the chamber volume and temperature. Both the magma compressibility β ($\sim 10^{-11}$ Pa^{-1}) and thermal expansivity α ($\sim 10^{-5}$ K^{-1}) depend on the major-element composition and volatile content of the magma, as well as depth (Dobran, 2001; Rivalta and Segall, 2008). Although the volume change during phase change varies with mineral species and magma composition, the net effect is a pressure increase during melting and a pressure decrease during solidification (Lange and Carmichael, 1990; Rushmer, 1995; Dobran, 2001). Melt buoyancy results from progressive crystallization of mineral phases from the melt (Marsh, 1996), which can progressively increase the density difference $\Delta\rho$ between magma and host rock. Magma buoyancy may be important for long-distance transport in dikes (Rubin, 1995; Roper and Lister, 2005; see Chapter 3), and for caldera-forming eruptions (Gudmundsson, 1998; McLeod, 1999; Pinel and Jaupart, 2005). Stresses may also accumulate as volatile species (primarily H_2O and CO_2) exsolve from the melt (Sparks *et al.*, 1977; Tait *et al.*, 1989). Volatile exsolution may also lead to anisotropic stresses as bubbles (super-critical fluid at these depths) rise toward the chamber roof (Woods and Cardoso, 1997).

The build-up of internal stress due to magma addition or withdrawal, volatile exsolution, and crystallization depends on the chamber depth and solubility relationships but generally can be described by (Woods and Huppert, 2003)

$$F\frac{dP}{dt}=\frac{1}{\rho_m}(Q_I-Q_o)+G\frac{dm_c}{dt}, \qquad (2.14)$$

where

$$F=V_g\frac{d\rho_g}{dP}+\rho_g\left(\frac{V_0}{\mu}+\frac{V_l}{K}\right)+\frac{\rho_l-\rho_g}{1-\phi_v}V_l\frac{d\phi_v}{dP}, \qquad (2.15)$$

and

$$G=\rho_g\left(\frac{1}{\rho_c}-\frac{1}{\rho_l}\right)+\frac{\rho_l-\rho_g}{\rho_l(1-\phi_v)}\phi_v. \qquad (2.16)$$

Here V_0 is the total volume of the magma chamber, Q_I and Q_o are magma mass flux into and out of the chamber, ϕ_v is volume fraction of a volatile species, dm_c/dt is the rate of magma crystallization, μ is the modulus of rigidity of the wall rock, K is the bulk modulus ($= 1/\beta$) of the magma, and the gas is assumed to behave as an ideal gas. The subscripts refer to the different phases, liquid, gas, or crystal (l, g, c), in the chamber.

Dynamic pressure evolution inside the magma chamber is of great interest, as it provides the source conditions for volcanic eruptions. Woods and Huppert (2003) analyzed a generalized form of Eq. (2.14) for a chamber containing an evolved silicic component and an intruding mafic phase. They found that chamber pressure evolution is strongly affected by the exsolution of volatiles, which increases the magma compressibility and therefore the volume and duration of eruptions from the chamber.

Of the sources of internal chamber stress mentioned above, accumulation of overpressure by the injection of melt is the most significant, followed by volatile exsolution and magma buoyancy. From Eq. (2.13), it is clear that melting of magma at depth to supply a magma chamber can in principle generate tremendous overpressures. The instantaneous melting of a 10^6 m³ body, for instance, or the injection of 10^6 m³ of magma into a 10^7 m³ reservoir generates ~10^{10} Pa of pressure. In comparison, pressures generated by exsolution of volatiles may be 10^4–10^6 Pa (Folch and Marti, 1998), and buoyancy stresses are on the order of 10^3–10^5 Pa because of density differences between magma and host rock

(although this scales with chamber size, e.g., Dobran, 2001). These are static stress changes, and do not account for compaction or viscous relaxation of stresses over time.

External stresses on magma chambers, occurring on both tectonic and eruptive time-scales, are also important and likely influence chamber construction (Jellinek and DePaolo, 2003; Gudmundsson, 2006). Tectonic processes provide a background stress field that may concentrate stresses on existing chambers and affect the ascent of magma rising towards the chamber. Extensional regional stresses promote vertical magma transport, whereas compressive stresses orient rising magma subvertically (Muller et al., 2001). This is quantified through the Anderson theory (Anderson, 1936) that posits that dike trajectories are perpendicular to the least compressive deviatoric principal stress at the dike tip. The total stress in the crust is given by

$$\sigma_{tot}=\sigma_{lith}+\sum\sigma_{ext}+\sum\sigma_{int}. \qquad (2.17)$$

Deviatoric stresses are $\sigma_d = \sigma_{tot} - \mathrm{Tr}(\sigma_{tot})$, where Tr is the trace operator. Equation (2.17) expresses the total crustal stress σ_{tot} as a sum of lithostatic stress σ_{lith}, far-field externally imposed stresses σ_{ext} (such as those due to tectonics), and stresses due to features internal to the transport system σ_{int} (such as propagating dikes, magma chamber pressure, edifice loading of the free surface, and structural heterogeneities).

According to this theory, tectonic stresses influence the propagation of dikes around a magma chamber (Muller and Pollard, 1977; Gudmundsson, 2006). External stresses may also come from other sources, such as loading by glaciers or volcanic edifices (Hieronymus and Bercovici, 1999; Jellinek et al., 2004), topography (McTigue and Mei, 1987), and from stress concentration due to structural heterogeneity such as pre-existing faults or material interfaces in the subsurface (see Chapter 3; Vigneresse and Tikoff, 1999; Kavanagh et al., 2006; Gaffney et al., 2007). Because internal magma chamber stresses contribute to the total background stress in Eq. (2.17), dikes emanating from a magma chamber are strongly mechanically coupled to the exterior stress trajectories (Meriaux and Lister, 2002), and may be focused towards it as they rise from greater depths (Karlstrom et al., 2009). Modeling stresses around magma chambers provides a means of predicting where chamber rupture, and hence volcanic eruption, is likely to occur (Muller and Pollard, 1977; Pinel and Jaupart, 2005). Conversely, knowledge of chamber-induced stresses at depth provides a means to estimate the spatial extent of a particular volcanic system (Muller et al., 2001; Karlstrom et al., 2009). The influence of chamber stresses on dike generation and focusing at depth thus has important consequences for magma chamber stability in the volcanic system as a whole.

To illustrate the role of magma chamber stress we consider some typical idealized models. Because of the large viscosity contrast between magma and rocks at mid to shallow crustal depths, magma chambers are commonly modeled as cavities in an elastic medium, solving the equations of Linear Elasticity (Fung, 1965):

$$\frac{\partial^2 \mathbf{d}_i}{\partial \mathbf{x}_j^2} + \frac{1}{1-2\upsilon}\frac{\partial}{\partial \mathbf{x}_i}\left(\frac{\partial \mathbf{d}_j}{\partial \mathbf{x}_j}\right) = \mathbf{f}_{bi}, \qquad (2.18)$$

where $\mathbf{d}$ is the displacement vector, $\mathbf{f}_b$ is the body force vector, and υ is Poisson's ratio. Application of Eq. (2.18) commonly assumes that no body forces exist in the elastic medium, which neglects free-surface effects. Most studies consider static stresses only, with some exceptions (Dragoni and Magnanensi, 1989; Bonafede, 1990).

Many three-dimensional boundary value problems in elasticity are far from trivial, so studies of magma chamber stresses, especially those that attempt to couple pure elasticity to heat transfer or fluid dynamics, tend to use very simplified chamber geometries and boundary conditions. It is of note, however, that a rich library of mathematical techniques for solving elasticity problems has been developed (Fung, 1965).

The simplest and most commonly used model for a magma chamber is a spherical inclusion in an infinite elastic medium. Displacements $\mathbf{d}$ and stresses σ for a pressurized and buoyant fluid-filled sphere of radius (R_d) are, in a spherical polar coordinate system (r, φ, ϑ) (Fung, 1965),

$$d_r = \frac{-\Delta P R_d^3(\upsilon+1)}{2Er^2}$$
$$+ \frac{2\Delta\rho g R_d^3}{E(13\upsilon-11)}\left(\frac{\upsilon^2-1}{r} - \frac{R_d^2(\upsilon+1)(4\upsilon-3)}{2r^3}\right)\cos\varphi,$$

$$d_\varphi = \left(\frac{\Delta\rho g R_d^3(r^2-R_d^2)(\upsilon+1)(4\upsilon-3)}{2E(13\upsilon-11)r^3}\right)\sin\varphi,$$

$$d_\vartheta = 0, \qquad (2.19)$$

and

$$\sigma_{rr} = \frac{\Delta P R_d^3}{r^3} + \frac{\Delta\rho g R_d^3(r^2(\upsilon-2)+3R_d^2(4\upsilon-3))\cos\varphi}{r^4(13\upsilon-11)},$$

$$\sigma_{\varphi\varphi}=\sigma_{\vartheta\vartheta}=\frac{-\Delta P R_d^3}{2r^3}$$
$$- \frac{\Delta\rho g R_d^3(r^2(\upsilon-2)+3R_d^2(4\upsilon-3))\cos\varphi}{2r^4(13\upsilon-11)},$$

$$\sigma_{r\varphi}=\frac{\Delta\rho g R_d^3(R_d^2-r^2)(4\upsilon-3)\sin\varphi}{2r^4(13\upsilon-11)},$$

$$\sigma_{r\vartheta}=\sigma_{\varphi\vartheta}=0, \qquad (2.20)$$

where ΔP is magma overpressure relative to lithostatic pressure, and $\Delta\rho g$ is the magma buoyancy. For isotropic overpressure alone, normal stresses decrease away from the chamber as $\Delta P/r^3$, while for buoyancy alone, stresses decrease as $\Delta\rho g/r^2$.

A spherical chamber in an infinite elastic medium approximates the qualitative aspects of static stresses, especially in the far field (Sartoris et al., 1990), but does not resolve higher-order moments if the surface boundary conditions become important. In particular, deviations from spherical geometry result in concentration of stresses in regions of high curvature around the chamber (Dieterich and Decker, 1975; Gudmundsson, 2006). Eshelby (1957) provides an analytic solution for stresses around a pressurized ellipsoidal inclusion in an infinite elastic medium that serves as a good example of

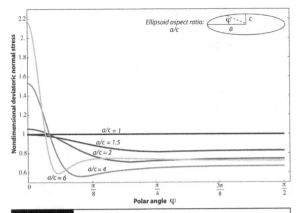

Figure 2.4 Stress concentration due to curvature of an oblate spheroid from the general solution of Eshelby (1957). Curves represent the magnitude of tensile deviatoric principal stress just outside the wall of a pressurized oblate spheroidal cavity in an infinite elastic medium, normalized by the deviatoric stress of a spherical chamber. Stresses plotted are compressive, and are oriented perpendicular to the wall of the chamber. Progressive flattening of the spheroid (increasing the aspect ratio a/c) results in the concentration of deviatoric stresses in areas of high curvature. In the limit that $a/c \rightarrow \infty$, the region near the medial plane of the ellipsoid develops a stress singularity.

such stress concentration. Figure 2.4 illustrates the stress concentration due to curvature by taking the particular case of an oblate spheroid ($a = b > c$), an appropriate approximate geometry for a sill (Fialko $et\ al.$, 2001), and gradually flattening it (decreasing c). It is evident that deviatoric stresses are progressively concentrated toward the medial plane. Indeed, the limit $c \rightarrow 0$ is a "penny-shaped" crack solution (Jaeger $et\ al.$, 2007), for which it can be shown that there is a $1/\sqrt{r}$ divergence of the normal stress at the edge of the ellipsoid. Other, more complicated chamber geometries have also been studied with numerical methods (Dieterich and Decker, 1975; Grosfils, 2007), but the qualitative features of solutions are the same.

Another case of interest to the evolution of the volcanic system occurs when the Earth's surface, a stress-free boundary, begins to affect concentration of stresses around magma chambers, such as during caldera formation (Gudmundsson, 1998). This occurs for shallow (< 20 km depth) chambers (McTigue, 1987). Although a general solution exists for a pressurized sphere in an elastic half-space (Tsuchida and Nakahara, 1970), it is sufficiently complicated that approximations are often employed. The most widely used model for chamber stresses in a half-space is a point-source model (Mogi, 1958), which corrects the (overpressured only) stresses from Eq. (2.20) on the free surface, by adding equal and opposite tractions. This gives rise to surface displacements (in a cylindrical coordinate system (r,θ,z) centered over a magma chamber at depth l below the surface)

$$d_z(r, z = 0) = \frac{\Delta P(1-\upsilon)R_d^3}{\mu} \frac{l}{(r^2+l^2)^{3/2}}$$
$$d_r(r, z = 0) = \frac{\Delta P(1-\upsilon)R_d^3}{\mu} \frac{r}{(r^2+l^2)^{3/2}}. \qquad (2.21)$$

This is equivalent to the first term in a power-series expansion of stress boundary conditions around a pressurized spherical cavity, matching boundary conditions on the sphere and the free surface in alternating, higher-order terms (McTigue, 1987). Two-dimensional solutions to this problem are algebraically more simple, and are given in the bipolar coordinate system by Jeffery (1921).

Other boundary conditions are also of interest in a careful treatment of magma chamber stresses. Depth-dependent density is neglected in most elastic treatments of chamber stresses, yet this can be important (Grosfils, 2007). Rheological change due to prolonged heating and partial melting of country rocks around a chamber is expected to produce a "shell" of material that behaves viscoelastically, and the presence of this material strongly affects chamber stresses (Bonafede $et\ al.$, 1986; Dragoni and Magnanensi, 1989; Jellinek and DePaolo, 2003; Newman $et\ al.$, 2006). In particular, the presence of a viscoelastic shell provides a mechanism by which high magma-chamber overpressures may be accommodated without large-scale chamber failure, and thus may explain how large, high melt-fraction chambers can grow in the first place (Jellinek and DePaolo, 2003). Thermoelastic stresses due to the heating or cooling of magma have also been addressed, and exert stresses comparable to other sources (e.g., Bonafede, 1990).

The fact that such varied boundary conditions affect the first-order mechanical evolution of magma chambers, both in theory and observation, is highly suggestive that chamber stresses are strongly coupled to the dynamic thermal and chemical evolution inside and around the chamber (Scandone et al., 2007). From a modeling perspective this requires coupling elastic, thermal, and fluid dynamic equations, and thus often requires numerical simulation (Gerya et al., 2004).

2.4 | Magma chamber convection

Convective motion in magma chambers helps determine the rate of magma differentiation, sets the length scale of compositional heterogeneities in these bodies, and redistributes melts with different physical properties, and thus may be responsible for the triggering and stalling of eruptions (Sparks et al., 1977; Snyder, 2000). Injection and storage of magma at upper or middle crustal depth can lead to a large thermal disequilibrium between the host rocks and the magma body. This thermal disequilibrium drives buoyant motion through crystallization, gas exsolution, or thermal expansion of the magma.

As magmatic systems cool below their liquidus temperatures, denser crystals generally form at the cooling boundaries and can be carried by gravitational instabilities from the roof into the bulk of the convecting body where they will remain in suspension or settle downwards (Sparks and Huppert, 1984; Martin and Nokes, 1989; Bergantz and Ni, 1999). The presence of this secondary phase can have an important effect on the convection style, even at low crystal fractions (< 10 vol.%; Martin and Nokes, 1989; Koyaguchi et al., 1990, 1993; Sparks et al., 1993). At higher crystal fractions (closer to the solidus), long-range (1/r) interactions between crystals will start to play a dominant role in the rheology of the mixture, as demonstrated by hindered settling and even the onset of yield strength beyond the percolation threshold, i.e., when crystals form a rigid connected framework (Vigneresse et al., 1996; Saar and Manga, 2002; Caricchi et al., 2007; Champallier et al., 2008).

Magmas also contain dissolved, and commonly exsolved, volatile phases, mostly H_2O, CO_2, and SO_2. The fractions and relative importance of these phases depend on the tectonic setting from which the magma originates. As a consequence of crystallization and cooling of the magma, saturation conditions change and the exsolved volatile volume fractions increase. This additional phase provides a new source of buoyancy and affects both the flow dynamics and the rheology of the mixture (Eichelberger, 1980; Huppert et al., 1982; Rust and Manga, 2002; Longo et al., 2006; Ruprecht et al., 2008).

When a magma with different physical properties is injected into the chamber, or when the residing magma undergoes differentiation by convective fractionation (Sparks and Huppert, 1984), this may induce large variations in transport coefficients (e.g., viscosity) and density. The amount of mixing or segregation between different magmas depends on the injection rate, the buoyancy of the injected magma, and the viscosity ratio (Turner and Campbell, 1986; Koyaguchi and Blake, 1991; Jellinek and Kerr, 1999). The presence of multiple sources of buoyancy, such as temperature and compositional differences between the magma batches, can lead to a range of fluid dynamical instabilities (Chen and Turner, 1980; Sparks and Huppert, 1984; Turner and Campbell, 1986).

2.4.1 Rayleigh–Bénard convection

Before addressing the complex dynamics of magma bodies or even their simpler laboratory analogs, we review Rayleigh–Bénard convection (Bénard, 1900; Rayleigh, 1916). Rayleigh–Bénard convection occurs in a fluid of constant thermal properties heated from the base and cooled from above. The set of equations describing the problem consists of mass, momentum, and energy conservation. We use the Boussinesq approximation, where the density of the fluid is assumed to vary linearly with temperature for the buoyancy force term and is assumed constant everywhere else. Assuming the flow is incompressible, mass conservation is given by

$$\frac{\partial \mathbf{u}_i}{\partial \mathbf{x}_i} = 0, \qquad (2.22)$$

where $\mathbf{u}$ is the fluid velocity. Momentum conservation for a Newtonian fluid is given by

$$\frac{\partial \mathbf{u}_i}{\partial t} = \mathbf{u}_j \left[\frac{\partial \mathbf{u}_i}{\partial \mathbf{x}_j} \right] = -\frac{1}{\rho_0} \frac{\partial P}{\partial \mathbf{x}_i} + v \left[\frac{\partial^2 \mathbf{u}_i}{\partial \mathbf{x}_j^2} \right] \\ -\frac{\Delta \rho}{\rho_0} g \delta_{ij}, \qquad (2.23)$$

where ρ_0 is the reference density for the fluid, v is the kinematic viscosity of the fluid ($v = \eta/\rho_0$), δ_{ij} is the Kronecker delta, and the last term represents thermal buoyancy, such that

$$\Delta \rho = \rho_0 \alpha (T - T_0). \qquad (2.24)$$

Here α is the coefficient of thermal expansion ($\sim 10^{-5}$ K^{-1} in most geological materials), and T_0 is the reference temperature corresponding to density ρ_0. Finally, when neglecting viscous dissipation effects, the energy conservation equation becomes (modified from Eq. (2.4) for the absence of phase transitions)

$$\frac{\partial T}{\partial t} + \mathbf{u}_j \frac{\partial T}{\partial \mathbf{x}_j} = \kappa \frac{\partial^2 T}{\partial \mathbf{x}_j^2} \qquad (2.25)$$

where κ is the thermal diffusivity. Introducing the length scale H as the distance between the lower hot boundary and the upper cold boundary, the diffusion timescale H^2/κ, and the temperature scale $\Delta T = T_{hot} - T_{cold}$ as the temperature difference between the two boundaries, Eqs. (2.22), (2.23), and (2.25) may be nondimensionalized to

$$\frac{\partial \mathbf{u}_j^*}{\partial \mathbf{x}_j^*} = 0, \qquad (2.26)$$

$$\frac{\partial \mathbf{u}_i^*}{\partial t^*} + \mathbf{u}_j^* \frac{\partial \mathbf{u}_i^*}{\partial \mathbf{x}_j^*} = \frac{1}{\rho_0^*} \frac{\partial P^*}{\partial \mathbf{x}_i^*} + \Pr \frac{\partial^2 \mathbf{u}_i^*}{\partial \mathbf{x}_j^{*2}} \\ - \Pr \operatorname{Ra} \delta_{ij}, \qquad (2.27)$$

$$\frac{\partial T^*}{\partial t^*} + \mathbf{u}_j^* \frac{\partial T^*}{\partial \mathbf{x}_j^*} = \frac{\partial^2 T^*}{\partial \mathbf{x}_j^{*2}}, \qquad (2.28)$$

where again the superscript (*) represents a dimensionless variable. This convection problem has two independent dimensionless numbers: the Prandtl number $\Pr = v/\kappa$, which is the ratio of momentum diffusivity to thermal diffusivity, and the Rayleigh number $\operatorname{Ra} = \alpha g \Delta T H^3/v\kappa$, which is the ratio of the diffusion timescale to the advection timescale for heat transfer. Above a critical value of ~1000, thermal boundary layers become unstable and the fluid starts to convect.

Any two systems sharing the same geometry and dimensionless numbers Ra and Pr will be dynamically similar. This is commonly known as the principle of similitude and is the basis for justifying the study of scaled laboratory experiments (Burgisser et al., 2005). It is important to note, however, that in the context of Rayleigh–Bénard convection, only two independent dimensionless numbers are required to fully describe the single-phase dynamics. As more complexity is introduced to the model, the number of dimensionless parameters required to ensure dynamic similitude between laboratory experiments and natural systems increases substantially.

2.4.2 Multiphase convection

The largest source of internal buoyancy in magma chambers comes from discrete phases such as crystals and bubbles. These phases may move with different velocities than the surrounding magma, and exert drag forces on the magmatic fluid. Although many magmatic problems can be approximated as a single fluid phase with modified properties due to crystallinity, others require a more detailed treatment. For instance, the gathering and dispersal of discrete phases has important implications for the chemical diversity that is preserved in bulk composition and on the micro-scale in phenocryst zoning (Hibbard, 1981; Speer et al., 1989; Castro and de la Rosa, 1994; Wallace and Bergantz, 2005).

Crystal-driven convection has been studied experimentally mainly at low crystal fractions (below a few volume percent) where the complexity introduced by crystal–crystal interactions can typically be neglected. Most crystalline phases follow convective motion of the magma closely, except in very low viscosity magma (Martin and Nokes, 1989; Koyaguchi et al., 1993; Sparks et al., 1993). Some particles can, however, still be removed from the flow when magma stagnates and crystals reside for an extended period of time. Martin and Nokes (1989) described this process with an exponential relationship assuming that, in the bulk of the chamber the crystals are well mixed, and only in the bottom boundary layer do crystals decouple from the flow. This behavior also explains the results of Koyaguchi et al. (1993), who described periodicity between three regimes: (1) clarified chamber where crystals are deposited at the bottom, (2) well-defined layers (crystal-poor overlying crystal-rich), and (3) wholesale overturn that mixes both layers.

At shallow depths, water can exsolve from the melt and form a dispersed bubble phase. The exsolution of volatiles strongly affects the fluid dynamics of magma chambers, generally resulting in large buoyancy contrasts within the melt (Huppert et al., 1982; Phillips and Woods, 2002). The role of exsolved volatiles has been investigated for triggering eruptions (Snyder, 2000; Huppert and Woods, 2002; Fowler and Spera, 2008) and for increasing the buoyancy of an unsegregated melt–bubble mixture (Bergantz and Breidenthal, 2001; Ruprecht et al., 2008).

A particular example of some of these effects is magma chamber recharge, where a mafic magma is injected at the base of a chamber filled with a more silicic magma (Eichelberger, 1980; Huppert et al., 1982). As the mafic intrusion cools, it partially crystallizes and exsolves bubbles. If enough bubbles exsolve and remain in suspension, this situation can lead to an unstable density stratification where the underplating magma becomes lighter than the host, causing overturn of the layer (Huppert et al., 1982; Bergantz and Breidenthal, 2001; Ruprecht et al., 2008). Conversely, if bubbles can separate from an underplating mafic melt, they will pond at density interfaces to form an unstable foam layer from which volatile-rich plumes originate and ascend through the less-dense magma. This

alternative process leads to (limited) mixing (Eichelberger, 1980; Bergantz and Breidenthal, 2001).

One approach in analyzing the motion of dispersed phases, such as crystals and bubbles in magma, is to examine the path of individual crystals as well as the velocity field of the magma. This approach, termed Eulerian–Lagrangian (here Eulerian refers to the continuous quantity, magma, and Lagrangian refers to the discrete quantity, crystals) is usually favored for dilute conditions when the coupling between the magma and crystal is essentially one-way (i.e., the magma imparts inertia and relative velocity to the crystals, but is not affected by drag due to the crystals).

Lagrangian analysis is especially useful in the analysis of the trajectory of individual particles and can be used to better understand the fluid volumes through which these particles have passed. However, the large numbers of individual particles in magmatic systems makes this approach computationally unattractive. An alternative that still preserves the ability to compute phase relative motion is to treat the discrete phase as an effective continuum, or Eulerian phase. In this approach, bubbles or crystals of the same diameter and density are averaged into a fluid and are represented locally by a volume fraction. In the averaging process, details of particle trajectories and histories are lost, but two-way interaction between the particles and magma can be computed for realistic volume fractions of discrete particles.

This approach is often referred to as Eulerian–Eulerian, multi-fluid, or multi-continuum. Each continuum (magma, crystals, bubbles) is represented by a conservation equation for mass, momentum, and thermal energy, along with constitutive relations that describe the density, rheology, and thermal properties of these continua. The set of equations is coupled through equal and opposite drag terms and terms for the transfer of thermal energy.

We illustrate this approach with an example of a crystal-driven instability in a magma composed of a melt and a denser crystal phase (Fig. 2.5). We assume that this basaltic chamber was intruded at its liquidus with a size of 400 ×

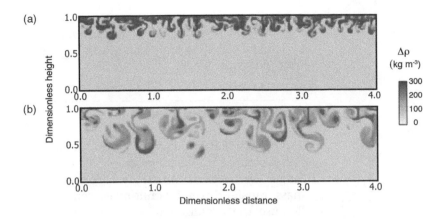

(a)

Dimensionless height

(b)

Dimensionless distance

$\Delta\rho$
(kg m⁻³)

300
200
100
0

Figure 2.5 Crystal-driven density instability. Crystals from the top margin drive fluid instabilities and mix the melt during their descent as plumes. The units shown are normalized by the magma chamber height, H. Panels (a) and (b) show the evolution of the density difference as the drips descend with time.

100 m, and has formed a 10 m thick layer of crystals (10% crystals by volume) at the top of the chamber due to cooling. The crystal-rich layer is denser than the magma ($\Delta\rho = 300$ kg m⁻³) and creates a density instability with dripping crystal plumes (Bergantz, 1999). Further information about the details of this approach, constitutive relations, and numerics can be found in Dufek and Bergantz (2005b, 2007) and Ruprecht *et al.* (2008).

We see from this example that density instabilities drive stirring in the entire chamber. This example also illustrates the value of the multi-fluid approach in predicting the spatial and temporal patterns of large-scale heterogeneity.

2.4.3 Convection and mixing

Mixing is a result of the stretching and twisting of flow paths by shear and normal strains integrated over time, and is important in determining compositional heterogeneity (or lack thereof) in magma chambers. For simplicity we focus on mixing of a single-phase magma. However, the concepts introduced here can be applied to more complex systems. Ottino *et al.* (1979) formally described how fluid motion at low Reynolds number leads to chaotic mixing. For large-wavelength heterogeneities, the stretching induced by shear strain ($\dot{\varepsilon}$) dominates, and leads to a deformation proportional to $\dot{\varepsilon}^{-1}$ (Olson *et al.*, 1984). Once the heterogeneities have been reduced to smaller sizes, normal strain accounts for the majority of the deformation, which becomes proportional to exp($-2\dot{\varepsilon}$).

Following Coltice and Schmaltzl (2006), for steady-state convection, we relate the stirring time to the strain rate

$$\tau_m = \frac{1}{2\dot{\varepsilon}}\log\left(\frac{\dot{\varepsilon}H^2}{D}\right), \tag{2.29}$$

where H^2/D is the diffusion timescale of interest in the mixing process. The dependence of the average strain rate on the Rayleigh number follows (Coltice and Schmalzl, 2006):

$$\dot{\varepsilon} = 0.023\frac{\kappa}{H^2}\text{Ra}^{0.685}. \tag{2.30}$$

Convection in magma chambers is, however, strongly time-dependent, and as cooling proceeds the mechanical properties of the mixture vary strongly because of temperature changes and the presence of crystal phases. One measure of cumulative mixing in unsteady convection (Ra = Ra(t)) is the number of overturns experienced by the fluid (Huber *et al.*, 2009). The product of the stirring time (Eq. (2.29)) and the strain-rate (Eq. 2.30) gives an estimate of the total strain experienced by the fluid to reach a point where molecular diffusion takes over to further homogenize the system. This quantity is largely invariant with respect to Ra (Figure 2.6(b)), and therefore independent of the dynamical history of the fluid (Huber *et al.*, 2009). This offers a convenient alternative to quantify the mixing efficiency of a convective fluid subjected to temporal changes. Newly created heterogeneities require about five overturns to be efficiently mixed.

(a)

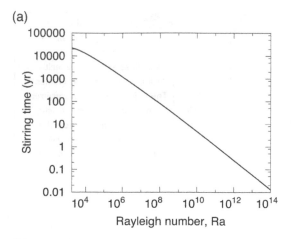

(b)

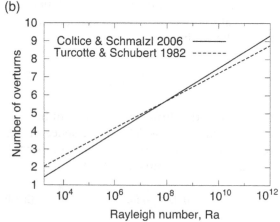

Figure 2.6 (a) Stirring time from Eq. (2.29). (b) Comparison of the total number of overturns required to achieve a well-stirred chamber.

For convection to occur in a magmatic system, the volume fraction of crystals in the magma must be below the rigid percolation threshold, i.e., where the crystallinity does not prevent convective motions of the magma (Vigneresse et al., 1996). There is extensive evidence that magmatic systems spend most of their suprasolidus time at high crystallinity (> 50%) (Marsh, 1981; Koyaguchi and Kaneko, 1999), and that large silicic systems such as Long Valley caldera and the Fish Canyon magma body can stay above their solidi for extended periods of time (> 10^5 years) owing to injections of new magma (Christensen and DePaolo, 1993; Reid et al., 1997; Bachmann et al., 2007b). Magmatic systems in which the bulk of the crystallization

occurs in a narrow temperature range close to the solidus experience strong thermal buffering by latent heat, thus prolonging their residence time (Bergantz, 1990).

2.5 | Future directions

The growth and evolution of magma chambers are dependent on a range of processes from long-range tectonic stresses to the heat and mass transfer driven by multiphase convection. Because the evolution of a magmatic system takes place over timescales that range over ten orders of magnitude, from minutes (volcanic eruptions) to hundreds of thousands of years (large chamber residence times), the multi-scale nature of magma transport cannot be over-emphasized. Much of the work in studying the dynamics of magma chambers has shown that almost all of these processes are coupled to some degree, even if traditionally they have been examined by different subdisciplines. One of the likely future directions in the study of magma chamber dynamics will be in improving our understanding of these different degrees of coupling. For instance, as magma chambers alter the surrounding stress state of the crust, they can potentially influence the flux of magma (and enthalpy) to the chamber thus prolonging its lifetime and providing new sources for compositional heterogeneity.

Further work on the relative motion of discrete phases, and in particular the transport of volatiles, is needed to better understand the role of volatiles. In addition to their role in dynamics as a source of buoyancy, volatiles and the relative concentration of H_2O versus CO_2 also affect the phase assemblage and stability of magmas. Coupling between exsolution of volatiles and crystallization can have a significant impact on the rheology of magmas and the separation of the different phases. Crystallization studies, as well as rheological experiments on mixtures containing both crystals and bubbles, will be important to constrain these effects in dynamic calculations.

Many fundamental aspects of magma transport physics are still poorly constrained. For example, the processes of magma chamber formation are still not well defined in space or time, and have not yet been actively observed. Similarly, the mechanisms by which a magma chamber is tapped are largely unknown. Dike propagation might dominate transport away from a magma chamber, but "leaking" and anelastic processes, such as sub-critical crack propagation (Chen and Jin, 2006) or channel flow through matrix compaction (Spiegelman and Kenyon, 1992) may be important. The formation of dikes from magma chambers, and the subsequent mechanical coupling between chambers and volcanic eruptions is still a largely outstanding problem.

The vastly different scales involved in magmatic problems also challenges our ability to describe these coupled systems numerically, even with foreseeable advances in computing power. Coupling between different types of modeling and development of subgrid models, or models of small-scale phenomena, will likely be necessary to span much of the behavior exhibited by natural systems. With the ongoing improvement of algorithm design, analytical and numerical methods, one of the most significant improvements in recent years has been to generate models that can be compared and tested against the diversity of observations provided by other techniques. One example of this is examining the path of crystals using a Lagrangian framework and then comparing predicted populations of crystals with those observed using microanalytical geochemical techniques. Another example is the application of multiphysics models, combined with high-resolution isotopic work, structural geology and geochronology of well-exposed field sites. Models of chamber dynamics developed with these applications in mind will be of more use to the broader earth science community, as the dynamics inferred from modeling can be ground-truthed in the field. Coupling between thermal, rheological, and stress modeling will also be helpful in interpreting both seismic and geodetic observations. Further advances in the study of magma chambers will likely continue to follow this course and take advantage of the rich and multidisciplinary observations on a range of scales.

2.6 | Summary

- Magma chambers play an important role in determining the composition of erupted magmas through a combination of fractionation and assimilation.
- In most crustal settings, crustal melting at magma chamber margins is an inefficient process.
- Magma chambers modify the local crustal stress environment influencing the eruptability of magmas, the influx of magmas from below, and ground deformation.
- Buoyancy from discrete phases (bubble and crystals) can drive multiphase convection in magma chambers and can result in large-scale homogeneity as a result of mixing processes.

2.7 | Notation

*	dimensionless value
a, b, c	dimensions of oblate spheroid (m); aspect ratio $= a/c$
c_p	specific heat capacity (J kg^{-1} K^{-1})
c_{pB}	specific heat of basalt (J kg^{-1} K^{-1})
c_{pC}	specific heat of crustal rocks (J kg^{-1} K^{-1})
$\mathbf{d}$	displacement vector (m)
D	generic diffusivity (m^2 s^{-1})
E	Young's modulus (Pa)
$E_\%$	efficiency of melting of crustal rocks (%)
f	melt volume fraction
$\mathbf{f_b}$	body force vector (kg m^{-2} s^{-2})
g	acceleration due to gravity (m s^{-2})
h	heat transfer coefficient (W K^{-1})
H	length scale (thickness) of convective layer (m)
H_T	enthalpy (J)
k	thermal conductivity (W m^{-1} K^{-1})
K	bulk modulus (= 1/β; Pa)
l	depth of magma chamber (m)
L	latent heat of fusion (J kg^{-1})
L_B	latent heat of basalt (J kg^{-1})
L_C	latent heat of crustal rocks (J kg^{-1})
m_c	mass of crystals (kg)
P	magma pressure (Pa)
ΔP	magma chamber overpressure (Pa)

Q_I magma mass flux into chamber (kg s^{-1})

Q_O magma mass flux out of chamber (kg s^{-1})

r, φ, ϑ spherical polar coordinates

R^* dimensionless rate of melt production

R_d magma chamber radius (m)

t time (s)

T temperature (K)

ΔT change in temperature (K)

T_0 reference temperature (K)

T_b temperature at magma chamber boundary (K)

T_{liq} liquidus temperature (K)

T_{liqB} liquidus temperature of basalt (K)

T_R background temperature of country rock (K)

T_{sol} solidus temperature (K)

T_{solC} solidus temperature of crustal rock (K)

u_0 reference velocity (m s^{-1})

V volume (m^3)

ΔV volume change (m^3)

V_0 total volume of the magma chamber (melt + crystals + bubbles; m^3)

V_g volume of gas phase (m^3)

V_l volume of liquid (melt) phase (m^3)

V_{Ceff} volume of crust melted at 100% efficient enthalpy transfer (m^3)

V_{Cmod} volume of crustal melt predicted in thermal model (m^3)

u   velocity field (m s^{-1})

x, x, y, z Cartesian coordinates (m)

α thermal expansion coefficient (K^{-1})

β compressibility (Pa)

δ thermal boundary layer thickness (m)

δ_{ij} Kronecker-delta (= 1 when $i = j$ and zero when $i \neq j$)

$\dot{\varepsilon}$ strain rate (s^{-1})

η dynamic viscosity (Pa s)

κ thermal diffusivity (m^2 s^{-1})

κ_{eff} effective thermal diffusivity (m^2 s^{-1})

μ modulus of rigidity of wall rock (Pa)

ν kinematic viscosity (= η/ρ; m^2 s^{-1})

ρ magma density (kg m^{-3})

$\Delta\rho$ density difference between magma and country rock (kg m^{-3})

ρ_0 reference density (kg m^{-3})

ρ_c density of crystal phase (kg m^{-3})

ρ_g density of gas phase (kg m^{-3})

ρ_l density of liquid (melt) phase (kg m^{-3})

ρ_B basalt density (kg m^{-3})

ρ_C crust density (kg m^{-3})

σ_{ij} stress tensor (Pa)

σ_d deviatoric stress (Pa)

σ_{ext} external (far-field) stress (Pa)

σ_{int} internal stress (Pa)

σ_{lith} lithostatic stress (Pa)

σ_{tot} total stress in crust (Pa)

τ_m stirring time (s)

υ Poisson's ratio

ϕ_v volume fraction of volatiles

Bi   Biot number, ratio of thermal resistance of magma chamber versus host rocks (= $h\delta/k$)

Pe   Peclet number, ratio of advective to diffusive heat transport (= $c_p\rho_0\delta\, u_0/k$)

Pr   Prandtl number, ratio of momentum diffusivity to thermal diffusivity (= ν/κ)

Ra   Rayleigh number, ratio of diffusion timescale to advection timescale (= $\alpha g\Delta T H^3/\nu\kappa$)

Ste   Stefan number, ratio of sensible to latent heat (= $c_p T_0/L$)

Acknowledgments

This work was supported in part by NASA 08-MFRP08–0073, NSF 0948532, and a Swiss Postdoctoral Fund Fellowship.

References

Ajakaiye, D. E. (1970). Gravity measurements over the Nigerian Younger granite province. *Nature*, **225**, 50–52.

Anderson, E. M. (1936). Dynamics of the formation of cone-sheets, ring-dykes, and cauldron-subsidences. *Proceedings of the Royal Society of Edinburgh*, **56**, 128–157.

Anderson, A. T. (1976). Magma mixing – Petrological process and volcanological tool. *Journal of Volcanology and Geothermal Research*, **1**(1), 3–33.

Annen, C. and Sparks, R. S. J. (2002). Effects of repetitive emplacement of basaltic intrusions on the thermal evolution and melt generation in the crust. *Earth and Planetary Science Letters*, **203**(3–4), 937–955.

Bachmann, O., Miller, C. F. and de Silva, S. L. (2007a). The volcanic-plutonic connection as a stage for understanding crustal magmatism. *Journal of Volcanology and Geothermal Research*, **167**, 1–23.

Bachmann, O., Oberli, F., Dungan, M. A. *et al.* (2007b). ^{40}Ar/^{39}Ar and U-Pb dating of the Fish Canyon magmatic system, San Juan Volcanic Field, Colorado: Evidence for an extended crystallization history. *Chemical Geology*, **236**, 134–166.

Barboza, S. A. and Berganz, G. W. (1996). Dynamic model of dehydration melting motivated by a natural analogue: applications to the Ivrea-Verbano zone, northern Italy. *Transactions of the Royal Society of Edinburgh*, **87**, 23–31.

Beard, J. S., Ragland, P. C. and Crawford, M. L. (2005). Reactive bulk assimilation: A model for crust-mantle mixing in silicic magmas. *Geology*, **33**(8), 681–684.

Bénard, H. (1900). Les tourbillons cellulaires dans une nappe liquide. *Revue générale des sciences pures et appliquées*, **11**, 1261–1271.

Berganz, G. W. (1989). Underplating and partial melting: Implications for melt generation and extraction. *Science*, **245**, 1093–1095.

Berganz, G. W. (1990). Melt fraction diagrams: the link between chemical and transport models. In *Modern Methods of Igneous Petrology: Understanding Magmatic Processes*, ed. J. Nicholls and J. K. Russell. *Reviews in Mineralogy*, **24**, Mineralogical Society of America, pp. 239–257.

Berganz, G. W. (1995). Changing paradigms and techniques for the evaluation of magmatic processes. *Journal of Geophysical Research*, **100**(B9), 17 603–17 613.

Berganz, G. W. and Breidenthal, R. E. (2001). Non-stationary entrainment and tunneling eruptions: A dynamic link between eruption processes and magma mixing. *Geophysical Research Letters*, **28**(16), 3075–3078.

Berganz, G. W. and Ni, J. (1999). A numerical study of sedimentation by dripping instabilities in viscous fluids. *International Journal of Multiphase Flow*, **25**, 307–320.

Berlo, K., Turner, S., Blundy, J., Black, S. and Hawkesworth, C. (2006). Tracing pre-eruptive magma degassing using (^{210}Pb/^{226}Ra) disequilibria in the volcanic deposits of the 1980–1986 eruption of Mount St. Helens. *Earth and Planetary Science Letters*, **249**(3–4), 337–349.

Bittner, D. and Schmeling, H. (1995). Numerical modelling of melting processes and induced diapirisms in the lower crust. *Geophysical Journal International*, **123**(1), 59–70.

Blundy, J. and Wood, B. (2003). Mineral-melt partitioning of uranium, thorium and their daughters. In *Uranium-Series Geochemistry*, ed. B. Bourdon, G. M. Henderson, C. C. Lundstrom and P. S. Turner. Washington, DC: Minerological Society of America and Geochemical Society, pp. 59–123.

Bonafede, M. (1990). Axi-symmetric deformation of a thermo-poro-elastic half-space: inflation of a magma chamber. *Geophysical Journal International*, **183**, 289–299.

Bonafede, M., Dragoni, M. and Quareni, F. (1986). Displacement and stress produced by a center of dilation and by a pressure source in a viscoelastic half-space: application to the study of ground deformation and seismicity at Campi Flegeri, Italy. *Geophysical Journal of the Royal Astronomical Society*, **87**(2), 455–485.

Burgisser, A., Berganz, G. W. and Breidenthal, R. E. (2005). Addressing complexity in laboratory experiments: the scaling of dilute multiphase flows in magmatic systems. *Journal of Volcanology and Geothermal Research*, **141**(3–4), 245–265.

Caricchi, L., Burlini, L., Ulmer, P. *et al.* (2007). Non-Newtonian rheology of crystal-bearing magmas and implications for magma ascent dynamics. *Earth and Planetary Science Letters*, **264**, 402–419.

Carrigan, C. R. (1988). Biot number and thermos bottle effect: Implications for magma-chamber convection. *Geology*, **16**, 771–774.

Castro, A. and de la Rosa, J. (1994). Nomarski study of zoned plagioclases from granitoids of the Seville Range batholith, SW Spain, Petrogenetic implications. *European Journal of Mineralogy*, **6**, 647–656.

Champallier, R., Bystricky, M. and Arbaret, L. (2008). Experimental investigation of magma rheology at 300 MPa: From pure hydrous melt to 76 vol % of crystals. *Earth and Planetary Science Letters*, **267**, 571–583.

Chapman, D. S. (1986). Thermal gradients in the continental crust. *Geological Society Special Publication*, **24**, 63–70.

Chapman, D. S. and Furlong, K. P. (1992). Thermal state of the continental lower crust. In *Continental Lower Crust*, ed. D. M. Fountain, R. Arculus, and R. W. Kay. Amsterdam: Elsevier, pp. 179–198.

Chen, C. F. and Turner, J. S. (1980). Crystallization in double-diffusive systems. *Journal of Geophysical Research*, **85**, 2573–2593.

Christensen, J. N. and DePaolo, D. J. (1993). Time scales of large volume silicic magma systems: Sr isotopic systematics of phenocrysts and glass from the Bishop Tuff, Long Valley, California. *Contributions to Mineralogy and Petrology*, **113**, 100–114.

Coltice, N. and Schmalzl, J. (2006). Mixing times in the mantle of the early Earth derived from 2-D and 3-D numerical simulations of convection. *Geophysical Research Letters*, **33**(L23304), doi:10.1029/2006GL027707.

Cooper, K. M. and Reid, M. R. (2003). Re-examination of crystal ages in recent Mount St. Helens lavas: implications for magma reservoir processes. *Earth and Planetary Science Letters*, **213**, 149–167.

Costa, F., Chakraborty, S. and Dohmen, R. (2003). Diffusion coupling between trace and major elements and a model for calculation of magma residence times using plagioclase. *Geochimica et Cosmochimica Acta*, **67**(12), 2189–2200.

Daly, R. A. (1914). *Igneous Rocks and Their Origin*. McGraw-Hill.

DePaolo, D. J., Perry, F. V. and Baldridge, W. S. (1992). Crustal versus mantle sources of granitic magmas: a two-parameter model based on Nd isotopic studies. *Transactions of the Royal Society of Edinburgh*, **83**, 439–446.

Dieterich, J. H. and Decker, R. W. (1975). Finite-element modeling of surface deformation associated with volcanism. *Journal of Geophysical Research*, **80**(29), 4094–4102.

Dobran, F. (2001). *Volcanic Processes, Mechanisms in Material Transport*. Kluwer Academic, 590pp.

Dragoni, M. and Magnanensi, C. (1989). Displacement and stress produced by a pressurized, spherical magma chamber, surrounded by a viscoelasitic shell. *Physics of the Earth and Planetary Interiors*, **56**, 316–328.

Druitt, T. H. and Francavigilia, V. (1992). Caldera formation on Santorini and physiography of the islands in the late Bronze Age. *Bulletin of Volcanology*, **54**, 484–493.

Dufek, J. D. and Bergantz, G. W. (2005a). Lower crustal magma genesis and preservation: A stochastic framework for the evaluation of basalt-crust interaction. *Journal of Petrology*, **46**, 2167–2195.

Dufek, J. D. and Bergantz, G. W. (2005b). Transient two-dimensional dynamics in the upper conduit of a rhyolitic eruption: A comparison of the closure models for the granular stress. *Journal of Volcanology and Geothermal Research*, **143**, 113–132.

Dufek, J. and Bergantz, G. W. (2007). The suspended-load and bed-load transport of particle laden gravity currents: Insight from pyroclastic flows that traverse water. *Journal of Theoretical and Computational Fluid Dynamics*, **21**(2), 119–145.

Dumond, G., Hahan, K., Williams, M. W. and Karlstrom, K. E. (2007). Metamorphism in middle continental crust, Upper Granite Gorge, Grand Canyon, Arizona: implications for segmented crustal architecture, processes at 25-km-deep levels, and unroofing of orogens. *Geological Society of America Bulletin*, **119**, 202–220.

Eichelberger, J. C. (1980). Vesiculation of mafic magma during replenishment of silicic magma reservoirs. *Nature*, **288**, 446–450.

Eshelby, J. D. (1957). The determination of the elastic field of an ellipsoidal inclusion, and related problems. *Proceedings of the Royal Society of London*, **A241**(1226), 376–396.

Feeley, T. C., Cosca, M. A. and Lindsay, C. R. (2002). Petrogenesis and implications of cryptic hybrid magmas from Washburn Volcano, Absaroka Volcanic Province, U.S.A. *Journal of Petrology*, **43**, 663–703.

Fialko, Y. A., Khazan, Y. and Simons, M. (2001). Deformation due to a pressurized horizontal circular crack in an elastic half-space, with applications to volcano geodesy. *Geophysical Journal International*, **146**, 181–190.

Folch, A. and Marti, J. (1998). The generation of overpressure in felsic magma chambers by replenishment. *Earth and Planetary Science Letters*, **163**, 301–314.

Fowler, S. J. and Spera, F. (2008). Phase equilibrium trigger for explosive volcanic eruptions. *Geophysical Research Letters*, **35**, doi:10.1029/2008GL03365.

Fung, Y. C. (1965). *Foundations of Solid Mechanics*. Englewood Cliffs, NJ: Prentice-Hall.

Gaffney, E. S., Damjanac, B. and Valentine, G. A. (2007). Localization of volcanic activity: 2. Effects of pre-existing structure. *Earth and Planetary Science Letters*, **263**, 323–338.

Gerya, T. V., Yuen, D. A. and Sevre, O. D. (2004). Dynamical causes for incipient magma chambers above slabs. *Geology*, **32**(1), 89–92.

Ghiorso, M. S. and Sack, R. O. (1995). Chemical mass transfer in magmatic processes IV. A revised and internally consistent thermodynamic model for the interpolation and extrapolation of liquid-solid equilibria in magmatic systems at elevated temperatures and pressures. *Contributions to Mineralogy and Petrology*, **119**, 197–212.

Green, D. H. (1973). Experimental melting studies on a model upper mantle composition at high pressure under water-saturated and water-undersaturated conditions. *Earth and Planetary Science Letters*, **19**(1), 37–53.

Grosfils, E. B. (2007). Magma reservoir failure on the terrestial planets: Assessing the importance of gravitational loading in simple elastic models. *Journal of Volcanology and Geothermal Research*, **166**, 47–75.

Grout, F. F. (1918). Two phase convection in igneous magmas. *Journal of Geology*, **26**, 481–499.

Grout, F. F. (1926). The use of calculations in petrology – A study for students. *Journal of Geology*, **34**(6), 512–558.

Grunder, A. L., Klemetti, E. W., Feeley, T. C. and McKee, C. M. (2006). Eleven million years of arc volcanism at the Aucanquilcha Volcanic Cluster, Northern Chilean Andes: implications for the life span and emplacement of plutons. *Transactions of the Royal Society of Edinburgh*, **97**, 415–436.

Gudmundsson, A. (1998). Formation and development of normal-fault calderas and the initiation of large explosive eruptions. *Bulletin of Volcanology*, **60**, 160–170.

Gudmundsson, A. (2006). How local stresses control magma-chamber ruptures, dyke injections, and eruptions in composite volcanoes. *Bulletin of Volcanology*, **79**, 1–31.

Gudmundsson, M. T. and Hognadottir, T. (2007). Volcanic systems and calderas in the Vatnajokull region, central Iceland: Constraints on crustal structure from gravity data. *Journal of Geodynamics*, **43**(1), 153–169.

Hart, G. L., Johnson, C. M., Shirey, S. B. and Clynne, M. A. (2002). Osmium isotope constraints on lower crustal recycling and pluton preservation at Lassen Volcanic Center, CA. *Earth and Planetary Science Letters*, **199**, 269–285.

Hibbard, M. J. (1981). The magma mixing origin of mantled feldspars. *Contributions to Mineralogy and Petrology*, **76**, 158–170.

Hieronymus, C. F. and Bercovici, D. (1999). Discrete alternating hotspot islands formed by interaction of magma transport and lithospheric flexure. *Nature*, **397**(18), 604–607.

Hildreth, W. and Moorbath, S. (1988). Crustal contributions to arc magmatism in the Andes of Central Chile. *Contributions to Mineralogy and Petrology*, **98**, 455–489.

Hopson, C. A. and Mattinson, J. M. (1994). Chelan Migmatite Complex, Washington: Field evidence for mafic magmatism, crustal anatexis, mixing and protodiapiric emplacement. In *Geologic Field Trips in the Pacific Northwest: 1994 Geological Society of America Annual Meeting*, ed. D. A. Swanson and R. A. Haugerud. Seattle: Department of Geological Sciences, University of Washington, pp. 1–21.

Huber, C., Bachmann, O. and Manga, M. (2009). Homogenization processes in silicic magma chambers by stirring and latent heat buffering. *Earth and Planetary Science Letters*, **283**, 38–47.

Huppert, H. E. and Sparks, R. S. J. (1988). The generation of granitic magmas by intrusion of basalt into continental crust. *Journal of Petrology*, **29**(3), 599–624.

Huppert, H. E. and Woods, A. W. (2002). The role of volatiles in magma chamber dynamics. *Nature*, **420**, 493–495.

Huppert, H. E., Sparks, R. S. J. and Turner, J. S. (1982). Effects of volatiles on mixing in calc-alkaline magma systems. *Nature*, **297**, 554–557.

Hutton, J. (1788). Theory of the Earth; or an investigation of the laws observable in the composition, dissolution and restoration of the land upon the globe. *Transactions of the Royal Society of Edinburgh*, **1**, 209–304.

Ilg, B., Karlstrom, K. E., Hawkins, D. and Williams, M. L. (1996). Tectonic evolution of paleoproterozoic rocks in Grand Canyon: Insights into middle crustal processes. *Geological Society of America Bulletin*, **108**, 1149–1166.

Jaeger, J., Cook, N. G. and Zimmerman, R. (2007). *Fundamentals of Rock Mechanics*, 4th edn., Wiley-Blackwell.

Jeffery, G. B. (1921). Plane stress and plane strain in bipolar co-ordinates. *Philosophical Transactions of the Royal Society of London A*, **221**, 265–293.

Jellinek, A. M. and DePaolo, D. J. (2003). A model for the origin of large silicic magma chambers: precursors of caldera-forming eruptions. *Bulletin of Volcanology*, **65**, 363–381.

Jellinek, A. M. and Kerr, R. C. (1999). Mixing and compositional stratification produced by natural convection 2. Applications to the differentiation of basaltic and silicic magma chambers and komatiite lava flows. *Journal of Geophysical Research*, **104**(B4), 7203–7218.

Jellinek, A. M., Manga, M. and Saar, M. O. (2004). Did melting glaciers cause volcanic eruptions in eastern, California? *Journal of Geophysical Research*, **109**(B09206), doi: 10.1029/2009JB002978.

Jessup, M. J., Jones, J. V., Karlstrom, K. E. *et al.* (2006). Three Proterozoic orogenic episodes and an intervening exhumation event in the Black Canyon of the Gunnison region, Colorado. *Journal of Geology*, **114**, 555–576.

Karlstrom, K. E. and Williams, M. L. (2006). Nature of the middle crust – Heterogeneity of structure and process due to pluton-enhanced tectonism: an example from Proterozoic rocks of the North American Southwest. In *Evolution and Differentiation of the Continental Crust*, ed. M. Brown and T. Rushmer. Cambridge University Press, pp. 268–295.

Karlstrom, L., Dufek, J. and Manga, M. (2009). Organization of volcanic plumbing through magmatic lensing by magma chambers and volcanic loads. *Journal of Geophysical Research*, **114**, B10204, doi:10.1029/2009JB006339.

Kavanagh, J. L., Menand, T. and Sparks, R. S. J. (2006). An experimental investigation of sill formation and propagation in layered elastic media. *Earth and Planetary Science Letters*, **245**, 799–813.

Koyaguchi, T. and Blake, S. (1991). Origin of mafic enclaves: Constraints on the magma mixing model from fluid dynamic experiments. In *Enclaves and Granite Petrology*. Amsterdam: Elsevier.

Koyaguchi, T. and Kaneko, K. (1999). A two-stage thermal evolution model of magmas in continental crust. *Journal of Petrology*, **40**, 241–254.

Koyaguchi, T., Hallworth, M. A., Huppert, H. H. and Sparks, R. S. J. (1990). Sedimentation of particles from a convecting fluid. *Nature*, **343**, 447–450.

Koyaguchi, T., Hallworth, M. A. and Huppert, H. E. (1993). An experimental study on the effects of phenocrysts on convection in magmas. *Journal of Volcanology and Geothermal Research*, **55**, 15–32.

Lange, R. L. and Carmichael, I. S. E. (1990). Thermodynamic properties of silicate liquids with emphasis on density, thermal expansion and compressibility. In *Modern Methods of Igneous Petrology: Understanding Magmatic Processes*, ed. J. Nicholls and J. K. Russell. Reviews in Mineralogy, Mineralogical Society of America, pp. 25–64.

Lees, J. M. (2007). Seismic tomography of magmatic systems. *Journal of Volcanology and Geothermal Research*, **167**, 37–56.

Longo, A., Vassalli, M., Papale, P. and Barsanti, M. (2006). Numerical simulation of convection and mixing in magma chambers replenished with CO_2-rich magma. *Geophysical Research Letters*, **33**, doi:10.1029/2006GL027750.

Manzella, A., Volpi, G. and Zaja, A. (2000). New magnetotelluric soundings in the Mt. Somma-Vesuvius volcanic complex: preliminary results. *Annali di Geofisica*, **43**(2), 259–270.

Marsh, B. D. (1981). On the crystallinity, probability of occurrence, and rheology of lava and magma. *Contributions to Mineralogy and Petrology*, **78**, 85–98.

Marsh, B. D. (1989). Magma chambers. *Annual Review of Earth and Planetary Sciences*, **17**, 439–474.

Marsh, B. D. (1996). Solidification fronts and magmatic evolution. *Mineralogical Magazine*, **60**, 5–40.

Martin, D. and Nokes, R. (1989). A fluid-dynamical study of crystal settling in convecting magmas. *Journal of Petrology*, **30**, 1471–1500.

McLeod, P. (1999). The role of magma buoyancy in caldera-forming eruptions. *Geophysical Research Letters*, **26**(15), 2299–2302.

McTigue, D. F. (1987). Elastic stress and deformation near a finite spherical magma body: resolution of the point-source paradox. *Journal of Geophysical Research*, **92**(B12), 12 931–12 940.

McTigue, D. F. and Mei, C. C. (1987). Gravity-induced stresses near axisymmetric topography of small slope. *International Journal of Numerical and Analytical Methods in Geomechanics*, **11**, 257–268.

Meriaux, C. and Lister, J. R. (2002). Calculation of dike trajectories from volcanic centers. *Journal of Geophysical Research*, **107**(B4), 2077–2087.

Mogi, K. (1958). Relationships between the eruptions of various volcanoes and the deformation of the ground surface around them. *Bulletin of the Earthquake Research Institute, University of Tokyo*, **36**, 99–134.

Muller, O. H. and Pollard, D. D. (1977). Stress state near Spanish Peaks. *Pure and Applied Geophysics*, **115**, 69–86.

Muller, J. R., Ito, G. and Martel, S. J. (2001). Effects of volcano loading on dike propagation in an elastic half-space. *Journal of Geophysical Research*, **106**(B6), 11 101–11 113.

Newman, A. V., Dixon, T. H. and Gourmelen, N. (2006). A four-dimensional viscoelastic deformation model for Long Valley Caldera, California, between 1995 and 2000. *Journal of Volcanology and Geothermal Research*, **150**, 244–269.

Olson, P., Yuen, D. A. and Balsiger, D. (1984). Convective mixing and the fine structure of mantle heterogeneity. *Physics of the Earth and Planetary Interiors*, **36**, 291–304.

Ottino, J. M., Ranz, W. E. and Macosko, W. (1979). A lamellar model for analysis of liquid-liquid mixing. *Chemical Engineering Science*, **34**, 877–890.

Parfitt, E. A., Wilson, L. and Head, J. W. (1993). Basaltic magma reservoirs: factors controlling their rupture characteristics and evolution. *Journal of Volcanology and Geothermal Research*, **55**, 1–14.

Patiño Douce, A. E. and Beard, J. (1995). Dehydration-melting of biotite gneiss and quartz amphibolite from 3 to 15 kbar. *Journal of Petrology*, **36**, 707–738.

Patiño Douce, A. E. and Harris, N. (1998). Experimental constraints on Himalayan anatexis. *Journal of Petrology*, **39**, 689–710.

Pedersen, T., Heeremens, M. and van der Beek, P. (1998). Models of crustal anatexis in volcanic rifts: applications to southern Finland and the Oslo Graben, southeast Norway. *Geophysical Journal International*, **132**(2), 239–255.

Petford, N. and Gallagher, K. (2001). Partial melting of mafic (amphibolitic) lower crust by periodic influx of basaltic magma. *Earth and Planetary Science Letters*, **193**, 483–499.

Phillips, J. C. and Woods, A. W. (2002). Suppression of large-scale magma mixing by melt-volatile separation. *Earth and Planetary Science Letters*, **204**, 47–60.

Pinel, R. and Jaupart, C. (2005). Caldera formation by magma withdrawal from a reservoir beneath a volcanic edifice. *Earth and Planetary Science Letters*, **230**, 273–287.

Poland, M., Hamburger, M. and Newman, A. V. (2006). The changing shape of active volcanoes: History, evolution and future challenges for volcano geodesy. *Journal of Volcanology and Geothermal Research*, **150**, 1–13.

Prakash, C. (1990). Two-phase model for binary solid-liquid phase change, part I: governing equations. *Numerical Heat Transfer*, **18**(B), 131–145.

Raia, F. and Spera, F. J. (1997). Simulations of crustal anatexis: Implications for the growth and differentiation of continental crust. *Journal of Geophysical Research*, **102**(B10), 22 629–22 648.

Rapp, R. P. (1995). Amphibole-out phase boundary in partially melted metabasalt, its conrol over liquid fraction and composition, and source permeability. *Journal of Geophysical Research*, **100**, 15 601–15 610.

Rapp, R. P. and Watson, E. B. (1995). Dehydration melting of metabasalt at 8–32 kbar: Implications for continental growth and crust–mantle recycling. *Journal of Petrology*, **36**(4), 891–931.

Rayleigh, Lord (1916). On convection currents in a horizontal layer of fluid when the higher temperature is on the under side. *Philosophical Magazine*, **32**, 529–546.

Reid, M. R. (2008). How long does it take to supersize an eruption? *Elements*, **4**(1), 23–28.

Reid, M. R., Coath, C. D., Harrison, T. M. and McKeegan, K. D. (1997). Prolonged residence times for the youngest rhyolites associated with the Long Valley Caldera. *Earth and Planetary Science Letters*, **150**, 27–39.

Rivalta, E. and Segall, P. (2008). Magma compressibility and the missing source for some dike intrusions. *Geophysical Research Letters*, **35**, L04306, doi:10.1029/2007GL032521.

Roper, S. M. and Lister, J. R. (2005). Buoyancy-driven crack propagation from an over-pressurized source. *Journal of Fluid Mechanics*, **536**, 79–98.

Rubin, A. M. (1995). Propogation of magma-filled cracks. *Annual Review of Earth and Planetary Sciences*, **23**, 287–336.

Rudnick, R. L., McDonough, W. F. and O'Connell, R. J. (1998). Thermal structure, thickness and composition of continental lithosphere. *Chemical Geology*, **145**, 395–411.

Ruprecht, P., Bergantz, G. W. and Dufek, J. (2008). Modeling of gas-driven magmatic overturn: Tracking of phenocryst dispersal and gathering during magma mixing. *Geochemistry Geophysics Geosystems*, **9**, doi:10.1029/2008GC002022.

Rushmer, T. (1991). Partial melting of two amphibolites: contrasting experimental results under fluid-absent conditions. *Contributions to Mineralogy and Petrology*, **107**, 41–59.

Rushmer, T. (1995). An experimental deformation of partially molten amphibolite: Application to low-melt fraction segregation. *Journal of Geophysical Research*, **100**, 15 681–15 695.

Rust, A. C. and Manga, M. (2002). The effects of bubble deformation on the viscosity of suspensions. *Journal of Non-Newtonian Fluid Mechanics*, **104**, 53–63.

Saar, M. O. and Manga, M. (2002). Continuum percolation for randomly oriented soft-core prisms. *Physical Review E*, **65**, doi:10.1103/PhysRevE.65.056131.

Sartoris, G., Pozzi, J. P., Phillippe, C. and Mouel, J. L. L. (1990). Mechanical stability of shallow magma chambers. *Journal of Geophysical Research*, **95**(B4), 5141–5151.

Scandone, R., Cashman, K. V. and Malone, S. D. (2007). Magma supply, magma ascent and the style of volcanic eruptions. *Earth and Planetary Science Letters*, **253**(3–4), 513–529.

Segall, P. and Davis, J. L. (1997). GPS applications for geodynamics and earthquake studies. *Annual Review of Earth and Planetary Sciences*, **25**, 301–336.

Simon, J. I. and Reid, M. R. (2005). The pace of rhyolite differentiation and storage in an 'archetypical' silicic magma system, Long Valley, California. *Earth and Planetary Science Letters*, **235**(1–2), 123–140.

Snyder, D. (2000). Thermal effects of the intrusion of basaltic magma into a more silicic magma chamber and implications for eruption triggering. *Earth and Planetary Science Letters*, **175**, 257–273.

Sparks, R. S. J. and Huppert, H. E. (1984). Density changes during fractional crystallization of basaltic magmas: Fluid dynamical implications. *Contributions to Mineraology and Petrology*, **85**, 300–309.

Sparks, R. S. J., Sigurdsson, H. and Wilson, L. (1977). Magma mixing: a mechanism for triggering acid explosive eruptions. *Nature*, **267**, 315–318.

Sparks, R. S. J., Huppert, H. E., Koyaguchi, T. and Hallworth, M. A. (1993). Origin of modal and rhythmic igneous layering by sedimentation in a convecting magma. *Nature*, **361**, 246–249.

Speer, J. A., Naeem, A. and Almohandis, A. A. (1989). Small-scale variations and subtle zoning in granitoid plutons: the Liberty Hill pluton, South Carolina, U.S.A. *Chemical Geology*, **75**, 153–181.

Spera, F. J. and Crisp, J. A. (1981). Eruption volume, periodicity, and caldera area: Relationships and inferences on developments of compositional zonation in silicic magma chambers. *Journal of Volcanology and Geothermal Research*, **11**, 169–187.

Spiegelman, M. and Kenyon, P. (1992). The requirements for chemical disequilibrium during magma migration. *Earth and Planetary Science Letters*, **109**, 611–620.

Tait, S., Jaupart, C. and Vergniolle, S. (1989). Pressure, gas content and eruption periodicity of a shallow, crystallizing magma chamber. *Geophysical Research Letters*, **25**(18), 3413–3416.

Touloukian, Y. S., Judd, W. R. and Roy, R.F. (ed.) (1981). *Physical Properties of Rocks and Minerals*. New York: McGraw-Hill.

Tsuchida, E. and Nakahara, I. (1970). Three-dimensional stress concentration around a spherical cavity in a semi-infinite elastic body. *Japan Society of Mechanical Engineering Bulletin*, **13**, 499–508.

Turcotte, D. L. and Schubert, G. (1982). *Geodynamics: Applications of Continuum Physics to Geologic Problems*. New York: Wiley.

Turner, J. S. and Campbell, I. H. (1986). Convection and mixing in magma chambers. *Earth Science Reviews*, **23**, 255–352.

Vazquez, J. A. (2004). Time scales of magma storage and differentiation beneath caldera volcanoes from uranium-238-thorium-230 disequilibrium dating of zircon and allanite. PhD Thesis, University of California, Los Angeles.

Vigneresse, J. L. and Tikoff, B. (1999). Strain partitioning during partial melting and crystallizing felsic magmas. *Tectonophysics*, **312**, 117–132.

Vigneresse, J. L., Barbey, P. and Cuney, M. (1996). Rheological transitions during partial melting and crystallization with application to the felsic magma segregation and transfer. *Journal of Petrology*, **37**, 1579–1600.

Voller, V. R. (1985). A heat balance integral method for estimating practical solidification parameters. *IMA Journal of Applied Mathematics*, **35**, 223–232.

Voller, V. R. and Cross, M. (1993). Solidification processes: Algorithms and codes. In *Mathematical Modelling for Materials Processing*, ed. M. Cross, J. F. T. Pittman and R. D. Wood. Oxford: Clarendon Press.

Voller, V. R. and Swaminathan, C. R. (1991). General source-based method for solidification phase change. *Numerical Heat Transfer*, **19B**, 175–189.

Wallace, G. and Bergantz, G. W. (2005). Reconciling heterogeneity in crystal zoning data: an application of shared characteristic diagrams at Chaos Crags, Lassen Volcanic Center, California. *Contributions to Mineralogy and Petrology*, **149**(1), 98–112.

Wells, P. R. A. (1980). Thermal models for the magmatic accretion and subsequent metamorphism of continental crust. *Earth and Planetary Science Letters*, **46**, 253–265.

Winter, J. (2001). *An Introduction to Igneous and Metamorphic Petrology*. Upper Saddle River, NJ: Prentice-Hall.

Wolf, M. B. and Wyllie, P. J. (1994). Dehydration-melting of amphibolite at 10 kbar: the effects of temperature and time. *Contributions to Mineralogy and Petrology*, **115**, 369–383.

Wolff, J. A., Balsley, S. D. and Gregory, R. T. (2002). Oxygen isotope disequilibrium between quartz and sanidine from the Bandelier Tuff, New

Mexico, consistent with a short residence time of phenocrysts in rhyolitc magma. *Journal of Volcanology and Geothermal Research*, **116**(1–2), 119–135.

Woods, A. W. (1995). The dynamics of explosive volcanic eruptions. *Reviews of Geophysics*, **33**(4), 495–539.

Woods, A. W. and Cardoso, S. S. S. (1997). Triggering of basaltic volcanic eruptions by bubble-melt separation. *Nature*, **385**, 518–520.

Woods, A. W. and Huppert, H. E. (2003). On magma chamber evolution during slow effusive eruptions. *Journal of Geophysical Research*, **108**(B8), 2403.

Wyllie, P. J. (1977). Crustal anatexis: An experimental review. *Tectonophysics*, **43**, 41–71.

Younker, L. W. and Vogel, T. A. (1976). Plutonism and plate tectonics: the origin of circum-Pacific batholiths. *Canadian Mineralogist*, **14**, 238–244.

Exercises

2.1 A basaltic sill (1280 °C) intrudes a kilometer beneath a water-saturated sedimentary layer (where the surrounding rocks are ~300 °C). Using simple scaling, estimate the amount of time that will elapse before the layer begins to be heated by the intrusion if heat transfer is only by conduction and no phase change occurs. Some of the following quantities may be useful for your estimate: The latent heat of the magma is 2×10^5 J kg^{-1}, the density of the rock is 3000 kg m^{-3}, heat capacity is 800 J kg^{-1} K^{-1}, and the conductivity is 5.3 W m^{-1}K^{-1}.

2.2 Calculate the critical thickness of magma required for convection. Assume a simplified magma chamber with a large aspect ratio (thickness $H \ll$ width W), and that the source of convection in the magma chamber is the density contrast between crystal-rich plumes sinking from the roof and the ambient magma, estimated here to be $\Delta \rho = 100$ kg m^{-3}. Taking the bulk magma viscosity of (a) basalt ($\eta = 10^2$ Pa s), (b) andesite ($\eta = 10^4$ Pa s), and (c) rhyolite ($\eta = 10^5$ Pa s), calculate the critical thickness of magma H_{cr} necessary for the onset of convection in the magma chamber, using

$$Ra_{cr} = \frac{\Delta \rho g H_{cr}}{\eta \kappa} = 10^3,$$

where $\kappa = 10^{-6}$ m^2 s^{-1} is the thermal diffusivity.

Repeat the calculations for convection driven by bubble plumes (density contrast $\Delta \rho = 500$ kg m^{-3}). Finally, compare these results with the case where convection is driven by thermal expansion alone, $\Delta \rho = \rho_0 \alpha \Delta T$, with $\rho_0 = 2500$ kg m^{-3}, $\alpha = 10^{-5}$ K^{-1} and $\Delta T = 10$ K.

Online resources available at www.cambridge.org/fagents
- Supplement 2A: Summary of dimensionless numbers relevant to volcanic processes
- Additional exercises
- Answers to exercises

Chapter 3

The dynamics of dike propagation

Steve Tait and Benoit Taisne

Overview

The aim of this chapter is to provide an overview of physical models of the dynamics of propagation of magmatic dikes. Experimental studies of fissure propagation provide an important way of validating hypotheses made in theoretical models, and hence provide a vital link between theory and field observations. Geological field observations of dikes can provide detailed information about rock and magma properties and post-emplacement dike geometry, but do not permit assessment of propagation rates. Conversely, geophysical studies can record dike emplacement as it takes place, allowing estimates of speed or demonstrating intermittency of propagation, but giving little information about dike geometry. We highlight some examples of field observations of dike dynamics that provide useful constraints for models. We discuss the main assumptions and key results of theoretical models that treat dike emplacement into both homogeneous and heterogeneous media. These models predict both the geometry of a propagating dike and the emplacement dynamics (e.g., speed), although they require assumptions about the source conditions, such as magma supply rates. There are open questions arising from shortcomings of current theory, which

can be addressed using laboratory experiments that permit detailed investigation of real-time geometric and dynamic information. Notable among recent studies are attempts to quantify factors that may lead to arrest of a dike before it reaches the surface, the potential for solidification to influence the dynamic regime of propagation, and the influence of three-dimensional fracturing effects on propagation.

3.1 | Introduction

A dike is a hydraulic fracture that brings magma through the brittle outer shell of the Earth from a zone of partial melting or storage to the site of a volcanic eruption or intrusion. The term dike originally applied to the solidified "fossil" fissure observed in field outcrops or detected in the subsurface by paleomagnetic studies (e.g., Ernst and Baragar, 1992), but is now also applied to the active fracture. The conditions leading to dike initiation are discussed in Chapter 2. This chapter focuses on the physics of propagation of individual dikes, which can be considered as the basis for a broader understanding of the significance of dikes in geologic processes. The dynamics of fissures can control, for example, whether magma produced in the mantle is erupted at

Modeling Volcanic Processes: The Physics and Mathematics of Volcanism, eds. Sarah A. Fagents, Tracy K. P. Gregg, and Rosaly M. C. Lopes. Published by Cambridge University Press. © Cambridge University Press 2013.

the surface or intruded as a horizontal or a vertical sheet. For this reason, the emplacement of dikes has a fundamental influence on the thermal and geochemical structure and evolution of the lithosphere. All volcanic eruptions involve the process of dike propagation as the means by which magma and associated gases reach the surface. The motion of dikes below the Earth's surface can be detected as dynamic geophysical signals such as seismic swarms and real-time strain measurements. Understanding the dynamics of fissure propagation is therefore of key importance for interpreting sequences of ground deformation and seismic swarms prior to eruptions. Ultimately, a good understanding of such precursory signals is required during volcanic crisis management. For example, advanced warning of the likely dynamics of an impending eruption, before a dike breaches the surface, would be highly beneficial. Dike propagation is a challenging physical problem, involving fluid flow, elasticity, fracturing, and heat transport. Strong rheologic gradients develop in the magma, which partially solidifies and degasses during propagation. Deformation and fracture of the host must be achieved in order for propagation to proceed. This combination of physics subdisciplines represents a complex mathematical problem, but equally, a rich phenomenology can be anticipated.

This chapter focuses on the fluid modeling aspects of dike propagation, emphasizing the relationships between experiment and theory. More detailed reviews of theoretical aspects of dike dynamics are given by Lister and Kerr (1991) and Rubin (1995). When faulting and dike injection take place simultaneously, the tectonic strain is absorbed both by the emplacement of magma and the faulting. Understanding this system therefore requires coupling of magma fluid dynamics and fault mechanics. While such studies are rare, there have been some interesting contributions at this research interface in recent years, notably building on intensive studies of mid-ocean ridge and rifting processes. Larger-scale models of rift zone evolution involving repeated dike injection events are largely concerned with explaining horizontal versus vertical motion, or extension and subsidence in

rift zones, for example, as responses to diking (Rubin, 1992; Buck, 2006). This class of models pertains to the "before and after" of a given dike event, without modeling dike propagation in detail. Although this chapter does not discuss these models in detail, the two approaches are complementary.

3.2 | Field observations

The most obvious geologic context in which to think of dikes is that of rifting, whether along a continental rift, such as in East Africa, or at ocean ridges. The significance of sheeted dike complexes in obducted oceanic crust, consisting of essentially 100% dikes, has been understood for several decades. Nevertheless, dikes are not limited to divergent plate boundaries – they are the fundamental means by which magma is transferred to the surface during volcanic activity, whatever the geodynamic context. The relatively rapid propagation and flow through a fissure is effective for preventing magma solidification during ascent. This section highlights some recent episodes of dike propagation that have been captured in real time by geophysical instrument arrays, and which illustrate the dynamics of dike emplacement. To date, these recorded episodes are neither sufficiently numerous nor sufficiently accurately captured to have thoroughly documented dike dynamics. Vertical propagation has been observed less frequently than horizontal propagation, for example. Furthermore, diking processes may vary among different tectonic stress regimes. Geologic evidence of swarms of mega-dikes (Halls and Fahrig, 1987) much larger than events observed by geophysical arrays, also suggests that gaps remain in the record of real-time observations. Before describing geophysically observed diking episodes, we briefly discuss insights into dike properties gleaned from geologic field observations.

3.2.1 Dike geometry as seen in the field
The geometry of dikes can be inspected directly in the field where they have been exposed

Figure 3.1 (a) Dike exposed by erosion on Piton des Neiges, Reunion Island (photograph courtesy of N. Villeneuve). The quasi two-dimensional geometry of dikes commonly adopted in theoretical models can be appreciated in this view. Pervasive horizontal jointing was caused by lateral cooling. (b) Dike emplaced just at the surface during the 2002 Nyiragongo eruption (photograph courtesy of J.-C. Komorowski). A solidified crust can be seen but the magma that occupied the interior of the dike was drained during the eruption. See color plates section.

by erosion. Observations made in the Tertiary Igneous Province in the UK (Jolly and Sanderson, 1995) related to continental rifting, or in the older, more eroded parts of Iceland (Gudmundsson, and Brenner, 2005) related to oceanic rifting, for example, indicate that basaltic dikes commonly have widths on the order of one to a few meters. The lengths and breadths (vertical and horizontal extents, respectively) of dikes are variable, but always much greater than dike widths, and thus it is natural to approximate dikes as two-dimensional sheets (Fig. 3.1(a)). Nevertheless, geologic exposures that allow examination of individual dikes over large vertical or horizontal extents are rare, so it is difficult to characterize dike geometry in greater detail. Moreover, when a dike is propagating, pressure changes in the magma and stress changes in the host will cause variations with time in the opening (width) of

the dike at a given location. Hence there is no guarantee that a solidified dike has the same thickness as it did during emplacement. Despite this complication, field observations provide a first-order estimate of dike geometry. Silica-rich dikes, in which the magma is typically several orders of magnitude more viscous than basalt, tend to be thicker than their basaltic counterparts by about an order of magnitude, although examples of thin silicic dikes also exist. However, in general there are fewer field studies of silicic dikes than of basaltic dikes. In this chapter, we do not emphasize magma composition as far as liquid properties are concerned, although the greater viscosity and volatile content are clearly important factors for dike emplacement and eruption of silica-rich magmas.

While the basic picture of dike geometry is that of a two-dimensional sheet, it is clear that dike surfaces are rarely smooth. It is common to observe damage in the host rocks in the form of joints and cracks, and pieces of host rock spalling off the wall, as well as irregular offsets of the dike profile when viewed in cross section. Currently it is not clear whether such features are important clues to propagation processes or simply secondary details, but they may be relevant in the context of characteristic seismicity accompanying propagation. In notable field studies on the Colorado Plateau, Delaney and Pollard (1981) and Delaney *et al.* (1986) examined the damage recorded in host rocks, and attempted to quantify *in situ* resistance to dike propagation. They concluded that a considerable amount of energy must be expended in irreversible damage to the host, and suggested that effective fracture toughness might be a lot higher than values measured on small rock samples in the laboratory. The underlying quantitative questions, related to use of Linear Elastic Fracture Mechanics (LEFM; see Section 3.3.2) in theoretical models, concern the size of the zone of irreversible damage and stress levels therein (e.g., Rubin, 1993a).

Field observations of chilled margins and vesicles in dikes indicate that magma properties vary both spatially and temporally during dike propagation, although it is not straightforward to relate the predictions of models of dynamic processes to these static observations, which are "frozen in" by solidification of the magma. For example, Figure 3.1(b) shows a dike just at the surface from which the magma drained laterally after formation of a chilled margin. In addition, heterogeneities in the host medium can affect dike propagation. Gudmundsson and Brenner (2001) and Gudmundsson (2005) described the possible effects on dike trajectory through a geologic formation in which successive layers have contrasting mechanical properties such as elastic stiffness, leading to stress refraction effects. Kavanagh and Sparks (2011) emphasized the importance of variable host rock properties in affecting the final geometry of a swarm of kimberlite dikes, and Baer and Hamiel (2010) interpreted a set of dikes in the Dead Sea region as having propagated horizontally under the influence of a density interface in the host rock.

3.2.2 Geophysically observed dike injection events

In this section, we discuss episodes of dike injection that have been recorded geophysically in real time, including the rifting event in Afar, Ethiopia, which started in 2005 and is ongoing at the time of writing, the 1997–98 event near the Izu Peninsula of Japan, and recent eruptions of Piton de la Fournaise. Valuable data have also been obtained for events elsewhere, notably in Iceland (e.g., Einarsson and Brandsdottir, 1980; Björnsson, 1985; Buck *et al.*, 2006; Pedersen *et al.*, 2007) and Hawaii (e.g., Pollard *et al.*, 1983; Rubin and Pollard, 1987). Here, we highlight the general features of such events that provide useful constraints for propagation models, including examples of rift-related, subduction-related, and hot-spot- (shield volcano-) related magmatism.

In 2005 a dike ~60 km long and 5–8 m wide, as constrained by satellite-borne radar interferometry, was injected along the Manda Harraro segment of the rift in Afar, Ethiopia (Wright *et al.*, 2006). Approximately a dozen earthquakes with magnitudes of 5 to 5.5 were recorded by regional networks over an interval of a few days. Although there was insufficient ground instrumentation close to the site of injection to observe accurately the duration of emplacement, an upper bound for the duration is ~1 week. A notable feature of this event is the considerable length of the first dike that was emplaced, which was somewhat longer than that of the eruptive fissure of Laki 1783, and comparable with the length of typical mid-ocean ridge segments. The overall volume of magma intruded in the September 2005 Afar event was 1.5 km^3 (Wright *et al.*, 2006; Grandin *et al.*, 2009), approximately an order of magnitude less than the volume erupted at Laki. In striking contrast to Laki, no basaltic magma reached the surface at Afar in September 2005. A dozen or so smaller dikes have been injected since the initial event (on average 1 dike every few months) from a central location on the rift segment (Ebinger *et al.*, 2010). A small proportion of these injections led to eruptions of basaltic magma, indicating that the volume of the dikes is in large part accommodating tectonic strain in the depth interval of 4–10 km, while the strain at shallower depths is being taken up by motion on faults.

At the time of writing, the Afar rifting episode appears to be nearing its end after more than six years of activity. The lack of vegetation in this region has allowed high-quality radar interferometry to constrain the volumes and final positions of the dikes. Moreover, some of the later injections were recorded on more proximal, albeit sparse, seismic networks (Ebinger *et al.*, 2008; Keir *et al.* 2009; Grandin *et al.*, 2011).

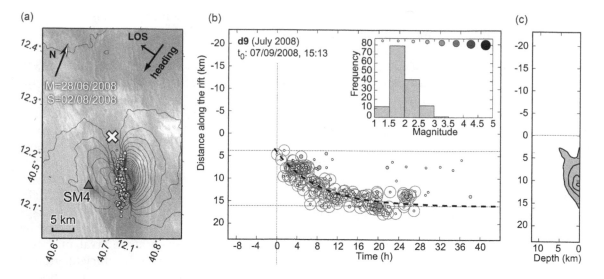

Figure 3.2 Data obtained during dike emplacement in the central section of the Manda Harraro rift segment in Afar, Ethopia, modified from Grandin *et al.* (2011). (a), (b) Earthquake locations deduced from data acquired by a local temporary seismic network; the white cross indicates the main source reservoir feeding the dikes. (c) Dike opening constrained by satellite-borne radar interferometry (Grandin *et al.*, 2009). Each contour represents an opening of 0.5 m. The dashed line shows the position of the dike tip based on a first-order interpretation that seismicity is centered on the propagating tip (Grandin *et al.*, 2011). However, it is clear in (b) that seismicity at any given time is distributed over a horizontal distance at least 5 km in width, i.e., considerably greater than the size of a process zone that would be consistent with the LEFM framework. The majority of this seismicity might therefore actually be located in front of or behind the tip.

In Figure 3.2, seismic recordings of one such injection clearly show the migration of events. A general picture emerges of a shallow magma source, positioned roughly centrally in the rift segment, from which magma migrates along the rift. This interpretation largely confirms earlier suggestions based on the pioneering use of microseismicity to follow dike injection during the Krafla rifting episode in the 1980s (Brandsdottir and Einarsson, 1979; Einarsson and Brandsdottir, 1980), and is similar to interpretations of events that have been recorded at submerged mid-ocean ridges (e.g., Blackman *et al.*, 2000; Tolstoy *et al.*, 2006). Seismic images

of crust at some mid-ocean ridges have identified magma storage/accumulation regions on the order of a kilometer in vertical and cross-ridge extent (Singh *et al.*, 2006). However, taking a broader view of magmatic activity in continents and oceans, observations that tightly constrain magma distribution prior to dike emplacement are still quite rare. One challenge is to record a sufficiently clear picture of dike dynamics that, via a physical model, robust inferences can be made about tectonic stress boundary conditions and the magma source pressure driving dike emplacement. For recent models that focus on how a sequence of dikes can be interpreted as a response to a given set of conditions prior to magmatic activity, we refer the reader to Buck (2006) and Buck *et al.* (2006), and here focus on the dynamics of individual injections.

In Figure 3.2(b), one can envisage that the dike tip is located at the barycenter of the events and fit a curve to dike tip position versus time, and hence deduce dike velocity by differentiation (Rivalta, 2010; Grandin *et al.*, 2011). The seismic events are sufficiently well located that one can probably improve on this first-order interpretation, but even if one locates the dike tip in this way, there is clearly seismicity occurring at locations well behind and beyond the tip. It is reasonable to expect some seismicity ahead of the dike tip, and to either side of the dike close behind the tip, depending on the stress distribution (Rubin and Gillard, 1998).

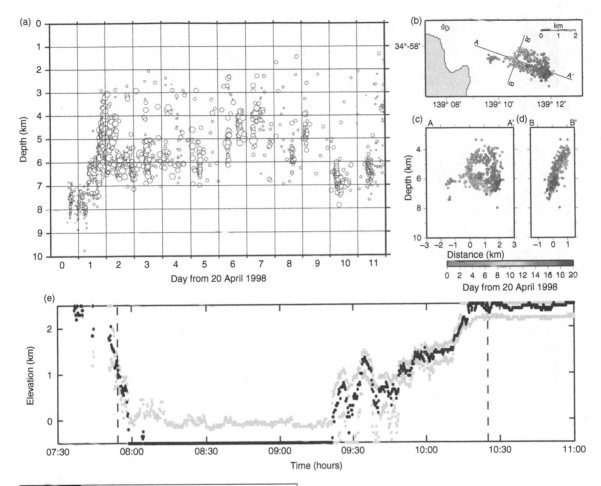

Figure 3.3 (a)–(d) Earthquake swarm activity occurring off the east coast of Izu Peninsula in April to May 1998; modified from Hayashi and Morita (2003) and Morita et al. (2006). (a) Hypocenters inferred from P- and S-wave arrivals. (b), (c) and (d) Distribution and migration of accurately relocated hypocenters (Hayashi and Morita, 2003). Symbol shades indicate dates on which the events occurred. Epicenters (b) and cross sections (c) and (d) along A–A' and B–B', respectively, with distances measured from the intersection of the two profiles. The hypocenters are clearly aligned on a nearly vertical plane, and migrate outward from the center of the distribution. (e) Location of radiated seismic energy using amplitude ratio analysis associated with the feeder dike of the January 2010 eruption at Piton de la Fournaise volcano, La Réunion, modified from Taisne et al. (2011a). Gray symbols indicate the error on location; the first dashed line indicates the beginning of the seismic crisis and the second represents the onset of eruption See online supplement for color version.

Therefore, identifying two lines bounding the distribution of seismicity permits a second-order interpretation, for example by locating the tip between these bounds where seismicity is maximum. In any case, we conclude that the seismically emitting zone covers several kilometers parallel to the dike's long axis, which is large compared to the size of a postulated mechanical process zone based on Linear Elastic Fracture Mechanics (LEFM; Section 3.3).

Another remarkable field data set was acquired during the 1997 and 1998 seismic swarms that occurred on the Izu peninsula, Japan. Geodetic and seismic networks were denser than the Afar example, allowing more accurate seismic locations. Hayashi and Morita (2003) present both classical location of the seismicity using P- and S-wave onset, and precise hypocenter determinations from waveform correlation (Figs. 3.3(a)–(d)). The geodetic data (including those from GPS receivers, a leveling survey, and tiltmeters) were successfully

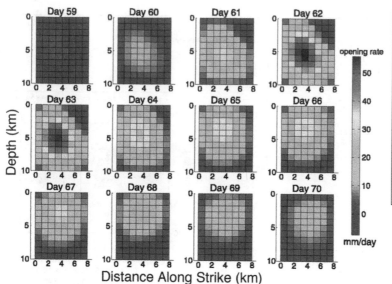

Figure 3.4 Temporal and spatial distribution of dike opening from time-dependent inversion of the 1997 seismic swarm at Izu Peninsula, Japan. Each panel represents the dike plane, viewed from the southwest, with depth in kilometers along the vertical axis, and along-strike distance in kilometers along the horizontal axis. The color represents the rate of dike opening, measured in millimeters per day. Each panel represents one day. From Aoki et al. (1999). See color plates section.

inverted using a model of an opening and moving dike by Aoki *et al.* (1999) for the 1997 seismic swarm (Fig. 3.4) and by Okada *et al.* (2000) for the 1998 seismic swarm. Morita *et al.* (2006) combined seismic and geodetic data using the former as a priori information while inverting the latter. These impressive studies start to provide a picture of dike geometry during propagation. We conclude that propagation proceeded in a somewhat discontinuous way with bursts of seismicity occurring around the margins of the dike rather than a continuous migration of a well-identified source. Seismic events took place over a depth interval of 3–8 km throughout the duration of the swarm (Fig. 3.3), which supports the idea that earthquakes also occur well away from the tip region.

Piton de la Fournaise, Réunion Island, is a highly active volcano in which seismic swarms preceding eruptions provide an opportunity to follow dike motion (Battaglia *et al.*, 2005; Lengliné *et al.*, 2008). Other methods have recently been derived to analyze seismic swarms too dense to allow a classical approach (Taisne *et al.*, 2011a). Figure 3.3(e) shows an example of migrating seismicity associated with a propagating dike recorded before it reached the surface at Piton de la Fournaise. Similarly to the relatively long-lived injection at Izu peninsula, this relatively short-lived

injection shows intermittent rather than steady seismic emission as the dike moves towards the surface. These observations demonstrate the importance of transient processes in propagation dynamics, which could arise due to host rock heterogeneities (e.g., in elastic properties or density), or variations in magma properties (e,g., due to cooling/solidification or degassing). Sections 3.3.2 and 3.4.4 investigate the role of variable magma and host rock properties from theoretical and experimental perspectives, in an effort to aid in the interpretation of these field observations.

This summary of three diking events that were recorded by geophysical arrays sets the scene for discussing modeling contributions that have been made over the last 20 years or so. The observational data available at present are biased towards predominantly horizontal rather than vertical propagation, whereas much theoretical work has explicitly tackled vertical buoyancy-driven propagation. Section 3.3 sets out the basic mathematical description of a propagating fissure, with an emphasis on lubrication theory, initially for situations in which magma and host rock properties are homogeneous. We then broaden the discussion to more recent studies involving heterogeneous magma and/or host properties, which attempt to more closely model natural conditions.

3.3 | Theoretical considerations

3.3.1 Constant magma and host properties

We start with the basic situation in which an isolated crack with some internal overpressure or buoyancy follows a linear trajectory within a homogeneous lithostatic environment. Adding a tensile opening stress field of tectonic origin does not fundamentally change the situation, because specifying the internal pressure and the direction of propagation is mathematically equivalent to specifying an opening stress, with the dike propagating normal to the direction of this opening stress (or in general normal to the direction of the least compressive stress). The presence of a more complex stress field in the elastic host would in general cause a dike to follow a curved trajectory. The crack is assumed to be isothermal and two dimensional. Muskhelishvili (1953) and Weertman (1971) established the application of theory describing dislocations in crystalline solids (i.e., when a regular solid lattice includes some offset, for example) to describe propagating fractures in geophysical contexts, in particular to handle mathematically the coupling between the pressure in the fluid within the fissure and the elastic host. Spence and Turcotte (1985), Spence et al. (1987), and Lister (1990a,b) described flow in the magma-filled crack using lubrication theory (i.e., assuming a geometry wherein one dimension is much smaller than the other two). The magma is assumed to be incompressible. To describe conditions in the vicinity of the tip, these studies adopted the approach of Linear Elastic Fracture Mechanics (LEFM), in which certain forms are assumed for the mathematical singularity in the elastic field just ahead of the crack tip, as well as the shape of the tip region. This reasoning leads to the following set of equations:

$$\frac{\partial h}{\partial t} = \frac{1}{3\eta}\frac{\partial}{\partial z}\left[h^3\left(\frac{\partial P_e}{\partial z} - \Delta\rho g\right)\right], \tag{3.1a}$$

$$P_e = \frac{-G}{(1-v)\pi}\int_{-\infty}^{z_f}\left(\frac{\partial h}{\partial \Gamma}\frac{\partial \Gamma}{\Gamma - z}\right), \tag{3.1b}$$

$$h(z,t) \sim \frac{1-v}{G}K_c\sqrt{\frac{2}{\pi}(l-z)}, \tag{3.1c}$$

$$P_e \sim \frac{K_c}{2\sqrt{z-z_f}} \quad \text{for } z > z_f, \tag{3.1d}$$

where h is the size of the fissure opening (thickness or width), η is the magma viscosity, P_e is the elastic pressure, $\Delta\rho$ is the difference between magma density and country rock density, g is acceleration due to gravity (such that $\Delta\rho g$ is a buoyancy term), t is time, z is the vertical coordinate, G is the shear modulus, v is the Poisson ratio, Γ is an integration variable, z_f is the position of the fissure tip, K_c is fracture toughness, and l is the fissure vertical length. All notation is defined in Section 3.7.

These equations are, respectively, momentum conservation in the form of the Reynolds lubrication equation (Eq. (3.1a)), the integral equation coupling fluid pressure to the elastic host medium (Eq. (3.1b)), the relation describing the shape of the crack in the tip region, derived by setting the stress intensity factor of the moving dike equal to the fracture toughness of the host (Eq. (3.1c)), and the form of the elastic stress singularity in the propagation plane ahead of the tip (Eq. (3.1d)). A more general mathematical framework can be found in Atkinson and Craster (1995).

This system was originally solved for steady propagation driven by magma buoyancy and for horizontal motion by specifying a constant flux of magma (Lister and Kerr, 1991; Rubin, 1995). The equations can be put into dimensionless form, such that, in the case of constant properties, one parameter – the dimensionless fracture toughness, K^* – is sufficient to describe the system (Lister, 1990b):

$$K^* = \frac{K_c}{\Delta\rho g V^{1/2}}, \tag{3.2}$$

where V is the volume of magma in the fissure.

Figure 3.5 shows results for the shape of a steadily propagating buoyant dike as a function of K^*. To first order, the dike has a swollen head region, where elastic pressure gradients are present, and the thickness of this head increases with increasing K^*. The head is followed by a

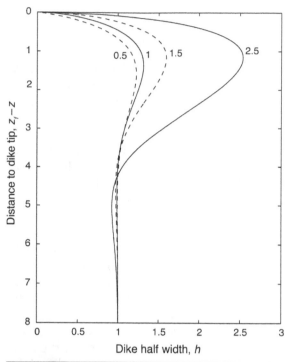

Figure 3.5 Dimensionless dike half-width plotted as a function of dimensionless distance behind the dike tip for different values of the dimensionless fracture toughness, K^* (0.5, 1, 1.5, 2.5, indicated on curves). An increase in dimensionless fracture toughness leads to a more swollen head. Note that horizontal and vertical distances are not scaled to the same reference length, which exaggerates the swelling. Redrawn from Taisne and Jaupart (2009).

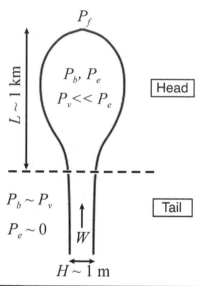

Figure 3.6 Schematic drawing of the head and tail structure of a buoyancy-driven dike (not to scale). The relative importance of the pressure scales defined in the text is indicated. In the head, viscous losses are neglected and buoyancy, elastic, and fracturing pressures are important. In the tail region, where elastic pressure gradients become negligible, the key balance is between buoyancy and viscous pressures. The order of magnitude of the tail width H and length L of the head region are indicated. The direction of the breadth dimension, B, is into the page.

tail region of constant thickness or half-width (Fig. 3.6) in which elastic pressure gradients are negligible. Flow in the tail region is a laminar viscous flow driven by buoyancy. It follows that the width of the dike tail h_∞ and the velocity of the propagating fissure, c, are related to the magma flux Q_{2D}, buoyancy and viscosity as:

$$h_\infty = \left(\frac{3\eta Q_{2D}}{2\Delta\rho g}\right)^{1/3}, \qquad (3.3a)$$

$$c = \frac{Q_{2D}}{2h_\infty}. \qquad (3.3b)$$

In this basic situation the host is continuously cracked by a combination of increased elastic pressure and buoyancy in the head of the fissure, whereas the speed of the crack is fixed by a balance

between fluid buoyancy and viscous pressures in the tail. This situation has been further developed in work motivated by oil reservoir applications (e.g., Garagash and Detournay, 2000).

This physical description is not limited to steady-state motion, and results have been obtained for transient problems (e.g., Mériaux and Jaupart, 1995; Roper and Lister, 2007). For transient calculations in which the fissure is connected to the magma source, the magma flux is calculated rather than specified a priori. For the case of buoyancy-driven propagation, Taisne and Jaupart (2009) showed that, starting from some initial shape, the fissure geometry adjusts to attain the steady solution.

Another important canonical case is that of the release of a finite volume of liquid. Because the dike is not fed continuously, propagation is not steady. Roper and Lister (2007) showed that, after adjusting from initial conditions, the propagation rate follows a power law such that

Table 3.1 | Fissure growth relationships for different conditions.

Model situation	Balance between resisting and driving forces	Form of length–time relation
2D – buoyant liquid	viscosity vs. buoyancy (P_v, P_b)	$t^{1/3}$
3D – neutrally buoyant, penny-shape	viscosity vs. elasticity (P_v, P_e)	$t^{1/9}$
3D – buoyant, penny-shape	viscosity vs. buoyancy (P_v, P_b)	$t^{1/5}$
3D – neutrally buoyant, constant breadth	viscosity vs. elasticity (P_v, P_e)	$t^{1/5}$
3D – buoyant, constant breadth	viscosity vs. buoyancy (P_v, P_b)	$t^{1/3}$

the fissure length follows $l \sim t^{1/3}$, and the crack thins with time as it extends in length.

The above theoretical work has inherent simplifications when compared with natural situations, perhaps the most obvious being the assumptions of two-dimensionality and of constant magma and host rock properties. Before discussing more recent results that have extended this basic theoretical approach to cases with variable magma or host properties, we introduce the approach of writing simple physical balances to obtain scaling laws. This gives a more intuitive picture of the basic physics involved. Four pressure scales can be defined (Lister and Kerr, 1991), which represent elasticity P_e, buoyancy P_b, viscous resistance P_v, and fracture toughness, P_f, respectively:

$$P_e \sim \frac{E}{2(1-v^2)} \frac{H}{\min[L,B]}, \tag{3.4a}$$

$$P_b \sim \Delta\rho g L, \tag{3.4b}$$

$$P_v \sim \frac{4\eta LC}{H^2}, \tag{3.4c}$$

$$P_f \sim \frac{K_c}{\sqrt{\min[L,B]}}. \tag{3.4d}$$

Here E is Young's modulus, C is the fissure velocity scale, and H, B, L are the fissure thickness, breadth, and length scales, respectively, where H and B represent dimensions normal to the direction of propagation (with $H<B$), and L is the dimension in the direction of propagation. The function $\min[L,B]$ means the smaller of the two variables is used. The volume of magma in the fissure can be specified conveniently in the form

$$V = At^\alpha, \tag{3.5}$$

where A is a constant whose dimensions depend on the value of the exponent α, and $\alpha = 1$ for constant magma flux and $\alpha = 0$ for constant magma volume. Table 3.1 summarizes fissure propagation relationships for various model situations represented by balances between different driving and resistive forces. As a fissure lengthens, it can pass from a regime defined by the balance between elastic and viscous pressures (P_e vs. P_v), to one between buoyancy and viscous pressures (P_b vs. P_v), and thus its rate of growth will vary with time. If the fissure is close to equidimensional in the fissure plane ($L \sim B$), and buoyancy is absent or small, cracks are usually referred to as "penny-shaped." Once buoyancy has become dominant in the vertical direction and L and B differ significantly ($L > B$), then the crack is inherently three dimensional, and the analysis becomes much harder (Taisne et al., 2011b). Exercises 3.1–3.3 (see end of chapter) explore the balances that lead to the results presented in Table 3.1, and expose some of consequences of assuming two-dimensional geometry in the theory.

3.3.2 Variable magma and host properties

To describe geological conditions more realistically, certain simplifying assumptions must be discarded. As regards magma properties, the most obvious issues are: (1) including heat transfer in the model will lead to viscosity increases and solidification of the magma, particularly in the tip region; and (2) volatiles initially dissolved in the magma will exsolve to form a gas phase as pressure decreases – again the greatest effects will be close to the propagating tip.

Volatile exsolution will also increase magma buoyancy and melt viscosity.

In addition, it is unlikely that the host rocks will be homogeneous for dikes propagating over distances of kilometers to tens of kilometers. There are large changes in confining pressure in the case of vertical propagation, and the dike is likely to encounter rock formations with different densities and elastic properties. Another complication is the possible existence of a regional tectonic stress such that the least compressive stress is not consistently orientated with respect to the initial propagation direction of the dike. These problems have not all been tackled comprehensively, but here we summarize some results obtained for situations involving variable properties.

Rubin (1993a) suggested, by using the model of Barenblatt (1962) to treat the tip region, that the LEFM approach may break down for high confining pressure. For LEFM to be valid, the "process zone," in which stresses exceed the elastic limit for rocks and inelastic processes take over, must be small. If LEFM breaks down, the resistance to propagation could vary as the dike moves and the process zone changes in size. Degassing of magma occurs as it rises towards the surface and experiences lower pressure, which introduces the major feature of magma compressibility into the requirements of a model. In models of dike propagation in the crust beneath a deep ocean, degassing can probably be justifiably ignored, but in the case of subaerial volcanism it is inevitably an issue. Lister (1990b) made a preliminary analysis of the effects of gas exsolution by including a gas pocket of fixed volume near the tip of the steady dike, and demonstrated the effects on the shape of the head. Taisne and Jaupart (2011) investigated the more realistic transient problem in which the gas pocket grows with time. Figure 3.7 shows results for the shape of the head region as a function of time: the fissure tends to narrow in the gas region just above the liquid–gas interface. Under real conditions, damage to the rocks in the tip region might allow gas to leak out of the dike and infiltrate the host (Rubin, 1993a).

The inclusion of heat conduction into the host and the consequent increase in magma

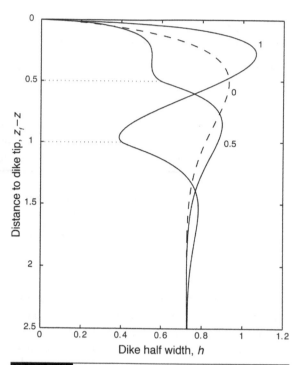

Figure 3.7 Dimensionless dike half-width plotted as a function of dimensionless distance behind the dike tip for a fixed dimensionless fracture toughness, but for a case with a growing gas pocket, with dimensionless lengths of 0, 0.5, and 1 indicated on the curves. The fracture narrows as the gas pocket increases in size, with the minimum width being attained just within the gas-filled region (dotted lines). Redrawn from Taisne and Jaupart (2011).

viscosity and potential solidification have not yet been thoroughly investigated. Delaney and Pollard (1982), Rubin (1993b) and Bolchover and Lister (1999) determined the conditions under which a dike may be thermally viable (i.e., able to propagate), or not, such that propagation is arrested by freezing. It is unlikely that a truly steady regime can be established when solidification or cooling-driven viscosity increases are significant, but for now a full theoretical treatment is lacking. We return to this problem in Section 3.4.3.

We now turn to consideration of heterogeneities in the host medium. Lister and Kerr (1991) considered propagation of a dike in a vertical density gradient, showing how propagation changed from vertical to horizontal, while the dike plane remained vertical. Taisne and

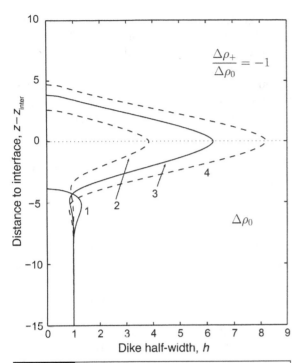

Figure 3.8 Time sequence of the evolution of dike shape at times 1 through 4 as it advances from country rocks with density greater than that of the magma to those with density less than that of the magma. The position of the density interface in the country rocks is indicated by the dotted line. As the dike encounters the interface it decelerates and evolves from vertical to horizontal propagation suggesting that a sill may form at the interface. Redrawn from Taisne and Jaupart (2009).

Jaupart (2009) analyzed the problem of a dike containing magma of fixed density, propagating vertically under its own buoyancy through a denser host, but which encounters a rock layer of density lower than that of the magma. The buoyancy inversion causes the dike to decelerate and swell horizontally. Figure 3.8 shows the shape of the head region calculated at consecutive times, demonstrating the potential for the dike plane to evolve to a horizontal orientation and form a sill. Note that in both studies, the dike does penetrate above the interface of buoyancy inversion. Another approach to the problem of sill formation has been to consider layers of contrasting elastic properties in the host. Gudmundsson and Brenner (2001) observed that large contrasts in elastic modulus or stiffness within a geologic formation that is traversed by

a dike should lead to stress refraction effects, such that the minimum compressive stress can become rotated, causing a vertically propagating dike to evolve into a horizontal orientation. At the time of writing, density inversions and stress refraction are both plausible explanations for sill formation, which can account for large volumes of magma being intruded into the crust rather than erupted. This is relevant to the thermal evolution of the continental crust, and the emplacement of thick sills of basaltic magma is one of the mechanisms that have been proposed to generate large volumes of silicic magmas by crustal melting (Huppert and Sparks, 1988).

The problem of stress refraction raises the general question of the existence of regional stress fields that may cause a dike to deviate from a linear trajectory or halt vertical progress. An important example of this situation is the stress field due to loading of the crust by a volcanic edifice. As volcanism builds an edifice, the increasing weight will progressively impose a stress field in the crust, which will ultimately affect the trajectory of rising dikes (Pinel and Jaupart, 2000, 2004).

3.3.3 Questions to resolve: an appeal to experiments

We end this summary of theoretical considerations by setting the scene for discussion of experimental simulations of dike propagation. It is not easy to compare the results of theoretical calculations directly with field observations, but laboratory experiments provide one way of linking the two. It is difficult to design experiments that reflect geologic reality, given the large length scales, pressure and rheologic changes involved. The general principle is that one must use materials at laboratory scale that enable us to respect the principle of dynamic similarity, i.e., that an appropriate set of dimensionless numbers should take similar values in both model and natural applications in order to reproduce the appropriate dynamic regime. Despite limitations, experiments can provide dynamic and geometric information during propagation, and hence help to verify aspects of theory that can then be applied with more confidence at the scale of geologic phenomena.

We have raised several issues that are addressed in the discussion of experimental studies in Section 3.4. (1) Natural fissures have finite breadth and therefore a three-dimensional shape that must be accounted for; only if the magma source is linear and very long may a two-dimensional approximation suffice. Of key interest are the factors that might limit dike breadth, and how they might affect fissure propagation. (2) Laboratory analyses can help to verify the scaling laws predicted by theory, such as the $l \sim t^{1/3}$ relationship obtained for the case of constant volume propagation. (3) Experiments can provide insights into propagation behavior in the case of significant variations in magma properties due to degassing and cooling. If a steady regime is not possible under these conditions, regimes of unsteady propagation may be defined.

3.4 | Experimental investigations

3.4.1 Experimental methods
Three types of experiments, involving a host elastic solid made of gelatin, provide insight into the dynamics of dike propagation. The first set of experiments involves continuous injection of buoyant fluid under isothermal conditions into the base of the gelatin, either by maintaining the source reservoir at constant overpressure or by imposing a constant injection flux. In a second set of experiments, also isothermal, a constant volume of positively or negatively buoyant fluid was injected into the gelatin. A third set of experiments investigated the effects of solidification on propagation, which was achieved by injecting hot paraffin at constant flux into gelatin at a temperature below the solidus of the paraffin. The results used for illustration are drawn from Taisne and Tait (2009, 2011) and Taisne et al. (2011b), but we also discuss work that has produced similar, contrasting, or complementary results.

Experimental techniques and conditions are described in detail by Taisne and Tait (2009, 2011), and references therein. In summary, the experiments were carried out in transparent acrylic tanks 50 cm high and either 30 × 30 cm or 30 × 45 cm in horizontal section. Fluid was injected from a reservoir into the base or top of a gelatin block. A pre-cut in the gelatin allowed clean initiation and orientation of the fissure. For solidification experiments, the reservoir was kept at a constant temperature above the solidus of the working fluid, while the gelatin temperature was below this solidus. Visual observations were made by dyeing the fluid and viewing normal to the crack plane, using the intensity of color to deduce crack thickness, and/or by exploiting the photo-elastic properties of the gelatin to visualize the elastic stress field around the fissure. Preparation of the gelatin followed the protocol of Menand and Tait (2001, 2002) – elastic modulus was varied via gelatin concentration, and measured in situ prior to each experiment. During solidification experiments the formation of solid caused the fissure to become opaque, hindering extraction of quantitative information about fissure thickness from the photographs. Nevertheless, fissure propagation was followed by taking a sequence of photographs at regular intervals, and extracting from the images the total area of the fissure as observed normal to the plane of the fissure.

In most natural geologic situations, it is reasonable to suppose that the elastic host medium is essentially infinite, whereas this is impossible to achieve in the laboratory. Experimental studies have generally assumed that the experimental tanks are sufficiently large that wall effects are negligible, although this does not appear to have been rigorously tested. Taisne and Tait (2009, 2011) devised a method of introducing a 2-cm-wide vertical layer of water between the gelatin and the tank walls parallel to the plane of the experimental fissure (Fig. 3.9(a)), in order to effectively minimize the influence of the finite dimensions of the tank on fissure propagation.

3.4.2 Steady propagation regime
The two-dimensional theory of fissure propagation was first established in the framework of constant imposed flux, but a more natural condition is constant overpressure in the source. Nevertheless, the calculations of Taisne and Jaupart (2009) and experiments of Menand and Tait (2002) showed that, after an initial transient, a steady-state regime is attained for a buoyant

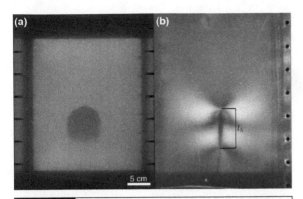

Figure 3.9 Photographs of an experiment in which a dike is supplied by a constant flux of liquid. The head and tail structure of the dike can be seen. (a) In transmitted light the dike is viewed normal to the propagation plane; the intensity of the color represents the thickness of the dike head and shows how it tapers to a thin tail. (b) In polarized light the photo-elastic properties of the gelatin reveal the stress field around the dike via birefringence, including the singularity centered on the tip, and the absence of fringes around the tail. The fringes appear to show that the elastic field does not interact with the vertical boundaries of the gelatin. From Taisne and Tait (2011). See color plates section.

liquid propagating from a source at constant overpressure. However, it remained uncertain whether effects of the three-dimensional experimental geometry influenced the relationship between source overpressure and liquid flux. Further experiments, similar to those of Menand and Tait (2002), used a computer-controlled injection system to impose a constant fluid flux (Taisne and Tait, 2011), and also investigated a greater range of fluid buoyancy. These improvements help to achieve more rigorous comparison with theory. Figure 3.9 shows representative photographs of one such experiment. Figure 3.9(a) shows the view perpendicular to the fissure plane, allowing the head to be identified by its stronger color in contrast to the tail. The photo-elasticity of the gelatin host (Fig. 3.9(b)) identifies strong stress concentrations at the tip of the fissure, and a head region surrounded by stress fringes. The head tapers into a tail that lacks stress fringes, indicating that there are no significant elastic pressure gradients there. This basic structure corresponds well to that predicted by the two-dimensional theory outlined in Section 3.3.1.

We now consider whether the experimental fissures are well described by two-dimensional theory. This theory uses a two-dimensional "volumetric" flux (Q_{2D} in m² s⁻¹; Eq. (3.3)), whereas the experimental flux is in m³ s⁻¹ and a physical balance determines the horizontal fissure breadth B in the plane of the fissure. Even if the breadth is assumed to be constant in the experiments so that a vertical pseudo-2D flux can be defined, this flux still depends on the third dimension, as well as the source. A truly two-dimensional fissure would have zero elastic pressure gradient parallel to the fissure in the horizontal plane. It is plausible that fracture resistance arrests horizontal spreading. We note that the experimental fissures initially tend to spread both vertically and horizontally, but then vertical propagation becomes dominant and horizontal spreading ceases. At this transition, we assume that vertical and horizontal elastic pressures are equal and the fissure aspect ratio is ~ 1 ($B \sim L$). Following the approach of writing physical balances (Eq. (3.4)), with $P_b \sim P_e \sim P_f$, we have

$$\Delta \rho g L \sim \Delta \rho g B \sim \frac{E}{2(1-v^2)} \frac{H}{B} \sim \frac{K_c}{\sqrt{B}}. \quad (3.6)$$

From this, and substituting $E = 2G(1+v)$, we obtain

$$HB \sim \Delta \rho g B^3 \frac{1-v}{G}. \quad (3.7)$$

Finally, because

$$C \sim \frac{Q_{3D}}{HB}, \quad (3.8)$$

we obtain

$$C \sim \left(\frac{Q_{3D} G \Delta \rho g}{1-v} \right) K_c^{-2}. \quad (3.9)$$

This three-dimensional velocity scale involves both the flux and the fracture toughness K_c because of its role in limiting lateral spreading and hence in determining B. Measurements of propagation velocity against this scale show a good correlation (Fig. 3.10).

Whether a propagating fissure is best described by two- or three-dimensional theory

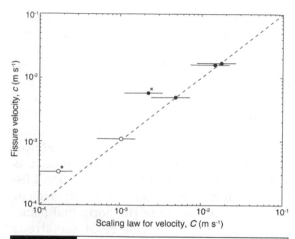

may largely depend on how the fissure is initiated. A fissure that initiates simultaneously over a long distance (as might occur at a spreading center where tectonic plates move apart) satisfies the requirement of two-dimensional theory that fluid is fed into a crack that is very long even at the source. However, it may be common for a crack to nucleate locally, such as at the wall of a magma storage region, and then propagate both laterally and vertically until appropriate physical balances are achieved. By analogy with laterally and vertically spreading experimental fissures, the three-dimensional theory (Eq. (3.9)) should describe this situation well.

3.4.3 Propagation (and arrest) of a constant volume fissure

Several experimental studies have investigated propagation of a fissure containing a fixed volume of fluid (e.g., Takada, 1990; Heimpel and Olson, 1994; Dahm, 2000), although interpretations and analyses have been diverse. More recent investigations (Taisne and Tait, 2009; Taisne *et al.*, 2011b) provide a basis for

comparison with these earlier studies. The experiments of Taisne and Tait (2009) involved injection of a fixed volume of fluid, typically a glucose solution, into a gelatin host medium. The glucose concentration was varied to achieve different degrees of buoyancy; in many cases the fluid density was greater than that of the gelatin. For buoyant propagation to occur, the buoyancy force (Eq. (3.4b)) must overcome the fracture resistance of the solid medium (Eq. (3.4d)), which, in a two-dimensional framework, requires that the vertical length of the fissure be greater than a "buoyancy length" (see Exercise 3.1) defined as:

$$L_b = \left(\frac{K_c}{\Delta \rho g} \right)^{2/3}. \tag{3.10}$$

In the experiments, a pre-cut of vertical extent slightly greater than L_b was made in the gelatin, so that buoyancy would be large enough to fracture the gelatin. Upon emplacement of the liquid in the pre-cut, the fracture initially spread faster horizontally than vertically, but after attaining a certain breadth (B_f), horizontal spreading ceased and vertical propagation dominated. The sequence of photographs in Figure 3.11(a) shows how experiments proceeded and Figure 3.11(b) shows the evolution of thickness determined by image analysis of color intensity of the dyed fluid (Taisne and Tait, 2009). As in the case of the constant flux experiments, the question of what limits the horizontal dimension of each fissure arises. Measurements of final fissure breadth B_f are plotted in Figure 3.12(a) as a function of a length scale B derived by assuming that horizontal spreading is ultimately halted by the fracture toughness of the medium. Equating P_e and P_b (Eqs. (3.4a) and (3.4b)) yields

$$B \sim \left(\frac{G}{(1-v)} \frac{V}{\Delta \rho g} \right)^{1/4}. \tag{3.11}$$

This analysis (Taisne *et al.*, 2011b) emphasizes that these fractures are fundamentally three-dimensional in character.

Figure 3.12(b) shows a typical example of the length of a fissure as a function of time. The initial phase involves horizontal spreading, dominated

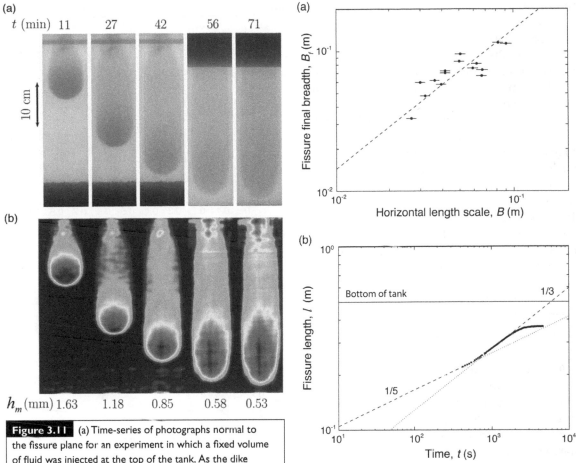

Figure 3.11 (a) Time-series of photographs normal to the fissure plane for an experiment in which a fixed volume of fluid was injected at the top of the tank. As the dike propagates downwards, the thickness of the dike slowly decreases; the head starts off with a strong color (i.e., is relatively thick) but this diminishes as fluid is lost to the trailing tail. (b) Dike thickness as a function of position, derived from the fissure color intensity. Here h_m is the maximum thickness of the fissure, corresponding to the dark red coloration. From Taisne et al. (2011b), reproduced with permission from Springer. See color plates section.

Figure 3.12 (a) Final breadth B_f of fixed-volume experimental fissures plotted as a function of the breadth scale B derived in Eq. (3.11). (b) Variation of fissure length with time for experimental fixed-volume fissures (dark symbols), showing initial vertical growth approximated by $l \sim t^{1/5}$, before increasing to $l \sim t^{1/3}$ after fissure stabilized at B_f. From Taisne et al. (2011b), reproduced with permission from Springer.

by a balance between elastic pressure and viscous resistance assuming that the fissure has an aspect ratio L/B of order 1, whereas the second phase involves vertical propagation, in which the dominant balance is between buoyancy and viscous resistance. The expected scaling laws (Table 3.1) indicate that the length evolves as $l \sim t^{1/5}$ and then as $l \sim t^{1/3}$ (dashed and dotted lines in Fig. 3.12(b)); agreement with the experimental data is good. This graph shows a general feature that is inconsistent with the two-dimensional theory, namely that the experimental fissures stop propagating altogether at some final length. This result again emphasizes the fundamentally three-dimensional nature of these cracks. A purely mechanical condition brings cracks propagating under buoyancy to a halt which, in the context of magmatic dikes, suggests that a minimum volume of magma is required for an eruption to take place from a given depth of origin of the dike (Taisne and Tait, 2009; Taisne et al., 2011b). Furthermore, as a constant volume crack propagates and hence thins, thermal effects will become progressively more important. Solidification of the fissure may therefore intervene before the mechanical effect halts propagation.

The results using glucose solutions show an important contrast with previous buoyant, fixed-volume experiments, in which the fissure adopted a steady-state shape and rose through the solid at constant velocity (Takada, 1990; Heimpel and Olson, 1994; Dahm, 2000). This difference may be attributed to the use of hydrophobic fluids in those studies, such that the fluid is unable to wet a thin tail region, and surface tension acts to retract fluid into the main crack (Taisne and Tait, 2009). Experiments with hydrophobic fluids are less relevant to the geological application of interest than studies using miscible fluids; in the natural case cooling will coat the wall of the magmatic fracture with a chilled margin which, being chemically identical to the magma in the fracture, will eliminate any strong surface energy effects between solid and liquid.

3.4.4 Propagation under conditions of variable properties

The seismic data presented in Section 3.2 show intermittent bursts and lulls, and a distribution of seismicity that was not just located in a narrow zone at the fracture tip. Experimental studies have shown that variable fluid or host properties can produce a variety of unsteady behaviors. For example, one set of experiments explored the behavior of a fissure undergoing fluid solidification as it propagates. The source conditions were kept fixed by injecting molten paraffin at a constant flux in order to determine whether steady regimes could become established. The introduction of solidification into the problem leads to the definition of two new dimensionless numbers (Taisne and Tait, 2011) defined as a dimensionless temperature

$$\Theta = \frac{T_s - T_\infty}{T_l - T_\infty} \tag{3.13a}$$

and dimensionless flux

$$\Phi = Q_{3D} \rho c_p \Delta T \frac{H}{\kappa \rho c_p \Delta T L B} = \frac{Q_{3D} H}{\kappa L B}. \tag{3.13b}$$

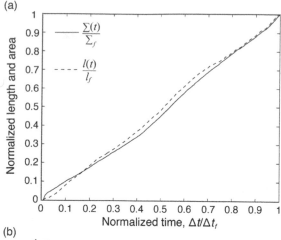

(a)

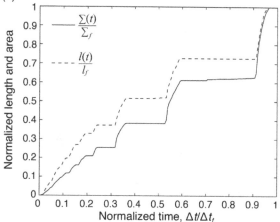

(b)

Figure 3.13 Normalized length (dashed lines) and normalized dike area viewed perpendicular to the fissure (solid lines) as a function of time for (a) an experiment in which intermittence was weak, and (b) an experiment in which intermittence was pronounced. From Taisne and Tait (2011).

Despite the constant supply of liquid, a dynamic regime of intermittent fracture propagation was observed. The fissure would stop propagating, i.e., the velocity at the tip would drop to zero, but then begin to swell without advancing, with the arriving liquid being stored in the fissure without increasing its surface area. This swelling inevitably led to an increase in the elastic pressure, and in some cases also buoyancy pressure, available to fracture the gelatin. In addition, a solid layer frozen against the cold fissure walls and particularly in the thin tip region thickened and presumably became stronger. As long as the

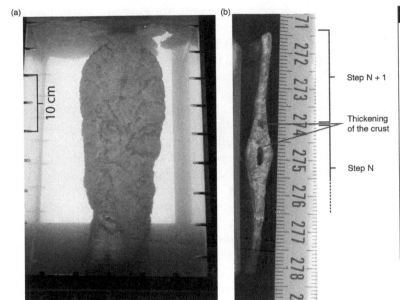

(a)

10 cm

(b)

Step N + 1

Thickening
of the crust

Step N

Figure 3.14 Photographs from a solidification experiment. (a) Frozen wax remaining in the tank just after the dike had reached the surface of the gelatin. (b) Cross section through the frozen wax after removal from the tank in an experiment with strong intermittent behavior. The sample brackets two large steps of progress, and shows a strong variation in the crust thickness: the thick crust that formed at the end of step N was breached and more wax flowed ahead to form the thinner crust in step N+1, which was last step of the experiment as the dike reached the top of the tank. See color plates section.

combined mechanical resistance of gelatin and frozen crust was sufficient to combat the increase in driving pressure, the fissure remained static. However, after a time interval that varied from experiment to experiment, and to a lesser extent during individual experiments, propagation resumed, accompanied by deflation of the swollen fissure. The fissure then grew rapidly only to be halted again by solidification.

Measuring the increase in fissure surface area by analyzing photographs taken at regular intervals allowed the rate of fissure propagation to be characterized. Figure 3.13 shows fissure length and surface area plotted as a function of time for two experiments with different initial conditions. In one case the behavior is quasi-continuous (Fig. 3.13(a)), but in the other case intermittent propagation is shown by the stepwise appearance of the curves (Fig. 3.13(b)).

Figure 3.14(a) shows a photograph of one of these experiments just after the dike had reached the surface of the gelatin, and Figure 3.14(b) shows a vertical cross section of the upper part of this solidified dike after removal from the tank. The section features a thick crust that formed at the tip during an interval in which the fissure was stationary, and then the thinner crust above that formed during the subsequent propagation event. The temporal variation of the crust thickness is the key to understanding the intermittency. Outbreaks commonly spread around a large part of the dike periphery, and did not remain confined to or typically even initiate at the tip. This mechanism of increasing the fissure surface area helps to explain the observed spatial and temporal distribution of seismicity accompanying dike injection events (Section 3.2.2; Figs. 3.2, 3.3): much of the seismicity is generated behind the dike tip during intermittent propagation.

Several other experimental studies also provide insights into non-steady behaviors observed in nature. Menand and Tait (2001) devised a way of accumulating gas in the tip region of a propagating liquid-filled crack, and found that once the gas pocket reached a critical size, it was able to fracture the gelatin with its own buoyancy. The gas pocket then propagated faster than the liquid and they proposed this as a possible explanation for gas-rich precursor eruptions. Muller et al. (2001) carried out experiments in which a weight was placed on the top of a gelatin block to simulate the presence of a volcanic edifice. This procedure induced a non-lithostatic stress field in the gelatin, which in turn caused dikes to deviate from their previously vertical trajectory. Acocella and Neri (2009) investigated propagation within an edifice itself, showing that the presence of the conical free surface causes dike trajectories to deviate. Kavanagh *et al.* (2006) examined propagation through a gelatin block consisting of two superposed layers having different elastic moduli, the upper layer being stiffer. They observed that the dike would deviate and propagate horizontally at the interface between the two layers, suggesting that this is one explanation for the formation of sills.

3.5 | Discussion and perspectives

Future research on dike propagation is likely to take a number of directions, including theoretical, field, and experimental studies, but also real-time geophysical data collection of active dike injection events. We have shown how laboratory experiments can be used to verify aspects of theoretical models of dike propagation dynamics. This step has been necessary because of the great difficulty of comparing model predictions directly with field observations. As an example of how experiments can also pose a challenge to theory, studies including solidification demonstrate an intermittent propagation regime governed by thermal effects. However, it remains unclear whether this mechanism is significant enough to account for the intermittency observed in nature.

Several dike propagation events have now been recorded by geophysical networks sufficiently dense as to provide quite detailed information (see Section 3.2.2), considerably enhancing the potential for making direct comparisons of theoretical predictions with the dynamic observations. Geologic observations of solidified dikes will continue to provide useful insights, but the integration of quantitative real-time information from dike-induced seismic and deformation measurements is a significant advance. Insights from these new data sets have already opened up a collaborative interface with researchers modeling larger-scale tectonic problems. Modeling of divergent boundary tectonics at timescales longer than individual dike

propagation events, for example, might benefit from incorporating some synthesis of theoretical models of dike dynamics. This merging of the shorter timescale picture into a medium to long-term model of rifts may not be too distant a prospect.

Two final remarks concern the process of dike initiation, the mechanisms of which are still not well understood. It is well known that rock undergoes partial rather than total melting, to produce magma, and the problem of how melts are extracted to generate crystal-poor magmas is an important one. In the theoretical and experimental work presented in this chapter, the boundary conditions concerning dike initiation do not take account of the fact that the dike is being fed from a region that has its own, perhaps slower, dynamic conditions. Therefore linking of realistic magma supply rates or availability with dynamic propagation models is also likely to be the subject of interesting future developments. A few pioneering contributions have touched on this matter (Sleep, 1974; Rubin, 1998), investigating dike behavior in a deformable porous medium, but there is a lot of interesting work yet to be done. The current view of magma chambers incorporated in dike propagation models is simplistic in both theory and experiment (McLeod and Tait, 1999; Buck *et al.*, 2006; Rivalta, 2010). It would therefore be of great interest to devise experiments allowing study of the coupling between a porous source and a dike, rather than simply treating an elastic interface.

3.6 | Summary

- The dynamics of dike emplacement involve complex interactions between the behavior of the enclosed fluid and the response of the host medium.
- Regimes of dike behavior can be defined by analyzing balances between the buoyancy and viscous pressures due to the fluid, and elastic and fracture pressures of the host.
- Experimental studies have provided insights into characteristics and behaviors observed in

nature, including dike geometry and trajectory, intermittency of propagation, and the distribution of seismicity around dikes during injection.

κ	magma thermal diffusivity (m^2 s^{-1})
v	Poisson's ratio
Σ	fissure area (m^2)
Σ_f	final fissure area (m^2)
Φ	dimensionless magma flux

3.7 | Notation

A	constant whose dimensions depend on the value of exponent α (m^3 for $\alpha = 0$; m^3 s^{-1} for $\alpha = 1$)
B	fissure breadth scale (m)
c	mean fissure propagation velocity (m s^{-1})
C	fissure velocity scale (m s^{-1})
c_p	specific heat of magma (J kg^{-1} K^{-1})
E	Young's modulus (Pa)
g	acceleration due to gravity (m s^{-2})
G	shear modulus (Pa)
h	fracture opening (thickness) (m)
h_m	maximum fissure thickness (mm)
h_∞	tail thickness (m)
H	fissure thickness scale (m)
K_c	fracture toughness (Pa $m^{1/2}$)
K^*	dimensionless fracture toughness
l	fissure length (m)
l_f	final fissure length (m)
L	fissure length scale (m)
L_b	buoyancy length scale (m)
P_b	buoyancy pressure scale (Pa)
P_e	elastic pressure scale (Pa)
P_f	fracture pressure scale (Pa)
P_v	viscous pressure scale (Pa)
Q_{2D}	magma volume flux per unit fissure breadth (m^2 s^{-1})
Q_{3D}	magma volume flux (m^3 s^{-1})
T_i	initial magma temperature (K)
T_s	magma solidus temperature (K)
T_∞	temperature of host medium far from the fissure (K)
V	volume of magma in fissure (m^3)
z	vertical coordinate (m)
z_f	position of fracture tip (m)
Γ	integration variable in Eq. (3.1b)
$\Delta\rho$	difference between densities of magma and country rock (kg m^{-3})
η	magma viscosity (Pa s)
Θ	dimensionless temperature

References

Acocella, V. and Neri, M. (2009). Dike propagation in volcanic edifices: Overview and possible developments. *Tectonophysics*, **471**, 67–77.

Aoki, Y., Segall, P., Kato, T., Cervelli, P. and Shimada, S. (1999). Imaging magma transport during the 1997 seismic swarm off the Izu Peninsula, Japan. *Science*, **286**, 927–930.

Atkinson, C. and Craster R. V. (1995). Theoretical aspects of fracture-mechanics. *Progress in Aerospace Sciences*, **31**, 1–83.

Baer, G. and Hamiel, Y. (2010). Form and growth of an embryonic continental rift: InSAR observations and modelling of the 2009 western Arabia rifting episode. *Geophysical Journal International*, **182**, 155–167.

Barenblatt, G. I. (1962). The mathematical theory of equilibrium cracks in brittle fracture. *Advances in Applied Mechanics*, **7**, 55–125.

Battaglia, J., Ferrazzini, V., Staudacher, T., Aki, K. and Cheminée, J.-L. (2005). Pre-eruptive migration of earthquakes at the Piton de la Fournaise volcano (Réunion Island). *Geophysical Journal International*, **161**, 549–558.

Björnsson, A. (1985). Dynamics of crustal rifting in NE Iceland. *Journal of Geophysical Research*, **90**(B12), 10 151–10 162.

Blackman, D. K., Nishimura, C. E. and Orcutt, J. A. (2000). Seismoacoustic recordings of a spreading episode on the Mohns Ridge. *Journal of Geophysical Research*, **105**, 10 961–10 974.

Bolchover, P. and Lister, J. R. (1999). The effect of solidification on fluid driven fracture, with application to bladed dykes. *Proceedings of the Royal Society of London, Series A*, **455**, 2389–2409.

Brandsdóttir, B. and Einarsson, P. (1979). Seismic activity associated with the September 1977 deflation of the Krafla central volcano in north-eastern Iceland. *Journal of Volcanology and Geothermal Research*, **6**, 197–212.

Buck, W. R. (2006). The role of magma in the development of the Afro-Arabian Rift System *Geological Society of London Special Publication*, **259**, 43–54, doi:10.1144/GSL.SP.2006.259.01.05

Buck, W.R., Einarsson, P. and Brandsdóttir, B. (2006). Tectonic stress and magma chamber size as controls on dike propagation: Constraints from the 1975–1984 Krafla rifting episode. *Journal of Geophysical Research*, **111**, B12404, doi:10.1029/2005JB003879.

Dahm, T. (2000). On the shape and velocity of fluid-filled fractures in the Earth. *Geophysical Journal International*, **142**, 181–192.

Delaney, P. T. and Pollard, D. D. (1981). Deformation of host rocks and flow of magma during growth of Minette dikes and breccia-bearing intrusions near Ship Rock, New Mexico. *United States Geological Survey Professional Paper* **1202**, 61pp.

Delaney, P. T. and Pollard, D. D. (1982). Solidification of basaltic magma during flow in a dike. *American Journal of Science*, **282**, 856–885.

Delaney, P. T., Pollard, D. D., Ziony, J. L. and McKnee, E. H. (1986). Field relations between dikes and joints: Emplacement processes and paleostress analysis. *Journal of Geophysical Research*, **91**, 4920–4938.

Ebinger, C. J., Keir, D., Ayele, A. *et al.* (2008). Capturing magma intrusion and faulting processes during continental rupture: seismicity of the Dabbahu (Afar) rift. *Geophysical Journal International*, **174**, 1138–1152.

Ebinger, C. J., Ayele, A., Keir, D. *et al.* (2010). Length and timescales of rift faulting and magma intrusion: The Afar rifting cycle from 2005 to present. *Annual Review of Earth and Planetary Sciences*, **38**, 439–466.

Einarsson, P. and Brandsdóttir, B. (1980). Seismological evidence for lateral magma intrusion during the July 1978 deflation of the Krafla volcano in NE-Iceland. *Journal of Volcanology and Geothermal Research*, **47**, 160–165.

Ernst, R. E. and Baragar, W. R. A. (1992). Evidence from magnetic fabric for the flow pattern of magma in the Mackenzie giant radiating dyke swarm. *Nature*, **356**, 511–513.

Garagash, D. and Detournay, E. (2000). The tip region of a fluid-driven crack in an elastic medium. *Journal of Applied Mechanics*, **67**, 183–192.

Grandin, R., Socquet, A., Binet, R. *et al.* (2009). September 2005 Manda Hararo-Dabbahu rifting event, Afar (Ethiopia): Constraints provided by geodetic data. *Journal of Geophysical Research*, **114**(B13), B08404, doi:10.1029/2008JB005843.

Grandin, R., Jacques, E., Nercessian, A. *et al.* (2011). Seismicity during lateral dike propagation: Insights from new data in the recent Manda Hararo–Dabbahu rifting episode (Afar, Ethiopia).

Geochemistry, Geophysics, Geosystems, **12**, Q0AB08, doi:10.1029/2010GC003434.

Gudmundsson, A. (2005). The effects of layering and local stresses in composite volcanoes on dyke emplacement and volcanic hazards. *Comptes Rendus Geosciences*, **337**, 1216–1222.

Gudmundsson, A. and Brenner, S. L. (2001). How hydrofractures become arrested. *Terra Nova*, **13**, 456–462.

Gudmundsson, A. and Brenner, S. L. (2005). On the conditions of sheet injections and eruptions in stratovolcanoes. *Bulletin of Volcanology*, **67**, 768–782.

Halls, H. C. and Fahrig, W. F., eds. (1987). *Mafic Dyke Swarms*. Geological Association of Canada Special Paper 34.

Hayashi, Y. and Morita, Y. (2003). An image of magma intrusion process inferred from precise hypocentral migration of the earthquake swarm east of the Izu Peninsula. *Geophysical Journal International*, **153**, 159–174.

Heimpel, M. and Olson, P. (1994). Buoyancy-driven fracture and magma transport through the lithosphere: Models and experiments. In *Magmatic Systems*, ed. M. P. Ryan. San Diego, CA: Academic Press, pp. 223–240.

Huppert, H. E. and Sparks, R. S. J (1988). The generation of granitic magmas by intrusion of basalt into continental crust. *Journal of Petrology*, **29**(3), 599–624.

Jolly, R. and Sanderson, D. (1995). Variation in the form and distribution of dykes in the Mull swarm, Scotland. *Journal of Structural Geology*, **17**, 1543–1557.

Kavanagh, J. L. and Sparks, R. S. J. (2011). Insights of dyke emplacement mechanics from detailed 3D dyke thickness datasets. *Journal of the Geological Society*, **168**, 965–978.

Kavanagh, J. L., Menand, T. and Sparks, R. S. J. (2006). An experimental investigation of sill formation and propagation in layered elastic media. *Earth and Planetary Science Letters*, **245**, 799–813.

Keir, D., Hamling, I. J., Ayele, A. *et al.* (2009). Evidence for focused magmatic accretion at segment centers from lateral dike injections captured beneath the Red Sea rift in Afar. *Geology*, **37**(1), 59–62.

Lengliné, O., Marsan, D., Got, J., Pinel, V., Ferrazzini, V. and Okubo, P. G. (2008). Seismicity and deformation induced by magma accumulation at three basaltic volcanoes. *Journal of Geophysical Research*, **113**, B12305, doi:10.1029/2008JB005937.

Lister, J. R. (1990a). Buoyancy-driven fluid fracture: similarity solutions for the horizontal and vertical propagation of fluid-filled cracks. *Journal of Fluid Mechanics*, **217**, 213–239.

Lister, J. R. (1990b). Buoyancy-driven fluid fracture: the effects of material toughness and of low-viscosity precursors. *Journal of Fluid Mechanics*, **210**, 263–280.

Lister, J. R. and Kerr, R. C. (1991). Fluid-mechanical models of crack propagation and their application to magma transport in dykes. *Journal of Geophysical Research*, **96**, 10 049–10 077.

McLeod, P. and Tait, S. R. (1999). The growth of dykes from magma chambers. *Journal of Volcanology and Geothermal Research*, **92**, 231–245.

Menand, T. and Tait, S. R. (2001). A phenomenological model for precursor volcanic eruptions. *Nature*, **411**, 678–680.

Menand, T. and Tait, S. R. (2002). The propagation of a buoyant liquid-filled fissure from a source under constant pressure: An experimental approach. *Journal of Geophysical Research*, **107**(B11), 8471–8481.

Mériaux, C. and Jaupart, C. (1995). Simple fluid dynamic models of volcanic rift zones. *Earth and Planetary Science Letters*, **136**, 223–240.

Morita, Y., Nakao, S. and Hayashi, Y. (2006). A quantitative approach to the dike intrusion process inferred from a joint analysis of geodetic and seismological data for the 1998 earthquake swarm off the east coast of Izu Peninsula, central Japan. *Journal of Geophysical Research*, **111**, B06208, doi:10.1029/2005JB003860.

Muller, J. R., Ito, G. and Martel, S. J. (2001). Effects of volcano loading on dike propagation in an elastic half-space. *Journal of Geophysical Research*, **106**(B6), 11 101–11 113.

Muskhelishvili, N. I. (1953). *Some Basic Problems of the Mathematical Theory of Elasticity*. Groningen: P. Noordhoff Ltd.

Okada, Y., Yamamoto, E. and Ohkubo, T. (2000). Coswarm and preswarm crustal deformation in the eastern Izu Peninsula, central Japan. *Journal of Geophysical Research*, **105**, 681–692.

Pedersen, R., Sigmundsson, F. and Einarsson, P. (2007). Controlling factors on earthquake swarms associated with magmatic intrusions: Constraints from Iceland. *Journal of Volcanology and Geothermal Research*, **162**, 73–80.

Pinel, V. and Jaupart, C. (2000). The effect of edifice load on magma ascent beneath a volcano. *Philosophical Transactions of the Royal Society of London, Series A*, **358**, 1515–1532.

Pinel, V. and Jaupart, C. (2004). Magma storage and horizontal dyke injection beneath a volcanic

edifice. *Earth and Planetary Science Letters*, **221**, 245–262.

Pollard, D. D., Delaney, P. T., Duffield, W. A., Endo, E. T. and Okamura, A. T. (1983). Surface deformation in volcanic rift zones. *Tectonophysics*, **94**, 541–584.

Rivalta, E. (2010). Evidence that coupling to magma chambers controls the volume history and velocity of laterally propagating intrusions. *Journal of Geophysical Research*, **115**, B07203, doi:10.1029/2009JB006922.

Roper, S. M. and Lister, J. R. (2007). Buoyancy-driven crack propagation: the limit of large fracture toughness. *Journal of Fluid Mechanics*, **580**, 359–380.

Rubin, A. M. (1992). Dike-induced faulting and graben subsidence in volcanic rift zones. *Journal of Geophysical Research*, **97**, 1839–1858.

Rubin, A. M. (1993a). Tensile fracture of rock at high confining pressure: Implications for dike propagation. *Journal of Geophysical Research*, **98**, 15 919–15 935.

Rubin, A. M. (1993b). On the thermal viability of dikes leaving magma chambers. *Geophysical Research Letters*, **20**(4), 257–260.

Rubin, A. M. (1995). Propagation of magma-filled cracks. *Annual Review of Earth and Planetary Sciences*, **23**, 287–336.

Rubin, A. M. (1998). Dike ascent in partially molten rock. *Journal of Geophysical Research*, **103**(B9), 20 901–20 919.

Rubin, A. M. and Gillard, D. (1998). Dike-induced earthquakes: Theoretical considerations. *Journal of Geophysical Research*, **103**, 10 017–10 030.

Rubin, A. M. and Pollard, D. D. (1987). Origins of blade-like dikes in volcanic rift zones. In *Volcanism in Hawaii*, ed. R. W. Decker, T. L. Wright and P. H. Stauffer, *United States Geological Survey Professional Paper 1350*, 1449–1470.

Singh, S. C., Crawford, W. C., Carton, H. *et al.* (2006). Discovery of a magma chamber and faults beneath a Mid-Atlantic Ridge hydrothermal field. *Nature*, **442**, 1029–1032.

Sleep, N. H. (1974). Segregation of magma from a mostly crystalline mush. *Geological Society of America Bulletin*, **85**, 1225–1232.

Spence, D. A. and Turcotte, D. L. (1985). Magma-driven propagation of cracks. *Journal of Geophysical Research*, **90**, 575–580.

Spence, D. A., Sharp, P. W. and Turcotte, D. L. (1987). Buoyancy-driven crack propagation: a mechanism for magma migration. *Journal of Fluid Mechanics*, **174**, 135–153.

Taisne, B. and Jaupart, C. (2009). Dike propagation through layered rocks. *Journal of Geophysical Research*, **114**, B09203, doi:10.1029/2008JB006228.

Taisne, B. and Jaupart, C. (2011). Magma expansion and fragmentation in a propagating dyke. *Earth and Planetary Science Letters*, **301**, 146–152.

Taisne, B. and Tait, S. (2009). Eruption versus intrusion? Arrest of propagation of constant volume, buoyant, liquid-filled cracks in an elastic, brittle host. *Journal of Geophysical Resarch*, **114**, B06202, doi:10.1029/2009JB006297.

Taisne, B. and Tait, S. (2011). The effect of solidification on a propagating dike. *Journal of Geophysical Research*, **116**, B01206, doi:10.1029/2009JB007058.

Taisne, B., Brenguier, F., Shapiro, N. M. and Ferrazzini, V. (2011a). Imaging the dynamics of magma propagation using radiated seismic intensity. *Geophysical Research Letters*, **38**, L04304, doi:10.1029/2010GL046068.

Taisne, B., Tait, S. and Jaupart, C. (2011b). Conditions for the arrest of a vertical propagating dyke. *Bulletin of Volcanology*, **73**(2), 191–204, doi: 10.1007/s00445-010-0440-1.

Takada, A. (1990). Experimental study on propagation of liquid-filled crack in gelatin: Shape and velocity in hydrostatic stress condition. *Journal of Geophysical Research*, **95**(B6), 8471–8481.

Tolstoy, M., Cowen, J. P., Baker, E. T. *et al.* (2006). A sea-floor spreading event captured by seismometers. *Science*, **314**, 1920–1922.

Weertman, J. (1971). Theory of water-filled crevasses in glaciers applied to vertical magma transport beneath oceanic ridges. *Journal of Geophysical Research*, **76**, 1171–1183.

Wright, T. J., Ebinger, C., Biggs, J. *et al.* (2006). Magma-maintained rift segmentation at continental rupture in the 2005 Afar dyking episode. *Nature*, **442**, 291–294.

Exercises

3.1 Consider an experiment in which a *fixed volume* of buoyant liquid is released in an elastic medium.

   (a) Derive the buoyancy length scale given by Eq. (3.10).

   (b) Derive the power laws that link length and time. (Hint: In the first phase of propagation make the penny-shape approximation, and in the second phase assume elongation at constant breadth.)

3.2 Consider a penny-shaped crack of fixed volume in an elastic solid of the same density. Starting from some initial size, the crack grows because the stress intensity factor exceeds the fracture toughness of the solid.

   (a) What is the time dependence of the evolution of the diameter? (Hint: you should derive the appropriate power-law given in Table 3.1.)

   (b) What will be the final static size of the injection? Take $V = 10^6$ m^3, $E = 10^9$ Pa, $K_c = 10^6$ Pa m$^{1/2}$, and $v = 0.5$.

3.3 Figure 3.6 shows the principal balances that exist within a propagating dike fed by a *constant flux*. From those simple balances you should be able to extract the geometrical characteristics of the problem. The thickness H of the dike, for instance, can be estimated using the equilibrium in the tail region, where the viscous pressure drop balances the buoyancy, which means that we can write $P_v \sim P_b$ and replace each pressure by their expressions.

   (a) What is the expression for H? Compare it with Eq. (3.3a).

   (b) Next find the typical length L, using the equilibrium that exists within the head region, $P_b \sim P_e$.

   (c) For evaluation of H and L, use reasonable order of magnitude values for each parameter: viscosity $\eta = 10$ Pa s, density difference $\Delta\rho = 100$ kg m^{-3}, volumetric flux $Q_{3D} = 10^6$ m^3 s^{-1}, horizontal extent (breadth) $B = 10$ km, Poisson ratio $v = 0.5$, and Young's modulus $E = 10$ GPa. Discuss the values obtained for H and L.

Online resources available at www.cambridge.org/fagents

- Additional figures
- Answers to exercises

Chapter 4

Dynamics of magma ascent in the volcanic conduit

Helge M. Gonnermann and Michael Manga

Overview

This chapter presents the various mechanisms and processes that come into play within the volcanic conduit for a broad range of effusive and dry explosive volcanic eruptions. Decompression during magma ascent causes volatiles to exsolve and form bubbles containing a supercritical fluid phase. Viscous magmas, such as rhyolite or crystal-rich magmas, do not allow bubbles to ascend buoyantly and may also hinder bubble growth. This can lead to significant gas overpressure and brittle magma fragmentation. During fragmentation in vulcanian, subplinian, and plinian eruptions, gas is released explosively into the atmosphere, carrying with it magma fragments. Alternatively, high viscosity may slow ascent to where permeable outgassing through the vesicular and perhaps fractured magma results in lava effusion to produce domes and flows. In low-viscosity magmas, typically basalts, bubbles may ascend buoyantly, allowing efficient magma outgassing and relatively quiescent magma effusion. Alternatively, bubbles may coalesce and accumulate to form meter-size gas slugs that rupture at the surface during strombolian eruptions. At fast magma ascent rates, even in low-viscosity magmas, melt and exsolved gas remain coupled, allowing for rapid acceleration and hydrodynamic fragmentation in hawaiian eruptions.

4.1 | Introduction

In the broadest sense, volcanic eruptions are either effusive or explosive. During explosive eruptions magma fragments and eruption intensity is ultimately related to the fragmentation mechanism and associated energy expenditure (Zimanowski et al., 2003). If the cause of fragmentation is the interaction of hot magma with external water, the ensuing eruption is called phreatomagmatic. Eruptions that do not involve external water are called "dry," in which case the abundance and fate of magmatic volatiles, predominantly H_2O and CO_2, as well as magma rheology and eruption rate, are the dominant controls on eruption style. Eruption styles are often correlated with magma composition and to some extent this relationship reflects differences in tectonic setting, which also influence magmatic volatile content and magma supply rate.

Modeling Volcanic Processes: The Physics and Mathematics of Volcanism, eds. Sarah A. Fagents, Tracy K. P. Gregg, and Rosaly M. C. Lopes. Published by Cambridge University Press. © Cambridge University Press 2013.

4.1.1 Getting magma to the surface

Volcanic eruptions represent the episodic or continuous surface discharge of magma from a storage region. We shall refer to this storage region as the magma chamber and to the pathway of magma ascent as the conduit. For magma to erupt, the chamber pressure has to exceed the sum of frictional and magma-static pressure losses, as well as losses associated with opening of the conduit. The latter may be due to tectonic forces and/or excess magma pressure, which in turn are a consequence of magma replenishment and/or volatile exsolution in a chamber within an elastically deforming wall rock. Magma-static pressure loss is strongly dependent on magma density, which is dependent on volatile exsolution and volume expansion as pressure decreases during ascent. Feedbacks between chamber and conduit can produce complex and unsteady eruptive behavior.

4.1.2 The volcanic conduit

Eruptions at basaltic volcanoes may begin as fissure eruptions, with typical dike widths of the order of one meter (Chapter 3; Wilson and Head, 1981). However, magma often erupts at the surface through structures that more closely resemble cylindrical conduits. The processes by which flow paths become focused into discrete cylindrical vents are not well understood. One possibility is that the temperature-dependent viscosity of magma results in a feedback between cooling, viscosity increase and reduction in flow velocity, so that the flow becomes more focused over time (Wylie-Lister, 1995). In addition, thermal erosion may also contribute to flow focusing (Bruce Huppert, 1989).

The conduit during silicic explosive eruptions is generally thought to be more or less cylindrical, at least at shallow depths. However, the conduit cross-sectional area can change with depth and time because of wall-rock erosion, caused by abrasion as pyroclasts collide with conduit walls, shear stress from the erupting magma, or wall collapse (Wilson et al., 1980; Macedonio et al., 1994; Kennedy et al., 2005; Mitchell, 2005).

4.2 | Volatiles

Volatiles play the central role in governing the ascent and eruption of magma, as summarized in Figure 4.1. Dissolved volatiles, in particular water, have large effects on melt viscosity. Exsolved volatiles form bubbles of supercritical fluids, which we will also refer to as gas or vapor bubbles. These bubbles allow for significant magma compressibility and buoyancy, ultimately making eruption possible (Pyle and Pyle, 1995; Woods and Cardoso, 1997). The decrease in pressure associated with magma ascent reduces volatile solubility, leading to bubble nucleation and growth, the latter a consequence of both volatile exsolution and expansion (Sparks, 1978). Volatile exsolution also promotes microlite crystallization, which in turn affects bubble growth and rheology – there is a complex feedback between decompression, exsolution, and crystallization (Tait et al., 1989). The occurrence and dynamics of explosive eruptions are mediated by the initial volatile content of the magma, and the ability of gases to escape from the ascending magma.

4.2.1 Solubility

Volatile solubility is primarily pressure dependent, with secondary dependence on temperature, melt composition, and volatile speciation (McMillan, 1994; Blank and Brooker, 1994; Zhang et al., 2007). Formulations for volatile solubility can be thermodynamic or empirical. In either case, calibration is achieved using measurements of dissolved volatile concentrations in quenched melts, equilibrated with a volatile phase of known composition at fixed pressures and temperatures. The principal species we consider here are H_2O and CO_2. Solubility of CO_2 in silicate melts is proportional to partial pressure, whereas for H_2O, dissociation into molecular H_2O and OH results in a square root dependence on partial pressure. An empirical formulation for combined solubility of H_2O and CO_2 in rhyolitic melts at volcanologically relevant conditions is given by Liu et al. (2005):

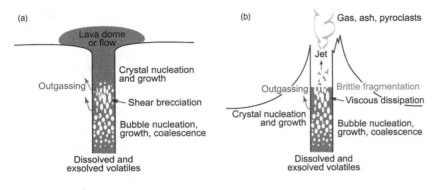

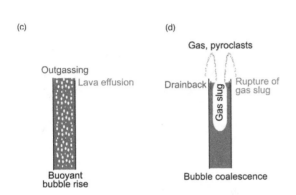

Figure 4.1 Schematic illustration of conduit processes. (a) In effusive eruptions of silicic magma (high viscosity, low ascent rate) gas may be lost by permeable flow through porous and/or fractured magma. (b) During (sub)plinian eruptions bubble walls rupture catastrophically at the fragmentation surface and the released gas expands rapidly as the flow changes from a viscous melt with suspended bubbles to gas with suspended pyroclasts. (c) Extensive loss of buoyantly rising bubbles occurs during effusive eruptions of low-volatile content, low-viscosity magma. (d) Coalescence and accumulation of buoyant bubbles, followed by their rupture at the surface, produces strombolian explosions in slowly ascending low-viscosity magmas. (e) Bubbles remain coupled to the melt in low-viscosity, hawaiian eruptions, which are characterized by relatively high ascent rates, hydrodynamic fragmentation, and sustained lava fountaining.

$$C_c = p_c \left[\frac{5668 - 55.99\, p_w}{T} + \left(0.4133\sqrt{p_w} + 0.002041\, p_w^{3/2}\right) \right].$$

(4.2)

Here C_w is total dissolved H_2O in wt.%, C_c is dissolved CO_2 in ppm and T is temperature in kelvin. p_w and p_c are the partial pressures in MPa of H_2O and CO_2, respectively. This formulation is also approximately applicable to other melt compositions (Zhang *et al.*, 2007), but more accurate models are available (Dixon, 1997; Newman and Lowenstern, 2002; Papale *et al.*, 2006). Equilibrium concentrations of dissolved CO_2 and H_2O, based on this formulation, are shown in Figure 4.2. Notice that almost all CO_2 exsolves at > 100 MPa for rhyolite and at > 25 MPa for basalt, whereas most H_2O exsolves at < 100 MPa for rhyolite and at < 25 MPa for basalt. Consequently, processes in the shallow conduit are predominantly affected by H_2O exsolution.

4.2.2 Diffusivity

Volatile diffusivities in silicate melts are best characterized for H_2O. A recent formulation for

$$C_w = 0.0012439\, p_w^{3/2}$$
$$+ \frac{354.94\sqrt{p_w} + 9.623\, p_w - 1.5223\, p_w^{3/2}}{T}$$
$$+ p_c\left(-1.084\times10^{-4}\sqrt{p_w} - 1.362\times10^{-5}\, p_w\right)$$

(4.1)

and

(a)

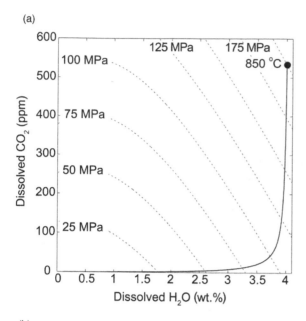

(b)

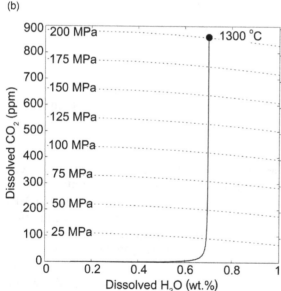

Figure 4.2 Solubility of CO_2 and H_2O (dashed contours) and equilibrium solubility degassing paths (solid lines) for typical rhyolitic and basaltic magmas undergoing closed-system degassing. (a) Rhyolite magma with no initial exsolved volatiles at 850 °C, initial pressure of 200 MPa, and initial dissolved H_2O content of 4 wt.%, using the solubility model of Liu *et al.* (2005). (b) Basaltic magma with no initial exsolved volatiles at 1300 °C, initial pressure of 200 MPa, and initial H_2O content of 0.7 wt.%, using the solubility model of Dixon (1997).

H_2O diffusivity (m² s⁻¹) in rhyolite with H_2O contents ≤2 wt. % is (Zhang and Behrens, 2000)

$$D_{wr} = C_w \exp\left(-17.14 - \frac{10661}{T} - 1.772\frac{p_m}{T}\right). \quad (4.3)$$

A formulation for higher H_2O concentrations is provided by Zhang *et al.* (2007), who also give the following equation for H_2O diffusivity in basalt

$$D_{wb} = C_w \exp\left(-8.56 - \frac{19110}{T}\right). \quad (4.4)$$

CO_2 diffusivity, D_c, is smaller than water diffusivity over a range of conditions (Watson, 1994). The most reliable formulation for D_c in silicate melts is based on argon diffusivity, which is essentially identical to CO_2 at volcanologically relevant conditions (Behrens and Zhang, 2001; Nowak *et al.*, 2004):

$$D_c = -18.239 - \frac{17367 + 1.0964\, p_m}{T} + \frac{(855.2 + 0.2712\, p_m)C_w}{T}. \quad (4.5)$$

4.2.3 Pre-eruptive volatile content of magmas

Water

Basaltic magmas show a wide range of water contents ranging from < 0.5 wt.% to 6–8 wt.% (Johnson *et al.*, 1994; Wallace, 2005), with non-arc basalts generally being considerably dryer than arc magmas, which have highly variable water contents. Silicic magmas also show a wide range of water contents up to 6 wt.%, and possibly higher (Carmichael, 2002).

Carbon dioxide

CO_2 content of magmas is more difficult to constrain, because of its low solubility at shallow depths (Fig. 4.2). Measurements of H_2O and CO_2 in melt inclusions, in conjunction with solubility relations, suggest that arc magmas contain several wt.% of CO_2, much of which exsolves pre-eruptively, for example during vapor-saturated fractional crystallization (Papale, 2005; Wallace, 2005). Thus, arc magmas may already contain, or have lost, significant amounts of exsolved

CO_2 prior to eruption. CO_2 contents of unde-gassed mid-ocean ridge basalts and ocean island basalts appear to be about 1 wt.% or less (Hauri, 2002). Because CO_2 exsolves at greater pressures than H_2O, relative proportions of CO_2 and H_2O in erupted gases, glasses and melt inclusions can provide constraints on depths of magma degassing. Because typical CO_2 contents of sili-cic magmas are considerably lower than H_2O, the latter dominates eruption dynamics.

Sulfur

H_2S and SO_2 are, after water and carbon diox-ide, the most abundant magmatic volatiles. S concentrations are, at a minimum, several thou-sand ppm in arc and back-arc magmas (Wallace, 2005). Volcanic sulfur emissions are spectro-graphically readily detectable and, therefore, play an important role in monitoring of active volcanoes via remote sensing. Emissions from active volcanoes, especially during voluminous explosive eruptions, have significant environ-mental and atmospheric impacts (Bluth *et al.*, 1993). Measured SO_2 fluxes from erupting volca-noes often exceed, by one to two orders of mag-nitude, the amounts thought to be dissolved in the magma prior to eruption. This "excess sulfur" problem is sometimes attributed to an exsolved S-bearing volatile phase accumulating within the magmatic system prior to eruption (Wallace *et al.*, 2003).

Chlorine and Fluorine

Measurable Cl and F are also present in volcanic gases, with Cl concentrations in the range of hundreds to thousands of ppm in arc and back-arc magmas (Wallace, 2005). Because S is less sol-uble than Cl, which in turn is less soluble than F (Carroll and Webster, 1994), these gases are also used to constrain depths of volatile exsolu-tion and magma degassing histories (Allard *et al.*, 2005).

4.3 | Bubbles

The "birth," "life," and "death" of bubbles are key to understanding the dynamics of volcanic eruptions and are the subject of this section.

4.3.1 Nucleation

Pressure decreases as magma ascends. If the concentration of dissolved volatiles exceeds the equilibrium solubility at a given pressure, the melt is supersaturated, a requirement for bub-ble nucleation. Supersaturation, Δp_s, is defined as the difference between actual pressure and the pressure at which the concentration of dis-solved volatiles would be in equilibrium with the co-existing vapor phase. The supersatur-ation required for nucleation corresponds to the energy that must be supplied to increase the sur-face area between two fluids. It is a consequence of surface tension, γ, a measure of the attractive molecular forces that produce a jump in pressure across a curved interface between two fluids. This so-called Laplace or capillary pressure is $2\gamma/R$ for a spherical bubble of radius R. Supersaturation is typically produced when fast magma ascent allows insufficient time for volatile diffusion into existing bubbles, which occurs when $\tau_{\text{dif}}/\tau_{\text{dec}} \gg 1$ (Toramaru, 1989). Here, $\tau_{\text{dec}} = p_m/\dot{p}_m$ is the char-acteristic decompression time, $\tau_{\text{dif}} = (S - R)^2/D$ is the characteristic time for volatile diffusion, S is the radial distance from bubble center to the midpoint between adjacent bubbles, D is volatile diffusivity in the melt, p_m is pressure of the melt, and $\dot{p}_m$ is the decompression rate.

Bubble nucleation can be heterogeneous or homogeneous. In heterogeneous nucleation, crystals provide substrates that facilitate nucle-ation, because of the lower interfacial energy between solid and vapor than between melt and vapor. Typically, Δp_s is a few MPa for heteroge-neous nucleation (Hurwitz and Navon, 1994; Gardner and Denis, 2004; Gardner, 2007) and ~10 to 100 MPa for homogeneous nucleation, in which nucleation sites are lacking (Mangan and Sisson, 2000).

Classical nucleation theory (Hirth *et al.*, 1970) predicts a very strong dependence of the nucle-ation rate J (number of bubbles per unit volume per unit time) on Δp_s and γ

$$J \propto \exp\left[-\frac{16\pi\gamma^3 \psi}{3k_B T \Delta p_s^2}\right]. \qquad (4.6)$$

Here k_B is the Boltzmann constant and γ is typ-ically 0.05 to 0.3 N m^{-1}, with a dependence on both temperature and composition of the melt (Bagdassarov et al., 2000; Mangan and Sisson, 2005). ψ is a geometrical factor that ranges between 0 and 1, depends on θ, the contact angle between bubble and crystal, and is given by

$$\psi = \frac{(2 - \cos\theta)(1 + \cos\theta)^2}{4}. \qquad (4.7)$$

J determines the number density of bubbles per unit volume of melt, N_d, and if bubbles nucleate at large Δp_s, they do so at a high rate.

4.3.2 Growth

As ambient pressure p_m decreases during magma ascent, bubbles may grow if volatiles exsolve and the vapor expands. The momentum balance for a growing bubble is obtained by neglecting inertial terms in the Rayleigh–Plesset Equation (Scriven, 1959). This is justified because large melt viscosities and short length scales result in Reynolds numbers for bubble growth $\ll 1$. The resulting equation for a bubble surrounded by a shell of melt with constant viscosity is (Proussevitch et al., 1993a)

$$p_g - p_m = \frac{2\gamma}{R} + 4\eta_0 v_R \left(\frac{1}{R} - \frac{R^2}{S^3} \right). \qquad (4.8)$$

Here, p_g is the pressure inside the bubble, η_0 is the Newtonian melt viscosity, and v_R is the radial velocity of the melt–vapor interface, i.e., bubble wall. An idealized uniform packing geometry is assumed, so that each bubble can be represented as a sphere of radius R, surrounded by a spherical melt shell of thickness $S - R$. In this formulation the time-dependent p_m couples bubble growth with magma ascent. Mass conservation of volatiles requires that

$$d\left(\rho_g R^3 \right) = 4R^2 \rho_m \sum_i D_i \left(\frac{\partial C_i}{\partial r} \right)_{r=R} dt, \qquad (4.9)$$

where r is radial distance from the center of the bubble, t is time, ρ_g is the pressure-dependent vapor density, ρ_m is melt density, D_i is the concentration-dependent diffusivity of dissolved volatile species i, and C_i is the mass fraction of dissolved volatile i. Volatile diffusion is governed by

$$\frac{\partial C_i}{\partial t} + v_r \frac{\partial C_i}{\partial r} = \frac{1}{r^2} \frac{\partial C_i}{\partial r} \left(D_i\, r^2 \frac{\partial C_i}{\partial r} \right), \qquad (4.10)$$

with v_r as the radial velocity of melt at radius r. The boundary condition at $r = S$ is $\partial C_i / \partial r = 0$. At $r = R$ the prescribed concentration is obtained from a suitable solubility model with the additional assumption that dissolved volatiles at the melt–vapor interface are locally in equilibrium with the vapor at p_g.

Under most conditions bubble growth is limited by (1) the rate of viscous flow of the melt to accommodate the ensuing volume increase; (2) diffusion rate of volatiles to the melt–vapor interface where they can exsolve; or (3) the change in solubility caused by decompression. For each growth limit it is possible to simplify the solution of the governing equations or to derive analytical growth laws.

Modeling of bubble growth

Explicit modeling of diffusive bubble growth during magma ascent was first presented by Sparks (1978), based on the equations of Scriven (1959) and Rosner and Epstein (1972). Subsequently, Proussevitch et al. (1993a) applied a model of bubble growth in magmas that included the diffusion of volatiles. This model was enhanced by Sahagian and Proussevitch (1996) and Proussevitch and Sahagian (1998) to include the thermal effects associated with the heat of volatile exsolution and the work done against the surrounding melt by bubble expansion. Figure 4.3 shows the results for diffusive bubble growth in a rhyolite containing both H_2O and CO_2.

Viscous limit

In the viscous limit $\tau_{\text{vis}} / \tau_{\text{dec}} \gg 1$ and bubble growth is retarded, following an exponential growth law (Lensky et al., 2004). Here $\tau_{\text{vis}} = \eta_0/(p_g - p_m)$ is the characteristic viscous timescale. The viscous limit to bubble growth is the consequence of fast decompression rates relative to the viscous deformation of melt by the growing bubbles, and is most significant at melt viscosities in excess of $10^9\,\text{Pa s}$ (Sparks et al., 1994), which in the absence of significant amounts

(a)

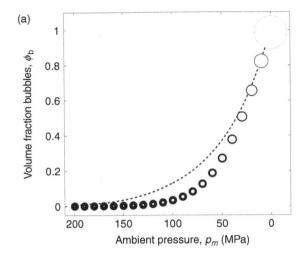

(b)

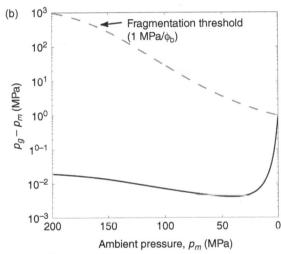

Figure 4.3 Model results for diffusive bubble growth in rhyolite with initial conditions of bubble volume fraction ϕ_0 = 0.001, bubble number density $N_d = 10^{11}$ m^{-3}, 5 wt.% H_2O, 268 ppm CO_2, T = 850 °C, magma pressure p_m = 200 MPa, and decompression rate $\Delta\dot{p}$ = 0.1 MPa s^{-1}. (a) Circles represent cross-sections through bubbles along the modeled trend of ϕ_b vs. p_m. For reference, the short-dashed black line represents the equilibrium closed-system degassing trend. Initially ϕ_b remains lower than equilibrium because volatile diffusion is slower than $\Delta\dot{p}$, and bubble overpressure $p_g - p_m \approx 2\gamma/R$. As bubble radius R increases and bubble separation, $S - R$, decreases, diffusion starts to keep pace with decompression and ϕ_b approaches equilibrium values. (b) Bubble overpressure, $p_g - p_m$, vs. p_m (solid curve), with fragmentation threshold, Δp_f (dashed curve) of Spieler et al. (2004). At p_m < 30 MPa, bubble growth becomes viscously limited, a consequence of H_2O exsolution and increasing melt viscosity. Consequently, magma pressure p_m decreases more rapidly than bubble vapor pressure pressure p_g until the bubble overpressure reaches the fragmentation threshold pressure, i.e., $p_g - p_m = \Delta p_f$.

of crystals, is achieved for silicic melts at shallow depths and low H_2O content. Because of retarded bubble growth, p_g decreases at a slower rate than p_m, resulting in the build-up of overpressure with $p_g - p_m \gg 2\gamma/R$ (Fig. 4.3). At the same time, a high p_g inhibits volatile exsolution, causing supersaturated conditions. This is referred to as "viscosity quench" (Thomas et al., 1994) and increases the potential for accelerated bubble nucleation and/or growth.

Diffusive limit

During decompression, the volatile concentration at the melt–vapor interface, based on the assumption of local equilibrium, decreases. This creates a concentration gradient for volatiles to diffuse to this interface and exsolve. If $\tau_{dec} \ll \tau_{dif}$, volatile concentrations will remain close to equilibrium throughout the melt. On the other hand, if $\tau_{dif}/\tau_{dec} \gg 1$ (i.e., at high decompression rates) the melt becomes supersaturated, which is a necessary condition for bubble nucleation (Lensky et al., 2004). Furthermore, in the diffusive limit, volatiles with different diffusivities may fractionate (Gonnermann and Manga, 2005b; Gonnermann and Mukhopadhyay, 2007).

Solubility limit

During solubility-limited growth, bubbles are close to mechanical and chemical equilibrium, that is, $\tau_{dif}/\tau_{dec} \ll 1$ and $\tau_{vis}/\tau_{dec} \ll 1$ (Lensky et al., 2004) so that both overpressure and supersaturation are small. Dissolved volatiles are near their equilibrium concentrations throughout the melt and bubble radius R can be directly calculated from mass balance, equilibrium solubility, and an equation of state. Conditions favorable to solubility-limited bubble growth are low melt viscosity and/or low magma ascent rates.

4.3.3 Coalescence

Coalescence takes place when bubbles come into increasingly close proximity and the melt film between bubbles ruptures. This may be the consequence of (1) gravitational or capillary drainage of interstitial liquid (Proussevitch et al., 1993b); (2) coalescence-induced coalescence, whereby the deformation of coalesced bubbles by capillary forces induces a flow field that brings nearby bubbles into proximity (Martula *et al.*, 2000); (3) bubble growth leading to stretching and thinning of the liquid film that separates individual bubbles (Borrell and Leal, 2008); and (4) advection and collision of bubbles by buoyancy or by magma flow (Manga and Stone, 1994).

Gravitational and capillary film drainage, as well as bubble collisions are most important in low viscosity magmas. A positive feedback between bubble coalescence, which increases bubble size, and bubble mobility due to buoyancy is expected (Section 4.3.5). The transition between capillary and gravitational drainage takes place at $R \sim \sqrt{\gamma / (p_m - p_g)g}$ where g is acceleration due to gravity. A characteristic velocity for capillary drainage can be obtained by balancing the capillary pressure gradient, γ/R^2, with viscous resistance to flow, $\eta_0 v_f/R^2$, where v_f is the velocity in the film. Because the bubble–melt surface is assumed to be a free-slip surface, the characteristic length scale is the bubble radius R rather than film thickness. The resultant scaling $v_f \sim \gamma/\eta_0$ highlights the importance of melt viscosity. Whether and at what rate bubble collisions result in coalescence depends on the Weber number, $We = 2\rho v^2 R/\gamma$, where v is the velocity at which two bubbles approach one another.

In high-viscosity magmas, bubbles remain essentially "frozen" in the melt and coalescence is principally a consequence of bubble growth. In this case, coalescence appears to create a permeable network of bubbles due to relatively persistent holes in the ruptured melt films (Klug and Cashman, 1996). Because the interfacial tension between melt and crystals tends to be lower than between gas and crystals, coalescence in the presence of crystals is expected to occur at greater film thickness than in the absence of crystals (Proussevitch *et al.*, 1993b).

4.3.4 Breakup

The breakup of bubbles into several smaller bubbles originates with their deformation by some combination of viscous stresses and inertial forces. In silicic magmas, the bubble Reynolds number, $Re_b = \dot{\varepsilon} R^2 \rho/\eta_0$, will always be small and inertia can be neglected. Here $\dot{\varepsilon}$ is the strain rate in the melt caused by buoyant

rise of the bubble or by magma flow. For small Re_b, deformation scales with the Capillary number, $Ca = \eta_0 \dot{\varepsilon} R/\gamma$, which characterizes the relative importance of viscous stresses that tend to deform bubbles, and surface tension stresses that act to keep bubbles spherical. If Ca exceeds some critical value, Ca_{cr}, bubble elongation by the flow becomes large enough that bubbles will break up. The value of the Ca_{cr} depends on the steadiness of the flow, melt viscosity, and flow type. For the viscosities of silicate melts, Ca_{cr} should range from ~1 to > 10^3. In both types of flows, bubbles will become highly elongated before breakup occurs. Capillary numbers will generally be large enough in conduits for large deformation and breakup to occur. However, large deformation also requires large strains, which may be limited to the sides of conduits. It is thought that tube pumice is a manifestation of these conditions (Marti *et al.*, 1999).

In basaltic magmas, inertial forces can no longer be neglected once bubbles become larger than a few centimeters. In this limit, velocity differences across bubbles can lead to breakup. The relative importance of inertial forces and surface tension forces is characterized by We, based on the velocity difference across the melt that surrounds the bubble. For large Re_b, breakup occurs if We exceeds a critical value, We_{cr}, which implies that there might be a maximum stable bubble size (Hinze, 1955). Although the actual value of We_{cr} depends on the origin of the velocity differences, it is generally between 1 and 5 for a wide range of flow conditions.

4.3.5 Bubble mobility

Bubbles are buoyant and their rise speed depends on size, volume fraction, and viscosity. If buoyant bubble rise is much slower than magma ascent, bubbles are dispersed throughout the continuous melt phase and move passively with the flow. Such flows are called "dispersed" and in the asymptotic limit of infinitesimally small dispersed bubbles the flow is termed "homogeneous." If the velocity of buoyant bubble rise is similar to or greater than magma ascent velocities, bubbles are

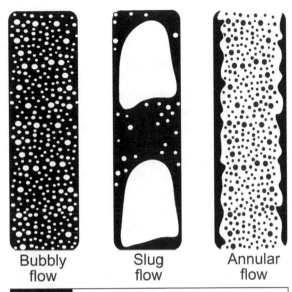

Bubbly
flow

Slug
flow

Annular
flow

Figure 4.4 Flow regimes observed in bubble column experiments (e.g., Wallis, 1969). The gas phase is shown in white and the liquid phase in black.

decoupled from the liquid phase and the flow is called "separated" (Brennen, 2005). During separated flow in laboratory experiments the topology of the gas and liquid phases is associated with different flow regimes (Fig. 4.4), which depend primarily on the relative volumetric flow rates of liquid and gas phases (Wallis, 1969). As this flux ratio increases to values of ~1, the flow transitions from "bubbly flow," where bubbles are randomly dispersed, to "slug flow," where coalesced bubbles form rapidly ascending gas slugs of diameter comparable to the conduit diameter. At flux ratios above ~10, the flow transitions to "annular flow," where the liquid phase forms an annular ring surrounding a cylindrical core of rapidly ascending gas. However, the extent to which these different flow regimes are applicable to volcanic eruptions remains controversial (Chapter 6).

The terminal velocity of a single bubble containing vapor of viscosity η_g and density ρ_g, rising in an infinite liquid of viscosity η_0 and density ρ_m is (Batchelor, 1967)

$$U_t = \frac{R^2 g\left(\rho_m - \rho_g\right)}{3\eta_0}\frac{\eta_0 + \eta_g}{\eta_0 + 3\eta_g/2}.$$

(4.11)

Therefore, bubbles within a silicic melt have negligible mobility, even at very low eruption rates, whereas significant bubble mobility can be expected for mafic melts. At finite volume fractions of bubbles, ϕ_b, hydrodynamic interactions reduce U_t by a factor $h(\phi_b)$, the hindering function. Various formulations exist for $h(\phi_b)$ and a commonly adopted one is the Richardson–Zaki equation (Richardson and Zaki, 1954):

$$h(\phi_b) = \left(1 - \phi_b\right)^n.$$

(4.12)

Here n is an empirical coefficient for the appropriate range of bubble size and flow conditions, typically with a value around 2.5 (Zenit et al., 2001). At flow regimes other than bubbly flow a different formulation for the relative velocity between gas and liquid phase has to be used (Dobran, 2001).

4.3.6 Bubbles and pressure loss

In general, mass discharge rate depends on the total pressure drop between magma chamber and vent. Within the conduit the pressure gradient is to first order the sum of magma-static pressure loss, $(dp/dz)_p$, and frictional pressure loss, $(dp/dz)_\eta$. Magma-static pressure depends on magma density and, hence, vesicularity as

$$\left(\frac{dp}{dz}\right)_\rho = \rho g \approx (1 - \phi_b)\rho_m g.$$

(4.13)

For dispersed bubbly flow the frictional pressure loss can be approximated on the basis of established single-phase flow correlations using the appropriate mixture viscosity (Ghiaasiaan, 2008):

$$\left(\frac{dp}{dz}\right)_\eta = \rho u^2 \frac{f}{a},$$

(4.14)

where a is conduit radius, u is magma velocity, and f is the friction factor given by

$$f = \frac{16}{\text{Re}} + f_0.$$

(4.15)

Here $\text{Re} = 2u\rho a/\eta$, where η is the magma viscosity and f_0 has values of about 0.002–0.02 (Mastin and Ghiorso, 2000). For highly viscous magmas the flow below the fragmentation level will, under most conditions, be laminar (Re < 10^3) and $f \approx 16/Re$. Above the fragmentation level the flow will be turbulent and $f \approx f_0$. For basaltic magma, frictional pressure loss over a wide range of ascent velocities is less than magma-static pressure loss and the flow dynamics will be more sensitive to changes in the density of the ascending magma than for eruptions where frictional pressure losses are larger. Consequently, under separated flow conditions there is potential for feedbacks between discharge rate, gas-to-melt flux ratios, and flow regime (Seyfried and Freundt, 2000; Guet and Ooms, 2006).

4.3.7 Permeable outgassing

The degassing history during ascent of an individual magma parcel may be complex and in many cases gas may separate from rising parcels of magma. We refer to this as outgassing, which may be a consequence of buoyant bubble rise, magma fragmentation, or permeable gas flow. Chapters 6 and 8 discuss outgassing associated with separated flow and high gas-to-melt flux ratios during strombolian and perhaps hawaiian eruptions. In some silicic eruptions high gas-to-melt flux ratios also suggest decoupled gas flow (Edmonds et al., 2003). However, high viscosities and negligible bubble mobility require that outgassing is associated with magma permeability, presumably a consequence of coalescing bubbles that form a permeable network (Eichelberger et al., 1986; Klug and Cashman, 1996) and perhaps cracks and fractures produced by brittle magma deformation (Gonnermann and Manga, 2003; Tuffen et al., 2003; Gonnermann and Manga, 2005a). It has been suggested that if conduit walls are permeable, which is controversial (Boudon et al., 1998), then volatiles may also escape laterally from ascending magma into the conduit walls and permeable outgassing has the potential to modulate eruptive behavior (Jaupart and Allegre, 1991; Woods and Koyaguchi, 1994; Eichelberger, 1995).

Vesicularity-permeability measurements for volcanic rocks, and by inference magmas, are usually analyzed using the Kozeny–Carman equation (Carman, 1956) where permeability, k, is given as

$$k(\phi_b) = \chi(\phi_b - \phi_{bp})^{\beta}. \qquad (4.16)$$

Here χ is an empirical constant and ϕ_{bp} is the volume fraction of bubbles that corresponds to percolation threshold (Saar and Manga, 1999), with typical values of $2 \leq \beta \leq 4$. An uncontroversial permeability model for vesicular magma remains elusive (Takeuchi et al., 2005; Wright et al., 2009). Furthermore, it appears that the nature of magma deformation plays a critical role in creating permeability (Okumura et al., 2008, 2010).

Permeable outgassing can be modeled as flow through a porous medium by combining the continuity equation for the exsolved volatile phase with Darcy's equation or, if the velocity of the gas phase becomes too large for inertial effects to be neglected, Forcheimer's equation (Rust and Cashman, 2004; Degruyter et al., 2012).

4.4 | Crystal nucleation and growth

Crystals nucleate due to undercooling, which is predominantly a consequence of H_2O exsolution. Nucleation rates determine the crystal number density and size distribution, which provide a record of the magma's ascent history. During crystallization the volatile content of the residual melt phase increases because volatiles preferentially partition into the melt phase. This in turn affects bubble nucleation and growth, as well as melt viscosity.

Undercooling, the thermodynamic driving force for crystallization, is the chemical potential of H_2O in the melt and can be expressed as ΔT_e, the difference between actual temperature, T, and the liquidus temperature, T_l. Because the latter depends on composition and volatile content, bubble nucleation and growth will increase T_l and can result in sufficient undercooling for crystallization (Hammer, 2004). The rate of homogeneous nucleation of critical nuclei, I (m^{-3} s^{-1}), is given by classical nucleation theory as

$$I = I_0 \exp\left(-\frac{\Delta G^* + \Delta G_D}{k_B T}\right). \qquad (4.17)$$

Here I_0 is a reference nucleation rate, k_B is the Boltzmann constant, and ΔG_D is the activation energy for diffusion (James, 1985). ΔG^* is the free energy required to form a spherical nucleus of critical size and depends on the free energy σ, associated with the crystal–liquid interface. Therefore nucleation rate increases with decreasing σ (Mueller et al., 2000). The rate of heterogeneous nucleation also follows Eq. (4.17), but with a modified σ to account for the lower interfacial energies (Spohn et al., 1988). Because σ is relatively poorly constrained, estimates of crystal nucleation rates are typically associated with large uncertainties. Moreover, the validity of classical nucleation theory and its assumptions that (1) the interface between nucleus and melt is sharp; (2) the critical nucleus has the thermodynamic properties of the bulk solid; and (3) σ can be treated as a macroscopic property equal to the value for a planar interface, remain controversial.

Once a crystal nucleates its growth rate, Y, is given by (Spohn et al., 1988)

$$Y = Y_0\left[1 - \exp\left(\frac{\Delta H \Delta T_e}{k_B T_l T}\right)\right]\exp\left(-\frac{\Delta G_D}{k_B T}\right), \qquad (4.18)$$

where Y_0 is a reference growth rate and ΔH is the change in enthalpy between melt and crystalline phases. If empirical parameters such as T_l and σ are known, the above equations can be used to model magma crystallization. Alternatively, parameterizations for I and Y can be used to explore the dependence of eruption behavior on syneruptive crystallization (Melnik and Sparks, 1999). The prediction of crystal size distributions is possible through the use of the Avrami equation (Marsh, 1998)

$$\phi_x = 1 - \exp\left(-k_v Y^3 t^4 I\right). \qquad (4.19)$$

Here a given crystal volume fraction, ϕ_x, is assumed to correspond to thermodynamic equilibrium after crystallization over time, t, and with k_v as a volumetric factor.

4.5 | Magma rheology

Silicate melts form a disordered network of interconnected SiO_4 tetrahedra in which the self-diffusive motion of atoms, called structural relaxation, results in a continuous unstructured rearrangement of the molecular structure. In the presence of an applied stress, this molecular rearrangement results in a directional motion of SiO_4 tetrahedra relative to one another and macroscopically manifests itself as viscous flow (Moynihan, 1995).

The intrinsic viscosity of silicate melts varies over orders of magnitude, even within a single volcanic eruption. It depends on the degree of polymerization, a function of chemical composition and volatile content. Within realistic ranges of compositional variability, eruptive temperature, and volatile content, the viscosity of basaltic melts may vary between about 10 and 10^3 Pa s, and between about 10^4 and 10^{12} Pa s or more for silicic melts (Fig. 4.5). Changes in viscosity during individual eruptions are especially pronounced at low pressures (depths < 1 km), where most of the H_2O exsolves (Figs. 4.2 and 4.6). Magma rheology is discussed further in the context of lava flow emplacement in Chapter 5.

4.5.1 The effect of dissolved volatiles and temperature

Dissolved water dissociates into molecular water and hydroxyl ions, thereby depolymerizing the melt. In rhyolitic melts the effect of water content can be tremendous and exceeds the effect of temperature (Fig. 4.5), but it is less pronounced for mafic magmas (Giordano and Dingwell, 2003). In contrast, viscosity is less affected by CO_2 (Bourgue and Richet, 2001). Recent viscosity formulations applicable over a range of compositions, temperature, and water contents are discussed in Hui and Zhang (2007), Zhang et al. (2007), and Giordano et al. (2008).

4.5.2 The effect of deformation rate

If the applied rate of deformation exceeds a certain threshold, the induced molecular motions of the melt are no longer compensated by random reordering of the melt structure. This is

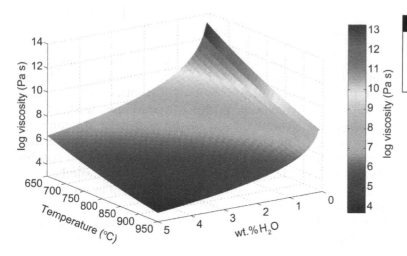

Figure 4.5 Dependence of rhyolitic melt viscosity on temperature and water content, after the empirical formulation of Hess and Dingwell (1996). See color plates section.

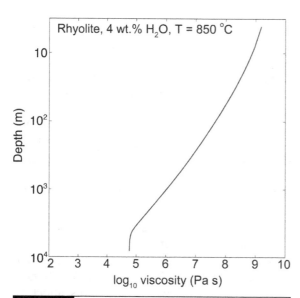

Figure 4.6 Dependence of rhyolitic melt viscosity (Hess and Dingwell, 1996) on depth, assuming lithostatic pressure and equilibrium volatile solubility. Note the wide range in viscosity and the tremendous viscosity increase at shallow depths, due to the exsolution of H_2O.

manifested in shear thinning, a decrease in viscosity from its "relaxed" or Newtonian value and will result in brittle failure if deformation rates exceed the onset of shear-thinning by approximately two orders of magnitude (Webb, 1997, and references therein). The strain rate at which shear thinning is first observed can be defined in terms of the viscous relaxation time, $\tau_r = \eta_0/G_\infty$, as

$$\dot{\varepsilon}_{st} \sim 10^{-3} \frac{G_\infty}{\eta_0}. \tag{4.20}$$

Here $G_\infty \sim 10^{10}$ Pa is the shear modulus and η_0 is the Newtonian melt viscosity, measured at $\dot{\varepsilon} \ll \dot{\varepsilon}_{st}$.

4.5.3 The effect of crystals

Crystals increase magma viscosity (Lejeune and Richet, 1995; Stevenson et al., 1996; Arbaret et al., 2007; Caricchi et al., 2007; Lavallee et al., 2007). For small crystal volume fractions, $\phi_x < 10–30\%$, and low strain rates, the viscosity of the crystalline magma relative to the crystal-free melt, $\eta_r = \eta/\eta_0$, can be approximated as

$$\eta_r = (1 - \phi_x/\phi_{xcr})^\alpha. \tag{4.21}$$

Here ϕ_{xcr} represents a critical volume fraction at which crystals start to impede the ability of the suspension to flow and $\alpha \approx -2.5$ (Llewellin and Manga, 2005). At ϕ_x greater than some value, typically in the range of 10–50%, crystals may come into contact, creating a framework that provides a finite yield strength. Shear thinning and shear localization may occur as melt and crystals redistribute themselves to decrease the flow disturbance. It most likely occurs in a narrow zone near the conduit wall and may produce a plug flow that can manifest itself at the surface as volcanic spine. A viscosity formulation that is calibrated to a broad range of crystal

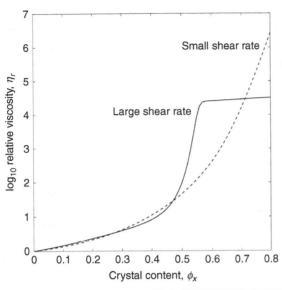

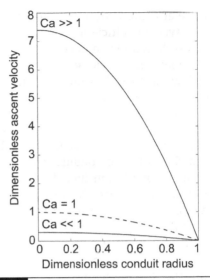

content and strain rates (Fig. 4.7) is provided by Caricchi *et al.* (2007);

$$\eta = \eta_0 \left[1 + \left(\frac{\phi_x}{\phi_{xcr}} \right)^{\delta} \right]$$
$$\left[1 - \alpha \operatorname{erf} \left(\frac{\sqrt{\pi}}{2\alpha} \frac{\phi_x}{\phi_{xcr}} \left[1 + \left(\frac{\phi_x}{\phi_{xcr}} \right)^{\beta} \right] \right) \right]^{-B\phi_{xcr}} . \quad (4.22)$$

Here $B = 2.5$ and the remaining empirical parameters δ, α, and β are provided in Figure 4.7.

4.5.4 The effect of bubbles

The effect of bubbles on viscosity depends on whether the bubbles are able to deform and is a function of Ca $= \eta_0 \dot{\varepsilon} R / \gamma$. If Ca $\ll 1$, bubbles remain nearly spherical and the viscosity is greater than that of the melt (Taylor, 1932). In contrast, for Ca $> O(1)$, bubbles can become elongated (Rust et al., 2003) and the viscosity decreases (Chapter 5; Bagdassarov and Dingwell, 1992; Manga et al., 1998; Lejeune et al., 1999; Stein and Spera,

2002). An expression for this dependence on Ca is given by Pal (2003) as

$$\eta_r \left(\frac{1 - 12\eta_r^2 \text{Ca}^2 / 5}{1 - 12\text{Ca}^2 / 5} \right)^{-4/5} = \left(1 - \frac{\phi_b}{\phi_{bcr}} \right)^{-\phi_{bcr}}, \quad (4.23)$$

where ϕ_{bcr} is an empirical parameter that is conceptually analogous to ϕ_{xcr}. The strain-rate dependent rheology of melt with suspended bubbles and crystals is derived from a linear superposition of the individual dependencies (Thies, 2002). As illustrated in Figure 4.8, during laminar flow at a given pressure gradient, the presence of bubbles can increase or decrease mass discharge rate by almost one order of magnitude. Chapter 5 further considers the combined effects of bubbles and crystals on magma rheology (Section 5.2.2).

4.6 | Magma fragmentation

During all explosive eruptions, pyroclasts are ejected from the volcanic vent. Here we focus on "dry" magma fragmentation, which does not involve any interaction with non-magmatic water. The type and efficiency of the fragmentation process determines pyroclast size and how much magmatic gas is released per unit mass of magma. This, in turn, has implications for the volcanic jet, column, and plume that are produced when the gas–pyroclast mixture exits the volcanic vent (Chapters 8 and 9). Fragmentation is thought to occur when specific conditions are reached within the conduit. The proposed criteria for fragmentation include: (1) a critical bubble volume fraction; (2) a stress criterion; (3) a strain-rate criterion; (4) a potential energy criterion; and (5) an inertial criterion.

The critical volume fraction criterion is thought to arise from some form of instability within the thin bubble walls, once $\phi_b \approx 0.75$ is reached (Verhoogen, 1951; Sparks, 1978). However, magma fragments have vesicularities that range from 0 (obsidian) to > 98% (reticulite) implying that a fragmentation criterion governed by a critical volume fraction cannot be generally applicable (though very high vesicularities may reflect post-fragmentation growth of bubbles).

The stress criterion is based on the view that fragmentation in high-viscosity magmas takes place when volatile overpressure, $\rho_g - \rho_m$, exceeds the tensile strength of the melt and ruptures bubble walls (Alidibirov, 1994; Zhang, 1999). Spieler et al. (2004) provide a formulation for this critical fragmentation overpressure Δp_f, which shows a good fit to a broad range of experimental data

$$\Delta p_f = 10^6 \, \text{Pa} \, / \, \phi_b, \qquad (4.24)$$

that has been modified by Mueller et al. (2008) to account for permeable gas flow.

The strain-rate criterion is based on the observation that silicate melts can fragment if deformation rates exceed the structural relaxation rate of the melt at $\dot{\varepsilon} \sim 10^2 \dot{\varepsilon}_{st}$ (Section 4.5.2). It is thought to be the consequence of a rapid decompression, which forces rapid bubble growth (Papale, 1999). Ittai et al. (2010) have argued that the stress criterion and the strain-rate criterion are equivalent, because of their mutual dependency on shear modulus and magma rheology.

Namiki and Manga (2005) suggested a potential energy criterion, based on the observation that during rapid decompression ϕ_b and $\rho_g - \rho_m$ determine the expansion velocity of a bubbly liquid. Potential energy depends on both ϕ_b and $\rho_g - \rho_m$ (Mastin, 1995), with fragmentation taking place above some threshold.

Hydrodynamic or inertial fragmentation should be predominantly associated with mafic magma, where rapid decompression results in rapid bubble growth, inertial stretching, and hydrodynamic breakup (Shimozuru, 1994; Zimanowski et al., 1997; Villermaux, 2007). A criterion for inertial fragmentation is based on the Reynolds number, Re_v, and Weber number, We_v, of the expanding magma (Namiki and Manga, 2008)

$$Re_v = \frac{\rho(1-\phi_b)v^2 L}{3\eta_0} > O(1) \qquad (4.25)$$

and

$$We_v = \frac{\rho(1-\phi_b)v^2 L}{\gamma} \gg 1. \qquad (4.26)$$

Here L is a characteristic length scale and v is the expansion velocity of the vesicular magma, which can be obtained from the bubble growth rate.

4.7 | Modeling of magma ascent

The past decade has seen much progress in modeling magma ascent (Sahagian, 2005). In contrast to the modeling of effusive eruptions, explosive eruptions require a criterion to estimate the depth of magma fragmentation, and merging of a model for the bubbly flow region below the fragmentation level with one for the gas–pyroclast region above. Flow in the gas–pyroclast flow region is compressible and is typically modeled as one-dimensional with a parameterization for wall friction and granular stresses during highly turbulent flow (Wilson and Head, 1981; Dobran, 1992; Koyaguchi, 2005).

Conduit models may be steady or time-dependent. The steady approximation is reasonable when the timescales over which changes in eruption rate and geometry exceed ascent times (Slezin, 2003). Time-dependent behavior becomes important when ascent times are similar to the timescale over which boundary conditions change. This can be a consequence of coupling between magma flow and (1) magma storage; (2) conduit wall elasticity; (3) magma compressibility; (4) magma outgassing; (5) viscous dissipation; and (6) magma rheology. A departure from one-dimensional models is necessary if lateral variations in properties are important, for example due to shear-rate dependent viscosity, shear localization, viscous dissipation, or permeable outgassing into conduit walls.

Dispersed two-phase flow below the fragmentation level is typically treated as homogeneous, using an appropriate magma viscosity. Above the fragmentation level, the homogenous flow assumption is sufficient if fragments are ≤ 10^{-4} m in diameter (Papale, 2001). However, treating the flow as separated for larger fragments results in a more accurate characterization of ascent velocity (Dufek and Bergantz, 2005). In the case of separated flow above or below the fragmentation level, there are two classes of models, those that explicitly incorporate interfacial momentum exchanges and those that do not (Dobran, 2001). Separated flow models with no interfacial momentum exchange treat the flow as single-phase, but incorporate closure laws for the gas volume fraction and frictional pressure loss of the two-phase mixture (Wilson and Head, 1981). Separated models with interface exchange allow for different gas and liquid (pyroclast) velocities and flow directions. However, a priori assumptions about the flow regime are required for these closure laws.

4.7.1 Steady homogeneous flow in one dimension

Below the fragmentation level homogeneous flow models advect bubbles passively with the melt. That is $U_t \ll U_m$ and $u = U_m = U_b$, where U_b is the bubble velocity. Together with the assumption of constant mass flux Q throughout the conduit, the equation of mass conservation becomes

$$\frac{dQ}{dz} = \frac{d}{dz}(\rho u A) = 0. \qquad (4.27)$$

Here A is conduit cross-sectional area and ρ is the bulk density of bubbly magma, given by $\phi_b \rho_g + (1 - \phi_b)\rho_m$. The value of ρ_g may be obtained from the ideal gas law, but use of another equation of state may be more appropriate at high pressures (Kerrick and Jacobs, 1981). ρ_m can be obtained from empirical formulations (Lange, 1994), but in many cases the melt phase is assumed to be incompressible. The bubble volume fraction ϕ_b depends on ρ_g and the amount of exsolved volatiles. If it is assumed that bubble growth is limited by solubility, the amount of exsolved volatiles per unit mass of melt is equal to the initial volatile concentration minus the equilibrium solubility at the given pressure. However, in many eruptions bubble growth is viscously and/or diffusively limited, requiring a concurrent calculation for bubble growth. If permeable outgassing is important, Eq. (4.27) may include a sink term based on a suitable porosity–permeability relation and Darcy's law or Forcheimer's equation for radial gas flow to the conduit walls and/or upward gas flow to the vent (Jaupart and Allegre, 1991).

Equation (4.27) states that mass flux is constant and volume expansion of the magma by bubble growth must be balanced by acceleration and/or by an increase in conduit cross-sectional area. Assuming a vertical conduit of cylindrical shape, conservation of momentum is governed by (Mastin and Ghiorso, 2000; Dobran, 2001)

$$\rho u \frac{du}{dz} = -\frac{dp}{dz} - \left(\frac{dp}{dz}\right)_\rho - \left(\frac{dp}{dz}\right)_\eta = -\frac{dp}{dz} - \rho g - \rho u^2 \frac{f}{a}.$$

$$(4.28)$$

Expanding Eq. (4.27) and substituting for du/dz in Eq. (4.28) gives

$$-\frac{dp}{dz} = \rho g + \rho u^2 \frac{f}{a} - \frac{\rho u^2}{A}\frac{dA}{dz} - u^2 \frac{d\rho}{dz}. \qquad (4.29)$$

Equation (4.29) indicates that the change in magma pressure with respect to height depends, from left to right, on magma-static pressure loss, frictional pressure loss, change in conduit diameter, and change in magma density. Assuming isentropic conditions (denoted by subscript S), that is an infinitesimal reversible pressure change and negligible time for heat transfer, then

$$\frac{d\rho}{dz} = \left(\frac{\partial \rho}{\partial p}\right)_S \frac{dp}{dz} = \frac{1}{c^2}\frac{dp}{dz}. \tag{4.30}$$

Here c is the speed of sound in the magma and using the Mach number, $M = u^2/c^2$, Eq. (4.29) becomes

$$-\frac{dp}{dz}(1 - M^2) = \rho g + \rho u^2 \frac{f}{a} - \frac{\rho u^2}{A}\frac{dA}{dz}. \tag{4.31}$$

Equations (4.27) and (4.31) are applicable to homogeneous flow above or below the fragmentation level and can be solved with standard numerical methods (Press et al., 1992). There are two solution approaches: (1) set dp/dz to be constant, for example lithostatic, and then dA/dz and u can be calculated; or (2) if A is known, then conduit entrance and exit pressures are specified and dp/dz and u can be calculated (Wilson and Head, 1981; Giberti and Wilson, 1990; Dobran, 1992; Mastin and Ghiorso, 2000). For $M < 0.3$ the effect of magma compressibility on the flow can be neglected, whereas for $M > 0.3$ magma compressibility contributes to the change in pressure of the ascending magma. If $p = 1$ atm at the vent exit, then the right-hand side of Eq. (4.29) has to go to zero to avoid a singular solution as M approaches 1. Therefore, the conduit flares upward with

$$\frac{da}{dz} = \frac{1}{2}\left(\frac{g\,a}{u^2} + f\right) > 0. \tag{4.32}$$

This result can be derived by setting the right-hand side of Eq. (4.29) to zero with $A = \pi a^2$ (Mastin and Ghiorso, 2000). If Eq. (4.32) is satisfied at $M < 1$ and the conduit above continues to flare, the magma will decelerate. On the other hand if $M = 1$, the magma will accelerate to supersonic velocities and p will continue to decrease, potentially to values below 1 atm. If that is the case, a shock wave will

adjust the flow back to 1 atm at the vent exit. If the exit pressure exceeds 1 atm, the flow is "choked," that is $M = 1$ and the exit velocity is equal to c, the speed of sound in the gas–pyroclast mixture.

In many cases the assumption of isothermal conditions is reasonable; if not, an equation for conservation of energy has to be added (Mastin and Ghiorso, 2000; Dobran, 2001):

$$dH + udu + gdz = 0. \tag{4.33}$$

Here H is the specific enthalpy of the melt–gas mixture from which the magma temperature can be calculated.

4.7.2 Steady separated flow in one dimension

As discussed in Section 4.3.5, the flow of the separated melt and gas phases below the fragmentation level may occur under different regimes. Criteria that determine the transitions between individual flow regimes, as well as formulations for interphase drag forces, F_{mg}, need to be specified. Existing models do not predict the occurrence of flow-regime transitions, but rather rely on a priori assumptions about the occurrence of flow transitions, equivalent to the flow transition associated with magma fragmentation. For example, the model of Dobran (1992) is based on the assumption of a constant mass flow rate through the conduit:

$$\frac{d}{dz}\left(Q_g + Q_{gd}\right) = 0 \tag{4.34}$$

and

$$\frac{d}{dz}\left(Q_m - Q_{gd}\right) = 0. \tag{4.35}$$

Here subscript g denotes the exsolved volatile phase, gd the dissolved volatile phase, and m the melt phase. As in the homogeneous flow model it is possible to account for permeable gas loss through an additional flux term on the right-hand side of Eq. (4.34). The momentum equations for the exsolved volatile and the melt phases are, respectively

$$\rho_g u_g \phi_b \frac{du_g}{dz} = -\phi_b \frac{dp}{dz} - F_{mg} - F_{wg} - \rho \phi_b g \qquad (4.36)$$

and

$$\rho_m u_m (1-\phi_b) \frac{du_m}{dz} = -(1-\phi_b)\frac{dp}{dz} - F_{mg} - F_{wm} \\ -\rho_m (1-\phi_b)g. \qquad (4.37)$$

Here F_{wg} and F_{wm} are the wall drag forces for the phase that is assumed to be in contact with the conduit wall. For example, Dobran (1992) used $F_{wm} = (dp/dz)_\eta$, $F_{wg} = 0$ below the fragmentation level and $F_{wm} = 0$, $F_{wg} = (dp/dz)_\eta$ above.

4.7.3 Two-dimensional flow

If magma flow is radially non-uniform, a two-dimensional modeling approach can provide additional insight. Cases where two-dimensional modeling has been important involve non-Newtonian magma rheology and shear heating, as well as permeable outgassing and heat loss through the conduit walls.

The direct approach is to solve the Navier–Stokes equations in two cartesian or axisymmetric cylindrical coordinates. For axisymmetric homogeneous flow, conservation of momentum is given by (Bird et al., 1960)

$$\rho \frac{\partial u_r}{\partial t} + \rho u_r \frac{\partial u_r}{\partial r} + \rho u_z \frac{\partial u_r}{\partial z} \\ = -\frac{\partial p}{\partial r} - \frac{1}{r}\frac{\partial}{\partial r}(r\tau_{rr}) - \frac{\tau_{rz}}{z} \qquad (4.38)$$

and

$$\rho \frac{\partial u_z}{\partial t} + \rho u_r \frac{\partial u_z}{\partial r} + \rho u_z \frac{\partial u_z}{\partial z} \\ = -\rho g - \frac{\partial p}{\partial z} - \frac{1}{r}\frac{\partial}{\partial r}(r\tau_{rz}) - \frac{\partial \tau_{zz}}{\partial z}, \qquad (4.39)$$

with stress tensors

$$\tau_{rr} = -2\eta \frac{\partial u_r}{\partial r} - \lambda \left(\frac{1}{r}\frac{\partial}{\partial r}(ru_r) + \frac{\partial u_z}{\partial z} \right), \qquad (4.40)$$

$$\tau_{zz} = -2\eta \frac{\partial u_z}{\partial z} - \lambda \left(\frac{1}{r}\frac{\partial}{\partial r}(ru_r) + \frac{\partial u_z}{\partial z} \right), \qquad (4.41)$$

and

$$\tau_{rz} = -\eta \left(\frac{\partial u_z}{\partial r} + \frac{\partial u_r}{\partial z} \right). \qquad (4.42)$$

Here $\lambda = 2\eta/3 - K$, where η is the viscosity of the magma. K is called the bulk viscosity, which accounts for the compressibility of the two-phase magma mixture (Massol et al., 2001) and is often assumed negligible. Various simplifications of Eqs. (4.38) and (4.39) can be employed. If the flow is laminar (Re < 10^3), which is usually the case below the fragmentation level, the second and third terms on the left-hand side can be neglected. Furthermore, if the flow is steady the time derivatives are zero. Assuming isothermal conditions, the momentum equations are complemented by an equation for the conservation of mass

$$\frac{\partial \rho}{\partial t} + \frac{1}{r}\frac{\partial}{\partial r}(\rho r u_r) + \frac{\partial}{\partial z}(\rho u_z) = 0. \qquad (4.43)$$

Examples of recent two-dimensional conduit models are given by Collier and Neuberg (2006); Costa et al. (2007a,b); Hale (2007); Hale and Muehlhaus (2007).

If vertical flow is a reasonable approximation ($u = u_z$, $u_r = 0$), Eqs. (4.38) and (4.39) become

$$\frac{\partial}{\partial r}(r\tau_{rz}) = -r\frac{\partial p}{\partial z}, \qquad (4.44)$$

and upon integration

$$\tau_{rz} = -\eta \frac{\partial u_z}{\partial r} = -\frac{r}{2}\frac{\partial p}{\partial z}. \qquad (4.45)$$

The relative ease with which Eq. (4.45) can be solved makes it easier to model more complex magma rheologies (Costa and Macedonio, 2005; Mastin, 2005; Vedeneeva et al., 2005), and to include subgrid scale calculations of bubble growth, where the model for bubble growth is discretized at a smaller scale than the conduit model itself (Fig. 4.9; Gonnermann and Manga, 2003, 2007).

To investigate the role of shear heating or magma cooling near the conduit walls, an energy equation has to be included. Neglecting thermal diffusion in the vertical direction, the energy equation is (Bird et al., 1960)

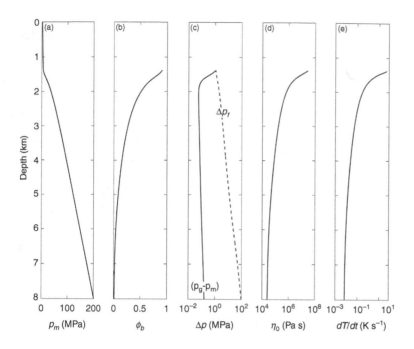

Figure 4.9 Illustrative model for rhyolite magma rising in a cylindrical conduit of 30 m diameter using Eq. (4.44) with sub-grid scale diffusive bubble growth at a constant bubble number density of 10^{15} m^{-3}, H$_2$O-content-dependent non-Newtonian rheology, viscous dissipation, and a constant mass discharge rate of $10^{7.5}$ kg s^{-1}. The vertical axis is depth on all plots. (a) The change in magma pressure p_m due to magma-static and frictional pressure loss below the fragmentation depth. Above the fragmentation depth pressure loss is assumed to be linear (Koyaguchi, 2005). (b) Bubble volume fraction ϕ_b at the center of the conduit. (c) Bubble overpressure $p_g - p_m$ (solid line) and fragmentation threshold Δp_f (dashed line). Note the large increase in bubble overpressure at shallow depths until $p_g - p_m = \Delta p_f$, when fragmentation is predicted to occur (Spieler et al., 2004). (d) Newtonian melt viscosity η_0 as a function of depth. Note the rapid viscosity increase at shallow depths due to H$_2$O exsolution, resulting in viscously limited bubble growth. (e) Viscous heating at the conduit wall becomes significant as viscosity increases, principally within a few hundred meters below the fragmentation depth.

$$\frac{\partial T}{\partial t} = \frac{\kappa}{\rho c_p}\left(\frac{\partial^2 T}{\partial r^2} + \frac{1}{r}\frac{\partial T}{\partial r}\right) - \frac{1}{\rho c_p}\left(\tau_{rz}\frac{\partial u_z}{\partial r}\right), \quad (4.46)$$

where c_p is the specific heat and κ is the thermal conductivity of the magma. The first term on the right-hand side accounts for diffusion of heat, and the last term represents viscous heating.

4.7.4 Coupling the conduit and the magma chamber

The boundary conditions for the momentum equations at the bottom of the conduit can be a specified magma flux or specified pressure. Either may be transient because of changes in chamber pressure, in turn a consequence of magma withdrawal or replenishment, as well as compressibility of the vesicular magma and surrounding rock. A simple equation for chamber pressure is obtained by combining conservation of mass with an equation of state (Denlinger and Hoblitt, 1999; Wylie et al, 1999)

$$\frac{dp}{dt} = \frac{1}{\rho \beta_e V}\left(Q_{in} - Q_{out}\right). \quad (4.47)$$

Here V is the volume of the magma chamber or conduit, Q_{in} is mass recharge rate, Q_{out} is mass discharge rate, and β_e is the effective compressibility. The latter incorporates both magma compressibility and wall-rock elasticity, and is usually much smaller than the compressibility of vesicular magma (Huppert and Woods, 2002; Woods et al., 2006). Typically it is the feedbacks between magma chamber and conduit that result in complex and unsteady eruptive

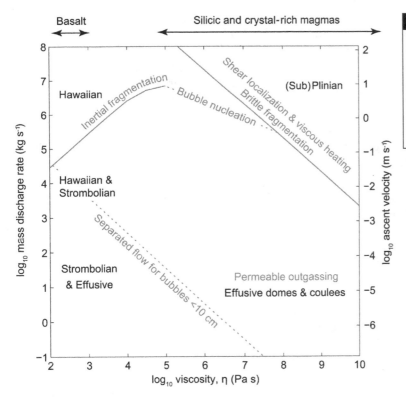

Figure 4.10 Summary of processes and mechanisms that govern magma flow in the volcanic conduit, based on modeling results and relations discussed in this chapter. Neglecting unsteady effects, the discharge rate and viscosity control which processes have sufficient time to become dominant and affect eruption style.

behavior (Melnik and Sparks, 1999; Macedonio et al., 2005).

4.8 | What conduit models have taught us

Figure 4.10 shows the relationship between magma composition (in essence a proxy for viscosity), eruption rate (or equivalently ascent rate), and eruption style. The boundaries shown are not "hard" in that uncertainties in thresholds for fragmentation, unsteady flow, variation in volatile content, conduit geometry, two-dimensional flow, and crystallization are not addressed. Overall these boundaries are meant to provide insight into the conditions under which certain processes are expected to dominate and to illustrate their relationships with eruption style.

4.8.1 Subplinian and plinian eruptions
Subplinian and plinian eruptions are characterized by brittle magma fragmentation at mass discharge rates on the order of 10^6 kg s^{-1} (subplinian) and 10^6–10^8 kg s^{-1} (plinian), sustained over several hours (Chapter 8). Syneruptive volatile exsolution and crystallization can lead to viscosity-limited bubble growth, large frictional pressure losses, and large decompression rates (Papale and Dobran, 1993, 1994), resulting in gas overpressure, supersaturation, and bubble nucleation (Massol and Koyaguchi, 2005), as well as fragmentation (Papale, 1999).

Whether fragmentation takes place depends on the conditions of magma ascent. However, fragmentation also feeds back on eruption dynamics. For example, the presence of pumice fragments after fragmentation, as opposed to complete fragmentation to ash, means that not all compressed gas contained in bubbles may be released, resulting in ascent velocities and pressures that are lower than if all gas were released. Incomplete outgassing of coarse fragments also increases the density of the erupting gas–pyroclast mixture, with consequences for the dynamics of the eruption column (Chapter 8; Papale, 2001). Furthermore, the granular stress induced by $\gtrsim 1$ cm-size pyroclasts decreases exit velocities and enhances the lateral inhomogeneity in clast-size distribution throughout the conduit (Dufek and Bergantz, 2005).

The strong dependence of viscosity on temperature in silicic magmas is of considerable importance for shear heating near the conduit walls (Rosi et al., 2004; Polacci et al., 2005). The resulting viscosity reduction will change the flow profile toward a more plug-like flow and reduce frictional pressure loss. Consequently, the discharge rate increases for a given pressure drop between chamber and surface, and hence affects the depth of fragmentation and allows for multiple stable discharge rates (Costa and Macedonio, 2002; Mastin, 2005; Vedeneeva et al., 2005; Costa et al., 2007c).

Eruption intensity may correlate with preeruptive H_2O content and magma composition (Papale et al., 1998; Polacci et al., 2004; Starostin et al., 2005). However, changes in CO_2 content can have the opposite effect of H_2O (Papale and Polacci, 1999; Ongaro et al., 2006). The transition from explosive to effusive eruptions in silicic magmas is thought to occur as magma ascent rates decrease, which allows more time for permeable outgassing, and bubbles grow at near equilibrium conditions (Jaupart and Allegre, 1991; Woods and Koyaguchi, 1994; Gonnermann and Manga, 2007).

4.8.2 Strombolian eruptions

Typical strombolian eruptions (Chapter 6) consist of prolonged periods of impulsive explosions that typically eject 0.01–10 m^3 of pyroclastic material and volcanic gases at about 100 m s^{-1}. Mass balance considerations imply that much of the gas is derived from a larger volume of magma than is erupted, a consequence of slow magma ascent rates (< 0.01–0.1 m s^{-1}). This permits significant bubble coalescence to form gas slugs and/or accumulations of bubbles, which decouple from the melt and rise to the surface where they burst. Aside from thermal modeling (Giberti et al., 1992), some numerical conduit models that have dealt directly with details of strombolian eruptions are Wilson and Head (1981); Parfitt and Wilson (1995); James et al. (2008); O'Brian and Bean (2008); D'Auria

and Martini (2009); James et al. (2009); Pioli et al. (2009).

4.8.3 Hawaiian eruptions

Hawaiian eruptions are characterized by gas-rich fountains of basaltic magma sustained for hours to days at mass discharge rates of the order of 10^4 to $> 10^6$ kg s^{-1} (Houghton and Gonnermann, 2008). Fountains reach heights that exceed hundreds of meters, producing ash and pyroclasts, presumably through hydrodynamic fragmentation (Namiki and Manga, 2008). Microtextural studies of pyroclasts indicate less open-system gas loss than for strombolian clasts (Polacci et al., 2006).

Conduit models by Wilson and Head (1981) predict that magma ascent in hawaiian eruptions is homogeneous, with dispersed bubbles, so that rapid decompression and shallow H_2O exsolution lead to expansion-driven acceleration and the formation of a sustained magma fountain above the vent (Parfitt et al., 1995; Parfitt, 2004). The transition from hawaiian to strombolian behavior is thought to occur at magma rise speeds of less than about 1 m s^{-1}, with a correlation between required rise speed and initial magmatic volatile content (Parfitt and Wilson, 1995). Alternatively, based on analog laboratory experiments it has been suggested that bubbles accumulate to form a magmatic foam, the collapse of which produces an annular gas–melt flow within the conduit (Chapter 6; Jaupart and Vergniolle, 1988).

4.8.4 Effusive eruptions

Large volumes of basaltic magma often erupt effusively over prolonged periods of time, at rates up to 10^4 kg s^{-1} (Wolfe et al., 1987). Silicic magmas of high viscosity may erupt effusively to form lava domes and coulees with time-averaged eruption rates ranging from 10^{-1} to 10^4 kg s^{-1} (Pyle, 2000). The resulting low ascent rates (≤ 0.01 m s^{-1}) are thought to facilitate permeable outgassing (Eichelberger et al., 1986; Jaupart and Allegre, 1991; Woods and Koyaguchi, 1994; Massol and Jaupart, 1999). Sparks (1997) attributed the occurrence of time-dependent behavior during dome-forming eruptions to feedbacks associated with volatile-dependent magma viscosity, magma degassing, outgassing, and crystallization. Conduit models suggest that large changes in discharge rate at relatively small changes in boundary conditions are the consequence of such feedbacks (Melnik and Sparks, 1999, 2005). They also demonstrate how nonlinear feedbacks produce different steady discharge rates for the same boundary conditions but different initial conditions, perhaps explaining how transitions between effusive and explosive behavior occur (Melnik et al., 2005; Clarke et al., 2007).

For long-lived dome-forming eruptions, magma recharge is often treated as constant over part of the eruption. However, transient eruptive behavior can ensue as a direct consequence of compressibility of the vesicular magma and/or surrounding wall rock (Meriaux and Jaupart, 1995; Denlinger, 1997; Huppert and Woods, 2002; Costa et al., 2007a). Feedbacks may be such that pressure build-up drives the system into a state where discharge rate exceeds recharge rate. With time, pressure decreases until the cycle repeats or is dynamically dampened toward a stable state (Barmin et al., 2002). High shear stress at the conduit walls, in conjunction with shear-thinning rheology, may result in shear localization and brittle deformation and this may in turn enhance permeability and outgassing. Conduit models have also explored how brittle deformation may be associated with stick-slip behavior at the conduit walls and shallow seismicity, especially for magmas with high crystal content (Denlinger and Hoblitt, 1999; Goto, 1999; Collier and Neuberg, 2006; Iverson et al., 2006; Hale, 2007).

4.9 | Summary

Conduit models of magma ascent help us understand which mechanisms and processes are important for a broad range of effusive and dry explosive volcanic eruptions (Fig. 4.10). Some important open questions are:

- What causes the observed unsteady behavior during sustained explosive eruptions? This will require models capable of exploring time-dependent dynamical feedbacks within the conduit, coupling between conduit and subaerial flow, and incorporation of bubble nucleation.

- The dynamics of magma flow within the volcanic conduit during strombolian and hawaiian eruptions remains controversial. The main challenge here is the modeling of the separated two-phase flow, where the topology of the flow can change drastically and rapidly in both space and time.

- Magma flow in the volcanic conduit is affected by processes that occur over a wide range of length scales, from the fluid dynamics of magma ascent over kilometers to crystal/bubble nucleation and growth, as well as fragmentation over millimeter to micrometer scales. Accounting for these different scales requires the development of subgrid-type models capable of coupling explicit modeling of the various small-scale processes with fluid dynamical modeling of magma flow.

- Eruption triggering is perhaps of foremost importance in terms of hazard mitigation. Integration of geodetic and remote-sensing observations with models of magma flow, supply, and storage is thus highly desirable. A major challenge is the incorporation of geological (tectonic and structural) complexities, which may require a departure from idealized one- or two-dimensional model geometries, as well as coupling between tectonically produced stresses and stresses within the magmatic system.

4.10 | Notation

a	conduit radius (m)
A	cross-sectional area of conduit (m^2)
B	empirical constant
c	speed of sound in magma (m s^{-1})
c_p	specific heat (J kg^{-1} K^{-1})
C_c	concentration of dissolved CO_2 (ppm)
C_i	mass fraction of dissolved volatile i
C_w	concentration of dissolved H_2O (wt.%)
D	diffusivity of dissolved volatile in melt (m^2 s^{-1})
D_c	diffusivity of CO_2 dissolved in melt (m^2 s^{-1})

D_i	diffusivity of dissolved volatile species i (m² s⁻¹)	Q_{in}	magma chamber mass recharge rate (kg s⁻¹)
D_{wb}	diffusivity of H_2O dissolved in basalt melt (m² s⁻¹)	Q_m	mass flow rate of melt phase (kg s⁻¹)
D_{wr}	diffusivity of H_2O dissolved in rhyolite melt (m² s⁻¹)	Q_{out}	magma chamber mass discharge rate (kg s⁻¹)
f	friction factor for flow in a cylindrical pipe	r	radial distance from center of bubble or from center of conduit (m)
F_{wg}	wall drag force of the gas phase (N)	R	bubble radius (m)
F_{wm}	wall drag force of the melt phase (N)	S	radial distance from bubble center to midpoint between adjacent bubbles (m)
g	acceleration due to gravity (m s⁻²)	t	time (s)
G_∞	shear modulus of melt (Pa)	T	temperature (K)
ΔG_D	activation energy for diffusion (J)	T_l	liquidus temperature (K)
ΔG^*	free energy for nucleus formation (J)	ΔT_e	undercooling of melt (K)
h	hindering function for Richardson–Zaki equation	u	magma velocity (m s⁻¹)
H	specific enthalpy (J kg⁻¹)	u_g	gas velocity (m s⁻¹)
ΔH	change in enthalpy between melt and crystal (J)	u_m	melt velocity (m s⁻¹)
		u_r	radial component of magma velocity (m s⁻¹)
I	homogeneous crystal nucleation rate (m⁻³ s⁻¹)	u_z	vertical component of magma velocity (m s⁻¹)
I_0	reference homogeneous crystal nucleation rate (m⁻³ s⁻¹)	U_b	bubble ascent velocity (m s⁻¹)
J	bubble nucleation rate (m⁻³ s⁻¹)	U_m	ascent velocity of the melt phase (m s⁻¹)
k	permeability (m²)	U_t	terminal rise velocity of a single bubble in infinite liquid (m s⁻¹)
k_v	volumetric factor for Avrami equation		
k_B	Boltzmann constant (J kg⁻¹)	v	approach velocity of two bubbles or expansion velocity of vesicular magma (m s⁻¹)
K	bulk viscosity (Pa s)		
L	characteristic length scale for inertial fragmentation (m)		
n	empirical coefficient	v_f	velocity in the melt film between bubbles (m s⁻¹)
N_d	number of bubbles per volume of melt (m⁻³)	v_r	radial velocity of the melt surrounding a growing bubble (m s⁻¹)
p	magma pressure (Pa)	v_R	radial velocity of the melt–vapor interface of growing bubble (m s⁻¹)
p_c	partial pressure of CO_2 (MPa)		
p_m	pressure of the melt (Pa)	V	volume of magma chamber (m³)
$\dot{p}_m$	melt decompression rate (Pa s⁻¹)	Y	crystal growth rate (m s⁻¹)
p_g	pressure of the vapor inside bubbles (Pa)	Y_0	reference crystal growth rate (m s⁻¹)
		z	vertical coordinate (m)
p_w	partial pressure of H_2O (MPa)	α	empirical constant
Δp_f	critical gas overpressure for brittle fragmentation (Pa)	β	empirical constant for Kozeny–Carman equation
Δp_s	supersaturation pressure of volatile in melt (Pa)	β_e	effective compressibility (m² N⁻¹)
		γ	vapor–melt surface tension (N m⁻¹)
Q	mass flow rate of magma (kg s⁻¹)	δ	empirical constant
Q_g	mass flow rate of gas phase (kg s⁻¹)	ε	strain rate (s⁻¹)
Q_{gd}	mass flow rate of dissolve volatile phase (kg s⁻¹)	$\dot{\varepsilon}_f$	critical strain rate for structural failure of the melt (s⁻¹)

$\dot{\varepsilon}_{st}$	strain rate at which shear thinning begins (s^{-1})
η	magma viscosity (Pa s)
η_0	Newtonian melt (liquid) viscosity (Pa s)
η_g	viscosity of vapor inside a bubble (Pa s)
η_r	relative viscosity
θ	contact angle
κ	thermal conductivity ($W\ K^{-1}\ m^{-1}$)
ρ	magma density ($kg\ m^{-3}$)
ρ_g	vapor density ($kg\ m^{-3}$)
ρ_m	melt density ($kg\ m^{-3}$)
σ	free energy of crystal–liquid interface (J)
τ_{dec}	characteristic decompression timescale (s)
τ_{dif}	characteristic diffusion time of volatile in melt (s)
τ_{ij}	components of the stress tensor (Pa)
τ_r	viscous relaxation time (s)
τ_{vis}	characteristic viscous timescale (s)
ϕ_b	volume fraction of bubbles
ϕ_{bcr}	critical volume fraction of bubbles
ϕ_{bp}	percolation threshold
ϕ_x	volume fraction of crystals
ϕ_{xcr}	critical volume fraction of crystals
χ	empirical constant for Kozeny–Carman equation
ψ	geometrical factor
Ca	Capillary number, $\eta_0 \dot{\varepsilon} R / \gamma$
Ca_{cr}	Capillary number at which bubbles break up, $\eta_0 \dot{\varepsilon} R / \gamma$
M	Mach number, u^2/c^2
Re	Reynolds number for flow in a cylindrical pipe, $a u_z / \eta$
Re_b	Reynolds number for a growing bubble, $\dot{\varepsilon} R^2 \rho / \eta_0$
Re_v	Reynolds number of expanding magma, $\rho (1 - \phi_b) v^2 L / (3\eta_0)$
We	Weber number, $2\rho v^2 R / \gamma$
We_{cr}	Weber number at which bubbles break up, $2\rho v^2 R / \gamma$
We_v	Weber number of expanding magma, $\rho (1 - \phi_b) v^2 L / \gamma$

Acknowledgments

The authors thank B. F. Houghton for stimulating discussions, as well as J. E. Hammer, L. Karlstrom, and M. Rudolph for constructive comments on the manuscript. Both authors were supported by NSF while preparing this review.

References

Alidibirov, M. (1994). A model for viscous magma fragmentation during volcanic blasts. *Bulletin of Volcanology*, **56**, 459–465.

Allard, P., Burton, M. and Mure, F. (2005). Spectroscopic evidence for a lava fountain driven by previously accumulated magmatic gas. *Nature*, **433**, 407–410.

Arbaret, L., Bystricky, M. and Champallier, R. (2007). Microstructures and rheology of hydrous synthetic magmatic suspensions deformed in torsion at high pressure. *Journal of Geophysical Research*, **112**, doi:10.1029/2006JB0048.

Bagdassarov, N. S. and Dingwell, D. B. (1992). A rheological investigation of vesicular rhyolite. *Journal of Volcanology and Geothermal Research*, **50**, 307–322.

Bagdassarov, N., Dorfman, A. and Dingwell, D. B. (2000). Effect of alkalis, phosphorus, and water on the surface tension of haplogranite melt. *American Mineralogist*, **85**, 33–40.

Barmin, A., Melnik, O. and Sparks, R. S. J. (2002). Periodic behavior in lava dome eruptions. *Earth and Planetary Science Letters*, **199**, 173–184.

Batchelor, G. K. (1967). *An Introduction to Fluid Dynamics*. Cambridge: Cambridge University Press.

Behrens, H. and Zhang, Y. X. (2001). Ar diffusion in hydrous silicic melts: implications for volatile diffusion mechanisms and fractionation. *Earth and Planetary Science Letters*, **192**, 363–376.

Bird, R. B., Steward, W. E. and Lightfoot, E. N. (1960). *Transport Phenomena*. New York: John Wiley.

Blank, J. G. and Brooker, R. A. (1994). Experimental studies of carbon-dioxide in silicate melts; solubility, speciation, and stable carbon-isotope behavior. In *Volatiles in Magmas*, ed. M. R. Carroll and J. R. Holloway, *Reviews in Mineralogy*, **30**, 157–186.

Bluth, G. J. S., Schnetzler, C. C., Krueger, A. J. and Walter, L. S. (1993). The contribution of explosive volcanism to global atmospheric sulfur-dioxide concentrations. *Nature*, **366**, 327–329.

Borrell, M. and Leal, L. G. (2008). Viscous coalescence of expanding low-viscosity drops; the dueling drops experiment. *Journal of Colloid and Interface Science*, **319**, 263–269.

Boudon, G., Villemant, B., Komorowski, J. C., Ildefonse, P. and Semet, M. P. (1998). The hydrothermal system at Soufriere Hills volcano, Montserrat (West Indies): Characterization and role in the on-going eruption. *Geophysical Research Letters*, **25**, 3693–3696.

Bourgue, E. and Richet, P. (2001). The effects of dissolved CO_2 on the density and viscosity of silicate melts: a preliminary study. *Earth and Planetary Science Letters*, **193**, 57–68.

Brennen, C. E. (2005). *Fundamentals of Multiphase Flow*. New York: Cambridge University Press, 2nd edition.

Bruce, P. M. and Huppert, H. E. (1989). Thermal control of basaltic fissure eruptions. *Nature*, **342**, 665–667.

Caricchi, L., Burlini, L., Ulmer, P. *et al.* (2007). Non-Newtonian rheology of crystal-bearing magmas and implications for magma ascent dynamics. *Earth and Planetary Science Letters*, **264**, 402–419.

Carman, P. C. (1956). *Flow of Gases Through Porous Media*. San Diego, CA: Academic Press.

Carmichael, I. S. E. (2002). The andesite aqueduct: perspectives on the evolution of intermediate magmatism in west-central (105–99 degrees W) Mexico. *Contributions to Mineralogy Petrology*, **143**, 641–663.

Carroll, M. R. and Webster, J. D. (1994). Solubilities of sulfur, noble-gases, nitrogen, chlorine, and fluorine in magmas. In *Volatiles in Magmas,* ed. M. R. Corroll and J. R. Holloway, *Reviews in mineralogy*, **30**, 231–279.

Clarke, A. B., Stephens, S., Teasdale, R., Sparks, R. S. J. and Diller, K. (2007). Petrologic constraints on the decompression history of magma prior to Vulcanian explosions at the Souffiere Hills volcano, Montserrat. *Journal of Volcanology and Geothermal Research*, **161**, 261–274.

Collier, L. and Neuberg, J. (2006). Incorporating seismic observations into 2D conduit flow modeling. *Journal of Volcanology and Geothermal Research*, **152**, 331–346.

Costa, A. and Macedonio, G. (2005). Viscous heating effects in fluids with temperature-dependent viscosity: Triggering of secondary flows. *Journal of Fluid Mechanics*, **540**, 21–38.

Costa, A., Melnik, O. and Sparks, R. S. J. (2007a). Controls of conduit geometry and wallrock elasticity on lava dome eruptions. *Earth and Planetary Science Letters*, **260**, 137–151.

Costa, A., Melnik, O., Sparks, R. S. J. and Voight, B. (2007b). Control of magma flow in dykes on cyclic lava dome extrusion. *Geophysical Research Letters*, **34**, doi: 10.1029/2006GL027466.

Costa, A., Melnik, O. and Vedeneeva, E. (2007c). Thermal effects during magma ascent in conduits. *Journal of Geophysical Research*, **112**, doi:10.1029/2007JB004985.

D'Auria, L. and Martini, M. (2009). Slug flow: Modeling in a conduit and associated elastic radiation. In *Monitoring and Mitigation of Volcano Hazards*, ed. R. A. Meyers. New York: Springer, pp. 8153–8168.

Degruyter, W., Bachmann, O., Burgisser, A. and Manga, M. (2012). The effects of outgassing on the transition between effusive and explosive silicic eruptions. *Earth and Planetary Science Letters*, **348–350**, 161–170.

Denlinger, R. P. (1997). A dynamic balance between magma supply and eruption rate at Kilauea volcano, Hawaii. *Journal of Geophysical Research*, **102**, 18 091–18 100.

Denlinger, R. P. and Hoblitt, R. P. (1999). Cyclic eruptive behavior of silicic volcanoes. *Geology*, **27**, 459–462.

Dixon, J. E. (1997). Degassing of alkalic basalts. *American Mineralogist*, **82**, 368–378.

Dobran, F. (1992). Nonequilibrium flow in volcanic conduits and application to the eruptions of Mt. St. Helens on May 18, 1980, and Vesuvius in A.D. 79. *Journal of Volcanology and Geothermal Research*, **49**, 285–311.

Dobran, F. (2001). *Volcanic Processes: Mechanisms in Material Transport*. New York, NY: Kluwer Academic / Plenum Publishers.

Dufek, J. and Berganz, G. W. (2005). Transient two-dimensional dynamics in the upper conduit of a rhyolitic eruption: A comparison of closure models for the granular stress. *Journal of Volcanology and Geothermal Research*, **143**, 113–132.

Edmonds, M., Oppenheimer, C., Pyle, D. M., Herd, R. A. and Thompson, G. (2003). SO_2 emissions from Soufriere Hills Volcano and their relationship to conduit permeability, hydrothermal interaction and degassing regime. *Journal of Volcanology and Geothermal Research*, **124**, 23–43.

Eichelberger, J. C. (1995). Silicic volcanism. *Annual Reviews in Earth and Planetary Science*, **23**, 41–63.

Eichelberger, J. C., Carrigan, C. R., Westrich, H. R. and Price, R. H. (1986). Non-explosive silicic volcanism. *Nature*, **323**, 598–602.

Esposti Ongaro, T., Papale, P., Neri, A. and Del Seppia, D. (2006). Influence of carbon dioxide on the large-scale dynamics of magmatic eruptions at Phlegrean Fields (Italy). *Geophysical Research Letters*, **33**, doi:10.1029/2005GL025528.

Gardner, J. E. (2007). Heterogeneous bubble nucleation in highly viscous silicate melts during instantaneous decompression from high pressure. *Chemical Geology*, **236**, 1–12.

Gardner, J. E. and Denis, M. H. (2004). Heterogeneous bubble nucleation on Fe-Ti oxide crystals in high-silica rhyolitic melts. *Geochimica et Cosmochimica Acta*, **68**, 3587–3597.

Ghiaasiaan, S. M. (2008). *Two-Phase Flow, Boiling, and Condensation in Conventional and Miniature Systems*. New York: Cambridge University Press.

Giberti, G. and Wilson, L. (1990). The influence of geometry on the ascent of magma in open fissures. *Bulletin of Volcanology*, **52**, 515–521.

Giberti, G., Jaupart, C. and Sartoris, G. (1992). Steady-state operation of Stromboli Volcano, Italy: constraints on the feeding system. *Bulletin of Volcanology*, **54**, 535–541.

Giordano, D. and Dingwell, D. B. (2003). Viscosity of hydrous Etna basalt: implications for Plinian-style basaltic eruptions. *Bulletin of Volcanology*, **65**, 8–14.

Giordano, D., Russell, J. K. and Dingwell, D. B. (2008). Viscosity of magmatic liquids: A model. *Earth and Planetary Science Letters*, **271**, 123–134.

Gonnermann, H. M. and Manga, M. (2003). Explosive volcanism may not be an inevitable consequence of magma fragmentation. *Nature*, **426**, 432–435.

Gonnermann, H. M. and Manga, M. (2005a). Flow banding in obsidian: A record of evolving textural heterogeneity during magma deformation. *Earth and Planetary Science Letters*, **236**, 135–147.

Gonnermann, H. M. and Manga, M. (2005b). Nonequilibrium magma degassing: Results from modeling of the ca.1340 AD eruption of Mono Craters, California. *Earth and Planetary Science Letters*, **238**, 1–16.

Gonnermann, H. M. and Manga, M. (2007). The fluid mechanics inside a volcano. *Annual Reviews in Fluid Mechanics*, **39**, 321–356.

Gonnermann, H. M. and Mukhopadhyay, S. (2007). Non-equilibrium degassing and a primordial source for helium in ocean-island volcanism. *Nature*, **449**, 1037–1040.

Goto, A. (1999). A new model for volcanic earthquake at Unzen Volcano: Melt rupture model. *Geophysical Research Letters*, **26**, 2541–2544.

Guet, S. and Ooms, G. (2006). Fluid mechanical aspects of the gas-lift technique. *Annual Reviews in Fluid Mechanics*, **38**, 225–249.

Hale, A. J. (2007). Magma flow instabilities in a volcanic conduit: Implications for long-period seismicity. *Physics of the Earth and Planetary Interiors*, **163**, 163–178.

Hale, A. J. and Muehlhaus, H. B. (2007). Modelling shear bands in a volcanic conduit: Implications for over-pressures and extrusion-rates. *Earth and Planetary Science Letters*, **263**, 74–87.

Hammer, J. E. (2004). Crystal nucleation in hydrous rhyolite: Experimental data applied to classical theory. *American Mineralogist*, **89**, 1673–1679.

Hauri, E. (2002). SIMS analysis of volatiles in silicate glasses, 2: isotopes and abundances in Hawaiian melt inclusions. *Chemical Geology*, **183**, 115–141.

Hinze, J. O. (1955). Fundamentals of the hydrodynamic mechanisms of splitting in dispersion processes. *American Institute of Chemical Engineers Journal*, **1**, 289–295.

Hirth, J. P., Pound, G. M. and St. Pierre, G. R. (1970). Bubble nucleation. *Metallurgical and Materials Transactions B*, **1**, 939–945.

Houghton, B. F. and Gonnermann, H. M. (2008). Basaltic explosive volcanism: constraints from deposits and models. *Chemie der Erde – Geochemistry*, **68**, 117–140.

Hui, H. J. and Zhang, Y. X. (2007). Toward a general viscosity equation for natural anhydrous and hydrous silicate melts. *Geochimica et Cosmochimica Acta*, **71**, 403–416.

Huppert, H. E. and Woods, A. W. (2002). The role of volatiles in magma chamber dynamics. *Nature*, **420**, 493–495.

Hurwitz, S. and Navon, O. (1994). Bubble nucleation in rhyolitic melts: Experiments at high-pressure, temperature, and water content. *Earth and Planetary Science Letters*, **122**, 267–280.

Ittai, K., Lyakhovsky, V. and Navon, O. (2010). Bubble growth in visco-elastic magma: Implications to magma fragmentation and bubble nucleation. *Bulletin of Volcanology*, **73**, 39–54.

Iverson, R. M., Dzurisin, D., Gardner, C. A. *et al.* (2006). Dynamics of seismogenic volcanic extrusion at Mount St Helens in 2004–05. *Nature*, **444**, 439–443.

James, M. R., Lane, S. J. and Corder, S. B. (2008). Modelling the rapid near-surface expansion of gas slugs in low viscosity magmas. In *Fluid Motion in Volcanic Conduits: A Source of Seismic and Acoustic Signals*, ed. S. J. Lane and J. S. Gilbert. Geological Society of London Special Publication, **307**, 147–167.

James, M. R., Lane, S. J., Wilson, L. and Corder, S. B. (2009). Degassing at low magma-viscosity volcanoes: Quantifying the transition between passive bubble-burst and Strombolian eruption. *Journal of Volcanology and Geothermal Research*, **180**, 81–88.

James, P. F. (1985). Kinetics of crystal nucleation in silicate glasses. *Journal of Non-Crystalline Solids*, **73**, 517–540.

Jaupart, C. and Allegre, C. J. (1991). Gas content, eruption rate and instabilities of eruption regime in silicic volcanoes. *Earth and Planetary Science Letters*, **102**, 413–429.

Jaupart, C. and Vergniolle, S. (1988). Laboratory models of Hawaiian and Strombolian eruptions. *Nature*, **331**, 58–60.

Johnson, M. C., Anderson, A. T. and Rutherford, M. J. (1994). Pre-eruptive volatile contents of magmas. In *Volatiles In Magmas*, ed. M. R. Carroll and J. R. Holloway, *Reviews In Mineralogy*, **30**, 281–330.

Kennedy, B., Spieler, O., Scheu, B. *et al.* (2005). Conduit implosion during Vulcanian eruptions. *Geology*, **33**, 581–584.

Kerrick, D. M. and Jacobs, G. K. (1981). A modified Redlich-Kwong Equation for H_2O, CO_2, and H_2O-CO_2 mixtures at elevated pressures and temperatures. *American Journal of Science*, **281**, 735–767.

Klug, C. and Cashman, K. V. (1996). Permeability development in vesiculating magmas: implications for fragmentation. *Bulletin of Volcanology*, **58**, 87–100.

Koyaguchi, T. (2005). An analytical study for 1-dimensional steady flow in volcanic conduits. *Journal of Volcanology and Geothermal Research*, **143**, 29–52.

Lange, R. A. (1994). The effect of H_2O, CO_2 and F on the density and viscosity of silicate melts. *Reviews in Mineralogy*, **30**, 331–369.

Lavallee, Y., Hess, K. U., Cordonnier, B. and Dingwell, D. B. (2007). Non-Newtonian rheological law for highly crystalline dome lavas. *Geology*, **35**, 843–846.

Lejeune, A. M. and Richet, P. (1995). Rheology of crystal-bearing silicate melts – An experimental study at high viscosities. *Journal of Geophysical Research*, **100**, 4215–4229.

Lejeune, A. M., Bottinga, Y., Trull, T. W. and Richet, P. (1999). Rheology of bubble-bearing magmas. *Earth and Planetary Science Letters*, **166**, 71–84.

Lensky, N. G., Navon, O. and Lyakhovsky, V. (2004). Bubble growth during decompression of magma: experimental and theoretical investigation. *Journal of Volcanology and Geothermal Research*, **129**, 7–22.

Liu, Y., Zhang, Y. X. and Behrens, H. (2005). Solubility of H_2O in rhyolitic melts at low pressures and a new empirical model for mixed H_2O-CO_2 solubility in rhyolitic melts. *Journal of Volcanology and Geothermal Research*, **143**, 219–235.

Llewellin, E. W. and Manga, A. (2005). Bubble suspension rheology and implications for conduit flow. *Journal of Volcanology and Geothermal Research*, **143**, 205–217.

Macedonio, G., Dobran, F. and Neri, A. (1994). Erosion processes in volcanic conduits and application to the AD 79 eruption of Vesuvius. *Earth and Planetary Science Letters*, **121**, 137–152.

Macedonio, G., Neri, A., Marti, J. and Folch, A. (2005). Temporal evolution of flow conditions in sustained magmatic explosive eruptions. *Journal of Volcanology and Geothermal Research*, **143**, 153–172.

Manga, M. and Stone, H. A. (1994). Interactions between bubbles in magmas and lavas: Effects of bubble deformation. *Journal of Volcanology and Geothermal Research*, **63**, 267–279.

Manga, M., Castro, J., Cashman, K. V. and Loewenberg, M. (1998). Rheology of bubble-bearing magmas: Theoretical results. *Journal of Volcanology and Geothermal Research*, **87**, 15–28.

Mangan, M. and Sisson, T. (2000). Delayed, disequilibrium degassing in rhyolite magma: Decompression experiments and implications for explosive volcanism. *Earth and Planetary Science Letters*, **183**, 441–455.

Mangan, M. and Sisson, T. (2005). Evolution of melt-vapor surface tension in silicic volcanic systems: Experiments with hydrous melts. *Journal of Geophysical Research*, **110**, doi:10.1029/2004JB003215.

Marsh, B. D. (1998). On the interpretation of crystal size distributions in magmatic systems. *Journal of Petrology*, **39**, 553–599.

Marti, J., Soriano, C. and Dingwell, D. B. (1999). Tube pumices as strain markers of the ductile-brittle transition during magma fragmentation. *Nature*, **402**, 650–653.

Martula, D. S., Hasegawa, T., Lloyd, D. R. and Bonnecaze, R. T. (2000). Coalescence-induced coalescence of inviscid droplets in a viscous fluid. *Journal of Colloid and Interface Science*, **232**, 241–253.

Massol, H. and Jaupart, C. (1999). The generation of gas overpressure in volcanic eruptions. *Earth and Planetary Science Letters*, **166**, 57–70.

Massol, H. and Koyaguchi, T. (2005). The effect of magma flow on nucleation of gas bubbles in a volcanic conduit. *Journal of Volcanology and Geothermal Research*, **143**, 69–88.

Massol, H., Jaupart, C. and Pepper, D. W. (2001). Ascent and decompression of viscous vesicular magma in a volcanic conduit. *Journal of Geophysical Research*, **106**, 16 223–16 240.

Mastin, L. G. (1995). Thermodynamics of gas and steam-blast eruptions. *Bulletin of Volcanology*, **57**, 85–98.

Mastin, L. G. (2005). The controlling effect of viscous dissipation on magma flow in silicic conduits. *Journal of Volcanology and Geothermal Research*, **143**, 17–28.

Mastin, L. G. and Ghiorso, M. S. (2000). A numerical program for steady-state flow of magma-gas mixtures through vertical eruptive conduits. United States Geological Survey, *Open-File Report* 00-209, **56** pp.

McMillan, P. F. (1994). Water solubility and speciation models. In *Volatiles in Magmas*, ed. M. R. Carroll and J. R. Holloway, *Reviews in Mineralogy*, **30**, 131–156.

Melnik, O. and Sparks, R. S. J. (1999). Nonlinear dynamics of lava dome extrusion. *Nature*, **402**, 37–41.

Melnik, O. and Sparks, R. S. J. (2005). Controls on conduit magma flow dynamics during lava dome building eruptions. *Journal of Geophysical Research*, **110**, doi: 10.1029/2004JB003183.

Melnik, O., Barmin, A. A. and Sparks, R. S. J. (2005). Dynamics of magma flow inside volcanic conduits with bubble overpressure buildup and gas loss through permeable magma. *Journal of Volcanology and Geothermal Research*, **143**, 53–68.

Meriaux, C. and Jaupart, C. (1995). Simple fluid dynamic models of volcanic rift zones. *Earth and Planetary Science Letters*, **136**, 223–240.

Mitchell, K. L. (2005). Coupled conduit flow and shape in explosive volcanic eruptions. *Journal of Volcanology and Geothermal Research*, **143**, 187–203.

Moynihan, C. T. (1995). Structural relaxation and the glass transition. *Reviews in Mineralogy and Geochemistry*, **32**, 1–19.

Mueller, S., Scheu, B., Spieler, O. and Dingwell, D. B. (2008). Permeability control on magma fragmentation. *Geology*, **36**, 399–402.

Muller, R., Zanotto, E. D. and Fokin, V. M. (2000). Surface crystallization of silicate glasses: nucleation sites and kinetics. *Journal of Non-Crystalline Solids*, **274**, 208–231.

Namiki, A. and Manga, M. (2005). Response of a bubble bearing viscoelastic fluid to rapid decompression: Implications for explosive volcanic eruptions. *Earth and Planetary Science Letters*, **236**, 269–284.

Namiki, A. and Manga, M. (2008). Transition between fragmentation and permeable outgassing of low viscosity magmas. *Journal of Volcanology and Geothermal Research*, **169**, 48–60.

Newman, S. and Lowenstern, J. B. (2002). VOLATILECALC: A siliate melt-H_2O-CO_2 solution model written in Visual Basic for Excel. *Computers and Geoscience*, **28**, 597–604.

Nowak, M., Schreen, D. and Spickenbom, K. (2004). Argon and CO_2 on the race track in silicate melts: A tool for the development of a CO_2 speciation and diffusion model. *Geochimica et Cosmochimica Acta*, **68**, 5127–5138.

O'Brian, G. S. and Bean, C. J. (2008). Seismicity on volcanoes generated by gas slug ascent. *Geophysical Research Letters*, **35**, L16308, doi: 10.1029/2008gl035001.

Okumura, S., Nakamura, M., Tsuchiyama, K., Nakano, T. and Uesugi, K. (2008). Evolution of bubble microstructure in sheared rhyolite: Formation of a channel-like bubble network. *Journal of Geophysical Research*, **113**, doi:10.1029/2007JB005362.

Okumura, S., Nakamura, M., Nakano, T., Uesugi, K. and Tsuchiyama, K. (2010). Shear deformation experiments on vesicular rhyolite: Implications for brittle fracturing, degassing, and compaction of magmas in volcanic conduits. *Journal of Geophysical Research*, **115**, doi:10.1029/2009JB006904.

Pal, R. (2003). Rheological behavior of bubble-bearing magmas. *Earth and Planetary Science Letters*, **207**, 165–179.

Papale, P. (1999). Strain-induced magma fragmentation in explosive eruptions. *Nature*, **397**, 425–428.

Papale, P. (2001). Dynamics of magma flow in volcanic conduits with variable fragmentation efficiency and nonequilibrium pumice degassing. *Journal of Geophysical Research*, **106**, 11 043–11 065.

Papale, P. (2005). Determination of total H_2O and CO_2 budgets in evolving magmas from melt inclusion data. *Journal of Geophysical Research*, **110**, doi:10.1029/2004JB003183.

Papale, P. and Dobran, F. (1993). Modeling of the ascent of magma during the Plinian eruption of Vesuvius in AD 79. *Journal of Volcanology and Geothermal Research*, **58**, 101–132.

Papale, P. and Dobran, F. (1994). Magma flow along the volcanic conduit during the Plinian and pyroclastic flow phases of the May 18, 1980, Mount St. Helens eruption. *Journal of Geophysical Research*, **99**, 4355–4373.

Papale, P. and Polacci, M. (1999). Role of carbon dioxide in the dynamics of magma ascent in explosive eruptions. *Bulletin of Volcanology*, **60**, 583–594.

Papale, P., Neri, A. and Macedonio, G. (1998). The role of magma composition and water content in explosive eruptions – 1. Conduit ascent dynamics. *Journal of Volcanology and Geothermal Research*, **87**, 75–93.

Papale, P., Moretti, R. and Barbato, D. (2006). The compositional dependence of the saturation surface of H_2O+CO_2 fluids in silicate melts. *Chemical Geology*, **229**, 78–95.

Parfitt, E. A. (2004). A discussion of the mechanisms of explosive basaltic eruptions. *Journal of Volcanology and Geothermal Research*, **134**, 77–107.

Parfitt, E. A. and Wilson, L. (1995). Explosive volcanic eruptions – IX. The transition between Hawaiian-style lava fountaining and Strombolian explosive activity. *Geophysical Journal International*, **121**, 226–232.

Parfitt, E. A., Wilson, L. and Neal, C. A. (1995). Factors influencing the height of Hawaiian lava fountains: Implications for the use of fountain height as an indicator of magma gas content. *Bulletin of Volcanology*, **57**, 440–450.

Pioli, L., Azzopardi, B. J. and Cashman, K. V. (2009). Controls on the explosivity of scoria cone eruptions: Magma segregation at conduit junctions. *Journal of Volcanology and Geothermal Research*, **186**, 407–415.

Polacci, M., Papale, P., Del Seppia, D., Giordano, D. and Romano, C. (2004). Dynamics of magma ascent and fragmentation in trachytic versus rhyolitic eruptions. *Journal of Volcanology and Geothermal Research*, **131**, 93–108.

Polacci, M., Rosi, M., Landi, P., Di Muro, A. and Papale, P. (2005). Novel interpretation for shift between eruptive styles in some volcanoes. *Eos, Transactions, American Geophysical Union*, **86**, 333–336.

Polacci, M., Corsaro, R. A. and Andronico, D. (2006). Coupled textural and compositional characterization of basaltic scoria: Insights into the transition from Strombolian to fire fountain activity at Mount Etna, Italy. *Geology*, **34**, 201–204.

Press, W. H., Teukolsky, S. A., Vetterling, W. T. and Flannery, B. P. (1992). *Numerical Recipes in C: The Art of Scientific Computing*. Cambridge: Cambridge University Press.

Proussevitch, A. A. and Sahagian, D. L. (1998). Dynamics and energetics of bubble growth in magmas: Analytical formulation and numerical modeling. *Journal of Geophysical Research*, **103**, 18 223–18 251.

Proussevitch, A. A., Sahagian, D. L. and Anderson, A. (1993a). Dynamics of diffusive bubble growth in magmas: Isothermal case. *Journal of Geophysical Research*, **98**, 22 283–22 307.

Proussevitch, A. A., Sahagian, D. L. and Kutolin, V. A. (1993b). Stability of foams in silicate melts. *Journal of Volcanology and Geothermal Research*, **59**, 161–178.

Pyle, D. M. (2000). Sizes of volcanic eruptions. In *Encyclopedia of Volcanoes*, ed. B. F. Houghton, H. Rymer, J. Stix, S. McNutt and H. Sigurdsson. San Diego: Academic Press, pp. 263–269.

Pyle, D. M. and Pyle, D. L. (1995). Bubble migration and the initiation of volcanic eruptions. *Journal of Volcanology and Geothermal Research*, **67**, 227–232.

Richardson, J. F. and Zaki, W. N. (1954). The sedimentation of a suspension of uniform spheres under conditions of viscous flow. *Chemical Engineering Science*, **3**, 65–73.

Rosi, M., Landi, P., Polacci, M., Di Muro, A. and Zandomeneghi, D. (2004). Role of conduit shear on ascent of the crystal-rich magma feeding the 800-year-BP Plinian eruption of Quilotoa Volcano (Ecuador). *Bulletin of Volcanology*, **66**, 307–321.

Rosner, D. E. and Epstein, M. (1972). Effects of interface kinetics, capillarity and solute diffusion on bubble growth rates in highly supersaturated liquids. *Chemical Engineering Science*, **27**, 69–88.

Rust, A. C. and Cashman, K. V. (2004). Permeability of vesicular silicic magma: Inertial and hysteresis effects. *Earth and Planetary Science Letters*, **228**, 93–107.

Rust, A. C., Manga, M. and Cashman, K. V. (2003). Determining flow type, shear rate and shear stress in magmas from bubble shapes and orientations. *Journal of Volcanology and Geothermal Research*, **122**, 111–132.

Saar, M. O. and Manga, M. (1999). Permeability-porosity relationship in vesicular basalts. *Geophysical Research Letters*, **26**, 111–114.

Sahagian, D. (2005). Volcanic eruption mechanisms: Insights from intercomparison of models of conduit processes. *Journal of Volcanology and Geothermal Research*, **143**, 1–15.

Sahagian, D. L. and Proussevitch, A. A. (1996). Thermal effects of magma degassing. *Journal of Volcanology and Geothermal Research*, **74**, 19–38.

Scriven, L. E. (1959). On the dynamics of phase growth. *Chemical Engineering Science*, **10**, 1–13.

Seyfried, R. and Freundt, A. (2000). Experiments on conduit flow and eruption behavior of basaltic volcanic eruptions. *Journal of Geophysical Research*, **105**, 23 727–23 740.

Shimozuru, D. (1994). Physical parameters governing the formation of Pele's hair and tears. *Bulletin of Volcanology*, **56**, 217–219.

Slezin, Y. B. (2003). The mechanism of volcanic eruptions (a steady state approach). *Journal of Volcanology and Geothermal Research*, **122**, 7–50.

Sparks, R. S. J. (1978). The dynamics of bubble formation and growth in magmas: A review and analysis. *Journal of Volcanology and Geothermal Research*, **3**, 1–37.

Sparks, R. S. J. (1997). Causes and consequences of pressurisation in lava dome eruptions. *Earth and Planetary Science Letters*, **150**, 177–189.

Sparks, R. S. J., Barclay, J., Jaupart, C., Mader, H. M. and Phillips, J. C. (1994). Physical aspects of magmatic degassing I. Experimental and theoretical constraints on vesiculation. In *Volatiles in Magmas*, ed. M. R. Carroll and J. R. Holloway, *Reviews in Mineralogy*, **30**, 415–445.

Spieler, O., Kennedy, B., Kueppers, U. *et al.* (2004). The fragmentation threshold of pyroclastic rocks. *Earth and Planetary Science Letters*, **226**, 139–148.

Spohn, T., Hort, M. and Fischer, H. (1988). Numerical simulation of the crystallization of multicomponent melts in thin dikes or sills. 1. The liquidus phase. *Journal of Geophysical Research*, **93**, 4880–4894.

Starostin, A. B., Barmin, A. A. and Melnik, O. E. (2005). A transient model for explosive and phreatomagmatic eruptions. *Journal of Volcanology and Geothermal Research*, **143**, 133–151.

Stein, D. J. and Spera, F. J. (2002). Shear viscosity of rhyolite-vapor emulsions at magmatic temperatures by concentric cylinder rheometry. *Journal of Volcanology and Geothermal Research*, **113**, 243–258.

Stevenson, R. J., Dingwell, D. B., Webb, S. L. and Sharp, T. G. (1996). Viscosity of microlite-bearing rhyolitic obsidians: An experimental study. *Bulletin of Volcanology*, **58**, 298–309.

Tait, S., Jaupart, C. and Vergniolle, S. (1989). Pressure, gas content and eruption periodicity of a shallow, crystallizing magma chamber. *Earth and Planetary Science Letters*, **92**, 107–123.

Takeuchi, S., Nakashima, S., Tomiya, A. and Shinohara, H. (2005). Experimental constraints on the low gas permeability of vesicular magma during decompression. *Geophysical Research Letters*, **32**. L10312, doi:10.1029/2006 GLO22491.

Taylor, G. I. (1932). The viscosity of a fluid containing small drops of another fluid. *Proceedings of the Royal Society of London A*, **138**, 41–48.

Thies, M. (2002). *Herstellung und rheologische Eigenschaften von porösen Kalk-Natron-Silicatschmelzen.* PhD dissertation, Technische Universität Berlin.

Thomas, N., Jaupart, C. and Vergniolle, S. (1994). On the vesicularity of pumice. *Journal of Geophysical Research*, **99**, 15 633–15 644.

Toramaru, A. (1989). Vesiculation process and bubble-size distributions in ascending magmas with constant velocities. *Journal of Geophysical Research*, **94**, 17 523–17 542.

Tuffen, H., Dingwell, D. B. and Pinkerton, H. (2003). Repeated fracture and healing of silicic magma generate flow banding and earthquakes? *Geology*, **31**, 1089–1092.

Vedeneeva, E. A., Melnik, O. E., Barmin, A. A. and Sparks, R. S. J. (2005). Viscous dissipation in explosive volcanic flows. *Geophysical Research Letters*, **32**, doi:10.1029/2004G02095.

Verhoogen, J. (1951). Mechanics of ash formation. *American Journal of Science*, **249**, 729–739.

Villermaux, E. (2007). Fragmentation. *Annual Reviews in Fluid Mechanics*, **39**, 419–446.

Wallace, P. J. (2005). Volatiles in subduction zone magmas: concentrations and fluxes based on melt inclusion and volcanic gas data. *Journal of Volcanology and Geothermal Research*, **140**, 217–240.

Wallace, P. J., Carn, S. A. R., William, I., Bluth, G. J. S. and Gerlach, T. M. (2003). Integrating petrologic and remote sensing perspectives on magmatic volatiles and volcanic degassing. *Eos, Transactions, American Geophysical Union*, **84**, 446–447.

Wallis, G. B. (1969). *One-Dimensional Two-phase Flow.* McGraw–Hill.

Watson, E. B. (1994). Diffusion in volatile-bearing magmas. *Reviews in Mineralogy*, **30**, 371–411.

Webb, S. (1997). Silicate melts: Relaxation, rheology, and the glass transition. *Reviews of Geophysics*, **35**, 191–218.

Wilson, L. and Head, J. W. (1981). Ascent and eruption of basaltic magma on the Earth and Moon. *Journal of Geophysical Research*, **86**, 2971–3001.

Wilson, L., Sparks, R. S. J. and Walker, G. P. L. (1980). Explosive volcanic eruptions – IX. The control of magma properties and conduit geometry on eruption column behavior. *Geophysical Journal of the Royal Astronomical Society*, **63**, 117–148.

Wolfe, E. W., Garcia, M. O., Jackson, D. B. *et al.* (1987). The Pu'u O'o eruption of Kilauea Volcano, episodes 1–20, January 3, 1983, to June 8, 1984. In *Volcanism in Hawaii*, ed. R. W. Decker, T. L. Wright and P. H. Stauffer. *United States Geological Survey Professional Paper*, **1350**, 471–508.

Woods, A. W. and Cardoso, S. S. S. (1997). Triggering basaltic volcanic eruptions by bubble-melt separation. *Nature*, **385**, 518–520.

Woods, A. W. and Koyaguchi, T. (1994). Transitions between explosive and effusive eruption of silicic magmas. *Nature*, **370**, 641–644.

Woods, A. W., Bokhove, O., de Boer, A. and Hill, B. E. (2006). Compressible magma flow in a two-dimensional elastic-walled dike. *Earth and Planetary Science Letters*, **246**, 241–250.

Wright, H. M. N., Cashman, K. V., Gottesfeld, E. H. and Roberts, J. J. (2009). Pore structure of volcanic clasts: Measurements of permeability and electrical conductivity. *Earth and Planetary Science Letters*, **280**, 93–104.

Wylie, J. J. and Lister, J. R. (1995). The effects of temperature-dependent viscosity on flow in a cooled channel with application to basaltic fissure eruptions. *Journal of Fluid Mechanics*, **305**, 239–261.

Wylie, J. J., Voight, B. and Whitehead, J. A. (1999). Instability of magma flow from volatile-dependent viscosity. *Science*, **285**, 1883–1885.

Zenit, R., Koch, D. L. and Sangani, A. S. (2001). Measurements of the average properties of a suspension of bubbles rising in a vertical channel. *Journal of Fluid Mechanics*, **429**, 307–342.

Zhang, Y. X. (1999). A criterion for the fragmentation of bubbly magma based on brittle failure theory. *Nature*, **402**, 648–650.

Zhang, Y. X. and Behrens, H. (2000). H_2O diffusion in rhyolitic melts and glasses. *Chemical Geology*, **169**, 243–262.

Zhang, Y. X., Xu, Z. J., Zhu, M. F. and Wang, H. Y. (2007). Silicate melt properties and volcanic eruptions. *Reviews of Geophysics*, **45**, doi:10.1029/2006RG000216.

Zimanowski, B., Buttner, R., Lorenz, V. and Hafele, H. G. (1997). Fragmentation of basaltic melt in the course of explosive volcanism. *Journal of Geophysical Research*, **102**, 803–814.

Zimanowski, B., Wohletz, K., Dellino, P. and Buttner, R. (2003). The volcanic ash problem. *Journal of Volcanology and Geothermal Research*, **122**, 1–5.

Exercises

Consider two magmas, a basalt ($\eta = 1$ Pa s, $\rho = 2800$ kg m^{-3}) and rhyolite ($\eta = 10^6$ Pa s, $\rho = 2600$ kg m^{-3}), rising in a conduit. Assume a conduit radius of 10 m and 1 m for the rhyolite and basalt, respectively. Assume a pressure gradient driving magma ascent of 500 Pa m^{-1}. For both cases:

4.1 Calculate the velocity as a function of radial position in the conduit.

4.2 Evaluate the strain rates at the conduit walls. Are the strain rates large enough that the melt might be shear-thinning or undergo structural failure?

4.3 Calculate the capillary number for a 1 cm radius bubble. Will the bubbles become deformed by the ascending magma?

4.4 Calculate the rise speed of a 1 cm bubble relative to the surrounding melt. Given this speed, what flow regime is likely to characterize the magma (see Fig. 4.4).

Online resources available at www.cambridge.org/fagents

• Answers to exercises

Chapter 5

Lava flows

Andrew J. L. Harris

Overview

There are many types of lava flow emplacement
models. One type of numeric model aims to
apply and link equations that describe lava flow
cooling rates, rheology, and dynamics to simulate lava flow spreading. Generally applied to
basaltic lava flow emplacement, it is this type
of model that we consider here. Within this
model type, we can define two simulation sub-
types: volume- and cooling-limited. By applying and linking laws governing the ability of a
Bingham fluid to spread under the influence of
gravity, both volume- and cooling-limited simulations spread control volumes across a digital
topography. Volume-limited approaches spread
and thin a finite erupted volume until a sheet
with a critical thickness defined by the lava
yield strength is attained. For a cooling-limited
model, heat losses due to radiation, convection,
and conduction are estimated to calculate core
cooling and crystallization rates. These can be
used to estimate the rheological properties (viscosity, yield strength) of the lava. Rheological
properties can be used to estimate flow thickness through a yield strength model, as well
as velocity using the Jeffreys or Navier–Stokes
equations. Such flow simulation models rely,
in turn, on well-constrained models that define
and link lava flow heat loss, cooling rates, crystallinity, rheology, velocity, and thickness.

5.1 | Lava flows and lava flow models

Following the classic lava channel morphological
model of Lipman and Banks (1987), an active
lava flow system can be split into two principal
components: a feeder system, which is either a
channel or tube, and a zone of dispersed lava
flow supplied by the feeder system. Within this
system, a range of lava types, defined by their
characteristic surface morphologies (including
'a'a, pahoehoe, transitional, blocky, and inflationary), can build a range of flow field architectures spanning simple, single-unit lava flows to
compound lava flow fields comprising hundreds
to hundreds of thousands of individual lava flow
units (e.g., Walker, 1972; Kilburn and Lopes,
1988; Mattox *et al.*, 1993). Each system can be
channel-, tube-, or channel-and-tube-fed, with
the feeder system often bifurcating to result in
multiple pathways which increase in number
with distance from the vent (e.g., Peterson *et al.*,
1994; Calvari and Pinkerton, 1998). It is the
challenge of the numerical modeler to simulate

Modeling Volcanic Processes: The Physics and Mathematics of Volcanism, eds. Sarah A. Fagents, Tracy K. P. Gregg, and Rosaly M. C.
Lopes. Published by Cambridge University Press. © Cambridge University Press 2013.

and link the complex interplays between topography, rheology, heat loss, and flow dynamics that lead to the emplacement of a specific lava flow system of given morphology, architecture, thickness, length, and width. By fitting lava flow models to real-world systems we can achieve two aims: (i) to understand the complex rheological, thermal, and dynamic processes involved in lava flow emplacement, and which give rise to specific flow emplacement styles, lava flow types architectures, morphologies, and dimensions; and (ii) to more accurately simulate lava flow emplacement events.

Consequently, lava flow numerical models can be generalized into two types: (I) those that improve our understanding of emplacement dynamics and lava flow morphology, and (II) those designed to simulate lava flow emplacement. Type I models (reviewed in online Supplement 5A – see end of chapter) can be split into theoretical or laboratory-based, with laboratory-based studies commonly using kaolin or "wax" as an analog for lava in scaled, tank-based simulations. Both theoretical and laboratory approaches typically attempt to parameterize links between lava flow heat loss, cooling and rheology, and the resulting flow dynamics, dimensions, and morphology. Many laboratory models relate flow dimensions or morphology to effusion rates by applying laws governing heat transfer to explain observed relationships (e.g., Fink and Griffiths, 1992; Blake and Bruno, 2000). Theoretical approaches can include empirical attempts to link flow dimensional properties, such as length and planform area, to volumetric effusion rates. Such models are reviewed by Kilburn et al. (1995) and Harris and Rowland (2009) but are not considered in Supplement 5A.

Type II models use fluid dynamic and heat transfer principles to predict flow emplacement by simulating likely flow direction and extent, as summarized in online Supplement 5B (see end of chapter). Type II studies commonly use findings from Type I modeling to evaluate cooling-induced changes in flow rheology, and hence to determine how flow forward motion and/or spreading varies in space and time. As such, Type II models are often used for hazard assessment, allowing generation of simulations to identify likely inundation zones during ongoing eruptions (e.g., Tedesco et al., 2007) or assessment of likely lava flow hazard zones for future effusive eruptions (e.g., Rowland et al., 2005), as well as evaluation of downflow changes in dynamic, thermal, and rheological properties of a spreading lava. This chapter focuses on the mechanics of Type II models, first reviewing the underlying fluid dynamic, rheological, and heat loss principles, many of which draw on the results of Type I modeling, before examining the way that Type II models function.

5.2 | Lava flow dynamics: fundamental principles and definitions

The principles governing lava flow dynamics were laid down by Hulme (1974) and later detailed by Chester et al. (1985). We begin by defining two fundamental terms: shear stress, τ (Pa or N m^{-2}), and strain rate $\dot{\varepsilon}$ (s^{-1}). Shear stress is the force per unit area acting on a fluid and depends on flow thickness, h, and density, ρ, as well as gravity, g, and the slope, θ, over which the fluid, in this case lava, is moving:

$$\tau = h\rho g \sin \theta. \qquad (5.1)$$

(All notation is given in Section 5.7.) Strain rate is the rate of deformation experienced by a fluid when a load stress is applied, and can be defined by the flow's velocity gradient (i.e., $\dot{\varepsilon} = du/dz$).

We next distinguish between Newtonian and Bingham fluids. A Newtonian fluid will flow (deform) when an infinitesimally small force (shear stress) is applied. In contrast, a Bingham fluid will only flow once sufficient shear stress has been applied. If a Bingham fluid is not sufficiently thick, shear stresses will not be high enough to cause a deformation response, so that strain rate will be zero. Upon attaining a critical thickness, h_0, shear stresses in Bingham fluids become sufficient to cause deformation or flow. For a Bingham fluid on an inclined plane, this point is defined by the yield strength of the fluid:

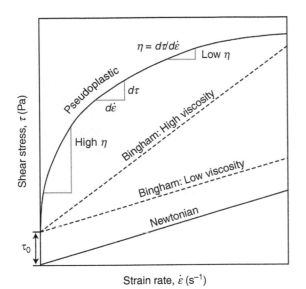

Figure 5.1 Shear-stress–strain-rate relationships for Newtonian, Bingham and pseudoplastic fluids. A Newtonian fluid will deform when placed under an infinitesimally small shear stress, τ. For a Bingham fluid, strain rate, $\dot{\varepsilon}$, will be zero (no deformation will occur) until a critical shear stress is applied, as defined by the yield strength, τ_0. In all cases, the slope of the τ–$\dot{\varepsilon}$ relationship defines the viscosity, η. Newtonian and Bingham fluids exhibit linear behavior, and have viscosities that do not vary with stress or strain rate. For pseudoplastic fluids, the slope of the relationship varies with strain rate, with shallower slopes (low viscosities) at high strain rates and steeper slopes (high viscosities) at low strain rates.

Once deformation is underway, the relation between shear stress and strain rate can follow a number of characteristic paths (Fig. 5.1). The slopes of the shear-stress–strain-rate relationships are dictated by the internal resistance of the fluid to flow, i.e., viscosity, η. If the internal resistance of the fluid is high, a greater increase in shear stress is required to achieve the same amount of change in strain rate than at lower viscosities, so that the slope of the relation is steeper for the high-viscosity case. For Newtonian and Bingham fluids, the relation between shear stress and strain rate is linear and viscosity is the same at all shear stresses and strain rates. For a pseudoplastic fluid, the rate of change in strain rate decreases as shear stress increases so that viscosity is lower at high shear stresses and strain rates than at low stress/strain rate conditions (Fig. 5.1).

Most lava flow models assume that the bulk behavior of a lava flow is Bingham rather than Newtonian (e.g., Dragoni *et al.*, 1986). The non-Newtonian behavior of basaltic lava (and magma) is borne out by the field data of Pinkerton and Sparks (1978), which show that lava has a yield strength and deforms in a pseudoplastic manner as shear stress increases. Pseudoplastic behavior is also apparent in the experimental data of Sonder *et al.* (2006), which reveal that the viscosity of basaltic magmas decreases with increased strain rates, even at temperatures of 1175–1225 °C.

$$\tau_0 = h_0 \rho g \sin\theta. \tag{5.2a}$$

Yield strength can also as be written in terms of the height-to-width ratio that will be attained by a fluid spreading under the influence of gravity on a flat surface (Blake, 1990). For a given volume, such a fluid will spread until it attains a critical thickness, h_0, and width, w, to produce a pile of stationary (non-deforming) fluid:

$$\tau_0 = g\rho h_0 / w. \tag{5.2b}$$

Bingham fluids will only spread as long as $\tau > \tau_0$; once $\tau \le \tau_0$, the flow will have reached its minimum thickness h_0, and spreading will stop. In contrast, Newtonian fluids have no yield strength and the fluid can spread to form an infinitesimally thin sheet.

5.2.1 Lava flow dynamics: velocity treatments

The fluid dynamics of viscous fluids are described extensively in engineering textbooks (e.g., White, 2006), as well as for geological applications by Furbish (1997), and many numerical treatments for viscous flow appear in the fluid mechanics literature. Mei and Yuhi (2001), for example, give one of the most complete solutions for three-dimensional flow of a Bingham fluid in a wide, shallow channel (see Balmforth *et al.* (2006) for review).

For lava flows, the relation of Jeffreys (1925) has become widely used. This provides a relation between fluid viscosity, η, and mean velocity, u, for gravity-driven Newtonian flow in one dimension:

$$\eta = \frac{h^2 g \rho \sin\theta}{Bu}, \qquad (5.3a)$$

in which h is flow depth, θ is the slope of the underlying surface, and B is a constant with a value of 8 for semi-circular channels and 3 for channels that are wider than they are deep. Equation (5.3a) was originally applied to estimate lava flow viscosity using velocity and slope data (e.g., Walker, 1967), but simple re-arrangement allows calculation of mean velocity if viscosity is known:

$$u = \frac{h^2 g \rho \sin\theta}{B\eta}. \qquad (5.3b)$$

In effect, Eq. (5.3b) considers the balance between gravitational forces (numerator) and resistive forces (denominator) on the flow. This equation is, however, only appropriate for flow of a Newtonian fluid in a channel (and so is not appropriate for calculating flow-front velocities).

Moore (1987) provides a treatment for a Bingham fluid that takes into account yield strength, where for flow in a semi-circular channel:

$$u = \frac{h^2 g \rho \sin\theta}{8\eta}\left[1 - \frac{4}{3}\frac{\tau_0}{\tau_b} + \frac{1}{3}\left(\frac{\tau_0}{\tau_b}\right)^4\right], \qquad (5.4a)$$

and for a channel that is wider than it is deep:

$$u = \frac{h^2 g \rho \sin\theta}{3\eta}\left[1 - \frac{3}{2}\frac{\tau_0}{\tau_b} + \frac{1}{2}\left(\frac{\tau_0}{\tau_b}\right)^3\right]. \qquad (5.4b)$$

In both cases the second term describes the damping effect of yield strength by using the ratio of yield strength, τ_0, to shear stress incident upon the flow, τ_b, i.e., the stress at the flow base found using the full flow thickness for h in Eq. (5.1). Multiplying Eqs. (5.4a) and (5.4b) by the width, w, and depth, h, of flow in the channel, allows the two equations to be written in terms of volume flux or effusion rate, E_r (m³ s⁻¹), e.g., for a wide channel,

$$E_r = \frac{w h^3 g \rho \sin\theta}{3\eta}\left[1 - \frac{3}{2}\frac{\tau_0}{\tau_b} + \frac{1}{2}\left(\frac{\tau_0}{\tau_b}\right)^3\right]. \qquad (5.5)$$

For flow in two dimensions, the equation of motion from Navier (1823) and Stokes (1845) can be applied. For steady state conditions, the Navier–Stokes equations can be used to assess the balance of forces incident upon the flow. As shown by Miyamoto and Sasaki (1997), the Navier–Stokes equations can be used to consider motion in the vertical (z) and downflow (x) dimensions:

$$\eta\frac{\partial^2 u}{\partial z^2} = \frac{\partial h}{\partial x}\rho g\cos\theta + \rho g\sin\theta. \qquad (5.6)$$

This describes a scenario where viscous forces (left-hand side of Eq. (5.6)) are balanced by gravity and pressure variations induced by changes in flow thickness, h (right-hand side of Eq. (5.6)). If the lava undergoes little or no variation in thickness, then the pressure term can be neglected so that Eq. (5.6) can be written solely in terms of viscous and gravitational forces (Dragoni et al., 1986):

$$\eta\frac{\partial^2 u}{\partial z^2} - \rho g\sin\theta = 0. \qquad (5.7)$$

Integration of this equation allows the vertical and horizontal profiles of flow velocity within and across a flow to be calculated (Fig. 5.2). Alternatively, the Navier–Stokes equations can be used to describe motion in both the down- and cross-flow (x and y) directions, as described by two velocity components, u and v, respectively (Del Negro et al., 2005):

$$\frac{\partial u}{\partial t} = \frac{\eta}{\rho}\left(\frac{\partial^2 u}{\partial x^2} + \frac{\partial^2 u}{\partial y^2}\right) - u\frac{\partial u}{\partial x} - v\frac{\partial u}{\partial y} - \frac{1}{\rho}\frac{\partial p}{\partial x} + F_x, \qquad (5.8a)$$

$$\frac{\partial v}{\partial t} = \frac{\eta}{\rho}\left(\frac{\partial^2 v}{\partial x^2} + \frac{\partial^2 v}{\partial y^2}\right) - u\frac{\partial v}{\partial x} - v\frac{\partial v}{\partial y} - \frac{1}{\rho}\frac{\partial p}{\partial y} + F_y \qquad (5.8b)$$

These equations account for viscosity effects (term 1 on the right-hand side), the velocity profile (terms 2 and 3), pressure (p) incident upon the flow (term 4), and external forces (F_x and F_y).

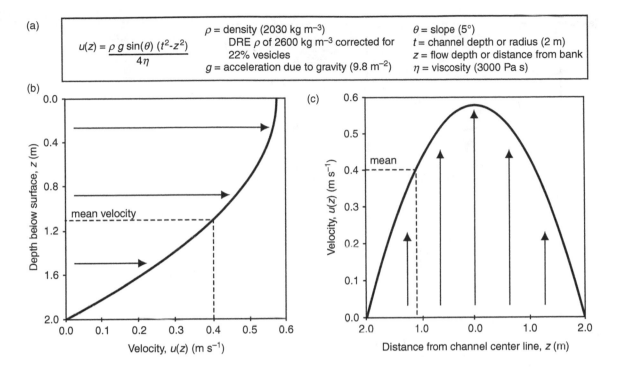

(a)

$$u(z) = \frac{\rho \, g \sin(\theta) \, (t^2 - z^2)}{4\eta}$$

ρ = density (2030 kg m^{-3})
DRE ρ of 2600 kg m^{-3} corrected for
22% vesicles
g = acceleration due to gravity (9.8 m^{-2})

θ = slope (5°)
t = channel depth or radius (2 m)
z = flow depth or distance from bank
η = viscosity (3000 Pa s)

Figure 5.2 Velocity profile equation obtained from integration of Eq. (5.7) and parameters used to derive the variation in flow velocity with depth and cross-channel distance as given in (b) and (c). Input parameters apply to a channel observed as active on Etna during 2001 by Bailey *et al.* (2006). Calculated maximum velocities compare well with values of ~0.6 m s^{-1} measured during brim-full flow.

Inspection of these equations reveals that defining and linking three sets of variables is crucial: (i) flow or channel dimensions (w, h), (ii) slope (θ), and (iii) rheology (η, τ_0). If we can understand the variation in these parameters in space and time, and numerically link them to the resulting flow dynamics, we can arrive at an effective Type II lava flow model. Of these, rheology is the most complicated variable to set, given that it depends on lava composition, temperature, crystallinity, and vesicularity. Thus, numerical relationships between, for example, viscosity and temperature need to be defined.

5.2.2 Controls on viscosity and yield strength

Lava is a polymer, i.e., a compound whose molecule is formed from a number of repeated units of one or more compounds (monomers). The presence of polymers leads to highly interconnected networks of molecules that pose internal resistance to flow. Because the strongest bonds in silicate liquids are the Si-O-Si bridging bonds, the internal resistance of the fluid to flow (i.e., viscosity) increases with silica content. However, network-modifying oxides and water serve to break up the bridging bonds, thereby depolymerizing the fluid and reducing flow viscosity. Temperature also influences viscosity: increased temperature provides increased energy, thereby weakening the strength of the bonds in the melt and lowering the viscosity. Below liquidus temperatures, crystallization also plays a role, whereby the presence of the rigid particles (crystals) impedes flow. The presence of bubbles in the melt also influences the internal resistance of magma. Viscosity relations for fluids, and fluids containing particles, are considered in numerous rheological texts, such as Larson (1999), and discussed in the context of conduit ascent models in Chapter 4. Here we provide a brief review of treatments relevant to silicate lava flow models.

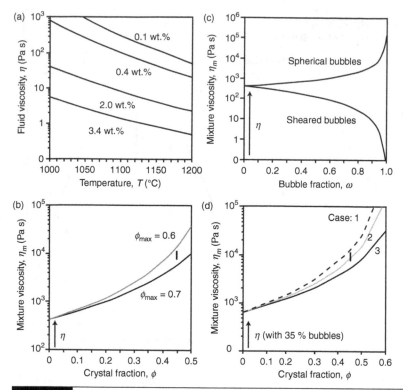

Figure 5.3 (a) Melt viscosity for Etna's alkali basalt as a function of temperature following the relation of Giordano and Dingwell (2003): $\log_{10}\eta = -4.643 + [5812.44 - 427.04 \times H_2O] / [T(K) - 499.31 + 28.74 \times \ln(H_2O)])$. The relation is given for the range of water contents associated with Etna's undegassed (2.0–3.4 wt.%) and degassed (0.1–0.4 wt.%) magma. (b) Relation between mixture viscosity and crystallinity defined by Eq. (5.11) using ϕ_{max} of 0.5 and 0.7. (c) Relation between mixture viscosity and bubble content defined by Eq. (5.12a) (spherical bubble relation) and Eq. (5.12b) (sheared bubble relation). (d) Three-phase mixture viscosity as a function of crystallinity defined using Eq. (5.13a) (Case 1), Eq. (5.13b) (Case 2), and Eq. (5.13c) (Case 3) using a bubble volume fraction of 0.35. Each mixture viscosity relation is set using fluid viscosity (η) of 425 Pa s. This is calculated using the relation of Giordano and Dingwell (2003) for a degassed lava (0.1 wt.% H_2O) erupted at a temperature of 1086 °C and defines the y-axis offset in (b), (c), and (d). Black bars in (b) and (d) indicate crystal content (45%) and viscosity range (9400 ± 1500 Pa s) measured for an Etnean lava by Pinkerton and Sparks (1978).

Viscosity: influence of composition and temperature

For lava, the compositional and temperature dependences of viscosity have traditionally been assumed to follow an Arrhenian relationship in which, for every 10 °C increase in temperature, the rate of chemical reaction doubles. Such a relation can be described by (Bottinga and Weill, 1972; Shaw, 1972):

$$\log \eta(T) = \log a + b/T, \qquad (5.9)$$

with a and b being two adjustable (best-fit) parameters that vary with melt composition. This equation describes a scenario in which viscosity decays exponentially with increasing temperature. Shaw (1972) provides a widely used method for defining a and b in Eq. (5.9) for a melt of given composition and water content.

More recent data show a non-Arrhenian temperature dependence for silicate melts (e.g., Hess and Dingwell, 1996; Russell et al., 2003; Getson and Whittington, 2007). To describe this scenario, the Vogel–Fulcher–Tammann (VFT) equation has been adopted as a purely empirical means of providing a best-fit to available data (Mano and Pereira, 2004):

$$\log \eta(T) = a + \frac{b}{T - c}, \qquad (5.10)$$

in which a, b, and c are adjustable (best-fit) parameters that depend on lava composition. Such relationships are derived on a case-by-case basis, with the fitting parameters depending strongly on composition. Thus, a relation must be selected that is appropriate for the melt under consideration. Currently VFT-based treatments are available for tephrite (Whittington et al., 2000) and Etna basalt (Giordano and Dingwell, 2003), as well as a range of compositions spanning basalt to rhyolite (Giordano et al., 2006). These relations define a situation whereby viscosity increases logarithmically with decreasing temperature (e.g., Fig. 5.3(a)).

Viscosity: influence of crystals and bubbles

Lava is a mixture of fluid (melt) and solids (crystals). Thus, to obtain the bulk viscosity of the lava mixture (η_m), the effect of crystals also needs to be taken into account. This can be achieved using the Einstein–Roscoe relationship (e.g., Marsh, 1981; Pinkerton and Stevenson, 1992),

$$\eta_m = \eta \, (1 - R\phi)^{-2.5}, \tag{5.11}$$

in which η is the fluid viscosity, ϕ is crystallinity (expressed as a volume fraction) and $R = 1/\phi_{max}$, where ϕ_{max} is the maximum crystal packing content. This defines a relation whereby mixture viscosity increases with crystal content following a power law (Fig. 5.3b).

A third phase is also present in the mixture, bubbles. These may serve to increase the mixture viscosity if the bubble population is spherical, or decrease mixture viscosity if the bubbles are sheared (e.g., Crisp et al., 1994; Manga et al., 1998; Section 4.5.4). Llewellin and Manga (2005) thus provide two relations for the mixture viscosity of a fluid containing bubbles, one for fluids containing spherical bubbles:

$$\eta_m = \eta \, [1 - \omega]^{-1} \tag{5.12a}$$

and one for fluids containing sheared bubbles

$$\eta_m = \eta \, [1 - \omega]^{5/3}, \tag{5.12b}$$

in which ω is bubble content. These relations are plotted in Figure 5.3(c), which shows the contrasting effects on mixture viscosity of spherical and sheared bubble populations.

To accurately characterize lava viscosity at subliquidus temperatures, a three-phase treatment is required which takes into account the viscosity of the fluid, as well as the effects of both crystals and bubbles. Although few such treatments have been applied to lava flows, the three-phase treatment of Phan-Thien and Pham (1997) shows promise in determining the three-component mixture viscosity for a basaltic lava with a variable amount of crystals and bubbles (Harris and Allen, 2008). The treatment simplifies to three cases:

Case (1): Crystals smaller than bubbles,

$$\eta_m = \eta \, [1 - \phi \, / \, (1 - \omega)]^{-5/2} \, (1 - \omega)^{-1}; \tag{5.13a}$$

Case (2): Crystals and bubbles in the same size range,

$$\eta_m = \eta \, (1 - \phi - \omega)^{-(5\phi + 2\omega)/2(\phi + \omega)}; \tag{5.13b}$$

Case (3): Crystals larger than bubbles,

$$\eta_m = \eta \, [1 - \omega \, / \, (1 - \phi)]^{-1} \, (1 - \phi)^{-5/2}. \tag{5.13c}$$

Although application of these equations requires exacting knowledge of the size and number of crystals and bubbles, when used with appropriate inputs, ω and ϕ, mixture viscosities in good agreement with field measurements can be obtained. Such a three-phase fit for Etna is given in Figure 5.3(d).

Yield strength

Yield strength can also be calculated as a function of temperature and crystallinity. Data in Chester et al. (1985) show that yield strength increases exponentially with $1/T$, so that

$$\tau_0(T) = a e^{b/T}. \tag{5.14}$$

As in Eq. (5.9), a and b are adjustable (best-fit) parameters that vary with lava composition. Chester et al. (1985) give values for Hawaiian (Kilauea) lava lake basalt and Columbia River basalt, and Dragoni (1989) gives values for Etnean basalt. However, such treatments apply to limited temperature ranges; Chester et al. (1985) suggest that τ_0 increases rapidly with the onset of crystallization, but trends towards a steady value before crystallization is complete. Pinkerton and Stevenson (1992) also give relationships that allow calculation of τ_0

on the basis of crystal concentration, size, and shape, with Ryerson *et al.* (1988) showing that data for picritic basalt from Kilauea Iki are best fitted with a power law expression whereby $\tau_0 = 6500\phi^{2.85}$.

The challenge with lava is that, unlike simple Bingham fluids, high temperature gradients mean that severe rheological gradients occur vertically within a lava flow, such that no single rheology applies. Lateral spreading, for example, will be influenced by the development of a cool, high viscosity, viscoelastic skin. This means that the fluid core will be confined within a strong outer shell (Hon *et al.*, 1994). As a result, one modeling direction has been to examine the ability of such rheologically defined outer shells to contain a steadily pressurizing or inflating core (e.g., Iverson, 1990) and to examine the role of this outer shell in influencing flow front break outs (e.g., Dragoni and Tallarico, 1996).

5.3 | Lava flow heat budget and cooling

Because temperature, and cooling-induced changes in crystallinity, play central roles in determining flow rheology, understanding the flow heat budget and cooling rates is an important step in achieving a Type II model. This involves the derivation of a complete lava flow heat budget that defines the balance between heat sinks and heat sources. While Keszthelyi (1995) provides a complete heat loss model for tube-confined lava, general texts such as Holman (1992) detail the principles of heat transfer. We here consider the heat budget for channel-confined lava and/or dispersed ('a'a, pahoehoe and blocky) lava flows whose surfaces are exposed to the atmosphere.

For an active channel-fed 'a'a, pahoehoe, or blocky lava flow, the main heat sinks will be radiation and convection from the flow surface (e.g., Keszthelyi *et al.*, 2003), with conduction causing heat loss through the flow levées (e.g., Quareni *et al.*, 2004) and base (e.g., Fagents and Greeley, 2001). Heat is conducted across a surface crust that forms over the molten core. This

heat is then lost from the surface by radiation and convection. Radiation from a lava surface at temperature T_{surf} (K) can be assessed using the Stefan–Boltzmann law. This describes the energy flux (q_{rad} in W m^{-2}) radiated directly from the surface of a gray-body, i.e., a body with an emissivity $\varepsilon < 1$:

$$q_{rad} = \varepsilon\sigma(T_{surf}^4 - T_a^4), \qquad (5.15)$$

in which σ is the Stefan–Boltzmann constant (5.67 × 10^{-8} W m^{-2} K^{-4}) and T_a is the temperature (K) of the environment into which radiation is being emitted. Multiplying by the area of the radiating surface (A) gives the total radiant heat loss, Q_{rad}, in watts. The first question is how to determine T_{surf}?

Crisp and Baloga (1990) suggested that the surface of an active lava is best modeled by a two-component thermal surface in which a chilled crust at temperature T_{crust} occupies fraction f of the surface, with cracks exposing the molten core at T_{core} occupying the remainder (Fig. 5.4(a)). As a result, T_{surf} is best described by the effective radiation temperature (T_e in K):

$$T_e = \left[fT_{crust}^4 + (1-f)T_{core}^4 \right]^{1/4}. \qquad (5.16)$$

In a stable case (when there is no destruction and renewal of the crust, so that it simply thickens with time), the cooler of these two thermal components can be determined on the basis of time since exposure (t). For Hawaiian pahoehoe, measurements by Hon *et al.* (1994) showed that surface temperature declined logarithmically with time ($T_{crust} = -140\log(t) + 303$) as the surface crust thickened with the square root of time ($D_{1070} = 0.0779\sqrt{t}$). Here, the 1070 subscript shows that this relation is for the depth of the 1070 °C isotherm, which is used to define the base of the visco-elastic surface crust. On the face of it, the hotter of the two components appears easy to set, being essentially equal to the lava core temperature. In reality, the maximum temperature of the flow core is rarely exposed at the surface. Instead, cracks penetrate a cooler viscoelastic layer that forms just below the brittle crust and which, for Hawaiian pahoehoe, has a maximum temperature of 1070 °C, as opposed to a

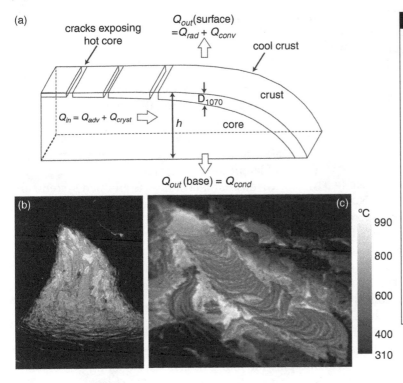

Figure 5.4 (a) Two-component model for the thermal structure of a lava flow surface, showing a chilled crust broken by cracks exposing the hot core, and the main heat sources and sinks (modified from Fig. 3 of Crisp and Baloga (1990)). (b) Photograph of a pahoehoe breakout (~1 m wide) showing at least three thermal components apparent from the different colors; yellow (~1090 °C), bright orange (~900 °C), dull orange (550–700 °C) and black (< 475 °C). (c) Thermal image of pahoehoe channel (central ropey slab is ~2 m wide and 5–6 m long) showing active surfaces ranging in temperature from ~400–1000 °C, with lower temperatures marking older, cooler crust developing on the flow surface as it advances and ages. See color plates section.

core temperature of 1140–1150 °C (Hon *et al.*, 1994), meaning that maximum surface temperatures measured at Hawaiian pahoehoe are typically ~1000 °C. Likewise, maximum surface temperatures measured at a lava channel on Mt. Etna (Italy) within 70 m of the vent were typically 730–1040 °C (mean of 880 °C) compared with a core temperature of 1065 °C (Harris *et al.*, 2005; Bailey *et al.*, 2006). Thus, use of a value for T_{core} that is offset from the true core temperature may yield more realistic values of T_e. In addition, more than one thermal component may be present on the flow surface (Figs. 5.4(b), 5.4(c)). Wright and Flynn (2003) found that active pahoehoe surfaces were best modeled using between five and seven thermal components at temperatures ranging between 200 and 1000 °C. In comparison, thermal data from active lava channels indicate that a four thermal component model may be used to describe such surfaces (see Exercise 5.1; online resources listed at end of chapter).

Convective heat loss from a flow surface involves heat transfer to an overlying gas or fluid. There are two cases. Free convection occurs in still conditions and can be conceptualized in terms of a heated plate exposed to ambient room air without an external source of motion (i.e., no wind). Heating of the air from beneath causes the air to expand. The subsequent reduction in density causes the heated air to become buoyant so that it rises, carrying heat with it and away from the surface. In contrast, forced convection is the cooling effect experienced when cool air blows over a hot surface. Both can be described in terms of the convective heat transfer coefficient, h_c:

$$q_{conv} = h_c (T_{surf} - T_a), \qquad (5.17)$$

where q_{conv} is the convective heat flux (in W m^{-2}), and multiplying by the area, A, of the hot surface gives the total convective heat loss, Q_{conv}, in watts. The convective heat transfer coefficient, h_c, can be calculated theoretically (as in Exercise 5.2; online resources listed at end of chapter), but differs for free and forced convection and varies with wind speed. However, values of between 5 and 150 W m^{-2} K^{-1} appear appropriate for subaerial active lavas (Keszthelyi and Denlinger, 1996; Neri, 1998; Keszthelyi *et al.*, 2003).

Across the flow base, heat will be transferred by molecular contact from the high-temperature core at T_{core}, through a basal boundary layer of thickness Δy, to the cold underlying country rock at T_a. This conduction heat flux (q_{cond}, in W m^{-2}) can be described using Fourier's Law:

$$q_{cond} = -k\frac{dT}{dy} = -k\frac{\Delta T}{\Delta y}, \qquad (5.18a)$$

where ΔT is the temperature difference ($= T_{core} - T_a$) across the basal boundary layer. Note that, because the temperature gradient across the base of a lava flow is negative, Fourier's Law has a minus sign on the right-hand side to make q_{cond} a positive quantity. The conductive heat flux will decrease with time as the basal crust thickens and as the substrate becomes heated (e.g., Kerr, 2001), so that the distance to the ambient isotherm increases. Following Turcotte and Schubert (2002), the basal boundary layer thickens at a rate on the order of $\sqrt{\kappa t}$, in which κ is thermal diffusivity and t is the characteristic time interval since flow emplacement (s), so that Eq. (5.18a) can be written in terms of time:

$$q_{cond} = -k\frac{\Delta T}{\sqrt{\kappa \pi t}}. \qquad (5.18b)$$

Heat sources within a lava flow include advection and crystallization (Crisp and Baloga, 1994; Fig. 5.4(a)). Advection describes heat brought into the control volume by a moving fluid, and can be written (q_{adv}, W m^{-2}):

$$q_{adv} = -uh\rho c_p\frac{dT}{dx}, \qquad (5.19)$$

in which c_p is the lava specific heat capacity, and dT/dx is the temperature change per unit distance downflow. Note there is a minus sign before dT/dx because the temperature decreases as distance increases. Writing the equation in terms of volume, V, and cooling per unit time ($dT/dt = u\,dT/dx$), i.e., $Q_{adv} = -V\rho c_p\,dT/dt$, gives the total advective heat gain in watts.

Heat generated by the phase change experienced during crystallization (q_{cryst}, W m^{-2}) can be described similarly:

$$q_{cryst} = uh\rho L\frac{d\phi}{dx}, \qquad (5.20)$$

in which L is latent heat of crystallization and ϕ is the mass fraction of crystals grown, so that $d\phi/dx$ is the crystallization rate per unit distance. Again, writing the equation in terms of volume and crystallization per unit time ($d\phi/dt = u\,d\phi/dx$), i.e., $Q_{cryst} = V\rho L\,d\phi/dt$, gives the total latent heat generated in watts.

Given that a lava flow is an open system, all heat flowing into the system (Q_{in}) must also flow out (Q_{out}), so that:

$$Q_{in} = Q_{out}. \qquad (5.21a)$$

If we consider all of the heat sources and sinks, Eq. (5.21a) can be expanded to

$$Q_{adv} + Q_{cryst} = Q_{rad} + Q_{conv} + Q_{cond}. \qquad (5.21b)$$

Because it allows calculation of core cooling and/or crystallization rates, this balance underpins modeling of lava core cooling and crystallization (e.g., Daneš, 1972; Crisp and Baloga, 1994; Keszthelyi and Self, 1998). This can most easily be shown by writing the left-hand side of Eq. (5.21b) in full, so that:

$$V\rho\,[c_p\,dT/dt + L\,d\phi/dt] = Q_{rad} + Q_{conv} + Q_{cond}. \quad (5.22a)$$

Rearranging Eq. (5.22a) yields the cooling rate per unit time (in K s^{-1}) from,

$$\frac{dT}{dt} = \frac{Q_{rad} + Q_{conv} + Q_{cond} - V\rho L\dfrac{d\phi}{dt}}{V\rho c_p}. \qquad (5.22b)$$

Alternatively, cooling rate per unit distance (K m^{-1}) can be obtained using the heat fluxes (in W m^{-2}),

$$\frac{dT}{dx} = \frac{q_{rad} + q_{conv} + q_{cond} - uhL\dfrac{d\phi}{dx}}{uh\rho c_p}. \qquad (5.22c)$$

Note that the use of heat fluxes, q, flow velocity, u, and depth, h, as opposed to the use of total heat losses, Q, and volume, V, in Eq. (5.22b), makes Eq. (5.22c) dimensionally correct if we require output in terms of change in temperature per unit distance. Thus, if (i) flow velocity and depth are known or measured, (ii) T_{surf} can be estimated to allow calculation of Q_{rad} and Q_{conv}, and (iii) a levée and basal conduction heat loss model is applied, then the core cooling rate can be calculated given reasonable assumptions for ρ, c_p, and crystallization rates. Exercises 5.3 and 5.4 (see online resources listed at end of chapter) illustrate this procedure.

5.3.1 Crystallinity

Viscosity, yield strength, and cooling rates also depend on crystallinity, which will increase with decreasing flow core temperature. Crystal content can be expressed as a function of time using the relationship defined by Crisp and Baloga (1994):

$$\phi(t) = \phi_{max} [1 - \exp(-t/t_{\phi max})], \qquad (5.23)$$

in which $t_{\phi max}$ is the time needed to reach the maximum crystallinity (ϕ_{max}). Alternatively, crystallization rates can be obtained as a function of distance using the cooling rate per unit distance (Harris and Rowland, 2001):

$$\frac{d\phi}{dx} = \frac{dT}{dx}\frac{d\phi}{dT}, \qquad (5.24)$$

in which $d\phi/dT$ is volume fraction of crystals grown ($d\phi$) over a given temperature drop (dT).

5.4 | Type II modeling

The aim of a Type I model is to draw links between any two of the following: (i) heat loss and core cooling, (ii) flow rheology, (iii) flow dynamics and dimensions, and/or (iv) lava surface morphology. The aim of the Type II model is to draw these findings together to generate a complete lava flow emplacement model. The main attempts to achieve this are summarized in Supplement 5B (online resource – see end of chapter).

The foundation of Type II modeling is either (a) volume-limited flow emplacement, or (b) cooling-limited flow emplacement. A cooling-limited flow is one whereby supply (lava effusion) is maintained until the full flow extent is realized. That is, the limits of the flow unit are defined by the processes of lava cooling and the

effect this has on inhibiting further flow advance through modifying the flow core rheology. In volume-limited cases, the flow stops once the available volume has been used up, irrespective of the cooling and/or rheological conditions. All Type II flow models also require a topographic model to determine the likely path that a fluid will follow. We thus begin with an assessment of flow path projection routines.

5.4.1 Flow paths

The simplest method to generate a flow path is to identify the course that water draining down a slope would follow (e.g., Fig. 5.5). Application of such flow path models are commonplace in hydrology and are available as part of many Geographical Information Systems (GIS). Many routines are based on the deterministic eight neighbors (D8) method (Mark, 1984; O'Callaghan and Mark, 1984). This routine finds the path of steepest descent across a gridded topography or Digital Elevation Model (DEM). To do this, the gradient between a pixel and all eight of its neighbors is calculated. The flow line is then projected from the host pixel in the direction of most negative gradient. This process is continued from pixel to pixel to define the flow line. The basic D8 algorithm has problems with long planar slopes, as well as flat areas and pits, i.e., cases where the host pixel will be lower than, or the same elevation as, all surrounding pixels (Fairfield and Leymarie, 1991; Turcotte et al., 2001; Jones, 2002). Pits may either be natural or due to noise in the DEM. To solve these problems, a number of modifications have been proposed to the D8 model including use of a stochastic function to solve flow path problems over planar slopes (Fairfield and Leymarie, 1991), use of maximum gradients calculated using eight triangular facets centered on the host pixel to improve flow line prediction (Tarboton, 1997), smoothing to remove pits (Hutchinson, 1989), or allowing the flow line to leave a pit through the lowest pixel on the pit rim (Fairfield and Leymarie, 1991; Tribe, 1992). In the latter case, once the pit outlet has been found, a flow line is projected across the pit from the entrance to exit point. This creates a flat area across the pit. To solve for problems in flat areas, Martz and Garbrecht (1998) proposed a solution

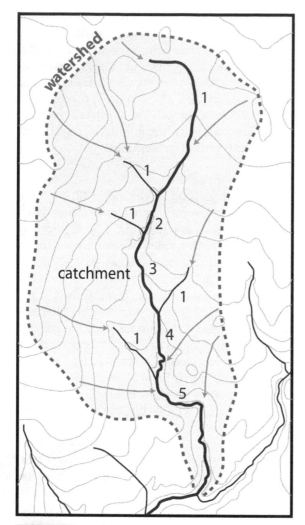

Figure 5.5 Drainage basin terminology and principles as applied to lava-shed modeling. In this case (modified from Fig. 2.1b of Gregory and Walling (1973)), the basin watershed (dashed line) is defined according to the contour information from a topographic map (gray lines). This encloses a basin that encompasses the drainage or catchment area (gray area) for all streams (black lines) within the watershed. The trunk stream (thick black line) is the stream that exits the catchment, and its order (in the system applied here and as numbered on the map) depends on the number of tributaries that feed it. Drainage lines from points along the watershed are given as gray arrows.

The D8 model is the basis of the Harris and Rowland (2001) FLOWGO model where, if a closed depression is encountered (e.g., a pit crater) then the program fills the depression and continues out of the depression from the rim pixel containing the maximum gradient. This approach defines a single line down which a lava flow erupting from a given point would drain (Fig. 5.6(a)). This does not mean that pixels through which drainage lines do not run are immune from inundation, because flow width is not considered, nor is the effect of flow emplacement upon topography. Thus, to determine flow area, as opposed to the most likely flow path given existing topography, a more sophisticated approach is required.

A simple but effective means of projecting flow areas is the stochastic slope-controlled model of Favalli *et al.* (2005). This projects a flow path down a DEM based on the maximum-gradient rule, and then adds random noise to the DEM in the $\pm\Delta h$ range, where Δh is the maximum variation in height allowed for a pixel. The flow path is then recalculated on the noise-modified DEM. This effectively represents a slightly new topography, so the flow path is not the same as in the preceding run. This procedure is repeated for each run, to predict an "envelope" of probable flow paths. This approach produces a spread of flow paths that provides a good approximation of the real flow field area (Figs. 5.6(b), 5.6(c)).

Kauahikaua *et al.* (1995) also used a hydrological approach to identify lava catchments. Use of a USGS 1:24 000 DEM with the ARC/INFO GRID module allowed identification of a drain point or polygon that was then used to compute a lava catchment, as defined in Figure 5.5. The catchment is defined as that region from which all surface water, or in this case lava, will flow to exit that region in a single, common, channel. In terms of drainage basin analysis, the exit channel will be the highest-order stream (Gregory and Walling, 1973). The DEST (Determination of Earth Surface Structures) algorithm of Favalli and Pareschi (2004) is fundamentally the same in that it projects maximum gradient paths down a DEM from multiple points, allowing catchment analysis by identifying drainage networks that

whereby elevation modifications are made across the flat area to ensure a flow line from the higher elevation to the lower elevation. Effectively, a slope is projected across the flat area from the entrance to the exit.

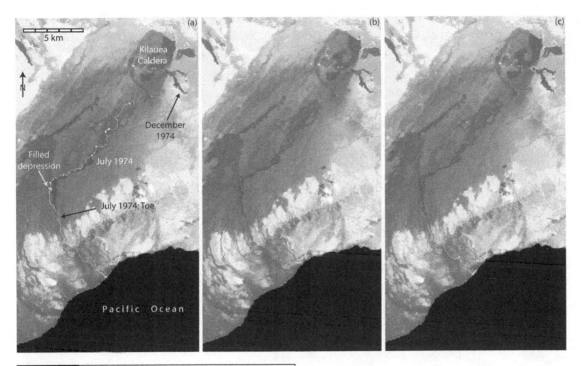

Figure 5.6 (a) Maximum gradient path (yellow line) from Kilauea's July 1974 vent location to the DEM edge (coast). (b) Stochastic flow path model from Kilauea's July 1974 vent location to the coast. Model applies DEM noise of 2 m and has been run for 20 iterations. (c) Twenty-iteration flow line runs with 2 m of DEM noise for Kilauea's July and December 1974 flows. Flow lines have been cut at the flow toe in each case to allow an assessment of the fit with reality, as revealed by the black flow areas in this Thematic Mapper (TM) color composite. Simulations have been projected onto a near true color (vegetation = bright green, lava = black) TM image of Kilauea. See color plates section.

define individual catchments. Multiple drainage lines, from points located around the rim of a catchment (i.e., along the watershed) will converge to exit the catchment at a single point, as shown in Figure 5.5. Thus, lava flows from multiple sources along a rift zone may be predicted to enter the ocean at the same place. Such an approach was used by Guest and Murray (1979) in assessing lava flow hazard on Etna, where all possible flow paths were generated by projecting downhill lines from all likely vent zones. These converged to define a smaller number of highest-order stream points at the base of Etna's cone at which flows would likely arrive.

DEM quality

In all flow path prediction algorithms, DEM quality is an issue; the DEM needs to be as smooth and noise-free as possible, but without losing topographic detail that may affect the flow path. On a small scale, flows moving over existing pahoehoe and 'a'a lavas will be influenced by inflation features, such as tumuli, or by channels, levées, and flow margins. It is not uncommon, for example, for a lava flow to follow a pre-existing flow margin (which represents a local topographic high), or even re-use a drained lava channel or tube (which represents a local topographic low). Also, care needs to be taken to remove the influence of trees from elevation data. If elevations are obtained from the tree canopy, DEM-projected flow paths may follow false boundaries induced by the apparent scarp caused by a forest edge bordering a lava flow field or road. The good fit for the modeled area of Kilauea's December 1974 flow in Figure 5.6(c), for example, is due to the simulation filling a false depression in the DEM caused by the emplacement of the real lava flow, which destroyed the trees in this otherwise forested area. Such false depressions due to tree clearance

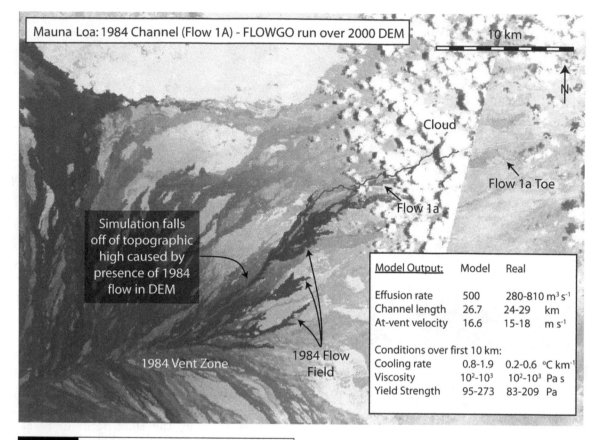

Mauna Loa: 1984 Channel (Flow 1A) - FLOWGO run over 2000 DEM

10 km

N

Cloud

Flow 1a Toe

Flow 1a

Simulation falls off of topographic high caused by presence of 1984 flow in DEM

1984 Flow Field

1984 Vent Zone

Model Output:	Model	Real	
Effusion rate	500	280-810	$m^3 s^{-1}$
Channel length	26.7	24-29	km
At-vent velocity	16.6	15-18	$m s^{-1}$
Conditions over first 10 km:			
Cooling rate	0.8-1.9	0.2-0.6	$°C \ km^{-1}$
Viscosity	10^2-10^3	10^2-10^3	Pa s
Yield Strength	95-273	83-209	Pa

Figure 5.7 FLOWGO simulation of Mauna Loa's 1984 flow 1A channel. Model run uses the input conditions for Mauna Loa's 1984 flow given in Table 5 of Harris and Rowland (2001). In executing the model the software first projects a maximum gradient path (in yellow) from the selected vent point to the edge of the DEM. The cooling-limited distance that the channel-contained control volume can travel down that flow path is then projected in red. Comparisons of model output and field measured parameters are given in the inset box. See color plates section.

are well-known artifacts in DEMs produced over forested volcanic areas (Stevens *et al.*, 1999).

On one hand, the DEM needs to be as recent as possible, because flow boundaries from recently emplaced flows will provide new topography that subsequent flows will follow. On the other hand, using current DEMs to model older flows is problematic because the topography over which the simulation is made will be modified by the presence of the flow itself. Paths projected to simulate the course of Mauna Loa's

1984 flow using a DEM generated from Shuttle Radar Topographic Mission data collected in February 2000, for example, invariably fall off of the edge of the 1984 flow (which is a local topographic high) and then follow the flow margin (Fig. 5.7), making the path of the 1984 flow itself impossible to simulate using the 2000 DEM.

5.4.2 Type IIa models: volume-limited flow emplacement

Volume-limited models take a volume of lava that is then allowed to spread across a surface until it thins to such an extent that flow ceases (e.g., Hulme, 1974; Blake, 1990). For a Bingham fluid, this will occur once the shear stress incident upon the fluid is insufficient to cause deformation or flow. Most volume-limited models thus assume a Bingham rheology and spread a finite lava volume until the critical thickness, defined by the yield strength, is attained. The FLOWFRONT model of Young and Wadge (1990) was one of the first examples of a volume-limited

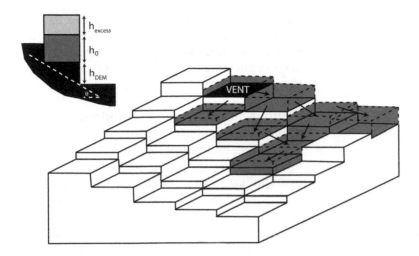

flow emplacement model, where lava from an "active" cell (i.e., a cell from which lava was moving) was divided proportionally between its eight adjacent cells depending on the slope between the source and target cell. "Fillable" cells were identified as those cells downslope from the active cell, so that lava was preferentially partitioned to those cells that were topographically lower than the active cell. For each cell, a critical thickness, h_0, of lava is left behind. The excess thickness is passed to each adjacent downslope cell (Fig. 5.8). If a volume (V_{crit}) equal to each cell area multiplied by h_0 is removed from the available volume (V_{erupt}) at each step, then a point will be reached where the entire erupted volume is used up. The model then stops and the flow field is defined on the basis of the most likely downslope inundation area defined by spreading the erupted volume under the influence of gravity.

5.4.3 Type IIb models: cooling-limited flow emplacement

In contrast to volume-limited spreading models, where the ability of the flow to spread is determined by the available volume, in cooling-limited models the extent of spreading is determined by cooling and rheology. The basic operation of a Type IIb model is summarized in Figure 5.9, and involves application of a heat loss model to allow calculation of downflow changes in core temperature and crystal content. This is used with an appropriate rheological model to estimate the variation in flow velocity and thickness down- and across-flow. Lava is spread or advanced until some stopping condition is met. This may involve application of the following steps:

(1) Initialize the model with an eruption temperature, crystal content, flow depth, underlying slope, and flow velocity estimate using Eqs. (5.3), (5.4), (5.6), (5.7) and/or (5.8);
(2) Calculate the heat flux for the current cell using Eqs. (5.15) to (5.20);
(3) Apply Eq. (5.22) to calculate cooling rate;
(4) Apply Eq. (5.23) or (5.24) to calculate crystallization rate;
(5) Estimate the new core temperature and crystallinity;

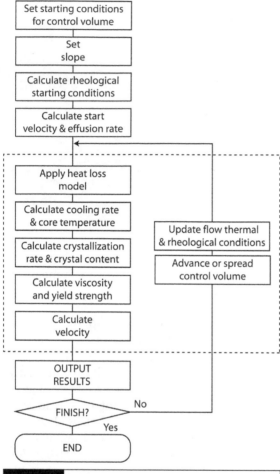

Figure 5.9 Flow chart for steps followed by a Type IIb (cooling-limited) lava flow simulation. Following Ishihara *et al.* (1990), steps enclosed by broken line are applied to all active cells in the cellular automata grid.

(6) Calculate the new temperature- and crystallinity-dependent viscosity using Eqs. (5.9) and (5.11), and yield strength using Eq. (5.14);

(7) Use the new viscosity and yield strength to recalculate velocity using Eq. (5.4);

(8) Estimate flow thickness using Eq. (5.2);

(9) Move the control volume, with its new thermal, rheological and velocity states, to the next cell;

(10) Loop until stopping condition is met.

This is the foundation of the FLOWGO-based model run for Mauna Loa's 1984 channel (Fig. 5.8), where a comparison between model-calculated and field-measured parameters, as well as runout, shows good agreement, suggesting that the model adequately represents the physics of flow emplacement.

Cellular automata cooling-limited emplacement models

The cooling-limited approach is the basis of cellular automata spreading models, which were first applied to lava flow emplacement by Crisci *et al.* (1986). A cellular automata (CA) model consists of a grid of cells, each of which has a number of states and the same rules for updating. Each time the rules are applied to the grid, a new generation of cells is created. In the case of a lava flow cellular automata model, each cell has an elevation defined by the DEM, and some rule(s) for updating the flow through, and volume residing in, each cell. As a result, lava is passed between cells depending on the rheological and thickness conditions of the lava. With each time step, the automata grid is updated, the rheological and thickness conditions of each active cell are modified, and the lava is spread across the DEM, leaving a layer behind in each inundated cell. Thus, the basic spreading and emplacement concept is much like FLOWFRONT in that a rheologically controlled thickness rule is used to spread volume from cell to cell (Fig. 5.8). The difference between FLOWFRONT and Type IIb CA models, though, is that the latter attempt to model the actual flow dynamics and place cooling limits on the ability of the flow to spread.

In the case of Navier–Stokes-based CA models, lava is passed from cell to cell with the thermal and rheological properties of lava in each active cell being updated at every time step to determine the cooling-limited extent of spreading. This approach forms the basis of several cellular automata models that build on the starting point of Ishihara *et al.* (1990) (as detailed in Supplement 5B). A progression from this foundation can be traced through the CA models of Miyamoto and Sasaki (1997) to the LavaSIM of Hidaka *et al.* (2005) and the MAGFLOW model of Vicari *et al.* (2007). In the cooling-limited CA model of Miyamoto and

Sasaki (1997), for example, yield strength, viscosity, thickness, and slope are varied from cell to cell as the lava spreads across the topographic grid obeying the equation of motion given in Eq. (5.6). The MAGFLOW model is instead built on the Cellular Nonlinear Network model of Del Negro *et al.* (2005), which applies Eq. (5.8) to also take into account cross-flow velocity. However, it too uses a cooling model to modify yield strength, viscosity, and flow thickness as the lava spreads from cell to cell across a topographic grid.

The SCIARA cellular automata model is different in that it is not based on the Navier–Stokes equations (Crisci *et al.*, 2004). Instead, the flow in and out of the cells is determined by hydrostatic pressure gradients due to differences in lava thickness within surrounding cells, and the rheology of lava in the cell (Barca *et al.*, 1993). Rather like FLOWFRONT, and all other CA models, it relies on an assessment of the critical thickness of lava that must remain in a given cell. Unlike FLOWFRONT, where a fixed yield strength (that does not vary downflow) is used to define this critical thickness, in SCIARA the critical thickness is set using an "adherence parameter," ϖ. This defines the thickness of lava that cannot flow out of a cell due to rheological resistance and varies with temperature following the form of the expected temperature-dependent relationship for yield strength ($\varpi = ae^{-bT}$). Thus, like all Type IIb models, SCIARA is driven by a heat budget based cooling model that is used to assess the temperature of lava in any given cell, and hence its rheological conditions.

5.5 | Discussion

Lava flow emplacement models are becoming increasingly sophisticated. Mostly this is because our ability to understand and parameterize the interplay among various flow emplacement dynamics has improved immensely in the last three decades, as witnessed by the weight of literature that comprises online Supplement 5A. As a result, flow emplacement simulations are providing increasingly better fits with the characteristics (e.g., effusion rate, rheological conditions, cooling, velocity, flow length, inundation area, and emplacement duration) of the real-world flows against which they are compared. However, one issue that has not been explored so far in this chapter is simulation run time. Cellular automata models are computationally complex, but allow for more complex flow geometries to be modeled than one-dimensional models, allowing flow spreading in two dimensions to be simulated. However, the complexity of the models makes run times quite long. Even the two-dimensional model of Miyamoto and Sasaki (1997), for example, has run times of up to two days when executing a 40-day simulation on a 3 GHz computer (~1.5 hours for each day of simulation). The FLOWGO model of Harris and Rowland (2001) is computationally much less complex, allowing run times for a single flow path of less than one second (e.g., Fig. 5.6(a)). Running multiple flow paths in a stochastic simulation increases run times but only to 30 seconds for a 20-iteration run (e.g., Fig. 5.6(b)).

If models are to be used for real-time hazard assessment we need rapid run times. For example, in the case of a rapidly advancing flow on a heavily populated volcano, run times need to be low so that results are available in a timely fashion. If a flow is moving towards a town at 10 km h^{-1}, even if the town is 25 km distant from the source vent, there is little time available to run a simulation to identify vulnerable infrastructure and zones in need of immediate evacuation. This was the case of Mauna Loa's 1950 eruption, which began at ~10:15 pm on 1 June 1950. Within 2 hours and 15 minutes the flow had extended 23 km to enter the town of Hookena, where it destroyed the main highway, post office, several houses and a gas filling station in short order (Finch and Macdonald, 1950). In such a scenario, there are less than 2.5 hours available to inform the decision-making process. Civil Protection authorities responding to such an event today would require information within minutes. Given reliable lava flow simulation models, one aim must now be to reduce run times to a minimum.

5.6 | Summary

- A large number of thermal, rheological and topographic treatments exist that allow construction of numerical models to simulate volume- and/or cooling-limited lava flow emplacement.
- Models presented here have tended to focus on basaltic scenarios, and describe emplacement properties of relatively simple flow types; they tend not to account for more complex emplacement processes such as channel blockage, overflow and diversion, inflationary growth, pahoehoe breakouts, and/or tube development.
- Good fits between model output and field data show that existing models do a reasonable job in describing real-world simple basaltic flow unit emplacement scenarios.
- Such off-the-shelf models can thus be selected depending on user requirements and computing capabilities, and may be used as a basis to achieve more complex models for basaltic flows.

5.7 | Notation

a, b, c	parameters in empirical equations
A	area (m^2)
B	constant in Jeffreys equation
c_p	specific heat capacity (J kg^{-1} K^{-1})
D_{1070}	thickness of lava surface crust (m)
E_r	effusion rate (m^3 s^{-1})
f	fraction of lava surface occupied by crust
F_x	external forces acting on flow in downflow direction, per unit mass (m s^{-2})
F_y	external forces acting on flow in crossflow direction, per unit mass (m s^{-2})
g	acceleration due to gravity (m s^{-2})
h	flow thickness (m)
h_0	critical thickness (m)
h_c	convective heat transfer coefficient (W m^{-2} K^{-1})

k	thermal conductivity (W m^{-1} K^{-1})
L	latent heat of crystallization (J kg^{-1})
p	pressure (Pa)
q_{adv}	advected heat flux (W m^{-2})
q_{cond}	conductive heat flux (W m^{-2})
q_{conv}	convective heat flux (W m^{-2})
q_{cryst}	heat flux due to crystallization (W m^{-2})
q_{rad}	radiative heat flux (W m^{-2})
Q_{adv}	total advected heat (W)
Q_{cond}	total conductive heat loss (W)
Q_{conv}	total convective heat loss (W)
Q_{cryst}	total heat due to crystallization (W)
Q_{in}	total heat entering system (W)
Q_{out}	total heat leaving system (W)
Q_{rad}	total radiative heat loss (W)
t	time (s)
$t_{\phi max}$	time taken to reach maximum crystallinity (s)
T_a	ambient temperature (K)
T_{core}	lava flow core temperature (K)
T_{crust}	lava flow crust temperature (K)
T_e	lava flow effective surface temperature (K)
T_{surf}	surface temperature (K)
ΔT	temperature difference across the basal boundary layer (K)
u, v	downflow, crossflow velocity component (m s^{-1})
V	volume (m^3)
w	channel/flow width (m)
x, y, z	downflow, crossflow, and vertical spatial coordinates (m)
Δy	thickness of basal boundary layer (m)
ε	emissivity
$\dot{\varepsilon}$	strain rate (s^{-1})
η	fluid viscosity (Pa s)
η_m	mixture bulk viscosity (Pa s)
θ	slope (degrees)
κ	thermal diffusivity (m^2 s^{-1})
ρ	density (kg m^{-3})
σ	Stefan–Boltzmann constant (W m^{-2} K^{-4})
τ	shear stress (Pa)
τ_b	basal shear stress (Pa)
τ_0	yield strength (Pa)
ϕ	crystallinity (% or volume fraction)
ϕ_{max}	maximum crystallinity (% or volume fraction)
ω	vesicularity (% or volume fraction)
ϖ	adherence parameter

References

Bailey, J. E., Harris, A. J. L, Dehn, J., Calvari, S. and Rowland, S. K. (2006). The changing morphology of an open lava channel on Mt. Etna. *Bulletin of Volcanology*, **68**, 497–515.

Balmforth, N., Craster, R., Rust, A. and Sassi, R. (2006). Viscoplastic flow over an inclined surface. *Journal of Non-Newtonian Fluid Mechanics*, **139**, 103–127.

Barca, D., Crisci, G. M., di Gregorio, S. and Nicoletta, F. (1993). Cellular automata methods for modeling lava flows: simulation of the 1986–1987 eruption, Mount Etna, Sicily. In *Active Lavas*, ed. C. R. J. Kilburn and G. Luongo. London: UCL Press, pp. 291–309.

Blake, S. (1990). Viscoplastic models of lava domes. In *Lava Flows and Domes*, ed. J. H. Fink. New York: Springer, pp. 88–128.

Blake, S. and Bruno, B. (2000). Modelling the emplacement of compound lava flows, *Earth and Planetary Science Letters*, **184**, 181–197.

Bottinga, Y. and Weill, D. F. (1972). The viscosity of magmatic silicate liquids: a model for calculation. *American Journal of Science*, **272**, 438–475.

Calvari, S. and Pinkerton, H. (1998). Formation of lava tubes and extensive flow field during the 1991–1993 eruption of Mount Etna. *Journal of Geophysical Research*, **103**(B11), 27 291–27 301.

Chester, D. K., Duncan, A. M., Guest, J. E. and Kilburn, C. R. J. (1985). *Mount Etna: The Anatomy of a Volcano*. London: Chapman and Hall, pp. 187–228.

Crisci, G. M., di Gregorio, S., Pindaro, O. and Ramieri, G. (1986). Lava flow simulation by a discrete cellular model: First implementation. *International Journal of Modeling and Simulations*, **6**, 137–140.

Crisci, G., Rongo, R., Gregorio, S. and Spataro, W. (2004). The simulation model SCIARA: the 1991 and 2001 lava flows at Mount Etna. *Journal of Volcanology and Geothermal Research*, **132**, 253–267.

Crisp, J. and Baloga, S. (1990). A model for lava flows with two thermal components. *Journal of Geophysical Research*, **95**(B2), 1255–1270.

Crisp, J. and Baloga, S. (1994). Influence of crystallization and entrainment of cooler material on the emplacement of basaltic aa lava flows. *Journal of Geophysical Research*, **99**(B6), 11 819–11 831.

Crisp, J., Cashman, K. V., Bonini, J. A., Hougen, S. B. and Pieri, D. C. (1994). Crystallization history of the 1984 Mauna Loa lava flow. *Journal of Geophysical Research*, **99**(B4), 7177–7198.

Daneš, Z. (1972). Dynamics of lava flows. *Journal of Geophysical Research*, **77**, 1430–1432.

Del Negro, C., Fortuna, L. and Vicari, A. (2005). Modelling lava flows by Cellular Nonlinear Networks (CNN): preliminary results. *Nonlinear Processes in Geophysics*, **12**, 505–513.

Dragoni, M. (1989). A dynamical model of lava flows cooling by radiation. *Bulletin of Volcanology*, **51**, 88–95.

Dragoni, M. and Tallarico, A. (1996). A model for the opening of ephemeral vents in a stationary lava flow. *Journal of Volcanology and Geothermal Research*, **74**, 39–47.

Dragoni, M., Bonafede, M. and Boschi, E. (1986). Downslope flow models of a Bingham liquid: Implications for lava flows. *Journal of Volcanology and Geothermal Research*, **30**, 305–325.

Fagents, S. A. and Greeley, R. (2001). Factors influencing lava-substrate heat transfer and implications for thermomechanical erosion. *Bulletin of Volcanology*, **62**, 519–532.

Fairfield, J. and Leymarie, P. (1991). Drainage networks from grid digital elevation models. *Water Resources Research*, **27**(5), 709–717.

Favalli, M. and Pareschi, T. (2004). Digital elevation model construction from structured topographic data: The DEST algorithm. *Journal of Geophysical Research*, **109**, F04004, doi:10.1029/2004JF000150.

Favalli, M., Pareschi, M., Neri, A. and Isola, I. (2005). Forecasting lava flow paths by a stochastic approach. *Geophysical Research Letters*, **32**, L03305, doi:10.1029/2004GL021718.

Finch, R. H. and Macdonald, G. A. (1950). The June 1950 eruption of Mauna Loa Part I: Narrative of the eruption. *Volcano Letter*, **508**, 1–12.

Fink, J. and Griffiths, R. (1992). A laboratory analog study of the surface morphology of lava flows extruded from point and line sources. *Journal of Volcanology and Geothermal Research*, **54**, 19–32.

Furbish, D. J. (1997). *Fluid Physics in Geology*. New York: Oxford University Press.

Getson, J. M. and Whittington, A. G. (2007). Liquid and magma viscosity in the anorthite-forsterite-diopside-quartz system and implications for the viscosity-temperature paths of cooling magmas. *Journal of Geophysical Research*, **112**, B10203, doi:10.1029/2006JB004812.

Giordano, D. and Dingwell, D. B. (2003). Viscosity of hydrous Etna basalt: implications for Plinian-syle basaltic eruptions. *Bulletin of Volcanology*, **65**, 8–14.

Giordano, D., Mangiacapra, A., Potuzak, M. *et al.* (2006). An expanded non-Arrhenian model for silicate melt viscosity: A treatment for metaluminous, peraluminous and peralkaline liquids. *Chemical Geology*, **229**, 42–56.

Gregory, K. J. and Walling, D. E. (1973). *Drainage Basin Form and Process*. London: Edward Arnold.

Guest, J. E. and Murray, J. B. (1979). An analysis of hazard from Mount Etna volcano. *Journal of the Geological Society of London*, **136**, 347–354.

Harris, A. J. L. and Allen III, J. S. (2008). One-, two- and three-phase viscosity treatments for basaltic lava flows. *Journal of Geophysical Research*, **113**, B09212, doi:10.1029/2007JB005035.

Harris, A. and Rowland, S. (2001). FLOWGO: a kinematic thermo-rheological model for lava flowing in a channel. *Bulletin of Volcanology*, **63**, 20–44.

Harris, A. J. L. and Rowland, S. K. (2009). Effusion rate controls on lava flow length and the role of heat loss: A review. In *Studies in Volcanology: The Legacy of George Walker*, ed. T. Thordarson, S. Self, G. Larsen, S. K. Rowland and A. Hoskuldsson, Special Publications of IAVCEI, **2**, 33–51.

Harris, A., Bailey, J., Calvari, S. and Dehn, J. (2005). Heat loss measured at a lava channel and its implications for down-channel cooling and rheology. *Geological Society of America Special Paper* **396**, 125–146.

Hess, K.-U. and Dingwell, D. B. (1996). Viscosities of hydrous leucogranitic melts: A non-Arrhenian model. *American Mineralogist*, **81**, 1297–1300.

Hidaka, M., Umino, S. and Fujita, E. (2005). VTFS project: Development of the lava flow simulation code LavaSIM with a model for three-dimensional convection, spreading, and solidification. *Geochemistry, Geophysics, Geosystems*, **6**, Q07008, doi:10.1029/2004GC000869.

Holman, J. P. (1992). *Heat Transfer*. London: McGraw-Hill.

Hon, K., Kauahikaua, J., Denlinger, R. and Mackay, K. (1994). Emplacement and inflation of pahoehoe sheet flows: Observations and measurements of active lava flows on Kilauea Volcano, Hawaii. *Geological Society of America Bulletin*, **106**, 351–370.

Hulme, G. (1974). The interpretation of lava flow morphology. *Geophysical Journal of the Royal Astronomical Society*, **39**, 361–383.

Hutchinson, M. F. (1989). A new procedure for gridding elevation and stream line data with automatic removal of spurious pits. *Journal of Hydrology*, **106**, 211–232.

Ishihara, K., Iguchi, M. and Kamo, K. (1990). Numerical simulation of lava flows on some volcanoes in Japan. In *Lava Flows and Domes*, ed. J. Fink. Berlin: Springer, pp. 184–207.

Iverson, R. M. (1990). Lava domes modeled as brittle shells that enclose pressurized magma, with application to Mount St. Helens. In *Lava Flows and Domes*, ed. J. Fink. Berlin: Springer, pp. 48–69.

Jeffreys, H. (1925). The flow of water in an inclined channel of rectangular section. *Philosophical Magazine*, **49**, 793–807.

Jones, R. (2002). Algorithms for using a DEM for mapping catchment areas of stream sediment samples. *Computers and Geosciences*, **28**, 1051–1060.

Kauahikaua, J., Margriter, S., Lockwood, J. and Trusdell, F. (1995). Applications of GIS to the estimation of lava flow hazards on Mauna Loa Volcano, Hawai'i. In *Mauna Loa Revealed: Structure, Composition, History and Hazards*, ed. J. M. Rhodes and J. P. Lockwood, American Geophysical Union, Geophysical Monograph 92, pp. 315–325.

Kerr, R. (2001). Thermal erosion by laminar lava flows. *Journal of Geophysical Research*, **106**(B11), 26 453–26 465.

Keszthelyi, L. (1995). A preliminary thermal budget for lava tubes on the Earth and planets. *Journal of Geophysical Research*, **100**(B10), 20 411–20 420.

Keszthelyi, L. and Denlinger, R. (1996). The initial cooling of pahoehoe flow lobes. *Bulletin of Volcanology*, **58**, 5–18.

Keszthelyi, L. and Self, S. (1998). Some physical requirements for the emplacement of long basaltic lava flows. *Journal of Geophysical Research*, **103**(B11), 27 447–27 464.

Keszthelyi L., Harris, A. and Dehn, J. (2003). Observations of the effect of wind on the cooling of active lava flows. *Geophysical Research Letters*, **30**(19), doi:10.1029/2003GL017994.

Kilburn, C. R. J. and Lopes, R. M. C. (1988). The growth of aa lava fields on Mount Etna, Sicily. *Journal of Geophysical Research*, **93**, 14 759–14 772.

Kilburn, C. R. J., Pinkerton, H. and Wilson, L. (1995). Forecasting the behaviour of lava flows. In *Monitoring Active Volcanoes*, ed. W. McGuire, C. R. J. Kilburn and J. B. Murray. London: UCL Press, pp. 346–368.

Larson, R. G. (1999). *The Structure and Rheology of Complex Fluids*. New York: Oxford University Press.

Llewellin, E. W. and Manga, M. (2005). Bubble suspension rheology and implications for conduit flow. *Journal of Volcanology and Geothermal Research*, **143**, 205–217.

Lipman, P. W. and Banks, N. G. (1987). Aa flow dynamics, Mauna Loa. *U.S. Geological Survey Professional Paper*, **1350**, 1527–1567.

Manga, M., Castro, J., Cashman, K. V. and Loewenberg, M. (1998). Rheology of bubble-bearing magmas. *Journal of Volcanology and Geothermal Research*, **87**, 15–28.

Mano, J. F. and Pereira, E. (2004). Data analysis with the Vogel-Fulcher-Tammann-Hesse equation. *Journal of Physical Chemistry*, **108**, 10824–10833.

Mark, D. M. (1984). Automated detection of drainage networks from digital elevation models. *Cartographica*, **21**, 168–178.

Marsh, B. D. (1981). On the crystallinity, probability of occurrence, and rheology of lava and magma. *Contributions to Mineralogy and Petrology*, **78**, 85–98.

Martz, L. W. and Garbrecht, J. (1998). The treatment of flat areas and depressions in automated drainage analysis of raster digital elevation models. *Hydrological Processes*, **12**, 843–855.

Mattox, T. N., Heliker, C., Kauahikaua, J. and Hon, K. (1993). Development of the 1990 Kalapana flow field, Kilauea Volcano, Hawaii. *Bulletin of Volcanology*, **55**, 407–413

Mei, C. and Yuhi, M. (2001). Slow flow of a Bingham fluid in a shallow channel of finite width. *Journal of Fluid Mechanics*, **431**, 135–159.

Miyamoto, H. and Sasaki, S. (1997). Simulation of lava flows by an improved cellular automata method. *Computers and Geosciences*, **23**, 283–292.

Moore, H. J. (1987). Preliminary estimates of the rheological properties of 1984 Mauna Loa lava. *U.S. Geological Survey Professional Paper*, **1350**, 1569–1588.

Navier, C. L. M. H. (1823). Mémoire sur les lois du mouvement des fluids. *Memoirs of the Academy Royale Paris*, **6**, 389–416.

Neri, A. (1998). A local heat transfer analysis of lava cooling in the atmosphere: application to thermal diffusion-dominated lava flows. *Journal of Volcanology and Geothermal Research*, **81**, 215–243.

O'Callaghan, J. F. and Mark, D. M. (1984) The extraction of drainage networks from digital elevation data. *Computer Vision, Graphics and Image Processing*, **28**, 323–344.

Peterson, D. W., Holcomb, R. T., Tilling, R. I. and R. L. Christiansen (1994). Development of lava tubes in the light of observations at Mauna Ulu, Kilauea volcano, Hawaii. *Bulletin of Volcanology*, **56**, 343–360.

Phan-Thien, N. and Pham, D. C. (1997). Differential multiphase models for polydispersed suspensions and particulate solids. *Journal of Non-Newtonian Fluid Mechanics*, **72**, 305–318.

Pinkerton, H. and Sparks, R. S. J. (1978). Field measurements of the rheology of lava. *Nature*, **276**, 383–385.

Pinkerton, H. and Stevenson, R. J. (1992). Methods of determining the rheological properties of magmas at sub-liquidus temperatures. *Journal of Volcanology and Geothermal Research*, **53**, 47–66.

Quareni, F., Tallarico, A. and Dragoni, M. (2004). Modeling of the steady-state temperature field in lava flow levees. *Journal of Volcanology and Geothermal Research*, **132**, 241–251.

Rowland, S., Garbeil, H. and Harris, A. (2005). Lengths and hazards from channel-fed lava flows on Mauna Loa, Hawai'i, determined from thermal and downslope modeling with FLOWGO. *Bulletin of Volcanology*, **67**, 634–647.

Russell, J. K., Giordano, D. and Dingwell, D. B. (2003). High-temperature limits on viscosity of non-Arrhenian silicate melts. *American Minerologist*, **88**, 1390–1394.

Ryerson, F. J., Weed, H. C. and Piwinskii, A. J. (1988). Rheology of subliquidus magmas 1. Picritic compositions. *Journal of Geophysical Research*, **93**(B4), 3421–3436.

Shaw, H. R. (1972). Viscosities of magmatic silicate liquids: an empirical method of prediction. *American Journal of Science*, **272**, 870–893.

Sonder, I., Zimanowski, B. and Buttner, R. (2006). Non-Newtonian viscosity of basaltic magma. *Geophysical Research Letters*, **33**, L02303, doi:10.1029/2005GL024240.

Stevens, N. F., Wadge, G. and Murray, J. B. (1999). Lava flow volume and morphology from digitised contour maps: a case study at Mount Etna, Sicily. *Geomorphology*, **28**, 251–261.

Stokes, G. G. (1845). On the theories of internal friction of fluids in motion. *Transactions of the Cambridge Philosophical Society*, **8**, 287–305.

Tarboton, D. G. (1997). A new method for the determination of flow determinations and upslope areas in grid digital elevation models. *Water Resources Research*, **33**(2), 309–319.

Tedesco, D., Badiali, L., Boschi, E. *et al.* (2007). Cooperation of Congo volcanic and environmental risks. *Eos, Transactions of the American Geophysical Union*, **88**(16), 177–181.

Tribe, A. (1992). Automated recognition of valley lines and drainage networks from grid digital elevation models: a review and a new method. *Journal of Hydrology*, **139**, 263–293.

Turcotte, D. L. and Schubert, G. (2002). *Geodynamics*. Cambridge: Cambridge University Press, 456 pp.

Turcotte, R., Fortin, J.-P., Rousseau, A. N., Massicotte, S. and Villeneuve, J.-P. (2001). Determination of the drainage structure of a watershed using a digital elevation model and a digital river and lake network. *Journal of Hydrology*, **240**, 225–242.

Vicari, A., Herault, A., del Negro, C. *et al.* (2007). Modeling of the 2001 lava flow at Etna volcano by a Cellular Automata approach. *Environmental Modelling and Software*, **22**, 1464–1471.

Walker, G. P. L. (1967). Thickness and viscosity of Etnean lavas. *Nature*, **213**, 484–485.

Walker, G. P. L. (1972). Compound and simple lava flows and flood basalts. *Bulletin of Volcanology*, **35**, 579–590.

White, F. M. (2006). *Viscous Fluid Flow*. New York: McGraw-Hill, 629 pp.

Whittington, A., Richet, P. and Holtz, F. (2000). Water and the viscosity of depolymerized aluminosilicate melts. *Geochimica et Cosmochimica Acta*, **64**(21), 3725–3736.

Wright, R. and Flynn, L. P. (2003). On the retrieval of lava-flow surface temperatures from infrared satellite data. *Geology*, **31**(10), 893–896.

Young, P. and Wadge, G. (1990). FLOWFRONT: Simulation of a lava flow. *Computers and Geosciences*, **16**, 1171–1191.

Online resources available at www.cambridge.org/fagents

- Supplement 5A: Review of Type I models
- Supplement 5B: Review of Type II models
- Exercises: Construction of a lava flow cooling model
- Answers to exercises

Chapter 6

Unsteady explosive activity: strombolian eruptions

Mike R. James, Steve J. Lane, and Bruce F. Houghton

Overview

During strombolian eruptions, large bubbles of exsolved magmatic gas, with sizes of meters or more, intermittently burst at the magma surface, spraying magma clots over distances of tens to hundreds of meters. This style of activity results from low magma viscosities allowing gas bubbles to move through the liquid magma phase. Relatively small bubbles rise and coalesce into bubbles with diameters similar to that of the conduit, at which point they are called gas slugs. This coalescence is responsible for converting the continuous degassing processes at depth into the observed intermittent surface activity, and may be controlled by the decompression expansion of the bubbles or by portions of non-vertical conduit geometry. Models of strombolian systems cover the bubble coalescence phase (slug generation), the slug ascent, and finally slug burst and the ejection of pyroclasts. A wide range of geophysical measurements, notably from Stromboli and Erebus volcanoes, are available to test these subsurface models. Nevertheless, key questions, such as the degree to which the activity is controlled by the geometry of the conduit, remain.

6.1 | Introduction

At volcanoes with relatively low-viscosity magmas such as basalt and basaltic andesite, large bubbles (with sizes of meters or greater) of exsolved gas can ascend rapidly through the melt and burst energetically at the surface, producing sprays of molten pyroclasts (Fig. 6.1). This type of intermittent explosion, in which a limited amount of magma is erupted with a relatively significant mass of gas, is known as "strombolian" after the characteristic activity at Stromboli volcano, where several such events usually occur every hour.

Typically, strombolian eruptions eject masses of a few tens of kilograms to 10^4 kg for the largest bubbles which, as observed at Heimaey in 1973, can have diameters of up to ~10 m (Blackburn et al., 1976). Pyroclast velocities are generally < 200 m s^{-1}, driven by a combination of high gas contents (with the discharge being between ~10 and 96% gas by mass) and bubble overpressures of several atmospheres. The fall deposits produced are confined close to the source vent and are some of the coarsest (Fig. 6.2), with typical median diameters of 16 to 64 mm, compared with samples from hawaiian-style eruptions at 2 to 16 mm and from basaltic plinian eruptions at

Modeling Volcanic Processes: The Physics and Mathematics of Volcanism, eds. Sarah A. Fagents, Tracy K. P. Gregg, and Rosaly M. C. Lopes. Published by Cambridge University Press. © Cambridge University Press 2013.

Figure 6.1 (a) A night view of a typical strombolian eruption at Stromboli, showing the ballistic trajectories of the large pyroclasts (field of view approximately 200 m across at the erupting vent). Photograph taken by B. Chouet, courtesy of the USGS. (b) Typical cinder cones associated with strombolian (and other) activity on Mount Etna, Sicily. In front of the La Montagnola (1761) cone that forms the skyline, is the 60-m-high 2001 "Laghetto" cinder cone on the Piano del Lago (Calvari and Pinkerton, 2004). The cones display similar eccentric morphologies that reflect the consistency of the prevailing winds. In 2001, the Laghetto cone was formed by a combination of phreatomagmatic and intense strombolian activity. Photograph B. Houghton.

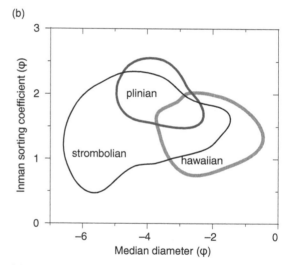

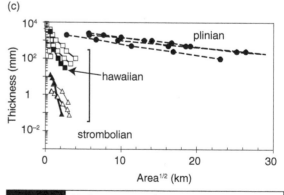

Figure 6.2 (a) Contrasting strombolian (lower, coarser, lighter-colored) and hawaiian (upper, darker, finer-grained) products of the eruption of Crater Hill volcano in the Auckland volcanic field, New Zealand. The better bedded nature of the strombolian phase reflects its origin in numerous discrete short-lived explosive events. The hawaiian deposits are dominated by microvesicular and highly fluidal lapilli whereas the strombolian clasts are more coarsely vesicular and ragged.

4 to 64 mm. This coarseness reflects the relatively inefficient fragmentation of the liquid magma by large gas bubbles, resulting from low liquid viscosities allowing relatively easy phase separation (Chapter 4). Nevertheless, a range of particle sizes is usually produced, a small proportion of which can be sufficiently fine-grained to be lofted into weakly buoyant thermal-like plumes.

The strombolian eruptive process thus represents the arrival, and subsequent bursting, of one or more large (sizes approaching that of the conduit diameter), overpressured gas bubbles at the magma free surface. During periods of activity, intervals between bursts are typically from tens of minutes to multiple hours. However, eruptive episodes can also involve periods of nearly continuous bubble bursts, with intervals of only seconds, producing persistent weak plumes of fine ejecta and cascades of spatter. This represents a transitional type of activity which, from a classification point of view, merges into hawaiian-style lava fountaining. In terms of landforms, the low dispersion and small volumes of coarse strombolian ejecta determine that small, relatively localized spatter- and scoria-dominated cones are formed around vents. However, in many cases, the style of activity varies between fountaining, transitional and pure strombolian, and significant cones (hundreds of meters high) can be built by the greater magma effusion rates involved in transitional and fountaining phases (Fig. 6.1(b)).

A key issue is what parameter or parameters can be used to distinguish between these eruptive styles. Unlike transitions between many other types of eruption, time averaged mass discharge rates cannot be used; for example, the 99 eruptions or eruption episodes at Kilauea since 1955 show a range of time-averaged discharge rates between 4×10^2 and 10^6 kg s^{-1}, overlapping completely with the rates displayed by strombolian behavior (Houghton and Gonnermann, 2008). Both strombolian and hawaiian eruptions produce cones composed of proximal pyroclast fall and, unlike Plinian falls, deposits do not follow simple exponential thinning relationships with distance from the vent (Fig. 6.2). This is perhaps predictable, because a simple exponential

thinning relationship requires, a priori, a constant release height of clasts, whereas the heights of strombolian jets tend to vary greatly. Consequently, strombolian and hawaiian eruptions cannot be distinguished by average discharge rate or ejecta dispersal; consideration of the steadiness of discharge rate is required, with the former being impulsive, short-lived, discrete events and the latter expressed as sustained fountaining, during which mass discharge rates are maintained for hours to days. So, although we concentrate here on classical discrete strombolian events involving individual large gas bubbles, the two-phase flow behavior involved is closely linked to that of hawaiian-style eruptions. Transitional behavior (Parfitt and Wilson, 1995), and the conditions leading up to it are discussed in Section 4.8.3.

The relatively high event frequency and low-hazard nature of strombolian activity has allowed intensive geophysical study of the processes involved. Many different data types, including seismic, imaging (both still and video, visible and thermal), infrasonic, Doppler radar and sodar (sound detection and ranging), have been collected in order to analyze ejecta velocities and masses, bubble overpressures and sizes, eruption intermittencies, and to model depths of slug acceleration and fragmentation. Consequently, when compared with other types of less frequent and more hazardous volcanic activity, there is a significant and rapidly growing volume of data from which models of strombolian activity can be developed and tested. Furthermore, in contrast to other scenarios in which magma rheological properties shift over many orders of magnitude (Chapter 4) and pyroclasts can exceed the sound speed and generate shocks (Chapter 7), strombolian eruptions are relatively straightforward to model. For example, the essential subsurface fluid flow processes can be reasonably represented using a liquid phase of constant viscosity. Nevertheless, in common with most two-phase flows, a complete understanding remains elusive, and current quantitative models are generally limited to distinct regions of the overall process; the initial formation of the large gas bubble, its ascent and burst, and

Caption for Figure 6.2 continued Photograph taken by Colin Wilson. (b) Plot of median diameter versus sorting coefficient for fall deposits from basaltic explosive eruptions modified after Walker and Croasdale (1972) and Houghton et al. (2000) to include plinian data from Carey et al. (2007). (c) Deposit thinning relationships indicated by plots of tephra thickness versus the square root of the area enclosed within a given isopach for selected basaltic pyroclastic deposits. Strombolian deposits are represented by open symbols (squares from Walker and Croasdale (1972); open triangles represent recent Arenal strombolian fall deposits from Cole et al. (2005)); closed triangles and bold line is the Etna 2003 paroxysm. The hawaiian-style 1959 Kīlauea Iki eruption is shown with closed squares and basaltic Plinian eruptions with solid circles. Where only isomass data were available, they were converted into thickness equivalents using field measurements of bulk density for the deposits.

the subsequent dispersion of the pyroclasts. These divisions are retained here and form the ensuing structure of the chapter. The value of such models is that they allow our understanding of the unobservable subsurface processes that control eruptive style to be tested, using a wide variety of surface measurements. Not only does this provide a starting point from which increasingly complex models of more hazardous eruptive phenomena can be derived but, with the volumetric importance of basaltic volcanism, it provides insight into the degassing processes that continue to shape much of our Earth and atmosphere.

6.2 | Slug-bubble formation

As magma ascends, decompression drives the process of gas exsolution and bubble expansion. Consequently, in order to understand the transition from continuous subsurface magma degassing to the sequence of discrete events observable at the surface, a fundamental aspect of modeling strombolian activity is to consider how individual, large bubbles can form from an evolving distribution of many small ones. In terms of two-phase flow, this represents a transition from a "bubbly" to a "slug" flow regime (although neither need be in a steady state), where a "slug" is a large bubble of diameter similar to, and length larger than, the width of the conduit (Wallis, 1969; see Chapter 4).

At depth, the first volumetrically important species to exsolve from the magma is CO_2 and, even at depths of several kilometers, most CO_2 will be exsolved (Burton *et al.*, 2007b; see Section 4.2.1). H_2O is the principal magmatic volatile by weight but, being more soluble than CO_2 it exsolves at shallower depths (e.g., for Stromboli magma at a depth of ~1.5 km, probably > 50% of the magma water content is exsolved (Burton *et al.*, 2007b)). The degassing process begins with the nucleation of small bubbles, at

number densities of ~10^{10} to 10^{11} m^{-3} which, in a Newtonian liquid, have buoyant rise velocities, u_b, that can be approximated by Stokes' law

$$u_b = \frac{2g\left(\rho - \rho_g\right)r_b^2}{9\eta},$$ (6.1)

where g is the acceleration due to gravity, ρ and ρ_g are the liquid magma and gas densities, respectively, r_b is the bubble diameter, and η is the magma viscosity. (All notation is given in Section 6.9.) For considering the terminal rise velocities of bubbles in basalt magma, note that Stokes' law is derived for rigid spheres and assumes creeping (laminar) flow within an infinite incompressible Newtonian fluid. Thus, it is only appropriate to use Eq. (6.1) for small (e.g., radii of order 1 cm or less), widely separated bubbles. At a pressure of 200 MPa and temperature of ~1300 K, exsolved CO_2 is a supercritical fluid[1] with a density of ~100 kg m^{-3}. So, for a magma viscosity of 100 Pa s, bubbles of radii between 10 μm and 1 mm have rise velocities of 0.01 to 10 μm s^{-1}, sufficiently slow that bubble motion will be dominated by the magma flow. Nevertheless, the bubbles will expand (and grow by diffusion) as they rise and depressurize and, when they reach radii of ~1 to 5 cm, ascent velocities will have increased to ~1 to 10 cm s^{-1}. These velocities are on the order of anticipated magma velocities, so that the flow can be considered to be separated, with bubbles having the chance to interact with each other or with the conduit walls.

Two models have been proposed for the bubble interactions that underpin explosive basaltic eruptions. Using different conduit geometries, the models differ in their consideration of where such interactions and the ensuing bubble coalescence occur (Fig. 6.3). First, in the "rise speed dependent" model (Wilson and Head, 1981), bubbles coalesce freely as they ascend a vertical conduit and eruptive style depends on, among other factors, the relative ascent rates of the bubbles and magma. In contrast, the "foam collapse" model (Jaupart and Vergniolle, 1989;

---

[1] A supercritical fluid is a fluid at a temperature and pressure above its critical point; for CO_2 the critical point is at 304.1 K and 7.38 MPa. Supercritical fluids tend to have physical properties in between those of a liquid and a gas.

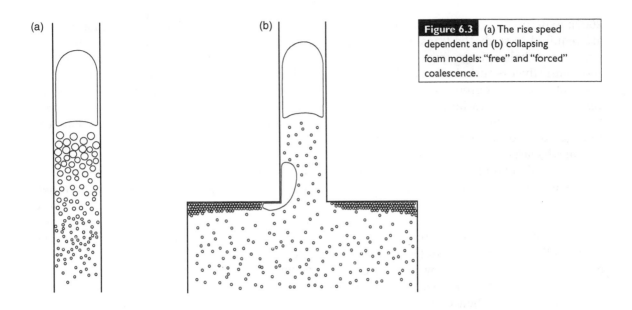

Figure 6.3 (a) The rise speed dependent and (b) collapsing foam models: "free" and "forced" coalescence.

Vergniolle and Jaupart, 1990) assumes bubbles collect and interact at major structural barriers such as magma chamber roofs and, under this scenario, the effects of any magma motion are not considered. The merits and drawbacks of both models are discussed in detail by Parfitt (2004).

6.2.1 Rise speed dependent model

The rise speed dependent model (Fig. 6.3(a)) considers the ascent rate of bubbles within magma with respect to the ascent velocity of the magma as a whole, within a long, vertical conduit. Larger bubbles ascend more rapidly (Eq. 6.1) and hence can encounter and absorb smaller bubbles at an increasingly fast rate. If the ascent velocity of the magma is low, then such bubbles have time for sufficient coalescence to form gas slugs before the magma free surface is reached, generating strombolian activity. Thus, a depressurizing (ascending) bubbly flow represents a transitional flow regime that will develop into slug flow if sufficient time is available and, in long conduits, depressurization expansion will enhance this process. In contrast, relatively high magma rise speeds inhibit significant bubble coalescence prior to arrival at the surface, at which point decompression-induced expansion may ultimately lead to fragmentation of the homogeneous, bubbly flow and the formation

of hawaiian fountains as the mixture of gas and pyroclasts exits the vent. One of the main applications of the rise speed model has been to address transitions between effusive, strombolian, and hawaiian activity (Parfitt and Wilson, 1995). For example, magma rise speed has a strong influence on eruptive style but, for a rise speed sufficiently low for strombolian activity to be maintained, this can change to hawaiian if the magma volatile content decreases, reflecting the fact that the lower gas content no longer allows sufficient coalescence to generate the slug flow regime, but still permits decompression-induced fragmentation and the formation of hawaiian fountains.

Although we address this model no further here, one point is worth discussion. In the development of the rise speed model, a relatively high gas volume fraction (0.75) was used to define the transition from bubbly to slug flow (Wilson and Head, 1981). In contrast, engineering research suggests that the bubble-to-slug transition generally occurs as the gas volume fraction exceeds ~0.25 (Taitel et al., 1980). This has yet to be experimentally verified in large diameter pipes (e.g., > 100 mm) because, in the low-viscosity hydrocarbons and aqueous solutions typically used in laboratory simulations ($\eta < 10^{-3}$ Pa s), turbulence disrupts large bubbles and hinders the development of a slug flow regime (Omebere-Iyari et al.,

2007); experiments using high-viscosity liquids are required to study the transition at volcanic scale. Nevertheless, large slugs can be observed transiently (before turbulence builds up) and, in magmatic systems, despite even larger diameter conduits, excessive turbulence is unlikely due to viscosities being up to six orders of magnitude greater. Consequently, there should be no reason why large gas bubbles will be unstable in magma-filled conduits and slug flow may well be possible at gas volume fractions much less than 0.75.

6.2.2 Collapsing foam model

In the collapsing foam model (Fig. 6.3(b)), bubble interactions are forced by bubbles collecting against a horizontal barrier, i.e., the roof of a magma reservoir, hence coalescence can be viewed as being directly driven by conduit geometry. Laboratory experiments in which small ascending bubbles formed a foam layer under a flat roof (vented by only a small vertical upward conduit) have shown that this system can exhibit either a steady or intermittent behavior (Jaupart and Vergniolle, 1989). As the foam grows, the thickness, h_m, of its deepest regions can be calculated by considering two gas fluxes: the input flux from bubbles from below and gas loss from the foam as it flows into the open conduit:

$$h_m = \Omega \left(\frac{\eta_f Q}{\varepsilon^2 \rho g} \right)^{\frac{1}{4}}, \qquad (6.2)$$

where ε is the gas volume fraction in the foam, η_f is the foam viscosity, Q is the volume flux of gas, and Ω is a factor determined by the roof area over which bubbles are collecting. However, a foam layer has a maximum potential thickness governed by the ability of surface tension to maintain the integrity of the foam against increasing buoyancy forces. At the point of failure, the critical foam thickness is given by

$$h_c = \frac{2\sigma}{\varepsilon \rho g r_b}, \qquad (6.3)$$

where σ is the surface tension. Experimental results were best represented when a value of 0.69 was used for ε, which, for magma, implies a foam thickness of < 10 cm for 1 mm bubbles, increasing to tens of meters for micron-sized bubbles (Jaupart and Vergniolle, 1989).

Equating these two foam thicknesses, h_m and h_c, relates the input gas flux to a foam breakdown criterion. At fluxes less than the critical flux, foam growth is steady and gas loss is maintained by foam flow out of the conduit. Above the critical gas flux, the system moves into a cyclical regime in which the foam grows to its maximum thickness then collapses into a large gas pocket which rises rapidly into the conduit, allowing the cycle to restart. The expression of this event at the free surface depends on the duration of the collapse itself. When the duration of the foam collapse is short with respect to foam flow timescales, a very large bubble can be produced and eruptions are deemed analogous to hawaiian-style fountaining. However, when foam collapse is relatively slow compared with foam flow, then numbers of small gas pockets are intermittently produced and released, in a manner equivalent to strombolian eruptions. In contrast to the rise speed model, in this model, increasing gas fluxes will convert strombolian-style activity to hawaiian as foam collapse becomes increasingly vigorous and continuous. Note that the large horizontal roof required for foam collection is highly unlikely to exist in the relatively near-surface, thus initiation of these eruptions will be controlled by CO_2-dominated foams, because significant H_2O is not exsolved at multiple-kilometer depths.

6.2.3 A combined model

The rise speed and collapsing foam models illustrate well the effects of different bubble coalescence scenarios, but it is likely that neither of the two idealized conduit geometries, invoked (a long, uniform vertical conduit or a flat, horizontal magma chamber roof) accurately represents a natural system. Furthermore, the foam collapse model does not involve the effects of bubble expansion and the rise speed model does not include any "forced" coalescence (such as is induced by a geometric trap).

The development of a combined model would allow both depressurization and more plausible complex geometries such as rough inclined planes representative of intersecting dikes, to be accounted for.

Within inclined conduits, rising bubbles collect at the upper conduit wall and then ascend along it. Hence, inclination leads to locally enhanced gas volume fractions and promotes bubble coalescence, with the effect of generating slug flow at much smaller overall gas volume fractions than in a vertical conduit (James et al., 2004). A similar effect can be produced by regions of constriction within the conduit (James et al., 2006) and by horizontal pathways branching from a vertical conduit (Menand and Phillips, 2007). However, unless conduits have completely smooth walls, relatively small irregularities and constrictions may also trap some bubbles and limited regions of foam are likely to accumulate. Laboratory experiments have suggested that the presence of many such gas-trapping regions can produce cascading foam collapse events, and are just as effective in converting a steady gas flux into a highly unsteady output flux as the one large trap in the collapsing foam model. Thus, for strombolian activity, a style of foam collapse mechanism can be effective without requiring large areas of foam envisaged by Jaupart and Vergniolle (1989).

The amount of gas trapping, foam collapse and bubble coalescence during bubble ascent will depend on the magma rise speed, the foam collapse mechanisms and the conduit geometry. If gas traps are generally small on the scale of the conduit width and magma ascent rates relatively high compared with the bubble rise velocity within the magma, then exsolved bubbles will be swept past traps and a rise-speed model will be most appropriate. Conversely, if traps are relatively large and magma rise speeds are relatively low, output may appear to be dominated by foam collapse events. Consequently, a natural conduit geometry will be likely to produce eruptions reliant on factors of both end-member scenarios, but a quantitative model of such a complex scenario has yet to be developed.

6.3 | Ascent of a gas slug

Once a large bubble is formed and begins to ascend to the surface, understanding of the ascent processes is required in order to interpret different geophysical signals (Chapters 15 and 16), as well as the development of the bubble overpressure, the effects of which are observable at the surface. Detailed modeling of bubble evolution is hampered by insufficient knowledge of the rheology of bubble- and crystal-rich magma and conduit geometries (cross-sectional shape, dimensions, and inclination) but applying several simplifying assumptions allows first-order results to be drawn based on previous fluid dynamic research on slug flow.

Firstly, the liquid phase is approximated as an initially stationary, homogenous, and incompressible Newtonian fluid. This assumes that the presence of crystals and small bubbles can be neglected and that inhomogeneities (such as those produced by cooling) are negligible. A more realistic stratified and cooling fluid with non-Newtonian rheology could be used, but uncertainties in the parameter values means that the results are unlikely to be any more realistic. For a meter-sized or larger bubble, the combination of its buoyancy and the relatively low magma viscosity promotes buoyant ascent rates of meters per second, significantly faster than the assumed background magma velocity. Consequently, initial background liquid motions are neglected and the gas is assumed to be ascending though initially stationary (stagnant) liquid.

Secondly, the conduit geometry is assumed to be adequately represented by a simple geometric cross section. At depth in a volcanic plumbing system, conduits will be dike-like and, for comparison with fluid dynamic literature, bubbles would best be represented as ascending in wide channels or under inclined planes. However, most of the important dynamics occur in the upper regions of the conduit where, for long-lived systems, the cross section will be increasingly controlled by thermodynamic constraints (which favor circular cross sections)

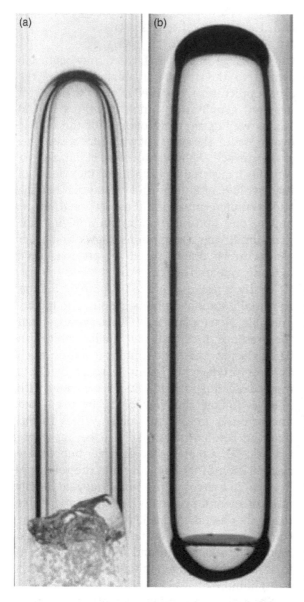

(a) (b)

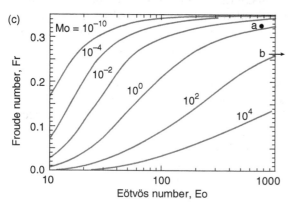

(c)

Figure 6.4 Ascending gas slugs or Taylor bubbles. In (a) and (b) gas slugs have been imaged ascending in a 76.2 mm internal diameter tube (the width of each image) within (a) water ($\eta = 10^{-3}$ Pa s) and (b) silicone oil ($\eta = 1.3$ Pa s). In the low-viscosity water, the slug has an unsteady, seiching base and is followed by a turbulent wake, illustrated by the entrained small bubbles; inertial forces control the slug ascent. In the higher viscosity oil, the bottom of the slug is convex; at intermediate viscosities, a stable concave base is observed and slug ascent is controlled by both inertial and viscous forces. Images reproduced from Viana et al. (2003) with permission from Cambridge University Press. (c) The dimensionless parameterization of ascent velocity for slugs in vertical tubes derived by White and Beardmore (1962). The Froude number, Fr, describing dimensionless velocity, is given as a function of the Eötvös number, Eo, for curves of different Morton number, Mo. The circle labeled "a" represents the slug shown in (a), the equivalent for (b) lies just off the plot at Eo = 2604, Fr = 0.264 as indicated by the arrow. Reprinted from James et al. (2008) with permission from the Geological Society of London.

rather than the subsurface stress field responsible for dike-like morphologies. Consequently, the ascent of gas slugs in low-viscosity volcanic systems has been based on investigations of slug flow in vertical cylindrical pipes (Fig. 6.4). Note that in engineering literature, this regime is often described as the motion of a "Taylor bubble" (the gas phase) and "liquid slug" for the connected, liquid phase.

6.3.1 Gas slug ascent velocities

For long slugs (where slug length is much greater than the slug diameter), ascent velocity is independent of the slug length and is determined by only the physical properties of the liquid and conduit (White and Beardmore, 1962). Compiling these into three dimensionless parameters gives the Froude, Morton, and Eötvös numbers which, in the case of gas–liquid flows where the gas density is sufficiently small that it can be neglected, are

$$\mathrm{Fr} = \frac{u_s}{\sqrt{gD}}, \tag{6.4}$$

$$\mathrm{Mo} = \frac{g\eta^4}{\rho\sigma^3}, \tag{6.5}$$

and

$$\text{Eo} = \frac{\rho g D^2}{\sigma}, \tag{6.6}$$

respectively. Here, u_s represents the terminal ascent velocity of the gas slug and D is the internal diameter of the conduit. The Froude number represents the ratio of inertial to gravitational forces, the Morton number is the ratio of viscous and surface tension forces, and the Eötvös number is the ratio of gravitational and surface tension forces. Figure 6.4(c) shows the Eo–Fr parameter space for curves of constant Mo, which allows Fr, and hence the slug velocity, to be determined for a system. Within this space, for Mo < 10^{-6} and Eo > 100, slugs are inertially controlled (viscous and surface tension forces are negligible) and rise at their maximum velocity in vertical tubes, given by Fr = 0.35 (White and Beardmore, 1962).

In basaltic systems with vertical conduits, typical ranges of these parameters are 0.1 < Fr < 0.35, 10^5 < Eo < 10^7, and 10^6 < Mo < 10^{11}, assuming 100 < η < 1000 Pa s, ρ = 2600 kg m^{-3} and σ = 0.4 N m^{-1} for basalt, and 1 < D < 10 m for the conduit. These ranges indicate that gas slugs usually ascend within a transitional regime with a velocity dependent on both inertial and viscous effects (Seyfried and Freundt, 2000). Note that, for these large bubbles in basaltic systems, interfacial (surface tension) effects are always negligible. The relationships shown in Figure 6.4 are difficult to parameterize but, over the region of combined viscous and inertial effects, Fr can be represented as

$$\text{Fr} = 0.345(1 - e^{-N_f/34.5}), \tag{6.7}$$

where N_f is a dimensionless inverse viscosity (or the buoyancy Reynolds number), given by

$$N_f = \left(\frac{\text{Eo}^3}{\text{Mo}}\right)^{1/4} = \frac{\rho D^{3/2} g^{1/2}}{\eta}. \tag{6.8}$$

In inclined pipes, the slug ascends along the upper region of the pipe wall, allowing a thicker return flow of liquid under the slug body. Consequently, ascent velocities in inclined pipes can exceed those in vertical pipes by up

to ~80% (Zukoski, 1966). The effect of inclination on ascent velocities has been assessed over only a relatively limited region of parameter space and several relationships have been proposed between Fr$_\theta$, the Froude number at an inclination angle, θ, from the horizontal and the Froude numbers in vertical and horizontal pipes, Fr$_{90}$ and Fr$_0$, respectively. For Mo < 10^4 and 5 < Eo < 20, maximum velocities occur close to inclinations of 45°. One parameterization proposed,

$$\text{Fr}_\theta = \text{Fr}_{90} \sin\theta + \text{Fr}_0 \cos\theta + K, \tag{6.9}$$

where $K = 1.37(\text{Fr}_{90} - \text{Fr}_0)^{2/3}$ for $\text{Fr}_{90} \geq \text{Fr}_0$ and $K = 0$ otherwise, has been shown to reproduce slug velocities reasonably well. However, for Mo > 10^4, maximum velocities occur at inclinations of around ~60°.

6.3.2 Gas slug expansion

The Fr–Eo–Mo parameterization (Fig. 6.4(c)) considers only the steady state ascent of gas slugs where decompressional gas expansion can be neglected. In volcanic scenarios, pressure changes and gas volumes are large, and decompressional expansion is a fundamental part of the process. Fortuitously, laboratory experiments carried out under reduced atmospheric pressures show that even rapidly expanding gas slugs maintain constant velocities at their base (James et al., 2008), which can be calculated from the Froude number (Eq. (6.7)).

Defining an ascent velocity for the slug base allows the slug expansion to be estimated by calculating the forces exerted on the liquid above the slug. Long slugs, for which curvature around the nose and base region make up only a small proportion of the slug length, can be adequately represented as gas cylinders of constant radius r_s, which can be related to the conduit radius, r_c, by the thickness of the falling liquid film surrounding the slug (Batchelor, 1967)

$$r_s = r_c - \left(\frac{3\eta r_c u_s}{2\rho g}\right)^{1/3}. \tag{6.10}$$

For a slug with initial length l_{s0}, within a vertical cylindrical conduit, conservation of liquid volume allows the slug length, l_s, at any time, t, to be given as

$$h_0 r_c^2 + l_{s0}\left(r_c^2 - r_s^2\right) = \left(h + u_s t\right) r_c^2 + l_s\left(r_c^2 - r_s^2\right), \quad (6.11)$$

where h_0 and h are the heights of liquid above the slug nose, initially and at time t, respectively, and the slug velocity, u_s, is defined at the slug base.

As the slug ascends, decompresses and expands, a cylindrical region of overlying liquid has to be accelerated upward to accommodate the increasing gas volume. A first-order approach can be used to define this acceleration in terms of the pressure, gravitational, and viscous forces acting on the liquid cylinder, given by $F_p = \pi r_s^2(P_s - P_a)$, $F_g = -\pi r_s^2 \rho h g$, and F_v, respectively, where P_s is the slug gas pressure and P_a is ambient surface pressure. For laminar flow in a pipe, the viscous drag force can be estimated from the pressure drop under Poiseuille flow (Batchelor, 1967) and, assuming that the liquid flux is equal to the gas expansion flux, this gives

$$F_v \approx -8\pi\,\eta h A' \frac{dl_s}{dt}, \quad (6.12)$$

where $A' = (r_s/r_c)^2$. Equating the forces to the product of mass and acceleration (of the center of mass of the liquid column directly over the gas slug) then gives

$$F_p + F_g + F_v = \pi r_s^2 \rho h \frac{d^2}{dt^2}\left(u_s t + l_s + \tfrac{1}{2}h\right). \quad (6.13)$$

If the slug is represented as a perfect gas (i.e., $P_s V_s^{\gamma} = $ constant for adiabatic conditions, where V_s is gas volume and γ the ratio of specific heats, in this case leading to $P_s = P_{s0}(l_{s0}/l_s)^{\gamma}$, for a constant radius cylinder and an initial slug pressure, P_{s0}), then

$$\tfrac{1}{2}\rho\left(1 + A'\right)\frac{d^2 l_s}{dt^2} + \frac{8\eta}{r_c^2}\frac{dl_s}{dt} - \frac{P_{s0}}{h}\left(\frac{l_{s0}}{l_s}\right)^{\gamma} + \rho g + \frac{P_a}{h} = 0.$$

$$(6.14)$$

Equation (6.14) can be solved numerically and Figure 6.5(a) demonstrates its application

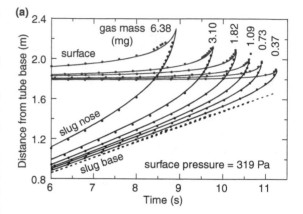

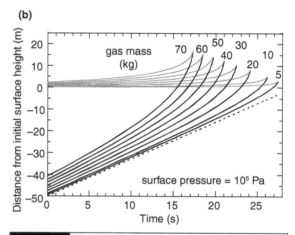

Figure 6.5 Modeling the positions of ascending, expanding gas slugs (lower black lines/symbols) and the liquid surface (upper gray lines/symbols) in (a) low-pressure laboratory environments and (b) a typical magmatic system. In both plots, the data and results are overlaid so that the position of the slug base is the same for each scenario. In (a) the symbols indicate slug position as determined from video data. Results are given for six different experiments carried out using different gas masses for the slug. The experiments used a silicone oil liquid phase with a density and viscosity of 862 kg m^{-3} and 0.124 Pa s, respectively, and were carried out in a tube of 1.3 cm internal radius. The best-fit model results shown were achieved using $\gamma = 1.0$, suggesting that the gas expansion was nearly isothermal rather than adiabatic. In (b) simulations are shown for a volcanic conduit of radius 1.5 m, with magma of density and viscosity of 2600 kg m^{-3} and 500 Pa s, respectively. Note the increase in the nose velocity as the slug approaches the surface. Reproduced from James et al. (2008) with permission from the Geological Society of London.

to laboratory data on expanding gas slugs. In Figure 6.5(b) the results from simulated volcanological scenarios illustrate the rapid acceleration of the slug nose towards the surface. For example, in a 1.5 m radius conduit and magma of density 2600 kg m^{-3} and viscosity 500 Pa s, the average slug nose velocity ascending through the last 10 m of magma is 4.4 m s^{-1}, over twice the steady slug base velocity of 1.7 m s^{-1} (Eq. 6.7). With Eq. (6.14) providing the slug length, its gas pressure can also be calculated; in Section 6.4 this is used to estimate burst overpressures.

6.3.3 Three-dimensional models

The simple slug ascent model discussed above is convenient for estimating ascent times and potential overpressures at burst but, because the dynamics of the fluid surrounding the slug are not included, forces and pressures exerted on the conduit cannot be calculated accurately. Thus, in order to interpret seismic data (Chapter 15), more sophisticated fluid dynamic models are required to fully describe the liquid motion. Some computational fluid dynamic (CFD) models of slug flow have been developed through engineering research (Mao and Dukler, 1990; Clarke and Issa, 1997; Taha and Cui, 2006); however, they only describe the local flow around non-expanding gas slugs and are thus restricted in their direct application to volcanological scenarios.

In order to simulate flow in a complete conduit, classic finite difference computational fluid dynamics (CFD) and Lattice Boltzman numerical models are currently being explored (O'Brien and Bean, 2008; D'Auria and Martini, 2009), with the results being linked to the generation of seismic signals. A similar approach has also been taken using the commercial computational fluid dynamics code Flow-3D®, and comparing the results with laboratory simulations (James et al., 2008). Flow-3D® solves the Navier–Stokes equations over a fixed grid and specializes in free-surface flows. For the ascent of a gas slug, the large density and viscosity contrasts between the gas and liquid phases allow the internal motion of the gas phase to be neglected; the position of the gas–liquid interface is controlled by the liquid and gas pressures, the liquid dynamics, and surface tension (if significant). The slug gas pressure can

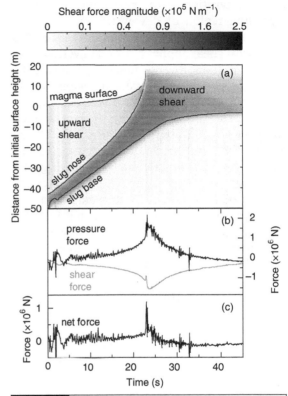

Figure 6.6 The vertical force components exerted on a conduit during the ascent and burst of a gas slug, as determined by 3D fluid dynamic modeling. The scenario simulated has a gas mass of 70 kg, a conduit radius of 1.5 m and a magma density and viscosity of 2600 kg m^{-3} and 500 Pa s, respectively. In (a) the force per vertical meter of the conduit exerted by viscous shear on the conduit wall is shown over the duration of slug ascent. The position of the magma surface and the slug nose and base are shown by the black curves, with the former two truncated ~1 s before slug burst for clarity. In (b) the total vertical forces over the domain due to pressure and shear forces are shown, and are summed in (c). The slug burst is accompanied by a sharp upward force on the domain. Redrawn from James et al. (2008) with permission from the Geological Society of London.

be adequately modeled by either P_sV_s or $P_sV_s^{\gamma} =$ constant, representing isothermal or adiabatic expansion, respectively.

A detailed discussion of the model is beyond the scope of this chapter but the results give insight into forces exerted on the conduit during slug expansion, in particular, those produced by viscous shear. Figure 6.6 illustrates the shear forces exerted on a vertical conduit, with the position of the slug indicated by the ascending region of downward shear exerted by the falling liquid film around the slug. This downward shear applies a region of vertical tension in the conduit walls immediately above the slug nose and a compressional region behind the slug base. As the slug begins to decompress rapidly near the surface, the upward acceleration of the liquid above the slug exerts a strong upward force on the conduit. Extension of this type of model to more complex conduit geometries may allow some seismic signals (e.g., very-long-period events) to be interpreted in terms of fluid dynamic processes (see Chapter 15). This raises the possibility that even in relatively low-viscosity magmas, the effect of viscous drag may be detectable, just as has been hypothesized for much more viscous magmas, for example at Asama, Soufrière Hills, and Mount St. Helens volcanoes.

6.3.4 Near-surface oscillations

As slugs approach the surface, seismic and acoustic data record a variety of signals (see Chapters 15 and 16), some of which have been interpreted in terms of pressure oscillations within the gas slugs (Vergniolle and Brandeis, 1996; Vergniolle et al., 1996). Different oscillatory modes have been proposed: longitudinal compression and dilation of the slug, with a frequency given by

$$f_L = \frac{1}{2\pi}\sqrt{\frac{\gamma P_s}{\rho l_s h}}, \qquad (6.15)$$

and kinematic waves forming at the slug nose, with a frequency

$$f_{NR_0} = \left(\frac{0.9^6 N^3 \rho g^2}{8\pi^3 \eta}\right)^{1/3}, \qquad (6.16)$$

where N is the mode ($N = 1$ being the fundamental mode), with odd modes being favored. In laboratory experiments involving low rates of slug expansion, pressure changes observed to increase in frequency (as $h^{-1/2}$) as the slug approached the surface have been attributed to longitudinal slug oscillations (James et al., 2004). However, such oscillations may be damped very quickly in viscous magma and may not be excited at all if the slug expansion occurs too rapidly to enable a resonance to become established. Note that these pressure changes are much lower in amplitude than the overall changes in magmastatic pressure due to the passage of a slug, but with frequency ranges (of around a few hertz) within the response bands of seismic and acoustic techniques, they lend themselves to being detected.

6.4 | Burst of a gas slug

When the bubble nose reaches the magma surface, a thinning magma membrane is produced which blisters upwards before bursting to produce the observable explosion. Some infrasonic waveforms have been interpreted as oscillations of the membrane prior to bursting (Vergniolle and Brandeis, 1996; Vergniolle et al., 1996). However, within conduits, viscous coupling to the conduit wall swiftly removes liquid from above rapidly expanding slugs, hence viable membranes will be extremely short lived transients, and oscillations are very unlikely. For unconfined gas bubbles bursting at the open surface of a lava lake, this wall interaction does not occur and membrane oscillations could be considered. Nevertheless, no such oscillations have been reported from video and radar data of large bubbles bursting in lava lakes (Dibble et al., 2008; Gerst et al., 2008) which, instead, indicates that bubble overpressures alone drive sufficient fluid accelerations for rapid disruption (fragmentation).

The criteria for predicting when the membrane will fail are difficult to quantify precisely, but the burst process is important for interpreting data such as ejecta velocity and infrasonic transients. A plausible and straightforward first-order burst criterion is to define a minimum thickness for the membrane before it ruptures and, with some of the ejecta being composed of parts of this membrane, it is reasonable to relate the minimum membrane thickness to the size of the observed ejecta. In the case of Stromboli, where characteristic particle diameters are often 1 to 4 cm, a membrane thickness of order

10 cm could be estimated. Bubble bursts can only be observed directly under conditions in which they are not confined by a conduit and these observations show considerable lateral expansion, with hemispherical membranes being produced. However, for bursts occurring within conduits, burst conditions must be inferred from the ejecta observed at the surface and, with little lateral expansion, jets are much more collimated. In conduits, bursts will be significantly influenced by the presence of the conduit wall and the associated viscous drag on the membrane.

6.4.1 Overpressures

Ascertaining the magnitude of bubble overpressure at burst has been the target of many measurements and is the source of much uncertainty. Early models for calculating bubble overpressure considered spherical bubbles in large bodies of liquid (Wilson, 1980); however, the slug ascent model given by Eq. (6.14) can be used to calculate more realistic anticipated overpressures for long gas bubbles in circular conduits. With a burst criterion given by a minimum membrane thickness of 10 cm, the surface overpressures for slugs of different gas masses are found to be 10^5–10^6 Pa (Fig. 6.7(a)), consistent with estimates made from field data (e.g., Vergniolle and Brandeis, 1996; Gerst et al., 2007a).

Note that in the model, the ascending slug is initially in equilibrium at magma-static pressure and the overpressure at the surface is generated only by the dynamics associated with the slug ascent. In order for the slug to expand within the conduit, the liquid above it must be accelerated and the forces required to overcome the liquid viscosity and inertia result in the slug overpressure. The absolute rate of gas expansion is strongly dependent on gas volume, so liquid accelerations and bubble overpressures increase rapidly near the surface where the bubble is largest. Hence, the development of significant overpressure should be considered a near-surface effect (i.e., at depths less than ~50 m), noting that in these large-bubble systems, any overpressure resulting from surface tension is entirely negligible.

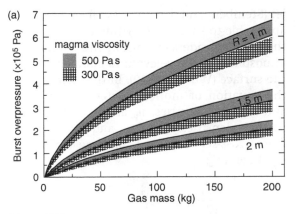

(a)

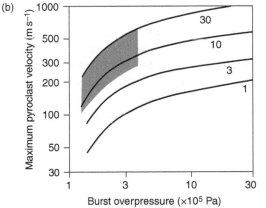

(b)

Figure 6.7 Burst pressures and ejecta velocities. (a) At-burst slug overpressures resulting from slug ascents calculated from Eq. (6.14). Each filled region represents the results for different conduit radii or magma viscosity, with the lower bound on each region representing the adiabatic case and the upper bound the isothermal case (i.e., $\gamma = 1.0$). Reprinted from James et al. (2008) with permission from the Geological Society of London. (b) Anticipated maximum ejecta velocities due to the burst of an initially overpressured bubble. The gray region indicates the anticipated region for strombolian eruptions. Redrawn from Wilson (1980).

6.4.2 Ejecta velocity

Ejecta velocities (for both the gas phase and the pyroclasts) have been determined by photographic, sodar, and Doppler radar techniques (Chouet et al., 1974; Weill et al., 1992; Dubosclard et al., 2004; Patrick et al., 2007; Gerst et al., 2008). Velocity measurements are used to estimate overpressures, in order to make inferences of the subsurface dynamics. Two approaches have been used to model ejecta velocities; one considers pyroclasts entrained within a confined jet (Steinberg and Babenko, 1978), and is hence applicable to slug bursts within a conduit, and the other considers expansion into a stationary atmosphere and is relevant to bubble bursts at the surface (Wilson, 1980). In the former case, the equation of motion relates the pyroclast and gas jet velocities, u_p and u_g respectively, by their densities ρ_p and ρ_g, and the pyroclast mass, cross-sectional area, and drag coefficient, m_p, A_p, and C_d;

$$m_p \frac{du_p}{dt} = \frac{\rho_g C_d A_p}{2C} \left(u_g - u_p\right)^2 - m_p \left(\frac{\rho_p - \rho_g}{\rho_p}\right) g. \quad (6.17)$$

As $t \to \infty$, this provides the steady state solution of a pyroclast falling at its terminal velocity within the gas jet. Assuming particles can be represented as spheres of diameter d_p allows simplification to

$$u_p = u_g - \sqrt{\left[\frac{4g}{3C_d}\left(\frac{\rho_p}{\rho_g} - 1\right)\right] d_p}. \quad (6.18)$$

Equation (6.17) considers only the pyroclasts and the entraining gas without involving any gas expansion or interaction with the surrounding atmosphere. Consequently, the results represent maximum possible clast velocities and should only be compared with field measurements made immediately above the vent.

The effects of atmospheric interaction and gas expansion can be considered using a general form of the energy equation for a fluid in motion

$$\int_{P_i}^{P_f} \frac{dP}{\rho} = \tfrac{1}{2}u_f^2 - \tfrac{1}{2}u_i^2 + gh + \text{friction terms}, \quad (6.19)$$

where P and ρ are the fluid pressure and density, respectively, and u_i, u_f, P_i, and P_f are the initial and final fluid velocities and pressures, respectively. Here, u_i is assumed to be negligible, u_f is taken as the maximum velocity prior to deceleration of the gas cloud, and P_f is taken to be equal to atmospheric pressure, P_a. For an ejecta cloud represented by a mixture of gas, and liquid or solid pyroclasts, the bulk density can be given by

$$\rho = \left(\frac{n}{\rho_g} + \frac{1-n}{\rho_p}\right)^{-1}, \quad (6.20)$$

where n is the gas weight fraction of the mixture. During the expansion of the cloud, an estimate of the friction resulting from air drag can be given by $\tfrac{1}{2}(m_a/m_c)u_f^2$, where m_c is the cloud mass and m_a is the mass of atmospheric air displaced. Wilson (1980) used this approach for the hemispherical and adiabatic expansion of a bubble-burst cloud, integrating Eq. (6.19) to yield

$$\tfrac{1}{2}u_f^2\left[1 + \left(\frac{RT_i\rho_a}{8P_i}\right)\left(\frac{P_i}{P_a}\right)^{1/\gamma}\right] = nRT_i\left(\frac{\gamma}{\gamma-1}\right)\left[1 - \left(\frac{P_a}{P_i}\right)^{\frac{\gamma-1}{\gamma}}\right]$$
$$+ \frac{(1-n)}{\rho_p}(P_i - P_a) - \tfrac{1}{2}gr_i\left(\frac{P_i}{P_a}\right)^{1/3\gamma},$$

$$(6.21)$$

which relates the pressure at bubble burst of an initially spherical bubble with initial radius and gas temperature r_i and T_i, respectively, to the maximum pyroclast velocity, where ρ_a is the atmospheric density. The results are shown in Figure 6.7(b) for different values of n. Note that here, both gas and the entrained pyroclasts are assumed to be initially stationary and then to accelerate to the same velocity.

6.4.3 Ejecta mass

Photographic and radar measurements have both been used to estimate the masses of gas and silicate material involved in strombolian eruptions (Chouet et al., 1974; Ripepe et al., 1993; Gerst et al., 2007b). Gas masses can be used to provide slug sizes for the slug ascent models and the gas-to-silicate mass ratio could be used to infer details of the burst process. Erupted silicate material is generally assumed to represent the burst bubble membrane, but the presence of material from the liquid film surrounding the slug (and then entrained into the gas jet) or even from the slug wake (which is exposed to a sudden depressurization during burst) should not be discounted. As yet, there are no models to predict ejecta masses and this will not be possible until burst processes are better understood.

6.5 | Example: Stromboli volcano, Aeolian Islands

Stromboli volcano rises 3000 m above the floor of the Tyrrhenian Sea to form part of the Aeolian Island volcanic arc. The summit region, at ~800 m above sea level, has three dominant craters (NE, Central, and SW), within which a variable number of vents (typically 5–15) are usually open and active with contrasting styles and intensities. Although interspersed with occasional larger explosions every year or so, normal activity consists of repeated mild strombolian events, one every ten minutes or so, each lasting up to a few tens of seconds, ejecting pyroclasts generally no more than 200 m above the surface. Stromboli also produces more powerful eruptions, called paroxysms (Barberi et al., 1993), every few decades. Major paroxysms occurred in 1930, 1950, 2003, and 2007, with weaker events recorded in September 1996, August 1998, and August 1999 (Metrich et al., 2005). During paroxysms, mass discharge rates reach values as large as 10^5–10^6 kg s^{-1} (Rosi et al., 2006) and a deeper, hotter, crystal-poor magma is erupted, largely bypassing the shallow storage region. These paroxysms are driven by episodic open-system escape of CO_2-rich gas from depths of multiple kilometers (Allard, 2010; Metrich et al., 2010), but are not considered further here.

Measurements at Stromboli made during typical activity have recorded ejecta velocities of up to ~110 m s^{-1}, with estimates of the mass of solid (or liquid) pyroclasts in explosions being between 16 and ~6000 kg, representing ~4 to 90 wt.% of the ejecta. The explosions can be classified by their duration (or complexity) and by their ejecta (Ripepe et al., 1993; Patrick et al., 2007). "Simple" explosions are generally relatively short in duration (~1 s) and associated with impulsive detonation sounds from the bursting of a single gas slug. Complex eruptions are longer in duration, often pulsed in nature, and sound more like a jet engine. They may be associated with either trains of multiple bubbles or much larger bubbles producing "organ-pipe" resonances within the conduit. Classification by ejecta type leads to

distinguishing lapilli-rich eruptions (type 1) from ash-rich (type 2, Fig. 6.8). The former represents the classical strombolian style described and modeled above, with an incandescent jet incorporating most or all of the coarse ejecta (Fig. 6.1). Often, type 1 explosions will dominate at one crater, and ash-rich type 2 explosions at an adjacent crater. Type 2 explosions generate weak ash-bearing convective plumes typically extending to a few hundred meters in height, and can be subdivided into events that may also be associated with an incandescent spatter-laden jet (Patrick et al., 2007).

Recent petrographic studies have shown considerable heterogeneity in ejecta at Stromboli, which must reflect complex dynamics for the melt phase in the shallow conduit (Lautze and Houghton, 2005; Polacci et al., 2008). In clast populations ejected by single explosions of typical strombolian activity, two different phenocryst-bearing melts are represented, with either sparse or relatively abundant populations of mm-sized and smaller bubbles. This textural diversity results from a combination of ongoing vesicle evolution in essentially stagnant melt within the shallow conduit (coalescence changing a population of abundant small bubbles to one in which small bubbles are relatively sparse, on timescales of hours to weeks), and a shorter timescale mingling process that is driven by passage of gas slugs and turbulent drainback of non-erupted melt following slug-bursts.

At Stromboli, very-long-period (VLP) seismic signals, with one source mechanism located at a consistent depth ~240 m below the surface, have been cited as resulting from the bubble coalescence forming the gas slug (Chouet et al., 2003). However, laboratory experiments have demonstrated that the magnitudes of the VLP signals are much more likely to be produced by the ascent of slugs through a change in conduit geometry (James et al., 2006; Chouet et al., 2008). This is consistent with gas geochemistry data indicating that slug gas geochemically decouples from the surrounding melt at depths of 800 to 3000 m (Burton et al., 2007a). With the positions of several sources of VLP signals now identified, it is clear that the plumbing system is best represented by multiple inclined and intersecting

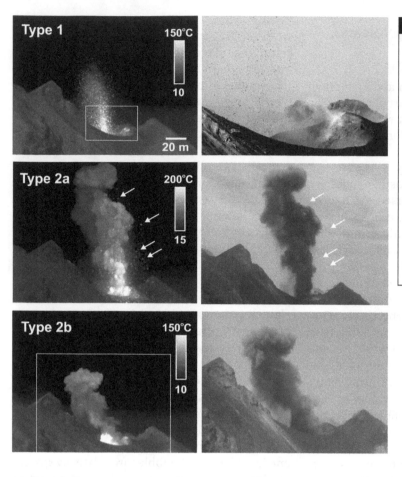

Type 1 150°C 10 20 m

Type 2a 200°C 15

Type 2b 150°C 10

Figure 6.8 Eruption styles at Stromboli as defined by thermal and visible images (the left and right columns, respectively). Type 1 explosions are lapilli rich with little or no visible ash. Type 2 explosions have a greatly increased ash component, and can be subdivided into 2a and 2b, representing events with and without a visible coarse (type 1 style) component (arrowed in the middle panels). The white boxes in the thermal images indicate the field of view in the associated visible image. Reproduced from Patrick *et al.* (2007, Fig. 3) with kind permission of Springer Science and Business Media. See color plates section.

dike segments, with dips ranging between 40° and ~70° (Chouet *et al.*, 2008).

The wealth of such geophysical data from Stromboli illustrates the complexity of the system with respect to the current first-order models of strombolian activity. Many key questions remain and are illustrated by inconsistencies between results derived from different techniques. For example, modeling oscillatory acoustic data has suggested low magma viscosities of ~300 Pa s (Vergniolle *et al.*, 1996). Although this is not inconsistent with values successfully used in fluid dynamic models of slug ascent, petrological modeling suggests that some magma could have a viscosity of up to 1.4×10^4 Pa s (Metrich *et al.*, 2001). A currently unexplored area is the extent to which a non-Newtonian and spatially varying magma rheology would affect our current understanding of slug ascent and burst processes. Further complexities such as the

role of post-burst drainback and rainfall (which has been observed to affect eruption velocities and pyroclast sizes as determined by Doppler radar) are also not understood. Even fundamental questions, such as where does the change from a dike-like morphology to a circular conduit occur, and what process is responsible for switching activity styles between vents, remain largely unaddressed.

6.6 | Example: Mount Erebus, Antarctica

A contrasting example is illustrated by activity at Mount Erebus, where strombolian activity is represented by bursts of large, observable bubbles within a small lava lake. Mount Erebus is a large stratovolcano located on Ross Island, at the western end of the West Antarctic rift system, with a summit at 3794 m. The summit cone hosts the Main Crater, within which the Inner Crater is approximately 100 m deep, with a radius of ~80 m. Over the last 30 years, the Inner Crater has contained an anorthoclase phonolite lava lake of between 5 and 15 m in radius that has been the site of variable strombolian activity. Events are generally simple rather than complex, and generally occur at a rate of several per day, with the occasional swarm of up to 900 per day. Erebus is now permanently geophysically monitored, with seismic, infrasonic, and video data being continuously collected unless instruments succumb to the harsh environmental conditions.

Early work described upwelling lava in the lake accompanied by seismic tremor preceding strombolian events by ~20 to 45 s. Erupting bubbles would have radii of up to 5 m, after which lake levels would drop by about 5 m before recovering over the following 15 minutes. This type of activity is indicative of slugs expanding as they ascend and, assuming a slug velocity between 1 and 2 m s^{-1}, suggests a minimum slug source depth of between 20 and 90 m.

More recently, VLP seismic events have been recorded and, although depths could be exaggerated by the effects of tilt, particle motions suggest VLP source depths of between 100 and 800 m (Rowe *et al.*, 1998, 2000). Gas geochemistry data are also consistent with gas being sourced at depths of up to 2 km (Oppenheimer and Kyle, 2008). However, the VLP events precede short-period signals (which accompany the surface explosion) by up to only 5 s, suggesting very short slug ascents, of order 10 m. This is in agreement with video data that indicate only the very smallest of disturbances to the lake level prior to each event and no significant period of decompression-related gas expansion. The relative timings of the seismicity and the polarity of the VLP signals vary, suggesting multiple slug ascent paths within a well-connected plumbing system. Furthermore, it is likely that the location of the VLP source reflects the region of greatest coupling between the varying fluid pressures and the surrounding rock rather than the locus of the initial fluid-dynamics driven pressure perturbation (Aster *et al.*, 2003; James *et al.*, 2006). Consequently, current models

for the subsurface plumbing at Erebus involve a relatively large, high-level storage region, possibly of the diameter of the crater itself, such that bubble coalescence would take place only in the upper few tens of meters (Aster *et al.*, 2003). The lack of significant upwelling of the lake immediately prior to slug burst supports this interpretation by suggesting there is only a brief period of slug ascent. Upwelling may be also suppressed by a strongly flared upper region of the conduit.

During bursts, the surface of the lava lake is observed to dome hemispherically upwards before being disrupted and showering the crater area with pyroclasts. Radar data collected in 2005–2006 suggest that pressures at burst could be ~3 × 10^5 to 8 × 10^5 Pa (Gerst *et al.*, 2007a). Radar (Gerst *et al.*, 2008) and infrasonic (Jones *et al.*, 2008) data indicate that the position (within the lava lake) and initial direction of the burst are random, strongly suggesting that, close to the surface, the conduit is likely to be vertical and that the burst process itself is controlled by instabilities or heterogeneities within the expanding magma membrane. With the membrane composed of magma that has been rapidly and recently cooling in the atmosphere, this is highly likely.

Erebus provides a good case for multiparameter (video, radar, infrasonic, seismic) measurements of bubble bursts. However, the subsurface geometries are poorly constrained and key questions remain, such as how to reconcile the contrasting evidence for the depth of slug coalescence.

6.7 | Future directions

Our understanding of strombolian activity has been advancing dramatically, most recently due to quantitative measurement of the gas species from spectroscopic data and improved imaging of conduit segments by inverted VLP seismic data. Such findings have been augmented by additional key interpretations drawn from an increasingly diverse array of other investigations, including laboratory experiments, computational fluid dynamic modeling, thermal imaging, and infrasonic data collection.

Nevertheless, major uncertainties still exist, as exemplified by the estimates of appropriate magma viscosities at Stromboli spanning two orders of magnitude. Uncertainty reduction will be driven by the increasing integration of different data types and the unprecedented data volumes from new continuous autonomous measurement techniques. A major challenge is to produce a similarly integrated model in which fluid dynamics, gas geochemistry, and petrology are all linked to measurables such as seismic and infrasonic data. Such a model will permit a much improved understanding of the physical processes recorded in remotely sensed data, and will consequently allow any hazardous changes within systems to be more readily identified and interpreted.

6.8 | Summary

- Strombolian activity is driven by exsolving gas bubbles within magma coalescing into large slug bubbles, with diameters on the order of the width of the conduit.
- At any significant depth, these gas slugs will ascend through the liquid magma at a few meters per second. Over the last few tens of meters to the surface, gas expansion can rapidly accelerate the slug nose to several tens of meters per second.
- On burst, coarse pyroclasts (mainly relatively dense clots of magma) are produced and are ejected at velocities of up to 100 m s^{-1} or so, travelling a few hundred meters.
- Current modeling approaches address the slug formation, slug ascent and slug burst processes separately. The models draw upon large volumes of geophysical data (e.g., seismic, radar, video, IR and infrasonic) to help understand the subsurface processes.

6.9 | Notation

A' — square of ratio of slug radius to conduit radius $(r_s/r_c)^2$

A_p — cross-sectional area of pyroclast (m^2)
C_d — pyroclast drag coefficient
d_p — pyroclast diameter (m)
D — internal conduit diameter (m)
f_L — longitudinal gas slug oscillation frequency (Hz)
f_{NR_0} — frequency of kinematic waves at slug nose (Hz)
F_g — vertical gravitational forces acting on a gas slug (N)
F_p — vertical pressure forces acting on a gas slug (N)
F_v — vertical viscous forces acting on a gas slug (N)
g — acceleration due to gravity (m s^{-2})
h — height of liquid above the slug nose (m)
h_0 — initial height of liquid above the slug nose (m)
h_c — critical foam thickness (m)
h_m — maximum foam thickness (m)
l_s — slug length (m)
l_{s0} — initial slug length (m)
m_a — mass of air displaced by expanding gas cloud (kg)
m_c — gas cloud mass (kg)
m_p — pyroclast mass (kg)
n — gas weight fraction of a gas–pyroclast mixture
N — kinematic wave mode
N_f — dimensionless inverse viscosity (or the buoyancy Reynolds number)
P_a — ambient surface pressure (Pa)
P_f — final fluid pressure (Pa)
P_i — initial fluid pressure (Pa)
P_s — slug gas pressure (Pa)
P_{s0} — initial slug gas pressure (Pa)
Q — gas volume flux (m^3 s^{-1})
R — molar gas constant (J K^{-1} mol^{-1})
r_b — gas bubble radius (m)
r_c — internal conduit radius (m)
r_i — initial gas bubble radius (m)
r_s — slug radius (m)
t — time (s)
T_i — initial gas temperature (K)
u_b — terminal bubble ascent velocity (m s^{-1})
u_f — final fluid velocity (m s^{-1})
u_g — gas jet velocity (m s^{-1})
u_i — initial fluid velocity (m s^{-1})
u_p — pyroclast velocity (m s^{-1})

u_s slug ascent velocity, defined at the slug base (m s^{-1})

V_s volume of a gas slug (m^3)

γ ratio of specific heats

ε foam gas volume fraction

η magma viscosity (Pa s)

η_f foam viscosity (Pa s)

ρ liquid magma density (kg m^{-3})

ρ_g gas density (kg m^{-3})

ρ_p pyroclast density (kg m^{-3})

σ magma surface tension (N m^{-1})

Ω area factor for bubble collection under a horizontal roof

Eo   Eötvös number

Fr   Froude number

Fr$_\theta$ Froude number at an inclination angle θ

Mo   Morton number

References

Allard, P. (2010). A CO_2-rich gas trigger of explosive paroxysms at Stromboli basaltic volcano, Italy. *Journal of Volcanology and Geothermal Research*, **189**, 363–374, doi:10.1016/j.jvolgeores.2009.11.018.

Aster, R., Mah, S., Kyle, P. *et al.* (2003). Very long period oscillations of Mount Erebus Volcano. *Journal of Geophysical Research*, **108**, 2522, doi:10.1029/2002JB002101.

Barberi, F., Rosi, M. and Sodi, A. (1993). Volcanic hazard assessment at Stromboli based on review of historical data. *Acta Vulcanologica*, **3**, 173–187.

Batchelor, G. K. (1967). *An Introduction to Fluid Dynamics*. Cambridge: Cambridge University Press.

Blackburn, E. A., Wilson, L. and Sparks, R. S. J. (1976). Mechanics and dynamics of Strombolian activity. *Journal of the Geological Society of London*, **132**, 429–440.

Burton, M., Allard, P., Mure, F. and La Spina, A. (2007a). Magmatic gas composition reveals the source depth of slug-driven Strombolian explosive activity. *Science*, **317**, 227–230.

Burton, M. R., Mader, H. M. and Polacci, M. (2007b). The role of gas percolation in quiescent degassing of persistently active basaltic volcanoes. *Earth and Planetary Science Letters*, **264**, 46–60, doi:10.1016/j. epsl.2007.08.028.

Calvari, S. and Pinkerton, H. (2004). Birth, growth and morphologic evolution of the "Laghetto" cinder cone during 2001 Etna eruption. *Journal of Volcanology and Geothermal Research*, **132**, 225–239.

Carey, R. J., Houghton, B. F., Sable, J. E. and Wilson, C. J. N. (2007). Contrasting grain size and componentry in complex proximal deposits of the 1886 Tarawera basaltic plinian eruption. *Bulletin of Volcanology*, **69**, 903–926.

Chouet, B., Hamisevicz, N. and McGetchin, T. R. (1974). Photoballistics of volcanic jet activity at Stromboli, Italy. *Journal of Geophysical Research*, **79**, 4961–4976.

Chouet, B., Dawson, P., Ohminato, T. *et al.* (2003). Source mechanisms of explosions at Stromboli Volcano, Italy, determined from moment-tensor inversions of very-long-period data. *Journal of Geophysical Research*, **108**, 2019, doi:10.1029/2002JB001919.

Chouet, B., Dawson, P. and Martini, M. (2008). Shallow-conduit dynamics at Stromboli Volcano, Italy, imaged from waveform inversions. In *Fluid Motion in Volcanic Conduits: A Source of Seismic and Acoustic Signals*, ed. S. J. Lane and J. S. Gilbert. Geological Society, London, Special Publication 307, pp. 57–84.

Clarke, A. and Issa, R. I. (1997). A numerical model of slug flow in vertical tubes. *Computers and Fluids*, **26**, 395–415.

Cole, P. D., Fernandez, E., Duarte, D. and Duncan, A. M. (2005). Explosive activity and generation mechanisms of pyroclastic flows at Arenal volcano, Costa Rica between 1987 and 2001. *Bulletin of Volcanology*, **67**, 695–716.

D'Auria, L. and Martini, M. (2009). Slug flow: Modeling in a conduit and associated elastic radiation. In *Encyclopedia of Complexity and Systems Science*, ed. R. A. Meyers, pp. 8153–8168.

Dibble, R. R., Kyle, P. R. and Rowe, C. A. (2008). Video and seismic observations of Strombolian eruptions at Erebus volcano, Antarctica. *Journal of Volcanology and Geothermal Research*, **177**, 619–634, doi:10.1016/j.jvolgeores.2008.07.020.

Dubosclard, G., Donnadieu, F., Allard, P. *et al.* (2004). Doppler radar sounding of volcanic eruption dynamics at Mount Etna. *Bulletin of Volcanology*, **66**, 443–456.

Gerst, A., Hort, M., Johnson, J. B. and Kyle, P. R. (2007a). The dynamics of a Strombolian bubble burst derived from Doppler radar. *Eos, Transactions of the American Geophysical Union*, **88**, Fall Meeting Supplement, Abstract V24B-06.

Gerst, A., Hort, M., Johnson, J. B. and Kyle, P. R. (2007b). The first second of a Strombolian eruption: Doppler radar and infrasound observations at Erebus volcano, Antarctica. *Geophysical Research Abstracts*, **9**, 07280.

Gerst, A., Hort, M., Kyle, P. R. and Vöge, M. (2008). 4D velocity of Strombolian eruptions and man-made explosions derived from multiple Doppler radar instruments. *Journal of Volcanology and Geothermal Research*, **177**, 648–660, doi:10.1016/j.jvolgeores.2008.05.022.

Houghton, B. F. and Gonnermann, H. M. (2008). Basaltic explosive volcanism: constraints from deposits and models. *Chemie der Erde*, **68**, 117–140.

Houghton, B. F., Wilson, C. J. N. and Smith, I. E. M. (2000). Shallow-seated controls on styles of explosive basaltic volcanism: a case study from New Zealand. *Journal of Volcanology and Geothermal Research*, **91**, 97–120.

James, M. R., Lane, S. J., Chouet, B. and Gilbert, J. S. (2004). Pressure changes associated with the ascent and bursting of gas slugs in liquid-filled vertical and inclined conduits. *Journal of Volcanology and Geothermal Research*, **129**, 61–82.

James, M. R., Lane, S. J. and Chouet, B. A. (2006). Gas slug ascent through changes in conduit diameter: Laboratory insights into a volcano-seismic source process in low-viscosity magmas. *Journal of Geophysical Research*, **111**, B05201, doi:10.1029/2005JB003718.

James, M. R., Lane, S. J. and Corder, S. B. (2008). Modelling the rapid near-surface expansion of gas slugs in low viscosity magmas. In *Fluid Motion in Volcanic Conduits: A Source of Seismic and Acoustic Signals*, ed. S. J. Lane and J. S. Gilbert. Geological Society, London, Special Publication 307, pp. 147–167.

Jaupart, C. and Vergniolle, S. (1989). The generation and collapse of a foam layer at the roof of a basaltic magma chamber. *Journal of Fluid Mechanics*, **203**, 347–380.

Jones, K. R., Johnson, J. B., Aster, R., Kyle, P. R. and McIntosh, W. C. (2008). Infrasonic tracking of large bubble bursts and ash venting at Erebus Volcano, Antarctica. *Journal of Volcanology and Geothermal Research*, **177**, 661–672, doi:10.1016/j.jvolgeores.2008.02.001.

Lautze, N. C. and Houghton, B. F. (2005). Physical mingling of magma and complex eruption dynamics in the shallow conduit at Stromboli volcano, Italy. *Geology*, **33**, 425–428.

Mao, Z. S. and Dukler, A. E. (1990). The motion of Taylor bubbles in vertical tubes 1. A numerical-simulation for the shape and rise velocity of Taylor bubbles in stagnant and flowing liquid. *Journal of Computational Physics*, **91**, 132–160.

Menand, T. and Phillips, J. C. (2007). Gas segregation in dykes and sills. *Journal of Volcanology and Geothermal Research*, **159**, 393–408.

Metrich, N., Bertagnini, A., Landi, P. and Rosi, M. (2001). Crystallization driven by decompression and water loss at Stromboli volcano (Aeolian Islands, Italy). *Journal of Petrology*, **42**, 1471–1490.

Metrich, N., Bertagnini, A., Landi, P., Rosi, M. and Belhadj, O. (2005). Triggering mechanism at the origin of paroxysms at Stromboli (Aeolian Archipelago, Italy): The 5 April 2003 eruption. *Geophysical Research Letters*, **32**, L10305.

Metrich, N., Bertagnini, A. and Di Muro, A. (2010). Conditions of magma storage, degassing and ascent at Stromboli: New insights into the volcano plumbing system with inferences on the eruptive dynamics. *Journal of Petrology*, **51**, 603–626, doi:10.1093/petrology/egp083.

O'Brien, G. S. and Bean, C. J. (2008). Seismicity on volcanoes generated by gas slug ascent. *Geophysical Research Letters*, **35**, L16308, doi:10.1029/2008gl035001.

Omebere-Iyari, N. K., Azzopardi, B. J. and Ladam, Y. (2007). Two-phase flow patterns in large diameter vertical pipes at high pressures. *AIChE Journal*, **53**, 2493–2504, doi: 10.1002/aic.11288.

Oppenheimer, C. and Kyle, P. R. (2008). Probing the magma plumbing of Erebus volcano, Antarctica, by open-path FTIR spectroscopy of gas emissions. *Journal of Volcanology and Geothermal Research*, **177**, 743–754, doi:10.1016/j.jvolgeores.2007.08.022.

Parfitt, E. A. (2004). A discussion of the mechanisms of explosive basaltic eruptions. *Journal of Volcanology and Geothermal Research*, **134**, 77–107.

Parfitt, E. A. and Wilson, L. (1995). Explosive volcanic eruptions IX. The transition between hawaiian-style lava fountaining and strombolian explosive activity. *Geophysical Journal International*, **121**, 226–232.

Patrick, M. R., Harris, A. J. L., Ripepe, M. *et al.* (2007). Strombolian explosive styles and source conditions: insights from thermal (FLIR) video. *Bulletin of Volcanology*, **69**, 769–784, doi:10.1007/s00445-006-0107-0.

Polacci, M., Baker, D. R., Bai, L. P. and Mancini, L. (2008). Large vesicles record pathways of degassing at basaltic volcanoes. *Bulletin of Volcanology*, **70**, 1023–1029, doi:10.1007/s00445-007-0184-8.

Ripepe, M., Rossi, M. and Saccorotti, G. (1993). Image-processing of explosive activity at Stromboli. *Journal of Volcanology and Geothermal Research*, **54**, 335–351.

Rosi, M., Bertagnini, A., Harris, A. J. L. *et al.* (2006). A case history of paroxysmal explosion at Stromboli: Timing and dynamics of the April 5, 2003 event. *Earth and Planetary Science Letters*, **243**, 594–606.

Rowe, C. A., Aster, R. C., Kyle, P. R., Schlue, J. W. and Dibble, R. R. (1998). Broadband recording of Strombolian explosions and associated very-long-period seismic signals on Mount Erebus volcano, Ross Island, Antarctica. *Geophysical Research Letters*, **25**, 2297–2300.

Rowe, C. A., Aster, R. C., Kyle, P. R., Dibble, R. R. and Schlue, J. W. (2000). Seismic and acoustic observations at Mount Erebus Volcano, Ross Island, Antarctica, 1994–1998. *Journal of Volcanology and Geothermal Research*, **101**, 105–128.

Seyfried, R. and Freundt, A. (2000). Experiments on conduit flow and eruption behavior of basaltic volcanic eruptions. *Journal of Geophysical Research*, **105**, 23 727–23 740.

Steinberg, G. S. and Babenko, J. J. (1978). Experimental velocity and density determination of volcanic gases during eruption. *Journal of Volcanology and Geothermal Research*, **3**, 89–98.

Taha, T. and Cui, Z. F. (2006). CFD modelling of slug flow in vertical tubes. *Chemical Engineering Science*, **61**, 676–687.

Taitel, Y., Barnea, D. and Dukler, A. E. (1980). Modelling flow pattern transitions for steady upward gas-liquid flow in vertical tubes. *AIChE Journal*, **26**, 345–354, doi:10.1002/aic.690260304

Vergniolle, S. and Brandeis, G. (1996). Strombolian explosions 1. A large bubble breaking at the surface of a lava column as a source of sound. *Journal of Geophysical Research*, **101**, 20 433–20 447.

Vergniolle, S. and Jaupart, C. (1990). Dynamics of degassing at Kilauea Volcano, Hawaii. *Journal of Geophysical Research*, **95**, 2793–2809.

Vergniolle, S., Brandeis, G. and Mareschal, J. C. (1996). Strombolian explosions 2. Eruption dynamics determined from acoustic measurements. *Journal of Geophysical Research*, **101**, 20 449–20 466.

Viana, F., Pardo, R., Yanez, R., Trallero, J. L. and Joseph, D. D. (2003). Universal correlation for the rise velocity of long gas bubbles in round pipes. *Journal of Fluid Mechanics*, **494**, 379–398.

Walker, G. P. L. and Croasdale, R. (1972). Characteristics of some basaltic pyroclastics. *Bulletin Volcanologique*, **35**, 303–317.

Wallis, G. B. (1969). *One-Dimensional Two-Phase Flow*. New York: McGraw-Hill.

Weill, A., Brandeis, G., Vergniolle, S. *et al.* (1992). Acoustic sounder measurements of the vertical velocity of volcanic jets at Stromboli Volcano. *Geophysical Research Letters*, **19**, 2357–2360.

White, E. R. and Beardmore, R. H. (1962). The velocity of rise of single cylindrical air bubbles through liquids contained in vertical tubes. *Chemical Engineering Science*, **17**, 351–361.

Wilson, L. (1980). Relationships between pressure, volatile content and ejecta velocity in three types of volcanic explosion. *Journal of Volcanology and Geothermal Research*, **8**, 297–313.

Wilson, L. and Head, J. W. (1981). Ascent and eruption of basaltic magma on the Earth and Moon. *Journal of Geophysical Research*, **86**, 2971–3001.

Zukoski, E. E. (1966). Influence of viscosity, surface tension and inclination angle on motion of long bubbles in closed tubes. *Journal of Fluid Mechanics*, **25**, 821–837.

Exercises

6.1 Consider the ascent of small bubbles of H_2O vapor within a basaltic magma conduit.

(a) Estimate the rise velocity through the fluid, u_b, of bubbles with diameters of 100 μm, 1 mm, 1 cm, and 10 cm, stating your assumptions.

(b) Stokes' law is valid only for laminar flow described by Reynolds numbers, Re, less than approximately 0.3, where Re $= \rho u_b L / \eta$, ρ and η are the fluid density and viscosity respectively, and L is a characteristic length (e.g., bubble diameter). In this case, comment on the validity of your answers to (a).

(c) Would anticipated rise velocities be less or greater than calculated? Why?

6.2 As gas ascends within conduits, it expands due to the decreasing pressure.

(a) For a bubble of volume 1 m³ at a depth of 1000 m in a basaltic conduit, estimate and plot the bubble radius against depth, assuming that the bubble could ascend to the surface in spherical form. What is the bubble volume on eruption?

(b) If the bubble ascends a conduit as a slug of near-cylindrical shape and constant radius 1.2 m, what length would the slug achieve at the surface?

(c) Discuss whether, if this slug were "real," it would reach this length or not; what is the evidence and what assumptions and omissions from the calculations result in the differences?

(d) Conditions appropriate to basaltic activity on Io or the early Moon would include very low atmospheric pressures and a much lower gravity compared to that on Earth. Consider these extraterrestrial environments by repeating your calculations for parts (a–c) using, for simplicity, a surface atmospheric pressure of 1 Pa and a gravitational acceleration of 1 m s^{-2}. Discuss the differences between the results and those for terrestrial conditions, commenting on the implications for eruptive style.

6.3 Figure 6.1(a) shows a long-exposure photograph of a strombolian eruption.

(a) As the gas slug arrived at the surface, what pressure conditions must have existed inside the slug in order to generate and eject the pyroclasts?

(b) With respect to the shape and position of the slug immediately prior to burst, where is the material that forms the pyroclasts sourced from?

(c) Considering the shape of the pyroclast tracks, what is the primary control on their trajectories and what other factors must also be influencing their motion?

(d) Given your answers to (c), explain why Strombolian deposits appear significantly less dispersed than Plinian deposits (Fig. 6.2(c)).

6.4 Data on the deposits from two well-documented historical eruptions are shown with filled symbols in Figure. 6.2(c), from the 1959 hawaiian fountaining eruption of Kilauea (filled squares) and the 2003 paroxysm at Stromboli (filled triangles).

(a) From the graph data alone, contrast these two eruptions in terms of their intensity (eruption rate) and volume.

(b) What do your interpretations imply about the relative duration of these eruptions?

(c) How then can you explain the thickness values given for the Strombolian deposits with open squares, which overlap with the data for the hawaiian fountaining eruption?

Answers to exercises available at www.cambridge.org/fagents

Chapter 7

Unsteady explosive activity: vulcanian eruptions

Amanda B. Clarke

Overview

Vulcanian eruptions are named for the 1888–90 eruptions of Vulcano, Aeolian Islands, Italy (Mercalli, 1907), and are defined here as short-lived, discrete explosions resulting from sudden decompression of a volcanic conduit caused by disruption of a sealing plug or dome. Resulting eruptions characteristically last only seconds to minutes and may produce buoyant columns, pyroclastic density currents, or both. They may occur as single events or in a sequence of discrete explosions. The short duration and unsteady vent conditions of vulcanian eruptions make them distinct from sustained plinian or subplinian eruptions. Pre-eruption pressures can reach 10 MPa, vent velocities may approach 400 m s^{-1}, eruption plumes typically rise to < 10 km, but in some cases may reach nearly 20 km, and the amount of magma erupted is typically < 10^{11} kg. This chapter reviews mechanisms associated with vulcanian eruptions and discusses several relevant conceptual and quantitative models. Topics include plug formation and disruption, magma fragmentation, calculation of vent flux, the production and propagation of shock waves, the dynamics of pyroclastic jets and plumes ascending from unsteady sources, and ballistic analysis. This chapter also addresses important questions regarding controls on the scale and duration of such short-lived explosions, as well as transitions in eruptive style.

7.1 | Introduction

Vulcanian eruptions are short-lived, discrete explosions resulting from sudden decompression of a volcanic conduit that commonly contains high-pressure, vesiculated magma (Fig. 7.1; Self et al., 1979; Alidibirov, 1994; Woods, 1995; Sparks et al., 1997; Stix et al., 1997; Morrissey and Mastin, 2000; Druitt et al., 2002). Explosions typically initiate when a competent sealing conduit plug or dome is disrupted because of increasing pressure in the underlying magma (Hoblitt, 1986; Tait et al., 1989; Stix et al., 1997) or as a result of groundwater vaporization (Schmincke, 1977). Pre-eruption pressures typically range from < 1 MPa to 10 MPa (Tait et al., 1989; Stix et al., 1997; Morrissey and Chouet, 1997; Voight et al., 1999; Formenti et al., 2003; Chojnicki et al., 2006). Upon plug disruption, a shock wave (pressure discontinuity) may travel away from the vent into the atmosphere and a decompression-induced fragmentation wave travels down the conduit (Kieffer, 1981; Turcotte et al., 1990; Woods, 1995). At the fragmentation

Modeling Volcanic Processes: The Physics and Mathematics of Volcanism, eds. Sarah A. Fagents, Tracy K. P. Gregg, and Rosaly M. C. Lopes. Published by Cambridge University Press. © Cambridge University Press 2013.

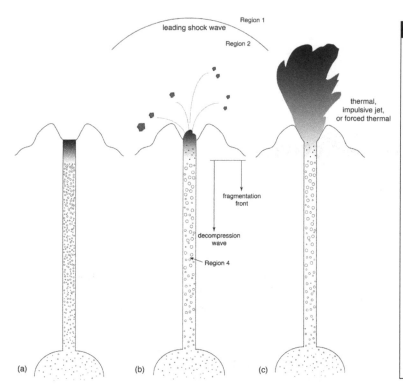

Figure 7.1 Schematic representation of vulcanian explosion sequence. (a) Initial conditions showing vesicular magma underlying a dense, sealing plug. (b) Vulcanian explosions initiate with plug disruption and launching of ballistic clasts; decompression causes simultaneous propagation of a leading shock wave into the atmosphere and propagation of a decompression wave into the conduit; a fragmentation wave follows closely behind the decompression wave. (c) A mixture of gas and fragmented magma propagates vertically into the atmosphere and the forces may be dominated by buoyancy (thermal), momentum (unsteady jet), or both (forced thermal). Region 1 is the atmosphere, region 2 is the zone immediately behind the shock but outside the conduit, and region 4 is the pre-explosion, high-pressure conduit; these designations are used in the equations of Section 7.5.

front, vesicular magma is disrupted into a gas–pyroclast mixture, propelled upward, and ejected from the vent at velocities up to 400 m s^{-1} (Self *et al.*, 1979). The resulting eruption may produce buoyant columns, pyroclastic density currents, or both. Typically, only a portion of the magma in the conduit is fragmented and evacuated (e.g., Druitt *et al.*, 2002), such that vulcanian eruptions characteristically last only seconds to minutes. They may occur as single events or as a sequence of explosions spaced sufficiently far apart in time to produce distinguishable, discrete, unsteady jets, and plumes.

The short duration and unsteady vent conditions of vulcanian eruptions make them distinct from quasi-steady plinian or subplinian eruptions, for which it is generally assumed that bubbly magma rises to meet a fragmentation wave and thus steadily feeds vent flux over long periods (hours to days). Vulcanian eruptions typically erupt < 0.1 km^3 DRE (dense rock equivalent), corresponding to a mass of ~10^{11} kg (Morrissey and Mastin, 2000), and produce columns < 20 km high (Nairn and Self, 1978),

although commonly plumes rise < 10 km (e.g., Rose *et al.*, 1978; Glaze *et al.*, 1989; Stix *et al.*, 1997; Druitt *et al.*, 2002; Wright *et al.*, 2006). For comparison, cruising altitude for commercial air traffic is generally 9–12 km. Vulcanian column heights typically exceed heights associated with strombolian eruptions (Chapter 6) and are less than those associated with plinian and subplinian eruptions (Chapters 8 and 9).

Vulcanian eruptions have a wide range of dispersal areas and degrees of fragmentation, making them difficult to classify from deposit characteristics alone (Fig. 7.2; Walker, 1973; Cas and Wright, 1987). Vulcanian fall deposits are typically thin, may consist of clasts ranging from fine ash to coarse lapilli, and contain variable proportions of lithics (non-juvenile) and juvenile material depending on the specifics of the eruption (Self *et al.*, 1979; Cas and Wright, 1987; Morrissey and Mastin, 2000). Juvenile clast densities vary even within a single eruption, ranging from dense to highly vesicular (e.g., Formenti and Druitt, 2003; Clarke *et al.*, 2007). Furthermore, unlike strombolian events, vulcanian eruptions

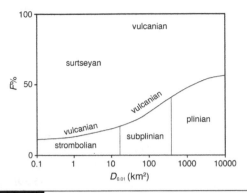

Figure 7.2 Classification of explosive volcanic eruptions in terms of F% and $D_{0.01}$. F% is the percentage of clasts < 1 mm in diameter on the main axis of dispersal where it crosses the $0.1T_{max}$ isopach; T_{max} is the maximum thickness of the fall deposit. $D_{0.01}$ is the area in km² enclosed by the $0.01T_{max}$ isopach. After Walker et al. (1973) and including vulcanian classification based on data from Cas and Wright (1987).

typically occur in intermediate-composition magmas with high viscosities (often due to high crystal contents), and therefore bubble coalescence is limited. Ballistic blocks or bombs, sourced from the displaced plug, are often associated with vulcanian deposits and range up to a few meters in diameter, are launched up to several kilometers from the vent, and may exhibit a breadcrusted texture indicating that interior gas expanded after the surface was quenched (Wilson, 1972; Fagents and Wilson, 1993; Waitt et al., 1995; Sparks et al., 1997; Druitt et al., 2002; Wright et al., 2007).

Vulcanian eruptions can precede large plinian eruptions, as at Mt. Pinatubo (Philippines) in 1991 (Hoblitt et al., 1996); produce dangerous pyroclastic flows, as at Mount St. Helens (USA) in the 1980s (Hoblitt, 1986) and at Soufrière Hills volcano (Montserrat, British West Indies) in 1997 (Druitt et al., 2002); and present a significant hazard to aircraft (e.g., Galunggung, Indonesia in 1982, http://lvo.wr.usgs.gov/zones/30410914–052_caption.html). At several volcanoes around the world, including Semeru, Indonesia (Carn and Oppenheimer, 2000), and Sakurajima, Japan (Ishihara, 1985), vulcanian eruptions occur regularly and can persist for years, potentially representing a large cumulative erupted mass, comparable perhaps to the mass erupted during a single plinian event (e.g., $\sim 10^{13}$ kg for the 1991

Pinatubo eruption; Scott et al., 1996). Ash and gases produced by vulcanian eruptions are not typically ejected into the stratosphere where they may have a global effect; nevertheless explosion products have devastating effects on local crops and nearby populations, especially during rainy seasons or spring snowmelt when lahars are commonly generated (Chapter 14).

Vulcanian eruptions occur much more frequently around the world than plinian eruptions, with some volcanoes producing multiple daily events, offering unparalleled opportunity for detailed quantitative field observations of explosive eruptions. To date, however, such observations are rare in the literature (Nairn and Self, 1978; Self et al., 1979; Ishihara, 1985; Sparks et al., 1997; Druitt et al., 2002; Clarke et al., 2002a,b; Formenti et al., 2003). Continued observation of vulcanian eruptions will play a critical role in advancing general theoretical understanding of explosive eruption dynamics.

7.2 | Eruption initiation: plug formation and disruption

The initial phase of a vulcanian eruption typically involves the disruption of a coherent magma plug or dome that sealed the conduit (Hoblitt, 1986; Hammer et al., 1999; Belousov et al., 2002; Druitt et al., 2002; Taddeucci et al., 2004; Cashman and McConnell, 2005; D'Oriano et al., 2005; Diller et al., 2006). Observational evidence of this initiation mechanism primarily consists of visually documented ballistic paths or ballistic clasts in deposits (Ishihara, 1985; Waitt et al., 1995; Druitt et al., 2002; Wright et al., 2007). Pressure in the underlying magma or gas must rise sufficiently to partially or fully dislodge this competent plug.

Two key processes of *plug formation* and *plug disruption* may be linked via magma degassing and subsequent microlite crystallization in the volcanic conduit (see Chapter 4). As magma rises, water exsolves from the melt due to decreasing ambient pressure. In response to the corresponding shift in liquidus, anhydrous phases crystallize (Geschwind and Rutherford, 1995; Stix et al.,

1997; Sparks, 1997; Moore and Carmichael, 1998; Hammer et al., 1999; Cashman and Blundy, 2000). Crystallization and degassing increase magma viscosity (Lejeune and Richet, 1995; Hess and Dingwell, 1996; Sparks, 1997; Stix et al., 1997; Melnik and Sparks, 1999) and, as a result of concentrating volatiles in the remaining melt, force further degassing. Bubble connections are thought to develop at some threshold vesicularity (Eichelberger et al., 1986; Takeuchi et al., 2005). However, simultaneous crystallization and degassing tends to concentrate vesicles into the interstices between crystals (Fig. 7.3(a)), and may enhance permeable connections among bubbles and lead to open-system degassing at lower-than-expected porosities. The consequent gas loss may cause vesicle collapse and magma densification, resulting in a dense, coherent plug that seals the conduit and may force underlying magma and gas to stagnate (Fig. 7.3; Hammer et al., 1999; Taddeucci et al., 2004; Cashman and McConnell, 2005; D'Oriano et al., 2005; Diller et al., 2006; Clarke et al., 2007).

Many of these same processes cause pressure to increase in the shallow conduit. Two mechanisms are particularly relevant: magma pressure may increase because of rheological stiffening of the ascending magma mixture (Sparks, 1997; Melnik and Sparks, 1999) and volatile pressure may increase because bubble growth is restricted by high-viscosity magma (Fig. 7.3(c); Sparks, 1978; Tait et al., 1989; Stix et al., 1997). Pressure in the shallow conduit increases until the plug is dislodged at pressures as high as 10 MPa, based on the strength of typical magmas. Disruption results in fragmentation of both the plug and underlying magma (Fig. 7.1(b)).

Magma overpressure develops in the shallow conduit because rheological stiffening concentrates most of the pressure gradient in the upper portions of the conduit (Sparks, 1997; Melnik and Sparks, 1999). The total pressure drop from magma chamber to surface drives flow upward from the chamber into the upper conduit between eruptions. For the one-dimensional steady-state, laminar flow case, and for slow ascent rates in a conduit of constant diameter, gravitational and viscous forces, which retard flow, must be balanced by the pressure gradient as follows:

$$\frac{dP}{dz} = -\rho g - \frac{32\eta u}{D^2}, \qquad (7.1)$$

where P is the magma pressure, z is the vertical coordinate increasing upward, ρ and η are the density and viscosity of the ascending magma, g is gravitational acceleration, D is the diameter of the conduit, and u is the ascent velocity of the magma (Melnik and Sparks, 1999). All parameters except D are allowed to vary with z. The first term on the right-hand side represents gravitational forces while the second term represents viscous forces according to assumptions associated with Poiseuille flow (steady, viscous flow through a pipe; Fox et al., 2006). The magma viscosity η is a function of composition, temperature, dissolved water content, crystal content, and bubble content (Chapters 4 and 5).

Because most of the increase in magma viscosity occurs near the surface (and because the increase in the viscous term outweighs the decrease in the gravitational term in Eq. (7.1)), most of the pressure drop must occur in the upper conduit precisely where viscosity increases most rapidly (Sparks, 1997). In other words, the pressure gradient dP/dz between the chamber and the upper portions of the conduit is gentle because the viscosity gradient is gentle in that region, whereas the pressure gradient in the top few hundred meters of the conduit is steep because the viscosity gradient is steep in that region. This complex pressure gradient allows high pressures to extend over a large portion of the conduit (Fig. 7.3(c)).

Bubble overpressure develops primarily because of viscosity-limited growth of bubbles for magma viscosities $> 10^7$ Pa s (Chapter 4; McBirney and Murase, 1971; Sparks, 1978; Proussevitch and Sahagian, 1996; Navon and Lyakhovsky, 1998). The gas phase expands in response to decreasing ambient magma pressure; however, bubble growth is resisted by surrounding high-viscosity melt, resulting in a trapped high-pressure gas phase. In general terms, bubble overpressure ΔP, defined as the difference between the pressure in the volatiles

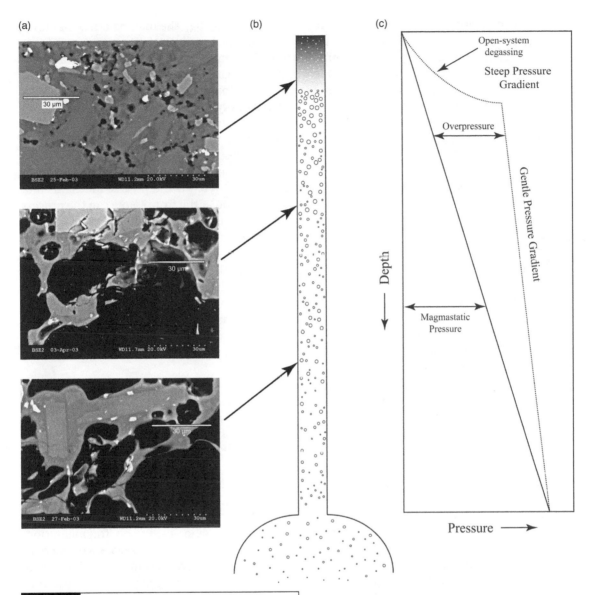

Figure 7.3 Schematic representation of plug formation, magma stalling, and pressure increase. (a) SEM images of vulcanian eruption products from the Soufrière Hills volcano (Clarke et al., 2007); crystals and glass appear as shades of gray, whereas bubbles appear black. Relative vertical source positions in the conduit for each sample are indicated by arrows pointed at (b), which is a schematic representation of conduit. (c) Pressure profile for the bubbly magma in both the ordinary static state (magma-static) and the overpressured state. Magma ascends, degasses, and crystallizes which can (1) increase magma viscosity promoting overpressure in the shallow conduit, and (2) enhance magma permeability, eventually forming a dense, sealing plug at the top of the conduit. After Clarke et al. (2007) and Mueller et al. (2008).

contained within the bubbles and the magma pressure, can be expressed by a form of the Rayleigh–Lamb equation:

$$\Delta P = 4\eta \frac{v_R}{r_b}, \qquad (7.2)$$

where v_R is the bubble growth rate and r_b is the bubble radius (Melnik et al., 2005; Mason et al., 2006). The basic concept of bubble overpressure is also relevant to magma fragmentation, as discussed in Section 7.3.

Some vulcanian eruptions are thought to initiate when groundwater comes into contact with magma or surrounding hot rock and, in turning to steam, increases pressure beneath the plug and dislodges it (Schmincke, 1977; Fagents and Wilson, 1993). This process is distinct from that of phreatomagmatic fragmentation. In the case of vulcanian eruptions, groundwater vaporization simply serves to dislodge the plug, which forces the underlying magma to fragment via a decompression-induced fragmentation wave.

7.3 | Decompression and fragmentation

Upon plug disruption, a decompression wave travels at the local sound speed into the conduit (Kieffer, 1981; Turcotte et al., 1990; Woods, 1995). The decompression wave is followed closely by a fragmentation wave that may travel more slowly through the bubbly magma (Melnik and Sparks, 2002; Spieler et al., 2004a; Fig. 7.1b). Behind the fragmentation front a mixture of expanding gases and freshly produced pyroclasts is projected upward and expelled from the conduit. The fragmentation wave is generally thought to fragment and quench the magma faster than dissolved gases can exsolve in response to the decompression. Therefore, to first order, exsolution of magmatic volatiles is assumed to be insignificant during fragmentation, and thus only volatiles that were already in the vapor phase prior to plug disruption participate in the eruption (Sparks, 1978; Woods, 1995; Melnik and Sparks, 2002; Clarke et al., 2002a,b, 2007). Eruptive products therefore preserve to some extent the pre-explosion state of magma vesiculation. However, up to several percent by volume bubble expansion can occur syn-explosively (i.e., during eruption), due to both bubble nucleation and growth (Melnik and Sparks, 2002; Formenti and Druitt, 2003; Giachetti et al., 2010) and thus erupted clasts may be more vesicular than the pre-eruptive magma. Furthermore, because the velocity of the fragmentation wave greatly exceeds magma ascent velocity (as the magma may have been stalled

by the sealing plug), the magma can be assumed to be stationary prior to fragmentation (Kieffer, 1981; Turcotte et al., 1990; Woods, 1995; Druitt et al., 2002; Clarke et al., 2002a,b). Therefore the fragmentation front meets magma with varying degrees of crystallinity, vesicularity, and viscosity, partly explaining the wide range of eruptive products found in vulcanian deposits.

Key questions about the specifics of fragmentation in vulcanian eruptions remain unanswered. What is the precise mechanism of fragmentation at the front? How can the process be quantified? What causes the fragmentation front to stop? Many models of fragmentation have been put forth (Chapter 4) that may help to answer these questions. The three most common and important are discussed here: (1) strain-rate threshold (Dingwell, 1996; Papale, 1999); (2) tensile stress threshold (Zhang, 1999; Melnik and Sparks, 2002); and (3) threshold volume fraction (Sparks, 1978).

7.3.1 Strain-rate threshold

High magma acceleration rates can generate large strain rates that may result in a rheological transition from ductile to brittle behavior (Dingwell, 1996, 2001; Papale, 1999). Essentially, under high acceleration or high strain-rate conditions, the magma relaxation or response time may exceed the time over which stresses are applied, causing the magma to behave in a brittle fashion, and leading to fragmentation. Magma relaxation time τ increases with increasing magma viscosity; therefore both increasing viscosity and increasing strain rate favor brittle fragmentation (Fig. 7.4). This concept can be expressed mathematically via the Maxwell relation

$$\frac{du}{dz} = \kappa\frac{1}{\tau} = \kappa\frac{G_\infty}{\eta} \text{ (Papale, 1999)} \qquad (7.3)$$

and

$$\frac{du}{dr} = \kappa\frac{1}{\tau} = \kappa\frac{G_\infty}{\eta} \text{ (Gonnermann and Manga, 2003),}$$

$$(7.4)$$

where du/dz is the spatial velocity gradient (or spatial acceleration) in the direction of flow

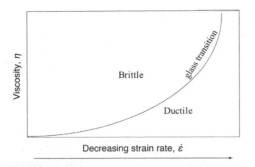

Figure 7.4 Magma fragments or fails brittlely under tension, at high viscosities and high strain rates. After Dingwell (1996).

(equivalent to elongational strain rate), du/dr is the spatial velocity gradient in the direction perpendicular to the direction of flow (equivalent to shear strain rate), κ is an empirical constant generally equal to 0.01 (Papale, 1999; Gonnermann and Manga, 2003), G_∞ is the elastic modulus of the magma, which ranges from 3 to 30 GPa depending on composition and temperature. The first equation (Eq. (7.3)) represents the fragmentation threshold associated with elongational strain, whereas the second (Eq. (7.4)) applies to shear strain. According to the elongational strain relationship (Eq. (7.3)) and over a wide range of eruptive conditions, the gas volume fraction (vesicularity) of the magma at fragmentation (for steady-state ascent) decreases linearly with the logarithm of the magma viscosity (Papale, 1999). This means that high-viscosity magmas can be fragmented at lower vesicularities than low-viscosity magmas. For example, magma with viscosity of 10^7 Pa s should fragment at ~85% vesicularity, whereas magma with viscosity of 10^8 Pa s should fragment at 60–65% vesicularity.

7.3.2 Tensile stress threshold

Brittle failure may also occur when a threshold tensile stress is exceeded; this may happen in a bubbly magma when the stress on bubble walls exceeds the tensile strength of the magma (Sparks, 1978; Alidibirov, 1994; Zhang, 1998, 1999). Fragmentation of this sort is thought to occur by disruption of bubbles near the free surface of a vesicular magma, where there can

be a significant pressure gradient between the ambient pressure and the bubble gas pressure (Sparks, 1978). This concept can be expressed quantitatively as follows (Zhang, 1998, 1999)

$$\frac{1+2\phi}{2(1-\phi)}\Delta P - P_{out} > \sigma_w, \qquad (7.5)$$

where ϕ is magma vesicularity or bubble volume fraction, P_{out} is the pressure on the outside surface of the bubble (or the bubble shell), and σ_w is the tensile strength of the bubble walls. This simplified relationship assumes that all bubbles are spherical, are of the same size and distributed uniformly throughout the melt such that the bubbly mixture is homogeneous, and that each bubble is surrounded by a spherical shell over which the stress in Eq. (7.5) is considered.

7.3.3 Bubble volume fraction

According to a series of numerical solutions (Sparks, 1978), bubbles should stop growing long before explosive fragmentation, primarily because of increasing melt viscosity. The solutions led to the concept that bubble volume fraction never exceeds 66–83% vesicularity. This range of vesicularity is often used as a fragmentation criterion in numerical models of magma ascent (e.g., Wilson and Head, 1981; Melnik and Sparks, 2002; Dufek and Bergantz, 2005; Koyaguchi, 2005).

7.3.4 Summary discussion: fragmentation

It is conceivable that each of these mechanisms contributes to fragmentation in vulcanian eruptions, depending on the specific circumstances. Upon plug disruption and decompression of the underlying bubbly magma, one or both of the strain-rate thresholds (Eqs. (7.3) and (7.4)) may be exceeded due to magma acceleration in response to the sudden decompression (increasing both du/dz and du/dr). The decompression also leads to rapid bubble expansion, which should produce very high strain rates within bubble walls such that bubble walls behave as a brittle solid and interbubble partitions are ruptured instead of stretched (Alidibirov, 1994). The second criterion may also apply because plug disruption suddenly exposes the underlying magma to a very low ambient pressure (P_{out}), which rapidly increases ΔP (Eq. (7.5)) and thus the magma exceeds the threshold overpressure criterion. Similarly, upon plug disruption, the volume-fraction criterion may be reached as bubbles expand in response to the decompression.

The effects of these three criteria on large-scale dynamics have been tested via a series of numerical solutions discussed in Section 7.9 (Melnik and Sparks, 2002). In general, the specific choice of fragmentation criterion does not significantly affect the resulting vent flux and velocity. However, strain-rate and overpressure criteria tend to produce pulsatory eruptions, whereas the volume-fraction criterion produces a single pulse. The numerical solutions also calculate the fragmentation front according to a particular criterion at each time step, and thus obtain the fragmentation front velocity as a function of time for each case. Maximum calculated fragmentation front velocities exceed 200 m s^{-1} for the first two criteria and are less than 50 m s^{-1} for the volume-fraction criterion.

7.3.5 Contributions from laboratory experiments

Laboratory experiments have been used to test fragmentation theories. The fragmentation threshold of a pressurized vesiculated magma under rapid (nearly instantaneous) decompression has been quantified as follows

$$\Delta P_{fr} = \frac{\sigma_m}{\phi}, \tag{7.6}$$

where σ_m is the effective tensile strength of the magma (~1 MPa; Spieler et al., 2004b; Mueller et al., 2008). This relationship holds for a wide range of magma compositions, crystallinities, and porosities. An interesting point to note is that for magmas with > 20% vesicularity, a sudden decompression of magnitude 5 MPa results in fragmentation, whereas low-vesicularity magmas ($\leq$ 10% vesicularity) may require a sudden pressure drop in excess of 15 to 30 MPa.

Permeability has been shown to relieve bubble pressure during propagation of a fragmentation wave (syn-fragmentation) by allowing high-pressure volatiles to escape via connected bubble pathways in magma below the fragmentation front. This effect serves to increase fragmentation threshold. Accordingly, Mueller et al. (2008) experimentally refine Eq. (7.6) as follows

$$\Delta P_{fr} = (1 + S)\frac{\sigma_m}{\phi}, \tag{7.7}$$

where $S = (a\sqrt{k})/\sigma_m$, a is a constant (8.21×10^5 MPa m^{-1}) and k is the permeability of the magma.

The corresponding propagation speed of the fragmentation front, as measured experimentally, generally falls between 2 and 70 m s^{-1} for magmas with 20–60% vesicularity and increases roughly linearly with the magnitude of the sudden decompression (ΔP; Spieler et al., 2004a). Maximum calculated fragmentation velocities (Section 7.3.4; Melnik and Sparks, 2002), greatly exceed experimental values, and the discrepancy is difficult to explain. Possible explanations include inappropriate model assumptions about fragmentation mechanisms, and simplifications required to perform fragmentation experiments.

Vulcanian explosions stop when the fragmentation front reaches magma that does not satisfy the appropriate fragmentation criteria. This may occur when the front reaches a depth in the conduit where: (1) the magma has insufficient vesicularity; (2) the bubbles are insufficiently overpressured; (3) the magma has sufficiently low viscosity allowing it to respond quickly to high strain rates; or (4) the front has weakened such that the pressure gradient is below the fragmentation threshold.

7.4 | Vent conditions

The subsequent acceleration of the pyroclastic mixture can be calculated by assuming that, upon decompression, available gas in the underlying magma expands as an ideal gas to atmospheric pressure and accelerates the pyroclasts to the same velocity as the gas itself (Self et al., 1979; Turcotte et al., 1990; Woods, 1995). Isothermal conditions can be assumed when heat is transferred from the clasts to the gas on timescales shorter than the duration of the explosion. This is thought to occur when most or all of the pyroclasts are < 1 mm in diameter (Woods, 1995). The resulting ascent of the pyroclastic mixture, assuming one-dimensional flow through a conduit of constant cross-sectional area, can be expressed in terms of the conservation of mass, momentum, and energy as follows (Turcotte et al., 1990; Woods, 1995):

$$\frac{\partial \rho_b}{\partial t} + \frac{\partial (\rho_b u)}{\partial z} = 0 \quad \text{(conservation of mass).} \quad (7.8)$$

The first term represents unsteadiness in the system, the second term is the inertial term (represented by the spatial gradient), and the zero on the right-hand side indicates that no mass enters or leaves the system. Here, ρ_b is the bulk density of the expanding mixture, t is time, z is the vertical coordinate, and u is the vertical velocity of the mixture.

$$\frac{\partial u}{\partial t} + u \frac{\partial u}{\partial z} = -\frac{1}{\rho_b} \frac{\partial P}{\partial z} - g - F \quad (7.9)$$

(conservation of momentum).

The first term on the left-hand side is the temporal acceleration of the flow, and the second is the remaining inertial term. The first term on the right-hand side is the pressure gradient (which drives the flow; flow moves in the direction of a negative pressure gradient), g is gravitational acceleration which retards flow, and F (approximated as 0.01; Wilson and Head, 1981) represents the frictional forces that also retard the flow. Typically, gravity and friction can be ignored because, during the first several seconds of the expansion, these two terms are small relative to the pressure, acceleration, and inertial terms.

$$\frac{dP}{\rho_b} + \left(c_m f (1-n) + c_g n\right) dT = 0 \quad (7.10)$$

(conservation of energy),

where f represents the proportion of solids in thermal equilibrium with the gas during decompression, n is the mass fraction of volatiles participating in the explosion, c_m is the specific heat of the magma ($\sim$1100 J kg^{-1} K^{-1}), c_g is the specific heat of the volatile phase at constant volume ($\sim$1000 J kg^{-1} K^{-1}).

Solutions to this system of equations (7.8)–(7.10) show that, in the isothermal limit (i.e., all particles in thermal equilibrium with the gas; $f = 1$), the corresponding velocity at the front of the expanding pyroclastic mixture can be approximated as follows (Woods, 1995):

$$u_f \cong \frac{1}{\gamma} (nRT\gamma)^{1/2} \ln \left(\frac{P_0}{P}\right), \quad (7.11)$$

where R is the gas constant for water vapor, T is the mixture temperature, P is the pressure of the mixture at any given time (atmospheric pressure P_a upon complete decompression), and P_0 is the initial (pre-explosion) pressure inside the conduit. The ratio of specific heats for the mixture, γ, approaches 1 for typical pyroclastic mixtures (Wohletz, 2001) and can be approximated as

$$\gamma = 1 + \frac{nR}{c_g n + c_m (1-n) f}. \quad (7.12)$$

Solutions for a wide range of initial gas mass fractions (0.01–0.1), initial temperatures (600–1400 K), and initial pressure ratios across the plug (0–100) reveal interesting trends (some solutions are shown in Fig. 7.5; Woods, 1995). Vent velocity increases with increasing initial pressure ratio and increasing volatile mass fraction. Mass flux per unit cross-sectional area of the conduit increases with increasing initial pressure ratio (due to an increase in vent velocity) and decreases with increasing volatile mass fraction (due to a decrease in mixture density).

As stated above, the isothermal assumption breaks down when a significant portion of the pyroclasts are > 1 mm, resulting in higher values of γ. Assuming dynamic equilibrium, the corresponding solution for flow-front velocity is (Woods, 1995)

$$u_f \cong \frac{2}{\gamma - 1} \left(\frac{nRT}{\gamma}\right)^{1/2} \left(\left(\frac{P_0}{P}\right)^{\frac{\gamma-1}{2}} - 1\right). \quad (7.13)$$

Mixture velocity decreases and mixture density increases with increasing levels of thermal disequilibrium (increasing proportion of large clasts; Fig. 7.5), pushing the system toward gravitational collapse and formation of pyroclastic density currents. These general trends

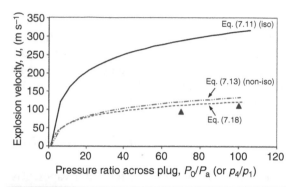

Figure 7.5 Calculated vent velocities for a range of initial pressure ratios, using four methods of calculation. The Woods (1995) isothermal solution (Eq. (7.11)) with 0.01 mass fraction volatiles at 1000 K (solid black line) significantly exceeds velocities calculated using shock-tube equations for pseudogas mixtures (Eq. (7.18); dashed line, where fluid properties in the high-pressure, pre-explosion conduit, region 4, are calculated to be $\gamma_4 = 1.004$ and $c_4 = 30$ m s^{-1}), and numerical solutions to full Navier–Stokes, multiphase equations (triangles; Clarke et al., 2002a). The non-isothermal solution (Eq. (7.13); Woods, 1995) for 0.01 mass fraction volatiles, initial temperature of 1000 K, and 3% of particles < 1 mm (dashed-dotted line) is very similar to the shock-tube and multiphase solutions.

are consistent with earlier solutions presented in Self et al. (1979). Note that this effect is independent of clast settling velocity, which is not considered here. In this regard, it is important to keep in mind that in reality large clasts may settle out of the rising mixture, contributing to the formation of pyroclastic density currents and/or reducing the density of the remaining mixture, allowing it to rise buoyantly.

Other approaches (e.g., Ishihara, 1985; Chojnicki et al., 2006) treat the system as a shock tube and calculate the velocity of the expanding mixture according to equations for shock-driven flow (Saad, 1985). In some cases, the mixture is treated as a thermally and dynamically perfectly coupled fluid, termed a pseudogas. The sound speed c and ratio of specific heats γ of a pseudogas are significantly less than values for the gas alone (Kieffer, 1981; Dobran et al., 1993; Wohletz, 2001). Shock-tube/ pseudogas methods for a given pre-explosion pressure ratio tend to predict lower vent velocities relative to solutions of Turcotte et al. (1990) and Woods (1995). Multiphase models that account for heat and momentum exchange

between phases have also been used to calculate vent velocities (Clarke et al., 2002a); multiphase solutions are reasonably consistent with the pseudogas/shock tube relationships. Details of the shock-tube relations are presented in Section 7.5 and the various model solutions are compared in Figure 7.5.

The vent flux associated with vulcanian eruptions is thought to be highly impulsive. The initial phase accelerates rapidly and quickly wanes. When the fragmentation wave reaches unfragmentable magma, the vent flux decays to near zero, although sustained gas exhalations following the main pulse have been documented for several eruptions (Hoblitt et al., 1996; Druitt et al., 2002). Seismic (Druitt et al., 2002) and computational evidence (Clarke et al., 2002a) of the impulsive nature of the vent flux is shown in Figure 7.6.

7.5 | Shock waves

Leading shock waves are a consequence of the pressure discontinuity between the high-pressure, gas-rich magma in the conduit and the atmosphere. They propagate ahead of and, according to shock-tube theory, drive motion of the pyroclastic mixture. Leading shock waves are sometimes visible because they condense atmospheric water vapor, allowing their properties to be documented and measured (Nairn, 1976; Ishihara, 1985; Morrissey and Chouet, 1997). Shock waves are characterized by a sharp increase in atmospheric pressure, followed by a dip to a pressure that is less than ambient, producing a characteristic N-shaped wave, which has been documented by stationary pressure sensors (Morrissey and Chouet, 1997). The amplitude of the wave decays nonlinearly with decreasing initial pressure ratio across the plug and with increasing distance from source.

For the simple case of adiabatic and inviscid flow of an ideal gas, shock characteristics have been derived in terms of the pre-explosion pressure ratio across the plug (p_4/p_1) by solving the conservation of mass, momentum, and energy equations over a control volume that encompasses and travels along with the shock wave

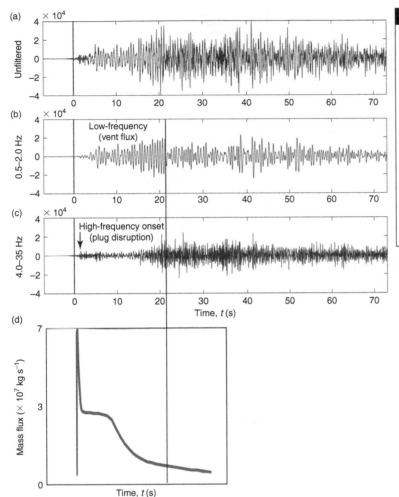

Figure 7.6 Seismic signal associated with the 7 August 1997 vulcanian eruption at the Soufrière Hills volcano, Montserrat. (a) Unfiltered signal. (b) Low-frequency portion of the signal (0.5–2 Hz). (c) High-frequency portion (> 4 Hz). (d) Mass flux vs. time calculated by multiphase models of vulcanian eruptions from Clarke et al. (2002a). The high-frequency onset (c) represents rock fracture associated with plug disruption. The first 20 s of low-frequency signal (bracketed by vertical lines) represents the main phase of mass flux. After Druitt et al. (2002).

(Saad, 1985). In Equations (7.14)–(7.18), subscript 1 represents atmospheric conditions, 2 represents conditions in the region behind the shock, but outside the conduit, and 4 represents properties in the pre-explosion, high-pressure conduit (Regions 1, 2 and 4 in Fig. 7.1). Assuming adiabatic and inviscid flow, the shock strength, or pressure ratio across the shock p_2/p_1 (Fig. 7.1), can be expressed in terms of the shock Mach number, $M_s = v_s/c$, where v_s is the shock velocity and c is the speed of sound in the atmosphere ($\sim$330 m s^{-1} for a standard terrestrial atmosphere; Saad, 1985):

$$\frac{p_2}{p_1} = 1 + \frac{2\gamma_1}{\gamma_1 + 1}(M_s^2 - 1), \qquad (7.14)$$

where γ_1 is the ratio of specific heats for the fluid into which the shock is propagating, in this case, the atmosphere. In turn, the initial pressure ratio across the plug p_4/p_1 can be expressed in terms of shock strength

$$\frac{p_4}{p_1} = \frac{p_2}{p_1}\left[1 - \frac{(\gamma_4 - 1)\left(\dfrac{c_1}{c_4}\right)\left(\dfrac{p_2}{p_1} - 1\right)}{\sqrt{2\gamma_1}\sqrt{2\gamma_1 + (\gamma_1 + 1)\left(\dfrac{p_2}{p_1} - 1\right)}}\right]^{-\frac{2\gamma_4}{(\gamma_4 - 1)}},$$

(7.15)

and in terms of M_s

$$\frac{p_4}{p_1} = \frac{\gamma_1 - 1}{\gamma_1 + 1}\left(\frac{2\gamma_1}{\gamma_1 - 1}M_s^2 - 1\right)\left[1 - \frac{\dfrac{\gamma_4 - 1}{\gamma_4 + 1}\left(\dfrac{c_1}{c_4}\right)(M_s^2 - 1)}{M_s}\right]^{-\frac{2\gamma_4}{(\gamma_4 - 1)}}.$$

(7.16)

The sound speed for a pseudogas can be approximated as

$$c_4 = \sqrt{\frac{RT}{n}\left(n + (1-n)\left(\frac{\rho_g}{\rho_m}\right)\right)}, \qquad (7.17)$$

where ρ_g and ρ_m are the densities of water vapor at high pressure and the magma particles, respectively (Dobran *et al.*, 1993). Equation (7.12) is an appropriate expression for the ratio of specific heats γ_4 for a pseudogas ($f=1$), while $\gamma_1 = 1.4$ for the atmospheric gas. Equations (7.14)–(7.16) must be solved by trial and error to calculate pressure ratios (see Exercise 7.1 at end of chapter).

Recent experimental work by Chojnicki *et al.* (2006) shows that the above relationships (Eqs. (7.14)–(7.17)), used with pseudogas fluid properties, under-predict one-dimensional laboratory shock strength and speed. This is in part due to the fact that, generally speaking, the shock wave is formed by the expanding gas alone, and particles hinder gas expansion, as well as hinder motion via an interphase drag force, rather than change the properties of the gas phase entirely. These findings are consistent with earlier work in which shock characteristics were calculated using solutions to the full Navier–Stokes equations for multiphase mixtures of gas and solid particles (Morrissey and Chouet, 1997). Comparisons are presented in Figure 7.7. The presence of particles and interphase drag reduce shock strength and velocity relative to shock waves created by an ideal gas alone; the magnitude of this reduction is independent of particle size (Morrissey and Chouet, 1997; Chojnicki *et al.*, 2006). Shock speed increases with increasing gas volume fraction in the magma beneath the plug.

Shock waves generated at Sakurajima volcano moved with velocities of 440–500 m s^{-1} (Ishihara, 1985); experiments (Chojnicki *et al.*, 2006) suggest corresponding pre-explosion conduit pressures of 1.5–10 MPa, which compares reasonably well with values of 0.2–5 MPa calculated from shock wave strength (Morrissey and Chouet, 1997).

The velocity of the fluid expanding behind the shock is

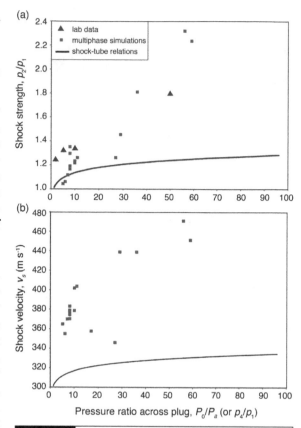

Figure 7.7 (a) Shock strength and (b) leading shock propagation velocity calculated using 1D pseudogas shock-tube relationships (Eqs. (7.14)–(7.16); solid lines), using axi-symmetric multiphase simulations (squares; Morrissey and Chouet, 1997), and measured in the laboratory under 1D conditions (triangles; Chojnicki *et al.*, 2006).

$$u_f = \frac{c_1}{\gamma_1}\left(\frac{p_2}{p_1} - 1\right)\left[\frac{\frac{2\gamma_1}{\gamma_1 + 1}}{\frac{p_2}{p_1} + \frac{\gamma_1 - 1}{\gamma_1 + 1}}\right]^{1/2}. \qquad (7.18)$$

Values calculated from Eq. (7.18) are compared against other models in Figure 7.5. Discussion of model differences is left to Exercise 7.2 (see end of chapter).

7.6 | Compressibility

Because vulcanian eruptions initiate via a sudden decompression of a high-pressure bubbly magma, the resulting pyroclastic mixture may enter the atmosphere in an overpressured state. This means that compressibility may be critical to understanding the subsequent propagation of the pyroclastic mixture. The combination of field observations from the Soufrière Hills volcano, Montserrat, British West Indies, and multiphase models for gas–particle mixtures (similar to that used in Morrissey and Chouet, 1997) shows that pressure inside the volcanic jets adjusts to atmospheric pressure within 50 m from the vent (less than two vent diameters) and that initially supersonic flows decelerate to subsonic conditions over similar distances (Formenti *et al.*, 2003). This finding is reasonably consistent with other field observations (Hoblitt, 1986; Sparks *et al.*, 1997; Clarke *et al.*, 2002a; Druitt *et al.*, 2002; Formenti *et al.*, 2003). To further address this issue, we can apply, with some assumptions, relationships derived from experiments on steady, supersonic, overpressured, ideal-gas jets (Yüceil and Ötügen, 2002) to the problem of vulcanian eruptions. For initial conduit pressures ranging from 1 to 10 MPa, fluid properties of volcanic gas–particle mixtures, and conduit diameters < 100 m, we find that vulcanian jets should reach atmospheric pressure by no more than 700 m from source in the most extreme cases, and by no more than 100 m from source in typical eruptions. During the decompression to atmospheric pressure, the jet expands in width by a factor of 2–8. Over this same distance, the jet accelerates, but only by as much as 10%. Simultaneously, the density of the mixture decreases significantly, by up to a factor of 70.

7.7 | Pyroclastic phase: vertical ascent

Key fluid dynamics principles that should be considered when describing the subsequent pyroclastic phase (beyond the vent) of vulcanian eruptions are Reynolds number (Re), reduced gravity (g'), Richardson number (Ri), total injected buoyancy (B), and total injected momentum (M).

Reynolds number, defined as the ratio of inertial to viscous forces, can be written for vulcanian jets as

$$\mathrm{Re} = \frac{\rho_b u L}{\eta_b}, \qquad (7.19)$$

where u is flow velocity, L is a characteristic length scale (usually the vent diameter), and η_b is the bulk dynamic viscosity of the pyroclastic mixture. Re values for typical vulcanian eruptions range from 10^5 to 10^7 and even as high as 10^8 for some large diameter jets; u can be as high as 400 m s^{-1}; L is on the order of tens of meters, although some vents can be hundreds of meters in diameter; and ρ_b can range from less than 1 kg m^{-3} (less than atmospheric density, and therefore buoyant) to as high as 100 kg m^{-3}; corresponding values of η_b are 10^{-4}–10^{-1} Pa s (Wohletz, 2001). Although some characteristics, such as the length scales of the smallest turbulent eddies, depend on Re, nearly all vulcanian eruptions are well within the turbulent regime, and thus inertial forces far outweigh viscous forces, and in general flow can be assumed to be inviscid, except near boundaries.

Richardson number (Ri) is defined as the ratio of buoyancy forces to inertial forces and can be expressed as

$$\mathrm{Ri} = \frac{g' L}{u^2}, \qquad (7.20)$$

where the reduced gravity, $g' = g(\rho_b - \rho_a)/\rho_a$, and ρ_a is the ambient (atmospheric) density at the vent. Ri expresses whether a vulcanian eruption is dominated by positive or negative buoyancy forces (large Ri), by momentum forces (small Ri), or whether both momentum and buoyancy are important (Ri ~ 1).

Total buoyancy injected B can be expressed as $B = g'V$, where V is the volume of fluid injected (time-integrated volume flux), and total momentum injected M can be expressed as $M = u_0 V$, where u_0 is the initial velocity at the vent.

Ascent of the pyroclastic mixture associated with a vulcanian eruption may be described by one of three simplified models: they may behave as *thermals* (short releases of a buoyant fluid), *impulsive jets* (short injections of fluid imparting

~400 m

Figure 7.8 (a) Recent eruption of Fuego volcano (Guatemala) showing release of a buoyant eruption cloud (thermal). (b) Streak lines illustrating velocity vectors for a laboratory experiment of an isolated thermal, showing upflow in center of thermal and downflow at the edges (modified from Turner, 1973). See color plates section.

initial momentum, but no buoyancy), or *forced thermals* (short injections of a buoyant fluid with both initial momentum and buoyancy). Note that the term forced thermals introduced here is modeled after the terminology of Morton (1959) which refers to a steady, sustained release of both buoyancy and momentum as a forced plume.

To address motion above the vent, the governing one-dimensional equations can be simplified according to the Boussinesq approximation, where density differences between fluids are ignored except in the buoyancy terms. In the one-dimensional case, we also ignore lateral variations in velocity and fluid properties. The resulting conservation of mass, momentum, and buoyancy are (summarized in Turner, 1973)

$$\frac{d}{dz}\left(r_p^2 u\right) = 2\varepsilon r_p u \quad \text{(conservation of mass)}, \quad (7.21)$$

$$\frac{d}{dz}\left(r_p^2 u^2\right) = r_p^2 g' \quad \text{(conservation of momentum)}, \quad (7.22)$$

$$\frac{d}{dz}\left(r_p^2 u g'\right) = -r_p^2 u N^2 \quad \text{(conservation of buoyancy)}. \quad (7.23)$$

The right-hand side of the conservation of mass is non-zero because mass is turbulently

entrained into the pyroclastic mixture as it propagates, where ε is the entrainment coefficient, and r_p is the radius of the pyroclastic jet. The pressure term is neglected in the momentum equation because it is small relative to the gravity term. Equation (7.23) simply states that buoyancy flux is conserved, while the right-hand side goes to zero in a uniform ambient environment. For a stably stratified environment, such as the atmosphere, N is the local Brunt–Väisälä frequency, which is defined as the natural oscillation frequency of a parcel of fluid of density ρ_0 displaced from its equilibrium position, quantified as $N^2 = (-g/\rho_0)(d\rho_a/dz)$. Similarity solutions (for thermals; Turner, 1973) and dimensional analysis (impulsive jets and forced thermals; Clarke et al., 2009) lead to the generalized relationships for flow-front velocity vs. time described below.

Thermals are defined as sudden (nearly instantaneous) releases of a finite volume of buoyant fluid, such that the timescale of release is much less than the timescale of flow propagation. Vertical motion of a thermal is controlled entirely by the total buoyancy injected B. Thermals are the unsteady/impulsive equivalent to *plumes*, which are steady, continuous releases of a buoyant fluid. Thermals commonly have spherical morphology and resemble a spherical vortex in which flow is non-uniform, with upflow in the center and downflow at the edges (Turner, 1973; Fig. 7.8). The vertical velocity of a thermal u_B varies as z^{-1}, where z is the height above the source, and as $t^{-1/2}$, where t is the time after the release (Fig. 7.9). Specific eruptions that behave as thermals are some of the ongoing eruptions

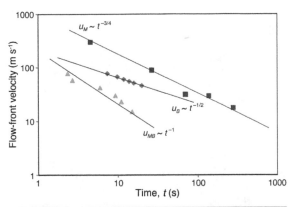

Figure 7.9 Vertical flow front velocity vs. time for three different vulcanian eruptions. The 1975 eruption of Ngauruhoe (New Zealand) was dominated by momentum M (impulsive jet) as shown by the relationship $u_M \sim t^{-3/4}$ (squares; Self et al., 1979). The 1982 eruption of Sakurajima (Japan) was dominated by buoyancy B (thermal) as shown by the relationship $u_B \sim t^{-1/2}$ (diamonds; Ishihara, 1985). The 7 August 1997 eruption of the Soufrière Hills volcano (Montserrat, British West Indies), was controlled by both momentum M and buoyancy B (forced thermal), as shown by the relationship $u_{MB} \sim t^{-1}$ triangles; Clarke et al., 2002a).

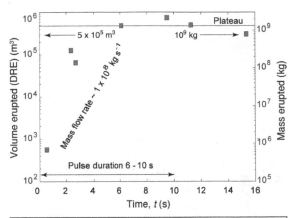

Figure 7.10 Total volume (DRE = 2500 kg m^{-3}) and total mass erupted vs. time (Eq. (7.24)) for the 7 August 1997 eruption of the Soufrière Hills volcano, Montserrat, calculated from the experimentally derived relationship of Clarke et al. (2009), which relates flow-front velocity to source conditions total momentum (M) and total buoyancy (B) injected. A value of 20 kg m^{-3} was assumed for ρ_b (Formenti et al., 2003). In this case the calculated eruption rate is 10^8 kg s^{-1}, lasting approximately 10 s, resulting in a total of 10^9 kg erupted. The calculated values are reasonably consistent with independent estimates of Druitt et al. (2002). (Modified from Clarke et al., 2009.)

of Fuego and Santiaguito volcanoes (Yamamoto et al., 2008), both in Guatemala, as well as a 1982 eruption of Sakurajima volcano, Japan (Fig. 7.9).

Analogous short injections of momentum only (i.e., impulsive jets) can be used to describe other vulcanian eruptions. In such cases, throughout most of the flow, the vertical velocity of the flow front u_M scales as $t^{-3/4}$ (Clarke et al., 2009). This scaling breaks down far from source, where buoyancy may become significant due to entrainment and heating of the ambient air during propagation. The well-documented eruptions of Ngauruhoe appear to be purely momentum-driven, as can be seen in Figure 7.9. Still other eruptions are best explained by a short injection of both momentum and buoyancy (i.e., forced thermals). These eruptions exhibit more rapid deceleration (u_{MB} scales as t^{-1}) than either a purely buoyant thermal (u_B scales as $t^{-1/2}$) or a purely momentum-driven unsteady jet (u_M scales as $t^{-3/4}$) (Clarke et al., 2009). Examples of this third type include: a February 1990 eruption of Lascar volcano in Chile (Sparks et al., 1997), a July 1980 eruption of Mount St. Helens in the western US (Hoblitt, 1986), and two eruptions of

the Soufrière Hills volcano (Fig. 7.9; Clarke et al., 2002a; Formenti et al., 2003).

For the mixed momentum and buoyancy (forced thermal) case, consolidation of experimental data reveals a relationship between two nondimensional terms, which upon rearranging allows one to calculate the total erupted mass as follows (Clarke et al., 2009):

$$Mass = \frac{\rho_b u_{MB}^4 t^4 g'}{u_0^2}. \qquad (7.24)$$

An example application of this relationship to field observations is presented in Figure 7.10. Vertical flow-front velocity vs. time is reasonably easy to measure using a video camera (Sparks and Wilson, 1982). However, the primary weakness of Eq. (7.24) is the difficulty in constraining ρ_b and the corresponding g', as pyroclastic mixtures can exit a vent with a wide range of solid particle concentrations; mixture densities can range over two orders of magnitude (< 1–100 kg m^{-3}).

The three flow categories described above are thought to have different atmospheric

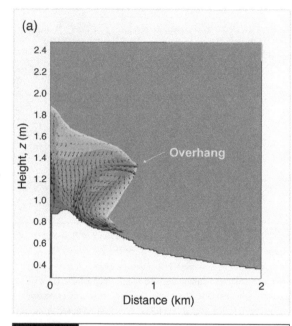

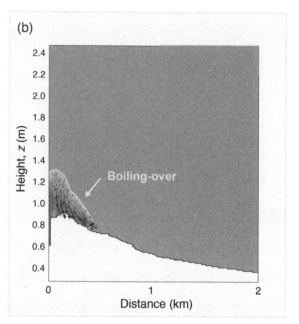

Figure 7.11 Pyroclastic density currents are generated when the pyroclastic mixture never becomes buoyant and collapses due to gravity. The collapse may initiate (a) hundreds of meters above the vent or (b) a few meters above the vent (boil-over), depending on the initial momentum (controlled by initial pressure and volatile content) and entrainment characteristics. See color plates section.

entrainment coefficients ε, as is the case for the well-documented steady equivalents (Chapter 8), which may partially explain their different deceleration patterns. Typical entrainment coefficients for steady, axisymmetric, momentum-driven jets are 0.065–0.116 (Turner, 1973; Sparks *et al.*, 1997), generally smaller than the entrainment coefficients for steady, axisymmetric plumes (0.12) and thermals (0.25). Entrainment constants for impulsive jets and forced thermals are not well documented. Furthermore, entrainment may be a function of local Richardson number (Kaminski *et al.*, 2005), overpressure, distance from vent (Solovitz and Mastin, 2009), and vent geometry, and thus coefficients may vary as the flow evolves, making the assumption of a constant ε an oversimplification. These complexities are important because column collapse is in part controlled by entrainment in the near-vent region.

For steady plume-forming eruptions, column collapse and pyroclastic density currents (Chapter 10) are favored for large vent radii, low vent velocities, high solid mass fractions at the vent (Wilson *et al.*, 1978, 1980; Chapter 8), and for large particles (Wohletz and Valentine, 1990; Woods, 1995). Depending on vent conditions and entrainment characteristics, vulcanian eruptions may also collapse and produce pyroclastic density currents (Fig. 7.11), as at Mount St. Helens in 1980 (Hoblitt, 1986), and at the Soufrière Hills volcano, Montserrat, in 1997 (Druitt *et al.*, 2002; Clarke *et al.*, 2002a).

Although the dynamics of pyroclastic ejection have been necessarily simplified to calculate key parameters such as exit velocity, mass flux, column height, and collapse characteristics, careful observations have led to documentation of interesting complex phenomenology. Detailed analysis of the 1997 vulcanian eruptions of the Soufrière Hills volcano reveals that explosion initiation involved multiple, individual finger-jets that have distinct characteristics and progressively increasing velocities (Formenti *et al.*, 2003). These observations indicate that there were pre-explosive gradients in volatiles in the shallow conduit, and that fragmentation, especially in the initial stages, was heterogeneous in time and space. Corresponding calculated entrainment coefficients are very low

(0.01) for individual jets, and this inefficient entrainment is thought to have contributed to collapse and pyroclastic flow formation.

7.8 | Ballistic analysis

Ballistic blocks or bombs are typically associated with vulcanian eruptions and are thought to represent the disrupted sealing plug. Ballistic block fields have been documented carefully for the 1977 phreatomagmatic eruptions of Ukinrek Maars, Alaska (Self *et al.*, 1980), the 1992 subplinian eruptions of Crater Peak Vent, Mount Spurr volcano, Alaska (Waitt *et al.*, 1995), the 1997 vulcanian eruptions of the Soufrière Hills volcano (Druitt *et al.*, 2002), and the 1999 vulcanian eruptions of Guagua Pichincha volcano, Ecuador (Wright *et al.*, 2007), among others. Blocks on the order of a half meter in diameter can be launched to 3 km from the vent, and in some cases smaller blocks can reach > 6 km from the vent. In general, ballistic size decreases with distance from the vent. However, in some cases the opposite is true because of complex drag interactions between the blocks and the expanding pyroclastic mixture and the surrounding air (Self *et al.*, 1980; Waitt *et al.*, 1995; de' Michieli Vitturi *et al.*, 2010). Their sizes, trajectories, ranges, and textures can be used to infer pre-explosion or vent conditions.

Ballistic trajectories can be calculated by solving Newton's second law of motion, $F = ma$ (force equals mass times acceleration) for a known launch velocity and angle. In the horizontal (x) direction, the ballistic clast is subject to drag forces alone:

$$\frac{dv_x}{dt} = \frac{-v_x \rho_a v' A C_d}{2m}. \tag{7.25}$$

Horizontal velocity v_x is equivalent to $v \cos \theta$, where v is the total velocity of the clast and θ is the trajectory angle relative to the horizontal; A is the cross-sectional area of the clast; v' is the clast velocity relative to the motion of the surrounding fluid (= $v - w$); and C_d is the drag coefficient (a function of particle Re, where clast diameter is the characteristic length scale in Eq. (7.19)) and may vary between 0.06 and 1 for typical volcanic conditions (Fagents and Wilson, 1993; Waitt *et al.*, 1995; Mastin, 2001).

In the vertical (z) direction, the ballistic block is subject to both drag and gravitational forces:

$$\frac{dv_z}{dt} = \frac{-v_z \rho_a v' A C_d}{2m} - g \frac{\rho_r - \rho_a}{\rho_r}. \tag{7.26}$$

The second term on the right accounts for the buoyancy of the clast in air; however, in most cases, ballistic block density ρ_r greatly exceeds atmospheric density and the second term goes to $-g$. Vertical velocity v_z can be expressed as $v \sin \theta$.

Solutions to Eqs. (7.25) and (7.26) can be used to interpret observed ballistic trajectories (range, size, and perhaps launch angle if available) in terms of initial vent velocity (Wilson, 1972; Fagents and Wilson, 1993; Waitt *et al.*, 1995; Mastin, 2001) and sometimes in terms of initial vent pressure (Ishihara, 1985). Because of the high velocity of the ejecting mixture surrounding the ballistic, one may ignore drag in the region immediately around the vent because the ejecting fluid is moving at (nearly) the same velocity as the ballistic clast ($v' = 0$) (Fagents and Wilson, 1993); or in some cases, the accelerating effect of a surrounding pyroclastic mixture must be considered (de' Michieli Vitturi *et al.*, 2010). Incorrect values of launch angle or inappropriate characterization of drag can lead to significant errors in estimates of initial ballistic velocity.

7.9 | Transitions in eruption style or scale

Vulcanian eruptions may transform into sustained, quasi-steady explosive eruptions or may end suddenly to be replaced by effusive dome-building. As stated in Section 7.3, it is reasonable to assume that gas does not diffuse from the melt into bubbles during propagation of the fragmentation front into the conduit. However, when volatile diffusion is fast and efficient, such as for high bubble number densities and large diffusion coefficients, this assumption becomes invalid. For these cases, numerical solutions to complex systems of equations have been used to explore the effects of syn-explosion gas diffusion on eruption characteristics (Melnik and Sparks, 2002; Mason et al., 2006). Results suggest that indeed some amount of syn-explosion gas diffusion may extend the scale and duration of a given vulcanian eruption, may explain repeated, pulsatory explosions, and may result in a transition in eruptive style, leading to a quasi-steady subplinian eruption.

The role of syn-fragmentation gas exsolution can be determined by examining the Peclet number (Pe), which is the ratio of the timescale for gas exsolution to the timescale of fragmentation (Navon and Lyakhovsky, 1998; Melnik and Sparks, 2002; Proussevitch and Sahagian, 2005; Koyaguchi and Mitani, 2005; Mason et al., 2006). The timescale of exsolution is determined by the rate of diffusion of a volatile in the melt of interest and the length scale in question. The timescale of fragmentation is a function of the distance between the decompression wave and the fragmentation front, and the velocity of the fragmentation front relative to the (possibly) ascending unfragmented magma; this relative velocity is simply the difference between the two velocity vectors (Melnik and Sparks, 2002). When Pe ≫ 1, syn-fragmentation gas exsolution can be ignored; when Pe ≪ 1, it must be considered. The degree of gas exsolution may be quantified according to a diffusion parameter that describes the intensity of gas exsolution during fragmentation; this parameter is a function of bubble number density per unit volume of magma and the diffusion coefficient of water in the melt (Melnik and Sparks, 2002; Mason et al., 2006). End-members are zero gas exsolution (the simplifying assumption in Section 7.3) and the equilibrium case in which gas exsolution instantaneously responds to decompression.

The system can be represented by one-dimensional conservation equations that account for the difference in pressure between the bubbles and the magma (Eq. (7.2)), allow gas exsolution during fragmentation (depending on Pe), and allow underlying magma to ascend in response to the decompression associated with conduit evacuation (Melnik and Sparks, 2002; Melnik et al., 2005; Mason et al.,

2006). Corresponding solutions show that no gas exsolution during fragmentation results in a single explosive pulse (until magma can slowly ascend in response to conduit evacuation and prepare for another explosion), intermediate syn-fragmentation gas exsolution (the disequilibrium case) leads to pulsatory eruptions, and the equilibrium case results in sustained eruption (Melnik and Sparks, 2002). Greater total volatile content for a given set of conditions pushes the system toward multiple pulses or quasi-steady behavior (Mason et al., 2006). The addition of crystals tends to increase the depth that the fragmentation front reaches for a single pulse, because of increased viscosity. However magmas with low crystal fractions (low viscosities) are more likely to stabilize into a steady-state eruption because the magma will more easily ascend to meet and feed the fragmentation wave (Mason et al., 2006).

Transition from periodic explosions to effusive activity may occur when permeability, rather than fragmentation, develops either throughout the magma (Mueller et al., 2008) or along conduit walls (Gonnermann and Manga, 2003; Tuffen et al., 2003), and therefore suppresses explosion. A particular sequence of periodic explosive eruptions at the Soufrière Hills volcano in 2003 is thought to have ended in this way via permeable escape of gas along conduit walls (Edmonds and Herd, 2007).

7.10 | Future directions

Vulcanian eruptions are very common and may present significant hazards to surrounding populations. Their impulsive nature, unsteady dynamics, as well as the potential importance of compressible fluid flow, makes theoretical understanding of their characteristics difficult. Furthermore, factors contributing to setting up conditions sufficient for explosion initiation remain less than fully understood, primarily because complex relationships exist among magma degassing and crystallization, magma permeability, and pressurization of the shallow conduit. Vulcanian eruptions may transform into sustained explosive volcanic eruptions, or may be replaced by effusion, offering an interesting focus of study for those interested in transitions in eruption style. Future observations and measurements of dynamic conditions will improve understanding. Useful measurements include temporal evolution of plume and pyroclastic flow morphology and flow-front velocity using simple stationary video cameras (e.g., Sparks and Wilson, 1982). These values can be compared to theoretical understanding of controls on flow propagation (thermal, impulsive jet, forced thermal), and be used to calculate total mass erupted (Clarke *et al.*, 2009) and entrainment coefficients (Formenti *et al.*, 2003). Profiles of plume and pyroclastic flow temperature, velocity, and particle concentration are also useful data. Currently such measurements are very difficult or impossible to make; however new ground-based imaging techniques (e.g., Doppler radar, thermal-IR, and UV cameras; Yamamoto *et al.*, 2008) may make them possible in the future, offering a potentially fruitful pathway for development and testing of new measurement techniques.

7.11 | Summary

- Vulcanian eruptions are characteristically short-lived and are fed by highly unsteady vent flux, distinguishing them from subplinian eruptions.
- Vulcanian eruptions represent a transitional eruption style, and may occur before or after long-lived explosive eruptions or extended periods of dome growth.
- Pyroclast grain size and dispersal area for vulcanian eruptions do not follow a particular trend; vulcanian deposit characteristics overlap with many other styles of explosive eruption.
- Plug formation and disruption are critical to generating vulcanian eruptions.
- During the main phase of a vulcanian eruption, a fragmentation front progressively accesses deeper and deeper magma, which in part explains the varied nature of eruption products.

- Leading shock waves may propagate through the atmosphere ahead of the eruption front.
- Ejection of ballistic blocks, which represent the former conduit plug, often characterizes the initial phases of vulcanian eruptions.
- Quantitative characteristics of shock waves, explosion flow front, and ballistic blocks and their trajectories can be used to constrain vent fluxes and pre-eruption conduit conditions.

7.12 | Notation

a constant equal to 8.21×10^5 (MPa m^{-1})

A cross-sectional area of ballistic block (m^2)

B total buoyancy injected/erupted, $= g'V$ (m^4 s^{-2})

c speed of sound in the atmosphere (m s^{-1})

c_1 speed of sound in fluid in low-pressure region (m s^{-1})

c_4 speed of sound in fluid in high-pressure region (m s^{-1})

c_m specific heat of the magma (J kg^{-1} K^{-1})

c_g specific heat of the volatile phase (J kg^{-1} K^{-1})

C_d drag coefficient for ballistic block

d diameter of plume/jet at the vent (m)

D conduit diameter (m)

f the proportion of solids in thermal equilibrium with the gas during eruption

F frictional forces per unit mass (N kg^{-1})

g gravitational acceleration (m s^{-2})

g' reduced gravity, $= g(\rho_b - \rho_a)/\rho_a$ (m s^{-2})

G_∞ elastic modulus of the magma (Pa)

k permeability of a bubbly magma (m^2)

m mass of ballistic block (kg)

L characteristic length scale (m)

M total momentum injected/erupted, $= v_0V$ (m^4 s^{-1})

$Mass$ total mass erupted (kg)

n mass fraction of volatiles in the erupting mixture

N Brunt–Väisälä (buoyancy) frequency (s^{-1})

p_2/p_1 shock strength, ratio of pressures behind and ahead of shock

p_4/p_1 ratio of pressures in high-pressure (pre-explosion conduit) region to low-pressure region above plug

P pressure of magma or pyroclastic mixture (Pa)

P_0 initial (pre-explosion) pressure inside the conduit (Pa)

P_a atmospheric pressure (Pa)

P_{out} pressure on outer surface of bubble shell (Pa)

ΔP pressure difference between bubble volatiles and magma (overpressure; Pa)

ΔP_{fr} fragmentation threshold for a pressurized magma (Pa)

r radial coordinate (m)

r_b bubble radius (m)

r_p plume/jet radius (m)

R gas constant for water vapor (461.5 J kg^{-1} K^{-1})

S parameter in Eq. (7.7), $= (a\sqrt{k})/\sigma_m$

t time (s)

T temperature (K)

u vertical velocity (m s^{-1})

u_0 initial velocity at the vent (m s^{-1})

u_B vertical flow front velocity for a thermal (m s^{-1})

u_f flow-front velocity of expanding gas–particle mixture (m s^{-1})

u_M vertical flow-front velocity for a short-lived momentum-driven jet (m s^{-1})

u_{MB} vertical flow-front velocity for a short-lived momentum- and buoyancy-driven forced thermal (m s^{-1})

v total velocity of ballistic block $= (v_x^2 + v_z^2)^{1/2}$ (m s^{-1})

v' velocity of ballistic block relative to ambient fluid motion (m s^{-1})

v_R bubble radial growth rate (m s^{-1})

v_s shock velocity (m s^{-1})

v_x horizontal velocity of ballistic block $= v \cos\theta$ (m s^{-1})

v_z vertical velocity of ballistic block $= v \sin\theta$ (m s^{-1})

V total volume injected/erupted (m^3)

w velocity of ambient fluid (air or volcanic gases) (m s^{-1})

z vertical spatial variable (m)

γ ratio of the specific heats at constant pressure and constant volume of the gas–particle mixture

γ_1 ratio of specific heats of fluid in low-pressure region

γ_4 ratio of specific heats of fluid in high-pressure region

ε entrainment coefficient

η magma viscosity (Pa s)

η_b bulk or effective viscosity of gas–particle mixture (Pa s)

θ angle of block trajectory measured from horizontal (°)

κ empirical constant

ρ magma density (kg m^{-3})

ρ_0 reference atmospheric density (kg m^{-3})

ρ_a atmospheric density (kg m^{-3})

ρ_b bulk density of the gas–particle mixture (kg m^{-3})

ρ_g gas density (kg m^{-3})

ρ_m density of magmatic particles (kg m^{-3})

ρ_r ballistic block density (kg m^{-3})

σ_m effective tensile strength of magma (Pa)

σ_w tensile strength of bubble walls (Pa)

τ magma relaxation timescale (s)

ϕ magma vesicularity

M_s Mach number of leading shock

Pe   Peclet number; ratio of the timescale for gas exsolution to the timescale of fragmentation

Re   Reynolds number; ratio of inertial to viscous forces

Ri   Richardson number; ratio of inertial to gravitational forces

References

Alidibirov, M. A. (1994). A model for viscous magma fragmentation during volcanic blasts. *Bulletin of Volcanology*, **56**(6–7), 459–465.

Belousov, A., Voight, B., Belousova, M. and Petukhin, A. (2002). Pyroclastic surges and flows from the 8–10 May 1997 explosive eruption of Bezymianny volcano, Kamchatka, Russia. *Bulletin of Volcanology*, **64**, 455–471.

Carn, S. A. and Oppenheimer, C. (2000). Remote monitoring of Indonesian volcanoes using satellite data from the Internet. *International Journal of Remote Sensing*, **21**, 873–910.

Cas, R. A. F. and Wright, J. V. (1987). *Volcanic Successions: Modern and Ancient*. London: Unwin Hyman Ltd.

Cashman, K. and Blundy, J. (2000). Degassing and crystallization of ascending andesite and dacite. *Philosophical Transactions of the Royal Society*, **358**, 1487–1513.

Cashman, K. V. and McConnell, S. M. (2005). Multiple levels of magma storage during the 1980 summer eruptions of Mount St. Helens, WA. *Bulletin of Volcanology*, **68**, 57–75.

Chojnicki, K., Clarke, A. B. and Phillips, J. C. (2006). A shock-tube investigation of the dynamics of gas-particle mixtures: Implications for explosive volcanic eruptions. *Geophysical Research Letters*, **33**, L15309, doi:10.1029/2006FL026414.

Clarke, A. B., Neri, A., Macedonio, G., Voight, B. and Druitt, T. H. (2002a). Computational modeling of the transient dynamics of August 1997 Vulcanian explosions at Soufrière Hills volcano, Montserrat: Influence of initial conduit conditions on near-vent pyroclastic dispersal. In *The Eruption of Soufrière Hills Volcano, Monserrat, from 1995 to 1999*, ed. T. H. Druitt and B. P. Kokelaar. Geological Society, London, Memoir 21, pp. 319–348

Clarke, A. B., Neri, A., Macedonio, G. and Voight, B. (2002b). Transient dynamics of Vulcanian explosions and column collapse. *Nature*, **415**, 897–901.

Clarke, A. B., Stephens, S., Teasdale, R., Sparks, R. S. J. and Diller, K. (2007). Petrological constraints on the decompression history of magma prior to Vulcanian explosions at the Soufrière Hills volcano, Montserrat. *Journal of Volcanology and Geothermal Research*, **161**, 261–274.

Clarke, A. B., Phillips J. C. , and Chojnicki K. N. (2009). An investigation of the dynamics of vulcanian eruptions using laboratory analogue experiments and scaling analysis. In *Studies in Volcanology: The Legacy of George Walker*, ed. T. Thordarson, S. Self, G. Larsen, S. K. Rowland and A. Hoskuldsson. Geological Society, London. Special Publications of IAVCEI, **1**, pp. 155–166.

de' Michieli Vitturi, M., Neri, A., Esposti Ongaro, T. Lo Savio, S. and Boschi, E. (2010). Lagrangian modeling of large volcanic particles: Application to Vulcanian explosions. *Journal of Geophysical Research*, **115**, B08206, doi:10.1029/2009JB007111.

Diller, K. D., Clarke A. B., Voight B. and Neri A. (2006). Mechanisms for conduit plug formation: implications for Vulcanian explosions. *Geophysical Research Letters*, **33**, L20302, doi:10.1029/2006GL027391.

Dingwell, D. B. (1996). Volcanic dilemma: flow or blow? *Science*, **273**, 1054–1055.

Dingwell, D. B. (2001). Magma degassing and fragmentation: Recent experimental advances. In *From Magma to Tephra*, ed. A. Freundt and M. Rosi. Amsterdam: Elsevier, pp. 1–23.

Dobran, F., Neri, A. and Macedonio, G. (1993). Numerical simulation of collapsing volcanic volumns. *Journal of Geophysical Research*, **98**, 4231–4259.

D'Oriano, C., Poggianti, E., Bertagnini, A. *et al.* (2005). Changes in eruptive style during the A.D. 1538 Monte Nuovo eruption (Phlegrean Fields, Italy): The role of syn-eruptive crystallization. *Bulletin of Volcanology*, **67**, 601–621.

Druitt, T. H., Young, S. R., Baptie, B. *et al.* (2002). Episodes of cyclic Vulcanian explosive activity with fountain collapse at Soufrière Hills volcano, Montserrat. In *The Eruption of Soufrière Hills Volcano, Monserrat, from 1995 to 1999*, ed. T. H. Druitt and B. P. Kokelaar. Geological Society, London, Memoir 21, pp. 281–306.

Dufek, J. and Berganzt, G. W. (2005). Transient two-dimensional dynamics in the upper conduit of a rhyolitic eruption: A comparison of closure models for the granular stress. *Journal of Volcanology and Geothermal Research*, **143**, 113–132.

Edmonds, M. and Herd, R. A. (2007). A volcanic degassing event at the explosive-effusive transition. *Geophysical Research Letters*, **34**, L21310, doi:10.1029/2007GL031379.

Eichelberger, J. C., Carrigan, C. R., Westrich, H. R. and Price, R. H. (1986). Non-explosive silicic volcanism. *Nature*, **323**, 598–602.

Fagents, S. A. and Wilson, L. (1993). Explosive volcanic eruptions – VII. The ranges of pyroclasts ejected in transient volcanic explosions. *Geophysical Journal International*, **113**, 359–370.

Formenti, Y. and Druitt, T. H. (2003). Vesicle connectivity in pyroclasts and implications for the fluidization of fountain-collapse pyroclastic flows, Montserrat (West Indies). *Earth and Planetary Science Letters*, **214**, 561–574.

Formenti, Y., Druitt, T. H. and Kelfoun, K. (2003). Characterization of the 1997 Vulcanian explosions of Soufrière Hills Volcano, Montserrat, by video analysis, I. *Bulletin of Volcanology*, **65**, 587–605.

Fox, R. W., McDonald, A. T. and Pritchard, P. J. (2006). *Introduction to Fluid Mechanics*, 6th edn. Hoboken, NJ: John Wiley and Sons.

Geschwind, C. H. and Rutherford, M. J. (1995). Crystallization of microlites during magma ascent: the fluid mechanics of 1980–1986 eruptions at Mount St. Helens. *Bulletin of Volcanology*, **57**, 356–370.

Giachetti, T., Druitt, T. H., Burgisser, A., Arbaret, L. and Galven, C. (2010). Bubble nucleation, growth and coalescence during the 1997 Vulcanian explosions of Soufrière Hills Volcano, Montserrat. *Journal of Volcanology and Geothermal Research*, **193**(3–4), 215–231.

Glaze, L. S., Francis, P., Self, S. and Rothery, D. A. (1989). The 16 September 1986 eruption of Lascar volcano, north Chile: satellite investigations. *Bulletin of Volcanology*, **51**, 149–160.

Gonnermann, H. M. and Manga, M. (2003). Explosive volcanism may not be an inevitable consequence of magma fragmentation. *Nature*, **426**, 432–435, doi:10.1038/nature02138.

Hammer, J. E., Cashman, K. V., Hoblitt, R. P. and Newman, S. (1999). Degassing and microlite crystallization during pre-climactic events of the 1991 eruption of Mt. Pinatubo, Philippines. *Bulletin of Volcanology*, **60**, 355–380.

Hess, K. U. and Dingwell, D. B. (1996). Viscosities of hydrous leucogranitic melts: A non-Arrhenian model. *American Mineralogist*, **81**, 1297–1300.

Hoblitt, R. (1986). *Observations of the Eruptions of July 22 and August 7, 1980, at Mount St. Helens, Washington.* United States Geological Survey Professional Paper 1335, 44 pp.

Hoblitt, R. P., Wolfe, E. W., Scott, W. E. *et al.* (1996). The preclimactic eruptions of Mount Pinatubo, June 1991. In *Fire and Mud: Eruptions and Lahars of Mount Pinatubo, Philippines*, ed. C. G. Newhall and R. S. Punongbayan. Seattle, WA: University of Washington Press, pp. 457–511.

Ishihara, K. (1985). Dynamical analysis of volcanic explosion. *Journal of Geodynamics*, **3**, 327–349.

Kaminski, E., Tait, S. and Carazzo, G. (2005). Turbulent entrainment into jets with arbitrary buoyancy. *Journal of Fluid Mechanics*, **526**, 361–376.

Koyaguchi, T. (2005). An analytical study for 1-dimensional steady flow in volcanic conduits. *Journal of Volcanology and Geothermal Research*, **143**, 29–52.

Koyaguchi T. and Mitani, N. K. (2005). A theoretical model for fragmentation of viscous bubbly magmas in shock tubes. *Journal of Geophysical Research*, **110**, B10202, doi:10.1029/2004JB003513.

Kieffer, S. W. (1981). Blast dynamics at Mount St. Helens on 18 May 1980. *Nature*, **291**, 568–570.

Lejeune, A.-M. and Richet, P. (1995). Rheology of crystal-bearing silicate melts; an experimental study at high viscosities. *Journal of Geophysical Research*, **100**(3), 4215–4229.

Mason, R. M., Starostin, A. B., Melnik, O. E. and Sparks, R. S. J. (2006). From Vulcanian explosions to sustained explosive eruptions: the role of diffusive mass transfer in conduit flow dynamics. *Journal of Volcanology and Geothermal Research*, **153**, 148–165.

Mastin, L. G. (2001). A simple calculator of ballistic trajectories for blocks ejected during volcanic eruptions. *United States Geological Survey Open-File Report 01–45*, 16pp., available at http://vulcan.wr.usgs.gov/Projects/Mastin/Publications/OFR01–45/framework.html.

McBirney, A. R. and Murase, T. (1971). Factors governing the formation of pyroclastic rocks, *Bulletin Volcanologique*, **34**(2), 372–384.

Melnik, O. and Sparks, R. S. J. (1999). Nonlinear dynamics of lava dome extrusion. *Nature*, **402**, 37–41.

Melnik, O. and Sparks, R. S. J. (2002). Modelling of conduit flow dynamics during explosive activity at Soufriere Hills Volcano, Montserrat. In *The Eruption of Soufrière Hills Volcano, Monsterrat, from 1995 to 1999*, ed. T. H. Druitt and B. P. Kokelaar. Geological Society, London, Memoir 21, pp. 307–317.

Melnik, O., Barmin, A. A. and Sparks, R. S. J. (2005). Dynamics of magma flow inside volcanic conduits with bubble overpressure buildup and gas loss through permeable magma. *Journal of Volcanology and Geothermal Research*, **143**, 53–68.

Mercalli, G. (1907). *I Volcani Attivi della Terra*. Milano: Ulrico Hoepli.

Moore, G. and Carmichael, I. S. E. (1998). The hydrous phase equilibria (to 3 kbar) of an andesite and basaltic andesite from western Mexico: constraints on water content and conditions of phenocryst growth. *Contributions to Mineralogy and Petrology*, **130**, 304–319.

Morrissey, M. M. and Chouet, B. A. (1997). Burst conditions of explosive volcanic eruptions recorded on microbarographs. *Science*, **275**, 1290–1293.

Morrissey, M. M. and Mastin, L. G. (2000). Vulcanian Eruptions. In *Encyclopedia of Volcanoes*, ed. H. Sigurdsson. San Diego, CA: Academic Press, pp. 463–475.

Morton, B. R. (1959). Forced plumes. *Journal of Fluid Mechanics*, **5**, 151–163.

Mueller, S., Scheu, B., Spieler, O. and Dingwell, D. B. (2008). Permeability control on magma fragmentation. *Geology*, **36**(5), 399–402. doi:10.1130/G24605A.

Nairn, I. A. (1976). Atmospheric shock waves and condensation clouds from Ngauruhoe explosive eruptions. *Nature*, **259**, 190–192.

Nairn, I. A. and Self, S. (1978). Explosive eruptions and pyroclastic avalanches from Ngauruhoe in February 1975. *Journal of Volcanology and Geothermal Research*, **3**, 39–60.

Navon, O. and Lyakhovsky, V. (1998). Vesiculation processes in silicic magmas. In *The Physics of Explosive Volcanic Eruptions*, ed. J. S. Gilbert and R. S. J. Sparks. Cambridge, UK: Cambridge University Press, pp. 27–50.

Papale, P. (1999). Strain-induced magma fragmentation in explosive eruptions. *Nature*, **397**, 425–428.

Proussevitch, A. A. and Sahagian, D. L. (1996). Dynamics of coupled diffusive and decompressive bubble growth in magmatic systems. *Journal of Geophysical Research*, **101**, 17,156–17,447.

Proussevitch, A. and Sahagian, D. (2005). Bubbledrive-1: A numerical model of volcanic eruption mechanisms driven by disequilibrium magma degassing. *Journal of Volcanology and Geothermal Research*, **143**, 89–111.

Rose, W. I. (1980). Gas and hydrogen isotopic analyses of volcanic eruptions clouds in Guatemala sampled by aircraft. *Journal of Volcanology and Geothermal Research*, **7**, 1–10.

Rose, W. I., Anderson, A. T., Woodruff, L. G. and Bonis, S. B. (1978). The October 1974 basaltic tephra from Fuego Volcano: description and history of the magma body. *Journal of Volcanology and Geothermal Research*, **4**, 3–53.

Saad, M. A. (1985). *Compressible Fluid Flow*. Upper Saddle River, NJ: Prentice-Hall, Inc.

Schmincke, H.-U. (1977). Eifel-Vulkanismus östlich des Gebiets Rieden-Mayen. *Fortschritte der Mineralogie*, **55**, 1–31.

Scott, W. E., Hoblitt, R. P., Torres, R. C. et al. (1996). Pyroclastic flows of the June 15, 1991, climactic eruption of Mount Pinatubo. In *Fire and Mud: Eruptions and Lahars of Mount Pinatubo, Philippines*, ed. C. G. Newhall and R. S. Punongbayan. Seattle, WA: University of Washington Press, pp. 545–570.

Self, S., Wilson, L. and Nairn, I. A. (1979). Vulcanian eruption mechanisms. *Nature*, **277**, 440–443.

Self, S., Kienle, J. and Huot, J. -P. (1980). Ukinrek Maars, Alaska, II. Deposits and formation of the 1977 craters. *Journal of Volcanology and Geothermal Research*, **7**, 39–65.

Solovitz, S. A. and Mastin, L. A. (2009). Experimental study of near-field air entrainment by subsonic volcanic jets. *Journal of Geophysical Research*, **114**, B10203, doi:10.1029/2009JB006298.

Sparks, R. S. J. (1978). The dynamics of bubble formation and growth in magmas: a review and analysis. *Journal of Volcanology and Geothermal Research*, **3**, 1–37.

Sparks, R. S. J. (1997). Causes and consequences of pressurization in lave dome eruptions. *Earth and Planetary Science Letters*, **150**, 177–189.

Sparks, R. S. J. and Wilson, L. (1982). Explosive volcanic eruptions – V. Observations of plume dynamics during the 1979 Soufrière eruptions, St. Vincent. *Geophysical Journal of the Royal Astronomical Society*, **69**, 551–570.

Sparks, R. S. J., Bursik, M. I., Carey, S. N. et al. (1997). *Volcanic Plumes*. New York, NY: John Wiley and Sons.

Spieler, O., Dingwell, D. B. and Alidibirov, M. (2004a). Magma fragmentation speed: an experimental determination. *Journal of Volcanology and Geothermal Research*, **129**, 109–123.

Spieler, O., Kennedy, B., Kueppers, U. et al. (2004b). The fragmentation threshold of pyroclastic rocks. *Earth and Planetary Science Letters*, **226**, 139–148.

Stix, J., Torres, R. C., Narváez M., L. et al. (1997). A model of Vulcanian eruptions at Galeras volcano, Colombia. *Journal of Volcanology and Geothermal Research*, **77**, 285–303.

Taddeucci, J., Pompilio, M. and Scarlato, P. (2004). Conduit processes during the July–August 2001 explosive activity of Mt. Etna (Italy): Inferences from glass chemistry and crystal size distribution of ash particles. *Journal of Volcanology and Geothermal Research*, **137**, 33–54.

Tait, S. R., Jaupart, C. and Vergniolle, S. (1989). Pressure, gas content and eruptive periodicity of a shallow crystallizing magma chamber. *Earth and Planetary Science Letters*, **92**, 107–123.

Takeuchi, S., Nakashima, S., Tomiya, A. and Shinohara, H. (2005). Experimental constraints on the low gas permeability of vesicular magma during decompression. *Geophysical Research Letters*, **32**, L10312, doi:10.1029/2005GL022491.

Tuffen, H., Dingwell, D. B. and Pinkerton, H. (2003). Repeated fracture and healing of silicic magma generate flow banding and earthquakes? *Geology*, **31**(12), 1089–1092.

Turcotte, D. L., Ockendon, H., Ockendon, J. R. and Cowley, S. J. (1990). A mathematical model of vulcanian eruptions. *Geophysical Journal International*, **103**, 211–217.

Turner, J. S. (1973). *Buoyancy Effects in Fluids*. New York: Cambridge University Press.

Voight, B., Sparks, R. S. J., Miller, A. D. *et al.* (1999). Magma flow instability and cyclic activity at Soufrière Hills Volcano, Montserrat. *Science*, **283**(5405), 1138–1142.

Waitt, R. B., Mastin, L. G. and Miller, T. P. (1995). Ballistic showers during Crater Peak eruptions of Mount Spurr Volcano, summer 1992. *United States Geological Survey Bulletin*, **2139**, 89–106.

Walker, G. P. L. (1973). Explosive volcanic eruptions – a new classification scheme. *Geologische Rundschau*, **62**(2), 431–446.

Wilson, L. (1972). Explosive volcanic eruptions – II. The atmospheric trajectories of pyroclasts. *Geophysical Journal of the Royal Astronomical Society*, **30**, 381–392.

Wilson, L. and Head, J. (1981). Ascent and erupton of basaltic magma on the Earth and Moon. *Journal of Geophysical Research*, **86**, 2971–3001.

Wilson, L., Sparks, R. S. J., Huang, T. C. and Watkins, N. D. (1978). The control of volcanic eruption column heights by eruption energetics and dynamics. *Journal of Geophysical Research*, **83**, 1829–1836.

Wilson, L., Sparks, R. S. J. and Walker, G. P. L. (1980). Explosive volcanic eruptions – IV. The control of magma properties and conduit geometry on eruption column behaviour. *Geophysical Journal of the Royal Astronomical Society*, **63**, 117–148.

Wohletz, K. H. (2001). Pyroclastic surges and two-phase compressible flows. In *From Magma to Tephra*, ed. A. Freundt and M. Rosi. Amsterdam: Elsevier, pp. 247–312.

Wohletz, K. H. and Valentine, G. A. (1990). Computer simulations of explosive volcanic eruptions. In *Magma Transport and Storage*, ed. M. P. Ryan. London: John Wiley and Sons, pp. 113–135.

Woods, A. W. (1995). A model of vulcanian explosions. *Nuclear Engineering and Design*, **155**, 345–357.

Wright, H. M. N., Cashman, K. V., Rosi, M. and Cioni, R. (2007). Breadcrust bombs as indicators of Vulcanian eruption dynamics at Guagua Pichincha volcano, Ecuador. *Bulletin of Volcanology*, **69**, 281–300, doi:10.1007/s00445-006-0073-6.

Yamamoto, H., Watson, I. M., Phillips, J. C. and Bluth, G. J. S. (2008). Rise dynamics and relative ash distribution in Vulcanian eruption plumes at Santiaguito Volcano, Guatemala, revealed using an ultraviolet imaging camera. *Geophysical Research Letters*, **35**, L08314, doi:10.1029/2007GL032008.

Yüceil, K. B. and Ötügen, M. V. (2002). Scaling parameters for underexpanded supersonic jets, *Physics of Fluids*, **14**, 4206–4215.

Zhang, Y. (1998). Experimental simulations of gas-driven eruptions: kinetics of bubble growth and effect of geometry. *Bulletin of Volcanology* **59**, 281–290.

Zhang, Y. (1999). A criterion for the fragmentation of bubbly magma based on brittle failure theory. *Nature*, **402**, 648–650.

Exercises

7.1 Given the shock velocity in air of 450 m s^{-1} associated with the eruption of Sakurajima volcano in 1982 (Ishihara, 1985), calculate:

(a) The initial pressure ratio across the plug by assuming that the mixture behaves as a pseudogas with $c_4 = 60$ m s^{-1}, $\gamma_4 = 1.04$, $c_1 = 330$ m s^{-1}, $\gamma_1 = 1.4$.

(b) The shock velocity and strength for an initial pressure ratio of 50, given the same fluid properties.

7.2 Given the values used in and calculated for the Sakurajima case above:

(a) Calculate the corresponding vent velocity using the isothermal relationship of Woods (1995) assuming $T = 1200$ K, $\gamma = 1.04$, $R = 460$ J kg^{-1} K^{-1}, and $n = 0.03$.

(b) Calculate the vent velocity using shock tube relationships.

(c) Calculate vent velocity using the non-isothermal relationship of Woods (1995) assuming that the gas temperature drops to 500 K upon decompression with corresponding $\gamma = 1.3$ (R and n are the same as in part (a)).

(d) Compare your solutions and discuss possible reasons for the differences.

Online resources available at www.cambridge.org/fagents

- Additional exercises and supporting materials
- Answers to exercises

Chapter 8

Sustained explosive activity: volcanic eruption columns and hawaiian fountains

Andrew W. Woods

Overview

This chapter reviews some of the physical processes responsible for the injection of large volumes of volcanic tephra and gas high into the atmosphere, following sustained explosive eruption from a volcanic vent. The resulting volcanic *plumes* or *columns* can disperse ash and aerosols over vast distances, and cause global perturbations to climate. In contrast, incandescent lava fountains are common manifestations of the low-intensity end of the spectrum of explosivity. This chapter presents a series of modeling approaches, from dimensional analysis to fully time-dependent, three-dimensional numerical treatments, in order to develop some quantitative understanding of these phenomena.

8.1 | Introduction

When sustained explosive eruptions discharge tephra from a volcanic vent, the erupting material may form a convecting plume that rises high into the atmosphere; these flows are known as volcanic eruption columns (Fig. 8.1(a)). At moderate to high discharge rates *subplinian* and *plinian* columns rise to 10–40 km, penetrating the stratosphere, and erupt ejecta volumes of ~0.1

to ~10 km³, whereas very high *ultraplinian* discharges produce columns > 40–50 km high and erupt volumes of ≫ 10 km³. At low discharge rates, *hawaiian* activity (named after the archetypal eruption style of the Hawaiian basaltic volcanoes) erupts volumes of 10^4–10^8 m³ as incandescent lava fountains tens to hundreds of meters high (Fig. 8.1(b)), with weak plumes of fine pyroclasts rising above the fountains to heights of order 1–5 km.

Plinian eruption columns are capable of dispersing tephra and volcanic gases over thousands of square kilometers, and provide the dominant mechanism by which volcanic eruptions can impact global climate, given that fine particles of ash and volcanic aerosols, such as sulfuric acid, impact the radiative balance of the atmosphere. Deposition of large volumes of tephra and fine particulates over such vast areas of land has a devastating impact on agriculture, infrastructure, and buildings (see Chapter 9). From a geological perspective, deposits from volcanic eruptions provide a unique insight into the scale and evolution of the eruption, through analysis of lateral and vertical variations in the deposits, in terms of particle-size distribution and composition. Models that quantify the relationship between eruption rate and tephra dispersal can be combined with detailed field studies to infer details of the eruption history,

Modeling Volcanic Processes: The Physics and Mathematics of Volcanism, eds. Sarah A. Fagents, Tracy K. P. Gregg, and Rosaly M. C. Lopes. Published by Cambridge University Press. © Cambridge University Press 2013.

Figure 8.1 (a) Photograph of an eruption column at Mount Redoubt, Alaska. Redoubt is an active stratovolcano that has produced primarily andesitic magma in recent eruptions. Columns in the 1989–90 eruption rose to 14 km, whereas the 2009 eruption produced plumes as high as 20 km. (b) Episode 3 of the 1959 Kilauea Iki eruption, Hawaii, produced lava fountains > 500 m high. Photograph taken 11/29/59, courtesy of the United States Geological Survey. Color versions of these photographs appear in the online supplement for this chapter.

which in turn provides important constraints on the evolution of the subsurface processes that ultimately control the eruption rate and the erupted mass.

Numerous historic eruptions have provided a wealth of information about the dynamics of eruption columns. Two of the best studied large explosive events are the 18 May 1980 eruption of Mount St. Helens in Washington State, USA, and the 15 June 1991 eruption of Mount Pinatubo in the Philippines. The Mount St. Helens event started with explosive decompression of a shallow reservoir of magma

below the summit, which produced an intense lateral blast flow. The lateral blast traveled along the ground for 10–15 km as an intense eroding flow, mixing with the air and becoming progressively less dense until it eventually became buoyant and lifted off to form a large coignimbrite cloud. Subsequently, magma sourced deeper in the subsurface continued to erupt from the vent, forming the main plinian phase of the eruption. A large sustained eruption column rose > 15 km into the atmosphere and persisted for many hours, generating a massive cloud that spread downwind and producing tephra fall deposits > 1000 km from the source. The Pinatubo eruption on 15 June 1991 was an extremely violent event, producing an eruption column that rose to 34 km above sea level at the peak of the eruption, and forming a huge umbrella cloud, > 500 km in radius, above the eruption column. This umbrella cloud was the source for the fine ash and aerosol cloud that was carried around the globe. Observations of these, and many other events, point to the tremendous energy in eruption columns, which can transport vast quantities of erupted material, as well as lower atmospheric air, high into the atmosphere. To provide constraints on plume models, pertinent data from historical eruptions include the volume erupted and the duration of the events, as well as estimates of the height of rise of the material into the atmosphere. These data enable comparison of model predictions of column rise height as a function of eruption rate with field observations.

The purpose of this chapter is to review some of the present models and understanding of the dynamics that control the evolution of volcanic eruption columns. We start with a description of the dominant physical processes, in order to provide context for the quantitative models that are subsequently discussed. We present some of the important predictions from these models and discuss some of the models' limitations. A brief review of numerical and experimental models is then given, illustrated with some field observations. By way of comparison, we also explore some of the controls on the dynamics of *phreatomagmatic* eruptions (discussed in greater detail in Chapter 11). At the low-intensity end of the explosive continuum, the dynamics of lava fountains are described, before we conclude with some suggestions for future modeling and field work.

8.2 | Physical processes

There is a range of styles of explosive volcanic eruption, ranging from short-lived transient explosions (see Chapters 6 and 7) to more sustained hawaiian lava fountains, and subplinian, plinian, and ultraplinian eruptions involving the ejection of very large volumes of ash and gas. In this chapter, we primarily consider plinian-style eruptions, saving discussion of hawaiian-style fountains to Section 8.4. Furthermore the description and models that follow apply primarily to *strong* plumes, which are sufficiently vigorous that they are unaffected by atmospheric

winds. Chapter 9 further discusses the differences between strong and weak plumes.

Figure 8.2 shows the constituent parts of a large eruption column. On exit from the vent, the gas–particle mixture emerges into the *gas-thrust region* or *momentum-driven jet*, which typically extends to 1–2 km above the vent. The mixture is hot and dense, with a temperature in the range 800–1200 °C, density up to several tens times that of the air, and with pressures as large as 10–100 times atmospheric pressure, depending on the geometry of the vent and the mass of exsolved gas in the erupting mixture. The mixture initially undergoes rapid decompression to atmospheric pressure, producing significant expansion. Subsequently, the material continues to ascend into the atmosphere as a highly turbulent flow, with a large mean speed typically in the range 100–150 m s^{-1}. The mixture of dense particles and gas decelerates as it ascends, and some of the larger particles may separate from the ascending mixture and follow ballistic trajectories to the ground, while the bulk of the flow continues upwards. The column commonly exhibits lateral gradients in density and momentum, with the material at the edges mixing very effectively with the surrounding air, which is then transferred to the inner core of the column over time through shear mixing. As the air is stirred into the ascending mixture by the turbulence, there is rapid heat transfer from the fine particles to the air. As a result, the gas content of the ascending mixture increases and this gas is heated to temperatures hundreds of degrees greater than that of the surrounding air. The gas component of the ascending mixture therefore becomes much less dense than the surrounding air, even though in the lower parts of the column, the bulk density of the gas and solid material is greater than that of the air.

If the initial upward speed is sufficiently high or the eruption rate is sufficiently small, then the continuing entrainment of air may eventually lead to the bulk density of a part of the flow becoming less than that of the surrounding air. At this point, the buoyancy will drive the mixture upwards, and the overall upward momentum of the flow will increase. As this buoyant mixture continues to rise convectively, more

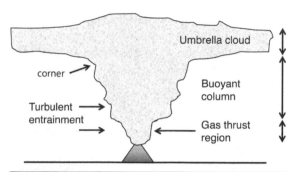

Figure 8.2 Schematic of an eruption plume, illustrating the dense jet above the volcanic vent, the buoyant column and the neutrally buoyant umbrella cloud that spreads out atop the column. The transition between the column and umbrella cloud is commonly called the corner.

air is entrained and the flow becomes progressively more buoyant as thermal energy stored in the large mass of solid material heats the air. In some cases, nearly all the erupted material then becomes buoyant to form the *buoyant column* region (Fig. 8.2), whereas in other cases, the outer parts of the flow continue upwards while some of the inner material falls back to the ground, producing an unstable fountain. With very large mass fluxes (see Section 8.2), only a small fraction of the material becomes buoyant while ascending, and the remainder of the erupted solids falls back to the ground.

The thermal energy of the erupting mixture is very much greater than the initial kinetic energy. The specific heat stored in the solid material is of order 10^6 J kg^{-1} while the kinetic energy per unit mass is of order 10^4 J kg^{-1}, for material with an initial temperature of order 800–1200 °C and upward speed 100–150 m s^{-1}. As a result, once the mixture becomes buoyant and is able to convert the thermal energy to potential energy, the mixture can ascend to heights of tens of km in the atmosphere, whereas rise heights of ballistic trajectories are only of order 1 km. The final rise height of the mixture is constrained by the stable density stratification of the atmosphere. As well as the erupting material, the eruption column carries a large mass of entrained air from the lower atmosphere high into the atmosphere. The initial thermal energy of the erupting mixture can only drive the column to a certain height into the atmosphere, at

which point the thermal energy of the erupted solids has been converted to potential energy. Subsequently, the mixture becomes denser than the surrounding air, overshooting the level of neutral buoyancy owing to its momentum, and decelerates until reaching a maximum height. The mixture then falls back to the neutral buoyancy height where it spreads laterally to form the *umbrella cloud* (Fig. 8.2), which represents the initial stages of the subsequent dispersal of the ash through the atmosphere (see Chapter 9).

If the dense mixture emerging from the volcanic vent is unable to mix with sufficient air to become buoyant before its upward momentum falls to zero, then the mixture will form a fountain that collapses back around the vent, producing pyroclastic density currents (PDCs) and associated deposits (see Chapter 10). In the region of transition between convecting and collapsing columns, a fraction of the material in the fountain may become buoyant and rise into the atmosphere, while the remainder forms a dense flow. This can lead to unstable flow, with intermittent fountaining inhibiting the ascent of the erupted material, and thereby leading to interspersed phases of buoyant plume or discrete thermal formation, and non-buoyant fountaining.

The phenomenological picture outlined above is somewhat simplified to identify the key processes, but there are a number of other effects which can have a major impact on plume dynamics. Important effects include *thermal disequilibrium* between the solid tephra and the air, in which conductive cooling rates of moderate to large particles limit heat transfer to the gas phase. This reduces the thermal energy available for heating the air, and therefore lowers the ultimate rise height of the column, but also may lead to formation of a collapsing column for a wider range of initial conditions. Secondly, the *re-entrainment* of particles back into the plume as they settle from the umbrella cloud can increase the solid load in the eruption column and thereby affect the buoyancy. This effect may drive the system to collapse even if the column would have been buoyant in the absence of particle recycling. Finally, if the erupting material interacts with external water, the tendency towards thorough fragmentation may increase the efficiency of heat transfer from fine particles to the gas phase, and hence drive higher columns (Chapter 11). However, when larger quantities of water are involved, the initial thermal energy of the solid may be depleted through heating and vaporization of the water, and this in turn may suppress the subsequent formation of a buoyant column, leading instead to vapor-rich or wet PDCs. We discuss some of these effects later in the chapter. We also describe some recent numerical modeling that has captured much of the dynamics of eruption columns, and provides new insights into the phenomena at the transition from plume-forming to flow-forming behavior.

8.3 | Quantitative modeling of eruption columns

In order to develop a quantitative description of the motion of volcanic eruption columns, which are highly turbulent, time-dependent phenomena, there is a range of possible modeling approaches. First, one can follow the principles of dimensional analysis and determine some of the key integral properties of the column in terms of the effective buoyancy flux in the column and the stratification of the environment (e.g., Morton et al., 1956). This provides a simplified but powerful means of assessing some of the main properties of the eruption, including the eruption rate and/or rise height of the eruption column.

Second, one can build a time-averaged model of the evolution of the properties of the column, accounting for the effects of turbulence by parameterizing the mixing rate between the ambient air and the ascending mixture in the column. In such models, integral properties of the mixture are related at each height in the column. In developing this modeling approach, it is helpful to examine the horizontally averaged properties of the column (Morton et al., 1956). The principles of mass, momentum, and energy conservation are then used to determine their variation with height (Woods, 1988; Sparks et al., 1997). This approach has proved very useful in exploring the dynamics of eruption columns and in assessing the different controls on the flow.

Third, it is possible to formulate a set of equations that govern the conservation of mass, momentum, and energy at each point in the flow. This includes a parameterization of the local effects of the turbulence, and the use of averaged quantities for the bulk properties of the multiphase flow at each point. These equations are then solved numerically, using an axisymmetric coordinate system (e.g., Neri and Dobran, 1994; Neri and Macedonio, 1996; Neri et al., 1998) or as a three-dimensional flow (e.g., Suzuki et al., 2005), by resolving the relevant dynamic scales to capture air entrainment and mixing, and thereby to examine the motion of the column.

This review focuses primarily on the first two approaches. However, Section 8.3.5 discusses some of the recent developments in numerical simulation of eruption columns, which replicate the predictions of the simpler integral models (e.g., Woods, 1988), but also provide new insights into the mixing and partitioning of the flow between plume- and PDC-forming regimes. These numerical simulations have evolved over the past ten years from axisymmetric steady-state models to the present time-dependent three-dimensional models.

8.3.1 Dimensional analysis

The motion of turbulent buoyant plumes has been studied in a variety of contexts over the past 50 years, following the pioneering work of Morton et al. (1956), in which some of the underlying physical principles of their motion were first presented. In that work, it was recognized that when a localized source of buoyancy produces a quasi-steady turbulent buoyant plume in a stratified environment, the height of rise of the plume depends on the buoyancy flux, B, defined as

$$B = \frac{\Delta\rho}{\rho_o} g\dot{V}, \qquad (8.1)$$

where $\dot{V}$ is the volume flux of the source, and $\Delta\rho$ is the difference between the density of the ambient fluid ρ_o at the level of the source, and the density of the source fluid. (Notation is summarized in Section 8.6.) The stratification is measured by the Brunt–Väisälä frequency, $N = \sqrt{-(g/\rho_o)(\partial\rho_a/\partial z)}$, where ρ_a is the density of the ambient fluid, and ρ_o is taken as 1 kg m^{-3}. By dimensional analysis the height of rise then takes the form

$$H = \lambda B^{1/4} N^{-3/4}, \tag{8.2}$$

where the constant λ is ~5, as deduced from experimental measurements. This model is based on an implicit assumption that the source flux is small compared to the flow rate in the column, which is typically the case sufficiently far from the source, as the ascending material entrains and convects air upwards. It also assumes that the buoyancy flux of the erupting material is conserved in a non-stratified, well-mixed environment, so that mixing of the source fluid with the ambient fluid causes the buoyancy to decrease at a rate that is inversely proportional to the volume flux.

This latter assumption does not apply to the motion of a volcanic eruption column however, because the buoyancy is not conserved. Indeed, once the material has emerged from the vent and decompressed to atmospheric pressure, the material is initially dense compared to the air, but through mixing and heat transfer, it becomes less dense than the air. However, in a uniform environment the energy flux is conserved, and this can be used to adapt Eq. (8.2) for application to a volcanic eruption column, as recognized by Wilson *et al.* (1978). For this purpose, it is necessary to define the effective buoyancy flux at the base of the plume in terms of the heat flux. If, at the volcanic vent, the source mass flux is Q, the specific heat is c and the temperature is T_o, then the heat flux measured relative to the temperature of the air at the ground T_a is

$$Q_H = Qc(T_o - T_a). \tag{8.3}$$

This heat flux evolves as the mixture entrains the relatively dense air from the lower atmosphere and carries it with the erupted material upwards thereby generating potential and kinetic energy. However, in the lower part of the column, we can define an effective buoyancy flux associated with this heat flux as

$$B = \frac{Q_H g}{\rho_a c T_a}. \tag{8.4}$$

We can then use Eq. (8.2) to determine the rise height of the eruption column (see Fig. 8.3). In evaluating the Brunt–Väisälä frequency N, some approximations are required because of the structure of the atmosphere. The lower atmosphere is relatively well mixed and so N is very small, whereas the stratosphere is much more stably stratified, and N is larger, which will arrest the motion of the plume. In applying the model, using a fixed value for N is a simplification likely to lead to errors in predicting the rise height of both small eruption columns that remain within the troposphere, and powerful eruption columns that penetrate well into the stratosphere. One of the benefits of using an integral model of the plume properties at each height in the plume (see Section 8.3.2) is that the variation with height of the atmospheric stratification can be included in the model of the turbulent buoyant plume.

There are a number of compilations of column rise height and eruption rate estimates for various historic eruptions (Sparks *et al.*, 1997). Some of the key eruptions include Mount Pinatubo, 1991, for which the eruption column rose over 34 km with an eruption rate of 4×10^6 m^3 s^{-1}; Vesuvius AD 79 (the "grey pumice phase"), for which the column rose 32 km with an estimated eruption rate of 6×10^5 m^3 s^{-1}; Quizapu, Chile, 1932, which generated a 27–30 km high column with an estimated eruption rate of 6×10^5 m^3 s^{-1}; and a number of smaller eruptions, including La Soufriere, St. Vincent, 1902, which produced a column 15.5–17 km high with an estimated eruption rate of 1.1–1.5×10^4 m^3 s^{-1}. Figure 8.3 shows a comparison of these data (symbols) with the dimensional scaling model (line), which indicates that the model is of considerable value in determining the overall

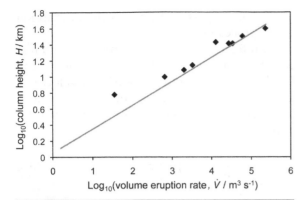

Figure 8.3 Plot of rise height of a turbulent buoyant plume vs. volumetric eruption rate, comparing the dimensional analysis scaling (gray line; Eq. 8.2) with data for some historical eruptions (symbols).

rise height of the column, and that the essential physics controlling ascent is consistent with the fundamental dimensional analysis.

8.3.2 Integral plume models

As indicated in the preceding section, the motion of eruption columns is more complex than suggested by Eq. (8.2), owing to the high initial density of the erupting mixture, which only becomes buoyant through the process of entrainment in the region near the vent. This balance is now explored further with the development of a time- and horizontally-averaged model for an eruption column, including a parameterization of the entrainment rate from the surrounding air, building from the original work of Morton *et al.* (1956) (Wilson, 1976; Sparks, 1986; Woods, 1988). In this model, the time averaging corresponds to averaging over the turnover time of the eddies, so that the model characterizes the mean properties of the flow rather than the detailed temporal fluctuations. It may be recognized from Section 8.3.1, and models of turbulent buoyant plumes in stratified environments, that the timescale of ascent of the material through an eruption column is of order $1/N$. In the atmosphere, this corresponds to several hundred seconds. Time averaging therefore corresponds to an average over a period of a few minutes. In many eruptions,

the eruption rate evolves with time, but in long-lived eruptions this typically occurs over periods much longer than a few minutes, and so the motion of the eruption column will evolve in a quasi-steady fashion. In order to model this evolution, it is possible to use integral models in conjunction with a slowly waning flux. In more short-lived, transient eruptions, the flow will not fully establish a turbulent plume, but will form a starting plume or discrete thermal (Chapter 7); the physics of these transient phenomena is also controlled by the entrainment of and heat transfer to the ambient air. The modeling approach described here treats the gas–particle mixture as a single phase in thermal and dynamic equilibrium; this is sometimes called the *dusty gas* or *pseudogas* approximation.

Figure 8.4 shows the simplified model of the eruption column of radius $b(z)$, upward mean speed $u(z)$, average temperature $T(z)$, gas mass fraction $n(z)$, and density $\rho(z)$, moving through an atmosphere having temperature $T_a(z)$ and density $\rho_a(z)$. The initial model assumption is that the mixture is all carried to the top of the plume, and that it is in thermal equilibrium with the air. This is valid when tephra are sufficiently fine grained for rapid heat transfer to the air (see Section 8.3.3 for further discussion of this condition), and when there is little fallout from or re-entrainment of particles into the plume. In this simplified case, the conservation of mass takes the form

$$\frac{d}{dz}\left(\rho u b^2\right) = 2\varepsilon b u \rho_a, \tag{8.5}$$

where ε is the coefficient of entrainment of the ambient air into the plume. In a purely buoyant plume, this takes a constant value of $\sim$0.1, whereas for more jet-like (momentum-driven) flow, the value of the entrainment coefficient is smaller, of order 0.06–0.07. However, with the large density differences between the volcanic plume and ambient air in the lower atmosphere, the entrainment is also modified. This can change the precise conditions for collapse in comparison with the predictions based on using an entrainment coefficient ε. Furthermore, in

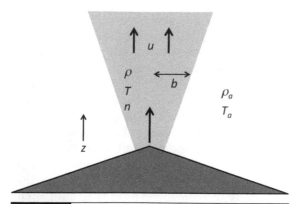

Figure 8.4 Schematic illustration of the variables used in the theoretical model of an eruption column. The plume of radius b, density ρ, temperature T, and gas mass fraction n rises with velocity u through an atmosphere of density ρ_a and temperature T_a. All parameters vary with height, z, but are assumed to be constant across the width of the plume at any given height.

flows in which there is internal generation of buoyancy through heating and expansion of the entrained air (Caulfield and Woods, 1995), the entrainment coefficient can change because of the evolving structure of the turbulence in the flow. This can be described in the modeling through inclusion of a variable entrainment coefficient (Woods, 1988; Kaminski et al., 2005; Suzuki et al., 2005). Changes in the entrainment rate lead to differences in the quantitative predictions of the model, but many of the key phenomena can be identified using a constant value for the entrainment coefficient. It is worth noting that there are many approximations in the model, including the assumptions of thermal equilibrium and momentum equilibration between the gas and particles, which may evolve with time if the particle-size distribution evolves, as well as the possible effects of particle fallout from the umbrella cloud and subsequent re-entrainment. An important role of integral models is to capture some of the key processes and explore how they may impact the flow dynamics and hence the overall evolution of the flow. It is also possible to replicate specific eruption histories with these models, in order to predict eruption rates or column rise heights. However, inferences from such simulations should include error estimates owing to

the simplifying assumptions in the description of the processes.

By integrating across the eruption column at each horizontal level, it is possible to define column properties (radius b, density ρ, velocity u, gas content n, and temperature T) as functions of height z. The conservation of momentum then takes the form

$$\frac{d}{dz}\left(\rho u^2 b^2\right) = b^2 g(\rho_a - \rho), \tag{8.6}$$

while the steady flow energy equation, which describes the conservation of enthalpy flux in the eruption column, takes the form

$$\frac{d}{dz}\left[\rho u b^2\left(c(T-T_o)+\frac{u^2}{2}+g z\right)\right]$$
$$= 2\varepsilon b u \rho_a\left(c_a(T_a-T_o)+g z\right), \tag{8.7}$$

where c is the specific heat, and subscripts a and o represent, respectively, properties of the ambient air and a reference or initial value (e.g., the ambient temperature at the ground). Equation (8.7) represents an expression for the rate of change of the averaged internal energy, kinetic energy, and potential energy of the ascending mixture (terms on left-hand side of Eq. 8.7), which results from the entrainment of internal energy and potential energy associated with the ambient fluid (terms on right-hand side). Equation (8.7) may be expressed in the form

$$\frac{d}{dz}\left[\rho u b^2 c(T-T_o)\right] = 2\varepsilon b u \rho_a c_a(T_a-T_o)$$
$$- g\rho u b^2 - \frac{d}{dz}\left(\rho u^3 b^2\right). \tag{8.8}$$

The specific heat of the material in the column is given by the mass average of the solid and gas specific heats, denoted by subscripts s and g,

$$c = \left((1-n)c_s + n c_g\right), \tag{8.9}$$

where specific heat of the gases, c_g, can be found from the weighted average of the specific heats of volcanic gas and air. The gas mass fraction in the column follows from the equation for conservation of solid material in the column

$$\frac{d}{dz}\left((1-n)\rho u b^2\right)=0 \qquad (8.10)$$

in the case of no particle fallout or re-entrainment into the column. The mass of air in the column is found by recording the mass of air entrained up to that height, from Eq. (8.5). Using these relations it may be shown that the average specific heat of the gas phase is

$$c_g = \left[(n-n_o)c_a + n_o(1-n)c_{vg}\right]\frac{1}{n(1-n_o)}, \qquad (8.11)$$

where c_a and c_{vg} are the specific heats of the air and volcanic gas, respectively, and n_o is the initial gas mass fraction in the erupting material. Finally, the bulk density of the mixture of gas and particles is given by

$$\rho = \left[\frac{n}{\rho_g}+\frac{1-n}{\rho_s}\right]^{-1}, \qquad (8.12)$$

where the gas density ρ_g is given by the mass average of the air and volcanic gases.

The above system of equations can be solved numerically to determine the motion through an atmosphere with given vertical temperature, pressure, and density profiles. In many simulations, including the calculations presented here, a simple model atmosphere is used, such as that described by Gill (1981). Solutions of the equations corresponding to a dry eruption column composed of fine ash, air, and volcanic gas, identify some of the dominant processes in operation in a volcanic plume. Although the model is simplified, especially with (i) the entrainment and mixing processes in the lower part of the column being approximated with a single coefficient, and (ii) the fact that in this region, the flow may divide into a dense PDC-forming component and a buoyant eruption column rather than remaining as a coherent flow, the model is able to delineate broadly the different styles of behavior observed in nature. Figure 8.5 illustrates the variation of plume velocity with height for three different initial speeds of 50, 75, and 200 m s⁻¹. At the lowest initial velocity, the erupting material falls to zero velocity not

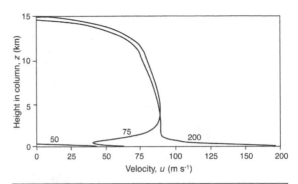

Figure 8.5 Model calculations showing the evolution of the ascent velocity of the plume as a function of height above the vent. Curves are given for initial speeds of 50, 75, and 200 m s⁻¹. The lowest speed does not allow the mixture to become buoyant before collapse of the plume occurs.

far above the vent and the column collapses to produce a dense fountain that sheds pyroclastic density currents that spread along the ground. At the larger eruption speed of 75 m s⁻¹, the material again decelerates under gravity in the region just above the source, but then becomes buoyant as a result of entrainment of air. At this point, the mixture begins to accelerate upwards because it has such a low speed. However, as it entrains progressively more air, the momentum associated with the entrainment of the air begins to dominate the upward buoyancy force, and the velocity decreases towards the maximum rise height. At the largest initial velocity of 200 m s⁻¹, the flow decelerates rapidly above the source, but then the mixture becomes buoyant and continues to ascend high into the atmosphere.

The critical velocity for column collapse can be calculated from the model. We note that there is some variation with such predictions depending on the precise model for entrainment in the lower part of the column. Smaller values of the entrainment coefficient lead to prediction of collapse in more cases, owing to the reduction of mixing and hence inhibition of buoyancy generation. Furthermore, the assumption that the flow remains well mixed and unidirectional in this lower part of the eruption column is somewhat simplified; in practice the flow may become unsteady as conditions approach those for collapse of the column. Therefore, calculations of

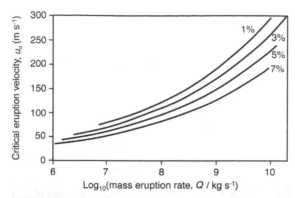

Figure 8.6 Critical eruption velocity at the vent as a function of the mass flux at which the erupting material is able to become buoyant and form an eruption column, rather than collapsing back to form a dense fountain. Curves are given for four different gas contents in the erupting material, 1, 3, 5, and 7 wt.%, and the calculations use a constant entrainment coefficient, ε, of 0.09.

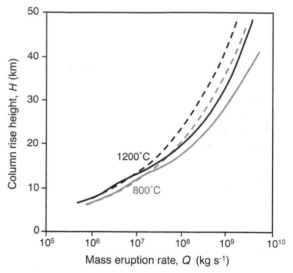

Figure 8.7 Total rise height of the column as a function of the mass eruption rate, as calculated from the model of the quasi-steady eruption column (solid lines), and from the dimensional analysis (dashed lines). Curves are given for eruption temperatures of 800 °C (gray) and 1200 °C (black) (after Woods, 1988; reprinted with kind permission from Springer Science+Business media).

column collapse conditions provide an approximate guide to the cases for which there may be transitions in flow behavior, rather than a precise prediction. Figure 8.6 shows one such calculation, which illustrates the critical eruption velocity as a function of the eruption mass flux, with curves given for volatile contents of 1, 3, 5, and 7 wt.%. For eruption velocities exceeding the critical value for collapse, a large convecting ash plume develops, whereas for lower eruption velocities, column collapse ensues (Bursik and Woods, 1991).

The other key result from integral plume models, in addition to recognition of column collapse, is the prediction of the column rise height, shown in Figure 8.7 as a function of the mass flux at the source, in which larger mass fluxes produce greater column heights. The simplified curves from the dimensional analysis (dashed curves) are in broad agreement with the rise height predictions from the theoretical model (solid curves), but differences arise owing to the increasing strength of the stratification with height in the atmosphere.

Glaze and Baloga (1996) further explored the sensitivity of column rise height to ambient atmospheric structure, and found that both seasonal and latitudinal differences can significantly influence column rise heights. Glaze et al. (1997)

investigated how eruption columns entrain water vapor from the lower atmosphere and carry it to much greater heights. Condensation of water vapor releases latent heat, and liquid water has a higher specific heat than the vapor phase; both factors act to further drive plume ascent, such that plumes erupting in relatively wet atmospheres can rise to significantly greater heights than those in dry atmospheres.

Experimental analogs

There have been some systematic experiments carried out to explore column dynamics using simplified analog systems. In one experiment (Woods and Caulfield, 1992) mixtures of MEG (methanol–ethylene glycol) were injected upwards into a tank of fresh water. The MEG is initially dense compared to the water, but as it mixes, it becomes less dense than the water and continues to rise through the tank, forming a buoyant plume. As with a volcanic eruption column, the buoyancy flux is not conserved in such a MEG plume, although a model can be developed based on the conservation of the MEG in the flow. A simplified version of the integral model presented above, including a constant entrainment coefficient, was shown to agree with these experiments (Woods and Caulfield, 1992). However, in the volcanic problem, which includes heat transfer and mixing, the picture is more complex. Woods and Caulfield (1992) also demonstrated that below a critical mass flux, the MEG is unable to mix with sufficient water to become buoyant, and instead it remains dense and forms a collapsing fountain. Near the conditions for collapse, the experimental flows became highly unsteady, with intermittent periods of plume- and flow-forming behavior; the exterior of the ascending mixture commonly entrained and mixed efficiently with the surrounding fluid, while the interior of the flow remained denser and less well-mixed, leading to these instabilities. As the flow collapsed asymmetrically, the continuing flow became buoyant and lifted off the horizontal surface in a manner analogous to co-PDC plumes.

8.3.3   Non-equilibrium effects

The model described in Section 8.3.2 is based on the assumption of equilibrium, in terms of both the dynamical and thermal coupling between the particles and the gas. For eruptions in which the erupting material does not fragment as efficiently, there will be some disequilibrium between the coarser-grained fragments and the gas. *Thermal disequilibrium* occurs because heat is transferred less efficiently between coarse particles and the gas. This reduces the supply of heat and hence may suppress the generation of buoyancy as the plume ascends. In turn, this may lead to lower column heights. *Dynamical disequilibrium* occurs as the larger particles, under the influence of gravity, decouple aerodynamically from the gas. This reduces the load of dense particles and so may assist in the buoyancy generation low in the column. However, higher in the column, the lack of solid material reduces the overall heat flux, and hence may lead to lower column heights, even though the mass of solid material is reduced.

Thermal equilibration between the solid particles and the air is ultimately controlled by the diffusion of heat within the solid fragments,

and this depends on the thermal diffusivity κ and the particle diameter, d. Typically, thermal equilibration occurs within a time of order d^2/κ. To understand the significance of this timescale compared to the timescale of the plume, we compare the conduction timescale with the timescale of ascent in the column, which is of order $1/N$. For the atmosphere, $1/N$ has a value of 100–200 s, and pumice clasts have a thermal diffusivity of order 10^{-7} m^2 s^{-1}, so that fragments smaller than 1 mm equilibrate in less than one-tenth of the ascent time of the column. This implies that particles larger than about 1 mm contribute progressively less of their thermal energy during ascent in the plume. In principle, 3 mm fragments are the largest that are able to fully equilibrate prior to reaching the plume top. In more coarse-grained eruptions, this lack of thermal equilibrium may have a profound impact on the plume ascent and also on the conditions for column collapse. Section 8.4 discusses some of the dynamics of lava fountains, and shows how the thermal and dynamical disequilibria of such coarse pyroclast populations lead to behavior very different from convecting columns.

To account for thermal disequilibrium effects, the ascent of the particles, including the vertical dispersion resulting from the turbulence, needs to be modeled along with the time-dependent heat transfer to the surrounding material. This involves a non-trivial development of the model to quantify the vertical dispersion within a turbulent plume and also to keep an inventory of the different particles in the plume in terms of their size and time of release, and hence their temperature. Although this is possible, in the context of the simplified integral models of an eruption column, a simpler heuristic approach has been developed to explore the impact of disequilibrium through introduction of a parameter, f, which reduces the fraction of the particles that are in good thermal contact with the gas. This approach essentially leads to a bimodal distribution of particles from a thermal perspective, with a fraction in equilibrium and the remainder in disequilibrium throughout their ascent in the column; this corresponds to particles smaller than ~0.5–1.0 mm and larger than ~5–10 mm,

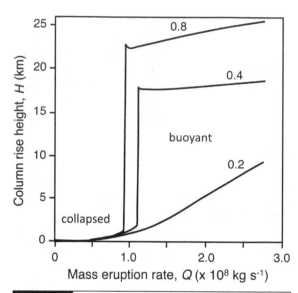

Figure 8.8 Variation of column rise height with eruption rate for a fixed vent radius and for different values of the thermal disequilibrium parameter f. For $f \sim 0.2$, a convecting column never forms. After Woods and Bursik (1991), reprinted with kind permission from Springer Science+Business media.

respectively. The specific heat used in calculating the thermal properties of the plume then becomes

$$c = \frac{c_a m_a + f c_s m_s + c_{vg} m_{vg}}{m_a + m_s + m_{vg}}. \tag{8.13}$$

A model of this form was investigated by Woods and Bursik (1991) and they found that, as the parameter f decreases from 1, the height of the column correspondingly decreases, because of the reduction in the thermal energy available to drive ascent and power the transition from a dense to a buoyant mixture (Fig. 8.8). In addition, eruption columns develop only with much larger mass fluxes as the thermal disequilibrium increases (i.e., for smaller f), and there is a critical eruption rate above which an eruption column is unlikely to form – instead the flow collapses to form a fountain feeding PDCs. For values of $f \sim 0.2$, a convecting column is unable to form at any mass flux.

A further impact of the particles in an eruption column can arise if the particles settle out

of the plume (Ernst *et al.*, 1996). If the settling is from the margin of the plume, then the material may form a coherent dense flow leading to near-field deposition. However, particles falling out from the spreading neutrally buoyant umbrella cloud may be re-entrained into the ascending column if they are sufficiently close to the plume. This process may occur on a timescale comparable to the fall time of the particles in the surrounding environment, and can lead to a substantial increase in the particle load of the column. To model the process, we first need to describe the spreading of the particles outwards in the umbrella cloud and the associated sedimentation from this cloud. The model of the eruption column then needs revision to account for this re-entrainment of particles, which impacts the mass balance in the column. Veitch and Woods (2002) presented a detailed model for particle re-entrainment in an idealized wind-free environment. Their model includes an equation for the rate of sedimentation from the spreading umbrella cloud in the vicinity of the column

$$\frac{\partial}{\partial t}(rH\rho\,\phi) + \frac{\partial}{\partial r}(ru_r H\rho\,\phi) = -ru_t\rho\phi, \tag{8.14}$$

where u_r is the outward radial velocity, u_t is the fall speed of the particles through the air, H is the column height, and ϕ is the volume fraction of particles in the umbrella cloud. This has solution

$$\phi(r) \approx \phi(0)\exp\left[-\frac{\rho u_t}{Q}\left(r^2 - r_o^2\right)\right] \tag{8.15}$$

for the concentration of particles in the cloud as a function of distance r from the axis of the eruption column, where r_o is the column radius at the point of transition to the umbrella plume (the *corner*; Fig. 8.2), $\phi(0)$ is the particle volume fraction at the corner, and Q is the mass flux, given by

$$Q = r\rho u_r H. \tag{8.16}$$

If the farthest distance from the column at which particles can be re-entrained is r_{me}, then the concentration of particles at the top of the column relative to the concentration at the base of the column is approximated by the relation

$$\phi(\text{top}) = \phi(\text{source}) \exp\left[\frac{\rho u_t}{Q}\left(r_{me}^2 - r_o^2\right)\right]. \quad (8.17)$$

This represents the amplification of the flux of particles as a result of the re-entrainment. Figure 8.9 shows the increasing particle flux with height, as well as the decrease in the upward column velocity resulting from re-entrainment, compared to the simple case with no re-entrainment. It can be seen that the additional particle loading in the column significantly slows the ascent rate. The critical conditions for collapse also change, with collapse occurring for higher eruption velocities (Fig. 8.10).

8.3.4 Phreatomagmatic effects

In the above description of volcanic eruption columns, the source conditions have been assumed to involve the ejection of a mixture of solid particles and volcanic gas, at high temperature. The material has been assumed to decompress in a region just above the vent, and then continue its ascent as a high-speed dense jet which mixes with air and becomes buoyant, thereby ascending up to several tens of kilometers into the atmosphere. However, in some situations the erupting material may issue into a crater lake, or interact with liquid water, leading to very rapid quenching of a proportion of the ejecta and rapid heat transfer to produce vapor (see Chapter 11). This lowers the residual thermal energy of the solid material that is available for heating any entrained air as the mixture of ejecta, air, water, and water vapor rises into the atmosphere. As a result, there will be less heat transfer to the mixture, and this itself may lead to column collapse and the formation of wet or vapor-rich, dense pyroclastic density currents. Koyaguchi and Woods (1996) explored the dynamics of such currents for different degrees of mixing in the erupting material, and this led to a series of predictions about the conditions for collapse as a function of the initial water content in the erupting mixture (Fig. 8.11).

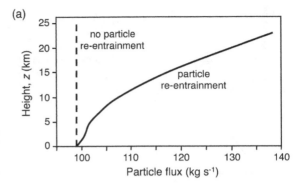

(a)

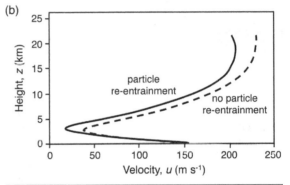

(b)

Figure 8.9 Illustration of the increase in particle load with height in an eruption column in which the effects of re-entrainment of particles are included: (a) variation of the particle flux with height; (b) variation of the column velocity with height. Magma mass flux is 10^8 kg s^{-1} and vent radius is 133 m; after Veitch and Woods (2002), reprinted with kind permission from Springer Science+Business media.

Each curve in Figure 8.11 represents a different mass fraction of water mixed into the erupting mixture, showing the transition in the critical eruption velocity required for formation of an eruption column vs. column collapse leading to pyroclastic flows. As the water content increases, the critical velocity initially decreases, because the water is all vaporized and therefore essentially provides an additional gas phase promoting buoyancy. However, as the water content increases to much larger values, the thermal energy required to heat and vaporize the water becomes more substantial and this reduces the temperature of the mixture and hence the ability for the mixture to become buoyant. In this case, wet ash flows or mud flows may develop unless the eruption velocity is high enough to allow for a much greater degree of atmospheric entrainment. Although this is a simplified

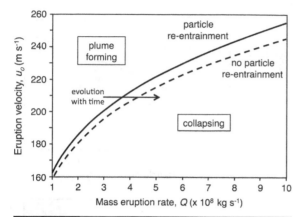

Figure 8.10 Illustration of the critical conditions for collapse of an eruption column, including the effect of the re-entrainment of particles (solid line), compared to the case with no re-entrainment (dashed line). If re-entrainment develops with time, then an initially buoyant column might later collapse. Magma temperature is 1000 K; after Veitch and Woods (2002), reprinted with kind permission from Springer Science+Business media.

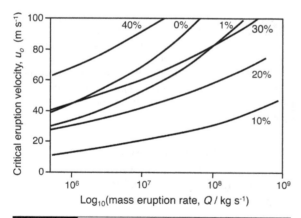

Figure 8.11 Minimum eruption velocity required for a buoyant column to develop, plotted as a function of mass flux, for eruptions in which there is significant mixing of water with the erupting material. The water fraction mixed into the erupting mixture is shown on the curves as a percentage. With sufficient water mixed into the flow, the material collapses to form a dense flow, even if the original eruption rate would otherwise have formed a buoyant column. Initial magma temperature is 1000 K and magmatic H_2O content is 3 wt.%. After Koyaguchi and Woods (1996).

picture of the full explosive interaction of water and ash, it does provide useful insight into the generation of buoyancy and the associated dynamics of phreatomagmatic explosions.

8.3.5 Numerical models

In addition to the one-dimensional, time-averaged models for steady-state eruption columns, a range of numerical models have been developed to model the dynamics of eruption columns with greater degrees of complexity. Early numerical simulations were based on an axisymmetric, steady-state treatment of the flow, and accounted for the motion of the gas and particles separately, initially including a single particle size but gradually evolving to account for a distribution of particle sizes. Flow dynamics were constrained by parameterizations of turbulent mixing in the column, and also by parameterizations of turbulent drag between the particles and air (Valentine and Wohletz, 1989; Dobran et al., 1993; Neri and Dobran, 1994). These models provided some fascinating insights into the formation of co-ignimbrite clouds above pyroclastic density currents generated from collapsing fountains, but the axisymmetric nature of the models represented a somewhat simplified picture of this complex phenomenon.

More recently, as computing power has increased, fully three-dimensional time-dependent models have been developed (e.g., Suzuki et al., 2005). These models provide interesting results concerning the time-dependent behavior of the eruption column, and the three-dimensional structures that develop in the flow, reminiscent of field observations. A critical feature of such numerical models is their ability to replicate the turbulent entrainment processes as measured in laboratory experiments (cf. Morton et al., 1956), through proper resolution of the engulfment of air by turbulent eddies and the subsequent mixing of this air with the material in the column. Suzuki et al. (2005) have shown that such resolution of entrainment requires a careful refinement of the computational grid size.

The three-dimensional time-dependent models have illustrated that the plume retains a dense inner core over the first 1–2 km above the vent. For the case in which a turbulent buoyant plume develops, the strong shear between this inner region and the outer, well-mixed zone disperses the inner dense region, and is key for the generation of bulk buoyancy (Fig. 8.12(a)). In

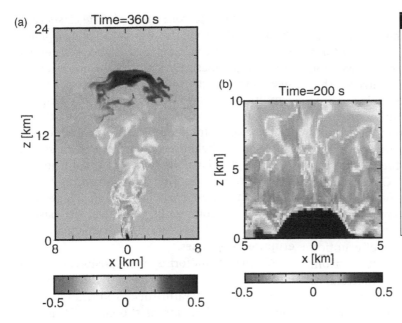

(a)

Time=360 s

(b)

Time=200 s

Figure 8.12 Results from the numerical simulations of Suzuki et al. (2005) illustrating the dynamics of (a) a buoyant eruption column (magma temperature 1000 K, mass flux = $10^{7.6}$ kg s^{-1}, and eruption velocity 150 m s^{-1}); and (b) a collapsing fountain (magma temperature 1000 K, mass flux = $10^{8.65}$ kg s^{-1}, and eruption velocity 150 m s^{-1}). The colors represent the density of the column relative to the ambient atmosphere, with blue being relatively buoyant and orange being relatively dense. See color plates section.

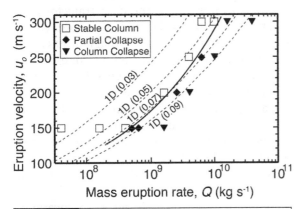

Figure 8.13 Comparison of numerical predictions of the conditions for collapse (symbols; Suzuki et al., 2005) with the predictions of the simpler one-dimensional (1D) integral models using a range of values for the entrainment coefficient as shown in parentheses on the individual curves. Square symbols represent a stable column regime, triangles indicate column collapse, and diamonds represent partial column collapse. From Suzuki et al. (2005), using initial temperature 1000 K and magma gas content 5 wt.%.

ground in co-ignimbrite (co-PDC) plumes. For even greater fluxes, the model predicts large-scale column collapse (Fig. 8.12(b)), as observed in experimental models of eruption columns (Woods and Caulfield, 1992).

One of the interesting features of these numerical simulations is that they do not assume any particular value for the entrainment coefficient that is a key ingredient of the averaged one-dimensional models presented in Sections 8.3.2–8.3.4. Suzuki et al. (2005) compared their numerical predictions of the conditions for collapse with earlier one-dimensional models and showed that the appropriate value for the entrainment coefficient in collapsing columns is of order 0.07 ± 0.01, whereas for buoyant columns, the appropriate entrainment coefficient is closer to 0.1 (Fig. 8.13). This difference in the entrainment coefficient arises from the suppression of mixing in a negatively buoyant jet, as compared to the turbulent buoyant plume, and is consistent with detailed experimental studies (e.g., Kaminski et al., 2005), and with field evidence that the entrainment of air into columns that collapse is much less than in columns that develop turbulent buoyant plumes (Kaminski and Jaupart, 2001).

larger eruption columns, some of the material in the column becomes buoyant in the same way, while other parts of the flow remain dense and fall back to the ground, leading to the formation of pyroclastic density currents which themselves then becomes buoyant and rise from the

8.3.6 Co-ignimbrite clouds

One of the other fascinating phenomena that have been observed to occur in large explosive eruptions is the formation of large convecting plumes above pyroclastic density currents that are generated by collapsing fountains. These plumes arise through the entrainment of air and sedimentation of particles from the ground-hugging flow, which results in the continuing flow becoming less dense than the air and lifting off to form buoyant plumes. Other plumes arise from the elutriation of the finer size fraction as the current continues to flow. The resultant dynamics of such co-ignimbrite clouds is analogous to the convective region of eruption columns, but the material may lift off far from the vent (Dobran *et al.*, 1993). If the material in these flows cools sufficiently as it spreads along the ground, for example as a result of mixing with water, then the continuing flow may be unable to generate buoyancy (Koyaguchi and Woods, 1993).

8.4 | Basaltic systems and lava-fountaining eruptions

Basaltic magma is less viscous and has lower volatile contents than the intermediate to felsic magmas that tend to produce convecting eruption columns. Because of the low viscosity, the exsolving gas phases can decouple more readily from the melt as it ascends to the surface, so that for low magma ascent speeds, intermittent emission of large gas slugs can produce strombolian explosions (Chapter 6). At greater magma rise rates, the bubbles and magma remain dynamically coupled, and the low viscosity prevents significant bubble overpressures from building up as the bubbly magma decompresses during ascent (Chapter 4, Section 4.3.2). The lack of significant bubble overpressure, together with the low total volatile content, means that fragmentation, if it occurs, is less vigorous than for more silicic systems. As a result, a coarser pyroclast size distribution is produced (with clasts typically ranging from millimeters to tens of centimeters), and gas expansion produces relatively low

speeds upon exit from the vent. The clasts readily decouple from the low-velocity gas phase, so instead of erupting as a high-speed, effectively single-phase flow as for the more explosive counterparts, the erupting material forms a multiphase flow. This can lead to a range of eruption phenomena from ballistic trajectories for the coarser fragments, forming incandescent *lava fountains* (when erupted from point source vents) or *curtains of fire* (when erupted from linear fissures), to buoyant plume-forming behavior for the finest particle fraction (Fig. 8.1(b)). Fountaining commonly gives way to passive lava effusion of less volatile-rich magma as the eruption progresses.

A range of landforms and deposits can be produced from lava fountaining activity, depending on the local temperature of clasts on landing and the accumulation rate of the clasts (Head and Wilson, 1989). Temperature and accumulation rate in turn depend on the clast trajectories (a function of fountain size and structure), which are ultimately related to the magma mass flux and volatile content (Fig. 8.14(a,b)). For example, relatively large volatile contents lead to high eruption velocities, tall fountains, and widespread dispersal from the vent. Particles landing far from the vent will be relatively cool on landing, whereas coarse clasts landing close to the vent will be hot and fluid. Therefore a range of deposits may be observed, from rheomorphic lava flows produced by coalescence of rapidly accumulating, hot clasts in proximal locations, to cone-forming agglutinated spatter and/or brittle scoria at increasing distances, and sheet-like ash fall deposits dispersed distally from the fine-grained component rising above the fountain (Fig. 8.14(c)). As an illustration, during the 1995 eruption of Cerro Negro (Hill *et al.*, 1998), an eruption column reached an altitude of about 5 km and transported small particles with median size of 0.7 mm to > 30 km downwind of the volcano, while also contributing to the growth of a near-source cone from larger volcanic blocks and bombs up to 2 m in diameter.

Lava fountains range from a few meters high to several hundred meters in height. For example, the 1959 eruption of Kilauea Iki in Hawaii (Fig. 8.1(b)) produced fountains close to

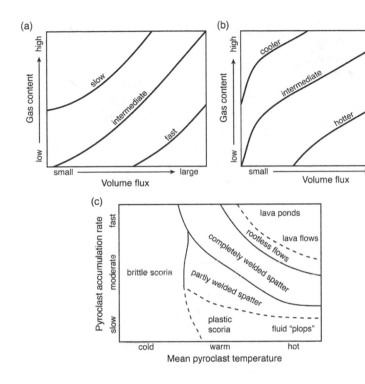

Figure. 8.14 Illustration of the influence of magma mass flux and volatile content on the (a) accumulation rate and (b) temperature of pyroclastic material on landing, and (c) the influence of accumulation rate and temperature on the nature of the resulting deposit (after Head and Wilson, 1989; reprinted with permission from Elsevier).

600 m high (Richter *et al.*, 1970). Fountain height, H_f, can be related to the vent speed, u_o, given in terms of a simple ballistic relation, neglecting air resistance for sufficiently large particles,

$$u_o = \sqrt{2gH_f}, \qquad (8.18)$$

which implies that pyroclasts are being ejected from the vent with speeds up to 110 m s^{-1}. Fountain heights and eruption velocities can in turn be linked to the exsolved magma volatile content driving the eruption (Head and Wilson, 1987). The coarse size of most clasts means that, upon eruption, they are slipping relative to the gas phase, which suggests that gas eruption speeds are significantly greater than the velocities of magma fragments calculated from fountain heights (Wilson, 1999).

While coarse clasts fall rapidly and thus are unable to contribute thermally to driving a buoyant column, the gas phase and fine particles may rise above the lava fountains to form a small ash plume. Stothers *et al.* (1986) estimated the height of rise of these plumes in terms of the mass flux supplied from basaltic

eruptions, and Woods (1993) extended this to account for the effects of the entrainment of moisture. Such calculations show that even if only a small fraction of the mass erupting in a basaltic fissure eruption leads to formation of an ash plume, the plume may rise several kilometers through the lower atmosphere because the air is only weakly stratified. The potential for such plumes to inject large quantities of aerosols high into the atmosphere has been proposed as one process whereby massive flood basalts can impact climate. Such plumes typically only carry a small flux of erupting material, so that in humid environments, the entrainment and vertical transport of water may be a key factor for increasing the height of rise of such plumes.

In considering sustained explosive activity, it is important to note that there is a range of phenomena, from eruption of very coarse-grained, fluid lava fountains to the highly fragmented mixture that forms ash plumes. This range may be considered as a continuum of activity, dependent on the degree of fragmentation (and hence efficiency of heat transfer between particle and gas phases) as well as the total volatile content of the erupting magma, both of which are largely dependent on magma composition. The different styles of behavior may be thought of as a spectrum from hawaiian-style lava fountaining, to subplinian activity, to vigorous plinian eruptions. Finally, we note that while plinian eruptions of basalt are rare, they are not unheard of, and may require rapid magma ascent, leading to delayed outgassing and the development of large supersaturations, and/or some degree of interaction with external water, to enhance fragmentation efficiency and drive a convecting plume (e.g., Williams, 1983; Walker *et al.*, 1984).

8.5 | Summary and outlook

This chapter summarizes research from the past 20 years on the dynamics of volcanic eruption columns. Many of the key principles are understood, and there is agreement emerging now between field observations, theoretical, and numerical models, concerning some of the key properties of eruption columns, including the rise height as a function of eruption rate and the principles governing the conditions for collapse. Recent numerical models have confirmed the quantitative predictions of the simpler integral models for column collapse and for total rise height, and detailed modeling has led to more accurate parameterizations of the entrainment and mixing of air into the flow. This opens up new possibilities to explore other features of column dynamics and motion which are beyond the simpler integral models. It will also be interesting to build more constraints on the models from field data, in terms of thermal equilibration, fallout of particles, and phreatomagmatic processes. However, models are already able to provide good predictions of column height for use in subsequent tephra dispersal models (see Chapter 9), and for constraining eruption rate and eruption histories from fall deposits (cf. Sparks *et al.*, 1997).

8.6 | Notation

b	column radius (m)
B	buoyancy flux (m^4 s^{-3})
c	specific heat capacity (J kg^{-1} K^{-1})
c_a	specific heat capacity of ambient air (J kg^{-1} K^{-1})
c_g	specific heat capacity of volcanic and atmospheric gases (J kg^{-1} K^{-1})
c_s	specific heat capacity of particles (J kg^{-1} K^{-1})
c_{vg}	specific heat capacity of volcanic gases (J kg^{-1} K^{-1})
d	particle diameter (m)
f	mass fraction of particles in good thermal contact with gas
g	gravitational acceleration (9.81 m s^{-2})
H	rise height of eruption column (m)
H_f	fountain height (m)
m_a	mass of air in column (kg)
m_s	mass of particles in column (kg)
m_{vg}	mass of volcanic gases in column (kg)
n	gas mass fraction
n_o	initial gas mass fraction
N	Brunt–Väisälä frequency (s^{-1})
Q	mass eruption rate (kg s^{-1})
Q_H	heat flux (J s^{-1})
r	radial coordinate in umbrella cloud (m)
r_{me}	maximum entrainment distance (m)
r_o	radius of column at corner (m)
T	temperature (K)
T_a	temperature of ambient air (K)
T_o	temperature of ambient air at source level (K)
u	plume average vertical speed (m s^{-1})
u_o	vent eruption velocity (m s^{-1})
u_r	radial velocity in umbrella cloud (m s^{-1})
u_t	fall velocity of particles (m s^{-1})
$\dot{V}$	volume flux (m^3 s^{-1})
z	distance above vent (m)
ε	entrainment coefficient
κ	thermal diffusivity (m^2 s^{-1})
λ	constant of proportionality
ρ	density (kg m^{-3})
$\Delta\rho$	density difference between ambient fluid at source level and the source fluid (kg m^{-3})

ρ_a density of ambient air (kg m^{-3})
ρ_g density of gas phase (kg m^{-3})
ρ_0 density of ambient fluid at source level (kg m^{-3})
ρ_s particle density (kg m^{-3})
ϕ volume fraction of solid particles in plume
$\phi(0)$ particle volume fraction at plume corner

References

Bursik, M. I. and Woods, A. W. (1991). Buoyant, superbuoyant and collapsing eruption columns. *Journal of Volcanology and Geothermal Research*, **5**, 347–350.

Caulfield, C. C. P. and Woods, A. W. (1995). Plumes with nonmonotonic mixing behavior. *Geophysical and Astrophysical Fluid Dynamics*, **79**, 173–199.

Dobran, F., Neri, A. and Macedonio, G. (1993). Numerical simulation of collapsing volcanic columns. *Journal of Geophysical Research*, **98**(B3), 4231–4259.

Ernst, G., Sparks, R. S. J., Carey, S. N. and Bursik, M. I. (1996). Sedimentation from turbulent jets and plumes. *Journal of Geophysical Research*, **101**, 5575–5589.

Gill, A. (1981). *Atmosphere–Ocean Dynamics.* International Geophysics Series. New York: Academic Press.

Glaze, L. S. and Baloga, S. M. (1996). Sensitivity of buoyant plume rise heights to ambient atmospheric conditions: Implications for volcanic eruption columns. *Journal of Geophysical Research*, **101**(D1), 1529–1540.

Glaze, L. S., Baloga, S. M. and Wilson, L. (1997). Transport of atmospheric water vapor by volcanic eruption columns. *Journal of Geophysical Research*, **102**(D5), 6099–6108.

Head, J. W. and Wilson, L. (1987). Lava fountain heights at Pu'u 'O'o, Kilauea, Hawaii: Indicators of amount and variations of exsolved magma volatiles. *Journal of Geophysical Research*, **92**, 13 715–13 719.

Head, J. and Wilson, L. (1989). Basaltic pyroclastic eruptions: Influence of gas-release patterns and volume fluxes on fountain structure, and the formation of cinder cones, spatter cones, rootless flows, lava ponds and lava flows. *Journal of Volcanology and Geothermal Research*, **37**, 261–271.

Hill, B. E., Connor, C. B., Jarzemba, M. S. *et al.* (1998). 1995 eruptions of Cerro Negro volcano, Nicaragua, and risk assessment for future eruptions. *Geological Society of America Bulletin*, **110**, 1231–1241.

Kaminski, E. and Jaupart, C. (2001). Marginal stability of atmospheric eruption columns and pyroclastic flow generation. *Journal of Geophysical Research*, **106**(B10), 21 785–21 798.

Kaminski, E., Tait, S. and Carazzo, G. (2005). Turbulent entrainment in jets with arbitrary buoyancy. *Journal of Fluid Mechanics*, **526**, 361–376.

Koyaguchi, T. and Woods, A. W. (1996). On the explosive interaction of magma and water, *Journal of Geophysical Research*, **101**, 5561–5574.

Morton, B., Taylor, G. I. and Turner, J. (1956). Turbulent gravitational convection from maintained and instantaneous sources. *Proceedings of the Royal Society of London*, **A231**, 1–23.

Neri, A. and Dobran, F. (1994). Influence of eruption parameters on the thermofluid dynamics of collapsing volcanic columns. *Journal of Geophysical Research*, **99**, 11 833–11 857.

Neri, A. and Macedonio, G. (1996). Numerical simulation of collapsing volcanic columns with particles of two sizes. *Journal of Geophysical Research*, **101**(B4), 8153–8174.

Neri, A., Papale, P. and Macedonio, G. (1998). The role of magma composition and water content in explosive eruptions: II. Pyroclastic dispersion dynamics. *Journal of Volcanology and Geothermal Research*, **87**, 95–115.

Richter, D. H., Eaton, J. P., Murata, K. J., Ault, W. U. and Krivoy, H. L. (1970). *Chronological Narrative of the 1959–60 Eruption of Kilauea Volcano, Hawaii.* United States Geological Survey Professional Paper 537-E.

Sparks, R. S. J. (1986). The dimensions and dynamics of eruption columns. *Bulletin of Volcanology*, **48**, 3–15.

Sparks, R. S. J., Bursik, M., Carey, S. *et al.* (1997). *Volcanic Plumes.* Chichester: John Wiley.

Stothers, R., Wolff, J. A., Self, S. and Rampino, M. (1986). Basaltic fissure eruptions, plume heights and atmospheric aerosols. *Geophysical Research Letters*, **13**, 725–728.

Suzuki, Y., Koyaguchi, T., Ogawa, M. and Hachisu, I. (2005). A numerical study of turbulent mixing in eruption clouds using a three dimensional fluid dynamics model. *Journal of Geophysical Research*, **110**(B08), 201–219.

Valentine, G. and Wohletz, K. (1989). Numerical models of Plinian eruption columns and pyroclastic flows. *Journal of Geophysical Research*, **94**(B2), 1867–1887.

Veitch, G. and Woods, A. W. (2002). Particle recycling in volcanic eruption columns. *Bulletin of Volcanology*, **64**, 31–39.

Walker, G. P. L., Self, S. and Wilson, L. (1984). Tarawera 1886, New Zealand – a basaltic plinian fissure eruption. *Journal of Volcanology and Geothermal Research*, **21**, 61–78.

Williams, S. N. (1983). Plinian airfall deposits of basaltic composition. *Geology*, **11**, 211–214.

Wilson, L. (1976). Explosive volcanic eruptions - III. Plinian eruption columns. *Geophysical Journal of the Royal Astronomical Society*, **45**, 543–556.

Wilson, L. (1999). Explosive volcanic eruptions – X. The influence of pyroclast size distributions and released magma gas content on the eruption velocities of pyroclasts and gas in Hawaiian and Plinian eruptions. *Geophysical Journal International*, **136**, 609–619.

Wilson, L., Sparks, R. S. J., Huang, T. C. and Watkins, N. D. (1978). The control of volcanic column heights by eruption energetics and dynamics. *Journal of Geophysical Research*, **83**(B4), 1829–1836.

Woods, A. W. (1988). The fluid dynamics and thermodynamics of eruption columns. *Bulletin of Volcanology*, **50**, 169–193.

Woods, A. W. (1993). A model of the plumes above basaltic fissure eruptions. *Geophysical Research Letters*, **20**, 1115–1118.

Woods, A. W. and Bursik, M. I. (1991). Particle fallout, thermal disequilibrium and volcanic plumes. *Bulletin of Volcanology*, **53**, 559–570.

Woods, A. W. and Caulfield, C. P. (1992). A laboratory study of explosive volcanic eruptions. *Journal of Geophysical Research*, **97**, 6699–6712.

Exercises

8.1 Estimate the height of rise of a plinian eruption column rising through an environment with a Brunt–Väisälä frequency of 0.01 s^{-1} if the eruption rate Q is 10^6 kg s^{-1}. Use typical values for atmospheric temperature and density at sea level, and assume a rhyolite eruption temperature. How will column height change if the particle size gradually increases with time through the eruption?

8.2 As basaltic fissure eruptions evolve, there is often some flow localization, leading to discrete axisymmetric fountains rising from one or more point-source vents. If 10% of the material from a basaltic eruption rises to form an ash plume above each lava fountain, and the plume has a height of 2.5 km, estimate the eruption rate for each fountain. Assume $N = 0.01$ s^{-1} and similar atmospheric characteristics to Exercise 8.1.

Online resources available at www.cambridge.org/fagents

- Answers to exercises
- Additional reading
- Additional figures

Chapter 9

Modeling tephra sedimentation from volcanic plumes

Costanza Bonadonna and Antonio Costa

Overview

Tephra erupted in volcanic plumes can be transported over distances of thousands of kilometers, causing respiratory problems to humans and animals, serious damage to buildings and infrastructure, and affecting economic sectors such as aviation, agriculture, and tourism. Models with different degrees of complexity have been developed over the last few decades to describe tephra dispersal. Depending on the application, different simplifications and assumptions can be introduced to make the problem tractable. Highly sophisticated models are not suited for the computationally expensive probabilistic calculations required by long-term hazard assessments. In contrast, the simplified models typically used for probabilistic assessments have to compromise the sophistication of the physical formulation for computational speed. A comprehensive understanding of tephra deposits and hazards can only result from a critical and synergistic application of models with different levels of sophistication, ranging from purely empirical to fully numerical. A review of the main approaches to tephra dispersal modeling is presented in this chapter.

9.1 | Introduction

Explosive volcanic eruptions have intrigued scientists because of their dramatic display of physical processes, their crucial role in the geological evolution of Earth, and their potentially catastrophic consequences for society. A key way of improving our understanding of explosive volcanism is to study the resulting pyroclastic deposits, which often represent the only direct evidence of explosive eruptions. Tephra deposits retain a considerable amount of information about the nature of the eruption, such as erupted mass, bulk grain-size distribution, and eruption intensity. However, tephra falls also represent significant hazards for people living close to active volcanoes. These hazards include collapse of buildings, disruption to water and electricity supplies, disruption to transportation networks, as well as health hazards from respirable ash, crop pollution, and lahar generation. Developing an understanding of tephra fall is crucial to public safety. In this chapter *tephra* is used in the original sense of Thorarinsson (1944) as a collective term for all particles ejected from volcanoes, irrespective of size, shape, and composition, whereas *tephra fall* indicates the process of particle fallout.

Modeling Volcanic Processes: The Physics and Mathematics of Volcanism, eds. Sarah A. Fagents, Tracy K. P. Gregg, and Rosaly M. C. Lopes. Published by Cambridge University Press. © Cambridge University Press 2013.

The preservation of tephra deposits is typically incomplete. In most eruptions proximal areas are buried or collapsed, and distal deposition occurs in the sea or becomes eroded. For small eruptions, the whole tephra deposit may be eroded away within a few years of deposition. As a result, empirical, analytical, and numerical models have been developed to allow quantitative interpretation of tephra deposits and to fully understand the nature of ancient eruptions that are incompletely preserved. Dedicated analytical and numerical models have also been produced to investigate plume dynamics and particle sedimentation, and to provide long-term assessments for land-use planning and rapid response during volcanic crises. Model validations have shown good agreement with field data, which justifies the use of these models for hazard applications.

This chapter describes: (1) empirical and analytical models used to determine eruption parameters, such as column height, eruption duration, magnitude, and intensity, (2) analytical and numerical models developed for the study of the dynamics of volcanic plumes and particle sedimentation, and (3) models commonly used for hazard assessments and forecast of plume spreading. Model assumptions and caveats are discussed, and a key case study is presented to facilitate comprehension of the application of the models described (the 22 July 1998 explosive eruption of Mt Etna, Italy; Coltelli et al., 2006; Scollo et al., 2008a).

9.2 | Plume dynamics and particle sedimentation

Before describing the models used to characterize tephra dispersal and deposits, it is important to understand some basic concepts of plume dynamics and particle sedimentation (see Chapter 8 for a detailed review of plume dynamics). Volcanic plumes are typically associated with explosive activity and consist of a mixture of lithics (wall rock), volcanic gas, and juvenile particles (fragmented magma), which, whether generated from a point source (i.e., single vent) or from an extended source (i.e., fissure eruption or pyroclastic density current), eventually develops into a turbulent buoyant current whose dynamics are strongly controlled by the degree of interaction with the atmosphere. If the plume upward velocity is much stronger than the wind velocity, the initial jet phase (gas thrust) evolves into a vertical buoyant column that then eventually spreads laterally as a gravity current (i.e., umbrella cloud) around the neutral buoyancy level H_b (i.e., strong plume; Fig. 9.1(a)). In contrast, if the wind velocity is much stronger than the plume upward velocity, the turbulent current will be bent over above the basal jet before spreading laterally around H_b (i.e., weak plume; Fig. 9.1(c)). It is important to distinguish between vigorous and low-energy weak plumes: both plumes are bent over by the wind but they are characterized by different energetics (e.g., steepness of plume trajectory relative to wind speed). Typically, vigorous weak plumes characterize the beginning of low-intensity sustained eruptions (e.g., Ruapehu, 17 June 1996; Bonadonna et al., 2005a), whereas low-energy weak plumes characterize the last phase of an eruption when wind eventually dominates and the cloud starts propagating as a lens of aerosol (e.g., Mount St. Helens, 22 July 1980; Sparks et al., 1997).

Volcanic clasts (juvenile and lithic fragment) are carried up within the turbulent current according to their settling velocity, which depends on both particle and atmospheric characteristics. When particle settling velocities are larger than the upwards component of the turbulent current, they fall out and are advected by local winds. Particles that are sufficiently small will typically aggregate into micron- to millimeter-sized clusters having greater settling velocities (Sparks et al., 1997). In addition, the deposition of fine particles is also enhanced by pronounced convective instabilities and mammatus that often form at the base of the sedimenting turbulent current (Bonadonna et al., 2002b; Durant et al., 2009). As a result, the characteristics of tephra deposits are the result of plume dynamics (e.g., plume height, velocity profile, weak-plume vorticity), particle parameters (e.g., size, density and shape), atmospheric characteristics (e.g., wind field, atmospheric density and viscosity) and sedimentation

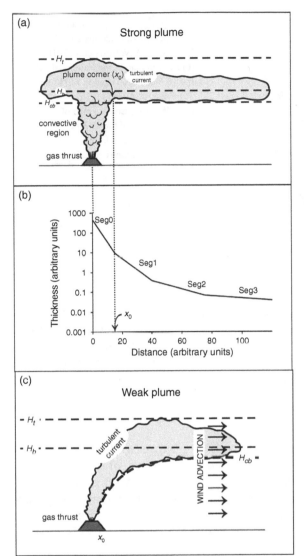

Diagrams showing (a) the main characteristics of a strong volcanic plume: H_b is the neutral buoyancy level of the plume, H_t and H_{cb} are the top and the base of the spreading turbulent, umbrella cloud (modified from Bonadonna and Phillips, 2003); (b) sedimentation from plume margins (Seg0) and umbrella cloud (Seg1 = turbulent regime, Reynolds number, Re > 500; Seg2 = intermediate regime, 6 < Re < 500; Seg3 = laminar regime, Re < 6) of a strong plume (fallout regimes defined as in Bonadonna and Phillips, 2003; (c) the main characteristics of a weak volcanic plume (modified from Bonadonna et al., 2005a).

processes (e.g., particle aggregation, convective instabilities, mammatus).

Tephra fall from strong plumes is characterized by four main sedimentation regimes: turbulent fallout from the plume margins and turbulent, intermediate, and laminar fallout from the umbrella cloud. These fallout regimes result in four exponential segments in semi-log plots of deposit thickness vs. distance from vent (Fig. 9.1(b)). Resulting isopach or isomass maps range from nearly concentric in the absence of wind (Fig. 9.2(a)) to strongly elongated in the case of significant wind advection (Figs. 9.2(b,c)). Because of the strong dependence of the settling velocity and Reynolds number on particle size, the final deposit morphology is also controlled by the initial grain-size distribution. In comparison to strong plumes, tephra fall from weak plumes is characterized by a lack of up-wind sedimentation, more pronounced proximal thinning due to the bent-over structure of the turbulent current, and narrower tephra deposits due to the presence of vortex structures underneath the plume. For example, blocks, lapilli, and a proportion of coarse ash were deposited from the rising phase of the 17 June 1996 Ruapehu weak plume (within ~30 km of the vent), whereas the horizontal spreading was characterized by deposition of coarse and fine ash (Bonadonna et al., 2005a). The resulting deposit (Fig. 9.2(c)) shows maximum accumulation along the dispersal axis, but as soon as the plume was no longer sustained the eruption lost its vigor and the spreading turbulent current started bifurcating due to vorticity conservation (~800 km from the vent).

9.3 | Empirical and analytical models used for the characterization of tephra deposits

Field investigations of tephra deposits are crucial for characterization of volcanic eruptions and their hazards. In particular, distributions of tephra thickness and mass per unit area (isopach and isomass maps; Fig. 9.2) are necessary for estimating erupted volume or mass (Pyle, 1989; Fierstein and Nathenson, 1992; Bonadonna and Houghton, 2005), whereas the distribution of largest clasts (isopleth maps) is typically used for estimating column height and wind speed at the

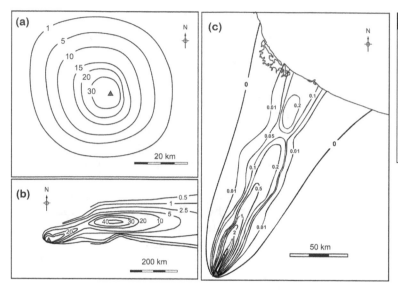

Figure 9.2 (a) Isopach map of the Basal Fall of Pululagua 2450 b.p., Ecuador (cm) (Volentik *et al.*, 2010); (b) isopach map of tephra deposit of the 18 May 1980 eruption of Mount St. Helens, USA (mm) (Sarna-Wojcicki *et al.*, 1981); (c) isomass map of the tephra deposit of the 17 June 1996 eruption of Ruapehu, New Zealand (kg m^{-2}) (Bonadonna *et al.*, 2005a).

time of the eruption (Carey and Sparks, 1986). Isopach, isomass and isopleth maps can also be used to determine vent location and to classify eruptive style (Walker, 1973; Walker, 1980; Pyle, 1989). Mass eruption rate and the duration of the sustained phase of the eruption can be calculated from plume height and erupted mass, respectively (Sparks, 1986; Wilson and Walker, 1987; Carey and Sigurdsson, 1989). Inferences of fragmentation mechanisms can also be made from the study of particle sizes (Kaminski and Jaupart, 1998; Neri *et al.*, 1998; Zimanowski *et al.*, 2003).

The empirical and analytical models used for these purposes, together with their assumptions and limitations, require thorough analysis to assess the variability of resulting eruption parameters. This is crucial not only because these eruption parameters are used to characterize volcanic eruptions, but also because they are used as input to numerical models and to construct potential activity scenarios for hazard assessment.

9.3.1 Determination of erupted volume based on the assumption of exponential thinning of tephra deposits

This approach was introduced by Pyle (1989), adopting the preliminary observation of Thorarinsson (1954) that both thickness and grain-size of tephra deposits follow an exponential decay with distance from the vent. As a result, the logarithm of tephra thickness can be described by straight lines (i.e., exponential segments) when plotted against distance from vent or square root of the area enclosed by each isopach ($\sqrt{A}$):

$$T = T_o \exp\left(-k\sqrt{A}\right) \tag{9.1}$$

where T_o is the maximum thickness of the deposit and k defines the rate of thinning of the deposit (i.e., slope of the associated exponential segment). All notation is summarized in Section 9.9. Assuming that isopachs have elliptical shapes, the volume of tephra deposit is:

$$V = 13.08\, T_o b_t^2 \tag{9.2}$$

where $b_t = \ln(2)/(k\sqrt{\pi})$.

Fierstein and Nathenson (1992), Pyle (1995), and Bonadonna and Houghton (2005) developed this method to account for abrupt changes in the rate of thinning of some tephra deposits:

$$
\begin{aligned}
V &= \frac{2T_{1_0}}{k_1^{\,2}} + 2T_{1_0}\left[\frac{k_2 S_1 + 1}{k_2^2} - \frac{k_1 S_1 + 1}{k_1^2}\right]\exp(-k_1 S_1) \\
&\quad + 2T_{2_0}\left[\frac{k_3 S_2 + 1}{k_3^2} - \frac{k_2 S_2 + 1}{k_2^2}\right]\exp(-k_2 S_2) + \ldots + 2T_{(n-1)_0} \\
&\quad \times \left[\frac{k_n S_{(n-1)} + 1}{k_n^2} - \frac{k_{(n-1)} S_{(n-1)} + 1}{k_{(n-1)}^2}\right]\exp\left(-k_{(n-1)} S_{(n-1)}\right)
\end{aligned}
\tag{9.3}
$$

where T_{n0}, k_n, and S_n are the intercept, slope, and position of the break in slope of line segment n. Their approach to estimating volume by defining several exponential segments (i.e., different values of k from proximal to distal portions of the deposit) is consistent with observations of well-preserved tephra deposits (Hildreth and Drake, 1992; Scasso et al., 1994) and with the results of some analytical models (Bursik et al., 1992a; Sparks et al., 1992; Bonadonna et al., 1998).

The approach of Pyle (1989) was also modified to estimate erupted volume for cases in which only one proximal isopach can be defined based on the available data (Legros, 2000). This technique was derived from empirical investigation of 74 tephra deposits and gives estimated minimum volumes of the same order of magnitude as for the case in which only the first two segments on semi-log plots of thickness T vs. square root of area $\sqrt{A}$ are available (Seg0 and Seg1 in Fig. 9.1(b)):

$$V = 3.69\, T_x A_x \qquad (9.4)$$

where A_x (m²) is the area enclosed within the isopach with thickness T_x (m). Sulpizio (2005) presented additional techniques for the determination of distal volume when most distal data are missing, based on extrapolation of the distribution of proximal deposits to distal areas (up to thickness > 1 cm). Finally, Mannen (2006) suggested an analytical method to derive the total erupted mass of relatively small eruptions by adopting the model of Bursik et al. (1992a) and integrating two exponential segments determined from isopleth maps.

9.3.2 Determination of erupted volume based on the assumption of power-law thinning of tephra deposits

Based on the results of analytical investigations of Bonadonna et al. (1998) and on observations of well-preserved deposits for which thinning can be described by a power-law fit on a semi-log plot of T vs. $\sqrt{A}$, Bonadonna and Houghton (2005) suggested deriving the total erupted volume by integrating the power-law best fit to

field data. In particular, the power-law best fit can be described as:

$$T = T_0 \left(\frac{\sqrt{A_0}}{\sqrt{A}} \right)^m = C_{pl} \left(\sqrt{A} \right)^{-m} \quad \text{with} \quad C_{pl} = T_0 A_0^{m/2}$$
$$(9.5)$$

where C_{pl} and m are the power-law coefficient and exponent respectively. The associated volume can be calculated as:

$$V = \frac{2C_{pl}}{2-m} \left(\sqrt{A_{dist}}^{2-m} - \sqrt{A_0}^{2-m} \right), \qquad (9.6)$$

which is equivalent to:

$$V = \frac{2}{2-m} T_0 A_0 \left(\sqrt{\frac{A_{dist}}{A_0}}^{2-m} - 1 \right), \qquad (9.7)$$

where $\sqrt{A_0}$ and $\sqrt{A_{dist}}$ are two arbitrary integration limits. In particular, A_{dist} ideally represents the area of the isoline of zero thickness, whereas A_0 is area enclosed by the isoline corresponding to the maximum deposit thickness. Note that, to guarantee convergence in the limit of large distances, m has to be > 2.

Caveats

Sensitivity analyses of volume calculations have shown that integration of less than three exponential segments can underestimate deposit volume when distal data are missing (Bonadonna and Houghton, 2005). For example, integration of only two exponential segments described by data within 10 km of the 1996 Ruapehu eruptive vent resulted in an underestimation by half of the actual deposit volume. Such an underestimation does not affect the classification of the eruption in terms of the volcanic explosivity index (VEI), but is significant when simulating tephra dispersal and compiling hazard assessments. In contrast, the power-law fit is a good approximation to well-preserved deposits and is consistent with theoretical models, but it is also problematic because integration limits have to be chosen. In particular, the volume of deposits having limited dispersal ($m > 2$) is very sensitive to the choice of $\sqrt{A_0}$ but not to the choice of $\sqrt{A_{dist}}$. In contrast, the volume of deposits having wide dispersal ($m < 2$) is very sensitive to $\sqrt{A_{dist}}$ but not to $\sqrt{A_0}$. Given that $\sqrt{A_{dist}}$ is difficult to constrain, the power-law method may be used for small deposits ($m > 2$) but is not recommended for widely dispersed deposits ($m < 2$) when the outer limit of the deposit cannot be accurately identified. Essentially, empirical fitting of poor data sets can be problematic, especially for large eruptions, even with the power-law method because there is no simple theoretical relationship between proximal and distal thinning. Proximal deposition is controlled by high-Reynolds-number particles, whereas distal deposition is controlled by low-Reynolds-number particles (Fig. 9.1(b)).

9.3.3 Determination of column height and wind speed

Even though a buoyant eruptive column is characterized by fluctuating vertical velocities, plume studies have shown that the horizontal profile of time-averaged vertical speed can be represented by a Gaussian distribution symmetrical with respect to the plume axis (Turner, 1979). From comparison between this Gaussian-distributed plume velocity and the settling velocities of volcanic particles, Carey and Sparks (1986) defined a series of theoretical "envelopes" that support the particles within plumes. When the settling velocity of the particles exceeds the vertical velocity of the plume (characteristic of a given envelope), particles will leave the plume and deposit on the ground at distances that depend on release height, and wind speed and direction. As a result, the column height and wind speed can be derived by plotting the maximum downwind range vs. the crosswind range of isopleth values (i.e., the length and half-width of individual isopleths). Carey and Sparks (1986) provide plots that are based on a specific wind profile (Shaw et al., 1974) and specific particle characteristics (i.e., size and density). In particular, the wind profile is assumed to be uni-directional at all heights, with a maximum velocity (5–50 m s^{-1}) at the tropopause level (11 km), a linear decay to zero at ground level, and a value above the tropopause of 3/4 of the maximum velocity.

Caveats

The method of Carey and Sparks (1986) provides a maximum column height from the distribution of largest clasts (pumice and lithic fragments). However, there are some important caveats to bear in mind when applying this technique. First, the distribution of largest clasts is not unique and is very sensitive to the techniques used to determine average clast size at a given outcrop and to contour maximum values (Carey and Sparks, 1986; Barberi et al., 1995). Second, the method of Carey and Sparks (1986) is based on two main assumptions: (1) a monotonic decrease in vertical plume velocity within a plinian column, and (2) a mid-latitude wind profile such as that described by Shaw et al. (1974). The first assumption requires that this method should be applied only to deposits from plinian and subplinian (i.e., sustained) eruptions generating strong plumes. In addition, Woods (1988) has shown that large plumes might be characterized by super-buoyancy (see Fig. 8.5), for which the vertical velocity profile does not decrease monotonically. Such an effect could result in a premature loss of large clasts and, therefore, higher plumes than predicted by Carey and Sparks (1986). The method of Carey and Sparks (1986) therefore gives the best results for small clasts (< 32 mm; Papale and Rosi, 1993; Barberi et al., 1995; Rosi, 1998). Finally, wind direction is typically very variable with height, and the profiles can show large discrepancies with that of Shaw et al. (1974). For example, Carey and Sigurdsson (1986) used a modified wind profile to account for the direction inversion above the tropopause that occurred during the 1982 eruption of El Chichon.

9.3.4 Determination of mass eruption rate and eruption duration

The mass eruption rate $\dot{M}$ (kg s^{-1}) can be derived from plume height H (m) by applying the semi-empirical formula of Wilson and Walker (1987)

$$H = C\dot{M}^{1/4} \tag{9.8}$$

where the empirical factor $C = 236$ m kg$^{-1/4}$ s$^{1/4}$. Eruption duration can be determined by dividing the total erupted mass by $\dot{M}$.

Mass eruption rate can also be derived from the column height H (m) using the analytical model of Sparks (1986), which was based on buoyant plume theory (BPT) (Morton *et al.*, 1956; Settle, 1978; Wilson *et al.*, 1978) and improved by accounting for a varying adiabatic lapse rate and atmospheric temperature. As a result, $\dot{M}$ and H show a nonlinear correlation, which strongly depends on eruption temperature.

Caveats

Equation (9.8) holds only for circular-vent plumes < 35 km high and is supported by theoretical investigations based on BPT (Morton *et al.*, 1956; Wilson and Walker, 1987) which show that maximum plume height is roughly proportional to the fourth root of the heat injection rate, and therefore to the fourth root of the mass eruption rate (see also Chapter 8, Section 8.3.1). However, there are several limitations in extending BPT to eruption plumes in a stratified atmosphere where buoyancy flux varies with height and crosswinds significantly affect plume entrainment (Bursik, 2001; Ishimine, 2006; Carazzo *et al.*, 2008). In addition, Eq. (9.8) is strictly valid for a plume temperature of ~800 °C, appropriate for andesitic magma. Basaltic magmas are typically hotter by at least 200 °C, and therefore, to achieve the same column height the corresponding mass discharge rates are lower for basaltic magmas (Carey and Sparks, 1986; Sparks, 1986; Woods, 1988). For example, Wehrmann *et al.* (2006) found that $C = 295$ m kg$^{-1/4}$ s$^{1/4}$ in Eq. (9.8) describes the relationship between $\dot{M}$ and H for a basaltic plinian eruption of Masaya volcano, Nicaragua (i.e., Fontana Lapilli). Scollo *et al.* (2007) and Andronico *et al.* (2008) found $C = 247$ m kg$^{-1/4}$ s$^{1/4}$ and $C = 244$ m kg$^{-1/4}$ s$^{1/4}$ for the 2001 and 2002 eruptions of Etna volcano, respectively. Finally, it is important to bear in mind that Eq. (9.8) is strictly valid only for values of maximum column height and therefore gives maximum values of $\dot{M}$. As a result, the corresponding eruption durations could be underestimated. In contrast, the model of Sparks (1986) can easily be applied by using dedicated diagrams compiled for sustained buoyant plumes with heights up to 35 km, eruption temperatures between 400 and 1000 °C, and for both tropical and temperate atmospheres. The model shows good agreement with observed data. However, to compile diagrams from more elaborate theory, Sparks (1986) made assumptions about tropopause height, surface temperature, temperature gradient, wind profile, and air-entrainment models that need to be carefully verified prior to application.

9.4 | Models based on the Advection–Diffusion–Sedimentation (ADS) equation

Models for tephra dispersal are based on the mass conservation equation with different degrees of simplicity, following either *Eulerian* or *Lagrangian* formulations. The Eulerian approach describes changes in the fluid at fixed points, whereas the Lagrangian approach describes changes by following a fluid parcel along its trajectory. Each approach is useful for different applications. For example, weather forecasting is based on the Eulerian approach (fixed measurement system) because it uses data from fixed stations around the world. The Lagrangian approach is more useful when describing the evolution of a given material as it moves within a certain fluid (e.g., chemical modeling). Tephra dispersal is often described using both approaches. In particular, models commonly defined as Lagrangian are based on an Eulerian–Lagrangian approach, which describes the dynamics of single particles within an Eulerian flow field. In contrast, Eulerian models consider the particle phase and the flow field as two continua. The governing equation derived from the mass conservation condition has the following form (Costa *et al.*, 2006):

$$\frac{\partial \bar{c}_j}{\partial t} + \frac{\partial \bar{u}_x \bar{c}_j}{\partial x} + \frac{\partial \bar{u}_y \bar{c}_j}{\partial y} + \frac{\partial \bar{u}_z \bar{c}_j}{\partial z} - \frac{\partial \bar{v}_j \bar{c}_j}{\partial z} =$$
$$-\frac{\partial \overline{u_x' c_j'}}{\partial x} - \frac{\partial \overline{u_y' c_j'}}{\partial y} - \frac{\partial \overline{u_z' c_j'}}{\partial z} + S, \tag{9.9}$$

where $c_j(x,y,z,t) = \bar{c}_j + c_j'$ is the concentration of particle class j (with c_j' the turbulent fluctuation and $\bar{c}_j$ the ensemble average), $\left(u_x, u_y, u_z\right) = \left(\bar{u}_x + u_x', \bar{u}_y + u_y', \bar{u}_z + u_z'\right)$ is the ambient fluid velocity, and $v_j(x,y,z)$ is the terminal velocity of a particle class j. The first term on the left-hand side of Eq. (9.9) represents the time rate of change of the average concentration $\bar{c}_j$ (i.e., the transient term), whereas the second, third, and fourth terms represent advection (i.e., wind transport) and the fifth term describes sedimentation. The first three terms on the right-hand side of Eq. (9.9) represent the diffusive transport due to the atmospheric turbulence, whereas the fourth term, $S(x,y,z,t)$, denotes the source (i.e., the mass flux of particle class j injected per unit volume and unit time). For most applications $v_j(x,y,z)$ can be considered a function of height z only. Turbulent fluxes are given by the product of the fluctuation terms on the right-hand side of Eq. (9.9). Typically, the simplest approach consists of expressing turbulent fluxes as proportional to the gradient of average concentration, e.g., for the x-direction:

$$\overline{u_x' c_j'} \approx -K_x \frac{\partial \overline{c_j}}{\partial x} \qquad (9.10)$$

where the coefficient K_x is the x-component of the turbulent diffusion tensor (typically in the free atmosphere, $K_V/K_H \ll 1$, where $K_H = K_x = K_y$ and $K_V = K_z$).

9.4.1 One-dimensional analytical sedimentation models

Depending on the application, if some approximations hold (e.g., horizontally uniform wind field, constant diffusion coefficient, negligible vertical motion and diffusion), Eq. (9.9) can be simplified and solved analytically. For instance, one-dimensional ADS models are typically used to investigate particle sedimentation along the dispersal axis, assuming negligible diffusion, steady plume conditions, and a single coordinate direction (s) along the direction of the carrying medium (Bursik *et al.*, 1992a,b; Sparks *et al.* 1992; Koyaguchi, 1994; Bonadonna *et al.*, 1998; Koyaguchi and Ohno, 2001a,b; Bonadonna and Phillips, 2003). As a result, Eq. (9.9) becomes:

$$\frac{\partial U c_j}{\partial s} - \frac{\partial v_j c_j}{\partial z} = 0, \qquad (9.11)$$

where U denotes the current velocity in the s-direction (e.g., direction along the vertical plume or the horizontal umbrella cloud). Under these conditions, and integrating for the sedimentation rate derived by Martin and Nokes (1988), the total mass of particles, M (kg), of a given size fraction carried by the spreading current at a certain distance x_1 is:

$$M = M_0 \exp\left\{-\int_{x_o}^{x_1} \frac{v_{Hb} w}{Q} dx\right\} \qquad (9.12)$$

where M_0 (kg) is the initial mass for a given grain size injected into the current, v_{Hb} (m s^{-1}) is the particle terminal velocity at the neutral buoyancy level H_b, w (m) is the maximum crosswind width of the current at the source, Q (m^3 s^{-1}) is the volumetric flow rate into the current at the neutral buoyancy level, and x_o (m) is the plume corner position (Fig. 9.1; Bursik *et al.*, 1992a). These models are supported by experimental data (Sparks *et al.*, 1991) and have given crucial insights into tephra deposit thinning. For example, the models of Bursik *et al.* (1992a) and Sparks *et al.* (1992) provided a theoretical basis for empirical methods (Pyle, 1989; Fierstein and Nathenson, 1992) of calculating erupted volumes. Furthermore, these analytical models showed that tephra fall from the plume margins and from the umbrella cloud can be described by two exponential segments on a semi-log plot of thickness vs. distance from vent, with a break in slope coinciding with the plume corner. Bonadonna *et al.* (1998) described sedimentation from the umbrella cloud by also accounting for particle Reynolds number and showed that the associated tephra deposits can be better described by three exponential segments (Fig. 9.1(b)). All of these results have crucial implications for the determination of the total erupted volume (Pyle, 1990; Rose, 1993). Koyaguchi and Ohno (2001a,b) also used this approach, in combination with dedicated inversion solutions of observed data, to determine the expansion rate of an umbrella cloud, the grain-size distribution at the top of the eruption column, and the amount of fine ash dispersed in the atmosphere, and found good agreement with satellite observations.

Further developments of these models have shown that the main processes affecting particle sedimentation are the effects of wind advection on the spreading of the umbrella cloud and on particle transport, and particle-aggregation processes in the case of ash-rich plumes and concentrated flows (Bursik *et al.*, 1992b; Bonadonna and Phillips, 2003). In particular, wind advection typically shifts the position of breaks in slope downwind, whereas aggregation processes allow fine particles to fall in the turbulent and intermediate fall-out regimes with the result that: (1) breaks in slope can be shifted either closer to the vent or further downwind depending on the size and density of the aggregates, (2) the thinning rate is controlled by the amount of aggregating particles, and (3) small aggregates are likely to generate secondary maxima of mass accumulation when advected (Bonadonna and Phillips, 2003). Bonadonna *et al.* (2005a) further developed Eq. (9.12) to account for the variation of volumetric flux with distance from the vent to describe sedimentation from a vigorous weak plume.

9.4.2 Two-dimensional analytical sedimentation models

Two-dimensional models are based on analytical solution of the ADS equation under the assumptions of constant and isotropic atmospheric diffusion and of negligible vertical wind velocity and vertical diffusion (Suzuki, 1983; Armienti *et al.*, 1988; Glaze and Self, 1991; Hurst and Turner, 1999; Connor *et al.*, 2001; Bonadonna *et al.*, 2002a, 2005b; Folch and Felpeto, 2005; Macedonio *et al.*, 2005; Pfeiffer *et al.*, 2005; Connor and Connor, 2006). The eruption column is typically described as a vertical line source that can be characterized by various mass distribution functions ranging from uniform to exponential (e.g., Suzuki, 1983). As a result, Eq. (9.9) becomes:

$$\frac{\partial c_j}{\partial t} + u_x \frac{\partial c_j}{\partial x} + u_y \frac{\partial c_j}{\partial y} - \frac{\partial v_j c_j}{\partial z} = K_H \frac{\partial^2 c_j}{\partial x^2} + K_H \frac{\partial^2 c_j}{\partial y^2} + S$$

(9.13)

where, for simplicity, we have eliminated the over-bar symbol to denote the average quantities. A solution of Eq. (9.13) is given by a Gaussian distribution (e.g., Suzuki, 1983; Pfeiffer *et al.*, 2005). The total mass on the ground is computed as the sum of the contributions of each of the point sources distributed above the ground and of each particle class.

Commonly, ADS models are based on empirical parameters, such as the diffusion coefficient K_H and column shape parameters introduced for describing the term S. As a result, they must be validated and calibrated with field data for specific eruptions before they can be used for a reliable hazard assessment. However, the advantage of these models is the simplicity of the physical parameterization and, therefore, the high computation speed. This allows for comprehensive probabilistic analysis of the associated inputs and outputs, and for solution of inverse problems to estimate eruption parameters, such as total erupted mass and column height. Key findings include the great sensitivity of ADS models to both erupted mass and column height, which justifies the use of inversion solutions for estimating these parameters (Connor and Connor, 2006; Scollo *et al.*, 2008a,b; Volentik *et al.*, 2010).

9.4.3 Three-dimensional numerical models for particle sedimentation

When the simplifying assumptions made to derive Eq. (9.13) are no longer valid, or when there is a need to describe three-dimensional dispersion of volcanic clouds within the atmosphere, fully numerical models must be adopted. One example requiring three-dimensional treatment is when most of the transport occurs within the atmospheric boundary layer (ABL), i.e., the part of the troposphere that is directly influenced by the presence of the Earth's surface and which responds to surface forcing with a timescale of ≤ 1 hour (Stull, 1988). Due to both topographic effects and rapid temporal variations of wind and temperature fields, turbulent tensor components are significantly more complex in the ABL than in the higher (i.e., free) atmosphere. Another case requiring a numerical approach is the tracking of volcanic ash clouds for diversion of aircraft flight paths routinely performed by the Volcanic Ash Advisory Centers (VAACs) (Witham et al., 2007).

However, depending on the application, several simplifying assumptions can be introduced to make the problem tractable for practical purposes, even for this category of models. For example, the VAAC particle-tracking models are mainly used to describe the atmospheric transport of volcanic ash for aviation safety, but, except for a few cases (Tanaka and Yamamoto, 2002), they are not used to calculate ground deposition. Other examples of VAAC three-dimensional, time-dependent Eulerian models are VAFTAD (Volcanic Ash Forecast Transport And Dispersion; Heffter and Stunder, 1993), CANERM (CANadian Emergency Response Model; D'Amours, 1998) and MEDIA (Eulerian Model for DIspersion in the Atmosphere; Sandu et al., 2003). The last model, used by the Toulouse VAAC, simulates the effects of advection by the average wind, diffusion of particles by thermal and dynamical turbulence, rainout (occurring inside the cloud) and washout (occurring below the cloud) of particles by precipitation, and sedimentation due to gravitational settling. Examples of VAAC models based on Lagrangian formulations are NAME (Numerical Atmospheric-dispersion Modelling Environment; Ryall and Maryon, 1998), PUFF (Searcy et al., 1998), and HYSPLIT (Hybrid Single Particle Lagrangian Integrated Trajectory Model; Draxler and Hess, 1998). Only a few models, such as the three-dimensional, time-dependent Eulerian FALL3D (Costa et al., 2006; Folch et al., 2009) and the Lagrangian VOL-CALPUFF (Barsotti and Neri, 2008; Barsotti et al., 2008) were designed for both particle transport and particle sedimentation.

In particular, FALL3D solves the ADS equation for each particle class concentration in a terrain-following coordinate system:

$$\frac{\partial c_j}{\partial t} + u_x \frac{\partial c_j}{\partial x} + u_y \frac{\partial c_j}{\partial y} + (u_z - v_j)\frac{\partial c_j}{\partial z} =$$
$$\frac{\partial}{\partial x}\left(K_H \frac{\partial}{\partial x} c_j\right) + \frac{\partial}{\partial y}\left(K_H \frac{\partial}{\partial y} c_j\right) + \qquad (9.14)$$
$$\frac{\partial}{\partial z}\left(K_V \frac{\partial}{\partial z} c_j\right) + c_j \frac{\partial v_j}{\partial z} - c_j \nabla \cdot \vec{u} + S.$$

The terrain-following wind speed components $\vec{u} = (u_x, u_y, u_z)$ can be passed through meteorological models such as CALMET (Scire et al., 2000), which is used for assimilating and interpolating short-term forecasts (or re-analysis) from mesoscale meteorological prognostic models. The vertical component of the turbulent diffusivity tensor K_V can be described on the basis of similarity theory in the ABL, and by the Richardson number (ratio of thermally-produced turbulence and turbulence generated by vertical shear) in the free atmosphere. The horizontal component K_H can be described following a large eddy simulation (LES) approach that consists of solving large scale motions of the flow and modeling the effect of the smaller universal scales using a sub-grid scale (SGS) model. The generic particle class j is defined by three values characterizing each particle (diameter, density, and shape factor). Several semi-empirical parameterizations can be used for calculating particle terminal settling velocities. The source term S (i.e., the amount of mass injected per unit volume and unit time) can be described as a point source, as a Suzuki distribution (Suzuki, 1983; Pfeiffer et al., 2005), or through a buoyant plume model (Bursik, 2001). The last option involves the solution of the one-dimensional, radially averaged, governing equations that describe the convective region of an eruption column, over which the mass is released and distributed in several layers according to the particle terminal velocity. These equations are intimately coupled with the wind field, which, for small plumes, may cause substantial bending over of the plume. The numerical solution is based on a second-order finite differences scheme. Particle sedimentation at the bottom of the computational domain (i.e., ground level) is calculated as the temporal integral of the outgoing mass flux from this surface (Costa *et al.*, 2006). In a recent study (Folch *et al.*, 2008), FALL3D was generalized to the mesoscale-synoptic domain and coupled with the Weather Research and Forecasting (WRF) meteorological model (Michalakes *et al.*, 2005; www.wrf-model.org).

Finally, it is worth mentioning another category of models designed to describe in detail the evolution of the eruption plume from its rise to its collapse. Given the high degree of complexity of such models, they are usually applied over relatively small horizontal domains (up to few tens of kilometers). Some of these models are focused on describing the dynamics and thermodynamics of the mixture of hot gases and particles (Dobran *et al.*, 1993; Esposti Ongaro *et al.*, 2007), whereas other models such as ATHAM (Active Tracer High Resolution Atmospheric Model; Herzog *et al.*, 1998; Oberhuber *et al.*, 1998) are more appropriate for dealing with the chemical interactions and microphysical processes of volcanic and cloud water, cloud ice, rain, and graupel.

9.5 | Limitations of input parameters and parameterizations adopted by ADS models

Regardless of the complexity of different sedimentation models, the reliability and uncertainties of the associated outputs strongly depend on the reliability and uncertainties of input parameters (i.e., erupted mass, column height, total grain-size distribution, meteorological data) and of the parameterizations used to describe critical sedimentation processes, such as particle aggregation, particle terminal velocity, and column dynamics.

9.5.1 Input parameters: erupted mass, grain-size distribution, plume height, wind profile

Erupted mass is the most important input parameter and is one of the most difficult to derive accurately from field data. In fact, volume (and mass) estimation strongly depends on

the technique used, on deposit exposure, and on data distribution and density. Most erupted volumes derived from field data should be considered minimum values unless the data sets extend hundreds of kilometers from the vent (the higher the plume, the larger the deposit to be investigated). A review of several methods can be found in Froggatt (1982), whereas the most recent techniques are summarized in Section 9.3. In addition, recent applications of inversion techniques to analytical models have shown promising results. Specifically, mass per unit area and particle-size data from individual outcrops are inverted through the use of two-dimensional analytical models to derive eruption parameters, such as erupted mass and column height (Connor and Connor, 2006; Scollo *et al.*, 2008a; Volentik *et al.*, 2010).

Column height is very important for defining the source term and is related to eruption intensity (i.e., mass flux). The best evaluation of column height comes from well-documented and calibrated direct observations. Column height can also be estimated through analysis of satellite images, based on geometry (cloud shadow clinometry), thermal infrared (IR) data (using a cloud-top temperature/temperature-profile method), and correlation of cloud trajectory with meteorological motion (cloud stereoscopy) (Holasek and Self, 1995; Prata and Turner, 1997; Glaze *et al.*, 1999; Prata and Grant, 2001). Field studies have also shown that the derivation of plume height using the method of Carey and Sparks (1986) gives fairly consistent results even for poorly exposed deposits (Wehrmann *et al.*, 2006). Inversion of two-dimensional analytical models using particle-size data also gives a good constraint on plume height (Volentik *et al.*, 2010). However, it is important to bear in mind that even a small uncertainty in plume height results in an uncertainty about four times larger in mass flux $\dot{M}$, because of the fourth power relationship between $\dot{M}$ and H (Eq. 9.8).

All tephra dispersal models are strongly dependent on the choice of initial grain-size distribution. Nonetheless, even though several methods have been proposed, including simple data averages, sectorization, Voronoi tessellation, and analytical models (Carey and Sigurdsson, 1982; Bonadonna and Houghton, 2005; Mannen, 2006; Volentik *et al.*, 2010), none of these techniques is able to reproduce missing field data. As a result, any extrapolation based on empirical fitting of poor data sets is likely to be problematic.

Tephra dispersal is significantly affected by local meteorological conditions. The required resolution of meteorological data depends mainly on the length scale and specific problem being considered. However, wind profile information is commonly difficult to include in simulations of particle sedimentation for two main reasons: (1) wind profile data are not always available for a given eruption, certainly not for prehistoric events; (2) many models for tephra dispersal do not capture the complexity of the local meteorological conditions (e.g., all analytical models described above). The only meteorological information for prehistoric eruptions is average wind direction, derived from the dispersal axis, and maximum wind speed at the tropopause, derived using the method of Carey and Sparks (1986); however, wind direction typically varies with height, especially in the lower part of the atmosphere. For recent eruptions, global meteorological data with a grid resolution of 2.5° are available from 1948 to the present (Kalnay *et al.*, 1996; www.cdc.noaa.gov/cdc/data. ncep.reanalysis.html, http://data.ecmwf.int/ data/d/era40_daily/). For recent years, other useful data come from direct soundings for many locations worldwide (http://weather.uwyo.edu/ upperair/sounding.html). Unfortunately, the grid resolution of these data sets is too coarse for many applications, so data derived from mesoscale meteorological models (e.g., the Fifth-Generation NCAR/Penn State Mesoscale Model (MM5), and the Weather Research and Forecasting (WRF) model) are more practical. For example, Byrne *et al.* (2007) showed that dispersal of tephra from the 1995 eruption of Cerro Negro (Nicaragua) can be described by accounting for small-scale wind variations during the eruption. For applications where a very fine scale is needed (≤ 1 km) diagnostic wind (mass consistent) models such as CALMET can be used (Scire *et al.*, 2000).

9.5.2 Parameterizations of particle aggregation and settling velocity

Particle aggregation is a fundamental process that typically occurs in ash-rich volcanic clouds for particle diameters < 100 μm, and results in the formation of coated crystals, dry aggregates, and accretionary lapilli, depending on the water content (Sparks *et al.*, 1997). Ash-coated crystals (typically particles > 200 μm coated by < 20-μm particles) are considered responsible for scavenging only small volumes of fine particles and the associated ash coating is not expected to change the particle terminal velocity significantly (James *et al.*, 2002). In contrast, both accretionary lapilli and dry aggregates can significantly affect sedimentation of ash-rich tephra, as shown by numerous field observations (Hobbs *et al.*, 1981; Brazier *et al.*, 1982; Sorem, 1982; Hildreth and Drake, 1992; Scasso *et al.*, 1994) and both analytical and numerical investigations (Carey and Sigurdsson, 1982; Cornell *et al.*, 1983; Wiesner *et al.*, 1995; Veitch and Woods, 2001; Bonadonna *et al.*, 2002a; Textor *et al.*, 2006). In particular: (1) aggregation processes can make fine particles fall closer to the vent than expected, generating secondary maxima of accumulation (Carey and Sigurdsson, 1982) (e.g., Fig. 9.2(b)); (2) aggregation processes affect tephra-deposit thinning and the position of thinning breaks in slope on semi-log plots of thickness vs. distance from vent (Bonadonna and Phillips, 2003); (3) deposition of co-pyroclastic-density-current ash (generated both from large ignimbrites and block-and-ash flows) cannot be described unless aggregation is accounted for (Cornell *et al.*, 1983; Bonadonna *et al.*, 2002a); (4) dry aggregation processes significantly affect health hazard assessments because a considerable number of particles with diameters < 10 μm are present as small aggregates in all volcanic plumes (James *et al.*, 2003). The complexity of aggregation processes explains why they are not accounted for in most dispersal models. Such a simplification, combined with the empirical nature of analytical models, is valid when sedimentation of ash-poor tephra is considered (e.g., eruption of Askja 1875 D; Bonadonna and Phillips, 2003). However, even analytical models fail to accurately reproduce tephra deposits characterized by ash-rich grain-size distributions.

Several studies have shown that particle-settling velocities strongly depend on particle shape (Wilson and Huang, 1979), although for simplicity particles are typically assumed to be spheres, for which terminal velocities can be determined using simple expressions (Kunii and Levenspiel, 1969; Arastoopour et al., 1982). The settling velocity v_j of particles of size d_j is obtained from the balance between gravity and air drag. The drag coefficient, C_d, is a function of the particle shape and the Reynolds number, Re = $d_j \rho_a v_j / \eta_a$, where η_a is the air dynamic viscosity (Pa s). The assumption of spherical particles is valid as a first-order approximation only; for non-spherical particles the determination of C_d is more complicated. Walker et al. (1971) showed that pumice clasts > 5 mm are better described by cylinders than spheres, and Wilson and Huang (1979) found that, for particle diameters between 30 and 500 μm, glass and feldspar fragments have a very high proportion of flattened particles, whereas pumice clasts have a greater variety of shapes, including equant particles.

Following a review of available methods for estimating the drag coefficient of non-spherical particles, Chhabra et al. (1999) showed that the best approach appears to be that of Ganser (1993), which uses the equal volume sphere diameter and the sphericity ψ of particles, with a resulting overall error within ~16% for Re ranging from 10^{-4} to 5×10^5. Unfortunately, expressions for terminal velocity that account for the complexity of irregular particles are commonly based on particle parameters that are impractical to measure. For example, particle surface area necessary to calculate ψ cannot be easily determined because this would imply a complicated integration over surface elements of an irregular particle. For this reason, Wadell (1933) and Aschenbrenner (1956) introduced the concepts of "operational" and "working sphericity," based on the determination of the volume and of the three dimensions of a particle respectively:

$$\psi_{work} = 12.8 \sqrt[3]{\frac{S^2}{IL}} \Bigg/ \left[1 + \frac{S}{I}\left(1 + \frac{I}{L}\right) + 6\sqrt{1 + \frac{S^2}{I^2}\left(1 + \frac{I^2}{L^2}\right)} \right],$$

(9.15)

where L is the longest particle dimension, I is the longest dimension perpendicular to L, and S is the dimension perpendicular to both L and I. In addition, Riley et al. (2003) considered the determination of particle sphericity from two-dimensional images, which permits analysis of small particles. However, these methods are all approximations that need to be tested thoroughly for application to calculating terminal velocities of volcanic particles. As a result, the effect of particle shape on terminal velocity is a critical factor that remains to be adequately described. However, the drag coefficient strongly depends on particle shape only for relatively large particles (Fig. 9.3). For example, the model of Ganser (1993) was used here to investigate the terminal velocities of irregular particles of known shape and diameter ranging between 1.5 and 7 cm, and with a shape factor ($F = (I + S) / 2L$; Wilson and Huang, 1979) ranging between 0.3 and 0.9. Figure 9.3 shows the associated drag coefficients for Reynolds numbers Re between 0.001 and 10^6. Note that the shape factor F does not uniquely constrain elongated and platy particles (i.e., $F < 0.7$), and that rounded particles can have similar drag coefficients to elongated particles. Furthermore, Figure 9.3 shows that the drag coefficient varies significantly with shape only for particles falling in the intermediate and turbulent regimes (Re > 1). As a result, future studies of the effects of particle shape on terminal velocity should focus on medium- and high-Re particles, i.e., particles with diameters > 63 μm and > 2 mm respectively, for heights > 10 km above sea level.

An accurate description of plume dynamics is crucial for both analytical and numerical models (Scollo et al., 2008b). Model results are very sensitive to the choice of velocity profile within the plume, which ultimately controls both the mass and the grain-size distribution within the eruptive column (e.g., Carey and Sparks, 1986; Bursik et al., 1992a). As a first approximation, models based on BPT consider a Gaussian distribution profile across the plume and a monotonic trend with height (Carey and Sparks, 1986; Bursik et al., 1992a; Sparks et al., 1992; Bonadonna and Phillips, 2003; Bonadonna et al., 2005a). However, numerical models show more

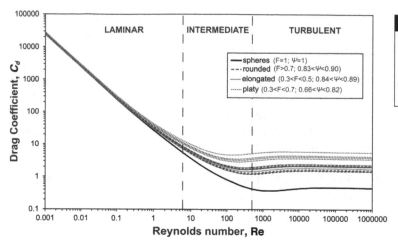

Figure 9.3 Plot of drag coefficient vs. Reynolds number for particles with different shapes. Drag coefficient was calculated analytically from the model of Ganser (1993). F denotes shape factor and ψ is sphericity.

complicated velocity profiles, and simulations from three-dimensional models of weak plumes show no Gaussian cross section at any time or even as a time-averaged property. The advantage of simple column models such as those used in the analytical models described above or the steady-state models based on BPT (Woods, 1988; Bursik, 2001; Ishimine, 2006) is their computational speed and flexibility compared to the complex three-dimensional time-dependent descriptions of plume dynamics (Dobran et al., 1993; Herzog et al., 1998; Oberhuber et al., 1998; Esposti Ongaro et al., 2007). As a result, for some regimes and applications, a challenge remaining for the volcanology community is to develop an accurate physical model for column dynamics that is also computationally fast.

9.6 | Case study

The 22 July 1998 paroxysmal event of Mt. Etna provides a useful illustration of the application of the main empirical, analytical, and numerical models described in Sections 9.3–9.5 (Coltelli et al., 2006; Scollo et al., 2008a). This was one of the strongest explosive events at Mt. Etna in the last century and produced a short-lived strong plume associated with hawaiitic magma (Corsaro and Pompilio, 2004) that rose 12 km above sea level (a.s.l.), ~9 km above the vent in Voragine crater,

leading to tephra dispersal to the southeast. The highest seismic tremor recorded for this plume-forming lava fountain lasted for about 25 minutes (Aloisi et al., 2002; Coltelli et al., 2006). The associated tephra blanket was sampled between ~3 and 30 km from the vent soon after deposition. As a result, a detailed isomass map was compiled (Fig. 9.4), and a maximum column height of 11 km a.s.l. and a maximum wind speed at the tropopause of 10–30 m s^{-1} were determined using the method of Carey and Sparks (1986) (Andronico et al., 1999; Table 9.1). The subplinian character of this eruption is suggested by comparison with other tephra deposits on a semi-log plot of T vs. $\sqrt{A}$ (Fig. 9.5). Plinian eruptions ranging from basaltic (e.g., Fontana Lapilli) to rhyolitic (e.g., Taupo) are characterized by larger maximum thicknesses and more gradual thinning with distance (i.e., larger b_t in Eq. (9.2)) than the 1998 Etna deposit, which instead plots with two other well-studied subplinian eruptions: the 17 June 1996 eruption of Ruapehu volcano (New Zealand) and the 22 July 1980 eruption of Mt. St. Helens (USA). In addition, sensitivity analyses based on time-series Meteosat images and theoretical modeling using the PUFF ash tracking model (Searcy et al., 1998) gave a best-fit value of column height of 13 km a.s.l., an eruption duration between 20 and 40 minutes, and horizontal and vertical diffusivity values of 5000 and 10 m^2 s^{-1}, respectively (Aloisi et al., 2002; Table 9.1).

Table 9.1 Comparison of eruption parameters obtained using different methods. See online supplement 9A for details of the methods used to derive parameters.

Parameter	Method								
	Observed data	Exponential method	Power law method	Carey and Sparks (1986)	Sparks (1986)	Wilson and Walker (1987)	Inversion: TEPHRA2	OAT: PUFF	Numerical solution: FALL3D
Erupted mass M (×10⁹ kg)	–	0.9 (1seg) 1.1 (2seg) 1.8 (3seg)	2.0	–	–	–	1.7	–	1.7
Column height H (km a.s.l.)	12	–	–	11	–	–	13	13	12
Mass eruption rate $\dot{M}$ (×10⁶ kg s⁻¹)	–	0.6 (1seg)[8] 0.7 (2seg) 1.2 (3seg)	–	–	0.6	1.8	–	–	2.4
Grain-size distribution	Md φ (0.8) STDV (1.8)	–	–	–	–	–	Mdφ (-0.6) STDV (2.2)	Mdφ (0) STDV (1.5)	–
Duration of sustained phase (min)	25 (total duration)	–	–	–	24 (1seg) 30 (2seg) 51 (3seg) 56 (PL) 47 (TEPHRA2)	8 (1seg) 10 (2seg) 17 (3seg) 19 (PL) 16 (TEPHRA2)	–	20–40	12
Maximum/ average wind speed (m s⁻¹)	11 / 6	–	–	10–30	–	–	6 / 6	–	–
K_H/K_V (m² s⁻¹)	–	–	–	–	–	–	0.5	5000 / 10	5000 / 0.004–600 (mean: 50)

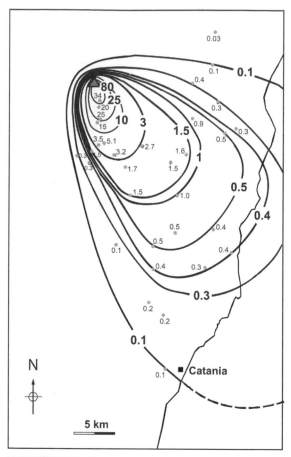

Figure 9.4 Isomass map of the 22 July 1998 eruption of Mt Etna (kg m^{-2}) based on data from Andronico (1999) and Coltelli *et al.* (2006).

9.6.1 Application of empirical models: erupted mass, mass eruption rate, and eruption duration

Because of the variation of deposit density with distance from vent, mass/area data were plotted against $\sqrt{A}$, and were fitted by three exponential segments and by a power-law trend (Fig. 9.6 and Eqs. (9.3) and (9.7)). For completeness, results for one (medial) and two (medial and distal) exponential segments are also shown in Table 9.1. Both power-law and three-exponential-segment fits give a total erupted mass of about 2×10^9 kg, which corresponds to a total erupted volume of 2×10^6 m^3 (assuming an average deposit density of 1050 kg m^{-3}) and to a dense rock equivalent (DRE) volume of 8×10^5 m^3 (assuming a magma

density of 2600 kg m^{-3}). For the power-law calculation, $\sqrt{A_0}$ and $\sqrt{A_{dist}}$ were chosen as 1.1 km (from Eq. (9.5)) and 500 km (based on the geometry of the deposit). However, the volume derived by integrating the power-law trend is not very sensitive to the choice of the outermost integration limit ($\sqrt{A_{dist}}$) because the associated power-law exponent $m > 2$, and therefore the deposit thins very rapidly (Table 9.1; Fig. 9.6(b)). In contrast, the derived volume is sensitive to the choice of $\sqrt{A_0}$ (Table 9.1), but this can be fixed using Eq. (9.5). A mass eruption rate of 0.6 $\times 10^6$ kg s^{-1} was determined from the model of Sparks (1986) for an eruption temperature of 1000 °C, as derived by Corsaro and Pompilio (2004). A mass eruption rate of 1.8×10^6 kg s^{-1} was calculated using Eq. (9.8) for a column height of 9 km and $C = 245$ m kg$^{-1/4}$ s$^{1/4}$ (averaged from Scollo *et al.* (2007) and Andronico *et al.* (2008)). Using the erupted volume determined from both the exponential and power-law methods, we obtain an eruption duration between 8 and 19 minutes (method of Wilson and Walker (1987) with $C = 245$ m kg$^{-1/4}$ s$^{1/4}$), and between 24 and 56 minutes (method of Sparks (1986) with an eruption temperature of 1000 °C; Table 9.1).

9.6.2 Application of 1D and 2D analytical sedimentation models

As described in Section 9.4.1, the (one-dimensional) model of Bonadonna and Phillips (2003) can be used to investigate the fallout dynamics and the thinning of tephra deposits (Fig. 9.7). The application of this model to the 22 July 1998 eruption of Mt. Etna explains the position of the breaks in slope shown in Figure 9.6 by accounting for the mass fractions of particles with different Reynolds numbers (Run 1, Table 9.2). The first field data point in Figures 9.6 and 9.7 corresponds to the position of the plume corner (~2 km from the vent for a plume height of ~9 km). As a result, all of the sampled deposit corresponds to fallout from the umbrella cloud, and the two breaks in slope correspond to the transitions between: (1) turbulent and intermediate sedimentation regimes (break in slope at 95% of intermediate-Re particles and 5% of high-Re particles), and (2) intermediate and laminar sedimentation

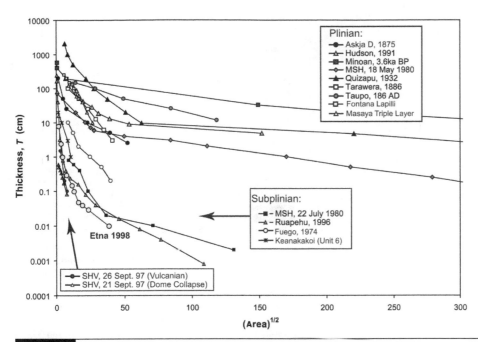

Figure 9.5 Semi-log plot of thickness vs. square root of isopach area for describing the thinning trend of eruptions of different styles. Vulcanian explosion and dome collapse of Soufrière Hills volcano, Montserrat, West Indies (Bonadonna et al., 2002b). Subplinian eruptions of Ruapehu volcano, New Zealand (17 June 1996; Bonadonna et al., 2005a); Mount St. Helens, USA (22 July 1980; Sarna-Wojcicki et al., 1981); Kilauea volcano, USA (Keanakakoi tephra, Unit 6; McPhie et al., 1990); Fuego 1974 (Rose et al., 2008). Plinian eruptions of Askja volcano, Iceland (Unit D; Sparks et al., 1981); Hudson volcano, Chile (12–15 August 1991; Scasso et al., 1994), Minoan eruption, Greece (Pyle, 1990), Mount St. Helens, USA (18 May 1980; Sarna-Wojcicki et al., 1981); Quizapu, Chile (1932; Hildreth and Drake, 1992); Tarawera, New Zealand (1886; Walker et al., 1984); Taupo, New Zealand (181 AD; Walker, 1980); Masaya volcano, Nicaragua (Fontana Lapilli; Costantini et al., 2008); Masaya volcano, Nicaragua (Triple Layer; Perez et al., 2009).

regimes (break in slope at 15% of intermediate-Re particles and 85% of low-Re particles). The discrepancy between the observed and computed tephra accumulation is probably due to the under-representation of fine particles in the grain-size distribution derived from field data, which results from the dominantly proximal exposure of the deposit. As a result, overestimates of terminal velocities will lead to overestimation of tephra accumulation (Eq. (9.12); Fig. 9.7).

The TEPHRA2 model can be used in combination with dedicated inversion techniques to determine eruption parameters. Connor and Connor (2006) applied the downhill simplex method to find the optimized set of eruption parameters corresponding to a given tephra deposit, based on comparison between observed and computed mass accumulation per unit area.

The goodness of fit is determined as the root mean square error (RMSE):

$$\text{RMSE} = \sqrt{\sum_{a=1}^{N} \frac{\left(Mc_a - Mo_a\right)^2}{Mo_a}}$$

(9.16)

where N is the number of observations and Mo_a and Mc_a are, respectively, the observed and computed deposit (i.e., mass per unit area) at sample location a respectively. In order to illustrate the distribution of minimum values of the goodness-of-fit measure, Figure 9.8 shows the RMSE corresponding to 0.2-log(mass) increments and 2-km-height increments (Run 2, Table 9.2). The wind direction was constrained based on the dispersal axis indicated by the isomass map (Fig. 9.4). Results show that this technique provides a very good constraint on the erupted mass but not on the column height. The optimized

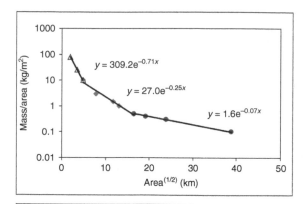

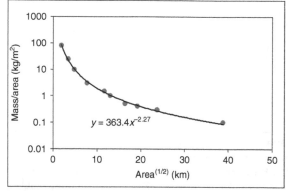

Figure 9.6 Semi-log plots of mass per unit area (kg m⁻²) vs. square root of area (km) for the Etna 1998 deposit, showing (a) the best fit of three exponential segments and (b) the power-law fitting. Best-fit equations are also shown.

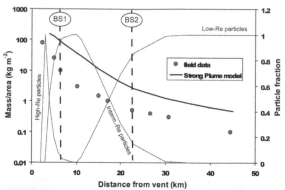

Figure 9.7 Semi-log plot of mass per unit area (kg m⁻²) vs. distance from vent (km) showing the comparison between Etna 1998 field data (from Figs. 9.4 and 9.5; symbols) and the results of the strong plume model of Bonadonna and Phillips (2003) modified to account for topography (bold curve; Run 1, Table 9.2). Fractions of particles with low, intermediate, and high Reynolds number are also shown (secondary axis; light curves). Vertical dashed lines indicate the position of the two breaks in slope shown in Fig. 9.6a (i.e., BS1 at ~6 km and BS2 at ~23 km, equivalent to BS1 at ~5 km and BS2 at ~16 km in units of $\sqrt{A}$). Inputs are summarized in Table 9.2.

erupted mass varies between 0.4×10^9 and 2.5×10^9 kg (corresponding to RMSE values of 1.5 ± 0.4 kg m⁻² for observed tephra accumulations between 0.1 and 84.0 kg m⁻²). The erupted mass is the only eruptive parameter that significantly affects model output, but not through interaction with other input parameters (Scollo et al., 2008a). However, given that for our case study the column height can be well constrained from observations, the inversion was run again for plume height from 10–15 km a.s.l. and erupted mass of $0.5 - 50 \times 10^9$ kg (a large enough interval for the algorithm to find a significant minimum). The best fit with field data was obtained for a plume height of 13 km a.s.l., a total erupted mass of 1.7×10^9 kg, and an average wind speed of 6 m s⁻¹ (Runs 3 and 4, Table 9.2). The low fall-time threshold (FTT) resulting from the inversion (i.e., 278 s) indicates that

most particles fell according to power-law diffusion (Eq. (8) in Bonadonna et al., 2005b). Figure 9.9 shows the forward solution of TEPHRA2 computed using these best-fit values and the quantitative comparison with field data.

9.6.3 Application of 3D numerical models for particle sedimentation

In order to apply the numerical model FALL3D (Costa et al., 2006; Folch et al., 2009), a pseudo-sounding profile was obtained for 37.5°N 15°E at 18:00 LT using the WRF model (Skamarock et al., 2005; Table 9.2, Run 5). The profile was used to initialize the meteorological processor CALMET (Scire et al., 2000), which produced a finer resolution wind and temperature field, also incorporating the effects of topography.

The observed column height of 9 km above the vent corresponds to a mass eruption rate $\dot{M}$ of 2.5×10^6 kg s⁻¹, derived using the model of Bursik (2001) with an exit velocity of 100 m s⁻¹, a temperature of 1000 °C, and a volatile content of 0.5 wt.% H_2O. In the FALL3D simulations, $\dot{M}$ was assumed to be constant for the eruption

Table 9.2	Parameters used in the simulations. See online supplement 9B for details of the methods used to derive parameters.				
	Run 1	Run 2	Run 3	Run 4	Run 5
Model	Strong Plume	TEPHRA2 (inversion)	TEPHRA2 (inversion)	TEPHRA2 (forward)	FALL3D (forward)
Grain-size distribution	GS Etna 1998	Mdϕ (-3–3) STDV (1–3)	Mdϕ (-3–3) STDV (1–3)	Mdϕ (-0.6) STDV (2.2)	GS Etna 1998
Plume height H (km a.s.l.)	12	4–20	10–15	13	Calculated within model
Wind speed w (m s^{-1})	3 (mean below base current)	1–20	1–20	6 (mean below plume height)	Wind Etna 1998
Wind direction (° from N)	–	90–200	90–200	110–184	Wind Etna 1998
Erupted mass M (×10^9 kg)	2	0.0001–100	0.5–50	1.7	Calculated within model (1.7)
Mass eruption rate (×10^6 kg s^{-1})	–	–	–	–	2.4
Particle density (kg m^{-3})	900–2600	900–2600	900–2600	900–2600	900–2600
K_H (m^2 s^{-1})	–	0.1–8000	0.1–8000	0.5	5000
Fall-time threshold FTT (s)	–	1–5000	1–5000	278	–

duration and the source terms (i.e., injected mass per second, column height, mass distribution in the plume) were determined through the model of Bursik (2001), based on BPT. In agreement with this model, a mass eruption rate $\dot{M}$ of 2.5 × 10^6 kg s^{-1} produces a column height of ~9 km above the vent (~12 km a.s.l.). Hence, taking this value for $\dot{M}$ and the duration of the climactic phase of 12 minutes (from best-fit analysis), the total mass erupted is found to be 1.7 × 10^9 kg. Figure 9.10 shows the deposit obtained by FALL3D using a grid of 51 × 51 × 18 km (1 km horizontal spacing) and the wind field refined by CALMET.

Terminal velocity was calculated using the model of Ganser (1993) with a sphericity of 0.93–0.95 (consistent with the data of Coltelli et al. (2008) for the 18 December 2002 eruption

of Etna volcano). The turbulent diffusivity tensor was described through similarity theory for the vertical component, and as both constant and using a large eddy simulation (LES) model for the horizontal component (Folch et al., 2009). Note that, when the wind field is derived from a single sounding, LES models can underestimate turbulent diffusion because there is no horizontal shear. In fact, fits to the data suggest a constant horizontal diffusion of $K_H \approx 5000$ m^2 s^{-1}, compared to the average value of $K_H \approx 2250$ m^2 s^{-1} predicted by the LES model. A comparison of the simulation results and observed deposit is shown in Fig. 9.10(b). Finally, FALL3D can also be used to assess ash concentration in the atmosphere and identify hazardous zones for air traffic (Folch et al., 2009).

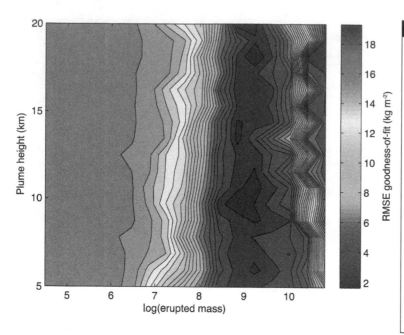

Figure 9.8 Plot of log(erupted mass) vs. plume height (km) showing the minimum values of the goodness-of-fit measure (RMSE; Eq. (9.16)) for the Etna 1998 tephra deposit. Erupted mass was varied between 10^4 and 10^{11} kg with 0.2-log(mass) increments and the plume height was varied between 4 and 20 km with 2-km-height increments (Run 2, Table 9.2). Resulting values were interpolated to produce 2D RMSE surfaces and the plot was cropped to show only the interpolated surfaces. Dark blue on the RMSE scale indicates the minimum values (i.e., best fit). RMSE values vary between 1 and 100 kg m^{-2} and are contoured with 0.8 kg m^{-2} contours (RMSE values > 20 kg m^{-2} were observed only for erupted masses > 3×10^{10} kg and were set equal to 20 kg m^{-2} for a better visualization of the minimum). See color plates section.

9.7 | Discussion

Integrated application of empirical, analytical, and numerical models to tephra deposits provides insights into the dynamics of the associated volcanic eruptions and allows for compilation of comprehensive hazard assessments.

9.7.1 Case study

The comprehensive data set of the 22 July 1998 Etna eruption provided the opportunity to investigate the uncertainties in eruption parameters derived from sedimentation models with different levels of complexity, and to apply these models to basaltic explosive eruptions. First, we stress that, even though magma composition does not seem to affect sedimentation dynamics (e.g., Fig. 9.5), it must be accounted for when deriving eruption parameters. For example, we have shown that inferred mass eruption rate strongly depends on eruption temperature (Sparks, 1986; Wilson and Walker, 1987).

Second, to assess the variation of eruption parameters derived using different modeling approaches, we have compared results for calculations of plume height, erupted mass, mass eruption rate, wind profile and diffusion

coefficients (Table 9.1). In particular, the standard deviation resulting from plume height calculations using four different models is 0.8 km, with discrepancies from the observed values between 0 and 8%. The standard deviation for erupted mass calculated using six different methods is 4×10^8 kg. Discrepancies with respect to the mass obtained with the inversion of TEPHRA2, and the numerical solution FALL3D (i.e., 1.7×10^9 kg) are 19% integrating the power-law fit, 49% integrating one exponential segment, 35% integrating two exponential segments, and 9% integrating three exponential segments.

Third, mass eruption rate is more difficult to constrain; we found a discrepancy of ~67% between the values obtained using the empirical model of Wilson and Walker (1987) and the model of Sparks (1986) for the same plume temperature (1000 °C). In addition, the mass eruption rate derived numerically using FALL3D is ~300% and 33% larger than the results from Sparks (1986) and Wilson and Walker (1987), respectively (Table 9.1). $\dot{M}$ derived by dividing the calculated erupted mass by the observed duration (i.e., 25 minutes) is 26–67% lower than the $\dot{M}$ of Wilson and Walker (1987) and up to 122% larger than that of Sparks (1986) (Table 9.1).

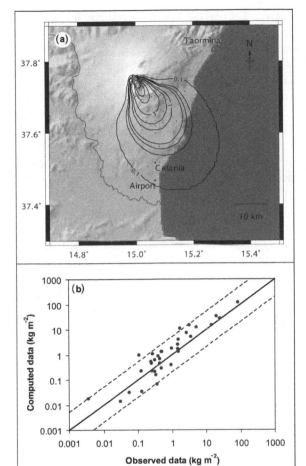

Figure 9.9 Forward solution of the TEPHRA2 model (Run 4 in Table 9.2) run with the best-fit values obtained from the inversion computation (Run 3, Table 9.2): (a) isomass map with contours of 0.1, 0.3, 0.4, 0.5, 1, 1.5, 3, 10, 25, and 80 kg m^{-2}, shown for comparison with Fig. 9.4. Locations of the volcano and the field data are indicated. (b) Comparison between computed (solid line) and observed (symbols) mass/area data. Dashed lines indicate over- or under-estimations of 1/5 and 5 times the observed values, respectively. TEPHRA2 was run using a Global Land One-km Base Elevation map (GLOBE; www.ngdc.noaa.gov/mgg/topo/globe.html).

Discrepancies in both mass eruption rate and erupted mass result in a discrepancy in the eruption duration ranging between 9 and 49% with respect to the duration obtained from the total mass derived using the inversion of TEPHRA2 and FALL3D (i.e., 16 and 47 minutes using the $\dot{M}$ of Wilson and Walker (1987) and Sparks (1986), respectively). In contrast, the duration derived

from the six different erupted mass values in Table 9.1 and the $\dot{M}$ of Sparks (1986) is always ~200% greater than when derived from the $\dot{M}$ of Wilson and Walker (1987). The duration derived using the $\dot{M}$ of Wilson and Walker (1987) and the mass inverted from TEPHRA2 is about 23% greater and 28% less than the minimum duration derived from the numerical solution of FALL3D and PUFF (Table 9.1). Discrepancies resulting from the calculation of wind velocity are within 50% of the observed value for all methods.

Finally, there is a large discrepancy observed for the diffusion coefficient calculated using FALL3D and PUFF, and the horizontal diffusion coefficient obtained using TEPHRA2. The diffusion coefficient used in analytical models is an empirical parameter that depends on various physical processes. In addition, TEPHRA2 describes particle diffusion using two diffusion laws for coarse and fine particles; a constant diffusion coefficient is used only for coarse particles (Fickian diffusion). As a result, the diffusion coefficient does not have the same meaning as for numerical models such as FALL3D and PUFF. However, the horizontal diffusion coefficients used in FALL3D and PUFF give the same result, which is about double the value derived from the LES model. The mean vertical diffusion coefficient obtained from similarity theory is also on the same order of magnitude as the coefficient derived from PUFF. In general, it is important to thoroughly understand parameters from different models before comparing them.

We conclude that for the 1998 Etna eruption, column height is the parameter that can be constrained with the least uncertainty. The total erupted mass can also be constrained within an acceptable range using empirical (exponential method applied to three segments and power-law fit), analytical (inversion of TEPHRA2), and numerical (FALL3D) models. In contrast, the derivations of $\dot{M}$, duration, wind profile, and diffusion coefficients are more problematic. This is probably due to the greater complexity associated with these parameters, given that they are strongly connected to atmospheric characteristics, and are also parameterized differently in different sedimentation models. In particular,

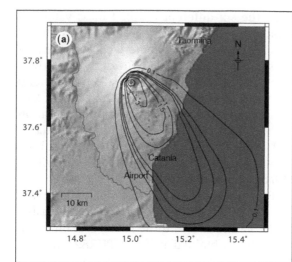

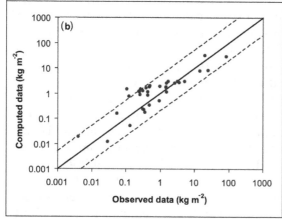

Figure 9.10 FALL3D simulation results for the Etna 22 July 1998 deposit (Run 5, Table 9.2): (a) Isomass map with contours of 0.1, 0.3, 0.4, 0.5, 1, 1.5, 3, 10, 25, and 80 kg m^{-2}, shown for comparison with Figs. 9.4 and 9.9. Locations of the volcano and the field data are indicated. (b) Comparison between computed (solid line) and observed (symbols) mass/area data. Dashed lines indicate over- or under-estimations of 1/5 and 5 times the observed values respectively.

discrepancies in the calculation of mass eruption rate are due to a combination of factors, including the more complex plume dynamics resulting from lava-fountaining fragmentation, and the effects of wind shear on air entrainment within the convective region described in FALL3D, which, for a given plume height, requires a larger $\dot{M}$ compared to standard analytical models.

9.7.2 Determination of eruption parameters

As described above, eruption parameters can be inferred by applying empirical, analytical, and numerical models, and through inversion solutions of analytical models. Our case study confirms that empirical extrapolation of poor data sets can be misleading (e.g., volume calculations based on the integration of only one or two exponential segments), but that integrations of at least three exponential segments and of the power-law fit give comparable results for moderate eruptions. However, this result is not necessarily expected to hold for more extensively dispersed deposits. In addition, sensitivity tests carried out for the integration of the power-law fit for the Etna deposit show larger discrepancies caused by the choice of $\sqrt{A_0}$ than of $\sqrt{A_{\mathrm{dist}}}$ (Table 9.1). This is consistent with the findings of Bonadonna and Houghton (2005) for deposits produced by relatively small eruptions characterized by a power-law exponent $m > 2$ (e.g., Ruapehu 1996). However, for widely dispersed deposits (i.e, $m < 2$) the power-law method is more sensitive to the choice of $\sqrt{A_{\mathrm{dist}}}$ than $\sqrt{A_0}$. As a result, poorly exposed deposits, in particular those of large eruptions, would be better described by applying inversion solutions of analytical models such as TEPHRA2 (Connor and Connor, 2006), ASHFALL (Hurst and Turner, 1999), and HAZMAP (Macedonio et al., 2005). Scollo et al. (2008a) have shown that the model TEPHRA can be used to determine erupted mass because it is very sensitive to this parameter, independent of other inputs, and that the erupted mass can be well constrained with at least 10 well-distributed field data points (the data set of our case study has 33 thickness measurements). These conclusions hold for the other analytical models. In contrast, determination of column height using TEPHRA is less straightforward because height is an input, along with the total grain-size distribution, that affects the model output through interaction with other input parameters (Scollo et al., 2008a). These observations are confirmed by application of inversion techniques using TEPHRA2, which shows that erupted mass is easier to constrain than column height (e.g., Fig. 9.8). However, Volentik *et al.* (2010) have shown that a better constraint on column height is obtained when tephra accumulation is inverted for individual grain-sizes. Further studies of the application of inversion techniques are needed to assess the minimum amount of field data required to provide a reliable characterization. In addition, inversion techniques are not always straightforward to apply; the choice of parameter ranges can significantly affect the final result and therefore requires critical analysis. We also stress that studies of poorly exposed deposits have shown that the method of Carey and Sparks (1986) gives good results even when the position of the vent is not well known (e.g., Wehrmann *et al.*, 2006). As a result, inversion techniques should always be used in combination with other models for constraining both erupted mass and plume height. In particular, granulometry data in proximal, medial, and distal areas should always be collected to better constrain plume height. Finally, models based on empirical observations should always be applied in their range of validation.

9.8 | Summary and outlook

Comprehensive characterization of tephra deposits and reliable hazard assessment can only result from critical and synergistic application of models with levels of sophistication ranging from purely empirical to fully numerical. First, tephra deposits need to be sampled accurately (for both mass/area and grain-size data) over an area proportional to the associated particle dispersal. Second, dedicated empirical and analytical models can be used for determination of plume height, erupted mass, initial grain-size distribution, mass eruption rate and duration. Inversion solutions of analytical models can also be used to obtain independent results for the same eruption parameters. Eruption parameters can be used to classify volcanic eruptions and build potential eruptive scenarios. Finally, following thorough model validation and calibration, analytical and numerical models can be applied to compile

probabilistic and/or real-time forecasting of tephra dispersal. Typically, analytical models are computationally fast and therefore can be used to compile fully probabilistic assessments, whereas numerical models are better suited to real-time forecasting to provide accurate estimates of ground sedimentation and of the position of the volcanic cloud with time. In particular, our case study confirms that the more sophisticated numerical models do not necessarily provide better accuracy in terms of ground sedimentation, especially given the uncertainties associated with the input parameters (e.g., Scollo *et al.*, 2008b; Figs. 9.9 and 9.10). However, numerical models can provide crucial information not possible with analytical models (e.g., a description of cloud movement with time). In general, verified and validated numerical models are also more appropriate for real-time forecasting because they require fewer empirical parameters (e.g., diffusion coefficient, mass distribution within the eruptive column) and, as a result, are simpler to apply. Models of all levels of sophistication would benefit significantly from better parameterization of critical sedimentation processes such as particle aggregation and settling, and from quantification of uncertainties associated with input parameters such as erupted mass, plume height, mass discharge rate, and total grain-size distribution.

9.9 | Notation

a	sample location in application of inversion techniques
A	area enclosed within isomass/isopach contours (m²)
A_x	area enclosed within isopach line of thickness T_x (m²)
$\sqrt{A_0}$	integration limit of power-law function, typically taken as distance of maximum deposit thickness (m)
$\sqrt{A_{dist}}$	integration limit of power-law function, typically taken as the downwind extent of deposit (m)
b_t	thickness half distance (m)

$c_j(x,y,z,t)$ — particle concentration (kg m^{-3})

c_j' — turbulent fluctuation of concentration of particle class j (kg m^{-3})

$\bar{c}_j$ — ensemble average of concentration of particle class j (kg m^{-3})

C — empirical factor (m kg$^{-1/4}$ s$^{1/4}$)

C_D — drag coefficient

C_{pl} — coefficient of power-law best fit (m$^{(1+m)}$)

d_j — diameter of particle of class j (m)

F — particle shape factor

H — maximum height of volcanic plume (m)

H_b — neutral buoyancy level of volcanic plume (m)

j — index of particle size

k — slope of deposit exponential best-fit curve (km^{-1})

k_n — slope of exponential segment n (km^{-1})

K_H — horizontal atmospheric diffusion coefficient ($K_H = K_x = K_y$) (m^2 s^{-1})

K_x — x-component of horizontal diffusion coefficient (m^2 s^{-1})

K_V — vertical atmospheric diffusion coefficient (m^2 s^{-1})

K_y — y-component of horizontal diffusion coefficient (m^2 s^{-1})

m — exponent of power-law best fit curve

L, I, S — dimensions of three perpendicular axes of particle (m)

M — total mass of particles of given size fraction carried by current at distance x_1 (kg)

$\dot{M}$ — mass eruption rate (kg s^{-1})

M_0 — initial mass for given grain size injected into current (kg)

Mo_a — observed mass per unit area at sample location a (kg m^{-2})

Mc_a — computed mass per unit area at sample location a (kg m^{-2})

n — number of exponential segments

N — number of field observations in application of inversion techniques

Q — volumetric flow rate into current at neutral buoyancy level (m^3 s^{-1})

s — spreading direction of current (m)

S_n — position of break in slope of exponential segment n (m)

$S(x,y,z,t)$ — source term (kg m^{-3} s^{-1})

t — time coordinate (s)

T — thickness of tephra deposit (m)

T_{n0} — intercept of exponential segment n (m)

T_0 — maximum thickness of tephra deposit (m)

T_x — thickness of given isopach x (m)

u_x, u_y, u_z — components of wind velocity vector (m s^{-1})

$\bar{u}_x, \bar{u}_y, \bar{u}_z$ — components of average wind velocity (m s^{-1})

u_x', u_y', u_z' — components of turbulent fluctuation of wind velocity (m s^{-1})

U — current velocity in s-direction of plume or umbrella cloud (m s^{-1})

v_{Hb} — particle terminal velocity at neutral buoyancy level H_b (m s^{-1})

v_j — particle terminal velocity of particle class j; $v_j = v_j(x,y,z)$ (m s^{-1})

V — erupted volume (m^3)

w — maximum crosswind width at source of spreading current (m)

x_0 — plume corner position (m)

x, y, z — spatial coordinates (m)

η_a — dynamic viscosity of air (Pa s)

ρ_a — density of air (kg m^{-3})

ψ — sphericity

ψ_{work} — "working" sphericity

Re — Reynolds number

List of acronyms

ABL — Atmospheric Boundary Layer
ADS — Advection Diffusion Sedimentation
ATHAM — Active Tracer High Resolution Atmospheric Model
BPT — Buoyant Plume Theory
CANERM — CANadian Emergency Response Model
DRE — Dense Rock Equivalent
FTT — Fall Time Threshold
GLOBE — Global Land One-km Base Elevation map
HYSPLIT — Hybrid Single Particle Lagrangian Integrated Trajectory Model
IR — thermal infrared
LES — Large Eddy Simulation
MEDIA — Eulerian Model for DIspersion in the Atmosphere

MM5   Fifth-Generation NCAR/Penn State Mesoscale Model

NAME   Numerical Atmospheric-dispersion Modelling Environment

OAT   One At a Time sensitivity tests

SGS   Sub-Grid Scale

VAAC   Volcanic Ash Advisory Centers

VAFTAD   Volcanic Ash Forecast Transport And Dispersion

WRF   Weather Research and Forecasting model

Acknowledgments

A. Costa was supported by the MIUR-FIRB Italian project "Sviluppo Nuove Tecnologie per la Protezione e Difesa del Territorio dai Rischi Naturali." The authors are grateful to D. Andronico, M. Coltelli, and P. Del Carlo for providing the field data of the 22 July 1998 eruption of Mt. Etna and to A. Folch for providing the pseudo-sounding used to run CALMET. S. Scollo and L. Connor are thanked for useful discussion.

References

Aloisi, M., D'Agostino, M., Dean, K. G., Mostaccio, A. and Neri, G. (2002). Satellite analysis and PUFF simulation of the eruptive cloud generated by the Mount Etna paroxysm of 22 July 1998. *Journal of Geophysical Research*, **107**(B12), 2373, doi:10.1029/2001JB000630.

Andronico, D., Del Carlo, P. and Coltelli, M. (1999). The 22 July 1998 fire fountain episode at Voragine Crater (Mt. Etna, Italy). *Volcanic and Magmatic Studies Group Annual Meeting, UK, 5–6 January*.

Andronico, D., Scollo, S., Caruso, S. and Cristaldi, A. (2008). The 2002–03 Etna explosive activity: Tephra dispersal and features of the deposits. *Journal of Geophysical Research*, **113**, B04209, doi:10.1029/2007JB005126.

Arastoopour, H., Wang, C. H. and Weil, S. A. (1982). Particle-particle interaction force in a dilute gas-solid system. *Chemical Engineering Sciences*, **37**, 1379–1386.

Armienti, P., Macedonio, G. and Pareschi, M. T. (1988). A numerical model for simulation of tephra transport and deposition – applications to May 18, 1980, Mount St. Helens eruption. *Journal of Geophysical Research*, **93**(B6), 6463–6476.

Aschenbrenner, B. C. (1956). A new method of expressing particle sphericity. *Journal of Sedimentary Petrology*, **26**, 15–31.

Barberi, F., Coltelli, M., Frullani, A., Rosi, M. and Almeida, E. (1995). Chronology and dispersal characteristics of recently (last 5000 years) erupted tephra of Cotopaxi (Ecuador): implications for long-term eruptive forecasting. *Journal of Volcanology and Geothermal Research*, **69**, 217–239.

Barsotti, S. and Neri, A. (2008). The VOL-CALPUFF model for atmospheric ash dispersal: 2. Application to the weak Mount Etna plume of July 2001. *Journal of Geophysical Research*, **113**, B03209, doi:10.1029/2006JB004624.

Barsotti, S., Neri, A. and Scire, J. S. (2008). The VOL-CALPUFF model for atmospheric ash dispersal: 1. Approach and physical formulation. *Journal of Geophysical Research*, **113**, B03208, doi:10.1029/2006JB004623.

Bonadonna, C. and Phillips, J. C. (2003). Sedimentation from strong volcanic plumes. *Journal of Geophysical Research*, **108**(B7), 2340–2368.

Bonadonna, C. and Houghton, B. F. (2005). Total grainsize distribution and volume of tephra-fall deposits. *Bulletin of Volcanology*, **67**, 441–456.

Bonadonna, C., Ernst, G. G. J. and Sparks, R. S. J. (1998). Thickness variations and volume estimates of tephra fall deposits: the importance of particle Reynolds number. *Journal of Volcanology and Geothermal Research*, **81**, 173–187.

Bonadonna, C., Macedonio, G. and Sparks, R. S. J. (2002a). Numerical modelling of tephra fallout associated with dome collapses and Vulcanian explosions: application to hazard assessment on Montserrat. In *The Eruption of Soufrière Hills Volcano, Montserrat, from 1995 to 1999*, ed. T. H. Druitt and B. P. Kokelaar. Geological Society London Memoir, **21**, 517–537.

Bonadonna, C., Mayberry, G. C., Calder, E. *et al.* (2002b). Tephra fallout in the eruption of Soufrière Hills Volcano, Montserrat. In *The Eruption of Soufrière Hills Volcano, Montserrat, from 1995 to 1999*, ed. T. H. Druitt and B. P. Kokelaar. Geological Society London Memoir, **21**, 483–516.

Bonadonna, C., Phillips, J. C. and Houghton, B. F. (2005a). Modeling tephra sedimentation from a Ruapehu weak plume eruption. *Journal of Geophysical Research*, **110**, B08209, doi:10.1029/2004JB003515.

Bonadonna, C., Connor, C. B., Houghton, B. F. *et al.* (2005b). Probabilistic modeling of tephra dispersion: hazard assessment of a multi-phase eruption at Tarawera, New Zealand. *Journal of Geophysical Research*, **110**, B03203, doi:10.1029/2003JB002896.

Brazier, S., Davis, A. N., Sigurdsson, H. and Sparks, R. S. J. (1982). Fallout and deposition of volcanic ash during the 1979 explosive eruption of the Soufrière of St. Vincent. *Journal of Volcanology and Geothermal Research*, **14**, 335–359.

Bursik, M. (2001). Effect of wind on the rise height of volcanic plumes. *Geophysical Research Letters*, **28**(18), 3621–3624.

Bursik, M. I., Sparks, R. S. J., Gilbert, J. S. and Carey, S. N. (1992a). Sedimentation of tephra by volcanic plumes: I. Theory and its comparison with a study of the Fogo A plinian deposit, Sao Miguel (Azores). *Bulletin of Volcanology*, **54**, 329–344.

Bursik, M. I., Carey, S. N. and Sparks, R. S. J. (1992b). A gravity current model for the May 18, 1980 Mount St. Helens plume. *Geophysical Research Letters*, **19**, 1663–1666.

Byrne, M. A., Laing, A. G. and Connor, C. (2007). Predicting tephra dispersion with a mesoscale atmospheric model and a particle fall model: Application to Cerro Negro volcano. *Journal of Applied Meteorology and Climatology*, **46**, 121–135.

Carazzo, G., Kaminski, E. and Tait, S. (2008). On the dynamics of volcanic columns: A comparison of field data with a new model of negatively buoyant jets. *Journal of Volcanology and Geothermal Research*, **178**, 94–103.

Carey, S. N. and Sigurdsson, H. (1982). Influence of particle aggregation on deposition of distal tephra from the May 18, 1980, eruption of Mount St. Helens volcano. *Journal of Geophysical Research*, **87**(B8), 7061–7072.

Carey, S. N. and Sigurdsson, H. (1986). The 1982 eruptions of El Chichon volcano, Mexico (2): observations and numerical modelling of tephra-fall distribution. *Bulletin of Volcanology*, **48**, 127–141.

Carey, S. and Sigurdsson, H. (1989). The intensity of Plinian eruptions. *Bulletin of Volcanology*, **51**, 28–40.

Carey, S. N. and Sparks, R. S. J. (1986). Quantitative models of the fallout and dispersal of tephra from volcanic eruption columns. *Bulletin of Volcanology*, **48**, 109–125.

Chhabra, R. P., Agarwal, L. and Sinha, N. K. (1999). Drag on non-spherical particles: an evaluation of available methods. *Powder Technology*, **101**, 288–295.

Coltelli, M., Puglisi, G., Guglielmino, F. and Palano, M. (2006). Application of differential SAR interferometry for studying eruptive event of 22 July 1998 at Mt. Etna. *Quaderni di Geofisica*, **43**, 15–20.

Coltelli, M., Miraglia, L. and Scollo, S. (2008). Characterization of shape and terminal velocity of tephra particles erupted during the 2002 eruption of Etna volcano, Italy. *Bulletin of Volcanology*, **70**, 1103–1112.

Connor, L. G. and Connor, C. B. (2006). Inversion is the key to dispersion: understanding eruption dynamics by inverting tephra fallout. In *Statistics in Volcanology*, ed. H. Mader, S. Cole, C. B. Connor and L. G. Connor. Special Publications of IAVCEI, **1**, pp. 231–242. London: Geological Society.

Connor, C. B., Hill, B. E., Winfrey, B., Franklin, N. M. and La Femina, P. C. (2001). Estimation of volcanic hazards from tephra fallout. *Natural Hazards Review*, **2**, 33–42.

Cornell, W., Carey, S. and Sigurdsson, H. (1983). Computer simulation of transport and deposition of the Campanian Y-5 Ash. *Journal of Volcanology and Geothermal Research*, **17**, 89–109.

Corsaro, R. A. and Pompilio, M. (2004). Magma dynamics in the shallow plumbing system of Mt. Etna as recorded by compositional variations in volcanics of recent summit activity (1995–1999). *Journal of Volcanology and Geothermal Research*, **137**, 55–71.

Costa, A., Macedonio, G. and Folch, A. (2006). A three-dimensional Eulerian model for transport and deposition of volcanic ashes. *Earth and Planetary Science Letters*, **241**, 634–647.

Costantini, L., Bonadonna, C., Houghton, B. F. and Wehrmann, H. (2008). New physical characterization of the Fontana Lapilli basaltic Plinian eruption, Nicaragua. *Bulletin of Volcanology*, **71**, 337–355.

D'amours, R. (1998). Modeling the ETEX plume dispersion with the Canadian emergency response model. *Atmospheric Environment*, **32**, 4335–4341.

Dobran, F., Neri, A. and Macedonio, G. (1993). Numerical simulation of collapsing volcanic columns. *Journal of Geophysical Research*, **98**, 4231–4259.

Draxler, R. R. and Hess, G. D. (1998). An overview of the HYSPLIT_4 modelling system for trajectories, dispersion and deposition. *Australian Meteorological Magazine*, **47**, 295–308.

Durant, A. J., Rose, W. I., Sarna-Wojcicki, A. M., Carey, S. and Volentik, A. C. M. (2009).

Hydrometeor-enhanced tephra sedimentation: Constraints from the 18 May 1980 eruption of Mount St. Helens. *Journal of Geophysical Research*, **114**, B03204, doi:10.1029/2008JB005756.

Esposti Ongaro, T., Cavazzoni, C., Erbacci, G., Neri, A. and Salvetti, M. V. (2007). A parallel multiphase flow code for the 3D simulation of explosive volcanic eruptions. *Parallel Computing*, **33**, 541–560.

Fierstein, J. and Nathenson, M. (1992). Another look at the calculation of fallout tephra volumes. *Bulletin of Volcanology*, **54**, 156–167.

Folch, A. and Felpeto, A. (2005). A coupled model for dispersal of tephra during sustained explosive eruptions. *Journal of Volcanology and Geothermal Research*, **145**, 337–349.

Folch, A., Jorba, O. and Viramonte, J. (2008). Volcanic ash forecast – application to the May 2008 Chaiten eruption. *Natural Hazards and Earth System Sciences*, **8**, 927–940.

Folch, A., Costa, A. and Macedonio, G. (2009). FALL3D: A computational model for transport and deposition of volcanic ash. *Computers and Geosciences*, **35**, 1334–1342

Froggatt, P. C. (1982). Review of methods estimating rhyolitic tephra volumes; applications to the Taupo Volcanic Zone, New Zealand. *Journal of Volcanology and Geothermal Research*, **14**, 1–56.

Ganser, G. H. (1993). A rational approach to drag prediction of spherical and nonspherical particles. *Powder Technology*, **77**, 143–152.

Glaze, L. S. and Self, S. (1991). Ashfall dispersal for the 16 September 1986, eruption of Lascar, Chile, calculated by a turbulent-diffusion model. *Geophysical Research Letters*, **18**, 1237–1240.

Glaze, L. S., Wilson, L. and Mouginis-Mark, P. J. (1999). Volcanic eruption plume top topography and heights as determined from photoclinometric analysis of satellite data. *Journal of Geophysical Research*, **104**(B2), 2989–3001.

Heffter, J. L. and Stunder, B. J. B. (1993). Volcanic Ash Forecast Transport and Dispersion (VAFTAD) Model. *Weather and Forecasting*, **8**, 533–541.

Herzog, M., Graf, H. F., Textor, C. and Oberhuber, J. M. (1998). The effect of phase changes of water on the development of volcanic plumes. *Journal of Volcanology and Geothermal Research*, **87**, 55–74.

Hildreth, W. and Drake, R. E. (1992). Volcano Quizapu, Chilean Andes. *Bulletin of Volcanology*, **54**, 93–125.

Hobbs, P. V., Lyons, J. H., Locatelli, J. D. *et al.* (1981). Radar detection of cloud-seeding effects. *Science*, **213**, 1250–1252.

Holasek, R. E. and Self, S. (1995). GOES weather-satellite observations and measurements of the May 18, 1980, Mount St. Helens eruption. *Journal of Geophysical Research*, **100**(B5), 8469–8487.

Hurst, A. W. and Turner, R. (1999). Performance of the program ASHFALL for forecasting ashfall during the 1995 and 1996 eruptions of Ruapehu volcano. *New Zealand Journal of Geology and Geophysics*, **42**, 615–622.

Ishimine, Y. (2006). Sensitivity of the dynamics of volcanic eruption columns to their shape. *Bulletin of Volcanology*, **68**(6), 516–537.

James, M. R., Gilbert, J. S. and Lane, S. J. (2002). Experimental investigation of volcanic particle aggregation in the absence of a liquid phase. *Journal of Geophysical Research*, **107**(B9), 2191, doi:10.1029/2001JB000950.

James, M. R., Lane, S. J. and Gilbert, J. S. (2003). Density, construction, and drag coefficient of electrostatic volcanic ash aggregates. *Journal of Geophysical Research*, **108**(B9), 2435, doi:10.1029/2002JB002011.

Kalnay, E., Kanamitsu, M., Kistler, R. *et al.* (1996). The NCEP/NCAR 40-year reanalysis project. *Bulletin of the American Meteorological Society*, **77**, 437–471.

Kaminski, E. and Jaupart, C. (1998). The size distribution of pyroclasts and the fragmentation sequence in explosive volcanic eruptions. *Journal of Geophysical Research*, **103**(B12), 29 759–29 779.

Koyaguchi, T. (1994). Grain-size variation of tephra derived from volcanic umbrella clouds. *Bulletin of Volcanology*, **56**, 1–9.

Koyaguchi, T. and Ohno, M. (2001a). Reconstruction of eruption column dynamics on the basis of grain size of tephra fall deposits. 1. Methods. *Journal of Geophysical Research*, **106**(B4), 6499–6512.

Koyaguchi, T. and Ohno, M. (2001b). Reconstruction of eruption column dynamics on the basis of grain size of tephra fall deposits. 2. Application to the Pinatubo 1991 euption. *Journal of Geophysical Research*, **106**(B4), 6513–6533.

Kunii, D. and Levenspiel, O. (1969). *Fluidization Engineering*. New York: Wiley and Sons.

Legros, F. (2000). Minimum volume of a tephra fallout deposit estimated from a single isopach. *Journal of Volcanology and Geothermal Resarch*, **96**, 25–32.

Macedonio, G., Costa, A. and Longo, A. (2005). A computer model for volcanic ash fallout and assessment of subsequent hazard. *Computers and Geosciences*, **31**, 837–845.

Mannen, K. (2006). Total grain size distribution of a mafic subplinian tephra, TB-2, from the 1986 Izu-Oshima eruption, Japan: An estimation based on a theoretical model of tephra dispersal. *Journal of Volcanology and Geothermal Research*, **155**, 1–17.

Martin, D. and Nokes, R. (1988). Crystal settling in a vigorously convecting magma chamber. *Nature*, **332**, 534–536.

McPhie, J., Walker, G. P. L. and Christiansen, R. L. (1990). Phreatomagmatic and phreatic fall and surge deposits from explosions at Kilauea volcano, Hawaii, 1790 A.D.: Keanakakoi Ash Member. *Bulletin of Volcanology*, **52**, 334–354.

Michalakes, J., Dudhia, J., Gill, D. et al. (2005). The weather research and forecast model: Software architecture and performance. *Use of High Performance Computing in Meteorology*, Proceedings of the Eleventh ECMWF Workshop, 156–168.

Morton, B., Taylor, G. L. and Turner, J. S. (1956). Turbulent gravitational convection from maintained and instantaneous source. *Proceedings of the Royal Society of London*, **A234**, 1–23.

Neri, A., Papale, P. and Macedonio, G. (1998). The role of magma composition and water content in explosive eruptions: 2. Pyroclastic dispersion dynamics. *Journal of Volcanology and Geothermal Research*, **87**, 95–115.

Oberhuber, J. M., Herzog, M., Graf, H. F. and Schwanke, K. (1998). Volcanic plume simulation on large scales. *Journal of Volcanology and Geothermal Research*, **87**, 29–53.

Papale, P. and Rosi, M. (1993). A case of no-wind plinian fallout at Pululagua caldera (Ecuador): implications for model of clast dispersal. *Bulletin of Volcanology*, **55**, 523–535.

Perez, W., Freundt, A., Kutterolf, S. and Schmincke, H.-U. (2009). The Masaya Triple Layer: A 2100 year old basaltic multi-episodic Plinian eruption from the Masaya Caldera Complex (Nicaragua). *Journal of Volcanology and Geothermal Research*, **179**, 191–205.

Pfeiffer, T., Costa, A. and Macedonio, G. (2005). A model for the numerical simulation of tephra fall deposits. *Journal of Volcanology and Geothermal Resarch*, **140**, 273–294.

Prata, A. J. and Grant, I. F. (2001). Retrieval of microphysical and morphological properties of volcanic ash plumes from satellite data: Application to Mt Ruapehu, New Zealand. *Quarterly Journal of the Royal Meteorological Society*, **127**, 2153–2179.

Prata, A. J. and Turner, P. J. (1997). Cloud-top height determination using ATSR data. *Remote Sensing of Environment*, **59**, 1–13.

Pyle, D. M. (1989). The thickness, volume and grainsize of tephra fall deposits. *Bulletin of Volcanology*, **51**, 1–15.

Pyle, D. M. (1990). New estimates for the volume of the Minoan eruption. In *Thera and the Aegean World*, ed. D. A. Hardy. London: The Thera Foundation, 113–121.

Pyle, D. M. (1995). Assessment of the minimum volume of tephra fall deposits. *Journal of Volcanology and Geothermal Research*, **69**, 379–382.

Riley, C. M., Rose, W. I. and Bluth, G. J. S. (2003). Quantitative shape measurements of distal volcanic ash. *Journal of Geophysical Research*, **108**, 2504, doi:10.1029/2001JB000818.

Rose, W. I. (1993). Comment on "Another look at the calculation of fallout tephra volumes". *Bulletin of Volcanology*, **55**, 372–374.

Rose, W. I., Self, S., Murrow, P. J. et al. (2008). Nature and significance of small volume fall deposits at composite volcanoes: Insights from the October 14, 1974 Fuego eruption, Guatemala. *Bulletin of Volcanology*, **70**, 1043–1067.

Rosi, M. (1998). Plinian eruption columns: particle transport and fallout. In *From Magma to Tephra: Modelling Physical Processes of Explosive Volcanic Eruptions*, ed. A. Freundt and M. Rosi. Elsevier, pp. 139–172.

Ryall, D. B. and Maryon, R. H. (1998). Validation of the UK Met. Office's NAME model against the ETEX dataset. *Atmospheric Environment*, **32**, 4265–4276.

Sandu, I., Bompay, F. and Stefan, S. (2003). Validation of atmospheric dispersion models using ETEX data. *International Journal of Environment and Pollution*, **19**, 367–389.

Sarna-Wojcicki, A. M., Shipley, S., Waitt, J. R., Dzurisin, D. and Wood, S. H. (1981). Areal distribution thickness, mass, volume, and grain-size of airfall ash from the six major eruptions of 1980. In *The 1980 Eruption of Mount St. Helens*, ed. W. P. Lipman and D. R. Mullineaux. Washington, D.C.: U.S. Geological Survey Professional Paper, 1250, 577–600.

Scasso, R., Corbella, H. and Tiberi, P. (1994). Sedimentological analysis of the tephra from 12–15 August 1991 eruption of Hudson Volcano. *Bulletin of Volcanology*, **56**, 121–132.

Scire, J. S., Robe, F. and Yamartino, R. (2000). *A User's Guide for the CALMET Meteorological Model*. Concord, MA: Earth Tech, Inc.

Scollo, S., Del Carlo, P. and Coltelli, M. (2007). Tephra fallout of 2001 Etna flank eruption: Analysis of the deposit and plume dispersion. *Journal of Volcanology and Geothermal Research*, **160**, 147–164.

Scollo, S., Folch, A. and Costa, A. (2008b). A parametric and comparative study of different tephra fallout models. *Journal of Volcanology and Geothermal Research*, **176**, 199–211.

Scollo, S., Tarantola, S., Bonadonna, C., Coltelli, M. and Saltelli, A. (2008a). Sensitivity analysis and uncertainty estimation for tephra dispersal models. *Journal of Geophysical Research*, **113**, B06202, doi:10.1029/2006JB004864.

Searcy, C., Dean, K. and Stringer, W. (1998). PUFF: A high-resolution volcanic ash tracking model. *Journal of Volcanology and Geothermal Research*, **80**, 1–16.

Settle, M. (1978). Volcanic eruption clouds and thermal power output of explosive eruptions. *Journal of Volcanological and Geothermal Research*, **3**, 309–324.

Shaw, D. M., Watkins, N. D. and Huang, T. C. (1974). Atmospherically transported volcanic glass in deep-sea sediments: Theoretical considerations. *Journal of Geophysical Research*, **79**, 3087–3094.

Skamarock, W., Klemp, J., Dudhia, J. *et al.* (2005). *A Description of the Advanced Research WRF Version 2.* Available online at: www.wrf-model.org.

Sorem, R. K. (1982). Volcanic ash clusters: tephra rafts and scavengers. *Journal of Volcanology and Geothermal Research*, **13**, 63–71.

Sparks, R. S. J. (1986). The dimensions and dynamics of volcanic eruption columns. *Bulletin of Volcanology*, **48**, 3–15.

Sparks, R. S. J., Wilson, L. and Sigurdsson, H. (1981). The pyroclastic deposits of the 1875 eruption of Askja, Iceland. *Philosophical Transactions of the Royal Society of London*, **229**, 241–273.

Sparks, R. S. J., Carey, S. N. and Sigurdsson, H. (1991). Sedimentation from gravity currents generated by turbulent plumes. *Sedimentology*, **38**, 839–856.

Sparks, R. S. J., Bursik, M. I., Ablay, G. J., Thomas, R. M. E. and Carey, S. N. (1992). Sedimentation of tephra by volcanic plumes. 2. Controls on thickness and grain-size variations of tephra fall deposits. *Bulletin of Volcanology*, **54**, 685–695.

Sparks, R. S. J., Bursik, M. I., Carey, S. N. *et al.* (1997). *Volcanic Plumes.* Chichester, UK: Wiley.

Stull, R. (1988). *An Introduction to Boundary Layer Meteorology.* Dordrecht: Kluwer Academic.

Sulpizio, R. (2005). Three empirical methods for the calculation of distal volume of tephra-fall deposits. *Journal of Volcanology and Geothermal Research*, **145**, 315–336.

Suzuki, T. (1983). A theoretical model for dispersion of tephra. In *Arc Volcanism, Physics and Tectonics*, ed. D. Shimozuru and I. Yokoyama. Tokyo: Terra Scientific, pp. 95–113.

Tanaka, H. L. and Yamamoto, K. (2002). Numerical simulation of volcanic plume dispersal from Usu volcano in Japan on 31 March 2000 using PUFF model. *Earth Planets Space*, **54**, 743–752.

Textor, C., Graf, H. F., Herzog, M. *et al.* (2006). Volcanic particle aggregation in explosive eruption columns. Part II: Numerical experiments. *Journal of Volcanology and Geothermal Research*, **150**, 378–394.

Thorarinsson, S. (1944). Petrokronologista Studier pa Island. *Geographes Annuales Stockholm*, **26**, 1–217.

Thorarinsson, S. (1954). The eruption of Hekla 1947–1948. In *The Tephra Fall from Hekla*. Reykjavik: Vis Islendinga.

Turner, J. S. (1979). *Buoyancy Effects in Fluids.* Cambridge: Cambridge University Press.

Veitch, G. and Woods, A. W. (2001). Particle aggregation in volcanic eruption columns. *Journal of Geophysical Research*, **106**(B11), 26 425–26 441.

Volentik, A., Bonadonna, C., Connor, C. B., Connor, L. J. and Rosi, M. (2010). Modeling tephra dispersal in absence of wind: insights from the climactic phase of the 2450BP Plinian eruption of Pululagua volcano (Ecuador). *Journal of Volcanology and Geothermal Research*, **193**, 117–136.

Wadell, H. (1933). Sphericity and roundness of rock particles. *Journal of Geology*, **41**, 310–331.

Walker, G. P. L. (1973). Explosive volcanic eruptions – a new classification scheme. *Geologische Rundschau*, **62**, 431–446.

Walker, G. P. L. (1980). The Taupo Pumice: product of the most powerful known (Ultraplinian) eruption? *Journal of Volcanology and Geothermal Research*, **8**, 69–94.

Walker, G. P. L., Wilson, L. and Bowell, E. L. G. (1971). Explosive volcanic eruptions – I. The rate of fall of pyroclasts. *Geophysical Journal of the Royal Astronomical Society*, **22**, 377–383.

Walker, G. P. L., Self, S. and Wilson, L. (1984). Tarawera, 1886, New Zealand – A basaltic Plinian fissure eruption. *Journal of Volcanology and Geothermal Research*, **21**, 61–78.

Wehrmann, H., Bonadonna, C., Freundt, A., Houghton, B. F. and Kutterolf, S. (2006). Fontana Tephra: A basaltic Plinian eruption in Nicaragua. In *Volcanic Hazards in Central America*, Geological Society of America Special Paper, 412, pp. 209–223.

Wiesner, M. G., Wang, Y. W. and Zheng, L. (1995). Fallout of volcanic ash to the deep South China Sea induced by the 1991 eruption of Mount Pinatubo (Philippines). *Geology*, **23**, 885–888.

Wilson, L. and Huang, T. C. (1979). The influence of shape on the atmospheric settling velocity of volcanic ash particles. *Earth and Planetary Sciences Letters*, **44**, 311–324.

Wilson, L. and Walker, G. P. L. (1987). Explosive volcanic eruptions – VI. Ejecta dispersal in plinian eruptions – the control of eruption conditions and atmospheric properties. *Geophysical Journal of the Royal Astronomical Society*, **89**, 657–679.

Wilson, L., Sparks, R. S. J., Huang, T. C. and Watkins, N. D. (1978). The control of volcanic column height by eruption energetics and dynamics. *Journal of Geophysical Research*, **83**, 1829–1836.

Witham, C. S., Hort, M. C., Potts, R. *et al.* (2007). Comparison of VAAC atmospheric dispersion models using the 1 November 2004 Grimsvotn eruption. *Meteorological Applications*, **14**, 27–38.

Woods, A. W. (1988). The fluid dynamics and thermodynamics of eruption columns. *Bulletin of Volcanology*, **50**, 169–193.

Zimanowski, B., Wohletz, K., Dellino, P. and Buttner, R. (2003). The volcanic ash problem. *Journal of Volcanology and Geothermal Research*, **122**, 1–5.

Exercises

9.1 Calculate the erupted mass of the 22 July 1998 eruption of Etna volcano, Italy, by plotting the data in the table below on a semilog plot of mass/area versus square root of area and fitting and integrating both

  (i) three exponential segments, and

(ii) a power-law curve.

Use a deposit density of 1050 kg m^{-3}

Values of square root of area and mass/area of isomass lines for the Etna 1998 tephra deposit.

Isomass-line $\sqrt{A}$ (km)	Isomass-line mass/ area (kg m^{-2})
1.9	80.0
3.5	25.0
4.9	10.0
7.7	3.0
11.7	1.5
13.1	1.0
16.4	0.5
19.2	0.4
23.8	0.3
38.8	0.1

Online resources available at www.cambridge.org/fagents

- Supplement 9A: Details for Table 9.1
- Supplement 9B: Details for Table 9.2
- Supplement 9C: Additional figures
- Additional exercises
- Answers to exercises

Chapter 10

Pyroclastic density currents

Olivier Roche, Jeremy C. Phillips, and Karim Kelfoun

Overview

This chapter summarizes the principal experimental and theoretical approaches used to investigate the physics of pyroclastic density currents (PDCs), which are gravity-driven hot gas–particle mixtures commonly generated during explosive volcanic eruptions. PDC behavior ranges from *pyroclastic surges*, which are dilute turbulent suspensions, to *pyroclastic flows*, which are dense (fluidized) granular avalanches. Most PDCs consist of a coupled basal flow and an overriding surge, which renders their physics particularly complex. Experiments and phenomenological theory have been used to characterize the propagation and deposition mechanisms of PDCs. Most work has used turbulent gravity currents as an analogue to dilute PDCs and has provided fundamental insight into propagation and deposition dynamics and mixing with their surroundings. Dense PDCs have been investigated as granular and fluidized flows, and these studies have provided insight into deposit levée-channel morphology typical of coarse-grained flows, shown that fines-rich flows may behave as inertial fluid currents, and suggested that deposits of PDCs may form by aggradation. Numerical formulations ranging from continuum depth-averaged to discrete element models have been used to simulate PDC emplacement on real topographies and are fundamental in the context of volcanic hazard assessment and mitigation.

10.1 | Principal characteristics of pyroclastic density currents

Pyroclastic density currents (PDCs) are common features of explosive volcanic eruptions. They are generated from the gravitational collapse of lava domes (Chapter 7) or eruptive columns (Chapter 8), by lateral explosions in the case of hydromagmatic activity (Chapter 11) or sudden decompression of a magma body (Fig. 10.1), as well as during the formation of collapse calderas. PDCs are hot (up to ~600–800°C), gravity-driven, gas–particle mixtures within which the interstitial fluid may control the flow dynamics. The pyroclasts result from magma fragmentation and their granulometry commonly ranges from micron-sized ash to centimeter-sized lapilli and sometimes meter-sized blocks. PDCs have typical volumes of ~10^4–10^8 m^3, though their accumulation during an eruptive event can form deposits > 10^3 km^3. They are highly mobile: they commonly travel on gentle slopes over distances of several kilometers to several tens of kilometers

Modeling Volcanic Processes: The Physics and Mathematics of Volcanism, eds. Sarah A. Fagents, Tracy K. P. Gregg, and Rosaly M. C. Lopes. Published by Cambridge University Press. © Cambridge University Press 2013.

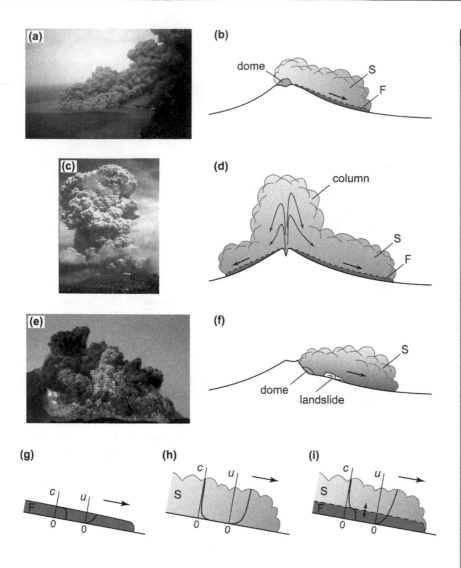

Figure 10.1 Photographs and schematic diagrams of PDCs generated by various mechanisms. The *dilute surge* and *dense flow* portions of the flow are labeled S and F, respectively (not to scale). (a) Lava dome collapse, Montserrat, 1997 (courtesy of R. S. J. Sparks), and (b) schematic diagram of gravitational or explosive lava dome collapse; (c) collapse of a vulcanian eruption column, Montserrat, 1997 (courtesy of A. B. Clarke), and (d) schematic diagram of gravitational collapse of discrete or continuous eruptive column; (e) lateral explosion, Mt. St. Helens, 18 May 1980 (courtesy of U.S.G.S), and (f) schematic diagram of lateral explosion caused by decompression of a cryptodome following landslide. (g–h) Schematic diagrams of the dense (F) and dilute (S) end-members of PDCs, with possible concentration (c) and velocity (u) profiles, whose curvature can be highly variable. (i) PDC consisting of a basal dense flow and an upper dilute surge; the double arrow and the dashed line indicate, respectively, that the transition between the basal and upper regions can be located at various heights, and be sharp or progressive.

at speeds of up to ~50–200 m s^{-1}, making them much more mobile than dry rock avalanches of equivalent volume. The principal characteristics of PDCs are summarized in Table 10.1.

As direct observations of PDCs are difficult and rare, their properties are inferred mainly from field analyses of their deposits. The characteristics of PDCs are presented in extensive reviews by Cas and Wright (1987), Druitt (1998), Freundt *et al.* (2000), Valentine and Fisher (2000), Branney and Kokelaar (2002), Sulpizio and Dellino (2008), and therefore detailed features will not be given here. PDCs represent a range of gravity-driven flows whose end-members are the *dilute pyroclastic surges* and the *dense pyroclastic flows*. Dilute surges are thought to be turbulent suspensions with solids concentrations of the order of 1 kg m^{-3} that increase downwards through the flow depth, and with mean internal horizontal velocities that increase upwards (Fisher, 1966; Valentine 1987; Dellino *et al.*, 2008). Particle interactions are negligible except at the basal boundary layer where the

Table 10.1 Selected microscopic and macroscopic parameters, and dimensionless numbers relevant to pyroclastic density currents.

Parameters	Symbol	Typical values
Particle diameter (m)	d	10^{-6}–1
Particle density (kg m^{-3})	ρ_p	10^2–10^3
Particle volume fraction	v_p	~0–0.8
Gas density (kg m^{-3})	ρ_g	~1
Gas dynamic viscosity (Pa s)	η_g	~10^{-5}
Current height (m)	h_c	1–10^2
Current length (m)	L	10^3–10^5
Current velocity (m s^{-1})	u_c	1–10^2
Current density (kg m^{-3})	ρ_c	1–10^3
Hydraulic permeability (m^2)^a	k	10^{-13}–10^{-9}

Numbers[b]	Symbol	Significance (ratio of quantities)
Mass	$\mathrm{Ma} = v_p\rho_p/(1-v_p)\rho_g$	solid inertia/fluid inertia
Froude	$\mathrm{Fr} = u_c/(g'h_c)^{1/2}$	current inertial stresses/gravitational stresses
Reynolds[c]	$\mathrm{Re} = \rho_c u_c h_c/\eta_g$	current inertial stresses/viscous stresses
Rouse[c]	$\mathrm{Pn} = u_s/(\kappa u_*)$	particle settling velocity/shear velocity
Bagnold[d]	$\mathrm{Ba} = \rho\,\dot\gamma d^2 v_p/\eta_g(1-v_p)$	collisional solid stresses/fluid viscous shear stresses
Darcy[d]	$\mathrm{Da} = \eta_g/v_p\rho_p k\dot\gamma$	solid–fluid interaction stresses/collisional solid stresses
Pressure[d]	$\mathrm{Pr} = (L/g')^{1/2}(D/h_c^2)$	current duration timescale/pore-pressure diffusion timescale
Savage[d,e]	$\mathrm{Sa} = \rho_p\dot\gamma^2 d^2/(\rho_p-\rho_g)gh_c$	collisional solid stresses/frictional solid stresses

$u_c/h_c = \dot\gamma$, the shear rate; D is a diffusion coefficient; κ is the von Kármán constant.
[a] Estimated
[b] See Iverson (1997); Iverson and Denlinger (2001).
[c] Applicable to dilute currents.
[d] Applicable to dense currents.
[e] Negligible role of the interstitial gas, modified from Savage (1984).

solid volume fraction can be high and from which deposition occurs. Surges are weakly influenced by topography as they are tens to hundreds of meters thick, and their deposits are commonly well sorted and laminated. On the other hand, dense flows (e.g., pumice-and-ash, block-and-ash, and scoria flows) have particle concentrations of the same order of magnitude as those of their deposits. They are a few meters to a few tens of meters thick, topographically controlled, pond in depressions, and form deposits that are commonly poorly sorted and massive. The nature of their internal velocity profile is poorly constrained. Two conceptual models for the deposition mechanisms of PDCs close to the dense end-member have been proposed: *en masse* deposition, in which the current freezes through its entire height (Sparks, 1976), and *progressive aggradation* in which the deposit builds up by accumulation of material (Fisher, 1966; Branney and Kokelaar, 1992). The dense and dilute end-members can coexist in most

PDCs, consisting of a coupled concentrated basal avalanche (the flow) and an overriding ash cloud (the surge) (Fig. 10.1), which results in complex internal particle concentration and velocity profiles through the depth of the current. Note also that a single eruption may produce PDCs that vary in a continuum between the end-members. When PDCs interact with topography, they can incorporate ambient air, whereas the overriding ash-cloud surge may generate secondary dense flows as it sediments particles out of suspension (Druitt *et al.*, 2002).

In the context of volcanic hazard assessment and mitigation, predictive theoretical models require improved understanding of the complex physics of PDCs, and experimental laboratory investigations can help provide fundamental insights. Dynamic similarity between the large-scale PDCs and their small-scale analogs is ensured when the dynamically relevant dimensional groupings are equivalent (see Table 10.1). In this chapter, we first describe the basic physics of gravity-driven motion of particle suspensions, and introduce simplified equations of motion for these currents. We then highlight important insights that have been obtained from laboratory experiments on the gravitational collapse of dense fluid suspensions, initially gas-fluidized particles, and granular columns, emphasizing the distinct effect on the frictional resistance to motion of these different initial states. We then describe numerical methods for solving the equations of motion relevant to emplacement following gravitational collapse, with particular focus on how to combine these dynamical models with complex natural topography for volcanic hazard prediction. Finally, we identify current and future research questions.

10.2 | Fundamental physics of gravity currents

10.2.1 Generalities

PDCs are an example of a predominantly horizontal, gravity-driven flow, more generally known as a *gravity current*. Gravity currents result whenever fluid of one density flows horizontally into a fluid of a different density, and are frequent occurrences in the natural world. Dust storms and sea-breeze fronts, thunderstorms and estuarine outflows, deep ocean turbidity flows and PDCs are just a few examples (Simpson, 1997). In these cases, the contrast in density between the two fluids can arise from compositional or thermal differences between the fluids, or the presence of suspended particles of a different density. The dynamics of gravity currents also depend strongly on whether the fluid that forms the current is released as a finite volume or continuously from a sustained source. An important paradigm for gravity current initiation is the gravitational collapse of a column of dense fluid or particles (Figs. 10.1(c), 10.1(d), and 10.2); recent experimental studies have led to the recognition that the dynamics of the collapse of a column of particles in air shares many common features with the fluid case (Lube *et al.*, 2004; Lajeunesse *et al.*, 2004; Balmforth and Kerswell, 2005; Roche *et al.*, 2008).

The dynamics of gravity currents are now generally well understood for a wide range of conditions (Simpson, 1997; Huppert, 2006). In this section, we introduce the fundamental dynamics of gravity currents, focusing in particular on the underlying physics and dynamical aspects that are relevant to PDCs. We start by considering the motion that develops between two fluids that have a small and constant density contrast, using energy balances to develop scaling relationships and simple theoretical models for gravity current motion. These approaches are then extended to apply to dilute turbulent PDCs (i.e., high Reynolds number; see Table 10.1), where the density contrast varies during the motion of the current, as a consequence of the sedimentation of particles initially held in suspension by turbulence (Valentine, 1987; Dellino *et al.*, 2008). We also consider situations in which the density contrast between the two fluids is large (i.e., dense PDCs), including granular flows, where the interstitial fluid phase can influence the flow dynamics to varying extents.

Figure 10.2 shows a series of images of the collapse of a column of a dense fluid and of initially gas-fluidized particles in a channel. The

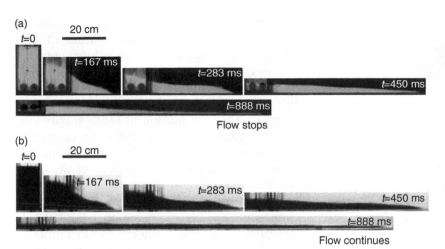

(a) t=0 20 cm

t=167 ms

t=283 ms

t=450 ms

t=888 ms

Flow stops

(b) t=0 20 cm

t=167 ms

t=283 ms

t=450 ms

t=888 ms

Flow continues

Figure 10.2 Photographs from laboratory experiments conducted in a horizontal channel showing the collapse in ambient air of (a) an initially air-fluidized granular column and (b) a water column. The currents have three distinct phases of emplacement: collapse (front acceleration, t = 167 ms), inertial (constant front velocity, t = 283 and 450 ms), and stopping (front deceleration of granular flows, t > 450 ms). Modified from Roche et al. (2008).

motion following this collapse can generally be divided into three distinct regimes (Huppert and Simpson, 1980; Roche et al., 2008). Initially, the flow dynamics are controlled by the release conditions and geometry. In the second regime, the flow dynamics are primarily controlled by the balance of inertia and buoyancy forces. In the final regime, the balance between buoyancy and resistance forces controls the stopping of the flow. Mass conservation in the collapsing flow requires that the mean depth of the flow decreases as its length increases (Fig. 10.2). In the case of fluid gravity currents, the stresses that resist motion arise from interactions at boundaries, internal turbulent stresses, and from the increasing influence of viscous stresses as the current depth decreases. In the case of granular flows, the stresses that resist motion arise as a consequence of interactions with the boundaries, and interactions between individual particles and between particles and the interstitial fluid. PDC motion is characterized by large inertial forces and PDCs typically propagate in a regime where inertia and buoyancy forces are in balance. In many cases, the motion of PDCs is approximately two-dimensional due to propagation on slopes (such that spreading of the current occurs predominantly in the downslope direction) or to confining topography such as valleys. We thus initially explore the dynamics of PDCs by considering two-dimensional, inertially dominated gravity currents.

10.2.2 Conservation of momentum

The driving force for gravity current motion is the buoyancy force that acts as a consequence of the density contrast between the current and the surrounding fluid. An appropriate way to express the density difference is as the *reduced gravity*, g', defined as

$$g' = g\frac{\rho_c - \rho_0}{\rho_{ref}} \quad (10.1)$$

where g is the acceleration due to gravity, ρ_c is the bulk density of the current, ρ_0 is the (uniform) density of the surrounding fluid, and ρ_{ref} is a reference density (all notation is given in Section 10.8). When the density of the current and the density of the surroundings are not very different in magnitude (i.e., dilute surges), the density of the surroundings is the appropriate reference density (e.g., Huppert and Simpson, 1980), so that $g' = g(\rho_c - \rho_0)/\rho_0$. The first theoretical study of gravity current motion by von Kármán (1940), and subsequent re-analysis by Benjamin (1968), considered the motion of a dense gravity current propagating under conditions in which inertia dominates, as shown in Figure 10.2 and schematically in Figure 10.3(a). A current of density ρ_c is shown flowing along a horizontal boundary and displacing a fluid of lower density ρ_0. Due to the weight of the dense fluid, a larger hydrostatic pressure exists inside the current as compared to the fluid ahead of it,

(a)

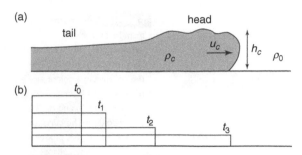

(b)

Figure 10.3 (a) Schematic diagram of an inertial gravity current having a depth h_c, flow front velocity u_c and density ρ_c in an ambient medium of density ρ_0. (b) Schematic diagram of mass conservation in a box model at times t_0 to t_3.

and this pressure difference provides the driving force for current motion. Applying conservation of energy at the current front leads to a relationship for the velocity of the front of a current, u_c, followed by a layer of fluid of depth, h_c, of the form (von Kármán, 1940; Benjamin, 1968)

$$\frac{u_c}{(g'h_c)^{1/2}} = \mathrm{Fr}, \tag{10.2}$$

where Fr is the Froude number, the ratio of inertial to buoyancy forces (Table 10.1). For a current flowing into deep surroundings ($h_c \ll H$) with no energy dissipation between the fluid layers, Fr is a constant with a theoretical value of $\sqrt{2}$ (von Kármán, 1940; Benjamin, 1968), compared with values of ~1.2 to 1.4 measured experimentally (Huppert and Simpson, 1980). In Eq. (10.2), all resistances to flow propagation, such as resistance to flow at the lower boundary, dissipation within the current, and dissipation against the surrounding fluid are parameterized in the value of the Froude number. Recent numerical studies have investigated the form of these resistances and, if the flow is depth-averaged so that all resistance is considered to operate at the lower boundary, an inertial form of resistance is found to best describe experimental data (Hogg and Pritchard, 2004). Dimensional considerations also lead inevitably to Eq. (10.2), as u_c, g', and h_c are the only variables in this problem, so the ratio $u_c/(g'h_c)^{1/2}$ has to be constant when the flow is controlled by a balance between inertia and buoyancy

(Huppert, 2006). Equation (10.2) is a simplified form of conservation of momentum (it is a solution to Euler's equation which approximates conservation of momentum for a turbulent flow) that has been widely used in studies of inertially controlled gravity currents, which suggests that the motion of these gravity currents is controlled at the front. An important consequence is that the dynamics of inertial gravity currents are independent of the slope of the underlying topography up to angles of ~30°, typical of the conditions for PDCs. It is important to note, however, that Eq. (10.2) is insufficient to specify the current dynamics as it is a single relationship between an unknown velocity and an unknown depth (Huppert and Simpson, 1980; Huppert, 2006).

10.2.3 Conservation of mass

A second relationship between gravity current depth and velocity can be derived from consideration of mass conservation for the current fluid. For a two-dimensional current, the shape of the spreading current can be approximated as a series of equal-area (i.e., volume in two dimensions) rectangles (Huppert and Simpson, 1980). This approach is illustrated schematically in Figure 10.3(b) and is a simplification of the detailed collapse dynamics shown in Figure 10.2. We can write the relationship for mass conservation for the current fluid as

$$h_c L = Q_0 t_c^\lambda, \tag{10.3}$$

where L is the length of the current, t_c is the duration of fluid release, and λ describes the style of fluid release, such that $\lambda = 0$ corresponds to the case of a release of finite volume Q_0, and $\lambda = 1$ corresponds to the case of a constant volume flux release. Equations (10.2) and (10.3) can be combined to form a simplified *box model* for gravity current motion, so called because of the approximation of mass conservation as equal-area rectangles. Since $u_c = dL/dt$, we can eliminate the current depth h_c between Eqs. (10.2) and (10.3) and integrate to find

$$L = \left(\frac{7}{12}\right)^{6/7} \left(\frac{6}{\lambda+6}\right)^{6/7} (g'^3 Q_0 H^2)^{1/7} t^{(\lambda+6)/7}, \tag{10.4}$$

in the region of initial collapse, and

$$L = \left(\frac{3}{2}\right)^{2/3}\left(\frac{2}{\lambda+2}\right)^{2/3} \mathrm{Fr}^{2/3}\left(g'Q_0\right)^{1/3} t^{(\lambda+2)/3} \quad (10.5)$$

for subsequent motion where inertia and buoyancy are in balance. For the special case of motion in deep surrounding fluid following a finite volume release, the propagation of a gravity current with constant density can be written,

$$L \approx 1.47(g'Q_0)^{1/3}t^{2/3}. \quad (10.6)$$

10.2.4 Fluid gravity currents with particles

In many natural situations, including PDCs, the density difference between gravity currents and the surrounding fluid depends on the concentration of suspended dense particles within the current. In this case, the bulk density of the current will vary in time if particles are sedimented at the current base. The framework of box models (or more sophisticated formulations of the conservation equations) can be extended in the case of relatively low particle concentrations to consider the case of varying current density by addition of a particle settling law (e.g., Bonnecaze et al., 1993; Dade and Huppert, 1995). The most common approach is to assume that the turbulent velocity scale within the current is sufficiently high to maintain a uniform concentration particle suspension. However, at the base of the flow, where the current fluid interacts with the underlying static boundary, the velocities in the current decrease below that of the settling speed of the particles, and sedimentation can occur. In this region, the particle-concentration profile and sedimentation process are governed by the balance of sedimentation and shear velocities, expressed as the Rouse number (Table 10.1),

$$\mathrm{Pn} = \frac{u_s}{\kappa u_*}, \quad (10.7)$$

where u_s is the settling velocity of the particles, κ is the von Kármán constant (~0.4), and u_* is the shear velocity (see Valentine, 1987, and Dellino et al., 2008, for applications to PDCs).

The sedimentation from a flow consisting of a deep turbulent layer above a narrow basal layer (where deposition can take place) can be described by an approximate particle settling law, sometimes known as Hazen's Law, and experimentally verified by Martin and Nokes (1988), of the form

$$c = c_0 \exp\left(-\frac{u_s t}{h_c}\right), \quad (10.8)$$

where c is the volume concentration of particles suspended in the current with initial value c_0. To complete the model, one further relationship is required to describe the change in current density with particle concentration. In the case where the excess density in the current arises solely from the presence of the particles (i.e., the density of the current fluid is equal to the density of the surrounding fluid), we can write

$$g' = \frac{g(\rho_p - \rho_0)c}{\rho_0} = g'_p c, \quad (10.9)$$

where ρ_p is the particle density. The solution for the system of equations (10.2), (10.3), (10.8), and (10.9) for a finite volume release can be written as

$$L^{5/2} = \frac{5\,\mathrm{Fr}(g'_p c_0 Q_0^3)^{1/2}}{u_s}\left(1 - \left(\frac{c}{c_0}\right)^{1/2}\right) \quad (10.10)$$

(Bonnecaze et al., 1993). If the excess density only arises from the presence of dense particles, the flow comes to a stop when $c = 0$, so the runout length, L_r, can be found as

$$L_r = \left(\frac{5\,\mathrm{Fr}(g'_p c_0 Q_0^3)^{1/2}}{u_s}\right)^{2/5}. \quad (10.11)$$

The particle settling speed, u_s, will vary with particle size (and density), so equations of the form of (10.10) and (10.11) are applicable to flows containing a single particle size, whose settling velocity can be found from standard settling laws such as Stokes' Law (e.g., Dade and Huppert, 1995) or Newton's impact law (Dellino et al., 2008). Extensions to polydispersed particle suspensions have been recently proposed by Harris et al. (2002).

An axisymmetric formulation of the box model for a particle-driven gravity current (Eqs. (10.2), (10.3), (10.8) and (10.9)) has been applied to investigate the emplacement of the 1800 B.P. Taupo ignimbrite (Dade and Huppert, 1996). The distribution of particle sizes in the ignimbrite is described using a probability density function for particle settling speed, and the model is fitted to the proximal deposit thickness in order to set the value of the input volume flux. Based on this initial calibration, the model shows good agreement with measurements of the deposit thickness, concentrations of different particle sizes in the deposit, and concentrations of particles of given size fractions, from the source to a distance of nearly 80 km where the PDC reached the coast. The flow conditions corresponding to these model predictions were a total volumetric flux of 40 km^3 s^{-1} for approximately 15 minutes, a flow depth of ~1 km, a temperature of 450°C, and a typical speed of 200 m s^{-1}. The good agreement between the model predictions and field observations (Wilson, 1985) led Dade and Huppert (1996) to suggest that large-volume ignimbrites could be emplaced by relatively dilute gravity currents. There is now, however, a developing consensus that most PDCs comprise a dense basal granular layer with an overlying dilute ash cloud (e.g., Druitt et al., 2002; see Table 10.1), and more complex two-layer models are emerging (Doyle et al., 2010; see Section 10.5).

With recent advances in computing power, numerical solutions of the complete governing equations can be used as the basis for models of gravity-driven flow (see Section 10.5). It is worth noting here that the two-dimensional equations for momentum and mass conservation (Eqs. (10.2) and (10.3)) can be obtained directly from the vector forms of the conservation equations presented in Section 10.5 (Eqs. (10.28) and (10.29)) when the flow internal stresses and boundary resistance are empirically approximated as a Froude number and the current is sufficiently shallow that the pressure within the flow can be assumed to be hydrostatic. One advantage of using simplified formulations to describe the dynamics of gravity currents is that these can be tested using laboratory experiments, as described in the next section.

10.3 | Experimental studies of fluid gravity currents

10.3.1 Lock-exchange experiments

Much of the basic physics of inertial gravity current motion described in Section 10.2 has been verified using laboratory experiments (Simpson, 1997; Huppert, 2006). The most widely used configuration is a lock-exchange experiment, in which the dense fluid is released into an adjacent lower density fluid (often of the same depth but larger volume) by removal of a rigid gate between them. The dense fluid collapses to the base of the tank, and propagates as a gravity current along a rigid boundary (Fig. 10.2). Early lock-exchange experiments in channels (summarized in Simpson, 1997) investigated the morphology of inertial gravity currents when the density of the current is close to that of the ambient fluid, which is applicable to dilute PDCs. The characteristic morphology is shown in Figure 10.3(a) (and online supplement OS 10.1; see end of chapter), with a well-defined flow front (commonly referred to as the *head*) of greater depth than the following gravity current fluid (the *tail*), and turbulent billows that result from fluid instability due to counterflowing motion of the dense and light fluid (Kelvin–Helmholtz instability). Simpson and Britter (1979) found that the speed of the main body of the current is about 15% greater than that of the front, which suggests that current fluid must be exchanged between these parts of the flow during motion. The propagation of a gravity current initiated in a lock-exchange experiment shows distinctive flow regimes, characterized by different time dependence of the flow velocity (cf. Eqs. (10.4)–(10.6)). The initial collapse of dense fluid, or *slumping phase*, was investigated by Huppert and Simpson (1980) who showed that the velocity of the flow was constant, and this has been interpreted as resulting from the counterflow of light fluid into the lock region at the channel top, required by mass conservation, while the slumping dense fluid propagates along the channel base. This constant velocity regime, however, could also result from the balance between inertia, buoyancy, and resistance to the flow (Hogg and Pritchard, 2004).

Recent experiments have considered the initial condition where the total depths of fluid on each side of the lock gate are equal, but where the depth of the dense fluid layer in the lock is some fraction of the depth of the lock, and is overlain by the same fluid as in the main body of the tank (Gladstone et al., 2004). In this case, the density stratification in the lock leads to streamwise stratification of the resulting flow, and the stratified currents are observed to propagate initially faster, then more slowly, than their unstratified counterparts. At early stages, gravity current propagation takes the form of Eq. (10.6) for a fixed volume release Q_0, and motion in the surrounding fluid plays a negligible role. Finally, as the gravity current energy is dissipated by displacing the surrounding fluid and by frictional interaction with the underlying surface, viscous dissipation becomes important and the current enters a flow regime where viscous and buoyancy forces are in balance (Huppert and Simpson, 1980).

10.3.2 Particle-laden gravity currents
Particle-driven gravity currents have also been widely studied using lock-exchange experiments. Bonnecaze et al. (1993) compared the predictions from a box model of the form developed in Section 10.2 with experimental measurements of the areal density of the deposit from a sedimenting particle current containing approximately monodispersed (i.e., uniform size) spherical particles. They found good agreement with the observed current dynamics and the distribution of deposit areal density with distance from the lock. This model has also been used to investigate the dynamics and deposition patterns of laboratory gravity currents in which the interstitial fluid is less dense than the fluid into which the current is propagating (Sparks et al., 1993).

The dynamics and sedimentation from particle-driven currents become much more complex when the current contains a range of particle sizes, as is typical of PDCs. Gladstone et al. (1998) conducted lock-exchange experiments using bidispersed and polydispersed particle mixtures and found that the effects of mixing different proportions of fine and coarse particles is strongly nonlinear. Adding a small amount of fine particles to a current containing coarse particles has a much larger influence on flow velocity and sedimentation patterns than adding a small amount of coarse particles to a current containing fine particles. Measurements of deposit areal density are not well reproduced by box models for bidispersed and polydispersed particle distributions (Dade and Huppert, 1995).

Particle concentration also influences gravity current dynamics and deposition (Choux and Druitt, 2002). Lock-exchange gravity currents containing bidisperse mixtures of dense and light particles with sizes chosen so as to be in hydrodynamic equivalence produce deposits that show normal grading of the dense particles, but the deposition of light particles depends strongly on the total particle concentration. The light particles are deposited in hydrodynamic equivalence (i.e., at the same settling velocity) in dilute flows, but are segregated efficiently in concentrated suspensions (up to 23% by volume; Choux and Druitt, 2002). The dynamics of gravity currents composed of high-concentration suspensions (up to 40% by volume) show an abrupt transition in deposition pattern with distance from their source (Hallworth and Huppert, 1998), which is very different from the deposition profile of a lower concentration current (e.g., Bonnecaze et al., 1993). Above a critical concentration of particles, the gravity currents stop abruptly and deposit the bulk of their sediment load as a relatively thick layer of constant thickness, with a much thinner layer of sediment being deposited from the residual low-concentration cloud.

10.3.3 Mixing processes
Experiments have also been used to investigate more complex phenomenology of gravity currents in order to develop simplified formulations that can be included in theoretical models. An important process relevant to dilute PDCs is the mixing of inertial gravity currents

with the surrounding fluid. The motion of turbulent billows along the upper surface of an inertial gravity current that contains a low or zero concentration of particles leads to incorporation of the surrounding fluid into the gravity current, or *entrainment*. Quantifying this process is important for PDCs as dilution of a current by lighter ambient fluid reduces the density contrast and hence the flow velocity and runout, even though the flow depth can increase.

Entrainment of fluid into a turbulent flow is difficult to calculate directly because the flow structure is three-dimensional and time-dependent, but entrainment can be measured in experiments. Hallworth *et al.* (1993) conducted experiments in which an alkaline inertial fluid gravity current was released into a two-dimensional channel containing an acidic ambient fluid, with the neutralization resulting from mixing visualized using universal indicator in the current. Different initial concentrations of alkali in the current resulted in neutralization at different distances from the source, so the proportion of current and entrained fluid could be determined with distance from the source. Entrainment was observed to take place primarily at the head of the gravity current. If the proportion of entrained fluid is defined as the ratio of the volume of entrained fluid to the total volume of the current (i.e., a dimensionless quantity), then dimensional considerations suggest that the proportion of entrained fluid depends only on the initial volume of the current and distance from the source, and is independent of the reduced gravity because g' is the only quantity including dimensions of time (Hallworth *et al.*, 1993). This result was confirmed in systematic experiments, and leads to a simplified description of entrainment by considering conservation of mass of the fluid in the gravity current head,

$$\frac{dQ}{dx} = (\alpha - \psi)Q^{1/2}, \qquad (10.12)$$

with

$$Q(0) = V_0, \qquad (10.13)$$

where Q is the volume of the current head (area in a two-dimensional current), V_0 is the initial

volume of the current, x is the distance from the current source, α is a constant that describes the amount of entrainment into the current head, and ψ is a constant representing the ratio of the height of the tail to the height of the head. The dependence on $Q^{1/2}$ indicates that the spatial rate of change of volume is proportional to the height of the head. The solution to this equation is

$$\frac{Q}{V_0} = \left(1 - \frac{1}{2}\frac{(\psi - \alpha)x}{V_0^{1/2}}\right)^2, \qquad (10.14)$$

with the best fit to experimental data suggesting values $\alpha = 0.078$ and $\psi = 0.147$ ($\pm 3\%$). Application of this simplified result suggests that dilution of the current can be significant (particle concentrations reducing from 40 vol.% to about 8 vol.% within the first quarter of the total run-out distance), and that entrainment provides an efficient mechanism for reducing the particle concentration and buoyancy of PDCs (Hallworth *et al.*, 1993). Furthermore, efficient mixing of fluid within the head of inertial gravity currents and between the head and the following fluid suggests that dilution will have an important effect throughout the depth and length of PDCs.

Whereas these experimental studies provide insights into the physics of dilute turbulent PDCs, their potential to address dense PDCs is uncertain. In this context, experimental studies on highly concentrated currents of dry granular material or of gas–particle mixtures are more relevant to the investigation of dense PDCs, as discussed in Section 10.4.

10.4 | Dynamics of granular flows

10.4.1 Fundamental physics of granular flows

This section describes key differences between the dynamics of dense granular currents propagating in air, hereafter called *granular flows*, and fluid gravity currents. Granular flows most resemble the dense end-member of PDCs, and are characterized by a particle volume fraction v_p

close to $v_{p,max}$ at loose packing, so that $\rho_{ref} = \rho_c$ (cf. Eq. (10.1)) and $g' = g(\rho_c - \rho_0)/\rho_c \sim g$. Furthermore, the mass number (Table 10.1) Ma $\gg$ 1, so that momentum is transferred almost entirely by the solid phase. We focus on end-member regimes of granular flows: *dry granular* and *gas–particle* flows, which represent negligible and dominant roles of the interstitial fluid phase, respectively.

Extensive reviews reveal the phenomenology and complexity of flowing *dry* granular matter (GDR MiDi, 2004; Forterre and Pouliquen, 2008). Studies on dry granular flows apply to dense PDCs in which fluidization effects are non-operant because limited sources of gas and/or large permeability (i.e., coarse-grained material) favor rapid pore-pressure diffusion (see Section 10.4.3). In dry media, energy dissipation is caused mainly by particle interactions, and flow dynamics depend on the Savage number (Sa; Table 10.1), which represents the ratio of inertial shear stresses resulting from particle collisions to quasi-static gravitational stresses associated with friction (Savage, 1984). Interparticle collisions govern the current dynamics at $v_p < v_{p,max}$ and/or high shear rate (Sa > 0.1), whereas frictional stresses dominate at $v_p \sim v_{p,max}$ and/or low shear rate (Sa < 0.1) (Savage and Hutter, 1989). The frictional, dense flow regime is achieved on rough substrates with inclinations close to the angle of repose of the material ($\sim \varphi_r$), and steady motion results from balance between driving gravitational and resisting frictional forces. The granular mass is commonly treated as a Coulomb material with constant, rate-independent interparticle friction coefficient (μ) and friction angle (φ), so that

$$\mu = \tan\varphi = \frac{\tau}{\sigma}, \qquad (10.15)$$

where τ and σ are the shear and normal stresses respectively. However, under steady flow conditions, the macroscopic friction coefficient, $\mu_{(I)}$, is shear-rate dependent and is a function of a dimensionless parameter called the inertial number, I, which represents the ratio of the microscopic timescale of particle rearrangement to the macroscopic timescale $(1/\dot\gamma)$ and is also the square root of the Savage number, so that

$$I = \frac{\dot\gamma d}{\sqrt{\sigma/\rho_p}}. \qquad (10.16)$$

Here $\dot\gamma$ is the shear rate, d is the particle diameter, and ρ_p is the particle density, and

$$\mu_{(I)} = \mu_1 + \frac{\mu_2 - \mu_1}{I_0/I + 1}. \qquad (10.17)$$

Values of typical friction coefficients for glass beads in experiments are $\mu_1 = \tan 21°$ and $\mu_2 = \tan 33°$, and $I_0 = 0.3$ is obtained from theoretical or experimental investigations (Forterre and Pouliquen, 2008).

Segregation according to particle size and density is an important process in polydisperse flows containing particles of different sizes as it may change their dynamics. Kinematic (dynamic) sieving is common and occurs when smaller particles fall into gaps beneath them and percolate downwards as force imbalances squeeze the large particles upwards, leading to reverse grading. In contrast, normal segregation may occur if large and dense particles can displace particles beneath them and move downwards under gravity, and this acts in opposition to dynamic sieving (the *mass effect*; Thomas, 2000).

Gas–particle flows have been less comprehensively studied than their dry counterparts though they probably represent most dense PDCs. Interactions between the gas and solid phases are likely to reduce particle interactions and to modify momentum transfer. For instance, granular flows that are continuously fluidized down inclines can propagate at slope angles below φ_r, and possibly on slopes close to horizontal, because of extreme friction reduction (Eames and Gilbertson, 2000).

10.4.2 Experiments on dry granular media

Particle interactions in steady flow

Particle interactions in *dry*, coarse-grained PDCs were investigated in shear-cell experiments by Cagnoli and Manga (2004, 2005). When a bed of pumice fragments is confined between two vertical and coaxial cylinders that rest on a rough horizontal rotating disk, energy dissipation occurs

(a)

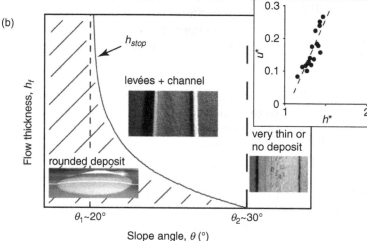

Figure 10.4 (a) 1993 pumice flow deposits with levée-channel and frontal lobe morphology, Lascar volcano, Chile. Persons for scale (photograph by O. Roche). (b) Deposit types reported in steady-state experiments, where h_{stop} is the deposit depth at given slope angle between θ_1 and θ_2 (equivalent to $h_{channel}$ for a channelized geometry). Inset: correlation between normalized flow velocity, $u^* = u_f/(gh_{levée})^{1/2}$, and thicknesses, $h^* = h_{levée}/h_{channel}$ (cf. Eq. (10.18), with $B \sim 0.5$); modified from Félix and Thomas (2004), reprinted with permission from Elsevier.

within a basal collisional layer at Sa up to ~0.4. The upper layer acts as a rigid raft but moves relative to the cylinders at a constant velocity, which is independent of the imposed shear rate, and this suggests a frictional Coulomb behavior. In this layer, reverse segregation of coarse light clasts and normal segregation of coarse dense clasts occurs. This is because the granular network expands and the coarse and fine components have contrasting inertia when they are pushed upwards by collisions originating at the basal layer.

Levées are common features of deposits of coarse-grained PDCs and are reproduced in experiments investigating steady finger-shaped flows down rough inclines (Fig. 10.4; Félix and Thomas, 2004). Levées form at the lateral static borders behind the flow front, and result from emptying of the central channel once source flux has decreased to zero and the flow is no longer steady. Static borders form when downslope gravitational forces are lower than frictional resistance at the flow margins because friction is depth-dependent (cf. Eqs. (10.15) and (10.16)), which suggests that grain-size segregation caused by polydispersity is not necessary for levée formation (Mangeney et al., 2007; see Section 10.5). The flow dynamics can be inferred from the deposit morphology according to Pouliquen's (1999) method, so that

$$\frac{u_f}{\sqrt{gh_{levée}}} = B\frac{h_{levée}}{h_{channel}},\qquad(10.18)$$

where u_f is the flow front velocity, $h_{levée}$ is the levée thickness, $h_{channel}$ thickness of the deposit in the channel, and B is an empirical constant ~0.5, typical of the granular material used by

(a)

(b)

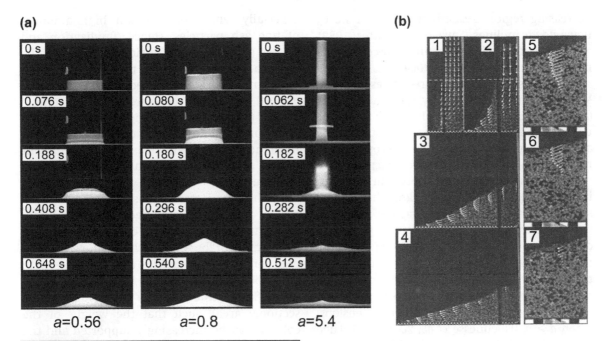

a=0.56 a=0.8 a=5.4

Figure 10.5 (a) Three-dimensional granular flows generated from the release of a cylindrical reservoir of aspect ratio a as a function of time; reprinted with permission from Lajeunesse et al. (2004), American Institute of Physics. (b) (1–4) Evolution of the velocity profile with time at a = 7, and (5–7) detailed view of the same experiment at a distance of one third the flow runout; reprinted with permission from Lube et al. (2005), copyright 2005 by the American Physical Society.

Félix and Thomas (2004). Although values of substrate slope and deposit height cannot be extrapolated directly to natural cases, these studies provide fundamental insights into the dynamics of coarse-grained PDCs.

Flow kinematics and runout in unsteady flow

Recent experimental studies have investigated the unsteady flows that result from the collapse of an initially static column of dry granular material, by analogy to lock-exchange gravity currents described in Section 10.3 (Figure 10.5). These simple experiments must be considered a first-step approach to investigating PDCs generated from gravitational collapse of a column or a dome, but despite their simplicity they provide important insights into the flow mechanisms. The kinematics and runout distance of monodisperse flows on a horizontal surface are controlled by the initial aspect ratio of the granular column

$$a = \frac{h_i}{x_i}, \qquad (10.19)$$

where h_i is the initial height and x_i is the initial length (in two dimensions) or radius (in three dimensions) of the column (Lajeunesse et al., 2004, 2005; Lube et al., 2004, 2005; Balmforth and Kerswell, 2005). Flank avalanches or spreading of the column generate a (truncated) conical deposit when $a < a_t$, where a_t is a critical aspect ratio ~1.7–3, whereas outward propagation of the base of the pile when $a > a_t$ triggers collapse of the upper portion with almost no deformation and this creates a low-angle conical deposit (Fig. 10.5(a)). Particles initially at the surface of the pile stay at superficial levels or are incorporated into a thin basal layer when overrun by the front (Lube et al., 2004). This layer is delimited by a dynamical interface that separates deposited and flowing particles, and propagates towards the upper free surface during emplacement. The velocity profile within the flowing layer consists of an upper low-shear, plug-like zone, a middle linear-gradient region, and a lower exponentially decreasing region, resembling those in steady flows down inclines (Savage and Hutter, 1989), and contrasting with those in shear-cell experiments (Cagnoli and Manga, 2004).

The flows propagate in three stages and their kinematics are controlled by the timescale $t_f = (h_i/g)^{1/2}$, proportional to that of free fall of the column (Lajeunesse et al., 2005). The front initially accelerates after release and then propagates at nearly constant velocity provided $a > a_t$; these two regimes are similar to those observed in fluid gravity currents (Section 10.3). The flows then enter a short stopping phase as they rapidly decelerate until motion ceases. There is a clear power-law dependence of the flow runout (x_f) on a because the normalized mobility $(x_f - x_i)/x_i$ is proportional to a when $a < a_t$ (in both two- and three-dimensional experiments), and is proportional to $a^{2/3}$ (in two dimensions) or $a^{1/2}$ (in three dimensions) when $a > a_t$ (Lajeunesse et al., 2004, 2005; Lube et al., 2004, 2005). Lube et al. (2004) concluded that the analysis is independent of any basal and internal friction parameter and questioned the role of Coulomb friction for most of the emplacement, until flows enter the stopping phase. In contrast, Balmforth and Kerswell (2005) showed that the normalized flow runout depends on the size and shape of the particles, which control the interparticle friction. Note that an erodible substratum has no influence on the flow kinematics and runout (Lajeunesse et al., 2004). In nature, the polydispersity of PDCs and the presence of moderately steep slopes may complicate the processes described above. Other experiments on granular column collapse have shown that the flow runout is strongly dependent on the proportions of fine and coarse particles for mixtures composed of two particle sizes (Phillips et al., 2006; Roche et al., 2005), and increases when an erodible substrate is mobilized by the passage of a flow at slope angle larger than $\sim \varphi_r/2$ for the material used (Mangeney et al., 2010).

10.4.3 Gas–particle flows

Characteristics of gas-fluidized pyroclastic materials

For most dense PDCs the interstitial gas phase may have a key influence on their dynamics, especially when they contain high amounts of fine ash particles. Hence, fluidization and related pore-pressure diffusion processes can be fundamental for governing their emplacement. Fluidization can be achieved when differential vertical motion between a (relatively ascending) gas and (relatively descending) particles is generated. For PDCs, this may occur as gas is released from the pyroclasts and/or as particles sediment during column collapse and subsequent flow propagation.

In the simple case of a gas flux injected at the base of a static non-expanded granular column, as in many engineering configurations (see Geldart, 1986, for review), the gas flow exerts a drag force on the particles that increases with the superficial gas velocity (u_g, defined as the mean flow rate divided by the column cross-sectional area), such that the weight of the column can be increasingly supported and the interparticle frictional stresses decrease. The dynamic pore-fluid pressure, P_d, across the column increases with u_g as described by the Ergun equation for a steady flow,

$$P_d = \left(\frac{150(1-\phi^2)\eta_g u_g}{\phi^3 d^2} + \frac{1.75(1-\phi)\rho_g u_g^2}{\phi^3 d} \right) h, \quad (10.20)$$

where h and ϕ are the height and porosity of the column, respectively, η_g the gas dynamic viscosity, ρ_g the gas density, and d the particle diameter. On the right-hand side of Eq. (10.20), the first and second terms are the laminar and turbulent components, respectively. The weight of the particles is fully supported at $u_g = u_{mf}$, the *minimum fluidization velocity*, when

$$P_{d,mf} = (\rho_p - \rho_g)(1-\phi)gh, \quad (10.21)$$

where ρ_p is the density of the particles. The total pore-fluid pressure at u_{mf} is then $P_{mf} = P_{d,mf} + P_h$, where the hydrostatic component $P_h = \rho_g gh$, so that

$$P_{mf} = \left[\rho_p(1-\phi) + \rho_g \phi \right] gh. \quad (10.22)$$

Note that Eq. (10.22) simplifies to $P_{mf} = \rho_p(1-\phi)gh$ if the gas is much less dense than the particles, as is the case for PDCs. At this stage, the granular material is *fluidized* (s.s.) and has a fluid-like behavior because stresses generated by interparticle frictional contacts, which vary inversely with the pore-fluid pressure, are negligible. One obtains u_{mf} by equating the laminar term of the Ergun equation (first term on right-hand side of Eq. (10.20)) and Eq. (10.22), so that

$$u_{mf} = \frac{k}{\eta_g} \frac{P_{mf}}{h},$$ (10.23)

where $k = \phi^3 d^2/[150(1-\phi)^2]$ is the hydraulic permeability. If the gas flux is no longer provided, the granular column defluidizes through a pore-pressure diffusion process and the particle frictional contacts rebuild. The pore pressure decreases with time according to

$$\frac{\partial P}{\partial t} = D \frac{\partial^2 P}{\partial h^2},$$ (10.24)

where $D = k/(\eta_g \phi \beta)$ is the hydraulic diffusion coefficient and β is the gas compressibility, and the duration of pressure diffusion is proportional to the timescale (cf. Iverson, 1997)

$$t_d = \frac{h^2}{D}.$$ (10.25)

For columns fluidized at $u_g \geq u_{mf}$, homogeneous expansion (i.e., no gas bubbles) is achieved for fine particles with grain sizes smaller than ~100 μm (at $\rho_p \sim 2500$ kg m^{-3}) to ~500 μm (at $\rho_p \sim 500$ kg m^{-3}) (Geldart, 1986), which are representative of the bulk mass of most PDCs. Once defluidization occurs, pore-pressure diffusion is accompanied by collapse of the column at a constant velocity called the *deaeration rate* (u_{de}), and a sedimentation interface that migrates upwards separates basal sedimented particles and settling ones above.

Experimental studies of gas-fluidization of static columns of pyroclastic materials provide insights into fluidization processes of PDCs. Wilson's (1980, 1984) seminal studies revealed that gas flow channeling readily occurs at room temperature due to interparticle cohesion in fines-rich, poorly sorted beds of ignimbrite. Later studies showed that high fines content, high temperature, and shear motion favor homogeneous fluidization with efficient support (Gravina *et al.*, 2004; Bareschino *et al.*, 2007;

Druitt *et al.*, 2007). High fines content strongly decreases u_{mf} and u_{de}, whereas temperatures above ~200 °C and shear both inhibit gas flow channeling, as the former eliminates moisture-derived interparticle cohesion and the latter breaks cohesive bonds and particle clusters. When these factors act together, ignimbritic materials are fluidized at u_{mf} as low as ~1 mm s^{-1}, exhibit homogeneous expansion, and deaerate slowly due to the high degree of bed expansion (up to 60–70% above loose packing) and low u_{de} ~ 0.5–1 cm s^{-1} (Druitt *et al.*, 2007). The duration and degree of fluidization of PDCs depend on the strength and longevity of gas sources but also, once gas supply has become ineffective, on the pore-pressure diffusion timescale, which depends on the current thickness (Eq. 10.25).

The dynamics of gas–particle flows

Several experimental studies have highlighted the fundamental role that the gas phase of PDCs can have on their emplacement dynamics through fluidization processes. Fluidization by air entrainment at the front of hot PDCs was proposed by McTaggart (1960) to explain increases in runout of laboratory-scale flows with temperature, by envelopment and heating of cold air by hot particles, and subsequent violent expansion. Air entrainment in flows of fine particles generated in a rotating drum is promoted by periodic and continuous projection of clusters of particles ahead of the front (Bareschino *et al.*, 2008). This may cause motion-induced self-fluidization (Salatino, 2005) and thus explain the weakly inclined free upper surface of the flows (Fig. 10.6). This mechanism occurs because the defluidization timescale is much larger than the periodicity of the avalanches.

Fluidization caused by the interstitial gas has also been considered in various experimental configurations. The emplacement of PDCs on moderate slopes (~5–25°) was investigated by Takahashi and Tsujimoto (2000), who found that continuously fluidized flows of fine ($d = 80$ μm) particles down inclines consist of a lower, concentrated layer ($v_p \sim 0.40$–0.45) and an upper, thinner, and more dilute layer. In contrast to *dry* flows, they have a nonlinear streamwise velocity profile whose gradient increases upwards, and a

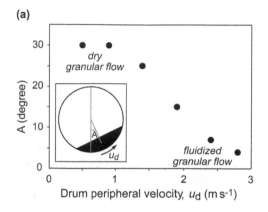

(a)

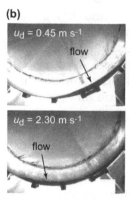

(b)

Figure 10.6 Flow of fine (< 100 μm) glass beads in a rotating drum, modified from Bareschino et al. (2008), reprinted with permission from Elsevier. (a) Mean central angle (A) as a function of the peripheral velocity (u_d) of the rotating drum of radius 0.9 m. (b) Photographs of the dry (top) and fluidized (bottom) flow regimes.

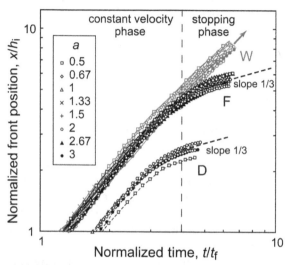

Figure 10.7 Normalized front position (x/h_i) as a function of normalized time (t/t_f) for dam-break flows of water (W), and of dry (D) or initially fluidized (F) fine particles ($d = 80$ μm), generated from the collapse of columns at various aspect ratios (a). The constant-velocity and stopping phases of the initially fluidized granular flows are indicated. During the stopping phase, they behave like the dry flows (slope = 1/3). Modified from Roche et al. (2008).

that the flow behavior depends on the diffusion timescale of the pore pressure imposed initially (see online supplement OS 10.2). Owing to the small permeability of the granular material, pore-pressure diffusion is slow compared to the typical flow duration timescale (i.e., low Pr values; Table 10.1) so that high pore pressure is maintained during emplacement of the air–particle mixture. Note that columns of high aspect ratio collapse vertically and are released continuously, suggesting that the resulting flows can be considered as simple analogs of PDCs generated from gravitational column collapse, though the initial velocity is zero. These experiments suggest that the behavior of dense PDCs may be two-fold: they can propagate as inertial fluid gravity currents for most of their emplacement (hence favoring high mobility) and then as frictional dry granular flows, as shown in Figures 10.2 and 10.7 (Roche et al., 2004, 2008). The air–particle flows first propagate at almost constant velocity (u_f) and height (h_f) after initial acceleration (i.e., slumping phase), and with Froude number values

$$\mathrm{Fr} = \frac{u_f}{(gh_f)^{1/2}},\qquad(10.26)$$

which are consistent with observations of inertial fluid gravity currents with large reduced gravities as Fr ~ 2√2 (Roche et al., 2004). Their morphologies and front kinematics are the same as those of inertial water flows, with a normalized front velocity (or initial Froude number),

$$\mathrm{Fr_i} = \frac{u_f}{(gh_i)^{1/2}},\qquad(10.27)$$

curvature that also increases with the mass flow rate (i.e., slope angle). *Dam-break* (equivalent to lock-exchange at high g′ ~ g) experiments on highly concentrated air–particle flows ($v_p \sim v_{p,max}$) were performed to investigate the propagation and deposition mechanisms of the dense end-member of PDCs (Roche et al., 2004, 2008). In this configuration, columns of fine ($d = 80$ μm) particles are fluidized in a reservoir, and then released into a horizontal channel which does not contain a source of fluidizing air, so

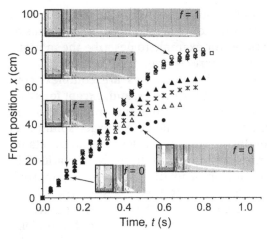

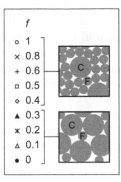

Figure 10.8 Flow-front kinematics for mixtures of fine (F, 80 μm) and coarse (C, 330 μm) particles generated from the release of an initially fluidized column (black rectangle, 10 × 15 cm). The inset shows the particle network arrangement as a function of the fine mass fraction (f) of the mixture. Modified from Roche et al. (2005), reprinted with permission from Elsevier.

that is independent of the aspect ratio of the initial column, and which has a constant value of $\sim\sqrt{2}$, as commonly reported in dam-break water flow experiments (Roche et al., 2008). The fluid-inertial behavior of these highly concentrated two-phase mixtures is consistent with the high pore-fluid pressure measured at the base (Roche et al., 2010), which may also result from significant air–particle viscous interactions (i.e., high Da values, see Table 10.1). Once pore pressure has decreased sufficiently by diffusion, and interparticle frictional contacts rebuild, the flows enter a stopping phase and behave like their dry counterparts as they rapidly decelerate and come to halt (Fig. 10.7). These flows consist of a sliding head followed by the body, at the base of which a deposit aggrades at nearly constant rate. Both parts are sheared pervasively as the internal velocity increases upwards (Roche et al., 2010; see online supplements OS 10.3 and OS 10.4). These results suggest that PDCs might have an erosive head and that their deposits are built by aggradation.

Further experimental studies have considered other characteristics of PDCs. In the dam-break configuration, polydisperse mixtures also propagate in a fluid-like inertial regime, provided fine particles form a continuous network (i.e., matrix) with embedded coarser components that are transported passively, such that very little segregation takes place (Fig. 10.8; Roche et al., 2005). This suggests that fines-rich (i.e., matrix-supported) dense PDCs are likely to propagate as inertial fluid gravity currents for most of their emplacement. Fluidized PDCs may be expanded so that their propagation is also controlled by the sedimentation rate of the particles. In this context, experimental laboratory investigations with pyroclastic material at high temperatures reveal a structure similar to that of dense flows of analog material, and deaeration (with a concomitant aggrading deposit) of the sheared mixtures occurs at a rate equivalent to that determined in static beds at the same initial expansion (Girolami et al., 2008, 2010). Other experiments using pyroclastic material at high temperature were conducted by Dellino et al. (2007) at larger scale in order to study the generation of PDCs by gravitational column collapse. Under these conditions, a basal concentrated flow and a dilute suspension of fine ash with frontal convective lobes, which are typical of PDCs, form readily. These large-scale experiments offer a means of investigating the impact dynamic pressure of the dilute component of PDCs (Dellino et al., 2008).

10.5 | Numerical modeling of PDCs

10.5.1 Principles

PDCs are transported with a velocity and thermal energy that may change with time. The basic concept of numerical simulation is to solve conservation equations of mass (sometimes density or thickness), momentum, and energy (sometimes enthalpy). We present below a simple formulation of the conservation equation of density (ρ) and volumetric momentum (ρu) of a compressible flow,

$$\underbrace{\frac{\partial}{\partial t}\rho}_{1} + \underbrace{\nabla \cdot (\rho \mathbf{u})}_{2} = 0, \tag{10.28}$$

$$\underbrace{\frac{\partial}{\partial t}\rho \mathbf{u}}_{3} + \underbrace{\nabla \cdot (\rho \mathbf{u}\mathbf{u})}_{4} = \underbrace{-\nabla P}_{5} + \underbrace{\nabla \cdot \tau}_{6} + \underbrace{\rho \mathbf{g}}_{7}. \tag{10.29}$$

Equation (10.28) means that, at a given position, the density only varies with time t (term 1), according to the flux of the density (term 2) that enters or leaves this point, i.e., the product of the spatial gradient of density (its variation in space) and the velocity at which it passes through the position. Mass displaces but is conserved. Adding terms to the right-hand side of Eq. (10.28) allows mass to be added to the system, for example by input from a vent into a PDC, or mass to be removed, for example by particle sedimentation during flow. The momentum equation (10.29) describes the conservation of momentum per unit volume, ρu, and is a little more complex because momentum is generated and dissipated by stresses. The velocity changes with time (term 3) according to the flux of momentum (term 4), but also due to the stresses that act on the flow at a given position. In this example, the stress is induced by pressure (term 5), mechanical shear stresses (term 6), and gravity (term 7). These stresses change according to the physics chosen for the simulated flow. To close this system of equations, we need to supply initial conditions (e.g., distribution of the density at the start of the simulation), boundary conditions (e.g., flux input, topography), and constitutive equations (e.g., equations relating stress and velocity, or equations of state relating density and temperature) that describe the physical behavior of the flow. Each model formulation is a simplified description of the natural phenomena, and a wide range of numerical models exists, from the simplest that deal only with a small subset of these equations, to the most complex that attempt to solve systems of several tens of equations, for example to describe the motions of multiple particle sizes.

Conservation equations are complex, containing partial derivatives, and, except for some idealized flows, solutions can only be obtained by numerical treatment. An exact solution is generally impossible to calculate and the standard methodology is to discretize equations in space and often in time, and obtain approximated solutions at the nodes of the discretized space (i.e., Toro, 2001). Different methods (finite differences, finite volumes, finite elements, cellular automata) and algorithms exist (Fig. 10.9). Their accuracy can be assessed by their ability to resolve shocks, which are discontinuous variations of properties (e.g., density) that can occur at the front of flows and also within them (e.g., sudden changes of pressure at the front of explosions, and of thickness at the front of dense flows). First-order resolution methods smooth the solutions of differential equations, meaning that shocks cannot be calculated correctly and changes in properties are more gradual than in reality. Consequently, gradients of variables at the position of the shock are incorrect, generating strong perturbations in the velocity field. For simulated PDCs, this induces large errors in velocity, thickness and runout, etc. Another indication of the quality of algorithms is their numerical stability. Some algorithms generate very strong unphysical oscillations that disturb the overall behavior. Another aspect of the numerical simulation of PDCs is that, whatever the quality of the model, there is always a compromise between accuracy and rapidity of calculations. The finer the computational mesh, the more accurate the solution (as is very clear for shock restitution) but also the longer the computational time required to solve the problem.

10.5.2 Kinetic models

The earliest numerical simulations of PDCs were made using the kinetic approach, which simulates trajectories only for the flow front, and considers the flow as a rigid block or a material point (Fig. 10.9(b)). Motion is calculated using

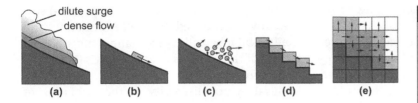

Figure 10.9 (a) Schematization of a natural PDC composed of a dense basal flow and an associated dilute ash-cloud surge. Four types of numerical approaches are illustrated: (b) kinetic, (c) discrete element, (d) depth-averaged, (e) multiphase models.

the fundamental equation of dynamics, which can be thought of as a drastic simplification of the momentum conservation equation:

$$\frac{d\mathbf{u}}{dt} = \frac{\sum \mathbf{F}}{m}, \qquad (10.30)$$

where $\sum \mathbf{F}$ represents the sum of forces that act on the block of mass m. The block is driven by gravity, following the topography, and by a resistive law. The resistive law is generally approximated using a basal friction term (with a constant value), a viscous stress term (which is a function of velocity), and a turbulent or a collisional term (which is a function of the square of the velocity), so that the resistive stress has the form $a_0 + a_1 u + a_2 u^2$ (e.g., Sheridan and Malin, 1983; McEwen and Malin, 1989; Saucedo et al., 2005). Wadge et al. (1998) used this approach to simulate the path of dense pyroclastic flows at Montserrat, adding an estimation of the areas that could be affected by a low-density surge generated from the dense flow.

Kinetic models significantly oversimplify the physical problem. The simulated flow cannot spread, and its depth, and any depth-dependent mechanical behavior, cannot be determined. This approach may be used to estimate the trajectories of dense flows but, as it cannot take into account density variations, sedimentation and air ingestion, it is too simplistic for dilute-flow simulations. However, this approach can generate results very rapidly. As it is generally very difficult to estimate volumes, rates, and initial velocity during a volcanic crisis, the accuracy of such a basic model is generally better than the knowledge of what will occur. It can thus be a useful tool for rapid evaluation of pyroclastic flow hazards.

10.5.3 Discrete element models

In this approach, Newton's equations of motion are solved for every particle in the flow, and the motion of each constituent grain is simulated (Fig. 10.9(c)). Behavior laws based on particle elasticity and deformation allow calculation of the dynamics of grain interaction. To our knowledge, the only direct application of this technique to PDCs was made by Mitani et al. (2004). They reproduced both normal and reverse grading in a granular medium (consisting of particles only, with no interstitial fluid) in motion on a slope and suggested that fluidizing gases are not required for the formation of coarse tail grading. This method cannot be used to simulate natural PDCs on real topography because of the huge number of particles that are required to obtain meaningful statistics. However, it is being increasingly used in other fields of science, and recent developments combining the discrete element method with a continuum method (such as the multiphase approach of Section 10.5.5) appear promising for the study of gas–particle interactions and their effects on PDC rheology at the particle scale.

10.5.4 Depth-averaged methods

As the emplacement of PDCs on natural topography cannot be calculated at the particle level, another solution consists of discretizing the space into a mesh and averaging the physical properties of particles on each mesh node. In the depth-averaged method, meshes can be thought as columns (Fig. 10.9(d)). All physical properties are vertically averaged and three-dimensional equations vertically integrated. The depth-averaged approximation requires that the flow length is much greater than its depth, so that vertical displacements are negligible. The mass of a column is considered to move either with the same velocity throughout the depth of the column, or with a fixed vertical velocity profile depending on the rheology chosen.

The depth-averaged method has been used to simulate granular flows in the laboratory (e.g., Savage and Hutter, 1989) and dense geophysical flows like mud flows, landslides, debris avalanches, or lahars (e.g., Heinrich et al., 2001; Pitman et al., 2003; Kelfoun and Druitt, 2005; Sheridan et al., 2005; Lucas and Mangeney, 2007; see Chapter 14), as well as pyroclastic flows (Patra et al., 2005; Capra et al., 2008; Kelfoun et al. 2009; Murcia et al., 2010). The depth-averaged method is very efficient for reproducing natural events if flow density can be assumed to be constant in time and space: for example, it was used successfully to reproduce a PDC from the 2006 eruption of Tungurahua volcano (Kelfoun et al., 2009; see online supplement OS 10.5). The depth-averaged approach is more suitable for the simulation of dense pyroclastic flows than for dilute surges that present strong vertical displacements at the end of their path and strong spatial variations of density. Nevertheless, Doyle et al. (2010) derived a depth-averaged model in which the two components of PDCs (dense and dilute) are coupled. This model reproduces the formation of the dense flow by sedimentation from the dilute ash cloud.

The main problem of the present models appears to be the rheology used. The complex and poorly understood physical behavior is generally approximated by first-order laws; a wide variety of rheological behavior has been used, including frictional ($\tau = \sigma \tan \varphi$), constant ($\tau = C$), viscous ($\tau = \eta\, \partial u/\partial y$), Bingham ($\tau = C + \eta\, \partial u/\partial y$), or Voellmy ($\tau = \sigma \tan \varphi + \rho g u^2/\xi$), where y is height within the flow, η is viscosity, and ξ is the Voellmy coefficient. The most common approximation uses a frictional behavior (e.g., Patra et al., 2005). Assuming that particle collisions are important in such granular flows, Itoh et al. (2000) used a collisional stress. Kelfoun et al. (2009) showed that a frictional behavior is not suitable for pyroclastic flow simulation and proposed a plastic rheology. Simulations of Mangeney et al. (2007) considered an empirical variable friction coefficient (i.e., Pouliquen and Forterre, 2002) to simulate flows over an inclined plane with a constant supply, and successfully reproduced the self-channeling and the frontal lobe and levée–channel morphology typical of the deposits of coarse PDCs (Fig. 10.4). They did not take into account segregation induced by polydispersity, or the formation of levées when frictional resistance overcomes downslope gravitational forces at the flow margins due to the depth-dependence of friction. The enormous advantage of the depth-averaged approach is the speed of calculation, because the third dimension is averaged. It is thus of great importance for real-time hazard assessment.

10.5.5 Multiphase approaches

To account for both temporal and spatial variations of the physical properties necessary to simulate dilute PDCs, the calculation domain is divided into a horizontal and vertical mesh (Fig. 10.9(e)). As a first approximation, it is possible to consider pyroclastic flows as a *dusty gas* (sometimes called a pseudogas) where particles and gas form a homogeneous phase, move at the same velocity, and are in thermal equilibrium (Cordoba, 2005; Ishimine, 2005). To develop more realistic simulations, gas and particle dynamics need to be distinguished and treated separately. Particles are considered as a continuous phase, their properties being averaged on each node in the same manner as the gas phase. All phases present (different size classes of particles, gases of various compositions) share the same mesh and interact with each other. The first application of this approach was made by Wohletz et al. (1984). The models have since become more complex as computational power has increased (e.g., Valentine and Wohletz, 1989; Wohletz and Valentine, 1990; Valentine et al., 1991; Dartevelle et al., 2004). Dobran et al. (1993) and Neri and Dobran (1994) introduced one solid phase and two gas phases; air and vapor water. These codes accounted for multiple particle sizes to allow for estimation of spatial particle segregation (e.g., Neri et al., 2003). Computational power limited those studies to a two-dimensional (or axisymmetric) approach and it is only recently that the first three-dimensional results (or four-dimensional, if time is considered) were obtained (Esposti

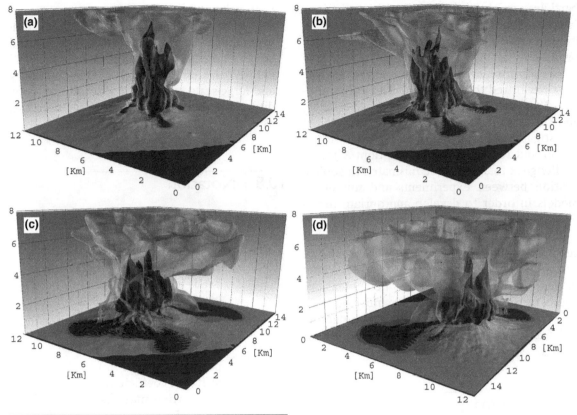

Figure 10.10 3D multiphase simulation of PDCs as a function of time (a) to (d). After Neri *et al.* (2007). Note that PDCs are strongly influenced by topography. See color plates section.

Ongaro *et al.*, 2007; Neri *et al.*, 2007), as shown in Figure 10.10. As for the depth-averaged method, the multiphase approach suffers from our incomplete understanding of the complex physics of PDCs; some codes are unable to form a deposit and can only be applied to dilute currents, for instance. However, results from multiphase models identify complex flow behavior such as vertical density stratification and progressive transitions between dense and dilute regions that cannot be resolved using other approaches. Detailed hazard assessments require future development of the multiphase approach, and the continued development of faster computational processing will enable these simulations to become more applicable to PDC motion.

10.6 | Future perspectives

PDCs are highly complex flows, and understanding the key controls on their dynamics and sedimentation remains a major challenge. Flow dynamics and particle transport are strongly time-dependent and vary over a wide range of scales, and the flow may undergo significant interactions with the underlying topography. In order to make progress in understanding PDC dynamics, idealized flows that share some of the observed features of PDCs have been investigated. Emerging methodologies are large-scale experiments and the use of unsteady (fluidized) granular collapse experiments, which build on the established fundamental dynamics of fluid collapses and gravity current motion. Outstanding research questions that can be studied in this way include investigation of the mechanisms of PDC propagation and deposition, the role of polydispersity as well as particle–particle and gas–particle interactions in controlling flow dynamics and resistance to motion, and measurement of internal velocity profiles.

Significant increases in computational power have permitted the development of fully multiphase codes that do not require depth-averaging, and these now offer the possibility of detailed investigation of mechanisms of particle interaction and resistance to flow motion. A major challenge is to address the mismatch in sophistication between experiments and numerical models in order to develop appropriate methodologies for model testing. One possibility is to explore the use of more complex experimental geometries that are closer to the variations in natural topography over which PDCs propagate. Ultimately, more sophisticated field measurements of the properties and dynamics of PDCs are required for testing numerical simulations and for better interpretation of experimental results.

10.7 | Summary

- PDCs cover a range of phenomena whose end-members are *dilute pyroclastic surges* and *dense pyroclastic flows*. These end-members coexist in most PDCs, which consist of a concentrated basal avalanche (the flow) and an overriding turbulent ash cloud (the surge).
- Lock-exchange laboratory-scale experiments on (particle-laden) fluid gravity currents are suitable for studying the physics of dilute PDCs, particularly the mixing processes with the ambient atmosphere and radial size distribution of the particles in the deposits.
- Laboratory experiments on dry or fluidized granular flows offer a means of investigating propagation and deposition processes of dense PDCs. In particular, dam-break gas–particle flows reveal front kinematics similar to those of inertial fluid gravity currents despite high particle concentrations, and deposits that form by aggradation.
- Numerical modeling taking into account a wide range of natural parameters is useful

to simulate PDC emplacement on real topographies. Available methods offer a good compromise between computational speed and accuracy of the simulated phenomena.
- Future research directions include laboratory and large-scale experimental studies on polydisperse (gas–particle) concentrated currents, as well as fully multiphase models.

10.8 | Notation

a	column aspect ratio
a_t	critical column aspect ratio
a_0, a_1, a_2	numerical constants
A	angle subtended by mass of glass beads in rotating drum (°)
B	empirical constant
c	particle concentration
c_0	initial particle concentration
C	constant stress (Pa)
d	particle diameter (m)
D	diffusion coefficient (m² s⁻¹)
f	fine particle mass fraction
$\mathbf{F}$	force (N)
g	acceleration due to gravity (m s⁻²)
g'	reduced gravity (m s⁻²)
g'_p	reduced gravity for particle-laden current (m s⁻²)
h	column height (m)
h_c	current height (m)
$h_{channel}$	thickness of deposit in channel (m)
h_f	flow height (m)
h_i	initial column height (m)
$h_{levée}$	height of deposit levée (m)
H	ambient medium height (m)
I	inertial number
k	hydraulic permeability (m²)
L	current length (m)
L_r	current runout length (m)
m	mass (kg)
P	pressure (Pa)
ΔP	pressure gradient (Pa)
P_d	dynamic pressure (Pa)
$P_{d,mf}$	dynamic pressure at u_{mf} (Pa)
P_{mf}	total pore-fluid pressure (Pa)
Q	finite 2D volume release (m²)
Q_0	2D volume of current head (m²)

t	time (s)
t_c	duration of current release (s)
t_d	pore-pressure diffusion timescale (s)
t_f	column free-fall timescale (s)
$\mathbf{u}$	local velocity (m s^{-1})
u_*	shear velocity (m s^{-1})
u_c	current front velocity (m s^{-1})
u_d	drum peripheral velocity (m s^{-1})
u_{de}	deaeration rate (m s^{-1})
u_g	superficial gas velocity (m s^{-1})
u_f	flow front velocity (m s^{-1})
u_{mf}	minimum fluidization velocity (m s^{-1})
u_s	particle settling velocity (m s^{-1})
V_0	initial 2D current volume (m^2)
w	width (m)
x	distance from source (m)
x_f	flow runout (m)
x_i	initial column length or radius (m)
y	height within flow (m)
α	entrainment constant
β	gas compressibility (Pa^{-1})
$\dot{\gamma}$	shear rate = u_c/h_c (s^{-1})
η	viscosity (Pa s)
η_g	gas dynamic viscosity (Pa s)
θ	slope angle (°)
κ	von Kármán constant
λ	exponent in Eq. (10.3)
μ	interparticle friction coefficient
$\mu_{(I)}$	macroscopic friction coefficient
μ_1, μ_2	friction coefficients
ν_p	particle volume fraction
$\nu_{p,max}$	maximum particle volume fraction
ξ	Voellmy coefficient
ρ	local density (kg m^{-3})
ρ_0	ambient density (kg m^{-3})
ρ_c	current density (kg m^{-3})
ρ_f	interstitial fluid density (kg m^{-3})
ρ_g	gas density (kg m^{-3})
ρ_p	particle density (kg m^{-3})
ρ_{ref}	reference density (kg m^{-3})
σ	normal stress (Pa)
τ	shear stress (Pa)
ϕ	bed porosity
ϕ_{mf}	bed porosity at u_{mf}
φ	friction angle (°)
φ_r	angle of repose (°)
ψ	current head/tail height ratio

References

Balmforth, N. J. and Kerswell, R. R. (2005). Granular collapse in two dimensions. *Journal of Fluid Mechanics*, **538**, 399–428.

Bareschino, P., Gravina, T., Lirer, L. et al. (2007). Fluidization and de-aeration of pyroclastic mixtures: the influence of fines content, polydispersity and shear flow. *Journal of Volcanology and Geothermal Research*, **164**, 284–292.

Bareschino, P., Marzocchella, A., Salatino, P., Lirer, L. and Petrosino, P. (2008). Self-fluidization of subaerial rapid granular flows. *Powder Technology*, **182**, 323–333.

Benjamin, T. B. (1968). Gravity currents and related phenomena. *Journal of Fluid Mechanics*, **31**, 209–248.

Bonnecaze, R. T., Huppert, H. E. and Lister, J. R. (1993). Particle-driven gravity currents. *Journal of Fluid Mechanics*, **250**, 339–369.

Branney, M. J. and Kokelaar, P. (1992). A reappraisal of ignimbrite emplacement: progressive aggradation and changes from particulate to non-particulate flow during emplacement of high-grade ignimbrite. *Bulletin of Volcanology*, **54**, 504–520.

Branney, M. J. and Kokelaar, P. (2002). *Pyroclastic Density Currents and the Sedimentation of Ignimbrites*. Geolocial Society of London, Memoirs, **27**.

Cagnoli, B. and Manga, M. (2004). Granular mass flows and Coulomb's friction in shear cell experiments: implications for geophysical flows. *Journal of Geophysical Research*, **109**, F04005, doi:10.1029/2004JF000177.

Cagnoli, B. and Manga, M. (2005). Vertical segregation in granular mass flows: a shear cell study. *Geophysical Research Letters*, **32**, L10402, doi:10.1029/2005GL023165.

Capra, L., Norini, G., Groppelli, G., Macías, J. L. and Arce, J. L. (2008). Volcanic hazard zonation of the Nevado de Toluca volcano, México. *Journal of Volcanology and Geothermal Research*, **176**, 469–484.

Cas, R. A. F. and Wright, J. V. (1987). *Volcanic Successions*. London: Chapman and Hall.

Choux, C. M. and Druitt T. H. (2002). Analogue study of particle segregation in pyroclastic density currents, with implications for the emplacement mechanisms of large ignimbrites. *Sedimentology*, **49**, 907–928.

Cordoba G. (2005). A numerical model for the dynamics of pyroclastic flows at Galeras Volcano, Colombia. *Journal of Volcanology and Geothermal Research*, **139**, 59–71.

Dade, W. B. and Huppert, H. E. (1995). A box model for non-entraining, suspension-driven gravity surges on horizontal surfaces. *Sedimentology*, **42**, 453–471.

Dade, W. B. and Huppert, H. E. (1996). Emplacement of the Taupo ignimbrite by a dilute turbulent flow. *Nature*, **381**, 509–512.

Dartevelle, S., Rose, W. I., Stix, J., Kelfoun, K. and Vallance, J. W. (2004). Numerical modeling of geophysical granular flows: 2. Computer simulations of plinian clouds and pyroclastic flows and surges. *Geochemistry, Geophysics, Geosystems*, **5**, Q08004, doi:10.1029/2003GC000637.

Dellino, P., Zimanowski, B., Büttner, R. *et al.* (2007). Large-scale experiments on the mechanics of pyroclastic flows: Design, engineering, and first results. *Journal of Geophysical Research*, **112**, B04202, doi: 10.129/2006JB004313.

Dellino, P., Mele, D., Sulpizio, R., La Volpe, L. and Braia, G. (2008). A method for the calculation of the impact parameters of dilute pyroclastic density currents based on deposit particle characteristics. *Journal of Geophysical Research*, **113**, B07206, doi:10.1029/2007JB005365.

Dobran, F., Neri, A. and Macedonio, G. (1993). Numerical simulation of collapsing volcanic columns. *Journal of Geophysical Research*, **94**, 1867–1887.

Doyle, E. E., Hogg, A. J., Mader, H. M. and Sparks, R. S. J. (2010). A two-layer model for the evolution and propagation of dense and dilute regions of pyroclastic currents. *Journal of Volcanology and Geothermal Research*, **190**, 365–378.

Druitt, T. H. (1998). Pyroclastic density currents. In *The Physics of Explosive Volcanic Eruptions*, ed. J. S. Gilbert and R. S. J. Sparks. *Geological Society of London Special Publication*, **145**, pp. 145–182.

Druitt, T. H., Calder, E. S., Cole, P. D. *et al.* (2002). Small volume, highly mobile pyroclastic flows formed by rapid sedimentation from pyroclastic surges at Soufrière Hills Volcano, Montserrat: An important volcanic hazard. In *The Eruption of Soufrière Hills Volcano, Montserrat, From 1995 to 1999*, ed. T. H. Druitt and B. P. Kokelaar. *Memoir of the Geological Society of London*, **21**, 263–279.

Druitt, T. H., Avard, G., Bruni, G., Lettieri, P. and Maez, F. (2007). Gas retention in fine-grained pyroclastic flow materials at high temperatures. *Bulletin of Volcanology*, **69**, 881–901.

Eames, I. and Gilbertson, M. A. (2000). Aerated granular flow over a horizontal rigid surface. *Journal of Fluid Mechanics*, **424**, 169–195.

Esposti Ongaro, T., Cavazzoni, C., Erbacci, G., Neri, A. and Salvetti, M. V. (2007). A parallel multiphase flow code for the 3D simulation of explosive volcanic eruptions. *Parallel Computing*, **33**, 541–560.

Félix, G. and Thomas, N. (2004). Relation between dry granular flow regimes and morphology of deposits: formation of levées in pyroclastic deposits. *Earth and Planetary Science Letters*, **221**, 197–213.

Fisher, R. V. (1966). Mechanism of deposition from pyroclastic flow. *American Journal of Science*, **264**, 350–363.

Forterre, Y. and Pouliquen, O. (2008). Flows of dense granular media. *Annual Reviews of Fluid Mechanics*, **40**, 1–24.

Freundt, A., Carey, S. and Wilson, C. J. N. (2000). Ignimbrites and block-and-ash flow deposits. In *Encyclopedia of Volcanoes*, ed. H. Sigurdsson *et al.* New York: Academic Press, pp. 581–600.

GDR MiDi (2004). On dense granular flows. *European Physical Journal E*, **14**, 341–365.

Geldart, D. (1986). *Gas Fluidization Technology*. Wiley.

Girolami, L., Druitt, T. H., Roche, O. and Khrabrykh, Z. (2008). Propagation and hindered settling of laboratory ash flows. *Journal of Geophysical Research*, **113**, B02202, doi:10.1029/2007JB005074.

Girolami, L., Roche, O., Druitt, T. H. and Corpetti, T. (2010). Velocity fields and depositional processes in laboratory ash flows. *Bulletin of Volcanology*, **72**, 747–759.

Gladstone, C., Phillips, J. C. and Sparks, R. S. J. (1998). Experiments on bidisperse, constant-volume gravity currents: propagation and sediment deposition. *Sedimentology*, **45**, 833–843.

Gladstone, C., Ritchie, L. J., Sparks, R. S. J. and Woods, A. W. (2004). An experimental investigation of density-stratified inertial gravity currents. *Sedimentology*, **51**, 767–789.

Gravina, T., Lirer, L., Marzocchella, A., Petrosino, P. and Salatino, P. (2004). Fluidization and attrition of pyroclastic granular solids. *Journal of Volcanology and Geothermal Research*, **138**, 27–42.

Hallworth, M. A. and Huppert, H. E. (1998). Abrupt transitions in high-concentration, particle-driven gravity currents. *Physics of Fluids*, **10**, 1083–1087.

Hallworth, M. A., Phillips, J., Huppert, H. E. and Sparks, R. S. J. (1993). Entrainment in turbulent gravity currents. *Nature*, **362**, 829–831.

Harris, T. C., Hogg, A. J. and Huppert, H. E. (2002). Polydisperse particle-driven gravity currents. *Journal of Fluid Mechanics*, **472**, 333–372.

Heinrich, P., Boudon, G., Komorowski, J. C. *et al.* (2001). Numerical simulation of the December

PYROCLASTIC DENSITY CURRENTS | 227

1997 debris avalanche in Montserrat, Lesser Antilles. *Geophysical Research Letters*, **28**, 2529–2532.

Hogg, A. J. and Pritchard, D. (2004). The effects of hydraulic resistance on dam-break and other shallow inertial flows. *Journal of Fluid Mechanics*, **501**, 179–212.

Huppert, H. E. (2006). Gravity currents: a personal perspective. *Journal of Fluid Mechanics*, **554**, 299–322.

Huppert, H. E. and Simpson, J. E. (1980). The slumping of gravity currents. *Journal of Fluid Mechanics*, **99**, 785–799.

Ishimine, Y. (2005). Numerical study of pyroclastic surges. *Journal of Volcanology and Geothermal Research*, **139**, 33–57.

Itoh, H., Takahama, J., Takahashi, M. and Miyamoto, K. (2000). Hazard estimation of the possible pyroclastic flow disasters using numerical simulation related to the 1994 activity at Merapi volcano. *Journal of Volcanology and Geothermal Research*, **100**, 503–516.

Iverson, R. M. (1997). The physics of debris flows. *Reviews of Geophysics*, **35**, 245–296.

Iverson, R. M. and Denlinger, R. P. (2001). Flow of variably fluidized granular masses across three-dimensional terrain 1. Coulomb mixture theory. *Journal of Geophysical Research*, **106**, 537–552.

Kelfoun, K. and Druitt, T. H. (2005). Numerical modeling of the emplacement of Socompa rock avalanche, Chile. *Journal of Geophysical Research*, **110**, B12202, doi:10.1029/2005JB003758.

Kelfoun, K., Samaniego, P., Palacios, P. and Barba, D. (2009). Testing the suitability of frictional behaviour for pyroclastic flow simulation by comparison with a well-constrained eruption at Tungurahua volcano (Ecuador). *Bulletin of Volcanology*, **71**, 1057–1075, doi: 10.1007/s00445-009-0286-6.

Lajeunesse, E., Mangeney-Castelnau, A. and Vilotte, J.-P. (2004). Spreading of a granular mass on a horizontal plane. *Physics of Fluids*, **16**, 2371–2381.

Lajeunesse, E., Monnier, J. B. and Homsy, G. M. (2005). Granular slumping on a horizontal surface. *Physics of Fluids*, **17**, 103302.

Lube, G., Huppert, H. E., Sparks, R. S. J. and Hallworth, M. A. (2004). Axisymmetric collapses of granular columns. *Journal of Fluid Mechanics*, **508**, 175–199.

Lube, G., Huppert, H. E., Sparks, R. S. J. and Freundt, A. (2005). Collapses of two-dimensional granular columns. *Physics Reviews E*, **72**, 041301, doi:10.1103/PhysRevE.72.041301.

Lucas, A. and Mangeney, A. (2007). Mobility and topographic effects for large Valles Marineris landslides on Mars. *Geophysical Research Letters*, **34**, L10201, doi:10.1029/2007GL029835.

Mangeney, A., Bouchut, F., Thomas, N., Vilotte, J.-P. and Bristeau, M. O. (2007). Numerical modeling of self-channeling granular flows and their levée-channel deposits. *Journal of Geophysical Research*, **112**, F02017, doi:10.1029/2006JF000469.

Mangeney A., Roche, O., Hungr, O. *et al.* (2010). Erosion and mobility in granular collapse over sloping beds. *Journal of Geophysical Research*, **115**, F03040, doi:10.1029/2009JF001462.

Martin, D. and Nokes, R. (1988). Crystal settling in a vigorously convecting magma chamber. *Nature*, **332**, 534–536.

McEwen, A. S. and Malin, M. C. (1989). Dynamics of Mount St. Helens' 1980 pyroclastic flows, rockslide-avalanche, lahars, and blast. *Journal of Volcanology and Geothermal Research*, **37**, 205–231.

McTaggart, K. C. (1960). The mobility of nuées ardentes. *American Journal of Science*, **258**, 369–382.

Mitani, N. K., Matuttis, H. G. and Kadono, T. (2004). Density and size segregation in deposits of pyroclastic flow. *Geophysical Research Letters*, **31**, L15606, doi:10.1029/2004GL020117.

Murcia H. F., Sheridan, M. F., Macías, J. L. and Cortés, G. P. (2010). TITAN2D simulations of pyroclastic flows at Cerro Machín Volcano, Colombia: Hazard implications. *Journal of South American Earth Sciences*, **29**, 161–170.

Neri, A. and Dobran, F. (1994). Influence of eruption parameters on the thermofluid dynamics of collapsing volcanic columns. *Journal of Geophysical Research*, **99**, 11 833–11 857.

Neri, A., Esposti Ongaro, T., Macedonio, G. and Gidaspow, D. (2003). Multiparticle simulation of collapsing volcanic columns and pyroclastic flows. *Journal of Geophysical Research*, **108**, 2202, doi:10.1029/2001JB000508.

Neri, A., Esposti Ongaro, T., Menconi, G. *et al.* (2007). 4D simulation of explosive eruption dynamics at Vesuvius. *Geophysical Research Letters*, **34**, L04309, doi:10.1029/2006GL028597

Patra, A. K., Bauer, A. C., Nichita, C. C. *et al.* (2005). Parallel adaptive numerical simulation of dry avalanches over natural terrain. *Journal of Volcanology and Geothermal Research*, **139**, 1–22.

Phillips, J. C., Hogg, A. J., Kerswell, R. R. and Thomas, N. H. (2006). Enhanced mobility of granular mixtures of fine and coarse particles. *Earth and Planetary Science Letters*, **246**, 466–480.

Pitman, E. B., Patra, A., Bauer, A., Sheridan, M. and Bursik, M. (2003). Computing debris flows and landslides. *Physics of Fluids*, **15**, 3638–3646.

Pouliquen, O. (1999). Scaling laws in granular flows down rough inclined planes. *Physics of Fluids*, **11**, 542–548.

Pouliquen, O. and Forterre, Y. (2002). Friction law for dense granular flows: application to the motion of a mass down a rough inclined plane. *Journal of Fluid Mechanics*, **453**, 133–151.

Roche, O., Gilbertson, M. A., Phillips, J. C. and Sparks, R. S. J. (2004). Experimental study of gas-fluidized granular flows with implications for pyroclastic flow emplacement. *Journal of Geophysical Research*, **109**, B10201, doi:10.1029/2003JB002916.

Roche, O., Gilbertson, M. A., Phillips, J. C. and Sparks, R. S. J. (2005). Inviscid behaviour of fines-rich pyroclastic flows inferred from experiments on gas-particle mixtures. *Earth and Planetary Science Letters*, **240**, 401–414.

Roche, O., Montserrat, S., Niño, Y. and Tamburrino, A. (2008). Experimental observations of water-like behavior of initially fluidized, dam break granular flows and their relevance for the propagation of ash-rich pyroclastic flows. *Journal of Geophysical Research*, **113**, B12203, doi:10.1029/2008JB005664.

Roche, O., Montserrat, S., Niño, Y. and Tamburrino, A. (2010). Pore fluid pressure and internal kinematics of gravitational laboratory air-particle flows: insights into the emplacement dynamics of pyroclastic flows. *Journal of Geophysical Research*, **115**, B09206, doi:10.1029/2009JB007133.

Salatino, P. (2005). Assessment of motion-induced fluidization of dense pyroclastic gravity currents. *Annals of Geophysics*, **48**, 843–852.

Saucedo, R., Macias, J. L., Sheridan, M. F., Bursik, M. I. and Komorowski, J. C. (2005). Modeling of pyroclastic flows of Colima Volcano, Mexico: implications for hazard assessment. *Journal of Volcanology and Geothermal Research*, **139**, 103–115.

Savage, S. B. (1984). The mechanics of rapid granular flows. *Advances in Applied Mechanics*, **24**, 289–366.

Savage, S. B. and Hutter, K. (1989). The motion of a finite mass of granular material down a rough incline. *Journal of Fluid Mechanics*, **199**, 177–215.

Sheridan, M. F. and Malin, M. C. (1983). Application of computer-assisted mapping to volcanic hazard evaluation of surge eruption: Vulcano, Lipari, and Vesuvius. *Journal of Volcanology and Geothermal Research*, **17**, 187–202.

Sheridan, M. F., Stinton, A. J., Patra, A. *et al.* (2005). Evaluating Titan2D mass-flow model using the 1963 Little Tahoma Peak avalanches, Mount Rainier, Washington. *Journal of Volcanology and Geothermal Research*, **139**, 89–102.

Simpson, J. E. (1997). *Gravity Currents in the Environment and the Laboratory*. Cambridge University Press.

Simpson, J. E. and Britter, R. E. (1979). The dynamics of the head of a gravity current advancing over a horizontal surface. *Journal of Fluid Mechanics*, **94**, 477–495.

Sparks, R. S. J. (1976). Grain size variations in ignimbrites and implications for the transport of pyroclastic flows. *Sedimentology*, **23**, 147–188.

Sparks, R. S. J., Bonnecaze, R. T., Huppert, H. E. *et al.* (1993). Sediment-laden gravity currents with reversing buoyancy. *Earth and Planetary Science Letters*, **114**, 243–257.

Sulpizio, R. and Dellino, P. (2008). Sedimentology, depositional mechanisms and pulsating behaviour of pyroclastic density currents. In *Caldera Volcanism: Analysis, Modelling, and Response*, ed. J. Gottsmann and J. Martí, Developments in Volcanology 10. Elsevier, pp. 57–96.

Takahashi, T. and Tsujimoto, H. (2000). A mechanical model for Merapi-type pyroclastic flow. *Journal of Volcanology and Geothermal Research*, **98**, 91–115.

Thomas, N. (2000). Reverse and intermediate segregation of large beads in dry granular media. *Physics Reviews E*, **62**, 961–974.

Toro, E. F. (2001). *Shock-Capturing Methods for Free-Surface Shallow Flows*. New York: Wiley.

Valentine, G. A. (1987). Stratified flow in pyroclastic surges. *Bulletin of Volcanology*, **49**, 616–630.

Valentine, G. A. and Fisher, R. V. (2000). Pyroclastic surges and blasts. In *Encyclopedia of Volcanoes*, ed. H. Sigurdsson *et al.* New York: Academic Press, pp. 571–580.

Valentine, G. A. and Wohletz, K. H. (1989). Numerical models of plinian eruption columns and pyroclastic flows. *Journal of Geophysical Research*, **94**, 1867–1887.

Valentine, G. A., Wohletz, K. H. and Kieffer, S. W. (1991). Sources of unsteady column dynamics in pyroclastic flow eruptions. *Journal of Geophysical Research*, **96**, 21 887–21 892.

von Kármán, T. (1940). The engineer grapples with nonlinear problems. *Bulletin of the American Mathematical Society*, **46**, 615–683.

Wadge, G., Jackson, P., Bower, S. M., Woods, A.W. and Calder, E. (1998). Computer simulations of pyroclastic flows from dome collapse. *Geophysical Research Letters*, **25**, 3677–3680.

Wilson, C. J. N. (1980). The role of fluidization in the emplacement of pyroclastic flows: An experimental approach. *Journal of Volcanology and Geothermal Research*, **8**, 231–249.

Wilson, C. J. N. (1984). The role of fluidization in the emplacement of pyroclastic flows, 2: Experimental results and their interpretation. *Journal of Volcanology and Geothermal Research*, **20**, 55–84.

Wilson, C. J. N. (1985). The Taupo eruption, New Zealand, 2. The Taupo ignimbrite. *Philosophical Transactions of the Royal Society of London A*, **314**, 229–310.

Wohletz, K. H. and Valentine, G. A. (1990). Computer simulations of explosive volcanic eruptions. In *Magma Transport and Storage*, ed. M.P. Ryan. London: Wiley, pp. 113–135.

Wohletz, K. H., McGetchin, T. R., Sandford, M. T. and Jones, E. M. (1984). Hydrodynamic aspects of caldera-forming eruptions: Numerical models. *Journal of Geophysical Research*, **89**, 8269–8285.

Exercises

10.1 Consider a dilute PDC (surge) of density ρ_c and height h_c propagating on a sub-horizontal surface with a front velocity u_c in a less dense atmosphere of density ρ_0 and height H (considered as semi-infinite).

(a) Calculate the Froude number at the front of the surge.

(b) Field observations reveal that $u_c = 65$ m s^{-1} and $h_c = 20$ m. Calculate the density of the surge ($\rho_0 = 1.2$ kg m^{-3}).

(c) Discuss the limitations of the analysis.

10.2 Laboratory experiments on granular flows in air ($\rho_0 = 1.2$ kg m^{-3}) are carried out to investigate concentrated, coarse-grained PDCs. The particles used are glass beads of grain size $d = 2$ mm and density $\rho_p = 2500$ kg m^{-3}. Flows are generated on an inclined rough substrate and have a typical thickness $h_f \sim$ 1 cm. Calculate the range of velocities u_f at which the flows will be in a frictional or in a collisional regime.

10.3 An axisymmetric constant-volume gravity current spreads radially from its source such that the position of the flow front r increases with time t, from an initial condition $r = 0$ at $t = 0$. The statement of mass conservation for this current is $Q = \pi r^2 h_c$, and the front condition when inertia and buoyancy forces are in balance is $u_c = \text{Fr}(g'h_c)^{1/2}$, where Q is the volume and h_c is the thickness of the current, u_c is the velocity of the current front, Fr is the Froude number, and g' is the reduced gravity. Using these formulae, show that the position of the front r varies as $t^{1/2}$ in the inertia-buoyancy regime. How does the position of the front depend on the volume of the current?

Online resources available at www.cambridge.org/fagents

• Movie files of PDC simulations
• Answers to exercises

Chapter 11

Magma–water interactions

Ken Wohletz, Bernd Zimanowski, and Ralf Büttner

Overview

Magma–water interaction is an unavoidable consequence of the hydrous nature of the Earth's crust, and may take place in environments ranging from submarine to desert regions, producing volcanic features ranging from passively effused lava to highly explosive events. Hydrovolcanism is the term that describes this interaction at or near the Earth's surface, and it encompasses the physical and chemical dynamics that determine the resulting intrusive or extrusive behavior, and the character of eruptive products and deposits. The development of physical theory describing the energetics and the hydrodynamics (dynamics of fluids and solids at high strain rates) of magma–water interaction relies on an understanding of the physics of water behavior in conditions of rapid heating, the physics of magma as a material of complex rheology, and the physics of the interaction between the two, as well as detailed field observations and interpretation of laboratory experiments. Of primary importance to address are the nature of heat exchange between the magma and water during interaction, the resulting fragmentation of the magma, and the constraints on system

energetics predicted by equilibrium and non-equilibrium thermodynamics. Taken together, these approaches elucidate the relationships among aqueous environment, interaction physics, and eruptive phenomena and landforms.

11.1 | Introduction: magma and the hydrosphere

The vast majority of volcanic eruptions take place under water because most volcanism concentrates at mid-oceanic ridges where new oceanic crust is produced. By definition, every kind of extrusive subaqueous volcanism on Earth is hydrovolcanic since some degree of water interaction must take place. The hydrosphere also exists in continental areas, as the consequence not only of lakes and rivers, but also of groundwater and hydrous fluids that circulate in joints and faults in the upper crust and fill pore space in sedimentary rocks. Such locations are typically referred to as geohydrological environments. As a consequence, subaerial volcanism is commonly influenced by magma–water interaction. Chapter 12 describes deep-sea eruptions in greater detail, whereas this chapter

Modeling Volcanic Processes: The Physics and Mathematics of Volcanism, eds. Sarah A. Fagents, Tracy K. P. Gregg, and Rosaly M. C. Lopes. Published by Cambridge University Press. © Cambridge University Press 2013.

focuses on magma–water interaction in surface and near-surface environments.

The explosive intensity of volcanic eruptions depends on the extrusion rate of magma and on the coupling of its thermal energy (i.e., heat flux) to the surroundings. Because the thermal conductivity of magma is very low, hydrovolcanic heat flux is mainly governed by the size of the interfacial area between magma and water and its growth rate during interaction. Thus, the key process that determines the thermal energy flux from the magma to its surroundings is magma fragmentation.

Water is the predominant thermodynamic working fluid on Earth, and practically every power plant in the world uses water for converting thermal energy into mechanical energy and finally into electricity. Where rising magma contacts water (in contrast to rocks or atmospheric gases) the major effect is an increase in thermal energy flux and, by analogy to commercial power production, in the efficacy of heat and power generation. For this reason there is a rich technical literature in science and engineering that can be applied to understanding magma–water interactions; this is the intent of this chapter.

11.2 | Hydrovolcanism: from pillow lava to maar tephra

11.2.1 History of hydrovolcanism

Magma–water interactions occur deep within the Earth (*hydromagmatism*) as well as at or near its surface (*hydrovolcanism*). These terms are roughly synonymous because in many cases the realm where interaction initiates and later manifests may span from deep within the Earth's crust to the surface. For this reason the term *phreatomagmatic* is used to designate interaction within the phreatic realm of the Earth's surface, which includes the zone of saturation where groundwater and surface water exist.

The topic of magma–water interaction may be considered to have its roots in the eighteenth-century Neptunists' theory about the origin of basaltic rocks in oceans (which was later formalized by Abraham Werner). Initial ideas about the role of ground and surface water in volcanism developed during the last century. These perceptions resulted largely from observations of unusually explosive periods of Hawaiian volcanism, during which groundwater entered rifts along which normal lava fountaining had previously occurred (Jaggar, 1949), as well as through examination of fragmental basalts found where lava had entered water (Fuller, 1931). Three well-documented eruptions during the late 1950s and early 1960s brought an increased awareness of explosive hydrovolcanism: Capelinhos, Azores (Tazieff, 1958; Servicos Geologicos de Portugal, 1959), Surtsey, Iceland (Thorarinsson, 1964), and Taal, Philippines (Moore *et al.*, 1966). Fisher and Waters (1970), Waters and Fisher (1971), and Heiken (1971) expanded the characterization of phreatomagmatic eruptions to include steam-rich eruption columns, base surges, and typical landforms such as maars, tuff rings, and tuff cones. As a result of this work, numerous twentieth-century phreatomagmatic eruptions are now recognized, many of which formed maar-like craters (e.g., Self *et al.*, 1980). After cinder cones, phreatomagmatic constructs (tuff rings, tuff cones, and maars) are perhaps the most abundant terrestrial volcanic landform. However, magma–water interaction is certainly not limited to explosive phreatomagmatic eruptions – the hydrologic environment plays a major role in determining the kind of interaction that can occur.

11.2.2 Hydrovolcanic environments

The wide variety of hydrovolcanic phenomena suggests that interaction between water and magma or magmatic heat may occur in any volcanic setting and geohydrological environment, and it is not restricted to a particular magma composition. From the formation of pillow lava and lineated, folded, and jumbled sheet flows in deep water, to the intrusion of breccia in dikes and sills deep in the crust, to the eruption of plumes of fine ash in desert, tropical, and shallow water environments, magma–water interaction includes both passive and dynamic phenomena. During ascent to the surface, magma commonly encounters

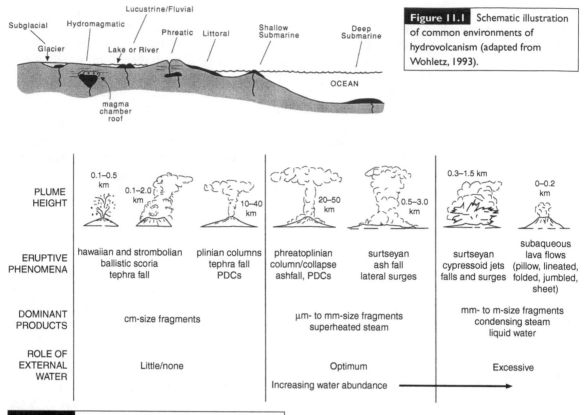

Figure 11.2 Relationship of typical eruptive phenomena and products to water abundance (adapted from Sheridan and Wohletz, 1983). Deposit features indicative of magma–water interaction may include planar to duneform bedding, impact sags or slumps and, particularly for greater water abundances, accretionary lapilli, soft sediment deformation structures, and cohesive, altered, or vesiculated tuff.

groundwater, connate (entrapped depositional) water, marine, fluvial, or lacustrine water, ice, or rain water (Fig. 11.1). From this diversity of hydrovolcanic environments comes a wide range of terminology. For example, the subaqueous environment includes all activity beneath a standing body of water (Kokelaar, 1986); products of this activity have been called *subaquatic* (Sigvaldason, 1968), *aquagene* (Carlisle, 1963), *hyaloclastite* (for deep marine; Bonatti, 1976), *hyalotuff* (for shallow marine; Honnorez and Kirst, 1975), and *littoral* (Wentworth, 1938). Volcanism that heats groundwater to produce steam explosions that do not eject juvenile

magma fragments is called *phreatic* (Ollier, 1974) or *hydrothermal* (Muffler *et al.*, 1971; Nairn and Solia, 1980). *Subglacial volcanism* (Noe-Nygaard, 1940; see Chapter 13) is best known from its products, including massive floods (*jökulhlaups*), table mountains (*tuyas* or *stapi*), and ridges (*tindars* or *mobergs*). In all these environments, a major factor determining the expression of hydrovolcanism is the abundance of water available to interact with the magma. Not only are the eruptive phenomena affected by water abundance, but so are the characteristics of the interaction products, their dispersal, and the resulting landforms (Fig. 11.2).

11.2.3 Hydrovolcanic eruption styles

A wide variety of eruption styles result from magma–water interactions, ranging from passive lava effusion to highly energetic thermohydraulic explosions, depending on ambient conditions and the proportions of water and magma interacting (Fig. 11.2). At the highest water abundances, explosive activity is generally suppressed, producing passive quenching and thermal granulation of lava flows (but see Chapter 12 for further discussion of deep-sea explosive activity). At somewhat more favorable water/magma mass ratios, surtseyan activity is a common expression of explosive hydrovolcanism in, for example, shallow marine environments. Dense, tephra-laden, cypressoid ("cock's tail") jets and steam clouds contain a significant component of liquid water, as both a result of vapor condensation and direct ejection from a shallow-water environment. Tephra is dispersed as fall or surges, which may be expressed as massive or planar-to-duneform bedded deposits. Phreatoplinian eruptions are the result of more intense interactions that produce a greater vapor-phase component at water–magma mass ratios approaching the optimum for energy conversion efficiency. Groundwater or shallow surface water is typically involved. Greater energy release allows for explosive vapor expansion, thorough magma fragmentation and the formation of convecting columns, and significantly more widespread tephra dispersal through fall and pyroclastic density currents. Duneform bedding becomes more common for energetic surge deposits. Transient explosive activity, i.e., vulcanian eruptions, may also be driven in part by pressurized vapor derived from meteoric water (Chapter 7). The deposits of phreatoplinian and vulcanian eruptions exhibit fewer indications of a liquid water phase in comparison to surtseyan deposits; the former are sometimes termed "dry" hydrovolcanic deposits, whereas the latter are termed "wet." Subglacial eruptions, discussed in detail in Chapter 13, commonly exhibit a range of hydrovolcanic styles (from pillow effusion to surtseyan and phreatoplinian activity) during their eruption sequence, and understanding of magma–ice interactions benefitted greatly from recognition of tephra and deposit characteristics first mapped in phreatomagmatic tuff cones and tuff rings.

11.2.4 Hydrovolcanic products

Hydrovolcanic solid products include tephra, blocks and bombs, explosion breccia, pillow or sheet lava, pillow breccia, and hyaloclastite. Posteruptive hydrothermal interaction with fragmental materials may produce palagonitic and zeolitic tuff, silica sinter, and travertine. Fragmental products are termed *hydroclasts* by Fisher and Schmincke (1984), instead of *pyroclasts*, a term that refers solely to the fragmental products of explosive eruptions driven by magmatic (juvenile) volatiles. Explosive hydrovolcanic products commonly contain significant abundances of lithic fragments derived from explosive fragmentation of country rock in various hydrovolcanic environments.

Petrographic studies of hydroclastic products involve the determination of particle size, shape, componentry (magmatic vs. lithic) and textural characteristics, and the chemical signatures caused by both rapid and slow alteration. These data are indicators of the degree and type of water interaction. For example, the grain size of hydroclasts is a function of the mass ratio of water and magma involved dynamically in the interaction; grain textures are indicative of the type of interaction – passive, explosive, brittle, or ductile. Field characterization of hydroclastic products focuses on analysis of deposit characteristics, including bedding, grading, sorting, lithification, and deposit thickness vs. distance from the vent. Variations of these characteristics within or among deposits can elucidate variability in eruptive intensity and style (e.g., fall vs. density current deposit), and degree of magma–water interaction.

Experimental and field research reveals a correlation between the median grain diameters of hydroclasts and the interacting water/magma mass ratio (Fig. 11.3). In general, hydrovolcanic tephra are distinguishable from magmatic tephra by their much finer grain size. Furthermore, for hydrovolcanic tephra there is an optimum water/magma mass ratio that produces the most thorough fragmentation; ratios less than or greater than the optimum value result in less finely fragmented tephra.

Microscopic examination of grain shapes and textures also reveals features indicative of hydrovolcanic origins; whereas the products of purely magmatic fragmentation are dominated

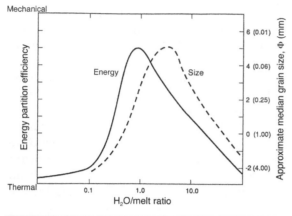

Figure 11.3 Correlation of grain size and thermal-to-mechanical energy partitioning to water/melt ratio (adapted from Frazzetta *et al.*, 1983, and Sheridan and Wohletz, 1983).

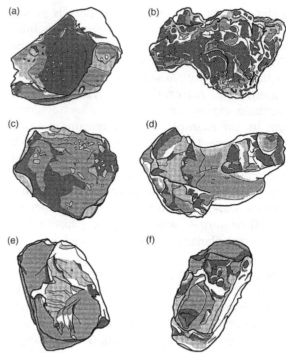

Figure 11.4 Sketches of particle textures found in hydrovolcanic deposits. (a) A characteristic blocky and equant glass shard, (b) a vesicular grain with cleaved vesicle surfaces, (c) a platy shard, (d) a drop-like or fused shard, (e) a blocky crystal with conchoidal fracture surfaces, and (f) a perfect crystal with layer of vesicular glass. In contrast, magmatically fragmented deposits are dominated by highly vesicular clasts, but may contain minor proportions of particle types shown here. (Adapted from Wohletz and Heiken, 1992.)

by highly vesicular clasts, hydroclast populations include poorly to non-vesicular clasts with a variety of textures (Fig. 11.4). Hydrovolcanic tephra may show features of both magmatic and hydrovolcanic origin; in such cases, detailed statistical study of tephra samples can help determine the relative proportions of the two end-member fragmentation processes (Büttner *et al.*, 1999; Dellino *et al.*, 2001). Some hydroclastic grain textures are also indicative of the type of magma–water interaction (e.g., whether a superheated vapor or liquid phase was involved; Wohletz, 1983). For example, quench cracks on grain surfaces indicate quenching in contact with excess liquid water, and chemical pitting is indicative of contact with corrosive fluids (Wohletz, 1987; Dellino *et al.*, 2001).

Field characteristics of deposits are also indicative of the proportion of liquid vs. vapor phases in phreatomagmatic eruptions. Broadly speaking, liquid water will produce more cohesive deposits that may contain accretionary lapilli, soft-sediment deformation structures, vesiculated tuffs (due to post-depositional vaporization of liquid water), and may produce steeper landforms (e.g., tuff cones) with some degree of hydrothermal alteration (e.g., palagonitization). In contrast, more energetic eruptions producing a superheated vapor phase may

result in broader, lower landform (e.g., maars, tuff rings) due to the greater mobility of the products (in falls or surges), with limited alteration of the hydroclasts. An individual eruption can evolve towards wetter or dryer conditions, which would be reflected in the deposit characteristics.

Explosive hydrovolcanic eruptions may occur as single, monogenetic events or at polygenetic centers (volcanoes constructed by multiple eruptions). The former commonly produce maars, tuff rings or tuff cones, depending on the environment of interaction and availability of water. The latter may simply introduce a phreatomagmatic deposit into the record of activity at that particular volcano.

11.3 | Magma–water interaction physics

11.3.1 Fuel–coolant interactions

Formulating an understanding of the physics of magma–water interactions requires consideration of the physical properties of magma, the thermodynamic behavior of water, and the physics of processes at the interface between the two fluids. Much of the physical understanding of hydrovolcanism has developed from the combined insights derived from studies of *molten fuel–coolant interaction* (MFCI or FCI) and laboratory experiments designed to replicate hydrovolcanic phenomena (e.g., Wohletz *et al.*, 1995a; Zimanowski *et al.*, 1997a). The field of FCI science arises from industrial applications, and concerns the interaction of two fluids for which the temperature of one (fuel) is above the vaporization temperature of the other (coolant) (Buchanan and Dullforce, 1973; Buchanan, 1974). Application of FCI theory to hydrovolcanic eruptions is described in Wohletz *et al.* (1995a), Zimanowski *et al.* (1991), and Zimanowski (1998). Figure 11.5 depicts a hypothetical geologic system in which magma (fuel) explosively interacts with water-saturated sediments (coolant). The stages of a FCI are as follows:

(1) Initial contact and coarse mixing of fuel and coolant, growth of vapor film;
(2) Quasi-coherent collapse of all vapor films in the premix caused by a triggering pressure pulse (seismically induced or by local over-expansion), leading to direct contact of fuel and coolant;
(3) Cycles of enhanced fuel–coolant heat transfer, rapid (< 1 ms) coolant expansion, fine fragmentation of fuel, producing superheated and pressurized water, and explosive energy release;
(4) Volumetric expansion of fuel–coolant mixture as superheated water transforms to superheated steam.

The process does not necessarily evolve through all these stages and may be arrested, for instance, before mixing or explosion.

Experimental studies have made it possible to quantify some controlling parameters by using field and laboratory measurements of hydrovolcanic products. The curves in Figure 11.3 were derived from early experiments (Wohletz, 1983; Wohletz and McQueen, 1984) that demonstrated that the explosive efficiency of the system (measured as the ratio of eruptive kinetic energy to magma thermal energy) is related to the mass ratio of interacting water and melt (thermite – a magma analog) and the confining pressure. These experiments displayed a variety of explosive and non-explosive behaviors that are analogous to natural volcanic activity, including classical strombolian, surtseyan, vulcanian, and plinian phenomena.

Experimental studies demonstrate that the thermodynamics of heat transfer (see Section 11.4.1) is a significant aspect of hydrovolcanic systems and their physical and chemical effects. The magmatic thermal energy transformed by interaction with external water is partitioned into many possible forms, including mechanical fragmentation of the magma and country rock, excavation of a crater, acceleration and dispersal of tephra, seismic and acoustic perturbations, and chemical processes such as solution, precipitation, and mass diffusion. Section 11.4.4 addresses energy partitioning further.

11.3.2 Physical properties of magma

Magma generally is treated as a three-phase system of melt, solids, and gases (see Chapter 4). The physical properties of magma are strongly controlled by chemical composition, the proportions and types of solids (phenocrysts, xenocrysts, and xenoliths), presence of exsolved and dissolved gases, flow speed, and temperature. Natural silicate melts have temperatures that exceed the solidus by ~100–200 K, such that the heat content of magma comprises a significant proportion of latent heat. However, during rapid cooling processes relevant to many magma–water interactions, the release of latent heat is negligible, because quenching (and the formation of glass) takes place, rather than crystallization.

Magma viscosity is the result of interaction between internal friction of the melt and

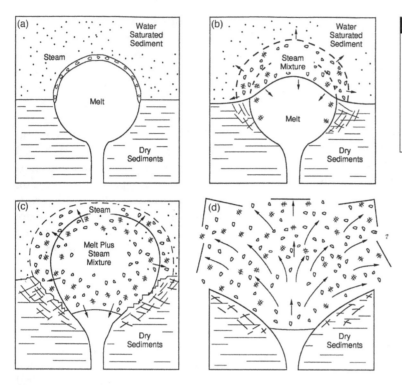

Figure 11.5 Hypothetical setting of subsurface hydrovolcanic activity, showing (a) initial contact of magma with water-saturated sediments, (b) vapor film growth, (c) mixing of magma with the sediments, and (d) expansion of the high-pressure steam in an explosion (adapted from Wohletz and Heiken, 1992).

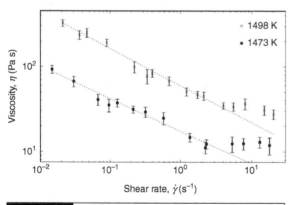

Figure 11.6 Shear stress dependent viscosity of a basaltic melt (< 5% crystals, < 2% bubbles), at two temperatures. The lines are viscosity calculations from a power-law model (modified from Sonder et al., 2006).

mechanically coupled compressible bubbles and incompressible crystals (see Chapter 4, Section 4.5). Crystals are not only characterized by various morphologies, but also by various interfacial couplings (e.g., wetting angles). It is therefore not surprising that the rheology of magma needs to be addressed by a non-Newtonian model. Viscosity is not only temperature dependent, but also strongly depends on the shear rate (Fig. 11.6). At high confining pressures (i.e., at depth within the crust or beneath water), magma viscosity depends primarily on the behavior of dissolved volatiles, and will generally decrease with increasing pressure. Although this pressure-dependence is not yet well characterized, it is the viscosity contrast between magma and water that primarily governs interaction dynamics, so little effect is expected on the hydrodynamics of the magma–water system at increasing pressure.

The intensity of magma–water interactions reflects the efficiency of heat transfer at the magma–water interface, which in turn reflects the degree of magma fragmentation. Fragmentation can generally be viewed as taking place in one of two regimes, depending upon whether the characteristic deformation time is greater or less than the mechanical relaxation time of the melt: (1) *hydrodynamic* fragmentation is restricted to deformation of two-dimensional interfacial areas (boundaries between the melt and gas) and is most efficient in systems subjected to high strain rates (rapid flow) at low viscosities, low interfacial tension, and high density contrast between the accelerating fluid and surrounding media; (2) *brittle* fragmentation is the result of three-dimensional crack growth caused by strain that exceeds the elastic properties of a medium (e.g., bulk modulus; cf. the stress and strain-rate criteria discussed in Chapter 4, Section 4.6). For hydrodynamic fragmentation to produce fine ash (< 63 μm), the required accelerations are extremely high. However, brittle fragmentation can readily produce fine ash particles through a steady increase in strain. The features of natural particles, experimental results, and theoretical models clearly show ash generation is dominated by brittle fragmentation (Zimanowski *et al.*, 2003). The material parameter that controls the brittleness of magma (i.e., the mechanical deformation energy needed to produce new crack surfaces) is the critical shear stress. The range of critical shear stress of magma is as large as the variability of viscosity, but in contrast to viscosity, the critical shear stress decreases with increasing silica content. Therefore, basaltic magma can be more than three orders of magnitude stronger than rhyolitic magma. Bubbles and crystals both weaken the structural strength and therefore reduce the critical shear stress. Further information on experimental and theoretical fragmentation studies can be found in Hermann and Roux (1990), Zimanowski *et al.* (1997a, 1997b, 2003), and Büttner *et al.* (1999, 2002, 2006).

11.3.3 Water physics

The thermodynamic behavior of water is well known from the steam-locomotion era, and many volcanological studies have approached the problem of magma–water interaction rather simplistically, by direct application of the first law of thermodynamics. While this approach does provide some limiting conditions, it ignores complex issues concerning the multiphase (steam, water, magmatic particles, melt) system. Consideration of multiphase fluid mechanics places further constraints on the application of thermodynamics to this problem, as described by Delaney (1982) and Wohletz (1986), and embodied in studies of MFCI (Section 11.3.1).

A crucial parameter for the behavior of water at a high-temperature interface is the phase transition temperature of water into steam. In steady-state, equilibrium thermodynamic conditions, the phase transition occurs at the boiling point, which depends on the ambient pressure. However, dynamic (short duration) events require application of quasi-steady-state thermodynamic models. Instead of the phase change occurring at the boiling point and spreading outward from discrete nucleation sites, the water is heated so rapidly that it greatly overshoots its boiling temperature, reaching the homogeneous nucleation temperature (HNT), where all of the water spontaneously changes state. The boiling regime of water under atmospheric conditions starts at 373 K and the HNT is reached at ~583 K. Little is known about the pressure dependence of the HNT. During extremely rapid heat transfer, as occurs in explosive MFCI, the critical point of water (22 MPa and 647 K) may be exceeded, so that water exists as a supercritical fluid, with no phase boundary separating vapor and liquid. Experimental observations indicate that water can remain at liquid state densities even at magmatic temperatures during such interactions (Zimanowski *et al.*, 1997b).

Given that > 60% of the Earth's surface is covered by oceans, magma–water interactions predominantly involve seawater. Seawater is a solution dominated by the presence of salts (mostly NaCl), and its thermodynamic behavior can be approximated by the two-component system of pure water and NaCl. Figure 11.7(a) illustrates a p–T phase diagram for the system NaCl–H_2O that shows phase boundaries of the pure components and projections of the phase boundaries for intermediate compositions. The salinity of seawater results in critical conditions occurring not at a single point but along a curve that connects the critical points of the two pure end-members. These phase relationships (Fig. 11.7(a)) show that at any temperature two fluid phases can coexist and a single critical point does not exist if solid NaCl is present. Critical behavior occurs at pressures and temperatures elevated from those of pure water to values approaching ~30 MPa and ~680 K for seawater with a salinity of 3.2 wt.% NaCl (Bischoff and

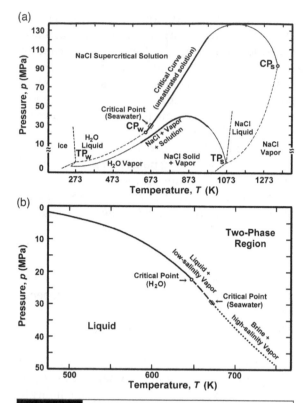

Figure 11.7 (a) Pressure–temperature diagram for the system NaCl-H$_2$O adapted from Krauskopf (1967) from experimental data of Sourirajan and Kennedy (1962). This diagram shows both pure H$_2$O and NaCl end-members (dashed lines). The solid lines (bold are experimental data) schematically represent the phase boundaries connecting the pure end-members. TP$_w$ and CP$_w$ denote the triple point and critical points of H$_2$O, respectively; TP$_s$ and CP$_s$ those points for NaCl. (b) The two-phase curve for standard seawater (3.2% NaCl) as a function of pressure and temperature, based on data from Bischoff and Rosenbauer (1984). Note that in this plot, pressure increases downward. The solid curve designates the boundary where pure water and seawater boundary are nearly coincidental. The boundary for pure water terminates at its critical point, whereas the boundary for seawater extends (dashed curve) to its respective critical point, along which it separates the stability regions of liquid and a mixture of low-salinity vapor and liquid. The phase boundary extends (dotted curve) from seawater's critical point to higher temperatures and pressures, separating the liquid region from that of a mixture of high-salinity vapor and brine.

is heated, solid NaCl is precipitated, which may greatly affect vapor nucleation (cf. White, 1996).

The two-phase boundary of seawater (Bischoff and Rosenbauer, 1984) is similar to that of pure water for subcritical conditions in pressure–temperature space (Fig. 11.7(b)), but unlike pure water, the critical point projects nearly linearly to higher pressures and temperatures (~680 K at 30 MPa to ~750 K at 50 MPa). Below the critical point the two-phase region of seawater consists of liquid and low-salinity vapor, and above the critical point it consists of brine and high-salinity vapor. These phase relationships indicate that a vapor phase can exist even at pressures above critical for seawater.

The dissolved solids in seawater also reduce the heat capacity of seawater by several percent relative to pure water. However, this effect is relatively small compared to the factor of ~2 variation in the heat capacity of pure water over the range of pressures and temperatures relevant to hydrovolcanism. For quantitative calculations, therefore, it is convenient to substitute pure water as a workable proxy for seawater because the heat capacity and phase transition of seawater appear to have offsetting effects in situations of rapid heating; however, one must bear in mind that details of heat transfer discussed in Section 11.3.4 will likely be greatly affected by the vapor nucleation and solid precipitation behavior of seawater.

The hydrodynamic properties of water, such as viscosity, heat content, and expansion coefficients, do not change drastically with increasing pressure, but they do with increasing temperature. Little is known about the effect of increasing pressure on interfacial tension. However, it can be expected that this influence is small over the range of pressures reflecting submarine volcanic scenarios. The viscosity of water increases slightly with increasing pressure but strongly decreases with increasing temperature. A salt content of 3–5% should have no significant influence on viscosity or interfacial tension.

Because magma–water interaction can result in a relatively high-pressure and high-temperature water phase, it is important to consider

Rosenbauer, 1988). Critical pressure is expected at a depth of ~3 km in the submarine environment. Figure 11.7(a) also shows that as seawater the variability of water properties near critical conditions. Figure 11.8 illustrates the variation of heat capacity, viscosity, and expansion coefficients at 30 and 60 MPa, analogous to deep submarine environments at about 3000 and 6000 m depth, respectively. Note that heat capacity, viscosity, and the isobaric expansion coefficient vary rapidly in the range 600–800 K, near the critical-point temperature (647 K). This means that small changes in temperature produce large changes in properties that determine the thermodynamic and hydrodynamic behavior of water. The large thermal gradients near the magma–water contact will produce large pressure and velocity gradients, and perturbations in water movement may result in hydrodynamic instability. For example, a supercritical fluid subjected to small pressure perturbations will tend to oscillate in density (Greer and Moldover, 1981) between liquid and vapor states (e.g., growth and collapse of vapor bubbles). The speed at which water flows from higher to lower pressure regimes depends not only on the magnitude of the pressure gradient but also on the viscosity; rapid viscosity fluctuation may enhance or dampen convective currents. These pronounced fluctuations in properties are critical to address with respect to magma–water interaction. This thermal-hydraulic system has received considerable attention for application to coolant flow stability in nuclear reactors (e.g., Ruggles et al., 1989, 1997).

11.3.4 Physics of the magma–water interface

Magma provides a high temperature heat source, such that the magma–water system is somewhat analogous to the thermal coupling and heat transfer to liquid coolants (water or aqueous liquids) used in power plant systems (e.g., Baierlein, 1999; Bejan and Eden, 1999). The major difference between magma–water interactions and power plant heat flow problems is that the interface conditions in a boiler are of a fixed geometry, whereas the magma–water interface is always dynamic. The heat flux from the magma–water interface to the water greatly exceeds the heat flux from the magma to the interface because of the relatively low

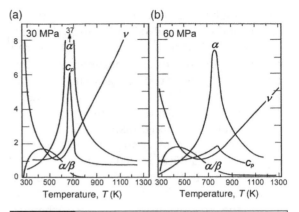

Figure 11.8 Variation of physical properties of water at (a) 30 MPa (~3000 m depth) and (b) 60 MPa (~6000 m depth) as a function of temperature. The properties are shown as curves designated by symbols and their corresponding SI units and scale factors are: α, isobaric expansion coefficient ($\times 10^3$ K^{-1}); α/β, pressure coefficient ((dp/dt)$_v$; $\times 10^{-2}$ MPa K^{-1}) where β is the isothermal expansion coefficient; v, kinematic viscosity ($\times 10^{-7}$ m^2 s^{-1}); and c_p, the constant pressure heat capacity (4.184 kJ kg^{-1} K^{-1}). Note the sharp inflections and discontinuities apparent near the critical temperature. (Adapted from Wohletz and Heiken, 1992.)

magma thermal conductivity (~1 W m^{-1} K^{-1}; Büttner et al., 1998, 2000; Ebert et al., 2003), and limited convection due to its high viscosity. Consequently, the temperature at the hot side of the interface drops rapidly, leading to solidification of the magma. Because the thermal conductivity of quenched melt is more than twice that of its liquid state, a cooling front migrates into the magma. Once a certain thickness has solidified, the strain caused by the 1–3% decrease in volume at the glass transition temperature leads to the formation of cooling cracks (thermal granulation) and thus to fragmentation of the magma into relatively coarse particles. This process has the potential to greatly increase the interfacial area between magma and water. The balance between quenching and surface area growth rates determines whether the heat transfer is relatively steady, or whether the heat transfer rate escalates. Section 11.4.1 discusses the thermodynamic regimes within which the heat transfer takes place. It is not only thermally induced fragmentation that modifies the heat flux, but also external hydro-mechanical forcing, such as differential movement between magma and water, and seismic impulses (earthquakes), which can affect the stability of the magma–water interface. Experimental studies suggest that external triggering is a major mechanism leading to the escalation of heat transfer into explosive conditions. For more details on magma–water mingling/mixing and fragmentation physics, see Morrissey *et al.* (2000) and Zimanowski and Büttner (2002, 2003).

In contrast to the hydrodynamics discussed in Section 11.3.3, the thermodynamic behavior of the magma–water interface changes dramatically with increasing temperature and hydrostatic pressure, and the effects of mineralization (e.g., seawater) are generally small by comparison. The thermal coupling between magma and water is largely dependent upon the interface temperature and development of a vapor film between the magma and the water. The vapor film consists of two regions: (1) a thicker hot layer of superheated steam at the same temperature as and directly in contact with the magma, and (2) a thinner cool layer (the *thermal boundary layer*) between the superheated steam layer and water, within which condensation and vaporization balance. The temperature of the thermal boundary layer can be only microscopically defined because its thickness may be much less than a few millimeters. Because the interface temperature is a hypothetical value, dependent upon thermal diffusivities of magma and vapor, it is useful to characterize the vapor film by its temperature contrast with the magma (i.e., the magma surface temperature minus water temperature). In so doing, the transfer of heat energy between substances, known as *thermal coupling*, can be classified in three regimes, with coupling increasing from (1) to (3):

(1) *Stable film boiling*: At a temperature contrast well above the water HNT and an ambient pressure well below the critical pressure of water, a macroscopic vapor film forms at the interface between magma and water and restricts the thermal coupling. The thickness of the vapor film is greater for high temperature contrasts and low ambient pressures.

(2) *Metastable film boiling*: At a temperature contrast below the HNT, a macroscopic vapor film does not form; however, the thermal coupling is modified by local formation of steam at the interface. At a temperature contrast well above the HNT, metastable film boiling can take place in two cases. In Case 1, high ambient pressure causes a reduction of the thickness of the superheated steam layer of the film so that it reaches the same thickness range as the thermal boundary layer. In Case 2, hydrodynamic disturbances cause interfacial instability waves with amplitudes of the same scale as the total film thickness.

(3) *Direct contact*: At a temperature contrast below the boiling point of water and/or at ambient pressures exceeding the critical point of water, the thermal coupling can be described using equilibrium thermodynamics, and it depends on the thermal properties of the liquids and the hydrodynamics at the interface. The thermal properties of magma are not strongly affected by increasing ambient pressure; however, the properties of water change markedly (Fig. 11.8). For magma and water in direct contact, the thermal coupling is one to two orders of magnitude higher than that of stable and metastable film boiling.

Considering the effect of increasing hydrostatic pressure on the thermal coupling of magma and water, the stable film boiling regime is restricted to ambient pressures below about 1 MPa where the vapor film thickness at the magmatic temperatures of the hot layer becomes critical. Above ambient pressures of 10 MPa, the moderating effect of metastable film boiling can practically be excluded. At thermal power plants, where an optimum heat flux (i.e., direct contact) is required, the water pressure of the heat exchange system usually is set to 15–20 MPa for temperature contrasts comparable to magmatic conditions. The thermal coupling at a constant magma–water interface at water depths exceeding ~1 km (equivalent to 10 MPa) can therefore be described well by the direct contact regime.

Magma fragmentation is an important aspect of the interface physics. Wohletz (1983) discusses a number of hydrovolcanic fragmentation mechanisms that create a complex interface geometry, but the actual fragmentation mechanisms involved are generally categorized as hydrodynamic (ductile) and brittle (Section 11.3.2). With increasing fragmentation, the interfacial surface area increases exponentially. This increase in surface area is often referred to as mixing. The resulting mixture evolves from a pre-expanded combination of magma fragments and high-pressure water and vapor to a post-expanded mixture of quenched fragments and steam, which typically forms an eruption column. Although Kokelaar (1986) distinguishes *contact-surface interaction* (dynamics along an interface between a free body of water and magma) from *bulk interaction* (dynamics of a volume of magma that confines water or water-rich clastic materials), both cases involve heat transfer along the interface between magma and water. Experimental evidence indicates that most of the fragmentation occurs during development of interface dynamics and prior to expansion and eruption. Nevertheless, Mastin (2007) develops a strong argument for the occurrence of *turbulent shedding* of glassy rinds from fragments in erupted jets of steam and ash. The physics of the dynamic interface are complex and poorly understood; however, experimental studies of shock waves associated with interface dynamics suggest the physics may involve detonation, i.e., formation of an exothermic shock front, the propagation of which creates an explosion.

Wohletz (2003) discusses how film boiling can be intimately linked to fragmentation and interface surface area increase. The interface not only becomes thermodynamically unstable with metastable film boiling, but it is also hydrodynamically unstable because of large pressure, density, sound speed, and conductivity gradients produced by the film. Such instability is prone to a kind of detonation, termed *thermal detonation*, especially if perturbed by some external pressure wave, such as that produced by volcanic seismicity. In a thermal detonation, the acceleration of the particle mixture by the shock wave must produce a relative velocity, u_r, high enough to satisfy the Chapman–Jouguet (C–J) condition: u_r is the speed of the shocked

material relative to the shock front, and u_r must equal the sonic velocity of the shocked material (Courant and Friedrichs, 1948; and Zel'dovich and Raizer, 1966).

The C–J condition can be evaluated on a pressure–volume diagram (Fig. 11.9) that shows the shock adiabat (termed the *Hugoniot* and defined as the locus of points representing pressure-volume states achievable by shocking a material from an initial state) and the release adiabat (called the *Rankine–Hugoniot curve* or *detonation adiabatic*). These adiabats are concave upward and the detonation adiabatic exists at higher volume states than the Hugoniot. The detonation adiabatic is defined by classical Rankine-Hugoniot jump conditions for conserving mass, momentum, and energy across a shock wave (Landau and Lifshitz, 1959; Zel'dovich and Kompaneets, 1960; and Zel'dovich and Raizer, 1966). These conditions can be stated as:

$$\rho_1 u_1 = \rho_2 u_2; \quad \text{(mass)} \tag{11.1}$$

$$p_1 + \rho_1 u_1^2 = p_2 + \rho_2 u_2^2; \quad \text{(momentum)} \tag{11.2}$$

$$E_2 - E_1 = \frac{1}{2}(p_1 + p_2)(V_1 - V_2); \quad \text{(energy)} \tag{11.3}$$

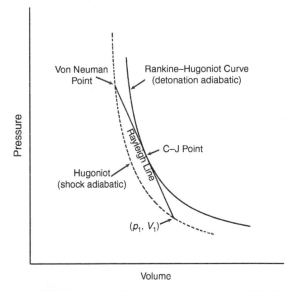

Figure 11.9 Pressure–volume diagram showing the relationships among the initial state of the magma–water mixture (p_1, V_1), the shock Hugoniot, the Rayleigh line, the Von Neumann and Chapman–Jouguet (C-J) points, and the detonation adiabatic.

where ρ is mixture density, V is mixture specific volume = $1/\rho$, p is pressure, u is mixture velocity, $E = E(p,V)$ is mixture internal energy, and subscripts 1 and 2 indicate the unshocked and shocked states. All notation is given in Section 11.6. For an inert material, the shocked thermodynamic states given by Eq. (11.3) define the Hugoniot, which is typically determined experimentally to derive the equation of state of a material. Two points on the Hugoniot, one at the initial pressure ((p_1,V_1) in Fig. 11.9) and the other at the pressure of the shock front (the *von Neumann point* or *spike*, since there is a transient pressure peak at this location), define the *Rayleigh line*. For a reactive material in which the shock induces vapor production in its wake (either chemically for common explosives, or physically for a magma–water mixture where nearly instantaneous heat transfer produces vapor), the energy equation (Eq. (11.3)) gives all possible states of the detonation adiabatic (Fig. 11.9).

By combining Eqs. (11.1) and (11.2), one obtains the value of $(p_2 - p_1)/(V_1 - V_2)$, which is a constant defined by the slope of the Rayleigh line. The square root of this slope is proportional to the velocity of the detonation wave. Zel'dovich and Kompaneets (1960) show that the only possible steady state (in which a detonation wave is sustained) is where the Rayleigh line is a tangent to the detonation adiabat at the C-J point. The points behind a propagating shock at which the C-J condition exists define a surface known as the *C-J plane* or the *detonation front* (not to be confused with the shock front).

Figure 11.10 shows a generalized conceptual view of thermal detonation that includes observed phenomena such as thermohydraulic fracturing and brittle reaction, described by Zimanowski *et al.* (1997b) and Büttner and Zimanowski (1998), which lead to enhanced heat transfer and catastrophic vapor expansion. Zimanowski *et al.* (1997a) document experiments that show development of intense shock waves in less than a millisecond under extreme rates of cooling (> 10^6 K s^{-1}) and stress (> 3 GPa m^{-2}). Yuen and Theofanous (1994) demonstrate that application of detonation theory successfully predicts results of many of

their MFCI experiments. In order to assess the role of ambient pressure in hydrovolcanism using detonation theory, the method of Board *et al.* (1975) can be used to calculate (using the Rankine–Hugoniot jump conditions) the speed of a propagating shock wave through a magma–water mixture. For mixture conditions of $R \sim 0.5$ (where R is the mass ratio of water to magma), a shock speed is ~300 m s^{-1}. This velocity defines a Rayleigh line slope, and then the Rankine–Hugoniot conditions must be specified in order to predict a detonation adiabatic that touches the Rayleigh line at a single point of tangency, the C-J condition of (p_{cj}, V_{cj}). To induce mechanisms for melt breakup (Fig. 11.10), the C-J condition requires that the slip velocity u_s between the shocked melt fragments and water is at least as large as u_r. Using the classical detonation theory described by Eqs. (11.1)–(11.3), the relative velocity of the shocked mixture leaving the front is a function of the mixture pressure and specific volume at initial and C-J conditions:

$$u_r = \sqrt{(p_{cj} - p_1)(V_1 - V_{cj})}. \qquad (11.4)$$

For an idealized thermal detonation in which $p_{cj} \approx 100$ MPa (Board *et al.*, 1975), Eq. (11.4) sets u_r at ~100 m s^{-1}. For volcanic MFCIs, the approach of Corradini (1981) yields a minimum u_r of 60 m s^{-1} (Wohletz, 1986). Drumheller (1979) combined the requirements for relative velocity and melt breakup time into a *critical Bond number* (the Bond number is ratio of body and surface tension forces). By assuming a constant p_{cj}, Wohletz (1986) evaluated the critical Bond number with respect to MFCI experimental data (Wohletz and McQueen, 1984) to predict the effects of R and ambient pressure on the development of relative velocities and magma particle sizes. Optimal conditions for thermal detonation exist at $0.5 < R < 2.0$, for ambient pressures ≤ 40 MPa (Wohletz, 2003). With increasing ambient pressure, predicted relative velocities fall to < 60 m s^{-1} (which is considered the lower limit for sustaining a detonation; Wohletz, 1986), and particle fragmentation decreases, meaning less thermal energy is released in the wake of the shock wave.

THERMAL DETONATION MODEL (1D)

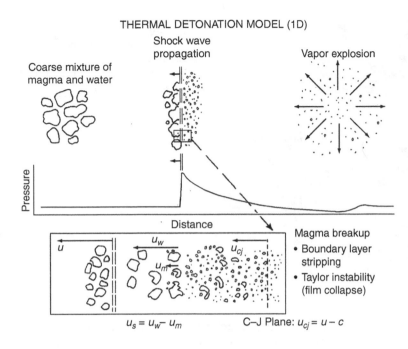

Figure 11.10 Schematic illustration of thermal detonation (adapted from Wohletz, 2003; Board *et al.*, 1975), showing propagation of a shock wave through a coarse mixture of magma fragments and water (vapor and liquid). The shock wave moves at velocity u and differentially accelerates the water and magma to velocities of u_w and u_m, respectively, resulting in a slip velocity u_s, which decays behind the shock. The slip velocity must be of sufficient magnitude to cause fine fragmentation of the magma fragments by mechanisms that shear away the boundary layer and cause interpenetration of the water and magma (Taylor instability) before the arrival of the C–J plane. At the C–J plane, the average mixture velocity is just sonic with respect to the shock wave. Fine fragmentation causes an exponential rise in heat transfer from the magma fragments to the water (characteristic conductive heat transfer times are ~ tens of μs for fine fragmentation) and catastrophic vapor expansion.

11.4 | Modeling magma–water interaction

11.4.1 Experimental insights into magma–water interaction

Experimental research on magma–water interaction has been strongly influenced by FCI studies in the context of nuclear safety research (e.g., Board *et al.*, 1975; Corradini, 1981; Henry and Fauske, 1981; Theofanous, 1995). Since the early 1980s, laboratory experiments have been conducted on magma–water interaction (e.g., Wohletz *et al.*, 1995a; Zimanowski *et al.*, 1997a; Büttner and Zimanowski, 1998), in which high-temperature melt (predominantly remelted volcanic rocks at ~1270 K) is brought into contact with water (Fig. 11.11). The philosophy of the experiments was to achieve a mesoscale dimension, minimizing danger, costs, and experimental effort, while fully representing the relevant physics. As discussed in Section 11.2.4, important information on natural explosive magma–water interaction is found in the characteristics of the products (tephra). In the experiments, the physical processes can be observed and

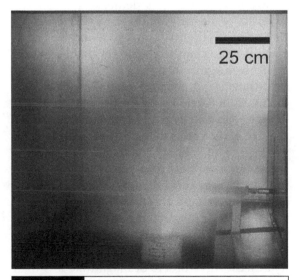

25 cm

Figure 11.11 Explosive interaction after injection of water into molten basaltic rock (Physikalisch-Vulkanologisches Labor, Universität Würzburg). See color plates section.

Figure 11.12 Products of thermal granulation experiments using remelted basalt (Physikalisch-Vulkanologisches Labor, Universität Würzburg).

measured directly. To verify the relevance to natural volcanic processes, the experiments were designed to produce artificial tephra in statistically relevant quantities. Furthermore, the products were nearly identical to the natural analogs in terms of grain size, morphology, and chemical composition (Fig. 11.12).

Experimental results demonstrate that the intensity of magma–water interaction, and thus the danger potential of hydrovolcanic eruptions, depends on the efficacy of the heat transfer from magma to water, i.e., the amount of heat transferred per unit volume and time. As discussed in Section 11.3.4, this heat transfer is directly correlated to the interfacial area (i.e., the size of the contact area between magma and water per unit volume) and the interfacial coupling conditions. Heat transport in the magma to the interface is controlled by the temperature-dependent magma thermal conductivity; convection can be neglected because of the high viscosity of magma and the short timescale. Heat transport in the water (having a viscosity at least three orders of magnitude less than magma) away from the interface is controlled by both conduction and convection, and therefore is in principle much more effective. The thermal coupling at the interface can be described in the three regimes of stable film boiling, metastable film boiling, and direct contact, with

direct contact producing about two orders of magnitude greater heat flux per unit area than stable film boiling (Section 11.3.4). Three heat flux regimes in the magma–water system can be defined to allow classification of magma–water interaction and facilitate theoretical description of the thermodynamics and hydrodynamics of the system (Wohletz, 1995; Büttner and Zimanowski, 1998; Zimanowski et al., 2003; Büttner et al., 2005):

(1) *Non-explosive interaction regime.* The heat flux in the system does not create a water overpressure (subcritical state of coolant). Steady thermodynamic sink conditions are maintained (i.e., water can always be described as a passive heat sink and steady-state thermodynamic models are applicable). Fragmentation of magma is governed by its rheology in an aqueous environment (e.g., pillow formation) and/or by thermal contraction of melt (thermal granulation), leading to passive fragmentation (Schmid et al., 2010; Sonder et al., 2011).

(2) *Subsonic explosive interaction regime.* The heat flux in the system creates a water overpressure (critical state of coolant). Superheated water is generated and complex phase transitions occur. Steady-state thermodynamic models need significant modifications, but still are generally applicable. Fragmentation

of magma is dominated by brittle processes caused by the hydraulic forcing of the coolant, but governed by the mechanical properties of the magma (i.e., mechanical deformation does not exceed critical shear conditions and propagation of cracks is subsonic with respect to the shear-wave velocity). This regime leads to subsonic active fragmentation (Austin-Erickson *et al.*, 2008).

(3) *Supersonic explosive interaction regime.* The heat flux in the system increases rapidly, due to a thermohydraulic feedback mechanism, and high overpressure is generated in the water (supercritical state of coolant). Non-equilibrium thermodynamic models are recommended. Fragmentation of magma is driven by hydraulic forcing of the coolant exceeding the mechanical properties, and propagation of cracks is supersonic. Consequently a significant proportion (up to 80%) of the mechanical energy is released as shock waves. This regime leads to supersonic active fragmentation.

The consequences for the eruptive behavior of the magma–water system depend on the interacting magma/coolant volume ratio, the rheological properties of the magma, the thermal and rheological properties of the coolant, the hydrodynamic mingling energy, the ambient pressure, and the geometry of the mingling space. In the following section, we discuss modeling techniques for steady state regimes 1 and 2, and we address the non-equilibrium conditions of regime 3 in Section 11.4.3.

11.4.2 Multiphase equilibrium thermodynamic models

Simple conservation of energy provides for a first-order assessment of magma–water interactions. For most cases, the magma temperature exceeds the water vaporization temperature, which varies somewhat with composition and ambient pressure. During interaction, the internal energy of the water increases by an amount equal to that lost by the magma. While most of that internal energy is involved in heating the water and quenching the magma, some of it may also be involved in phase transitions

(e.g., water vaporization and magma crystallization). During explosive interaction the energy exchange happens rapidly, resulting in a minimum of energy going towards phase transitions and most going towards raising the temperature of the water and cooling the magma. For this simple situation, the equilibrium temperature, T_e, is between the temperature of the magma, T_m, and that of the water, T_w. Therefore, simple energy conservation can be expressed as:

$$m_w c_w (T_e - T_w) = m_m c_m (T_m - T_e), \qquad (11.5)$$

where m is the mass and c is the specific heat capacity of the water (subscript w) and the magma (subscript m), respectively. Equation (11.5) shows that T_e varies with the mass ratio of water to magma, m_w/m_m, denoted as R. In addition, the specific heat ratio of water to magma is typically in the range of 3.0 to 4.0, depending upon magma composition and the range of typical water states at the surface of the Earth. For most cases, it is assumed that the ratio $\xi = c_w/c_m$ ~ 3.5. Accordingly, we can rearrange Eq. (11.5) to predict T_e:

$$T_e = \frac{\xi R T_w + T_m}{1 + \xi R}. \qquad (11.6)$$

The value of T_e can be thought of as an idealized initial thermal equilibrium prior to explosive vaporization of the water, which allows thermodynamic prediction of the resulting mechanical energy of the interaction. Figure 11.13(a) shows some predictions of T_e as a function of R. However, this thermal equilibrium is just an idealization and many factors can cause initial interaction temperatures to be higher or lower, one of which is the composition of the water.

Water at the surface of the Earth may contain dissolved or suspended constituents (e.g., muddy water), or it may be contained within pores of rock or sediments. The effect on the thermal equilibrium can be evaluated as follows. For interactions between magma and saturated rocks or impure water (White, 1996), rock and impurities act as heat sinks, and the mass ratio, R_r, is given as

$$R_r = \frac{m_r}{m_m}, \qquad (11.7)$$

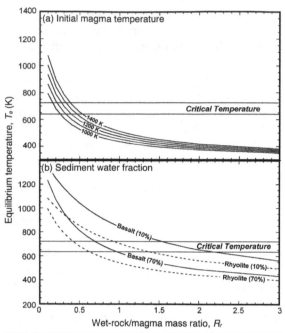

Figure 11.13 Thermal equilibrium (T_e) is the idealized temperature that water (in saturated rock) can reach during interaction with magma, and it is shown as a function of R_r (mass ratio of wet-rock/magma). These plots show cases where the initial water temperature is 298 K and the magma heat capacity c_m is ~1 kJ kg^{-1}K^{-1}. The range in critical temperature reflects the effect of dissolved solids. (a) T_e is shown for different initial temperatures of magma interacting with water (100% volume fraction, where $R_r = R$). (b) For basalt (solid curve; 1473 K) and rhyolite (dashed curve; 1173 K) interacting with wet rocks, T_e is shown as a function of rock water volume fraction, for which 10 and 70% bound a range centered on 40% ($x_w \approx 0.2$; Eq. (11.9)), representing a porous, fully saturated sandstone.

where m_r is the mass of rock and pore water (or water plus impurities). In these situations, the water/magma mass ratio is

$$R = x_w R_r = x_w \frac{\rho_r V_r}{\rho_m V_m},$$ (11.8)

where x_w is the water mass fraction in a rock (or impure water) volume V_r having bulk density ρ_r, interacting with a magma volume V_m having density ρ_m. For magma interacting with pure water alone, $x_w = 1$, $\rho_r = \rho_w$, and $V_r = V_w$. In order to cast this relationship into terms of rock porosity and saturation, which are commonly measured quantities, we write the following equation:

$$x_w = \frac{S_w \phi \rho_w}{\rho_r},$$ (11.9)

where S_w is the rock saturation (the volume fraction of pores filled by water), and ϕ is the rock porosity. For impure water with suspended rock particles one can consider $S_w = 1$ and ϕ somewhat less than unity (depending on the volume fractions of water and particles). From Eqs. (11.8) and (11.9), R may be approximated as:

$$R \approx S_w \phi \left(\frac{\rho_w}{\rho_m} \right) \left(\frac{V_r}{V_m} \right),$$ (11.10)

such that R can be constrained by measuring the volume ratio of lithic constituents and magma in samples of tephra or consolidated sediments such as peperite (a sedimentary rock containing igneous fragments formed during contact of magma and wet sediments).

Besides the effect on the value of R, one must also consider the heat capacity of water impurities or the porous rock containing the water. In Eq. (11.6), the heat capacity ratio ξ is replaced by the ratio involving wet rock instead of pure water, c_r/c_m. Here c_r is the effective heat capacity of the impure water or water–rock mixture, which can be approximated as a function of x_w by:

$$c_r = x_w c_w + (1 - x_w) c_s,$$ (11.11)

where c_s denotes the heat capacity of the solid constituents, and $c_s \approx c_w / 4$.

For a typical magma in contact with water-saturated rock at 298 K, T_e decreases with R_r, as shown in Figure 11.13(b). It is evident that, for magma interacting with wet rocks, T_e can exceed critical temperature (647 K (pure) to 720 K (5 wt.% dissolved solids)) where $R_r < 1.0$ (basalt) and $R_r < 0.5$ (rhyolite). Where critical temperature is exceeded during interaction prior to explosive expansion, supercritical pressures will be created.

Thermal equilibrium is probably never reached during the time span of interaction because of the insulating property of the vapor film that forms at the magma–wet-rock interface. Because of its relatively low thermal conductivity, a vapor film can greatly decrease the rate of heat transfer from the magma to the wet

Figure 2.3 Three scales of melt–crust interaction. (a) Dike network of 1.43 Ga Vernal Mesa pegmatite cross-cutting 1.73 Ga gneisses on the 670 m high Painted Wall, Black Canyon of the Gunnison, Colorado. Dikes were emplaced at about 10 km depth in tensional opening during dextral/oblique shear on the Black Canyon shear zone (Jessup et al., 2006). (b) Orthogonal dike network of ~1.69 Ga granite and pegmatite cross-cutting 1.73 Ga granodiorite within Tuna Canyon, tributary to Upper Granite Gorge of Grand Canyon. Dike networks in Grand Canyon reflect late-syntectonic granite dike swarms emplaced at about 20 km depth (Dumond et al., 2007) during late stages of accretionary assembly of southwestern North American lithosphere (note geologist for scale) (Ilg et al., 1996); photograph courtesy of Laurie Crossey. (c) Four generations of ~1.69 Ga granite dikes in Spaghetti Canyon of Lower Granite Gorge of Grand Canyon, note pen for scale. Earlier dikes underwent ductile deformation before later dikes were emplaced, suggesting an interplay of diking and ductile flow at middle crustal depths (Karlstrom and Williams, 2006); photograph courtesy of Laurie Crossey.

Figure 3.1 (a) Dike exposed by erosion on Piton des Neiges, Reunion Island (photograph courtesy of N. Villeneuve). The quasi two-dimensional geometry of dikes commonly adopted in theoretical models can be appreciated in this view. Pervasive horizontal jointing was caused by lateral cooling. (b) Dike emplaced just at the surface during the 2002 Nyiragongo eruption (photograph courtesy of J.-C. Komorowski). A solidified crust can be seen but the magma that occupied the interior of the dike was drained during the eruption.

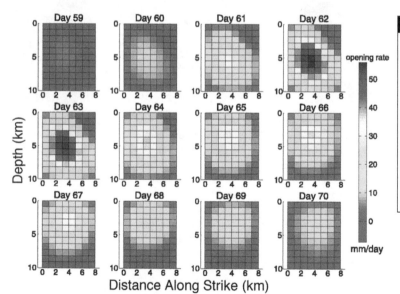

Figure 3.4 Temporal and spatial distribution of dike opening from time-dependent inversion of the 1997 seismic swarm at Izu Peninsula, Japan. Each panel represents the dike plane, viewed from the southwest, with depth in kilometers along the vertical axis, and along-strike distance in kilometers along the horizontal axis. The color represents the rate of dike opening, measured in millimeters per day. Each panel represents one day. From Aoki et al. (1999).

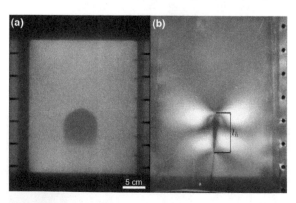

Figure 3.9 Photographs of an experiment in which a dike is supplied by a constant flux of liquid. The head and tail structure of the dike can be seen. (a) In transmitted light the dike is viewed normal to the propagation plane; the intensity of the color represents the thickness of the dike head and shows how it tapers to a thin tail. (b) In polarized light the photo-elastic properties of the gelatin reveal the stress field around the dike via birefringence, including the singularity centered on the tip, and the absence of fringes around the tail. The fringes appear to show that the elastic field does not interact with the vertical boundaries of the gelatin. From Taisne and Tait (2011).

(a)

t (min)   11   27   42   56   71

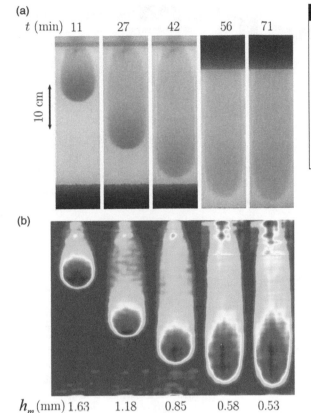

10 cm

(b)

h_m (mm) 1.63   1.18   0.85   0.58   0.53

Figure 3.11 (a) Time-series of photographs normal to the fissure plane for an experiment in which a fixed volume of fluid was injected at the top of the tank. As the dike propagates downwards, the thickness of the dike slowly decreases; the head starts off with a strong color (i.e., is relatively thick) but this diminishes as fluid is lost to the trailing tail. (b) Dike thickness as a function of position, derived from the fissure color intensity. Here h_m is the maximum thickness of the fissure, corresponding to the dark red coloration. From Taisne et al. (2011b), reproduced with permission from Springer.

(a)

10 cm

(b)

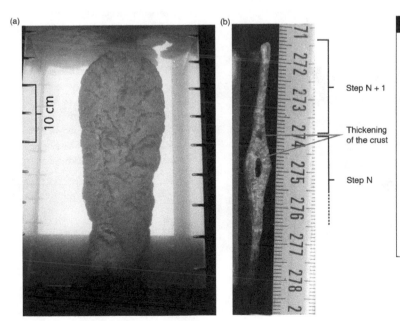

Step N + 1

Thickening of the crust

Step N

Figure 3.14 Photographs from a solidification experiment. (a) Frozen wax remaining in the tank just after the dike had reached the surface of the gelatin. (b) Cross section through the frozen wax after removal from the tank in an experiment with strong intermittent behavior. The sample brackets two large steps of progress, and shows a strong variation in the crust thickness: the thick crust that formed at the end of step N was breached and more wax flowed ahead to form the thinner crust in step N+1, which was last step of the experiment as the dike reached the top of the tank.

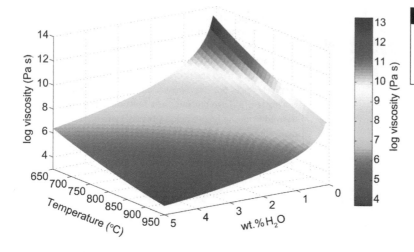

Figure 4.5 Dependence of rhyolitic melt viscosity on temperature and water content, after the empirical formulation of Hess and Dingwell (1996).

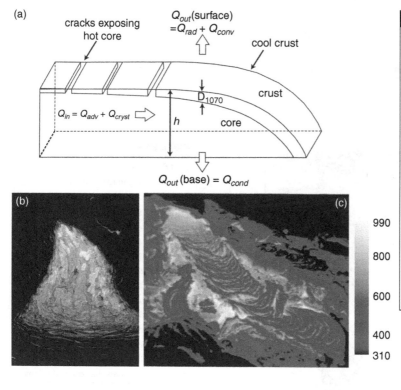

(a)

cracks exposing hot core

Q_{out}(surface) $=Q_{rad} + Q_{conv}$

cool crust

crust

D_{1070}

$Q_{in} = Q_{adv} + Q_{cryst}$

h

core

Q_{out} (base) $= Q_{cond}$

(b)

(c)

Figure 5.4 (a) Two-component model for the thermal structure of a lava flow surface, showing a chilled crust broken by cracks exposing the hot core, and the main heat sources and sinks (modified from Fig. 3 of Crisp and Baloga (1990)). (b) Photograph of a pahoehoe breakout (~1 m wide) showing at least three thermal components apparent from the different colors; yellow (~1090 °C), bright orange (~900 °C), dull orange (550–700 °C) and black (< 475 °C). (c) Thermal image of pahoehoe channel (central ropey slab is ~2 m wide and 5–6 m long) showing active surfaces ranging in temperature from ~400–1000 °C, with lower temperatures marking older, cooler crust developing on the flow surface as it advances and ages.

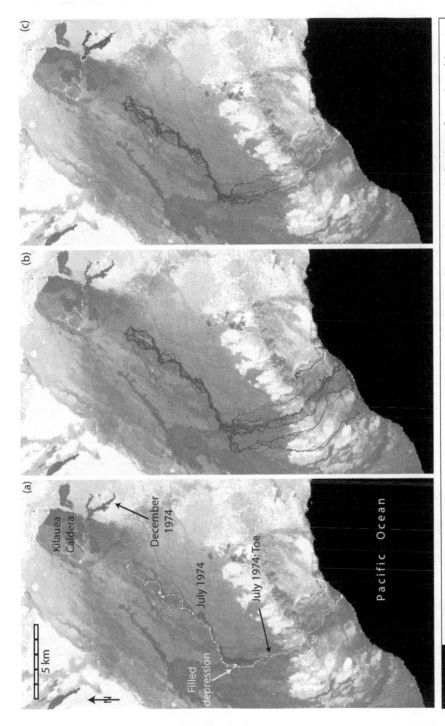

Figure 5.6 (a) Maximum gradient path (yellow line) from Kilauea's July 1974 vent location to the DEM edge (coast). (b) Stochastic flow path model from Kilauea's July 1974 vent location to the coast. Model applies DEM noise of 2 m and has been run for 20 iterations. (c) Twenty-iteration flow line runs with 2 m of DEM noise for Kilauea's July and December 1974 flows. Flow lines have been cut at the flow toe in each case to allow an assessment of the fit with reality, as revealed by the black flow areas in this Thematic Mapper (TM) color composite. Simulations have been projected onto a near true color (vegetation = bright green, lava = black) TM image of Kilauea.

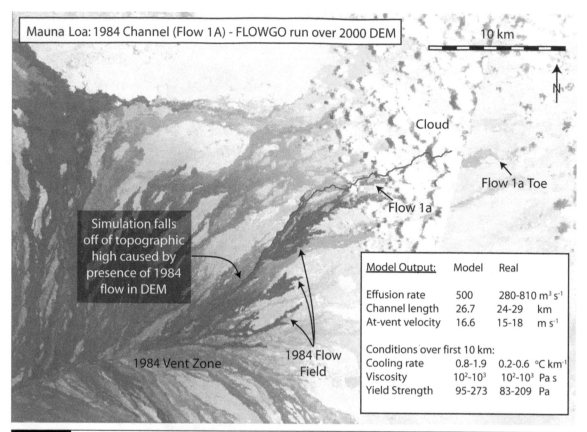

Figure 5.7 FLOWGO simulation of Mauna Loa's 1984 flow 1A channel. Model run uses the input conditions for Mauna Loa's 1984 flow given in Table 5 of Harris and Rowland (2001). In executing the model the software first projects a maximum gradient path (in yellow) from the selected vent point to the edge of the DEM. The cooling-limited distance that the channel-contained control volume can travel down that flow path is then projected in red. Comparisons of model output and field measured parameters are given in the inset box.

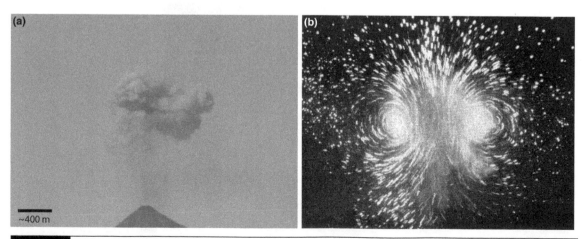

Figure 7.8 (a) Recent eruption of Fuego volcano (Guatemala) showing release of a buoyant eruption cloud (thermal). (b) Streak lines illustrating velocity vectors for a laboratory experiment of an isolated thermal, showing upflow in center of thermal and downflow at the edges (modified from Turner, 1973).

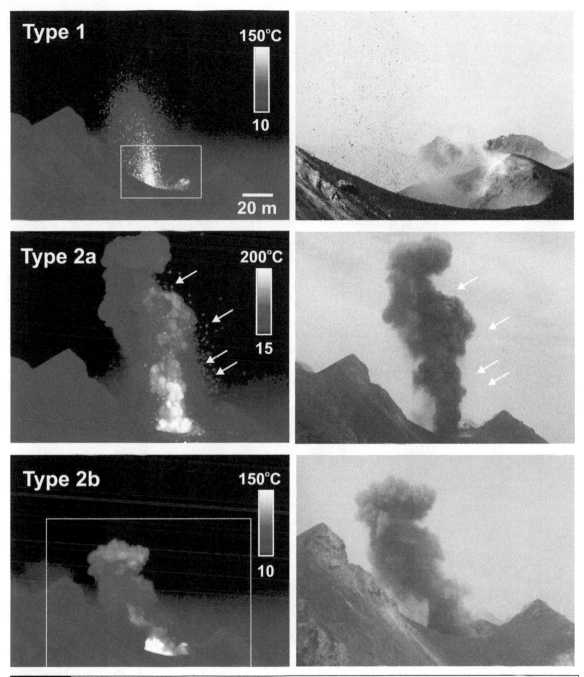

Figure 6.8 Eruption styles at Stromboli as defined by thermal and visible images (the left and right columns, respectively). Type 1 explosions are lapilli rich with little or no visible ash. Type 2 explosions have a greatly increased ash component, and can be subdivided into 2a and 2b, representing events with and without a visible coarse (type 1 style) component (arrowed in the middle panels). The white boxes in the thermal images indicate the field of view in the associated visible image. Reproduced from Patrick *et al.* (2007, Fig. 3) with kind permission of Springer Science and Business Media.

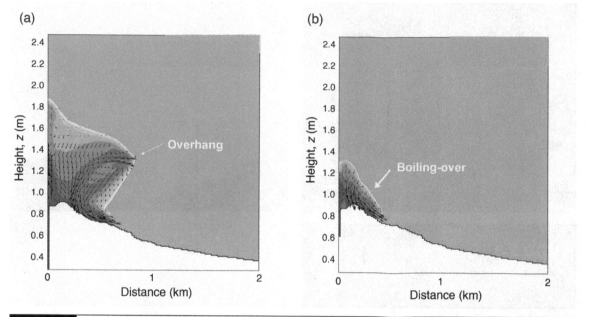

Figure 7.11 Pyroclastic density currents are generated when the pyroclastic mixture never becomes buoyant and collapses due to gravity. The collapse may initiate (a) hundreds of meters above the vent or (b) a few meters above the vent (boil-over), depending on the initial momentum (controlled by initial pressure and volatile content) and entrainment characteristics.

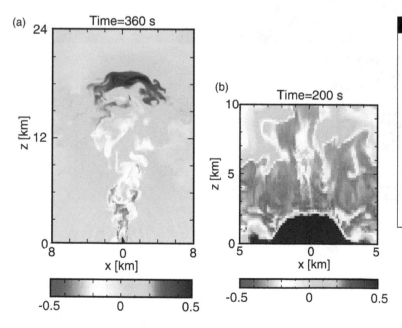

Figure 8.12 Results from the numerical simulations of Suzuki et al. (2005) illustrating the dynamics of (a) a buoyant eruption column (magma temperature 1000 K, mass flux = $10^{7.6}$ kg s^{-1}, and eruption velocity 150 m s^{-1}); and (b) a collapsing fountain (magma temperature 1000 K, mass flux = $10^{8.65}$ kg s^{-1}, and eruption velocity 150 m s^{-1}). The colors represent the density of the column relative to the ambient atmosphere, with blue being relatively buoyant and orange being relatively dense.

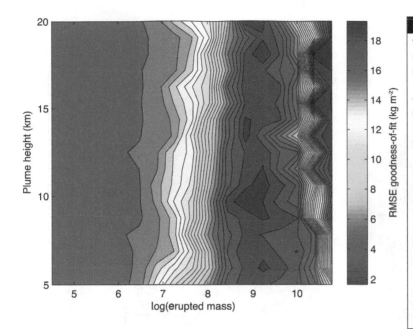

Figure 9.8 Plot of log(erupted mass) vs. plume height (km) showing the minimum values of the goodness-of-fit measure (RMSE; Eq. (9.16)) for the Etna 1998 tephra deposit. Erupted mass was varied between 10^4 and 10^{11} kg with 0.2-log(mass) increments and the plume height was varied between 4 and 20 km with 2-km-height increments (Run 2, Table 9.2). Resulting values were interpolated to produce 2D RMSE surfaces and the plot was cropped to show only the interpolated surfaces. Dark blue on the RMSE scale indicates the minimum values (i.e., best fit). RMSE values vary between 1 and 100 kg m^{-2} and are contoured with 0.8 kg m^{-2} contours (RMSE values > 20 kg m^{-2} were observed only for erupted masses > 3 × 10^{10} kg and were set equal to 20 kg m^{-2} for a better visualization of the minimum).

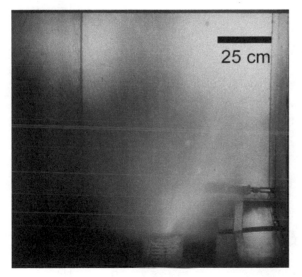

Figure 11.11 Explosive interaction after injection of water into molten basaltic rock (Physikalisch-Vulkanologisches Labor, Universität Würzburg).

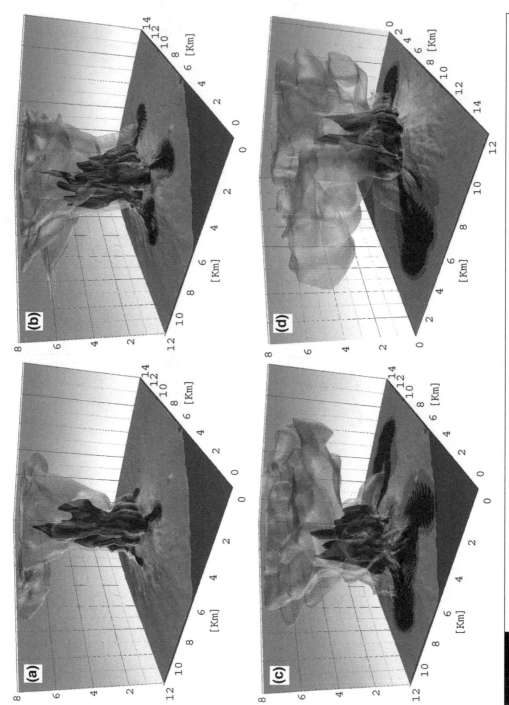

Figure 10.10 3D multiphase simulation of PDCs as a function of time (a) to (d). After Neri *et al.* (2007). Note that PDCs are strongly influenced by topography.

(a)

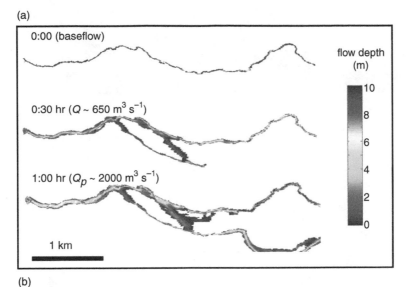

0:00 (baseflow)

0:30 hr ($Q \sim 650$ m³ s⁻¹)

1:00 hr ($Q_p \sim 2000$ m³ s⁻¹)

flow depth
(m)

10

8

6

4

2

0

1 km

(b)

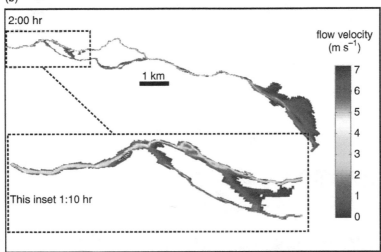

2:00 hr

1 km

This inset 1:10 hr

flow velocity
(m s⁻¹)

7

6

5

4

3

2

1

0

Figure 14.13 Sample outputs from Delft3D™ simulations of the March 2007 Crater Lake break-out lahar at Mt. Ruapehu, New Zealand. (a) Snapshots of flow depth at different time intervals showing downstream propagation of lahar and flooding of channel bifurcations. (b) Depth-averaged flow velocity illustrating higher flow speeds in deeper parts of the channel. Courtesy of J. Carrivick, unpublished data.

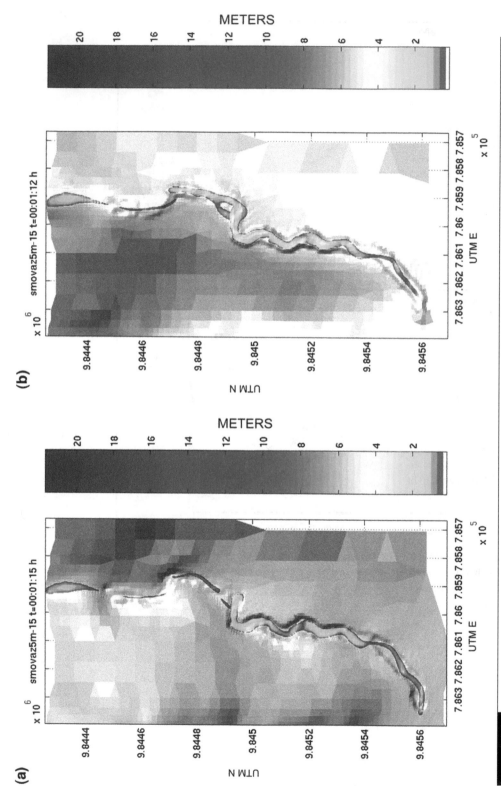

Figure 14.14 Simulations of the 2005 Vazcún Valley lahar, Ecuador, generated using the two-phase version of TITAN2D. Flow depth distributions at final time-steps for simulations of a 70 000 m³ volume for (a) 40 vol.% solids, and (b) 60 vol.% solids. Reprinted from Williams et al. (2008) with permission from Elsevier.

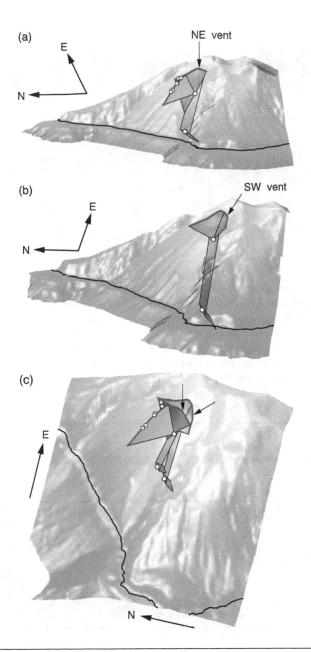

Figure 15.10 (a) Geometry of the upper 1 km of conduit underlying the northern vent area of Stromboli. A semi-transparent perspective cut-away view of the northwest quadrant of the edifice including the sector graben of the Sciara Del Fuoco, provides the reference for the location and geometry of the conduit, which is derived from the seismic source mechanisms obtained from inversions of VLP signals associated with explosions. Surface illumination from an external light source provides color contrasts, emphasizing topographical features. A black line indicates sea level. The summit of the volcano is 924 m above sea level (no vertical exaggeration). The two flow disruption sites that are sources of VLP elastic radiation are indicated by small white circles. The irregular red and blue lines, respectively, mark the surface traces of the dominant and subdominant dike segments constituting the shallowest portions of the conduit system. The eruptive vent is marked by a vertical arrow, and temporary vents active during the flank eruption in 2002–2003 are indicated by green squares. The lateral extents of individual dike segments are unknown and are shown for illustrative purpose only. (b) Same as (a) for conduit underlying the southern vent area. A slanted arrow points to the southern vent. (c) Plunging view of the Sciara Del Fuoco showing the two dike systems underlying the summit crater. The conduit structure underlying the northern vent (marked by a vertical arrow) is shaded green and that underlying the southern vent (marked by a slanted arrow) is colored blue (see (a) for details).

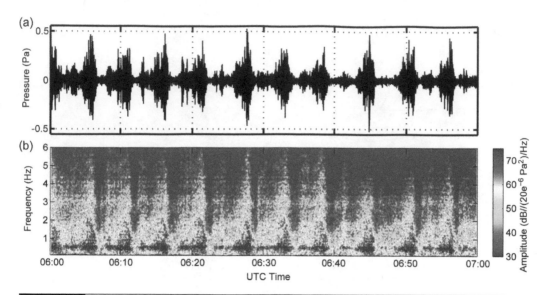

Figure 16.2 Episodic tremor (a) pressure signal and (b) PSD as a function of time (spectrogram) recorded ~7 km from the Halemaumau Vent (HV). The episodic tremor is characterized by cyclical temporal variations caused by filling and draining of the cavity at Halemaumau Vent, Kilauea Volcano. The variations are coincident with a capped surface, or "gas piston," rising and releasing an accumulated amount of gas, as well as with the rising and lowering of an exposed, degassing lava surface. The degassing portions of this episodic tremor are consistent with elevated infrasound, while the "capped" or quiescent lava surface correlates with little to no infrasound. The spectral peak around 0.5 Hz is related to the cavity being excited into Helmholtz resonance by the release of accumulated gas.

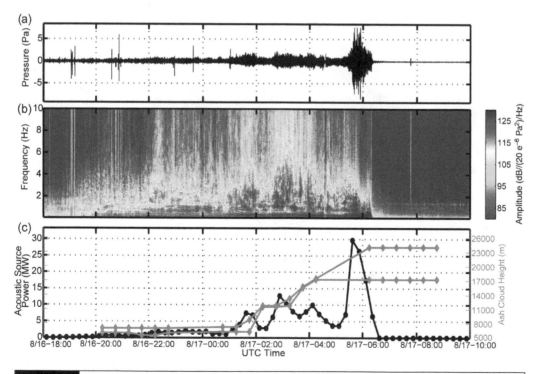

Figure 16.5 (a) Waveforms, (b) spectrogram, and (c) radiated acoustic power (in black) and ash cloud top (in green) for the suplinian to plinian Tungurahua eruption of 17 August 2006. The eruption was characterized by continuous jetting for >10 hours, with some explosions as well. Acoustic power broadly scales with ash cloud height for this eruption, particularly during the plinian phase of the eruption around 17 August 0600 UTC. From Fee *et al.* (2010b).

(a)

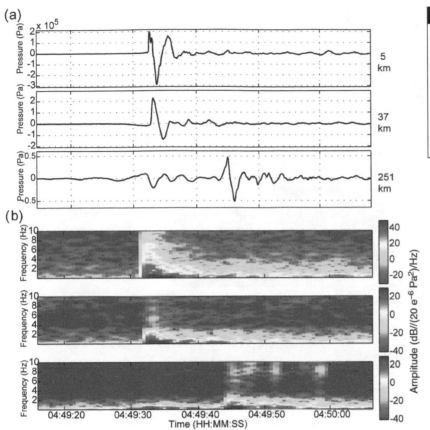

Figure 16.4 Time-corrected Tungurahua (a) explosion waveforms and (b) spectra recorded at distances of 5 km, 37 km, and 251 km. As the distance is increased, the signal loses high-frequency energy and its duration is extended. Nearest station data courtesy of H. Kumagai.

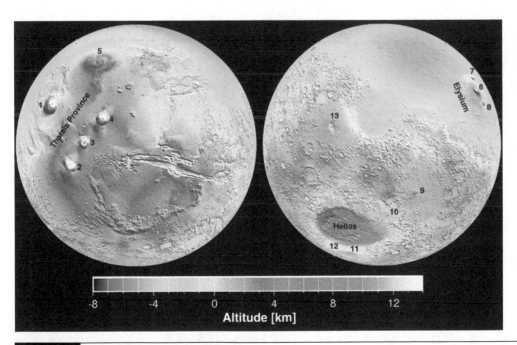

Figure 17.2 False color view of Mars topography with locations of key volcanic features labeled: 1 Olympus Mons; 2 Arsia Mons; 3 Pavonis Mons; 4 Ascraeus Mons; 5 Alba Mons; 6 Elysium Mons; 7 Hecatus Tholus; 8 Albor Tholus; 9 Tyrrhenus Mons; 10 Hadriacus Mons; 11 Amphitrites Patera; 12 Peneus Patera; 13 Nili Patera. Image: NASA/Goddard Space Flight Center.

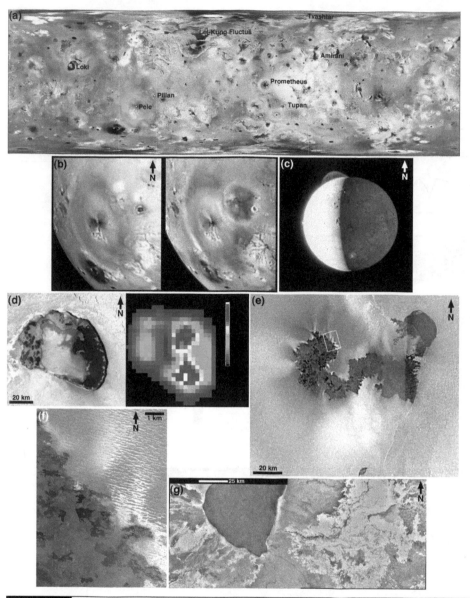

Figure 17.6 Volcanic features on Io. (a) Map of the Ionian surface showing key volcanic features. Image NASA/JPL/USGS; merged Voyager and Galileo mosaic available at http://astrogeology.usgs.gov/Projects/JupiterSatellites/io.html. (b) These two Galileo images show the appearance of a new, dark deposit 400 km in diameter that erupted from Pillan Patera in 1997, and partly covered the large red deposit from Pele Patera. Image PIA00744, NASA/JPL/University of Arizona. (c) The New Horizons Long Range Reconnaissance Imager captured this view of three of Io's plumes on 28 February 2007. In the far north is the 390-km-high Pele-type plume from Tvashtar volcano. In addition, a 60-km-high plume from Prometheus can be seen on the western limb, and the top of Masubi's plume can be seen poking into the sunlight from Io's night side, to the south of the image. Image PIA09248, NASA/Johns Hopkins Applied Physics Laboratory/Southwest Research Institute. (d) The floor of Tupan Patera (75 km across) is partly covered with dark material, presumably lava, that appears significantly warmer in Near Infrared Mapping Spectrometer data than the adjacent orange-colored "island," parts of which are cool enough for SO_2 to condense. Image PIA03601, NASA/JPL. (e) The source of the Prometheus plume lies at the distal (westernmost) end of a compound pahoehoe flow field. Image PIA02565, NASA/JPL/University of Arizona. Inset box shows the location of (f), which is a high-resolution view of the active flow front. Dark areas on the flow show recent lava breakouts, older flow surfaces brighten with time due to deposition of SO_2 from the plume. Fresh SO_2 frost is also seen on the surrounding terrain. Image PIA02557, NASA/JPL/University of Arizona. (g) Bright lobate features to the east of Emakong Patera have been interpreted as fresh sulfur flows. Image PIA02539, NASA/JPL/University of Arizona.

rock, allowing gradual quenching. With gradual quenching, the film slowly heats water in the rocks near the magma at a rate balanced by the heat transfer away from it by the convective movement of pore water, provided that pores are interconnected. However, such passive quenching is not always the case. Consider the hypothetical instantaneous interface temperature, T_I, attained by the initial contact of magma with water that can be estimated (Cronenburg, 1980) by:

$$T_I = \frac{T_m\left(k_m/\sqrt{\kappa_m}\right) + T_w\left(k_w/\sqrt{\kappa_w}\right)}{\left(k_m/\sqrt{\kappa_m}\right) + \left(k_w/\sqrt{\kappa_w}\right)}, \qquad (11.12)$$

where k and κ are thermal conductivity and diffusivity, respectively, and subscripts m and w refer to magma and water, respectively. For the contact of a typical basalt magma with pure water, T_I approaches 1000 K. Because of the rapidity of heat exchange, water may exist in the metastable state of superheating in which it is a liquid well above its vaporization temperature. A consequence of this superheated state is that it continues to absorb heat at a high rate, reaching temperatures well in excess of its spontaneous nucleation temperature (i.e., HNT ~583 K). As temperatures approach the critical temperature (647 K), instantaneous vaporization by homogeneous nucleation produces a vapor film that expands rapidly and is highly unstable – that is, it can expand well beyond the thermodynamic equilibrium thickness. In so doing, the vapor becomes supercooled, leading to spontaneous condensation. The condensation then leads to a rapid collapse such that liquid water impacts the magma surface with a finite amount of kinetic energy, leading to a second spontaneous vaporization event. This cyclic vapor film growth and collapse is repeated continuously, typically with a frequency of up to 1 kHz (analogous to the *Leidenfrost* phenomenon of a drop of water vibrating on a hot metal surface). Vapor film instability can generate enough kinetic energy to distort the interface between the magma and wet rock, as well as cause failure of the host rock. In some cases, film collapse can lead to jets of water-saturated rock fragments penetrating the magma surface (White, 1996).

In addition, the rapid heat loss from the magma by this continued vapor film instability leads to magma quenching and possible granulation. With these interface phenomena, the magma becomes increasingly fragmented, leading to larger surface areas for heat transfer and larger volumes of superheated water and vapor.

Further thermodynamic predictions about magma–water interactions can be derived from analysis of the thermodynamic work done by the expansion of water from its initial thermal equilibrium. We have assumed that nothing is known about the mechanism of the contact between magma and water, but that it results in production of high-temperature and high-pressure water that may explosively decompress. Thermodynamic work is manifested in the fracture and excavation of country rock to form a crater, fragmentation of the magma into fine-grained debris, and ejection of these fragments in an expanding jet of steam. For a hydrovolcanic eruption, it is necessary to find the work potential for expansion of the steam to atmospheric pressure. Calculation of the true potential requires determination of a complex set of boundary conditions unique to each eruption, but for simplicity and generality, one can make some standard assumptions that allow analytical calculation of maximum potential. For the system consisting of a mixture of magmatic particles and water and the surroundings being the volcano vent structure and the atmosphere, we assume that:

(1) All heat lost by the magma during the eruption is transferred to external water;
(2) Liquid water and magma are incompressible, so that for each the specific heats remain constant with changing pressure and volume;
(3) The specific volume (reciprocal of density) of liquid water is small compared to that of its vapor;
(4) Water vapor behaves as a perfect gas and the magma volume does not change during the eruption (i.e., it does not vesiculate or contract during cooling).

Some heat is lost from the magma to both the country rock and the atmosphere during eruption, and magma does exhibit a finite volume change on quenching. The combined effect of these caveats may account for energy discrepancies of several percent. The other assumptions concern the exact thermodynamic properties of water. At supercritical states a deviation from ideality of $\geq 10\%$ is expected, but since most volume change occurs at subcritical states for adiabatic expansion, the assumptions introduce very little error into calculations, especially when extended or extrapolated steam table data are considered, such as by using a modified Redlich–Kwong equation of state (Burnham *et al.*, 1969; Holloway, 1977; Kieffer and Delany, 1979).

Online Supplement 11A (see end of chapter) provides a derivation of the thermodynamic work of the interaction. From this analysis, the following predictions can be made for magma interacting with water or water-saturated rock. The conversion ratio of a magma–water interaction is the fraction of magmatic heat converted to thermodynamic work, and is a measure of the mechanical energy released by the interaction. Figure 11.14(a) shows conversion ratios as a function of the wet-rock/magma mass ratio R_r for the interaction of a basaltic magma with saturated rocks (with 40% porosity and $x_w = 0.2$). Also shown are curves depicting pure water–magma interaction (i.e., $x_w = 1$). Both *isentropic* expansion (in which steam separates from fragmented magma) and *pseudo-isothermal* expansion (in which steam and fragments remain at the same temperature, denoted in figures as *isothermal*) are calculated. Note that, compared to pure water–magma interactions, wet-rock–magma interactions have lower conversion ratios (are less energetic) with optimum conversion ratios near $R_r = 1.0$. Figure 11.14(b) shows the fraction of vaporized water that condenses back to liquid during expansion. For interactions at low R_r values, most of the water remains in the vapor state after expansion, leading to the likelihood of explosive behavior. In contrast, for $R_r > 3.0$ (wet rock), a dominant portion of the steam condenses during expansion, producing a liquid–particle system. This latter behavior allows convective heat transfer that promotes passive cooling, which is less likely to be explosive.

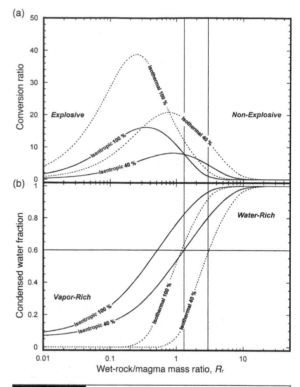

Figure 11.14 (a) Conversion ratio and (b) condensed water fraction vs. wet-rock/magma mass ratio, R_r. Solid curves are isentropic values and dashed curves are pseudo-isothermal values (labeled *isothermal*). Conversion ratio is the percentage of the magma's heat energy that is converted to thermodynamic work during interaction with water. This plot shows results for a basaltic magma at 1473 K interacting with water at 298 K (solid curves, 100% water by volume; $x_w = 1$) compared with those for water-saturated sediments (dashed curves; 40% porosity, 100% saturated; $x_w \approx 0.2$; Eq. (11.9)). The condensed water fraction represents the fraction of interacting water that condenses to a liquid state after expansion. For water-saturated sediment interactions having $R_r > 1.3$ (isentropic) to 3 (pseudo-isothermal), a dominant fraction (> 0.6) of vaporized water will condense to liquid during expansion to ambient pressures, and wet sediments have the ability to convectively carry heat from the magma, behaving as fluid substances rather than explosive vapor-rich ones. Using this criterion, an arbitrary region, separating explosive from non-explosive behavior, may be drawn over the range $1.3 \leq R_r \leq 3.0$.

Wet-rock–magma interaction may also occur at depths below the Earth's surface in hydrothermal systems of elevated temperature and pressure. Figure 11.15(a) shows that when pore water is at elevated temperatures (e.g., 358 K), a

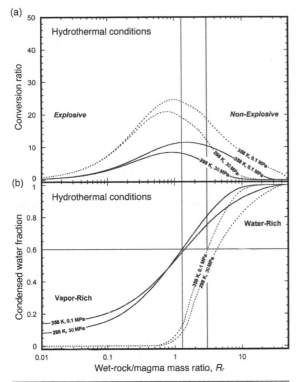

Figure 11.15 (a) Conversion ratio and (b) condensed water fraction vs. wet-rock/magma mass ratio R_r for hydrothermal conditions of elevated pore-water temperature (358 K) and elevated hydrostatic pressure (30 MPa). These curves represent a saturated sediment containing 40% by volume water ($x_w \approx 0.2$); solid curves are for isentropic expansion and dashed curves are for isothermal expansion.

slightly greater fraction of the magma's thermal energy is converted to thermodynamic work and optimum peaks occur at slightly higher values of R_r when compared to values shown in Figure 11.14(a). However, elevated hydrostatic pressure (e.g., 30 MPa) does not have much effect. Still, conversion ratios at hydrostatic pressures exceeding critical pressure are sufficiently large that such interactions could be explosive.

To further illustrate these calculations, Figure 11.16 depicts water phase diagrams with pressure-temperature-volume-entropy (p-T-V-S) relationships for theoretical initial equilibrium and final states. Both isentropic (Fig. 11.16(a)) and pseudo-isothermal (Figs. 11.16(b,c)) paths are shown. Isentropic expansion follows paths of constant entropy (Fig. 11.16(a)), whereas the

pseudo-isothermal expansion of the vapor–fragment mixture follows paths intermediate to those of isentropic and pure isothermal expansions (Kieffer and Delany, 1979). The phase diagrams also illustrate the three regimes of heat flux described in Section 11.4.1. The non-explosive interaction regime generally corresponds to interaction ratios $R_r > 10.0$, for which isentropic expansion is simply a vertical path in Figure 11.16(b) and water expansion is always subcritical. The subsonic explosive interaction regime approximately corresponds to interactions where $1.0 < R_r < 10.0$. For adiabatic and pseudo-isothermal expansions in this regime, water is converted to steam at high pressure but then some of it condenses during expansion. The supersonic explosive interaction regime ($R_r < 1.0$) involves significant expansion in supercritical states, especially for pseudo-isothermal conditions. The supercritical state is considered further in Section 11.4.3. It is important to note that the correspondence of these heat-flux regimes to the range of water/magma mass ratios and modeled thermodynamic states is very approximate because other factors contribute to determining the actual heat-flux regime, namely the rheology of the magma, as well as conditions of contact and subsequent mixing and fragmentation.

Figure 11.16 also illustrates an important point that experimental studies confirm: the critical point of water is not necessarily a limiting factor in vapor explosions. It has been commonly assumed that interactions occurring at confining pressures above the critical point of water (22 MPa) cannot result in explosions, because water exists as a supercritical fluid for which there is no liquid–vapor phase boundary. Considering the thermodynamic paths illustrated in Figure 11.16(c), where interactions are nearly an order of magnitude in excess of critical pressure, release of the interaction pressure involves a large volume increase, especially for pseudo-isothermal expansion, at rates determined by local sound speeds (Kieffer and Delany, 1979). Although the interactions extend to pressures exceeding critical, it is not yet clear if explosive expansion can initiate at these high pressures (equivalent to

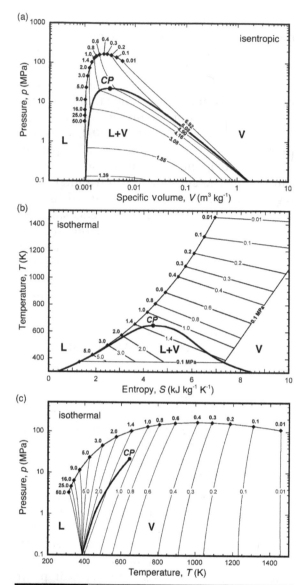

Figure 11.16 Phase diagrams illustrating calculations of wet-rock–magma interaction, using the method of Wohletz (1986). Labeled points (diamonds) are the theoretical initial equilibrium condition for interactions of various R_r (from 0.01 to 50.0; $x_w \approx 0.2$). The critical point (CP), liquid (L), vapor (V), and liquid plus vapor two-phase (L+V) regions are shown. (a) A p–V diagram shows expansion volumes and release isentropes (kJ kg^{-1} K^{-1}) followed during isentropic expansion of vapor. For all interactions, water expands into the two-phase region. (b) A T–S diagram for pseudo-isothermal expansion shows the increase in entropy as water stays in thermal equilibrium with magma fragments. For R_r < 1.3 water expands into the vapor field. (c) A p–T diagram shows the variation in water temperature during pseudo-isothermal expansion. For more detail on these diagrams see Kieffer and Delany (1979).

~2.2 km ocean depth). However, the presence of other pressure perturbations, such as those caused by seismicity, host media failure, and the vapor-film dynamics, add to the likelihood that expansion will lead to thermohydraulic explosion.

11.4.3 A non-equilibrium thermodynamic model

In the supersonic explosive interaction regime (regime 3; Section 11.4.1), water reacts as a supercritical fluid that undergoes complex phase transitions that are difficult to model. Consequently, the steady-state models described in Section 11.4.2 do not fully capture some important physics. Experimental observations reveal that magma quenches with little or no phase change and can therefore be described as a supercooled liquid (i.e., glass; Section 11.3.2), the physical properties of which are well known and/or can easily be measured. It is therefore much simpler to model the non-equilibrium behavior of the magma rather than the complex phase reactions of the water.

The basis of the model is the heat conduction equation of standard, non-equilibrium thermodynamics applied to the individual magma fragments:

$$\frac{\partial T_{(x,y,z,t)}}{\partial t} = k_m \nabla^2 T_{(x,y,z,t)}, \qquad (11.13)$$

where $T_{(x,y,z,t)}$ is the time-dependent temperature-field inside the fragment and k_m is the thermal conductivity (approximately constant for supercooled liquids). The solution of this partial differential equation can be found in Büttner et al. (2005). Inserting the result into the heat flux equation yields:

$$\frac{\Delta Q_{(t)}}{\Delta t} = m c_m \frac{\Delta T_{(x,y,z,t)}}{\Delta t}, \qquad (11.14)$$

where $\Delta Q_{(t)}/\Delta t$ represents the time-dependent heat flux, m is the mass of the fragment, c_m is the specific heat capacity, and $\Delta T_{(x,y,z,t)}/\Delta t$ is the variation of the temperature field in the fragment with time. Analysis of experimentally produced particles yields the size, shape and number of particles produced for a given starting mass. Summing the heat transferred from all particles

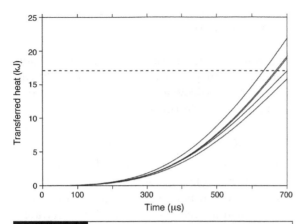

Figure 11.17 Results of heat flow modeling of a MFCI experiment, using the non-equilibrium thermodynamic approach. The different curves represent end-members of fragment morphology (e.g., cubes, spheres), which show the limited influence of shape on the results. The dashed line represents the total measured heat transfer during the experiment, which ended in explosion after 700 μs. (After Büttner et al., 2005.)

allows calculation of the total energy release and therefore the explosivity of this regime. Figure 11.17 shows the results of modeling the heat transfer during an MFCI experiment. The model curves predict well the total heat flow measured during the experiment (dashed line), confirming that this technique is robust with respect to the magma properties and fragment size and morphology. It is possible to extend this treatment to natural eruptions to evaluate the total energy release during explosive events by using grain-size and shape analyses of natural tephra samples, together with appropriate magma properties and an assessment of the total mass of the deposit.

11.4.4 A brittle fragmentation model

The above non-equilibrium behavior is intimately tied to magma fragmentation by application of the Gibbs thermodynamic function (Yew and Taylor, 1994), which is expressed as:

$$G = U + B - TS, \qquad (11.15)$$

where G is the Gibbs free energy, U is the pre-fragmentation internal energy of the magma, B is

the surface energy consumed by fragmentation, and S is the entropy generated by creation of new surfaces during fragmentation. In order to apply this general function, one must assume:

(1) Magma breaks up due to internal stress (i.e., in a brittle regime);
(2) Thermodynamic equilibrium is not achieved during crack growth within the magma until its fragmentation into a number, n, of spherical particles;
(3) Isothermal conditions exist during fragmentation;
(4) Surface energy is not recoverable, and the entropy of the system increases; and
(5) During increments of time the stress is treated as constant.

Following the methods of Grady (1982, 1985), each term in Eq. (11.15) can be evaluated as follows. The value of U is taken as the strain energy in the magma just prior to failure, and it is a function of n, the number of particles formed by fragmentation:

$$U = \frac{\mu}{8}\left(\frac{\dot{\varepsilon}}{c}\right)^2 \left(\frac{6}{\pi}\right)^{2/3} n^{-2/3} V^{5/3}, \qquad (11.16)$$

where μ is the magma bulk modulus ($\mu = \rho_m c^2$; ρ_m is density and c is sound speed), $\dot{\varepsilon}$ is the strain rate, and V is the total volume. Yew and Taylor (1994) give a simple geometric argument to constrain B, the total energy dissipated by fragmentation:

$$B = \left(\frac{\pi}{6}\right)^{1/3} \left(\frac{3K_c^2}{\rho_m c^2}\right) n^{1/3} V^{2/3}, \qquad (11.17)$$

where K_c is the stress intensity factor. The entropy S of new surfaces can be evaluated by the configuration entropy of n cracks that could occupy $N + n$ sites in a volume V (Varotsos and Alexopoulos, 1986):

$$S = k_B \ln\left\{\frac{(N+n)!}{N!n!}\right\}, \qquad (11.18)$$

where k_B is the Boltzmann constant and N is the total number of normalized new surfaces.

In order to apply Eq. (11.15) to the state where fragmentation has occurred, thermodynamic equilibrium is assumed at a constant temperature and pressure where the Gibbs function is at a minimum:

$$\left(\frac{\partial G}{\partial n}\right)_{T,p} = 0. \tag{11.19}$$

By taking the derivative of Eq. (11.15), setting the average fragment size (s; Grady, 1982, 1985; Yew and Taylor, 1994) to:

$$s = \left(\frac{6V}{\pi n}\right)^{1/3} \approx 2\left[\frac{\sqrt{3}\,K_c}{\sqrt{2}\,\rho_m c\dot{\varepsilon}}\right]^{2/3}, \tag{11.20}$$

the strain rate associated with fragmentation that produces particulates of an average size s is expressed by:

$$\dot{\varepsilon} = \frac{K_c}{\rho c}\sqrt{\frac{12}{s^3}}. \tag{11.21}$$

The value of s can be constrained by inspection of tephra sample subpopulation modes, distinguished either by shapes and textures (e.g., Heiken and Wohletz, 1985, 1991) or by mode dispersion (e.g., Wohletz et al., 1995b). Using an estimated stress intensity value K_c, fragmentation strain rates can be determined from Eq. (11.21). The strain rates are then related to E_f, the fragmentation energy, by:

$$E_f \approx \frac{1}{120}\left(\frac{6}{\pi}\right)^{2/3}\rho_m\,\dot{\varepsilon}n^{-2/3}V^{5/3}. \tag{11.22}$$

This treatment can be applied to natural hydrovolcanic deposits if the particle density and whole deposit volume and grain-size distribution are known (see Büttner et al., 2006, for details). Summation of the specific fragmentation energy for each grain-size fraction and for the total deposit volume allows calculation of the total fragmentation energy for the eruption. Using an assessment of the partitioning of total eruption energy among fragmentation (50–75%), kinetic (15–30%), seismic (10–20%), and acoustic (< 5%) energies, Büttner et al. (2006) calculated total eruption energies on the order 10^{14} J (~ 25 Mt of high explosive equivalent) for two eruptions at Campi Flegrei, Italy.

11.5 | Summary

The following points summarize a basic understanding of magma–water interactions during volcanism (hydrovolcanism). The numerous references cited in this chapter will permit the reader to gain a deeper appreciation of the physics of magma–water interaction.

- Magma–water interaction is not a rare process, and is relevant for nearly every volcanic system. Earth is a water planet and most of its volcanic activity is found underwater, especially at the mid-oceanic ridges (see Chapter 12).
- A well-developed catalog exists of diagnostics that are useful to characterize magma–water interaction from analysis of hydrovolcanic products, including vent/construct morphology, deposit dispersal, tephra bedding sequences, and ash analyses. However, a quantitative analysis in terms of energy balances and hazard mitigation needs experimental data and theoretical concepts.
- Whereas understanding of water physics is highly evolved because of the rich tradition of power plant technology, magma physics are less well understood, and volcanologists are just beginning to develop the complex physics needed to handle this multiphase, multicomponent, subliquidus system. Further data are required for basic physical parameters such as magma heat conductivity and viscosity, without which the dynamic heat transfer at the magma–water interface cannot be reliably quantified.
- The key process for understanding the energy balance of hydrovolcanism is magma fragmentation. To make the thermal energy available for conversion to mechanical energy via the working fluid water, heat must be transferred. Because heat transfer is dominated by conduction, the energy transfer rate increases with the increased surface area produced by fragmentation. Energy release on a short timescale (explosion) during hydrovolcanic processes is likely a result of a positive feedback mechanism in which heat exchange drives

fragmentation, which in turn drives escalating heat transfer rates.

- Three heat-flux regimes can be defined to cover the range of magma–water interaction dynamics from non-explosive passive cooling to supersonic thermal detonation. Modeling techniques comprise steady-state thermodynamics for the non-explosive to explosive interaction regimes, based on thermal power plant physics. Modeling of the supersonic explosive regime, however, benefits from application of non-equilibrium thermodynamics, which develops a strong link between explosion energy and fragment size distributions.

11.6 Notation

B	surface energy consumed by fragmentation (J)
c	sound speed (m s^{-1})
c_m	specific heat capacity of magma (J kg^{-1} K^{-1})
c_p	specific heat capacity at constant pressure (J kg^{-1} K^{-1})
c_r	specific heat capacity of saturated rock (J kg^{-1} K^{-1})
c_s	specific heat capacity of solids dissolved in water (J kg^{-1} K^{-1})
c_w	specific heat capacity of pure water (J kg^{-1} K^{-1})
E	mixture internal energy (J)
E_f	fragmentation energy (J)
G	Gibbs free energy (J)
k_B	Boltzmann constant ($1.3806488 \times 10^{-23}$ J kg^{-1})
k_m	thermal conductivity of magma (W m^{-1} K^{-1})
k_w	thermal conductivity of water (W m^{-1} K^{-1})
K_c	stress intensity factor (Pa m$^{1/2}$)
m	mass of particle (kg)
m_m	mass of magma (kg)
m_r	mass of porous rock (or impure water) (kg)
m_w	mass of water (kg)
n	number of particles produced by fragmentation event
N	number of new surfaces produced by fragmentation event
p	pressure (Pa)
p_{cj}	pressure at Chapman–Jouguet point (Pa)
R	mass ratio of water to magma
R_r	mass ratio of saturated rock (or impure water) to magma
Q	heat (J)
s	average fragment size (m)
S	entropy generated by fragmentation (J)
S_w	rock saturation (pore volume fraction filled with water)
t	time (s)
T	temperature (K)
T_e	equilibrium temperature (K)
T_I	instantaneous interface temperature (K)
T_m	magma temperature (K)
T_w	water temperature (K)
u	mixture velocity (m s^{-1})
u_r	velocity of shocked material relative to shock wave (m s^{-1})
U	internal energy of pre-fragmentation magma (J)
V	mixture volume (m^3) or specific volume (m^3 kg^{-1})
V_{cj}	volume at Chapman–Jouguet point (m^3)
V_m	volume of magma (m^3)
V_r	volume of porous rock or impure water (m^3)
V_w	volume of water (m^3)
W	thermodynamic work (J)
x_w	mass fraction of pure water
α	isobaric expansion coefficient (K^{-1})
β	isothermal expansion coefficient (Pa^{-1})
$\dot{\gamma}$	shear rate (s^{-1})
$\dot{\varepsilon}$	strain rate (s^{-1})
η	dynamic viscosity (Pa s)
κ_m	thermal diffusivity of magma (m^2 s^{-1})
κ_w	thermal diffusivity of water (m^2 s^{-1})
μ	magma bulk modulus (Pa)
ν	dynamic viscosity (m^2 s^{-1})
ξ	ratio of water and magma specific heats
ρ	mixture density (kg m^{-3})
ρ_m	magma density (kg m^{-3})
ρ_r	rock density (kg m^{-3})
ρ_w	water density (kg m^{-3})
ϕ	porosity

Acknowledgments

Much of this work was performed under the auspices of Los Alamos National Laboratory, operated by Los Alamos National Security, LLC for the National Nuclear Security Administration of the U.S. Department of Energy under contract DE-AC52-06NA25396.

References

Austin-Erickson, A., Büttner, R., Dellino, P., Ort, M. H., and Zimanowski, B. (2008). Phreatomagmatic explosions of rhyolitic magma: experimental and field evidence. *Journal of Geophysical Research*, **113**, B11201, doi:10.1029/2008JB005731.

Baierlein, R. (1999). *Thermal Physics*. Cambridge University Press, 456 pp.

Bejan, A. and Eden, M. (1999). *Thermodynamic Optimization of Complex Energy Systems*. Springer.

Bischoff, J. L. and Rosenbauer, R. J. (1988). Liquid-vapor relations in the critical region of the system NaCl-H_2O from 380 to 415°C: a refined determination of the critical point and two-phase boundary of seawater. *Geochimica et Cosmochimica Acta*, **52**, 2121–2126.

Board, S. J., Farmer, C. L. and Poole, D. H. (1975). Fragmentation in thermal explosions. *Journal of Heat and Mass Transfer*, **17**, 331–339.

Bonatti, E. (1976). Mechanisms of deep sea volcanism in the South Pacific. In *Researches in Geochemistry 2*, ed. P. H. Ableson. New York: John Wiley & Sons, pp. 453–491.

Buchanan, D. J. (1974). A model for fuel-coolant interaction. *Journal of Physics D: Applied Physics*, **7**, 1441–1457.

Buchanan, D. J. and Dullforce, T. A. (1973). Mechanism for vapor explosions. *Nature*, **245**, 32–34.

Burnham, C. W., Holloway, J. R. and Davies, N. F. (1969). Thermodynamic properties of water to 1000°C and 10 000 bars. *Geological Society of America Special Paper*, **132**, 1–96.

Büttner, R. and Zimanowski, B. (1998). Physics of thermohydraulic explosions. *Physics Reviews E*, **57**(5), 5726–5729.

Büttner, R., Zimanowski, B., Blumm, J. and Hagemann, L. (1998). Thermal conductivity of a volcanic rock material (olivine-melilitite) in the temperature range between 288 and 1470 K. *Journal of Volcanology and Geothermal Research*, **80**, 293–302.

Büttner, R., Dellino, P. and Zimanowski, B. (1999). Identifying modes of magma/water interaction from the surface features of ash particles. *Nature*, **401**, 688–690.

Büttner, R., Zimanowski, B., Lenk, C., Koopmann, A. and Lorenz, V. (2000). Determination of thermal conductivity of natural silicate melts. *Applied Physics Letters*, **77**, 1810–1812.

Büttner, R., Dellino, P., LaVolpe, L., Lorenz, V. and Zimanowski, B. (2002). Thermohydraulic explosions in phreatomagmatic eruptions as evidenced by the comparison between pyroclasts and products from Molten Fuel Coolant Interaction experiments. *Journal of Geophysical Research*, **107**, doi: 10.1029/2001JB000792.

Büttner, R., Zimanowski, B., Mohrholz, C.-O. and Kümmel, R. (2005). Analysis of thermohydraulic explosion energetics. *Journal of Applied Physics*, **98**, 043524, doi:10.1063/1.2033149.

Büttner, R., Dellino, P., Raue, H., Sonder, I., and Zimanowski, B. (2006). Stress induced brittle fragmentation of magmatic melts: Theory and experiments. *Journal of Geophysical Research*, **111**, B08204, doi:101029/2005JB003958.

Carlisle, D. (1963). Pillow breccias and their aquagene tuffs: Quadra Island, British Columbia. *American Journal of Science*, **71**, 48–71.

Corradini, M. L. (1981). Phenomenological modelling of the triggering phase of small-scale steam explosion experiments. *Nuclear Science and Engineering*, **78**, 154–170.

Courant, R. and Friedrichs, K. O. (1948). *Supersonic Flow and Shock Waves*. New York: Springer, 464 pp.

Cronenberg, A. W. (1980). Recent developments in the understanding of energetic molten fuel-coolant interactions. *Nuclear Safety*, **21**, 319–337.

Delaney, P. T. (1982). Rapid intrusion of magma into wet rock: ground water flow due to pressure increases. *Journal of Geophysical Research*, **87**, 7739–7756.

Dellino, P., Isaia, R., La Volpe, L. and Orsi, G. (2001). Statistical analysis of textural data from complex pyroclastic sequences: Implications for fragmentation processes of the Agnano Monte Spina Tephra (4.1 ka), Phlegraean Fields, southern Italy. *Bulletin of Volcanology*, **63**, 443–461.

Drumheller, D. S. (1979). The initiation of melt fragmentation in fuel-coolant interactions. *Nuclear Science and Engineering*, **72**, 347–356.

Ebert, H.-P., Hernberger, F., Fricke, J. *et al.* (2003). Thermophysical properties of a volcanic rock material. *High Temperatures – High Pressures*, **34**, 561–668.

Fisher, R. V. and Schmincke, H.-U. (1984). *Pyroclastic Rocks*. Berlin: Springer-Verlag, 472 pp.

Fisher, R. V. and Waters, A. C. (1970). Base-surge bed forms in maar volcanoes. *American Journal of Science*, **268**, 157–180.

Frazzetta, G., La Volpe, L. and Sheridan, M. F. (1983). Evolution of the Fossa cone, Vulcano. *Journal of Volcanology and Geothermal Research*, **17**, 329–360.

Fuller, R. E. (1931). The aqueous chilling of basaltic lava on the Columbia River Plateau. *American Journal of Science*, **21**, 281–300.

Grady, D. E. (1982). Local inertial effects in dynamic fragmentation. *Journal of Applied Physics*, **53**, 322–523.

Grady, D. E. (1985). Fragmentation under impulsive stress loading. In *Fragmentation by Blasting*, ed. W. L. Fourney and R. R. Boade. Society for Experimental Mechanics, Brookfield Center, pp. 63–72.

Greer, S. C. and Moldover, M. R. (1981). Thermodynamic anomalies at critical points of fluids. *Annual Reviews of Physical Chemistry*, **32**, 233–265.

Heiken, G. (1971). Tuff rings: examples from the Fort Rock-Christmas Lake Valley, south-central Oregon. *Journal of Geophysical Research*, **76**, 5615–5626.

Heiken, G. and Wohletz, K. H. (1985). *Volcanic Ash*. Berkeley, CA: University of California Press, 245 pp.

Heiken, G. and Wohletz, K. (1991). Fragmentation processes in explosive volcanic eruptions. In *Sedimentation in Volcanic Settings*, ed. R. V. Fisher and G. Smith, Society of Sedimentary Geology Special Publication, **45**, 19–26.

Henry, R. E. and Fauske, H. K. (1981). Required initial conditions for energetic steam explosions. In *Fuel-Coolant Interactions*. New York: American Society of Mechanical Engineers, Report HTD-V19.

Herrmann, H. J. and Roux, S. (1990). *Statistical Models for the Fracture of Disordered Media*, Amsterdam: North-Holland.

Holloway, J. R. (1977). Fugacity and activity of molecular species in supercritical fluids. In *Thermodynamics in Geology*, ed. D. G. Fraser. Dordrecht, Holland: Reidel, pp. 161–181.

Honnorez, H. and Kirst, P. (1975). Submarine basaltic volcanism: morphometric parameters for discriminating hyaloclastites from hyalotuffs. *Bulletin of Volcanology*, **39**, 441–465.

Jaggar, T. A. (1949). *Steam Blast Volcanic Eruptions; a Study of Mount Pelée in Martinique as a Type Volcano*. Hawaii Volcano Observatory, 4th Special Report. Honolulu: The Hawaiian Volcano Research Association, **137**.

Kieffer, S. W. and Delany, J. M. (1979). Isentropic decompression of fluids from crustal and mantle pressures. *Journal of Geophysical Research*, **84**, 1611–1620.

Kokelaar, P. (1986). Magma-water interactions in subaqueous and emergent basaltic volcanism. *Bulletin of Volcanology*, **48**, 275–290.

Krauskopf, K. B. (1967). *Introduction to Geochemistry*. New York: McGraw-Hill.

Landau, L. D. and Lifshitz, B. M. (1959). *Fluid Mechanics: Volume 6, Course of Theoretical Physics*. New York: Pergamon.

Mastin, L. G. (2007). Generation of fine hydromagmatic ash by growth and disintegration of glassy rinds. *Journal of Geophysical Research*, **112**, B02203, doi:10.1029/2005JB003883.

Moore, J. G., Nakamura, K. and Alcaraz, A. (1966). The 1965 eruption of Taal Volcano. *Science*, **151**, 995–960.

Morrissey, M., Zimanowski, B., Wohletz, K., and Büttner, R. (2000). Phreatomagmatic fragmentation. In *Encyclopedia of Volcanism*, ed. H. Sigurdsson. London: Academic Press, pp. 431–445.

Muffler, L. J. P., White, D. E. and Truesdell, A. H. (1971). Hydrothermal explosion craters in Yellowstone National Park. *Geological Society of America Bulletin*, **82**, 723–740.

Nairn, I. A. and Solia, W. (1980). Late Quaternary hydrothermal explosion breccias at Kawerau geothermal field, New Zealand. *Bulletin of Volcanology*, **43**, 1–13.

Noe-Nygaard, A. (1940). Subglacial volcanic activity in ancient and recent times. *Folia Geographica Danica*, **1**, no. 2.

Ollier, C. D. (1974). Phreatic eruptions and maars. In *Physical Volcanology*, ed. L. Civett, P. Gasparini, G. Luongo, and A. Rapolla. New York: Elsevier, pp. 289–310.

Ruggles, A E., Drew, D. A., Lahey, R. T., Jr. and Scarton, H. A. (1989). The relationship between standing waves, pressure pulse propagation and the critical flow rate in two-phase mixtures. *Journal of Heat Transfer*, **111**, 467–473.

Ruggles, A. E., Vasiliev, A. D., Brown, N. W. and Wendel, M. W. (1997). Role of heater thermal response in reactor thermal limits during oscillatory two-phase flows. *Nuclear Science and Engineering*, **125**, 75–83.

Schmid, A., Sonder, I., Seegelken, R. et al. (2010). Experiments on the heat discharge at the dynamic magma-water-interface. *Geophysical Research Letters*, **37**, L20311, doi: 10.1029/2010GL044963

Self, S., Kienle, J. and Huot, J.-P. (1980). Ukinrek maars, Alaska, II. Deposits and formation of the 1977 craters. *Journal of Volcanology and Geothermal Research*, **7**, 39–65.

Servicos Geologicos de Portugal (1959). Le volcanisme de l'isle de Faial et l'éruption de volcan de Capelinhos. *Memorandum* **4**, 100 pp.

Sheridan, M. F. and Wohletz, K. H. (1983). Hydrovolcanism: basic considerations and review. *Journal of Volcanology and Geothermal Research*, **17**, 1–29.

Sigvaldason, G. (1968). Structure and products of subaquatic volcanoes in Iceland. *Contributions to Mineralogy and Petrology*, **18**, 1–16.

Sonder, I., Büttner, R. and Zimanowski, B. (2006). Non-Newtonian viscosity of basaltic magma. *Geophysical Research Letters*, **33**, L02303, doi:10.1029/2005GL024240.

Sonder, I., Schmid, A., Seegelken, R., Zimanowski, B. and Büttner, R. (2011). Heat source or heat sink: What dominates behavior of non-explosive magma-water interaction? *Journal of Geophysical Research*, **116**, B09203, doi:10.1029/2011JB008280.

Sourirajan, S. and Kennedy, G. C. (1962). The system H_2O-NaCl at elevated temperatures and pressures. *American Journal of Science*, **260**, 115–141.

Tazieff, H. K. (1958). L'éruption 1957–1958 et la tectonique de Faial (Azores). *Annales de la Sociéte Géologique de Belgique*, **67**, 14–49.

Theofanous, T. G. (1995). The study of steam explosion in nuclear systems. *Nuclear Engineering Design*, **155**, 1–26.

Thorarinsson, S. (1964). *Surtsey, the New Island in the North Atlantic*. Reykjavik: Almenna Bokofelagid.

Turekian, K. K. (1968). *Oceans*. New Jersey: Prentice-Hall, 150 pp.

Varotsos, P. A. and Alexopoulos, K. D. (1986). *Thermodynamics of Point Defects and Their Relation with Bulk Properties*. Amsterdam: North-Holland.

Waters, A. C. and Fisher, R. V. (1971). Base surges and their deposits: Capelinhos and Taal Volcanoes. *Journal of Geophysical Research*, **76**, 5596–5614.

Wentworth, C. K. (1938). *Ash Formations of the Island of Hawaii*. Hawaii Volcano Observatory, 3rd Special Report. Honolulu: Hawaiian Volcano Research Association, 183 pp.

White, J. D. L. (1996). Impure coolants and interaction dynamics of phreatomagmatic eruptions. *Journal of Volcanology and Geothermal Research*, **74**, 155–170.

Wohletz, K. H. (1983). Mechanisms of hydrovolcanic pyroclast formation: grain-size, scanning electron microscopy, and experimental results. *Journal of Volcanology and Geothermal Research*, **17**, 31–63.

Wohletz, K. H. (1986). Explosive magma-water interactions: thermodynamics, explosion mechanisms, and field studies. *Bulletin of Volcanology*, **48**, 245–264.

Wohletz, K. H. (1987). Chemical and textural surface features of pyroclasts from hydrovolcanic eruption sequences. In *Clastic Particles*, ed. J. R. Marshall. New York: Van Nostrand Reinhold, pp. 79–97.

Wohletz, K. H. (1993). Hidrovolcanismo. In *Nuevas Tendencias: La Volcanologia Actual*, ed. J. Martí and V. Araña. Madrid: Consejo Superior de Investigaciones Cientificas, pp. 99–196.

Wohletz, K. H. (2003). Water/magma interaction: physical considerations for the deep submarine environment. In *Submarine Explosive Volcanism*, American Geophysical Union Monograph, **140**, 25–49.

Wohletz, K. H. and Heiken, G. (1992). *Volcanology and Geothermal Energy*. Berkeley, CA: University of California Press.

Wohletz, K. H. and McQueen R. G. (1984). Experimental studies of hydromagmatic volcanism. In *Explosive Volcanism: Inception, Evolution, and Hazards*. Washington: National Academy Press, pp. 158–169.

Wohletz, K. H. and Sheridan, M. F. (1983). Hydrovolcanic explosions II. Evolution of basaltic tuff rings and tuff cones. *American Journal of Science* **283**, 385–413.

Wohletz, K. H., McQueen, R. G. and Morrissey, M. (1995a). Analysis of fuel-coolant interaction experimental analogs of hydrovolcanism. In *Intense Multiphase Interactions*, ed. T. G. Theofanous and M. Akiyama. Proceedings of US (NSF) Japan (JSPS) Joint Seminar, Santa Barbara, CA, pp. 287–317.

Wohletz, K. H., Orsi, G. and de Vita, S. (1995b). Eruptive mechanisms of the Neapolitan Yellow Tuff interpreted from stratigraphy, chemistry, and granulometry. *Journal of Volcanology and Geothermal Research*, **67**, 263–290.

Yew, C. H. and Taylor, P. A. (1994). A thermodynamic theory of dynamic fragmentation. *International Journal of Impact Engineering*, **15**, 385–394.

Yuen, W. W. and Theofanous, T. G. (1994). The prediction of 2D thermal detonations and resulting damage potential. *Nuclear Engineering Design*, **155**, 289–309.

Zel'dovich, Ya. B. and Kompaneets A. S. (1960). *Theory of Detonation*. London: Academic Press.

Zel'dovich, Ya. B. and Raizer, Yu. P. (1966). *Physics of Shock Waves and High-Temperature Hydrodynamic Phenomena*, Volumes I and II. New York: Academic Press.

Zimanowski, B. (1998). Phreatomagmatic explosions. In *From Magma to Tephra*, ed. A. Freundt and M. Rosi. Developments in Volcanology, **4**. Amsterdam: Elsevier, pp. 25–54.

Zimanowski, B. and Büttner, R. (2002). Dynamic mingling of magma and liquefied sediments. In *Peperite*, ed. I. Skilling, J. D. L. White and J. McPhie. Amsterdam: Elsevier, pp. 37–44.

Zimanowski, B. and Büttner, R. (2003). Phreatomagmatic explosions in subaqueous eruptions. In *Explosive Subaqueous Volcanism*, ed. J. D. L. White, J. L. Smellie and D. Clague. American Geophysical Union Monograph, **140**. Washington, pp. 51–60.

Zimanowski, B., Fröhlich, G. and Lorenz, V. (1991). Quantitative experiments on phreatomagmatic explosions. *Journal of Volcanology and Geothermal Research*, **48**, 341–358.

Zimanowski, B., Büttner, R., Lorenz, V. and Häfele, H.-G. (1997a). Fragmentation of basaltic melt in the course of explosive volcanism. *Journal of Geophysical Research*, **107**, 803–814.

Zimanowski, B., Büttner, R. and Nestler, J. (1997b). Brittle reaction of a high temperature ion melt. *Europhysics Letters*, **38**, 285–289.

Zimanowski, B., Wohletz, K. H., Büttner, R. and Dellino, P. (2003). The volcanic ash problem. *Journal of Volcanology and Geothermal Research*, **122**, 1–5.

Exercises

The interested reader is encouraged to visit the following software websites to explore a range of hydrovolcanic and magmatic eruption styles:

(1) geodynamics.lanl.gov/Wohletz/PHM.htm. KWare PHM is designed to make thermodynamic calculations to illustrate the wide range of explosive potential that develops depending upon the initial conditions of water/magma contact.

(2) geodynamics.lanl.gov/Wohletz/Erupt-User.htm Erupt3 is a graphical program that simulates various volcanic eruption types, including strombolian, plinian, vulcanian/surtseyan, pyroclastic flows and surges, hawaiian fluid lava flows, fumarolic activity, and peleean viscous lava dome emplacement.

Online resources available at www.cambridge.org/fagents

- Supplement 11A: Calculation of interaction thermodynamic work
- Supplement 11B: Video of MFCI experiments
- Links to websites of interest
- Additional reading

Chapter 12

Deep-sea eruptions

Tracy K. P. Gregg

Overview

The most abundant and widespread volcanic behavior on Earth is the effusive eruption of basaltic lava on the ocean floors – most of it from mid-ocean ridge eruptions. The inaccessibility of the deep (> 500 m below sea level) sea floor limits direct observations of submarine volcanic phenomena. Observations of volcanic products are therefore the primary means by which the dynamics of mid-ocean ridge eruptions are constrained and evaluated. Submarine lava flow morphology can be used to constrain local flow rate if lava viscosity is known. Recent deep (> 500 m) explosive eruptions of seamounts have been directly observed.

12.1 | Introduction

The majority of Earth's surface is covered with oceans, and the oceanic floors are underlain by basaltic lavas erupted at mid-ocean ridges. The most common volcanic eruption on Earth is a submarine, *mid-ocean ridge* eruption, and understanding the processes that occur during these volcanic events provides a strong foundation for understanding other submarine eruptions.

Although Iceland, located at the intersection of a mid-ocean ridge and a mantle plume, provides an opportunity to examine mid-ocean ridge eruptions on land, it is important to remember that Iceland is not a typical mid-ocean ridge volcanic system. Throughout this chapter, then, the phrase "mid-ocean ridge" will refer to the submarine variety unless otherwise stated.

The behavior of deep submarine eruptions depends upon the magma viscosity, volatile content, effusion rate, and water depth. Before the 1990s, it was commonly accepted that deep submarine eruptions must be effusive because the overlying pressure of seawater would preclude hydromagmatic explosions. Only recently has there been a paradigm shift toward accepting that explosive deep submarine eruptions occur (see White *et al.*, 2003, and references therein). For example, there is evidence that Loihi (an active seamount located off the coast of Kilauea volcano, Hawaii, and under 2000 m of seawater) has erupted both effusively and explosively (Schipper *et al.*, 2010). Recent studies have suggested that infrequent explosive eruptions may occur at mid-ocean ridges, despite significant water depths, low lava viscosity, and low initial volatile contents (Sohn *et al.*, 2008; Clague *et al.*, 2009). In the past decade, researchers have been able to directly observe multiple

Modeling Volcanic Processes: The Physics and Mathematics of Volcanism, eds. Sarah A. Fagents, Tracy K. P. Gregg, and Rosaly M. C. Lopes. Published by Cambridge University Press. © Cambridge University Press 2013.

seamount eruptions for the first time, associated with seamounts in the submarine Pacific "Ring of Fire," including West Mata volcano (15.10°S, 173.75°E, with a summit ~1200 m below sea level; Baker *et al.*, 2010; Chadwick *et al.*, 2012) and Northwest Rota-1 volcano (14.60°S, 155.78°E, with a summit ~500 m below sea level; Embley *et al.*, 2006, 2007; Chadwick *et al.*, 2008a).

Chapter 11 discusses magma–water interaction in shallow crustal or surface environments, and the role that external water plays in producing vigorously explosive subaerial eruptions. This chapter examines eruptions under great depths of water. The tectonic settings in which submarine volcanism occurs are first presented, followed by a discussion of how the ambient conditions of the deep ocean floor affect submarine eruptive processes. Inferences gleaned from deep-sea eruptions can provide useful insights into eruptive processes in other high-pressure environments, such as on Venus (Chapter 17), which has a surface atmospheric pressure equivalent to the pressure beneath ~900 m of water.

12.2 | Settings for submarine volcanic deposits

12.2.1 Volcanic seamounts

Seamounts are submarine volcanic mountains taller than ~100 m. No one knows the precise number of seamounts in Earth's oceans because most of the ocean floor has yet to be surveyed at sufficiently high resolution, but there could be as many as 25 million of them (Wessel *et al.*, 2010, and references therein). Thus, not all deep submarine volcanism is constrained to mid-ocean ridges. Seamounts can be classified as belonging to one of the following groups: (1) part of a linear or arcuate seamount chain (the Hawaiian–Emperor seamount chain being among the best known) that are related to intra-plate mantle plumes or subduction (such as the Marianas volcanic arc); (2) associated with mid-ocean ridges (e.g., Smith and Cann, 1992; Cochran, 2008); (3) associated with flexure of

the oceanic lithosphere by a larger volcanic pile (such as those around the Hawaiian arch; Torresan *et al.*, 1991; Bridges, 1997; http://walrus.wr.usgs.gov/gloria/hwgloria/); or (4) linear seamount chains associated with, and oriented roughly perpendicular to, mid-ocean ridges (e.g., the Lamont seamount chain near 10°N off the East Pacific Rise; Allan *et al.*, 1989).

Although seamounts associated with intra-plate activity and with mid-ocean ridges tend to be basaltic (Batiza, 1982), seamounts formed in subduction zones and in back-arc basins display a range of eruptive products and compositions, including, but not limited to, tholeiite basalt, alkalic basalt, boninite (Embley *et al.*, 2007; Tamura *et al.*, 2010), and lavas containing as much as 70 wt.% SiO_2 (Stern *et al.*, 2010). This range of compositions can affect eruption behavior and the volcanic products (Deardorff *et al.*, 2011).

Seamounts are perhaps most familiarly found as part of seamount chains such as the Hawaiian–Emperor seamount chain that extends west and northwest of the Big Island of Hawaii. Wilson (1963) proposed that seamount chains are formed by the passage of Earth's lithospheric plates over a stationary mantle plume, or *hot spot*. Although there is not complete consensus in the scientific community about the existence or behavior of these mantle plumes, their presence most simply explains those seamount chains that display a well-constrained age progression. Linear to arcuate seamount chains without a clear age progression (e.g., the Wolf–Darwin lineament, located on the southern flank of the Galapagos spreading center) require other mechanisms (Ito *et al.*, 2003).

Seamounts clustered around mid-ocean ridges tend to have compositions similar to MOR lavas. Fast-spreading centers display more off-axis seamounts than do slow-spreading centers (Fig. 12.1; see online supplement OS 12A for color versions of figures). They tend to be small in volume (< 3 km³), and are likely the result of "leaky" MOR magma chambers (e.g., Fornari *et al.*, 1987; Perfit and Chadwick, 1998).

Seamount fields are associated with extensional flexure of the oceanic lithosphere, both around regions of volcanic loading (such as

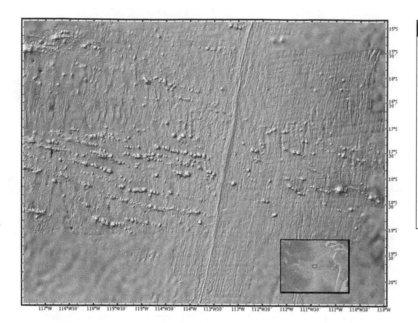

Figure 12.1 Bathymetry from the East Pacific Rise near 17°S, with north at the top; false illumination from upper left corner. Note the seamounts that are found off-axis, and many are in "chains" that were probably fed from a leaky axial magma chamber and rifted off-axis. See http://media.marine-geo.org; modified from Ryan *et al.* (2009). Color versions of this and other figures are found in the online supplement to Chapter 12 (see end of chapter).

around the Hawaiian arch) and subduction zones (such as in the northwest Pacific; Valentine and Hirano, 2010). Although research into these types of seamounts is limited, they may hold some similarities to subaerial monogenetic basalt fields (Valentine and Hirano, 2010).

Note that the formation and evolution of seamounts is a rich area of research; please see the online supplementary material (listed at end of chapter) for further reading on this topic.

12.2.2 Oceanic plateaus and large igneous provinces

Large igneous provinces (LIPs) are accumulations of mafic (typically basaltic) lavas that: (1) are emplaced in a short period of time (a few million years); (2) cover large areas (> 10^5 km²); and (3) comprise 10^5 to 10^6 km³ (Coffin and Eldholm, 1994). Submarine LIPs have been studied primarily by drilling (Ocean Drilling Program, 1991, 2001) and through shipboard sonar mapping (Neal *et al.*, 1997). The submarine Ontong Java Plateau is an Alaska-sized feature centered around 2.08°S, 158.09°E, and is one of the largest LIPs yet identified (Coffin and Eldholm, 1994; Neal *et al.*, 1997). As the descriptor "plateau" suggests, these features are typically 1–3 km higher than the surrounding oceanic floor; geophysical data suggest that

the total crustal thickness of the Ontong Java Plateau is between 30 and 43 km, giving a total volume of > 5×10^7 km³ (Neal *et al.*, 1997, and references therein).

An active eruption of this type has not yet been observed – on sea or on land – so eruption processes must be inferred. Drilling into the Ontong Java plateau reveals that the plateau is composed of extrusive lava flows, with no record of explosive deposits: roughly half of the returned cores were composed of pillowed flows, and half of massive (probably inflated sheet) flows (Neal *et al.*, 1997). Given that the lavas were emplaced on a relatively level surface, pillowed lavas suggest a lower effusion rate than both sheet lavas (Fink and Griffiths, 1992; Gregg and Fink, 1995) and the large-scale inflated pahoehoe flows identified in subaerial LIPs (Self *et al.*, 1998). Similar to most subaerial LIPs, vents and vent regions have not been identified in the submarine oceanic plateaus. Coffin and Eldholm (1994) identified 34 oceanic plateaus and oceanic flood basalt provinces, each of which is composed of countless individual submarine lava flows. The lack of hyaloclastites within these oceanic plateaus suggests that neither the lava volatile contents nor the effusion rates were sufficiently high to generate explosions during their genesis. The extraordinary

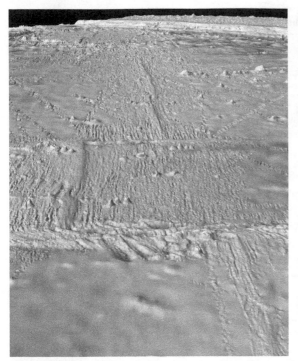

Figure 12.2 A perspective view north from 7°N along the East Pacific Rise. Note how the two east–west trending transform faults offset the ridge near the center of the image. See http://media.marine-geo.org; modified from Ryan et al. (2009).

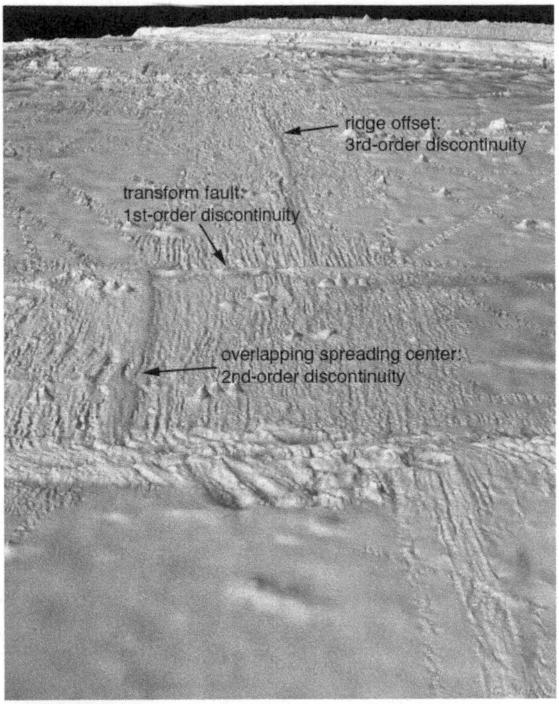

Figure 12.3 Same bathymetric map as shown in Fig. 12.2, with first-, second-, and third-order discontinuities labeled. See http://media.marine-geo.org; modified from Ryan et al. (2009).

volumes of these LIPs suggest a combination of long-lived eruptions and steady supply rates.

12.2.3 Mid-ocean ridges

The MOR system is the largest, most active volcanic system on Earth. Although < 5% of the global MOR system has been studied in detail, we have more compositional and time-series (i.e., repeated data set collection over time from the same location) data for MORs than for other submarine volcanic provinces. Available data suggest that products of a specific eruption along the MOR system tend to be relatively homogeneous in composition (Perfit and Chadwick, 1998; Rubin et al., 2001; Sinton et al., 2001), which simplifies the task of modeling these eruptions.

MORs are not continuous, but are divided into segments, or first-order discontinuities, by transform faults (e.g., Macdonald, 1998;

Fig. 12.2). The first-order discontinuities are further broken into second-, third-, and fourth-order discontinuities (Fig. 12.3). It is likely that the third-order discontinuities mark the boundaries of individual volcanic systems (Macdonald, 1998; Perfit and Chadwick, 1998; White et al., 2000, 2009; Soule et al., 2009). Between these discontinuities, lava composition and eruption styles are similar, consistent with this interpretation (Macdonald, 1998). Compositional variability is most noticeable at segment boundaries (Perfit and Chadwick, 1998), and segment centers tend to be more volcanically active than segment ends (Sinton et al., 2001). It is likely that these segment boundaries are dynamic, however, as suggested by observations that indicators of volcanic accretion over longer timescales are consistent across third- and fourth-order discontinuities (Soule et al., 2009).

Basalts comprise the majority of mid-ocean ridge lavas. There are localized exceptions; for

Table 12.1 | Average major element analyses of individual lava flows from mid-ocean ridges.

Weight %	EPR 9°50'N (BBQ)	EPR 18°38'S (Animal Farm)	MAR 22°55'N (Serocki Volcano)	JFR 44°56'N (N. Cleft 1993 flow)
No. samples	21	60	22	20
SiO_2	44.88	50.8	50.03	50.04
TiO_2	1.29	1.65	1.67	1.53
Al_2O_3	15.52	14.5	15.87	14.78
FeO	9.31	10.6	9.92	10.52
MnO	0.17	0.20	Not determined	0.20
MgO	8.55	7.29	7.51	7.58
CaO	12.26	11.7	11.21	11.85
Na_2O	2.56	2.96	2.93	2.47
K_2O	0.09	0.09	0.14	0.12
P_2O_5	0.12	0.14	0.17	0.14
Citation	Gregg et al. (1996)	Sinton et al. (2001)	Humphris et al. (1990)	Embley et al. (1991)

Table 12.2 | Typical spreading rates of mid-ocean ridges.

Spreading rate	Spreading rate (cm yr⁻¹ full rate)	Type locale	Citation
Ultra-slow	0.6–1.3	Gakkel Ridge	Sohn et al. (2008)
Slow	1.0–4.0	Mid-Atlantic Ridge	Perfit and Chadwick (1998)
Intermediate	4.0–8.0	Juan de Fuca Ridge	Perfit and Chadwick (1998)
Fast	8.0–16.0	East Pacific Rise, 9°50'N	Perfit and Chadwick (1998)
Superfast	> 16.0	East Pacific Rise, 30°S	Korenaga and Hey (1996)

example, dacites have been found along the submarine MOR system (Stakes et al., 2006; Wanless et al., 2010), but basalt – specifically, mid-ocean ridge basalt (MORB) – is the most abundant lava erupting from submarine mid-ocean ridges. Mid-ocean ridge basalts are classified by their major- and trace-element abundances. The range of Mg abundance in MOR basalts can vary considerably in time and space, as can Ti and others (see Table 12.1). Closely spaced (1–2 km) sampling along and across MORs and subsequent analyses reveal that ridge discontinuities can coincide with significant changes in underlying magma chemistry (Langmuir et al., 1986; Sinton et al., 1991, 2001; Perfit and Chadwick, 1998).

Eruption parameters, including erupted volume, eruption duration, and effusion rate (here, defined as volumetric rate, or m³ s⁻¹), are believed to correlate with spreading rate. Herein, the spreading rates given are full spreading rates: if you were to stand on one plate, the full spreading rate is how fast the plate on the other side of the MOR would appear to be moving away from you. Spreading rates vary from *ultraslow* (< 1.5 cm yr⁻¹) to *superfast* (> 16 cm yr⁻¹) (Table 12.2). The Southern East Pacific Rise (SEPR) near 17°30' S is the type locale for a superfast spreading center (e.g., Sinton et al., 2001); the Gakkel Ridge in the Arctic Ocean is the type locale for an ultraslow spreading center (Edwards et al., 2001).

Eruptions along fast-spreading MORs appear to be small volume (~10^6–10^7 m³), frequent (approximately every 5–10 years per kilometer of ridge length) and short duration (hours) (Gregg et al., 1996; Sinton et al., 2001; Chadwick, 2003; Soule et al., 2007). In contrast, eruptions at slow spreading centers are thought to be larger volume and longer-lived (Perfit and Chadwick, 1998; Sinton *et al.*, 2001).

12.3 | Effusive submarine eruptions

In 1977, R. Ballard and T. Van Andel led an expedition to the Mid-Atlantic Ridge near 37°N (Ballard and van Andel, 1977). In addition to scholarly journal articles, Ballard and Moore (1977) published a "coffee-table" book on the expedition, complete with striking full-color photographs of the deep ocean floor. By chance, the region they explored was covered with pillow lavas (Ballard and van Andel, 1977; Ballard and Moore, 1977); thereafter, most introductory geology textbooks stated that the ocean floor was dominated by pillow lavas (Fig. 12.4(a)). Furthermore, pillowed lavas were observed to form underwater where subaerial Hawaiian lavas flowed into the ocean (Tepley and Moore, 1974; Tribble, 1991; see online supplement OS 12B, "Pele Meets the Sea"). In addition, subglacial lavas also commonly display pillowed morphologies (e.g., Jones, 1970; Jakobsson and Gudmundsson, 2008). The presence of lava pillows, therefore, requires an abundance of water, but an abundance of water does not produce only lava pillows (cf. Embley *et al.*, 1991): most of the Pacific Ocean is likely floored by a type of sheet flow known as *lobate* (Fox *et al.*, 1988; Perfit and Chadwick, 1998; Fig. 12.4(b)).

On the deep sea floor (most MORs have a water depth of ~2500 m; Perfit and Chadwick, 1998), steam formation and explosions driven by magmatic gases are minor or rare (cf. Perfit *et al.*, 2003; Clague *et al.*, 2009). At a water depth of 2500 m, the ambient pressure is 250 MPa. A normal MORB would theoretically require a water content of 4.5 wt.% before a magmatic explosion could occur, and most MORs do not contain sufficient volatiles (Table 12.1) for magma fragmentation (Head and Wilson, 2003). Therefore, most MOR eruptions are effusive.

Submarine lava flow morphologies are similar to, but distinct from, subaerial basaltic lava morphologies because of the different ambient conditions. The high ambient pressure at MORs (250 MPa) keeps most volatiles dissolved within the magma; a typical vesicle content for MOR basalts is < 2 vol.% (e.g., Perfit and Chadwick, 1998), although volumes as high as 35 vol.% have been reported (Hekinian *et al.*, 2000). Commonly, however, volatile contents of MORB magmas are thought to be much less than 0.5 wt.% (Byers *et al.*, 1986). For example, it is unlikely that true 'a'a lavas are produced on the deep sea floor because the required volatile exsolution (and accompanied increase in lava viscosity) will not occur on the deep sea floor as it does on land. Different descriptive terms are therefore used for submarine lava morphologies, even though they may look similar to subaerial basalt morphologies. Submarine lava flows are described as pillowed, lobate, lineated, ropy (folded), or jumbled (hackly) (Fig. 12.4).

12.3.1 Submarine lava flow inflation

Observations suggest that almost all submarine lava flows are inflated (Hon *et al.*, 1994) to some extent, including pillow lavas. If lava inflation is defined as "molten lava injected beneath a solidified lava crust," then the rapidity with which submarine basalts form solidified crusts helps to explain the abundance of lava inflation features on the sea floor (Applegate and Embley, 1992; Gregg and Chadwick, 1996; Chadwick *et al.*, 1999). Submarine inflation features similar to those observed on land – such as tumuli – have been identified (Applegate and Embley, 1992).

Submarine *lava pillars* (Fig. 12.5) are also indicative of lava inflation. They are hollow cylinders of basalt, morphologically similar to subaerial *lava trees* or *tree molds*. They are commonly < 20 m tall and < 2 m in diameter, and form during the emplacement, inflation, and drainback of lobate sheet flows in submarine environments (Gregg and Chadwick, 1996; Chadwick *et al.*, 1999; Chadwick, 2003). Gregg and Chadwick (1996) state that pillars grow as heated seawater rises through gaps between adjacent lava lobes of an advancing submarine lava flow. Alternatively, it has been proposed that pillars form after the lava has been emplaced, when buoyant water rises through the overlying lava (Francheteau *et al.*, 1979; Perfit *et al.*, 2003). Cooling at the flow

Figure 12.4 (a) Lava pillows (foreground, light-toned) and small lobes (right side of the image; dark-toned) from the floor of Axial Volcano caldera in the Juan de Fuca Ridge. Image width is ~2 m. Image from www.pmel.noaa.gov/vents/nemo/logbook/images/aug09-lava.html. (b) Lobate lava flow from the Galapagos spreading center; light-toned sediment appears in pockets on the surface. Image width is ~2 m. Image #cTOW-20100319–073119–0000253.jpg collected using TowCam (Fornari, 2003) during the GRUVEE cruise; courtesy of J. Sinton. (c) Lineated sheet flow from the Juan de Fuca Ridge; image width ~1 m. Individual lineations have a 1–3 cm wavelength; light-toned sediment fills the troughs between adjacent lineations. Image courtesy of the PMEL NOAA Vents Program www.pmel.noaa.gov/vents/nemo/explorer/concepts/sheetflow.html. (d) Jumbled submarine lava flow from the Galapagos spreading center; low-lying areas are filled in with light-toned sediment. Image width ~2 m. Image # cTOW-20100319–071039–0000129.jpg collected using TowCam (Fornari, 2003) during the GRUVEE cruise; courtesy of J. Sinton.

margins or topographic obstruction causes the lava flow to pond, and if lava continues to erupt, the flow will inflate. As long as there is sufficient heated water flowing through the gaps between adjacent lobes, the gaps will remain as the lava flow thickens. Seawater funneling through these gaps cools the lava surrounding the gap. Once the lava drains away (either back into the vent, or downslope), the lava pillars remain (Chadwick, 2003). Pillar height reflects the maximum thickness of the lava flow.

In 1998, an instrument designed to measure the ambient pressure was caught up within a lobate sheet flow that erupted within the summit caldera of Axial Volcano along the Juan de Fuca Ridge (Chadwick, 2003). During

Figure 12.5 Lava pillar (< 2 m tall) from the Axial Volcano caldera on the Juan de Fuca Ridge. This pillar was generated during an eruption in 1998. Image courtesy of PMEL NOAA Vents Program: www.pmel.noaa.gov/vents/nemo1999/logbook/cal062799/.

the eruption, the instrument recorded being uplifted ~3 m, and then being lowered ~2 m. The interpretation is that initially, the lava flow was thin enough to flow beneath the 50-cm-tall legs of the instrument. Subsequently, the lava flow inflated and then drained away, leaving a lava flow only ~1 m thick. The entire inflation/deflation event took only 153 minutes (Fox *et al.*, 2001). Lava pillars about 2 m tall were subsequently found within the lava flow (Chadwick, 2003; Fig. 12.5).

Similarly, lobate lava flows along the East Pacific Rise are commonly pock-marked with collapse pits, revealing hollow interiors locally supported by lava pillars (Fornari *et al.*, 1998; Sinton *et al.*, 2001; Perfit *et al.*, 2003; Soule *et al.*, 2006; Fundis *et al.*, 2010). Observations therefore indicate that lobate flows are emplaced by lava inflation processes.

12.3.2 | Modeling effusive submarine eruptions

An active, effusive mid-ocean ridge eruption has not yet been observed using visible wavelengths – that is, we have yet to *see* one. Ocean-bottom seismometers and submarine

microphones known as *hydrophones* have recorded the vibrations from mid-ocean ridge eruptions (Fox, 1999; Fox *et al.*, 2001; Tolstoy *et al.*, 2006), and a portion of the East Pacific Rise near 9°50'N was visited within weeks of an eruption in 1991 (Haymon *et al.*, 1993). However, details such as effusion rate and emplacement style must be interpreted from the resulting volcanic morphology.

Chapter 5 discusses thermal modeling as one way to understand lava flow behavior (and thereby morphology); convective, radiative, and conductive cooling are the primary mechanisms by which subaerial lavas transfer heat to their environment. On the deep sea floor, convective cooling by water is the dominant cooling mechanism, and it is much more efficient than radiative, convective, and conductive cooling combined on land (Griffiths and Fink, 1992). Convective heat flux from the surface of a submarine lava flow is given by (Fink and Griffiths, 1990):

$$F_{conv} = \rho_a c_a \gamma \left(\frac{g' \alpha_a \kappa_a^2}{v_a} \right)^{\frac{1}{3}} \left(T_c - T_a \right)^{\frac{4}{3}}, \qquad (12.1)$$

where F_{conv} is the convective heat flux from the lava flow surface (W m^{-2}); ρ_a is density (kg m^{-3}) of the ambient medium (here, seawater at 2500 m depth and 4 °C); c_a is seawater specific heat (J kg^{-1} K^{-1}); γ is a dimensionless constant equal to 0.1; g' is reduced gravitational acceleration (m s^{-2}); α_a is the coefficient of thermal expansion of seawater (K^{-1}); κ_a is seawater thermal diffusivity (m^2 s^{-1}); v_a is the seawater kinematic viscosity (m^2 s^{-1}); and T_a and T_c are the temperatures of the seawater and lava crust (K). (Notation is summarized in Section 12.8.) Reduced gravity, g', takes into account the density of the surrounding seawater, and is given by (Fink and Griffiths, 1990):

$$g' = g(\rho - \rho_a) / \rho_a, \qquad (12.2)$$

in which ρ is the lava density. Calculations indicate that the heat flux from a typical MORB erupting on the mid-ocean ridge, with an eruption temperature of 1100 °C (Perfit *et al.*, 1995, 2003) and a viscosity of 100 Pa s (Gregg *et al.*, 1996), would form a solid crust in 0.7 s (Griffiths and Fink, 1992). Videos taken of subaerially erupted lavas on Kilauea volcano that flowed into the ocean confirm these calculations (Tepley and Moore, 1974; see OS 12B.1). Additionally, MOR lavas are typically covered with a glassy rind ≤ 2 cm thick, confirming rapid cooling to the glass transition temperature (Gregg and Chadwick, 1996, and references therein).

The glassy rind on MOR lavas acts to insulate the interior of the lava flow from additional heat loss. Heat from the lava interior can only be transferred to the lava flow surface via conduction through this glassy rind, and conduction is a slow mechanism of heat transfer (see Chapter 5). Once the glass rind forms, the lava flows erupted from MORs tend to be volume-limited rather than cooling-limited (Guest et al., 1987; Gregg and Fornari, 1998). Thus, in comparison with subaerial basalt flows, submarine lavas: (1) contain fewer vesicles; (2) display a glassy rind ≥ 2 cm thick formed by seawater rapidly quenching their surface; and (3) have interiors that cool more slowly because they are insulated by the glassy rind.

Although the pressure exerted at the great depths of MORs suppresses steam explosions, there is still evidence for vapor pockets existing in contact with submarine lavas (Perfit et al., 2003; Chadwick, 2003; Soule et al., 2006; Schiffman et al., 2010). These pockets are small (tens of cm across or smaller) and short-lived, but nonetheless contribute to the formation of nm- to cm-scale features observed on and in submarine lava flow crusts.

Fink and Griffiths (1990) developed a dimensionless parameter, Ψ, that can be related to lava flow morphology if the eruption rate, Q, and lava kinematic viscosity, v_l, are known. Ψ is essentially the ratio of the timescale required for the surface of a lava flow to solidify (t_s) to the timescale for heat to be advected along the flow (t_a):

$$\Psi = \frac{t_s}{t_a}. \tag{12.4}$$

The timescale for solidification is proportional to the heat flux from the lava flow surface; on the sea floor, this is dominated by convection (Eq. (12.1)). The timescale for advection is given

by (Fink and Griffiths, 1990; Griffiths and Fink, 1992)

$$t_a = \left(\frac{v_l}{g'}\right)^{3/4} Q^{-1/4} \tag{12.5a}$$

for a point-source eruption, and by

$$t_a = \left(\frac{v_l}{g'}\right)^{2/3} q^{-1/3} \tag{12.5b}$$

for a line-source (fissure) eruption in which q (m^2 s^{-1}) is the volumetric effusion rate per meter of fissure length. Most mid-ocean ridge eruptions appear to be fissure eruptions (Chadwick and Embley, 1998; Fornari et al., 1998). Laboratory simulations, in which polyethylene glycol (PEG) was extruded at a constant effusion rate with a controlled cooling rate, allowed Ψ values to be quantitatively attached to PEG flow morphologies. These morphologies were related to those seen at mid-ocean ridges (Gregg and Fink, 1995) to constrain eruption rates there (Fig. 12.6). Using this method, Gregg et al. (1996) estimated eruption durations for MOR eruptions; the 1998 eruption at Axial Volcano on the Juan de Fuca Ridge confirmed this methodology (Chadwick, 2003).

If eruption viscosity is known or can be estimated, flow morphology is assigned to one of the categories identified by Fink and Griffiths (1990) or Gregg and Fink (1995), and cooling rates calculated, then effusion rates can be constrained. Griffiths and Fink (1992), Gregg and Greeley (1993) and Gregg and Fink (1996) demonstrated the applicability of this technique for extraterrestrial lava flows, for example.

Klingelhofer et al. (1999) examined the fluid dynamic behavior of cooling basalt with a temperature-dependent viscosity to constrain the formation of submarine lava pillows. They modeled pillows as cylindrical "tubes" with known dimensions, and concluded that the cooling rate must be balanced by the advection of hot lava, which in turn is controlled by a pressure gradient. The critical pressure gradient can be related to the radius (r) of the pillow tube raised to the power 3.85 (i.e., $r^{3.85}$). Consistent with the laboratory simulations of Fink and Griffiths (1990), Klingelhofer et al. (1999) agree

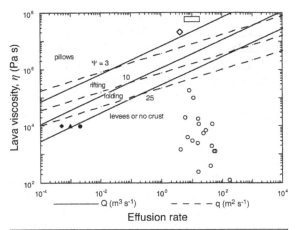

Figure 12.6 Relationship of flow morphology to effusion rate and lava viscosity. Dashed lines are for linear (fissure) sources, solid lines are for point sources. Symbols represent the 1984 Mauna Loa flow (open circles); the 1979 andesite dome at La Soufriere (open diamond); the Mount St. Helens dacite lobes erupted between October 1981 and August 1982 (open rectangle); and the 1975 basaltic andesite eruption at Mount Etna, which included channelized flows (solid circle), ropy pahoehoe tongues (solid triangle), and smooth lobes and toes (solid diamond). From Gregg and Fink (1996).

that lava viscosity also exerts a strong influence on flow morphology.

Submarine lava flow morphologies have subsequently been used to qualitatively and quantitatively constrain flow rates and eruption dynamics (e.g., White et al., 2000; Soule et al., 2005, 2007; Fundis et al., 2010). Where there are independent temporal data about eruption duration – collected by hydrophones (Dziak and Fox, 1999), seismometers (Tolstoy et al., 2006), or pressure sensors (Fox et al., 2001) – estimates of eruption duration based on lava morphology match well with the measured duration of activity (Chadwick, 2003; Fundis et al., 2010). However, there are problems in this method. Gregg and Fink (2000) point out that the results of analog experiments predict that on slopes steeper than 30° channeled flows should dominate. In fact, steep, constructional submarine slopes are dominated by elongate pillows and lobes. The likely problem in this case is that the solid analog material (PEG) has a higher tensile strength than does basalt (Soule and Cashman, 2004). Gregg and Fink (2000) explain that on

steep slopes, the basalt is pulled apart by gravitational forces, reducing the local flow rate and resulting in pillows or lobes. Thus, in ideal situations, lava flow morphology reflects local flow rates, but other factors (such as underlying slope, pre-existing roughness, or variations in lava crystal content or rheology) also exert local controls on lava morphologies.

In general, eruption volume and duration can be correlated with full spreading rate (Perfit and Chadwick, 1998; Sinton et al., 2001). Eruption volumes are estimated for only a handful of well-mapped mid-ocean ridge eruptions; eruption frequency is crudely calculated by using assumptions about the crustal thickness and known spreading rates (Sinton et al., 2001). The results of these calculations correlate well with known eruption frequencies along the Juan de Fuca Ridge and northern East Pacific Rise (e.g., Sinton et al., 2001, 2009). Fast-spreading MORs produce small ($\sim 10^6$ m³), short-lived (hours), frequent (every ~ 5 years) eruptions. The most common lava flow morphologies observed at fast-spreading MORs are lobate flows and sheet flows. Slow-spreading MORs are thought to generate larger, longer-lived, and less frequent eruptions dominated by pillow lavas. Erupted lavas typically have higher temperatures and lower viscosities at fast-spreading centers than at slow-spreading centers (Sinton and Detrick, 1992).

12.4 | Deep explosive submarine eruptions

Prior to the 1990s, the high ambient pressures exerted over most of Earth's MOR system were thought to preclude explosive eruptions (e.g., White et al., 2003; Wohletz, 2003). However, Haymon et al. (1993) found hyaloclastites generated during the 1991 eruption of the East Pacific Rise near 9°50′N. They interpret that these glassy shards were generated when lava flowed over an actively venting hydrothermal chimney, and that the continued flow of heated water into the lava flow eventually burst through and fragmented any solidified crust. Clague et al. (2009) reported on bubble wall glass shards (known as *limu o Pele* or *Pele's seaweed*) along the East Pacific Rise, the Juan de Fuca and Gorda Ridges, at Loihi seamount, along the Hawaiian arch, and in the Fiji back-arc basin. Sohn *et al.* (2008) discovered limu o Pele deposits at 4000 m depth along the Gakkel ridge. Clague *et al.* (2009) found a correlation between the abundance of limu o Pele and "high-rate eruptions," identified on the basis of lava flow morphologies: higher-effusion rate lava flow morphologies are associated with a greater abundance of bubble wall fragments than are lower-effusion rate morphologies. They assert that the morphology and composition of these limu o Pele require that they formed from violent expansion of magmatic gases (probably CO_2; Head and Wilson, 2003; Sohn *et al.*, 2008), rather than from explosions caused by expanding seawater, and suggest that the fragmentation and dispersal of limu o Pele are consistent with a strombolian eruption style. Similar to models presented by Head and Wilson (2003), Clague *et al.* (2009) propose that the submarine strombolian activity was generated by the addition of a magmatic foam from the top of the magma reservoir to the resident magma.

Schipper *et al.* (2010) collected lapilli and bombs from near the summit of Loihi seamount, at water depths of 1100–1291 m. They examined vesicle textures, as well as the geochemistry of the lavas, and the H_2O and CO_2 contents of the lapilli glasses, ash, and glass inclusions. Vesicles composed 34–45 vol.% of the samples examined – well below the fragmentation volume for subaerial eruptions (Wilson, 1980). Vesicles were poorly connected, in spite of their abundance. These characteristics are similar to basaltic scoria generated from subaerial hawaiian-style (lava fountain) eruptions. The ~1 km water depth corresponds to an ambient pressure of ~10 MPa. At this pressure, the presence of H_2O can affect the solubility of CO_2 in basalt (see Chapter 4), and this volatile coupling affects the resulting eruption style. Schipper *et al.* (2010) assert that hydromagmatic fragmentation likely occurs below the vent level of most submarine volcanoes, and that a strong coupling of volatile-phase bubbles with rising magma may be a requirement for efficient magma fragmentation.

Embley *et al.* (2006) report on the discovery of eruptive activity at NW Rota-1, a basaltic to basaltic andesite volcano located 60 km northwest of Rota island. The volcano's summit is ~500 m below sea level. This find marks the first direct observations and sampling of a submarine arc eruption, and provides opportunities for investigating the behavior of deep submarine volcanic explosions (Chadwick *et al.*, 2008a, 2012). During explosive events, the volcano emitted sulfur-rich eruption plumes, bubbles of CO_2, and basaltic andesite scoria (Embley *et al.*, 2006). Deardorff *et al.* (2011) report on a week's worth of explosive and effusive activity at NW Rota-1 in 2006. Activity ranged from diffuse gas venting to energetic explosions (see online supplement OS 12B).

Eruption plumes at NW Rota-1 had both momentum-driven and buoyancy-driven components. Video analyses show that momentum-driven plumes rose with velocities of ~1–4 m s^{-1}, whereas the buoyancy-driven plumes rose at a steady ascent rate of ~0.35 m s^{-1}. Based on Wilson and Self's (1980) model, Deardorff *et al.* (2011) estimated the efficiency of plume cooling by seawater by considering the plume ascent rate:

$$u = \sqrt{\frac{8rg\Delta\rho}{3\rho_a C_d}} \qquad (12.6)$$

in which u is the velocity of plume rise (m s^{-1}); r is the plume radius (measured from video); C_d is the drag coefficient (assumed to be 0.5 for a spherical plume); and $\Delta\rho$ is the density contrast between plume and seawater (kg m^{-3}). Deardorff *et al.* (2011) used measured velocities and plume radii, combined with known thermophysical properties of seawater (Fofonoff and Millard, 1983), to calculate the density contrast $\Delta\rho$, and then to relate $\Delta\rho$ to a required temperature contrast (neglecting the effect of volcanic particles on plume density). Observations of incandescence at the vent requires temperatures of at least 700 °C (Chadwick *et al.*, 2008a), and the calculations require temperatures on the order of 30 °C within the plume at ~1 m above the vent. This implies that efficient, rapid cooling (> 200 °C s^{-1}) of the plume occurs.

Interestingly, observations at NW Rota-1 clearly show that lava fragments can be repeatedly recycled within the erupting vent: individual fragments were observed to fall out of the plume and then become entrained again during the next explosion (cf. Section 8.3.3). Deardorff *et al.* (2011) estimate as much as 15% of the erupted fragments are repeatedly recycled in this way.

Direct observations of submarine explosive eruptions have revealed cyclic activity at both NW Rota-1 and at W Mata (15.10°S, 73.75°W; summit is 1174 m below sea level) (Rubin *et al.*, 2009; Smithsonian Institution, 2011; see OS 12B). It may be that this eruption cyclicity is related to the exsolution of magmatic gases, and accumulation of vesicles into a magmatic foam at the top of a magma chamber (cf. Wilson and Head, 2003; Chadwick *et al.*, 2008b, 2012). Additionally, these submarine volcanoes have experienced landslides or shallow sector collapses; it is likely that these landslides are caused by the generation of an unstable pile of pyroclastic materials near the volcanoes' summits (Chadwick *et al.*, 2008b, 2012). As in subaerial environments, the generation of pyroclasts can also lead to instabilities and gravity-driven flows on the sea floor.

Hydroacoustic monitoring of known active submarine volcanoes (such as Monowai; Chadwick *et al.*, 2008b) provides additional information about volcano behavior. These devices "listen" for volcanic activity, allowing scientists to target direct observations (using submersibles or remotely operated vehicles) on specific volcanoes shortly after – or even during – eruptions (e.g., Fox *et al.*, 2001; Chadwick *et al.*, 2008b, 2012). Additional deployments of these monitoring instruments will continue to provide important insights into the behavior of deep submarine explosive eruptions.

12.5 | Conclusions

Extreme explosive events are suppressed on the sea floor because of the high ambient pressure, although given the appropriate combination of parameters (both intrinsic and extrinsic), volcanic explosions can occur. The sea floor is dominated, however, by basaltic lavas that erupted effusively from mid-ocean ridges or from seamounts. High ambient pressures help keep volatiles dissolved in the lavas, precluding the formation of true 'a'a and other familiar lava morphologies on the sea floor. Furthermore, the efficient mechanism of convective heat transfer creates an outer, insulating glassy rind on any part of the lava flow exposed to seawater. This rind, if it remains more or less intact, acts to insulate the interior of the lava flow from additional heat loss, theoretically allowing it to travel farther underwater than it would if it were emplaced on land.

There are more questions than answers, however, regarding mid-ocean ridge volcanic eruptions. We do not know with certainty, for example, what the eruption frequencies and volumes are for most of the mid-ocean ridge system. Successful monitoring has begun to answer these questions in select locations: Axial Volcano along the Juan de Fuca Ridge, for example, and the East Pacific Rise between 9° and 10°N. We remain ignorant of how well the behaviors observed at these locations reflect volcanic behavior elsewhere along the ridge system.

12.6 | Summary

- Most of Earth's ocean crust is composed of basalt erupted from mid-ocean ridges; seamounts and oceanic plateaus also contribute to the formation of oceanic crust.
- Once believed to be impossible, explosive eruptions commonly occur on the deep sea floor, even with basaltic magma.
- Submarine explosive eruptions have been directly observed at NW Rota-1 and W Mata volcanoes in the Pacific Ocean (see OS 12B)
- Most submarine eruptions are basaltic and effusive, and demonstrate some amount of lava inflation.
- If eruption viscosity is known, submarine lava flow morphology can be used to constrain volumetric effusion rate.

12.7 | Notation

c_a	specific heat of seawater (J kg^{-1} K^{-1})
C_d	drag coefficient
F_{conv}	convective heat flux (J m^{-2} s^{-1})
g	gravitational acceleration (m s^{-1})
g'	reduced gravitational acceleration (m s^{-2})
q	linear effusion rate (m^2 s^{-1})
Q	volumetric effusion rate (m^3 s^{-1})
r	plume radius (m)
t_a	heat advection timescale (s)
t_s	lava solidification timescale (s)
T_a	temperature of ambient seawater (K)
T_c	temperature of lava crust (K)
u	velocity (m s^{-1})
α_a	coefficient of thermal expansion of seawater (K^{-1})
γ	dimensionless constant
κ_a	thermal diffusivity of seawater (m^2 s^{-1})
ν_a	kinematic viscosity of seawater (m^2 s^{-1})
ν_l	kinematic viscosity of lava (m^2 s^{-1})
ρ	density of lava (kg m^{-3})
$\Delta\rho$	density contrast between plume and seawater (kg m^{-3})
ρ_a	density of seawater (kg m^{-3})
Ψ	ratio of t_s to t_a

References

Allan, J. F., Batiza, R., Perfit, M. R., Fornari, D. J. and Sack, R. O. (1989). Petrology of lavas from the Lamont Seamount Chain and adjacent East Pacific Rise, 10°N. *Journal of Petrology*, **30**(5), 1245–1298.

Applegate, T. B. and Embley, R. W. (1992). Submarine tumuli and inflated tube-fed lava flows on Axial Volcano, Juan de Fuca Ridge. *Bulletin of Volcanology*, **54**, 447–458.

Baker, E. T., Resing, J. A., Lupton, J. E. *et al.* (2010). Multiple active volcanoes in the northeast Lau Basin, Abstract #T13B-2188, presented at *2010 Fall American Geophysical Union Meeting, San Francisco, CA, 13–17 December.*

Ballard, R. D. and Moore, J. G. (1977). *Photographic Atlas of the Mid-Atlantic Ridge Rift Valley.* New York: Springer, 114 pp.

Ballard, R. D. and van Andel, T. H. (1977). Morphology and tectonics of the inner rift valley at lat 36°50'N on the Mid-Atlantic Ridge. *Bulletin of the Geological Society of America*, **88**, 507–530.

Batiza, R. (1982). Abundances, distribution, and sizes of volcanoes in the Pacific Ocean and implications for the origin of non-hotspot volcanoes. *Earth and Planetary Science Letters*, **60**, 195–206.

Bridges, N. T. (1997). Characteristics of seamounts near Hawaii as viewed by GLORIA. *Marine Geology*, **138**, 273–301.

Byers, C. D., Garcia, M. O. and Muenow, D. W. (1986). Volatiles in basaltic glasses from the East Pacific Rise at 21°N: Implications for MORB sources and submarine lava flow morphology. *Earth and Planetary Science Letters*, **79**, 9–20.

Chadwick, W. W., Jr. (2003). Quantitative constraints on the growth of submarine lava pillars from a monitoring instrument that was caught in a lava flow. *Journal of Geophysical Research*, **108**, doi:10.1029/2003JB002422.

Chadwick, W. W., Jr. and Embley, R. W. (1998). Graben formation associated with recent dike intrusions and volcanic eruptions on the mid-ocean ridge. *Journal of Geophysical Research*, **103**, 9807–9825.

Chadwick, W. W., Jr., Gregg, T. K. P. and Embley, R. W. (1999). Submarine lineated sheet flows: A unique lava morphology formed on subsiding lava ponds. *Bulletin of Volcanology*, **61**, 194–206.

Chadwick, W. W., Jr., Cashman, K. V., Embley, R. W. *et al.* (2008a). Direct video and hydrophone observations of submarine explosive eruptions at NW Rota-1 Volcano, Mariana Arc. *Journal of Geophysical Research*, **113**, B08S10, doi:10.1029/2007JB005215.

Chadwick, W. W., Jr., Wright, I. C., Schwarz-Schampera, U. *et al.* (2008b). Cyclic eruptions and sector collapses at Monowai submarine volcano, Kermadec arc: 1998–2007. *Geochemistry Geophysics Geosystems*, **9**, doi:10.1029/2008GC002113.

Chadwick, W. W., Jr., Dziak, R. P., Haxel, J. H., Embley, R. W. and Matsumoto, H. (2012). Submarine landslide triggered by volcanic eruption recorded by in situ hydrophone. *Geology*, **40**, 51–54.

Clague, D. A., Paduan J. B., and Davis, A. S. (2009). Widespread strombolian eruptions of mid-ocean ridge basalt. *Journal of Volcanology and Geothermal Research*, **180**, 171–188.

Cochran, J. R. (2008). Seamount volcanism along the Gakkel Ridge, Arctic Ocean. *Geophysical Journal International*, **174**, 1153–1173.

Coffin, M. F. and O. Eldholm (1994). Large igneous provinces: Crustal structure, dimensions, and external consequences. *Reviews of Geophysics*, **32**, 1–36.

Deardorff, N. D., Cashman, K. V. and Chadwick, W. W., Jr. (2011). Observations of eruptive plume dynamics and pyroclastic deposits from submarine eruptions at NW Rota-1, Mariana arc. *Journal of Volcanology and Geothermal Research*, **202**, 47–59.

Dziak, R. P. and Fox, C. G. (1999). The January 1998 earthquake swarm at Axial Volcano, Juan de Fuca Ridge: Hydroacoustic evidence of seafloor volcanic activity. *Geophysical Research Letters*, **26**, 3429–3432.

Edwards, M. H., Kurras, G. J., Tolstoy, M. *et al.* (2001). Evidence of recent volcanic activity on the ultraslow-spreading Gakkel ridge. *Nature*, **409**, 808–812.

Embley, R. W., Chadwick, W. W., Jr., Perfit, M. R. and Baker, E. T. (1991). Geology of the northern Cleft segment, Juan de Fuca Ridge: Recent lava flows, sea-floor spreading, and the formation of megaplumes. *Geology*, **17**, 771–775.

Embley, R. W., Chadwick, W. W., Jr., Baker, E. T. *et al.* (2006). Long-term eruptive activity at a submarine arc volcano. *Nature*, **441**, 494–497.

Embley, R. W., Baker, E. T., Butterfield, D. A. *et al.* (2007). Exploring the submarine Ring of Fire. *Oceanography*, **20**, 68–79.

Fink, J. H. and Griffiths, R. W. (1990). Radial spreading of viscous-gravity currents with solidifying crust. *Journal of Fluid Dynamics*, **221**, 485–509.

Fink, J. H. and Griffiths, R. W. (1992). A laboratory analog study of the surface morphology of lava flows extruded from point and line sources. *Journal of Volcanology and Geothermal Research*, **54**, 19–32.

Fofonoff, P. and Millard, R. C., Jr. (1983). Algorithms for computation of fundamental properties of seawater. *UNESCO Technical Papers in Marine Science*, **44**, 1–53, http://www.es.flinders.edu.au/~mattom/Utilities/density.html.

Fornari, D. J. (2003). A deep-sea towed digital camera and multi-rock coring system. *Eos, Transactions of the American Geophysical Union*, **84**(8), 69–76.

Fornari, D. J., Batiza, R. and Luckmann, M. A. (1987). Seamount abundances and distribution near the East Pacific Rise 0°–24°N based on beam data. In *Seamounts, Islands and Atolls*, ed. B. H. Keating, P. Fryer, R. Batiza and G. W. Boehlert. American Geophysical Union Geophysical Monograph 43, Washington, DC, pp. 13–24.

Fornari, D. J., Haymon, R. M., Perfit, M. R., Gregg, T. K. P. and Edwards, M. H. (1998). Axial summit trough of the East Pacific Rise 9°–10°N; Geological characteristics and evolution of the axial zone on fast spreading mid-ocean ridge. *Journal of Geophysical Research*, **103**(B5), 9827–9855.

Fox, C. G. (1999). In situ ground deformation measurements from the summit of Axial Volcano during the 1998 volcanic episode. *Geophysical Research Letters*, **26**, 3437–3440.

Fox, C. G., Murphy, K. M. and Embley, R. W. (1988). Automated display and statistical analysis of interpreted deep-sea bottom photographs. *Marine Geology*, **78**, 199–216.

Fox, C. G., Chadwick, W. W., Jr. and Embley, R. W. (2001). Direct observation of a submarine volcanic eruption from a sea-floor instrument caught in a lava flow. *Nature*, **412**, 727–729.

Francheteau, J., Juteau, T. and Rangan, C. (1979). Basaltic pillars in collapsed lava-pools on the deep ocean floor. *Nature*, **281**, 209–211.

Fundis, A. T., Soule, S. A., Fornari, D. J. and Perfit, M. R. (2010). Paving the seafloor: Volcanic emplacement processes during the 2005–2006 eruptions at the fast spreading East Pacific Rise, 9°50'N. *Geochemistry Geophysics Geosystems*, **11**, doi:10.1029/2010GC003058.

Gregg, T. K. P. and Chadwick, W. W., Jr. (1996). Submarine lava-flow inflation: A model for the formation of lava pillars. *Geology*, **24**, 981–984.

Gregg, T. K. P. and Fink, J. H. (1995). Quantification of submarine lava flow morphology through analog experiments. *Geology*, **23**, 73–76.

Gregg, T. K. P. and Fink, J. H. (1996). Quantification of extraterrestrial lava flow effusion rates through laboratory simulations. *Journal of Geophysical Research*, **101**(E7), 16 891–16 900.

Gregg, T. K. P. and Fink, J. H. (2000). A laboratory investigation into the effects of slope on lava flow morphology. *Journal of Volcanology and Geothermal Research*, **96**, 281–292.

Gregg, T. K. P. and Fornari, D.J. (1998). Long submarine lava flows: Observations and results from numerical modeling. *Journal of Geophysical Research*, **103**(B11), 27 517–27 531.

Gregg, T. K. P. and Greeley, R. (1993). Formation of venusian canali: Considerations of lava types and their thermal behaviors. *Journal of Geophysical Research*, **98**(E6), 10 873–10 882.

Gregg, T. K. P., Fornari, D. J., Perfit, M. R., Haymon, R. M. and Fink, J. H. (1996). Rapid emplacement of a mid-ocean ridge lava flow on the East Pacific Rise at 9°46'–51'N. *Earth and Planetary Science Letters*, **144**, E1–E7.

Griffiths, R. W. and Fink, J. H. (1992). Solidification and morphology of submarine lavas: A dependence on extrusion rate. *Journal of Geophysical Research*, **97**(B13), 19 729–19 737.

Guest, J. E., Kilburn, C. R. J., Pinkerton, H. and Duncan, A. M. (1987). The evolution of lava flow-fields: Observations of the 1981 and 1983 eruptions of Mount Etna, Sicily. *Bulletin of Volcanology*, **49**, 527–540.

Haymon, R. M., Fornari, D. J., von Damm, K. L. *et al.* (1993). Volcanic eruption of the mid-ocean ridge along the East Pacific Rise crest at 9°45'–52'N: Direct submersible observation of seafloor phenomena associated with an eruption event in April 1991. *Earth and Planetary Science Letters*, **119**, 85–101.

Head, J. W. and Wilson, L. (2003). Deep submarine pyroclastic eruptions: Theory and predicted landforms and deposits. *Journal of Volcanology and Geothermal Research*, **121**, 155–193.

Hekinian, R., Pineau, F., Shilobreeva, S. *et al.* (2000). Deep sea explosive activity on the Mid-Atlantic Ridge near 34°50'N: Magma composition, vesicularity, and volatile content. *Journal of Volcanology and Geothermal Research*, **98**, 49–77.

Hon, K., Kauahikaua, J., Denlinger, R. and Mackay, K. (1994). Emplacement and inflation of pahoehoe sheet flows: Observations and measurements of active lava flows on Kilauea Volcano, Hawaii. *Geological Society of America Bulletin*, **106**, 351–370.

Humphris, S. E., Bryan, W. B., Thompson, G. and Autio, L. K. (1990). Morphology, geochemistry and evolution of Serocki volcano. In *Proceedings of the Ocean Drilling Program, Scientific Results, 106/109*, ed. R. S. Detrick, J. Honnorez, W. B. Juteau *et al.* College Station, Texas: Ocean Drilling Program, pp. 67–84.

Ito, G., Lin, J. and Graham, D. (2003). Observational and theoretical studies of the dynamics of mantle plume–mid-ocean ridge interaction. *Reviews of Geophysics*, **41**, 1017, doi:10.1029/2002RG000117.

Jakobsson, S. P. and Gudmundsson, M. T. (2008). Subglacial and intraglacial volcanic formations in Iceland. *Jokull*, **58**, 179–196.

Jones, J. G. (1970). Intraglacial volcanoes of the Laugarvatn region, southwest Iceland II. *Journal of Geology*, **78**, 127–140.

Klingelhofer, F., Hort, M., Kumple, J.-J. and Schmincke, H.-U. (1999). Constraints on the formation of submarine lava flows from numerical model calculations. *Journal of Volcanology and Geothermal Research*, **92**, 215–229.

Korenaga, J. and Hey, R. (1996). Recent dueling propagation history at the fastest spreading center, the East Pacific Rise, 26°–32°S. *Journal of Geophysical Research*, **101**(B8), 18 023–18 041.

Langmuir, C. H., Bender, J. F. and Batiza, R. (1986). Petrological and tectonic segmentation of the East Pacific Rise, 5°30'–14°30'N. *Nature*, **322**, 422–429.

Macdonald, K. C. (1998). Faulting, volcanism, hydrothermal activity on fast spreading centers. In *Faulting and Magmatism at Mid-ocean Ridges*. Washington, D.C.: American Geophysical Union Geophysical Monograph 106, pp. 27–58.

Neal, C. R., Mahoney, J. J., Kroenke, L. W., Duncan, R. A. and Petterson, M. G. (1997). The Ontong Java Plateau. In *Large Igneous Provinces: Continental, Oceanic, and Planetary Flood Volcanism*, ed. J. J. Mahoney and M. F. Coffin. Washington, DC: American Geophysical Union Geophysical Monograph 100, pp. 183–216.

Ocean Drilling Program (1991). Ontong Java Plateau, *Scientific Results of the Ocean Drilling Program*, **130**.

Ocean Drilling Program (2001). Basement drilling of the Ontong Java Plateau, *Scientific Results of the Ocean Drilling Program*, **192**.

Perfit, M. R. and Chadwick, W. W., Jr. (1998). Magmatism at mid-ocean ridges: Constraints from volcanological and geochemical investigations. In *Faulting and Magmatism at Mid-ocean Ridges*. Washington, DC: American Geophysical Union Geophysical Monograph 106, pp. 59–115.

Perfit, M. R., Fornari, D. J., Smith, M. C. *et al.* (1995). Small-scale spatial and temporal variations in mid-ocean ridge crest magmatic processes. *Geology*, **22**, 375–379.

Perfit, M. R., Cann, J. R., Fornari, D. J. *et al.* (2003). Interaction of sea water and lava during submarine eruptions at mid-ocean ridges. *Nature*, **426**, 62–65.

Rubin, K. H., Smith, M. C., Bergmanis, E. C. *et al.* (2001). Magmatic history and volcanological insights from individual lava flows erupted on the sea floor. *Earth and Planetary Science Letters*, **188**, 349–367.

Rubin, K. H., Embley, R. W., Clague, D. A. *et al.* (2009). Lavas from active boninite and very recent basalt eruptions at two submarine NE Lau Basin sites. *Eos, Transactions of the American Geophysical Union*, **89**, *Fall Meeting Supplement*, abstract V43I-05.

Ryan, W. B. F., Carbotte, S. M., Coplan, J. *et al.* (2009). Global multi-resolution topography (GMRT) synthesis data set. *Geochemistry Geophysics and Geosystems*, **10**, Q03014, doi:10.1029/2008GC002332.

Schiffman, P., Zierenberg, R., Chadwick, W. W., Jr., Clague, D. A. and Lowenstern, J. (2010). Contamination of basaltic lava by seawater: Evidence found in a lava pillar from Axial Seamount, Juan de Fuca Ridge. *Geochemistry Geophysics and Geosytems*, **11**, doi:10.1029/2009GC003009.

Schipper, C. I., White, J. D. L., Houghton, B. F., Shimizu, N. and Stewart, R. B. (2010). "Poseidic" explosive eruptions at Loihi Seamount, Hawaii. *Geology*, **38**, 291–294.

Self, S., Keszthelyi, L. P. and Thordarson, Th. (1998). The importance of pahoehoe. *Annual Review of Earth and Planetary Sciences*, **26**, 81–110.

Sinton, J. M. and Detrick, R. S. (1992). Mid-ocean ridge magma chambers. *Journal of Geophysical Research*, **97**(B1), 197–216.

Sinton, J. M., Smaglik, S. M., Mahoney, J. J. and Macdonald, K. C. (1991). Magmatic processes at superfast spreading mid-ocean ridges: glass compositional variations along the East Pacific Rise, 13°–23°S. *Journal Geophysical Research*, **96**, 6133–6155.

Sinton, J. M., Bergmanis, E., Rubin, K. *et al.* (2001). Volcanic eruptions on mid-ocean ridges: New evidence from the superfast spreading East Pacific Rise, 17°S–19°S. *Journal of Geophysical Research*, **107**(B6), doi:10.1029/2000JB000090.

Smith, D. K. and Cann, J.R. (1992). The role of seamount volcanism in crustal construction at the mid-Atlantic Ridge. *Journal of Geophysical Research*, **97**(B2), 1645–1658.

Smithsonian Institution (2011). Global Volcanism Program, www.volcano.si.edu/info/about/about_votw.cfm.

Sohn, R. A., Willis, C., Humphris, S. *et al.* (2008). Explosive volcanism on the ultraslow-spreading Gakkel Ridge, Arctic Ocean. *Nature*, **453**, doi:10.1038/nature07075.

Soule, S. A. and Cashman, K. V. (2004). The mechanical properties of solidified polyethylene glycol 600, an analog for lava crust. *Journal of Volcanology and Geothermal Research*, **129**, 139–153.

Soule, S. A., Fornari, D. J., Perfit, M. R. *et al.* (2005). Channelized lava flows at the East Pacific Rise Crest 9°–10°N: The importance of off-axis lava transport in developing the architecture of young oceanic crust. *Geochemistry Geophysics and Geosystems*, **16**, doi:10.1029/2005GC000912.

Soule, S. A., Fornari, D. J., Perfit, M. R. *et al.* (2006). Incorporation of seawater into mid-ocean ridge lava flows during emplacement. *Earth and Planetary Science Letters*, **252**, 289–307.

Soule, S. A., Fornari, D. J., Perfit, M. R. and Rubin, K. (2007). New insights into mid-ocean ridge volcanic processes from the 2005–2006 eruption of the East Pacific Rise, 9°46'N–9°56'N. *Geology*, **35**, 1079–1082.

Soule, S. A., Escartin, J. and Fornari, D. J. (2009). A record of eruption and intrusion at a fast spreading ridge axis: Axial summit trough of the East Pacific Rise at 9–10°N. *Geochemistry Geophysics and Geosystems*, **10**, doi:10.1029/2008GC002354.

Stakes, D. S., Perfit, M. R., Tivey, M. A. *et al.* (2006). The Cleft revealed: Geologic, magnetic, and morphologic evidence for construction of upper oceanic crust along the southern Juan de Fuca Ridge. *Geochemistry Geophysics and Geosystems*, **7**, doi:1029/2005GC001038.

Stern, R. J., Tamura, Y., Leybourne, M. I. *et al.* (2010). Felsic magmatism in intra-oceanic arcs: The Diamante Cross-chain in the Southern Mariana Arc, Abstract #V14A-02, presented at *2010 American Geophysical Union Fall Meeting, San Francisco, CA, 13–17 December.*

Tamura, Y., Ishizuka, O., Stern, R. J. *et al.* (2010). Two primary magma types from Northwest Rota-1 Volcano, Mariana Arc, Abstract #T13B-2197, presented at *2010 American Geophysical Union Fall Meeting, San Francisco, CA, 13–17 December.*

Tepley, L. and Moore, J. G. (1974). *Fire Under the Sea: The Origin of Pillow Lava*. Mountain View, CA: Moonlight Productions.

Tolstoy, M., Cowen, J. P., Baker, E. T. *et al.* (2006). A seafloor spreading event captured by seismometers. *Science*, **314**, 1920–1922.

Torresan, M. E., Clague, D. A. and Jacobs, C. L. (1991). Cruise report, Hawaiian GLORIA Cruise F12-89-HW. United States Geological Survey, Open-File Report 67.

Tribble, G. W. (1991). Underwater observations of active lava flows from Kilauea volcano. Hawaii. *Geology*, **19**, 633–636.

Valentine, G. A. and Hirano, N. (2010). Mechanisms of low-flux intraplate volcanic fields – Basin and Range (North America) and northwest Pacific Ocean. *Geology*, **38**, 55–58.

Wanless, V. D., Perfit, M. R., Ridley, W. I. and Klein, E. (2010). Dacite petrogenesis on mid-ocean ridges: Evidence for oceanic crustal melting and assimilation. *Journal of Petrology*, **15**, 2377–2410.

Wessel, P., Sandwell, D. T. and Kim, S.-S. (2010). The global seamount census. *Oceanography*, **23**, 24–33.

White, J. D. L., Smellie, J. L. and Clague, D. A. (2003). Introduction. In *Explosive Subaqueous Volcansim*, ed. J. D. L. White, J. L. Smellie and D. A. Clague. Washington, D.C.: American Geophysical Union Monograph 140, pp. 1–25.

White, S. M., Macdonald, K. C. and Haymon, R. M. (2000). Basaltic lava domes, lava lakes, and volcanic segmentation on the southern East Pacific Rise. *Journal of Geophysical Research*, **105**(B10), 23 519–23 536.

White, S. M., Mason, J. L., Macdonald, K. C. *et al.* (2009). Significance of widespread low effusion rate eruptions over the past two million years for delivery of magma to the overlapping spreading centers at 9°N. *Earth and Planetary Science Letters*, **280**, 175–184.

Wilson, J. T. (1963). A possible origin of the Hawaiian islands. *Canadian Journal of Physics*, **41**, 863–870.

Wilson, L. (1980). Relationships between pressure, volatile content and ejecta velocity in three types of volcanic explosion. *Journal of Volcanology and Geothermal Research*, **8**, 297–313.

Wohletz, K. H. (2003). Water/magma interaction: Physical considerations for the deep submarine environment. In *Explosive Subaqueous Volcanism*, ed. J. D. L. White, J. L. Smellie and D. A. Clague. Washington, D.C: American Geophysical Union Monograph 140, pp. 26–50.

Exercises

12.1 Calculate the heat flux from a submarine and subaerial lava flow, using information from this chapter and Chapter 5 (specifically, Eq. (5.15) for radiative cooling). Demonstrate which mechanism of heat transfer is most efficient in each environment.

12.2 Use Eq. (12.6) to calculate the density contrast present in a buoyancy-driven submarine eruption plume that has a rise velocity of 0.35 m s^{-1} and radius of 1 m.

12.3 Expand Figure 12.6 to incorporate a typical EPR eruption viscosity of 100 Pa s (Perfit and Chadwick, 1998; Gregg *et al.*, 2006). What effusion rates are required to generate a pillowed lava flow? A rifted lava flow?

Online resources available at www.cambridge.org/fagents

- OS 12A Additional figures
- OS 12B Video of submarine eruptions
- Additional reading
- Links to websites of interest
- Answers to exercises

Chapter 13

Volcano–ice interactions

Lionel Wilson, John L. Smellie, and James W. Head

Overview

This chapter reviews basic physical processes controlling interactions between silicate magmas and surface ice and snow layers, focusing on subglacial, englacial, and supraglacial interactions. Where possible, theoretical considerations are linked with observations of the lithofacies and sequence characteristics of the deposits expected as a result of these various interactions, with particular focus on the products of mafic eruptions. The range of possible interactions is large, resulting in a correspondingly diverse group of resulting landforms. These predictions are made for the environment of the Earth, but with suitable changes to atmospheric temperature and pressure and acceleration due to gravity are readily applicable on Mars. Numerous putative examples of volcano–ice interaction features on Mars have already been documented and this chapter provides a comprehensive unifying theoretical framework for further interpretation of features on both planets.

13.1 | Introduction

Magma–ice interactions can occur in a number of ways and can produce a range of products

and landforms (e.g., Lescinsky and Fink, 2000; Mee *et al.*, 2006; Komatsu *et al.*, 2007; Larsen and Eiriksson, 2008; Smellie, 2009), the details depending on the geometry and timescale of the interaction. No subglacial rhyolite eruptions have ever been observed. A "typical" mafic volcanic eruption progresses from initial rapid subsidence and collapse of the overlying ice surface to form a pit, simultaneous with subglacial emplacement of volcanic products (often but not always pillow lava, forming a pillow mound or ridge) in a water-filled cavity. Many eruptions might cease at this point but, commonly, as the volcanic edifice grows upward and the vent becomes shallower, the magma interacts explosively with the surrounding meltwater and a high subaerial eruption column is generated, accompanied by deposition of abundant ash. This results in the construction of a subaqueous tuff cone or ridge, the latter known as a *tindar* (Jones, 1969). If the edifice grows above the surface of the surrounding ice sheet and the vent dries out, the eruption may change to lava effusion and/or lava fountaining. Lava-fed deltas then form where lava enters and advances across the surrounding meltwater lake, resulting in distinctive flat-topped volcanic constructs known as a *tuyas* (Mathews, 1947) or *stapi*. In the latest stages, the ice cover is often floated, and a large volume of meltwater

Modeling Volcanic Processes: The Physics and Mathematics of Volcanism, eds. Sarah A. Fagents, Tracy K. P. Gregg, and Rosaly M. C. Lopes. Published by Cambridge University Press. © Cambridge University Press 2013.

escapes in a sudden, rapid, and environmentally devastating discharge known as a glacial outburst flood or *jökulhlaup*. Several of these features were displayed during the exceptionally well-documented eruption at Gjálp, Iceland, in 1996, which took place under the 700-m-thick Vatnajökull ice cap (Gudmundsson *et al.*, 1997, 2002, 2004; Jarosch *et al.*, 2008). Although the spectacular April–May 2010 eruption at Eyjafjallajökull (also Iceland) took place under much thinner ice (just 200–300 m), it also demonstrated many similar aspects of volcano–ice interactions, in particular a high, explosively generated tephra-bearing eruption column and meltwater floods that washed away the local highway. Unlike the Gjálp eruption, the effects of which were limited to Iceland and included the creation of new land as well as serious and costly damage to the local transport infrastructure, the Eyjafjallajökull eruption distributed ash over much of northern Europe. It is estimated that about $1.7 billion was lost to the EU economy as a result principally of flights that had to be cancelled (see Eyjafjallajökull links at end of chapter). In this chapter the human impact of volcano–ice eruptions is not described. Rather, the focus is on providing a theoretical framework of how mafic magmas might interact with ice under different situations, and examples are provided (where available) that show what the resulting rock record might look like.

13.1.1 Three generic scenarios for volcano–ice interactions

Ice, in the form of glaciers, snow-fall or frozen lakes, commonly occurs as a sheet overlying a silicate rock surface. Magma penetrating the shallow crust travels in brittle fractures, i.e., dikes and sills. Possible initial geometries of the interaction between ice and magma are (a) injection of a magmatic dike into a layer of ice; (b) formation of what is initially a sill-like magma intrusion beneath ice at the ice–rock interface, though this may ultimately develop into what might be thought of as a subglacial lava flow; and (c) emplacement on top of ice of a lava flow or pyroclastic material originating elsewhere (Wilson and Head, 2002, 2007, 2009).

Case (a), injection of magma into an ice layer as a dike, is predicted to be a possibility on theoretical grounds (Wilson and Head, 2002). The inherent instability of the resulting geometry means that a fragmental deposit surrounding the vent (fissure) at the base of the ice sheet is likely to be produced at the end of the interaction. The subsequent evolution of the subglacial volcanic mass might involve construction of a pillow mound, then an explosively generated subaqueous tuff cone and ultimately a subaerial edifice with or without lava-fed deltas, as described for many glaciovolcanic edifices (e.g., Jones, 1969, 1970; Smellie, 2000). Some of these eruptions might have been associated with jökulhlaups.

In case (b), melting of ice overlying a sill intrusion at the ice base can produce a large volume of water (Höskuldsson and Sparks, 1997; Wilson and Head, 2002, 2007; Tuffen, 2007). The water may accumulate relatively slowly, but in the case of a large body of water accumulated under a glacier it may eventually escape very rapidly in the form of a jökulhlaup. If the water escapes at the glacier edge and is replaced by air, and the overlying ice does not deform too quickly, the sill may evolve into what is effectively a subaerial lava flow in an ice cave, resulting in a range of possible features (Wilson and Head, 2002, 2007). Key issues then are the relative values of the rate of advance of the magma sheet beneath the ice, the rate of advance of thermal waves penetrating both the unmelted ice and the magma itself, and the rate of drainage of the meltwater produced. The evolution of such an eruption is less well known but at least some might be associated with construction of pillow mounds or ridges (Smellie, 2008).

In case (c), the presence of the ice plays no part in the magmatic eruption. If the eruption involves deposition of a pyroclastic fall deposit onto ice, the consequences depend on the rate of thickening of the deposit and its final total thickness (e.g., Wilson and Head, 2007, 2009). Except very close to the vent, fall deposits are likely to consist of clasts with temperatures close to the ambient atmospheric temperature, and so immediate heating of underlying ice will be minimal. Longer term, the fact that the pyroclasts will have a lower albedo than the ice, and hence will reach higher daytime temperatures during the diurnal solar heating cycle, may be important. If the volcanic material is hot on emplacement, e.g., a lava flow or a pyroclastic density current deposit, the effects on the underlying ice will depend on both the rate of advance and thickness of the flowing material, and the rate at which water can migrate beneath the advancing deposit to escape at its edges.

We now examine the above processes theoretically, and suggest examples of the deposits that might form from them, to facilitate the kinds of field observations needed to constrain the models, and to identify issues so far overlooked.

13.2 | Englacial dike emplacement

13.2.1 Theoretical issues

Magma propagation at shallow depths in planetary bodies takes place in brittle fractures (dikes and sills), held open by a combination of magma pressure and the stress field in the host rocks (Pollard, 1987; Rubin and Pollard, 1987; see Chapter 3). Some reach the surface to feed eruptions, but others stall as intrusions, and these may cause surface manifestations such as bedrock graben or ice fractures (crevasses) if their tops are sufficiently close to the surface (Mastin and Pollard, 1988; Rubin, 1992; Gudmundsson et al., 2004). If the least principal stress changes from horizontal to vertical, a dike will cease to propagate upward but may still propagate laterally to form a sill. A dike approaching the surface in a location capped by a substantial thickness of ice is likely to encounter such a stress change, which, coupled with a discontinuity in material properties, may encourage sill injection at the ice–rock interface. Such sills are expected to evolve rapidly into subglacial lava flows (Wilson and Head, 2002; Smellie, 2008; see Section 13.3).

The density structure of the crust in volcanic provinces is commonly such that mafic magma generated in the mantle is positively buoyant in its source zone but negatively buoyant near the surface. Magma reservoirs are likely to form from the accumulation of multiple stalled intrusions at intermediate levels of neutral magma buoyancy (Ryan, 1987). Dikes are driven upward from these reservoirs by excess pressures, due to the positive buoyancy of the melts feeding them from below, that are typically at least several MPa (Parfitt, 1991). Thus for reservoir depths of ~3 km the pressure gradient in excess of the weight of the magma driving magma motion is ~1000 Pa m^{-1}. Where mafic magma reaches the surface directly from lower crustal or mantle depths, net density differences between magma and host rocks are generally ~100 kg m^{-3} (Parfitt et al., 1993). This value, combined with the acceleration due to gravity of ~10 m s^{-2}, also leads to pressure gradients of order 1000 Pa m^{-1}. Widths of mafic dikes propagating from shallow reservoirs are ~1–3 m (Parfitt, 1991), and pressure gradients of this order drive mafic magma rise speeds of ~1 m s^{-1} (Wilson and Head, 1981). The strain rates near the dike tips implied by these speeds are ~1 s^{-1}, about seven orders of magnitude larger than the strain rates at which the surrounding ice can flow plastically given the rheological models (a pseudoplastic power-law fluid with a yield strength) proposed by Glen (1952), Nye (1953) and Paterson (1994). Thus a mafic dike can easily overshoot an ice–rock interface because the ice appears to the propagating crack as a brittle, low-density rock with elastic properties similar to those of the substrate (Wilson and Head, 2002). The time required for a dike containing magma propagating at 1 m s^{-1} to penetrate tens to hundreds of meters into ice is tens to hundreds of seconds, and with both ice and rock having thermal diffusivities of ~10^{-6} m^2 s^{-1}, the distance that heat can be conducted, chilling the magma and warming and then melting the ice, is ~20 mm. Thus the initial penetration of mafic dikes into ice may be a stable process, though subsequent, more extensive ice melting may lead to collapse of the dike.

Consider a mafic dike that has propagated from a magma reservoir whose top is at a depth z below an ice–rock interface (a list of all notation is given in Section 13.6). We adopt $z = 3$ km as typical of many mafic reservoirs (Ryan, 1987). The ice thickness is y (Fig. 13.1) and the top of

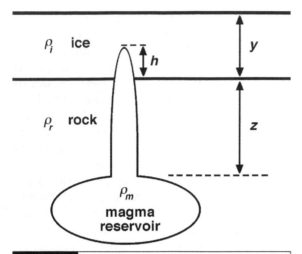

Figure 13.1 Geometry of a dike rising from a magma reservoir with its top located at a depth z below a rock surface on which an ice layer of thickness y is present. The dike tip penetrates to a distance h into the ice. The densities of ice, crustal rock, and magma are indicated.

the dike has come to rest having penetrated a distance h into the ice. We assume that y may be as large as 4 km. This is based on ice-cap thicknesses in Iceland measured under current conditions (up to ~900 m; Sigmundsson and Einarsson, 1992; Einarsson, 1994) and estimated for glacial periods (1000–1500 m; Einarsson and Albertsson, 1988; Geirsdóttir and Eiríksson, 1994; Bourgeois et al., 1998), and on thicknesses in the volcanically active West Antarctic Rift System (2–4 km; Blankenship et al., 1993; Behrendt et al., 2002; Corr and Vaughan, 2008). The crustal rock is assumed to be accumulated volcanics with a bulk density, ρ_r, of 2300 kg m^{-3} and the ice density, ρ_i, is taken as 900 kg m^{-3} to allow for the presence of a few tens of meters of low-density firn on top of compacted ice with density 917 kg m^{-3}. The excess pressure, ΔP, in the magma reservoir is initially assumed to be 5 MPa, a typical value inferred for the shallow mafic magma reservoir of Kilauea volcano (Parfitt, 1991). The pressure distribution in propagating dikes adjusts to maximize the pressure gradient driving flow of the magma against wall friction (Rubin, 1992). As a result, the pressure in the dike tip, P_t, is buffered at a low value determined by exsolution of volatiles that accumulate as a free gas phase in a cavity at the

upper tip of the dike. Various ways of estimating the gas pressure have been suggested (Lister and Kerr, 1991; Rubin, 1995) but the control is likely to be the requirement that the volume fraction, f, of gas bubbles at the base of the region of free gas is ~0.75–0.85, the value at which the underlying bubble foam becomes unstable and drains, causing magma fragmentation (Jaupart and Vergniolle, 1989; see also Chapter 4). If n_t is the total mass fraction of volatiles with average molecular weight m that have exsolved from the magma, the relative volumes of gas, V_g, and magma, V_m, in the collapsing foam are $V_g = (n_t / \rho_g)$ and $V_m = [(1 - n_t) / \rho_m]$, where ρ_g is the gas density and ρ_m is the density of the magmatic liquid. Assuming perfect gas behavior, $\rho_g = (m\,P_t)/(R\,T_m)$ where T_m is the magma temperature. By definition $f = V_g/(V_g + V_m)$ and so

$$P_t = \frac{n_t(1-f)\rho_m R T_m}{mf(1-n_t)}. \tag{13.1}$$

The data presented in Chapter 4 suggest that, if the pressure is low enough that most of the common volatiles, ~0.7 wt.% H_2O and ~900 ppm CO_2, have been released, then $n_t \approx 0.008$, and $m \approx$ 21 kg kmol^{-1}. Then with $f = 0.8$, $\rho_m = 2700$ kg m^{-3}, T_m for a mafic magma = 1473 K and $R = 8.314$ kJ kmol^{-1} K^{-1}, we find $P_t \approx 3.2$ MPa. When the magma comes to rest, this pressure will still be present in the gas in the dike tip. The distance, h, that the dike penetrates into the ice can be found by equating the pressure at the base of the magma column, equal to the tip pressure plus the weight of the magma, $P_t + (h + z)\,g\,\rho_b$, to the pressure in the magma reservoir, equal to the local lithostatic pressure plus the excess internal pressure, $P_a + y\,g\,\rho_i + z\,g\,\rho_r + \Delta P$, where P_a is the atmospheric pressure, ~ 0.1 MPa. The bulk density, ρ_b, of the magma in the dike will be a little less than the magmatic liquid density, ρ_m, due to the gas bubble foam below the gas tip cavity; examples in Wilson and Head (2002) suggest $\rho_b \approx 2500$ kg m^{-3}. Solving for h:

$$h = \frac{z(\rho_r - \rho_b)}{\rho_b} + y\frac{\rho_i}{\rho_b} + \frac{\Delta P + P_a - P_t}{g\rho_b}. \tag{13.2}$$

The relative controls on h can be assessed by comparing the sizes of the three terms on the

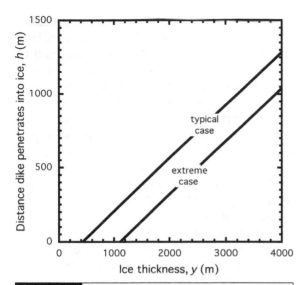

Figure 13.2 Variation of h, the distance that a dike tip can penetrate into a surface ice layer, as a function of the ice layer thickness, y, for typical and extreme conditions discussed in the text.

right of this equation. Since ρ_r is less than ρ_b, $[z\,(\rho_r - \rho_b)\,/\,\rho_b]$ is negative, and with $z = 3$ km, this term is of order 240 m. ΔP and P_t are of similar size and their difference will be a few MPa, so $[(\Delta P + P_a - P_t)/(g\,\rho_b)]$ may be positive or negative and of order 100 m. In the extreme case, with the sum of these terms equal to, say, –400 m, the ice thickness y will need to be at least ~1100 m before dike penetration into the ice is likely to occur. Figure 13.2 shows the variation of h with y for this extreme scenario and for a more plausible case where $\Delta P \approx 5$ MPa, $P_t \approx 3$ MPa, and so $[(\Delta P + P_a - P_t)/(g\,\rho_b)]$ is positive and ~80 m. In this case, ice penetration is possible when y is greater than ~430 m, and for very great ice thicknesses the penetration distance could in principle be many hundreds of meters. Also, the value $\Delta P \approx 5$ MPa may be conservative. For example, a magma reservoir being fed from a depth of 10 km by mantle magma that is buoyant relative to the host rocks by 200 kg m^{-3} will produce an excess pressure of ~20 MPa, not 5 MPa, allowing an additional 600 m of ice penetration in both cases. Finally, a rhyolitic magma, or a mafic magma that has exsolved significant CO_2 in its reservoir, may have ρ_b significantly less than the 2500 kg m^{-3} used above.

If, say, 2000 kg m^{-3} is used, then a further 210 m of penetration is predicted. Clearly, complete penetration of ice layers up to ~2000 m thick is possible given favorable circumstances.

Magma composition will influence the above results due to the effects of magma rheology on the timescales involved. Dike widths for high-viscosity magmas will be at least a few times larger than for mafic dikes, but magma viscosities may be several orders of magnitude larger (e.g., Wilson et al., 2007). The corresponding magma rise speed and dike-tip strain rate for a rhyolite with viscosity 10^6 Pa s rising through a 3-m-wide dike under a similar pressure gradient are ~10^{-3} m s^{-1} and 3×10^{-4} s^{-1}, so completely elastic deformation of the ice is not guaranteed. Also, thermal interaction times are so much longer that significant ice melting, with its consequent volume reduction, will change the stress distribution around the dike tip. However, Tuffen and Castro (2009) described a rhyolitic dike that penetrated a 35–55-m-thick layer of glacial ice and firn to build an 80-m-high obsidian ridge.

13.2.2 Phreatomagmatic activity from mafic englacial dikes

When mafic dikes do not initially penetrate completely through ice layers, explosive activity may still develop shortly after emplacement ceases (Wilson and Head, 2002). The important parameter is the difference, P_d, between the dike tip pressure P_t and the ambient stress in the surrounding ice. If the stress in the ice is isotropic, $P_a = [P_t - g\rho_i(y - h)]$, and so Eq. (13.2) can be used to find the variation of P_d with penetration distance h for various values of ice thickness y. Figure 13.3 gives the results for the two cases shown in Figure 13.2. The stress changes from compressive (positive) to tensile when the penetration distance exceeds ~290 m in the typical case and is tensile under all conditions in the extreme case. Furthermore, the tensile strength of ice at the strain rates relevant here is ~1.6 MPa (Lange and Ahrens, 1983). Therefore, the stress conditions shown in Figure 13.3 render it very likely that, if the penetration distance exceeds ~575 m in the typical case or ~190 m in the extreme case, the ice around the dike tip

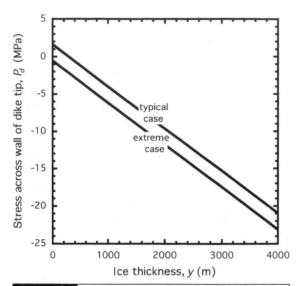

Figure 13.3 Variation of the stress difference, P_d, acting across the walls of a dike tip as a function of the distance that a dike tip penetrates into a surface ice layer, h, for the same conditions as Fig. 13.2.

will fracture in tension. Blocks of ice will fall through the gas-filled cavity onto the top of the magma column, initiating thermal and mechanical interactions. If progressive collapse occurs to the extent that a pressure pathway to the surface of the ice is formed, the excess water vapor pressure in the dike tip will be vented to the atmosphere and the consequent unloading of the magma will lead to further magma vesiculation as more volatiles exsolve, with the possible onset of explosive activity reaching the surface of the ice. Such activity would almost certainly be phreatomagmatic because of the intimate contact between magma, water, and spalled ice blocks.

The most extreme activity would occur if intimate mechanical mixing of ice blocks and magma took place so rapidly that no change in the total volume was possible. This process, involving the mixing of equal volumes of ice and magma, would probably be less extreme than some explosive interactions between surface lava flows and surface water, where the maximum energy release occurs at a melt/water ratio of ~2 (Wohletz, 1986; see Chapter 11). The magma would cool from its initial temperature,

T_m, to approach some equilibrium temperature, T_e (Chapter 11, Section 11.4.2). The ice, initially at the triple point $T_i = 273.15$ K, would melt to form a fluid heated toward the same equilibrium temperature. The fluid could be liquid water, water vapor, a mixture of the two, or, if the equilibrium pressure and temperature were above the critical point, a single-phase supercritical fluid. When equal volumes V of ice and magma mix, they reach equilibrium by sharing the total available thermal energy. The heat lost by the magma is $[V\rho_m c_m(T_m - T_e)]$, where c_m is its specific heat, ~900 J kg⁻¹ K⁻¹, averaged over the relevant temperatures (Dobran, 2001). The heat lost by the magma components can be equated to the energy required to heat the water fluid to T_e, equal to $[V\rho_i(L_i + E(T_e, P_e)]$, where L_i is the latent heat of fusion of the ice, 3.33×10^5 J kg⁻¹, and $E(T_e, P_e)$ is the heat energy (enthalpy) of the resulting H_2O fluid at the final equilibrium temperature and pressure, resulting in

$$\rho_i\left(L_i + E(T_e, P_e)\right) = \rho_m c_m(T_m - T_e). \tag{13.3}$$

This treatment neglects the heat contributed by the magmatic volatiles, but as they represent only ~1% of the magma mass, and hence heat budget, this is a small error. Values of $E(T_e, P_e)$ as a function of pressure and of temperature in excess of the triple point can be obtained from standard tables (ASME, 2000), and the equation can be solved recursively from an initial estimate of P_e.

Figure 13.4 shows values of the final equilibrium temperatures, T_e, and pressures, P_e, for a range of pre-mixing dike tip pressures, P_t, from 2 to 6 MPa, which Eq. (13.1) shows correspond to total magma volatile contents from 0.005 to 0.015, i.e., 0.5 to 1.5 wt.%, covering most of the range suggested for mafic melts in Chapter 4. Comparison of the values in Figure 13.4 with the saturation curve for H_2O shows that at values of T_e greater than ~590 K, corresponding to $P_t >$ 3.25 MPa and $n_t > $ ~0.8 wt.% total volatiles, the fluid is a liquid, whereas at lower temperatures it is a vapor. There is a much greater volume change on expanding hot liquid rather than hot vapor to atmospheric pressure. This implies that if the pressure pulse suggested by the above treatment causes a fracture to propagate to the

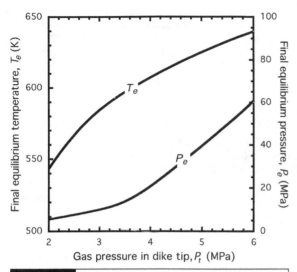

Figure 13.4 Variation of the final equilibrium temperature, T_e, and pressure, P_e, in an intimate mixture of magma and ice after the ice has melted to form a fluid, shown as a function of the pressure P_t in the propagating dike tip, an indicator of the magma H_2O content.

surface of the ice, the most violent phreatomagmatic activity will result from the intrusion of a volatile-rich magma. Whatever the details of the physical interaction between dike magma and surrounding ice, the initially high rate of heat transfer decreases with time. The absolute maximum amount of ice that might eventually be melted by a given dike can be found by equating (i) the heat lost by the magma in cooling to the ice melting point with (ii) the heat needed to melt the ice; the resulting water volume is ~10 times the magma volume (Wilson and Head, 2007).

13.2.3 Englacial dike intrusion products

Unambiguous evidence for dikes intruding into glaciers is sparse. However, during the 1996 eruption of Gjálp, Iceland, Gudmundsson et al. (2004) described an en echelon fracture in > 500-m-thick ice overlying the erupting fissure, in a location previously free of crevasses. The fracture was interpreted to be a consequence of the ice rupturing ahead of an englacial dike. Though there are no published examples of the lithofacies that might be created following dike intrusion into an ice sheet and its subsequent

collapse, the characteristics of those lithofacies can be predicted (Wilson and Head, 2002, 2007) from consideration of the likely sequence of events involved (Fig. 13.5). Initially after stalling, the dike will be flanked on both sides by a narrow but slowly expanding water-filled layer (Fig. 13.5(a)). Although all the heat available in a 2-m-wide dike is capable, ultimately, of melting up to 10 m of ice on either side of the dike (i.e., 20 m in total), that is unlikely to happen since the water cannot support the chilled and fractured, mechanically weak (thus gravitationally unstable) dike. The dike will quickly collapse by slumping under its own weight, displacing meltwater and filling the space available with coarse, chilled (glassy) dike rubble (hyaloclastite) up to a lower elevation than that reached by the dike tip when it stalled (Fig. 13.5(b)).

Magma will continue to extrude into the narrow overlying water-filled englacial space ("vault"), probably as pillow lava (Fig. 13.5(c)). With its coarse pore spaces, some of the dike rubble will also be intruded by magma in the form of subsurface breakouts, particularly immediately after collapse when the original dike pathway is disrupted. The pillow lava will spread laterally until confined by the vault ice walls, but continued melt-back of the walls will remove support from the tall rubble pile, causing further collapses and repeated lowering of what is now an unusual type of early-stage volcanic edifice (Figs. 13.5(d), 13.5(e)). Collapses will continue until the pile rests on the underlying bedrock and has stably graded flanks, thus finally forming a mound or ridge (Fig. 13.5(f)).

The lithofacies forming the mound or ridge is likely to be quite distinctive, comprising a heterogeneous pile of coarse, essentially fines-free hyaloclastite locally cemented by chilled dike material, and numerous fragmented and intact lava pillows. It will also likely show crude radially outward-dipping bedding and possible grading as a consequence of multiple avalanching events of the coarse debris (as grain flows) down the cone flanks. Stratification will probably only affect the latest stages of growth of the breccia pile, when the mound has formed a free upper surface. Further eruption will result in the basal breccia mound/ridge being draped and buried

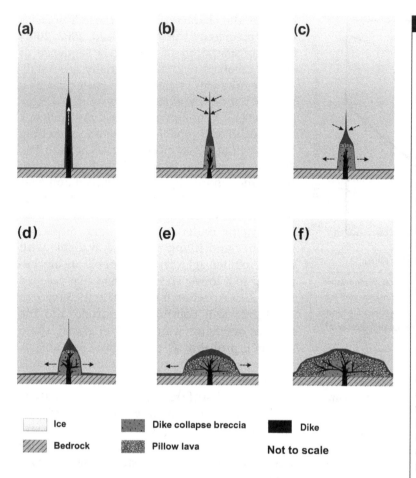

(a) **(b)** **(c)**

(d) **(e)** **(f)**

Legend:
- Ice
- Bedrock
- Dike collapse breccia
- Pillow lava
- Dike

Not to scale

Figure 13.5 Series of schematic diagrams showing development of englacially emplaced mafic dike-collapse breccias and subsequent edifice growth. (a) Dike stalls after intrusion in ice sheet; (b) lack of support at the dike base (by ice melt-back) causes dike to collapse and form hyaloclastite-rubble breccia that fills the available space by displacing meltwater; initial ice fracture caused by dike injection pinches off by stress relaxation and ice deformation; (c) chaotic dike re-injection causes pillow lava effusion on top of dike-collapse breccia, further melt-back of supporting ice walls, and further collapses as unstable steep pile repeatedly fills any gap between the breccia mass and the retreating confining ice walls; incorporation of lava pillows creates pillow-fragment breccia; (d) further collapses of the breccia–pillow-lava pile as confining ice walls melt back; (e) breccia–pillow-lava pile now draped entirely by pillow lava, with breccia core, as the edifice transforms into (f) pillow volcano; the breccia core will rapidly become volumetrically very minor compared with pillow lava, and hard to discern in outcrop; if the eruption is long-lived and voluminous, the edifice may ultimately evolve into a subaqueous tuff cone (tindar) or even a tuya. Not to scale.

by pillow lava, thus forming a small and probably inconspicuous core to the subsequent pillow volcano.

The recognition in the field of a dike-collapse lithofacies might prove difficult since it relies on the exposure of a section eroded fortuitously directly into the former vent region. Although there are no published descriptions of lithofacies interpreted in this way, a basal facies observed in a Pliocene glaciovolcanic sequence at Mussorgsky Peaks, Alexander Island (Antarctic Peninsula) might be an example. There, a poorly stratified outcrop of intrusive "pods" and broken and intact pillow lavas is mingled with up to 40% fines-free glassy breccia (Smellie and Hole, 1997). However, the outcrop is isolated and the field relationships are unclear, although the lithological characteristics are plausible. In another possible example, Wright (1978) described unusually highly fractured massive

basalt surrounded by a pile of basalt talus in the Royal Society Range, Antarctica. Because of its situation, she interpreted the outcrop speculatively as the remnants of an englacial dike, but again the field relationships are inconclusive. Moreover, according to our analysis of englacial dike intrusion (above), the preservation of the dike itself is unlikely (it should wholly collapse to form a breccia mound).

After the initial interactions described above, and depending on the duration and volume of the eruption, the subglacial volcano might evolve up into a subaqueous tuff cone or ridge (tindar; formed of stratified phreatomagmatic tuffs) and ultimately into laterally prograding lava-fed delta(s), which together comprise a tuya volcano. There is extensive literature on the lithofacies and construction of glaciovolcanic mafic pillow volcanoes, tindars, and tuyas (e.g., Jones, 1969, 1970; Skilling, 1994, 2002; Smellie and Hole, 1997; Werner and Schmincke, 1999; Smellie, 2000, 2006; Schopka et al., 2006).

13.3 | Magma intrusion at the base of an ice layer

13.3.1 Theoretical issues

If a dike reaches an ice–rock interface and feeds a sill intruding along the interface, the minimum requirement is that the pressure in the dike magma at the sill inlet must exceed the pressure due to the weight of the overlying glacial ice, P_g (Smellie, 2008). Thus, using Figure 13.1, we require

$$P_a + g\,(\rho_i y + \rho_r z) + \Delta P - (g\rho_b z) > P_a + g\rho_i y \qquad (13.4)$$

which simplifies to

$$\Delta P > gz\,(\rho_b - \rho_r). \qquad (13.5)$$

Thus the required value of the excess magma pressure ΔP increases with the depth of the reservoir z and the difference between the densities of the dike magma, ρ_b, and the crustal rocks, ρ_r. The magma in the sill inlet will contain some exsolved CO_2, but, except for a short time immediately after the intrusion starts, the pressure at the inlet will not be as low as in the magma just behind the propagating sill tip where significant H_2O has exsolved, and so a suitable magma density is ~2600 kg m^{-3} (Wilson and Head, 2002). With the value of $\rho_r = 2300$ kg m^{-3} adopted here, Eq. (13.5) then predicts that ΔP must be greater than ~9 MPa. The required value will be less than this in situations with more volatile-rich or lower-density magma, denser crustal rocks, or shallower reservoirs.

A subglacial sill may evolve in a number of possible ways. If the glacier is cold-based, i.e., the ice temperature at the base of the ice layer is less than the ice freezing point, the ice–rock interface will be strong enough to support tensile stresses and the intruding sill will propagate in the same way that a sill grows at a rock–rock interface, with both the ice and rock initially behaving as elastic solids. The eruption may cease while the system still has this configuration, but heat will continue to be transferred from magma to ice, and the H_2O volume decrease due to ice melting will at least partially relax any excess pressure in the magma. If the glacier is warm-based, water at the interface between the glacier and the underlying rock will prevent the support of tensile stresses. The requirement that the magma pressure at the dike-sill connection exceeds the weight of the glacier means that there is the potential for the glacial ice to be lifted by the magma. However, this can only occur locally: ice has a finite shear modulus, and a thick and laterally extensive slab of ice (the glacier) supported over an initially small region beneath it (around the dike-sill connection) will bend, remaining in contact with the underlying rock beyond some finite distance from the supported region. As a result, in the early stages of the intrusion the resulting stress in the ice will interact with the magma pressure in the same way as in the cold-based case, and the pressure in the intruding magma will be equal to or greater than the weight of the overlying ice.

However, as the sill margin approaches the edge of the glacier, a stage will be reached when the glacial margin is lifted, providing a local pressure connection to the atmosphere and an escape route for water, water vapor, and exsolved volcanic gases. Even in cold-based glaciers this condition must eventually be reached as the elastically propagating sill tip approaches the ice margin. Water loss is likely to happen faster than the overlying ice can deform plastically (Tuffen, 2007), especially if the water release is on the scale of a jökulhlaup (e.g., Björnsson, 1992), so unless wholesale disintegration of the glacier occurs, air will progressively replace the water overlying the magma. The pressure acting on the magma will be much closer to atmospheric pressure than to the weight of the overlying glacier, and a wave of magma fragmentation may propagate from the end of the sill near the ice margin both back toward the dike–sill inlet and downward into the sill as existing gas bubbles expand and additional volatiles exsolve from the magma. A possible example analogous to this is documented in Iceland by Höskuldsson et al. (2006), in which early-formed vesicular pillows underwent secondary vesiculation of the still-molten pillow cores, attributed to release of a jökulhlaup and associated sudden decompression of the pillows, although fragmentation of the lava did not occur. In extreme cases, enough volatiles may be released, especially from the least-cooled magma at the point where the feeder dike reaches the ice–rock interface, that a lava fountain forms over what becomes a subglacial fissure eruption. A sufficiently high fountain will radiate heat to the overlying ice (indeed, pyroclasts may even come into physical contact with it), thus greatly enhancing ice melting in this region. We now elaborate on each of these three stages.

13.3.2 Elastic intrusion

We argued above that the pressure in a sill intruded beneath an ice layer must be at least equal to the pressure due to the weight of the overlying ice, and that the pressure in a propagating dike or sill tip would probably not decrease below the pressure at which magma fragmentation occurs (a function of the total magma volatile content and of order 3 MPa for a typical mafic magma). Given that volatile exsolution in such a magma would be significant at pressures less that ~10 MPa (Chapter 4, Fig. 4.2), it is clear that a large range of combinations of magma reservoir pressure and magma water and CO_2 contents are consistent with the injection of vesicular magmatic liquids at ice–rock interfaces. Various aspects of the resulting growth and evolution of the system have been explored by Höskuldsson and Sparks (1997), Wilson and Head (2002, 2007) and Tuffen (2007).

Wilson and Head (2002) showed quantitatively that major morphological differences are expected between subaerial lava flows and elastically emplaced subglacial intrusions. The most obvious of these is the relative reduction of overall volatile exsolution due to the high ambient pressure, in a manner analogous to deep submarine eruptions (Head and Wilson, 2003; see Chapter 12). Other differences are more subtle. A subaerial flow has an upper free surface and acquires a cross-sectional shape (channel-and-levée, sheet-like, or compound pahoehoe flow field) dictated by the slope on

which it is emplaced, its volume effusion rate, and its evolving rheological properties. The absence of a free surface at the top of a subglacial magma body, together with the stress applied by the elastic properties of the ice above and rock beneath, will lead to it spreading into a sheet-like structure likely to be initially much wider, thinner, and more slowly advancing than a subaerial lava flow erupted at the same volume flux. Other differences between subglacial and subaerial behavior are similarly driven by the stress field. Thus, a subglacial intrusion will get thicker as its edges advance, this process offsetting the inward migration of the effects of cooling at its upper and lower faces and allowing it to advance to a much greater distance from the vent than a subaerial flow before cooling limits its growth. The predicted thickening process has some similarities to the inflation of subaerial lavas (Hon et al., 1994) but, whereas subaerial flow units inflate and thicken after their margins come to rest as a result of cooling, subglacial intrusions thicken continuously as their edges advance. These results will be broadly independent of magma composition, with caveats similar to those listed in Section 13.2.1.

Heat transfer rates from subglacial magma to melted ice have been predicted by Höskuldsson and Sparks (1997), Wilson and Head (2002, 2007), and Tuffen (2007), and measured in a few well-studied cases (e.g., Jarosch et al., 2008). The theoretical calculations differ somewhat in the values used for material properties such as the specific heat of the magma, which is typically averaged over the temperature ranges involved, and the magma density, a function of composition and volatile content. Wilson and Head (2007) found that the volume of water that can be produced by ice melting is ~6.5 times as much as the intruded sill magma volume, which contrasts with a volume ratio of ~10 if all of the magmatic heat is transferred to the water. Thus a 1-m-thick sill would generate a water layer ~6.5 m deep by melting a ~[(1000/900) × 6.5 =] ~7.2-m-thick layer of ice, the volume change thus accommodating (7.2 − 6.5 =) ~0.7 m of the 1 m sill thickness. As suggested by Dixon et al. (2002) and by Figure 4.2 in Chapter 4, as long as they do not make pressure contact with the atmosphere, subglacial intrusions are expected to exsolve a significant fraction of their pre-eruption CO_2 and at most a minor fraction of their pre-eruption H_2O. This suggests that examining residual volatile contents of subglacially intruded rocks is a potential diagnostic tool, in addition to purely morphological examination, for distinguishing such intrusions from subaerial lavas after all of the overlying ice has been removed by climate change. Also, volatile concentrations found in exposed subglacially erupted glasses can be used (with many caveats) to reconstruct quenching processes and minimum ice thicknesses (Dixon et al., 2002; Tuffen et al., 2010).

13.3.3 Subglacial eruption after atmospheric connection

When the margin of a growing subglacial sill approaches sufficiently close to the edge of an ice sheet that a connection to the atmosphere is made, the first consequence will be the escape of the pressurized water vapor in the sill tip. This will be followed by progressive water leakage and replacement by air, almost certainly leading to a wave of decompression advancing into the sill from its distal margin towards the inlet from the feeder dike. This can only be avoided completely if the ice overlying the sill can deform fast enough to replace the water and stay in close proximity to the top of the sill magma. For conditions similar to those proposed here, Höskuldsson and Sparks (1997) calculated an ice deformation rate of ~1 mm s^{-1}, so if the rate of thinning of the water layer exceeds this value, some pressure decrease will inevitably occur.

Initially, the magma will remain a vesicular foam: existing carbon dioxide bubbles will expand and new bubbles of both CO_2 and H_2O will form at a rate that may allow the magma to stay in physical contact with the overlying ice. If so, this will lead to continuing water production, and the system will tend toward an equilibrium where the pressure in the water is greatest near the dike and least at the edge of the ice sheet, with the resulting pressure gradient driving the water toward the exit. If water drainage becomes efficient enough, it is possible that the pressure at the sill–ice contact will eventually

decrease to atmospheric pressure, and then magma fragmentation must occur unless the total magma volatile content is extremely small (Wilson and Head, 1981).

Magma fragmentation will begin at the margin of the sill closest to the connection to the atmosphere. As the pressure in the space above the chilled magma crust decreases, the crust may temporarily prevent any response from the underlying magma. However, the crust is likely to be pervaded by cooling cracks and have little strength. When the crust fails, an expansion wave will propagate down into the sill at a large fraction of the local speed of sound which, in a vesicular liquid, will be ~100 m s^{-1} (Kieffer, 1977; Wilson and Head, 1981). Thus for a sill a few meters thick, this timescale will be less than a tenth of a second. The expansion wave will fragment the magma (Scheu et al., 2006), and expansion of the released gas will accelerate disrupted magma clasts to impact the overlying ice. Formulae given by Wilson (1980) for transient explosions show that, even if the magma contains no water, a pressure reduction from several MPa to 0.1 MPa in a magma still containing, say ~300 ppm of total dissolved volatiles will generate speeds in the hot vesiculated pyroclasts of up to ~25 m s^{-1}, thus projecting them upward to heights of ~30 m. If the vertical distance to the base of the overlying ice is less than this, the impact of these hot magma clasts will locally enhance ice melting; sufficiently violent interaction of magma and ice may be enough to trigger a sustained, violent molten fuel–coolant type of interaction (MFCI; Wohletz and McQueen, 1984; Zimanowski et al., 1991; see Chapter 11). The explosion products would be propelled toward the exit to the atmosphere as the wave of magma vesiculation and fragmentation propagated back toward the feeder dike. The propagation speed of the interaction would be a balance between the speed of the wave into the as-yet unaffected sill magma (again some large fraction of the ~100 m s^{-1} local speed of sound) and the speed at which water and fragmented magma could be discharged from under the ice. Such an explosive fragmentation process is an excellent candidate for the origin of sudden jökulhlaup production. Some fraction of the fragmented magma would be washed out with the escaping water and the rest would be left behind to form a vitroclastic (phreatomagmatic) deposit.

When the wave of magma disruption reaches the feeder dike, the system will behave like a subaerial eruption, with the formation of a chain of lava fountains. Depending on the efficiency with which water is drained from the vicinity of the vent, the lava fountains may be entirely magmatic or may be phreatomagmatic. Initially the fountains will "drill" into the overlying ice. A cavity will grow upward until the fountains reach the subaerial height corresponding to the total magma volatile content (Head and Wilson, 1987). Thereafter heat will only be transferred to the ice by radiation from pyroclasts in the fountains, and the larger clasts falling from the fountains will begin to form rootless lava flows (Head and Wilson, 1989) spreading away from the vent over the top of the disrupted sill material. A new balance will be approached between ice subsidence and ice melting, and if the eruption continues for long enough, the explosive activity at the vent may eventually emerge through the ice. Alternatively, the ice layer above the fragmented sill residue might fracture in a brittle fashion and collapse rather than slowly deforming plastically. The pressure at the dike vent would still be close to atmospheric, provided there was a reasonably high porosity and permeability in the collapsed ice block pile, but interactions between magma and ice would be more vigorous because of their greater proximity.

A hybrid situation could occur if the magma volatile content is very low. If water drainage is inefficient, the pressure reduction rate in the water above the sill will be less dramatic, but the pressure acting on the upper sill surface will still be greatly reduced. The distal part of the still-spreading magma body may then begin to evolve into a thicker, narrower morphology more like that expected for a subaerial lava flow. Meltwater will be channeled along the side(s) of the flow, and the system will remain stable as long as the pressure in the water above the flow is maintained high enough to suppress magma vesiculation to the point of fragmentation. One

or more discrete lava flow units may eventually emerge from beneath the ice. After ice removal by climate change, it should be possible to identify this kind of flow by its changing shape as a function of distance from its vent (taking due account of pre-existing topography), and to distinguish it from a flow generated by lava fountain activity in an ice cavern after sill disruption.

13.3.4 Geological examples of subglacial eruptions

Eruptions under thin ice, in cases where the pressure is not enough to suppress magma fragmentation, will be explosive and result in the abrupt removal of overlying ice and construction of pyroclastic cones, as scoria or tuff cones or tuff rings. Subglacial sills will not be formed. The generation of extreme conditions such as formation of a pyroclastic density current and its emplacement by intrusion beneath an ice cover is also unlikely in these cases. Once magma fragmentation is achieved by high levels of vesiculation, in the presence of abundant meltwater, a violent fuel–coolant interaction with meltwater will result in an abrupt transition to an explosive phreatomagmatic eruption (Chapter 11). This will form stratified tuff cones or tuff rings composed of interbedded fall and pyroclastic density current tephra. However, no examples have been observed, possibly because the edifice, formed of unconsolidated tephra, is easily eroded by ice. Additionally, some of the edifice will be formed on the ice itself and will be destroyed as the ice flows downslope and melts. The pyroclastic cones thus have a very low preservation potential. However, distinctive glaciovolcanic sequences, known as sheetlike sequences of Mount Pinafore type (after the Antarctic edifice), appear to be the products of eruptions under thin ice (probably < 150–200 m thick; Smellie et al., 1993; Smellie and Skilling, 1994; Loughlin, 2002; Smellie, 2008). They are outflow deposits (i.e., they accumulated away from the source edifice) and are characterized by prominent basal beds of fluvially deposited stratified sandstones rich in vitric ash clasts, for which an explosive phreatomagmatic source has been inferred. The sequences are typically composed of (from base up) diamict (i.e., *in*

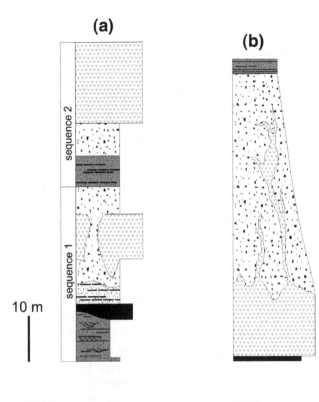

(a)

(b)

sequence 2

sequence 1

10 m

Figure 13.6 Schematic vertical profile sections showing "standard sequences" characteristic of (a) Mt. Pinafore- and (b) Dalsheidi-type mafic subglacial sheet-like eruptions. Products of two discrete Mt. Pinafore-type eruptive events are shown in (a), resulting in a repetition of most lithofacies, whereas products of just a single Dalsheidi-type eruptive episode are shown in (b). The Mt. Pinafore-type sequences in (a) are both incomplete as they lack capping subaerial lithofacies. See text for details. Adapted from Smellie et al. (1993) and Smellie (2008).

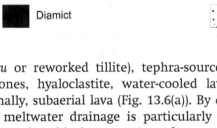

Stratified volcanic sandstones (a) & mudstones (b)

Diamict

Lava/sill, columnar- and/or blocky-jointed

Hyaloclastite, *in situ* & redeposited

situ or reworked tillite), tephra-sourced sandstones, hyaloclastite, water-cooled lava, and, finally, subaerial lava (Fig. 13.6(a)). By contrast, if meltwater drainage is particularly efficient beneath a thin ice cover (e.g., for eruptions on relatively steep bedrock), the vent might dry out and result in magmatic eruptions of strombolian or hawaiian type. This was observed during the 1969 subglacial fissure eruption of basaltic andesite to andesite magma at Deception Island, Antarctica, during which a line of cinder cones was generated along a fissure in ice ~100 m thick and which were partly constructed on the ice surface (Smellie, 2002).

Eruptions under much thicker ice might have quite different consequences for the lithofacies formed. Unusual lava–hyaloclastite sequences of Plio-Pleistocene age at Dalsheidi in southern Iceland, which were ascribed to lava extrusion in a shallow-marine (shelf) setting (Bergh

and Sigvaldason, 1991), were reinterpreted as products of subglacial eruptions, possibly multiple subglacial sill intrusions and associated meltwater floods (jökulhlaups; Smellie, 2008). The Icelandic outcrops, which were named subglacial sheet-like sequences of Dalsheidi type, are products of voluminous subglacial fissure eruptions individually up to > 30 km³. The products can be traced up to 30 km along strike and at least 14 km down-dip and are separated by sharp, undulating, likely glacial erosion surfaces. Each sequence is formed during a single eruption. A "standard sequence" of four major lithofacies has been identified, i.e., diamictite, lava (sill), hyaloclastite, and mudstone (from base up; Fig. 13.6(b)). Although broadly resembling sheet-like sequences of Mount Pinafore type, the Dalsheidi-type sequences are wider and usually much thicker (up to 300 m), and they lack basal explosively generated phreatomagmatic tuffs and subaerial capping lava, which are significant for indicating their formation in a subglacial environment (contrast Figs. 13.6(a) and 13.6(b)).

In Dalsheidi-type sequences, diamictite occurs above a basal unconformity. It probably represents a combination of tillite and melted-out basal glacier debris, and is succeeded by laterally extensive sheet lava, called an interface sill by Smellie (2008), showing evidence for basal loading and interaction with the diamictite, which must have been relatively soft, wet, and unconsolidated at the time. The sill shows spectacular water-induced columnar cooling joints. Its upper surface is locally deformed into structures resembling flow folds, some planed off by coeval erosion, but it is also conspicuously characterized by prominent apophyses, usually a few tens of meters long, that intrude overlying massive to faintly planar stratified hyaloclastite. Thick hyaloclastite dominates each eruptive unit. It is monomict, relatively fine grained (mainly coarse sand to granule grade) and fine-ash poor/free, with dispersed broken and intact pillows that apparently increase in proportion up-dip toward the source. The glassy upper surface of the sill has been locally stripped off by rapidly moving, partly turbulent, sediment-charged fluidal currents (known as hyperconcentrated flows) that deposited the hyaloclastite as they slowed down. The hyperconcentrated flow events represent periods of major meltwater outflows (jökulhlaups) that occurred towards the end of each eruption. The sequences are capped by thinly stratified mudstone and fine sandstone, representing finer-grained hyaloclastite detritus deposited by a waning flood. Some of the finer-grained capping sediments are formed from explosively generated lapilli tuff indicating that some eruptions were explosive (phreatomagmatic) in the final stages.

A genetic model for the emplacement of the Dalsheidi-type sequences was suggested by Smellie (2008), comprising: (1) sill emplacement at the base of an ice sheet; (2) sill stagnation caused by eruptive overpressures becoming insufficient to lift the overlying ice cover; (3) transformation to pillow lava effusion at the erupting fissure, with construction of a pillow ridge and simultaneous creation of a meltwater-filled vault; (4) floating of the overlying ice and release of meltwater in a major jökulhlaup that destabilized and destroyed much of the pre-existing pillow edifice, redepositing it as thick beds of hyaloclastite; and (5) the high-energy flood event locally eroding the surface of the sill. Finally, (6) finer-grained beds were laid down as the flood waned. The final stage was also associated with intrusion of apophyses of magma derived from the still-molten sill interior, which were injected up into the sluggishly moving flow and associated hyaloclastite pile.

Similar to eruptions of sheet-like sequences of Mount Pinafore type, the source edifices responsible for Dalsheidi-type eruptions have never been observed. In part, at least, they are thought to have been pillow ridges that were extensively removed during the late-stage jökulhlaups. Since explosive activity may have been absent or else confined to the final stages of eruptions, the sills and pillows are inferred to have been undegassed during emplacement because volatile exsolution was suppressed by high ambient pressures associated with unusually thick ice sheet conditions (empirically calculated to be at least 1000 m). An entirely subglacial setting is also suggested by the apparent absence of any capping subaerial lavas, indicating that the overlying ice was never completely melted through.

It was suggested in Section 13.3.3 that a combination of thicker-ice conditions (initially suppressing significant vesiculation) and decompression caused by sudden late stage pressure reduction, e.g., as a subglacial sill nears the margins of an ice sheet and connects with atmospheric or near-atmospheric pressures, might cause major changes within the sill. Many variations on this theme appear to be possible. Thus Höskuldsson et al. (2006) described mafic pillows inferred to have been erupted under 1.5–2.0 km of ice that have outer zones with 15–20% vesicularity surrounding cores with 40–60% vesicularity, interpreted to be the result of a ~4.5 MPa pressure decrease as a jökulhlaup abruptly removed much of the water that had been produced by ice melting. In contrast, Schopka et al. (2006) described a mafic "hyaloclastite" ridge formed under ~500 m of ice where the activity was explosive throughout; pressure conditions were maintained low enough for magma fragmentation by continuous drainage of meltwater at a high enough rate that the overlying ice could not deform fast enough to maintain contact with the erupting magma. Finally, Tuffen (2007) has stressed the fine balance that may exist between magma intrusion rate, water production and drainage rate, and ice deformation rate. Tuffen *et al.* (2008) described a rhyolitic eruption that took place beneath ~150 m of ice, beginning as an explosive event but changing to the intrusion of vesicular magma into the fragmental deposits from the previous explosive phase. Clearly a low-pressure pathway to the atmosphere never formed in this case.

The most extreme version of late-stage pressure reduction is proposed (Section 13.3.3) to be the explosive decompression of an intruded sill. The loci of the explosions will migrate rapidly back toward source (the vent), potentially generating subglacial pyroclastic density currents that will simultaneously mix with the abundant ambient meltwater and rapidly transform into hyperconcentrated flood-flows, resulting in a major jökulhlaup where the floods exit from the ice sheet. There are currently no known examples of those deposits, although they might easily have been missed. The deposits will probably be massive to weakly stratified and largely formed of angular, highly vesicular hyaloclasts in abundant fine ash matrix, and there will be evidence for lateral (down-dip) and vertical transitions to fluvial deposits as floods wane and normal stream-flow conditions become reestablished. Because they are generated from the destruction of the original sill, they will not overlie any coeval sill rock, which is a major distinction from "standard" Dalsheidi-type sheet-like sequences. The deposits will also be composed of phreatomagmatic lapilli tuffs, rather than the redeposited hyaloclastite (*sensu* White and Houghton, 2006) seen in the Dalsheidi-type sequences. Deposits that failed to escape from the ice will rest on a glacially eroded surface and/or glacial diamict, whereas proglacial deposits of the jökulhlaup will lie on outwash in front of the glacier. This relationship raises the possibility of using these features to identify the geographical limits of past ice sheets.

13.4 | Supraglacial eruptions

Eruptions onto glacial ice may involve either the advance onto the ice of a lava flow from a vent located off the glacier or the deposition onto the ice of pyroclasts from an explosive eruption (e.g., Wilson and Head, 2009). The vent for the latter may be located off the glacier or may lie within the glacier in cases where a dike penetrates into, or at least fractures, the ice leading to a phreatomagmatic eruption. To advance onto a glacier, the upper surface of an encroaching lava flow must be substantially higher than the glacier so that it can override the outwardly inclined glacier margins. Such events are probably rare, though lava flow advance over thin snow and ice deposits is common. Where a lower lava encounters a higher glacier surface, it will "pond" by thickening against the ice barrier; examples are common, both for basalts and more evolved lava flows on Earth (e.g., Lescinsky and Sisson, 1998; Lescinsky and Fink, 2000; Mee *et al.*, 2006; Stevenson *et al.*, 2006; Harder and Russell, 2007) and Mars (Shean *et al.*, 2005; Kadish *et al.*, 2008). Another way is for lava to spill over onto a glacier from an adjacent ice-free topographic high, although subsequent shear through flow of the glacier will separate the bedrock- and glacier-covering lava outcrops as the glacier moves down-valley. Examples of truncated lavas and scoria cones are present at relatively high elevations along the margins of some of the Dry Valleys in Antarctica that were formerly ice filled.

13.4.1 Lava flowing onto glacial ice
Wilson and Head (2007) treated the advance of a lava flow of a given thickness over ice in a region where the ambient temperature is slightly below the ice melting temperature. Their treatment, and that followed here, ignores the presence and likely insulating effects of a basal autobreccia layer, so the

calculations of extent of ice melting (below) should be regarded as optima. The entire core of the flow is assumed to be initially at the lava eruption temperature and the upper lava surface is assumed to rapidly cool to the ambient temperature. Thus the ice melting rates calculated will be maxima, corresponding to a flow encountering ice almost immediately after leaving the vent. The base of the flow melts ice to water at the triple point temperature 273.15 K and it is assumed that the water produced can drain efficiently. In that case, the temperature of both the water and the base of the flow remain at 273.15 K. The evolving temperature profile within the flow can then be evaluated as a function of time using the series expansion solution given by Carslaw and Jaeger (1959; Article 3.4, Eq. (1)). The temperature profile can be differentiated to yield the temperature gradient and this, multiplied by the thermal conductivity of the lava, provides the rate of heat transfer to the ice. If the heat transfer rate is divided by the ice density and latent heat of fusion, the result is the rate of decrease of ice thickness beneath the flow, and hence the water production rate per unit area of the flow base. Figure 13.7 shows examples of the ice melting rate as a function of time after the start of melting at any given location for flows 1, 3, 10, and 30 m thick using the computer program used by Wilson and Head (2007) to implement the above procedure.

Integration of the water production rate over the growing area of the base of an advancing flow yields the total water production rate as a function of time. After one day, the absolute thicknesses of ice melted by flows 1, 3, 10, and 30 m thick are respectively 4.95, 4.65, 4.0, and 1.8 m, corresponding to 99%, 31%, 8%, and 1.2% of the total melting that is ultimately achievable. The maximum value of the ratio (total ice thickness melted)/(lava flow thickness) is predicted to be independent of flow thickness and very close to 5 (Wilson and Head, 2007). The overall timescale for melting depends very strongly on the thickness of the flow: for flows 1, 3, 10, and 30 m thick, 90% melting is achieved after about 20 hours, one week, 3 months,

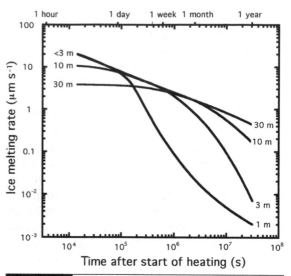

Figure 13.7 Variation of the ice melting rate beneath a lava flow near its vent as a function of time after the onset of melting. Curves are labeled for 4 flow thicknesses, 1, 3, 10, and 30 m. The curves for 1 and 3 m are indistinguishable ("< 3 m") at small times.

and 3 years, respectively. These lengthy delays partly explain observations (Einarsson, 1948; Kjartansson, 1948) that advancing flows cause negligible amounts of snow melting, apparently at odds with the above predictions. Additionally, the heat content of the flow available to cause melting will decrease steadily with increasing distance from the vent.

13.4.2 Pyroclasts emplaced onto glacial ice

When a layer of pyroclasts accumulates on a glacier, there are many possible outcomes (e.g., Wilson and Head, 2009). If the vent is nearby and the activity is not too energetic, e.g., as in a mafic lava fountain eruption, it is possible that many of the larger clasts will still be very hot on landing: Head and Wilson (1989) showed the trends of mean deposit temperature variation with magma volatile content and discharge rate. In these circumstances rapid melting of ice may occur, but the consequences will be very localized. Of more wide-ranging importance will be more energetic (i.e., high mass flux) eruptions leading to the emplacement of hot pyroclastic density currents onto glaciers, where a sufficient thickness of deposit may trigger a vigorous interaction, especially when scouring of the ice and mixing with the pyroclasts occurs, ultimately forming a lahar (Walder, 2000a). Chapter 14 discusses lahar generation and runout in more detail.

In contrast, a relatively high mass-flux, high volatile-content eruption (e.g., a rhyolitic plinian eruption) will generate an eruption plume from which the bulk of the clasts will have fallen from a great enough height that they will land at the ambient atmospheric temperature. If the glacier onto which they fall is stable (i.e., not in the process of melting) then they will not immediately begin to melt it. However, pyroclasts are likely to have a significantly lower albedo than ice, and will therefore absorb solar insolation more efficiently than the ice, raising the surface temperature of the pyroclast layer, possibly above the ice melting point. If this temperature increase is communicated by conduction through the pyroclast layer to the ice below, it will initiate ice melting. Field observations of tephra on glaciers suggest that a tephra layer < 2 cm thick acts to accelerate ice/snow melting, while thicker layers seem to insulate the snow surface (e.g., Manville *et al.*, 2000, and personal observations of the authors). However, such observations do not take time into account. On diurnal and annual timescales, melting is enhanced overall, to varying degrees (i.e., more for thinner layers), and the entire surface draped in ash is lowered at rates above "normal" (i.e., under ash-free conditions). However, much thicker ash layers do insulate snow/ice surfaces.

The warming effect can be quantified by considering the balance between the incoming solar heat flux and the heat flux radiated by the surface, equal to ($\varepsilon \sigma T^4$), where ε is the emissivity, σ is the Stefan-Boltzman constant, and T is the absolute surface temperature. Clearly, the mean surface temperature of an exposed surface is inversely proportional to the fourth root of its emissivity. The albedo of glacial ice ranges up to 0.4 (Paterson, 1994), whereas mafic silicates have albedos as low as 0.1 (Farrand and Singer, 1992). The corresponding emissivities are 0.6 and 0.9, and so the effect of emplacing a layer of pyroclasts will be to increase the surface temperature by a factor of up to $(0.9/0.6)^{1/4} \approx 1.11$. Thus ice at a temperature as low as (273/1.11 =) ~247 K (−26 °C) could be heated to the melting point by this process. Melting will not be instantaneous. A thermal wave will penetrate a layer of pyroclasts of thickness λ in a time τ equal to ~2.32 $(\kappa \tau)^{1/2}$, where κ is the thermal diffusivity, ~10^{-6} m^2 s^{-1}. The relevant timescales for diurnal and seasonal temperature fluctuations are one day (~9×10^4 s) and one year (~3×10^7 s), for which λ = 0.58 m and 10.6 m, respectively. Thus a low-albedo pyroclast layer much less than half a meter thick overlying ice in a region where the mean diurnal temperature is only a few degrees below the melting temperature may cause a significant amount of ice melting beneath it during each daily temperature cycle. Conversely, a pyroclast layer a few meters thick will delay the onset of ice melting for several days, even if the surface of the pyroclast layer warms above 273 K every day. However, a pyroclast layer more than about 20 m thick would be needed to protect the ice against an annual heating cycle.

13.4.3 Geological examples of supraglacial emplacement

Detailed descriptions of lava that flowed onto glaciers are rare and focus mainly on the secondary generation of lahar and meltwater flood events (e.g., summary by Major and Newhall, 1989; also Khrenov *et al.*, 1988; Vinogradov and Murav'ev, 1988). Smellie (2007, 2009) postulated that lava emplaced on a glacier surface would be redeposited immediately down-dip of the source edifice as breccia during rapid *in situ* downwasting of the glacier (e.g., at the glacial termination). With coeval glacier flow, however, it also seems likely that most of the lava will be carried away from its source by the glacier and deposited as lava clast-dominated breccia in ice-marginal moraines in much the same way as cool pyroclastic deposits laid down on moving ice. The only published

Figure 13.8 Close view of an andesite lava flow originally emplaced on snow. The lava shows distinctive curviplanar primary cooling fractures (solid arrows), as well as smaller secondary fractures (dashed arrows) orientated perpendicular to the primary set. Although such fractures are not diagnostic of lavas emplaced on snow, they are a distinctive and conspicuous feature and indicate thermal contraction during quenching of the lava probably by steam derived from melted underlying snow (see Mee et al., 2006; photograph provided by K. Mee).

description of *in situ* lava characteristics specifically attributed to snow interaction is by Mee *et al.* (2006), who described an andesite lava from a winter eruption in Chile that flowed over snow. The evidence comprised (1) a flow-front ~5 m wide and 5 m high composed of blocky glassy breccia, that formed as a talus apron during post-emplacement gravitational instability; and (2) a 20-m-wide zone behind the flow-front forming a basal layer several meters thick, that consisted of glassy andesite lava showing distinctive, cross-cutting, curviplanar "pseudopillow" fractures and abundant perpendicular small-diameter secondary fractures (Fig. 13.8). Both sets of fractures were regarded as cooling joints thought to have been caused by the overridden snow melting and flashing to steam. The lava in the snow-interaction zone was subsequently overridden during the same emplacement event by crystalline subaerial lava with a blocky surface autobreccia showing no evidence for water chilling.

Pyroclasts deposited on glaciers, whether derived from fallout or pyroclastic density currents, are reworked and redeposited by either eolian activity or in lahars generated by melting related to pyroclastic density currents (e.g., Thouret, 1990; Walder, 2000b). They will also be advected en masse to marginal locations by ice flow, where they will be dumped in moraines and/or extensively reworked as further mass flows or by proglacial or later non-glacial weathering and fluvial processes, leaving a broad zone centered on the erupting vent(s) swept free of pyroclasts (e.g., Manville *et al.*, 2000; Höskuldsson, 2001). The off-ice deposits will not be preserved in their original position, but will be remobilized, reworked, and redeposited. They are likely to be mixtures of pyroclasts, ice, snow, and water (perhaps as much as 65–90% snow and ice particles in some cases; Manville *et al.*, 2000; Fig. 13.9). The snow/ice components will melt out, leaving behind a "lag" deposit comprising mainly porous ashy material that is very susceptible to being washed away by rain or winnowed by wind, thus destroying any original textures. When mixed with blocks of accidental material, these lag deposits will simply resemble deposits of lahars. Of the many factors that may influence redistribution of pyroclasts after emplacement on a snow and ice-covered volcano, slope angle and aspect, ice thermal regime (warm-based vs. cold-based), pyroclast grain componentry and size distribution, and a variety of local climate parameters, such as mean temperature, diurnal temperature range, insolation, and precipitation, are inferred to be most important (Manville *et al.*, 2000).

(a)

(b)

Figure 13.9 Lahar deposits composed mainly of dark mafic pyroclasts (ash and lapilli) produced during explosive eruptive activity and abundant ice blocks deposited in (a) proximal supraglacial and (b) more distal proglacial (i.e., in front of the glacier) locations relative to the eruptive vents. Both deposits are products of the 1969 subglacial fissure eruption on Deception Island photographed about one month after the eruption. See Baker et al. (1975) and Smellie (2002) for descriptions of the eruption. The prominent ice blocks in (a) are mainly ~30–50 cm in diameter but some exceed 1 m. The long-term preservation potential of supraglacial lahar deposits such as those shown in (a) is negligible, whereas those deposited in a proglacial setting are more likely to be preserved, although also subjected to local fluvial reworking.

13.5 | Summary

We have described some basic physical principles that underlie the nature of volcanic eruptions taking place under, into, and onto glacial ice, and some candidate examples of such deposits and edifices encountered in the field. The likely modes of magma–ice interaction can be predicted theoretically, and form the basis for further testing these predictions in the field. Direct observations of magma–ice interactions beneath thick ice covers have not been made, but could be studied in the future with properly geophysically instrumented sites. Although the immediate products of such interactions are commonly extensively modified by the flow of the water that is inevitably produced, these specific theoretical predictions form a paradigm on which initial conditions can be visualized, and interpretations can be based. The range of possible interactions is large, and the potential clearly exists for the formation of a very diverse suite of deposits and landforms, many of which we have described. Furthermore, although our understanding of the relevant mechanisms has evolved from studies of eruptions and landforms on Earth, the same basic principles can also be applied to volcano–ice interactions on Mars, where examples of dike and sill intrusions into glacial ice, as well as related phreatomagmatic eruptions and tephra deposition have been documented (e.g., Head and Wilson, 2002, 2006; Shean *et al.*, 2005; Kadish *et al.*, 2008; Wilson and Head, 2009).

13.6 | Notation

c_m	specific heat of magma, 900 (J kg^{-1} K^{-1})
E	enthalpy of H_2O fluid (J kg^{-1})
f	volume fraction of gas bubbles in magma
g	acceleration due to gravity (m s^{-2})
h	distance dike tip penetrates into ice layer (m)
L_i	latent heat of fusion of ice, 333 (kJ kg^{-1})
m	molecular mass of volatile (kg kmol^{-1})
n_t	total mass fraction of exsolved magma volatiles
P_a	atmospheric pressure,10^5 (Pa)
P_d	dike tip pressure minus ambient stress in ice (Pa)
P_e	equilibrium pressure of ice–magma mixture (Pa)
P_g	weight of glacial ice layer (Pa)
P_t	final pressure in gas in dike tip cavity (Pa)
ΔP	super-lithostatic pressure in magma reservoir (Pa)
R	universal gas constant, 8.314 (kJ kmol^{-1} K^{-1})
T	absolute temperature of surface materials (K)
T_e	equilibrium temperature of ice–magma mixture (K)
T_i	triple point temperature of H_2O, 273.15 (K)
T_m	magma temperature, 1473 (K)
V	volumes of ice and magma mixing (m^3)
V_g	relative volume of gas (m^3)
V_m	relative volume of magma (m^3)
y	thickness of ice layer (m)
z	depth of magma reservoir top beneath ice–rock interface (m)
ε	emissivity of surface materials
κ	thermal diffusivity of ice and dike magma, ~10^{-6} (m^2 s^{-1})
λ	distance thermal wave propagates into surface (m)
ρ_b	bulk density of magma in dike, 2500–2600 (kg m^{-3})
ρ_g	density of gas (kg m^{-3})
ρ_i	mean density of ice, 900 (kg m^{-3})
ρ_m	density of magma in reservoir, 2700 (kg m^{-3})
ρ_r	crustal rock bulk density, 2300 (kg m^{-3})
σ	Stefan–Boltzman constant, 5.67 × 10^{-8} (W m^{-2} K^{-4})
τ	timescale for thermal wave penetration (s)

Acknowledgements

This chapter is a contribution to the British Antarctic Survey's GEACEP programme (ISODYN Project), which sought to investigate climate change over geological timescales and to develop appropriate novel climate change proxies, in this case glaciovolcanism. JWH gratefully acknowledges grants from the NASA Mars Data Analysis Program, NNG04G99G and NNX07AN95G, and the Mars Express High Resolution Stereo Camera Co-Investigator Program (DTM-3250-05). We are grateful to Katy Mee for permission to use one of her photographs.

References

ASME (2000). *ASME International Steam Tables for Industrial Use*. New York, NY: American Society of Mechanical Engineers.

Behrendt, J.C., Blankenship, S. D., Morse, D. L., Finn, C. A. and Bell, R. A. (2002). Subglacial volcanic features beneath the West Antarctic Ice Sheet interpreted from aeromagnetic and radar ice sounding. In *Volcano-Ice Interaction on Earth and Mars*, ed. J. L. Smellie and M. G. Chapman. Geological Society of London Special Publication, **202**, pp. 337–355.

Bergh, S. and Sigvaldason, G. E. (1991). Pleistocene mass-flow deposits of basaltic hyaloclastite on a shallow submarine shelf, South Iceland. *Bulletin of Volcanology*, **53**, 597–611.

Björnsson, H. (1992). Jokulhlaups in Iceland: Prediction, characteristics and simulation. *Annals of Glaciology*, **16**, 95–106.

Blankenship, D. D., Bell, R. E., Hodge, S. M. *et al.* (1993). Active volcanism beneath the West Antarctic Ice Sheet and implications for ice-sheet stability. *Nature*, **361**, 526–529.

Bourgeois, O., Dauteuil, O. and Van Vliet-Lanoe, B. (1998). Pleistocene subglacial volcanism in Iceland: Tectonic implications. *Earth and Planetary Science Letters*, **164**, 165–178.

Carslaw, H. S. and Jaeger, J. C. (1959). *Conduction of Heat in Solids*, 2nd edn. Oxford; Clarendon Press.

Corr, H. F. and Vaughan, D. G. (2008). A recent volcanic eruption beneath the West Antarctic Ice Sheet. *Nature Geoscience*, **1**, 122–125.

Dixon, J. E., Filiberto, J. R., Moore, J. G. and Hickson, C. J. (2002). Volatiles in basaltic glasses from a subglacial volcano in northern British Columbia (Canada): implication for ice sheet thickness and mantle volatiles. In *Volcano-Ice Interaction on Earth and Mars*, ed. J. L. Smellie and M. G. Chapman. Geological Society of London Special Publication, **202**, pp. 255–271.

Dobran, F. (2001). *Volcanic Processes: Mechanisms in Material Transport*. New York, NY: Kluwer Academic/Plenum.

Einarsson, T. (1948). The flowing lava. Studies of its main physical and chemical characteristics, *Societas Scientiarum Islandica*, **IV**(3), 1–70.

Einarsson, T. (1994). *Geology of Iceland. Rocks and Landscape*. Reykjavik: Málog Menning Publishing Company, 309 pp.

Einarsson, T. and Albertsson, K. J. (1988). The glacial history of Iceland during the past three million years. *Philosophical Transactions of the Royal Society London Series A*, **318**, 637–644.

Farrand, W. H. and Singer, R. B. (1992). Alteration of hydrovolcanic basaltic ash – observations with visible and near-infrared spectrometry. *Journal of Geophysical Research*, **97** (B12), 17 393–17 408.

Geirsdóttir, A., and Eiríksson, J. (1994). Growth of intermittent ice sheet in Iceland during the Late Pliocene and Early Pleistocene. *Quaternary Research*, **42**, 115–130.

Glen, J. W. (1952). Experiments on the deformation of ice. *Journal of Glaciology*, **2**, 111–114.

Gudmundsson, M. T., Sigmundsson, F. and Björnsson, H. (1997). Ice–volcano interaction of the 1996 Gjálp subglacial eruption, Vatnajökull, Iceland. *Nature*, **389**, 954–957.

Gudmundsson, M. T., Pálsson, F., Björnsson, H. and Högnadóttir, T. (2002). The hyaloclastite ridge formed in the subglacial 1996 eruption of Gjálp, Vatnajökull, Iceland: present day shape and future preservation. In *Volcano-Ice Interaction on Earth and Mars*, ed. J. L. Smellie and M. G. Chapman. Geological Society of London Special Publication, **202**, pp. 319–335.

Gudmundsson, M. T., Sigmundsson, F., Björnsson, H. and Högnadóttir, T. (2004). The 1996 eruption at Gjálp, Vatnajökull ice cap, Iceland: efficiency of heat transfer, ice deformation and subglacial water pressure. *Bulletin of Volcanology*, **66**, 46–65.

Harder, M. and Russell, J. K. (2007). Basanite glaciovolcanism at Langorse mountain, northern British Columbia, Canada. *Bulletin of Volcanology*, **69**, 329–340.

Head, J. W. and Wilson, L. (1987). Lava fountain heights at Pu'u 'O'o, Kilauea, Hawai'i: indicators of amount and variations of exsolved magma volatiles. *Journal of Geophysical Research*, **92**, 13 715–13 719.

Head, J. W. and Wilson, L. (1989). Basaltic pyroclastic eruptions: Influence of gas-release patterns and volume fluxes on fountain structure and the formation of cinder cones, spatter cones, rootless flows, lava ponds and lava flows. *Journal of Volcanology and Geothermal Research*, **37**, 261–271.

Head, J. W. and Wilson, L. (2002). Mars: A review and synthesis of general environments and geological settings of magma-H_2O interactions. In *Volcano-Ice Interaction on Earth and Mars*, ed. J. L. Smellie and M. G. Chapman. Geological Society of London Special Publication, **202**, pp. 27–57.

Head, J. W. and Wilson, L. (2003). Deep submarine pyroclastic eruptions: theory and predicted landforms and deposits. *Journal of Volcanology and Geothermal Research*, **121**, 155–193.

Head, J. W. and Wilson, L. (2006). Heat transfer in volcano-ice interactions on Mars: Synthesis of environments and implications for processes and landforms. *Annals of Glaciology*, **45**, 1–13.

Hon, K., Kauahikaua, J., Denlinger, R. and Mackay, K. (1994). Emplacement and inflation of pahoehoe sheet flows: Observations and measurements of active lava flows on Kilauea Volcano, Hawaii. *Geological Society of America Bulletin*, **106**, 351–370.

Höskuldsson, A. (2001). Late Pleistocene subglacial caldera formation at Cerro las Cumbres, eastern Mexico. *Jökull*, **50**, 49–64.

Höskuldsson, A. and Sparks, R. S. J. (1997). Thermodynamics and fluid dynamics of effusive subglacial eruptions. *Bulletin of Volcanology*, **59**, 219–230.

Höskuldsson, A., Sparks, R. S. J. and Carroll, M. R. (2006). Constraints on the dynamics of subglacial basalt eruptions from geological and geochemical observations at Kverkfjöll, NE-Iceland. *Bulletin of Volcanology*, **68**, 689–701.

Jarosch, A., Gudmundsson, M. T., Högnadóttir, T. and Axelsson, G. (2008). Progressive cooling of the hyaloclastite ridge at Gjálp, Iceland, 1996–2005. *Journal of Volcanology and Geothermal Research*, **170**, 218–229.

Jaupart, C. and Vergniolle, S. (1989). The generation and collapse of a foam layer at the roof of a basaltic magma chamber. *Journal of Fluid Mechanics*, **203**, 347–380.

Jones, J. G. (1969). Intraglacial volcanoes of the Laugarvatn region, south-west Iceland – I. *Journal of the Geological Society of London*, **124**, 197–211.

Jones, J. G. (1970). Intraglacial volcanoes of the Laugarvatn region, south-west Iceland – II. *Journal of Geology*, **78**, 127–140.

Kadish, S. J., Head, J. W., Parsons, R. L. and Marchant, D. R. (2008). The Ascraeus Mons fan-shaped deposit: Volcano-ice interactions and the climatic implications of cold-based tropical mountain glaciation. *Icarus*, **197**, 84–109, doi:10.1016/j.icarus.2008.03.019.

Khrenov, A. P., Ozerov, A. Yu., Litasov, N. E. *et al.* (1988). Parasitic eruption of Kluchevskoi volcano (Predskazannyi eruption 1983). *Volcanology and Seismology*, **7**, 1–24.

Kieffer, S. W. (1977). Sound speed in liquid-gas mixtures: water-air and water-steam. *Journal of Geophysical Research*, **82**, 2895–2904.

Kjartansson, G. (1948). Water flood and mud flows. In *The Eruption of Hekla (1947–1948). Societas Scientiarum Islandica*, **II**(4).

Komatsu, G., Arzhannikov, S. G., Arzhannikova, A. V. and Ershov, K. (2007). Geomorphology of subglacial volcanoes in the Azas Plateau, the Tuva Republic, Russia. *Geomorphology*, **88**, 312–328.

Lange, M. and Ahrens, T. (1983). The dynamic tensile strength of ice and ice-silicate mixtures. *Journal of Geophysical Research*, **88**(B2), 1197–1208.

Larsen, G. and Eiríksson, J. (2008). Late-quaternary terrestrial tephrochronology of Iceland – frequency of explosive eruptions, type and volume of deposits. *Journal of Quaternary Science*, **23**(2), 109–120.

Lescinsky, D. T. and Fink, J. H. (2000). Lava and ice interactions at stratovolcanoes: Use of characteristic features to determine past glacial extents and future volcano hazards. *Journal of Geophysical Research*, **105**(B10), 23 711–23 726.

Lescinsky, D. T. and Sisson, T. W. (1998). Ridge-forming, ice-bounded lava flows at Mount Rainier, Washington. *Geology*, **26**, 351–354.

Lister, J. R. and Kerr, R. C. (1991). Fluid-mechanical models of crack propagation and their application to magma transport in dykes. *Journal of Geophysical Research*, **96**, 10 049–10 077.

Loughlin, S. C. (2002). Facies analysis of proximal subglacial and proglacial volcaniclastic successions at the Eyjafjallajokull central volcano, southern Iceland. In *Volcano-Ice Interaction on Earth and Mars*, ed. J. L. Smellie and M. G. Chapman. Geological Society of London Special Publication, **202**, pp. 149–178.

Major, J. J. and Newhall, C. G. (1989). Snow and ice perturbation during historical volcanic eruptions and the formation of lahars and floods. *Bulletin of Volcanology*, **52**, 1–27.

Manville, V., Hodgson, K. A., Houghton, B. F., Keys, J. R. and White, J. D. L. (2000). Tephra, snow and water: complex sedimentary responses at an active snow-capped stratovolcano, Ruapehu, New Zealand. *Bulletin of Volcanology*, **62**, 278–293.

Mastin, L. G. and Pollard, D. D. (1988). Surface deformation and shallow dike intrusion processes at Inyo Craters, Long Valley, California. *Journal of Geophysical Research*, **93**, 13 221–13 235.

Mathews, W. H. (1947). "Tuyas," flat-topped volcanoes in Northern British Columbia. *American Journal of Science*, **245**, 560–570.

Mee, K., Tuffen, H. and Gilbert, J. (2006). Snow-contact volcanic facies and their use in determining past eruptive environments at Nevados de Chillán volcano, Chile. *Bulletin of Volcanology*, **68**, 363–376.

Nye, J. F. (1953). The flow law of ice from measurements in glacier tunnels, laboratory experiments, and the Jungfraufirn borehole expedition. *Proceedings of the Royal Society of London Series A*, **219**, 477–489.

Parfitt, E. A. (1991). The role of rift zone storage in controlling the site and timing of eruptions and intrusions of Kilauea volcano, Hawai'i. *Journal of Geophysical Research*, **96**, 10 101–10 112.

Parfitt, E. A., Wilson, L. and Head, J. W. (1993). Basaltic magma reservoirs: factors controlling their rupture characteristics and evolution. *Journal of Volcanology and Geothermal Research*, **55**, 1–14.

Paterson, W. S. B. (1994). *The Physics of Glaciers*, 3rd. edn. Oxford: Pergamon Press.

Pollard, D. D. (1987). Elementary fracture mechanics applied to the structural interpretation of dikes. In *Mafic Dike Swarms*, ed. H. C. Halls and W. F. Fahrig. Geological Association of Canada Special Paper, **34**, 5–24.

Rubin, A. M. (1992). Dike-induced faulting and graben subsidence in volcanic rift zones. *Journal of Geophysical Research*, **97**, 1839–1858.

Rubin, A. M. (1995). Propagation of magma-filled cracks. *Annual Review of Earth and Planetary Science*, **23**, 287–336.

Rubin, A. M. and Pollard, D. D. (1987). Origins of blade-like dikes in volcanic rift zones. In *Volcanism in Hawaii*. United States Geological Survey Professional Paper, **1350**, pp. 1449–1470.

Ryan, M. P. (1987). Neutral buoyancy and the mechanical evolution of magmatic systems. In *Magmatic Processes: Physico-chemical Principles*, ed. B. O. Mysen. *Geochemical Society Special Publication*, **1**, pp. 259–287.

Scheu, B., Spieler, O. and Dingwell, D. B. (2006). Dynamics of explosive volcanism at Unzen volcano: an experimental contribution. *Bulletin of Volcanology*, **69**, 175–187.

Schopka, H. H., Gudmundsson, M. T. and Tuffen, H. (2006). The formation of Helgafell, southwest Iceland, a monogenetic subglacial hyaloclastite ridge: sedimentology, hydrology and volcano–ice interaction. *Journal of Volcanology and Geothermal Research*, **152**, 359–377.

Shean, D. E., Head, J. W. and Marchant, D. R. (2005). Origin and evolution of a cold-based tropical mountain glacier on Mars: The Pavonis Mons fan-shaped deposit. *Journal of Geophysical Research*, **110**, E05001, doi:10.1029/2004JE002360.

Sigmundsson, F. and Einarsson, P. (1992). Glacio-isostatic crustal movements caused by historical volume change of the Vatnajökull ice cap, Iceland. *Geophysical Research Letters*, **19**, 2123–2126.

Skilling, I. P. (1994). Evolution of an englacial volcano: Brown Bluff, Antarctica. *Bulletin of Volcanology*, **56**, 573–591.

Skilling, I. P. (2002). Basaltic pahoehoe lava-fed deltas: large-scale characteristics, clast generation, emplacement processes and environmental discrimination. In *Volcano-Ice Interaction on Earth and Mars*, ed. J. L. Smellie and M. G. Chapman. Geological Society of London Special Publication, **202**, pp. 91–113.

Smellie, J. L. (2000). Subglacial Eruptions. In *Encyclopedia of Volcanoes*, ed. H. Sigurdsson. San Diego: Academic Press, pp. 403–418.

Smellie, J. L. (2002). The 1969 subglacial eruption on Deception Island (Antarctica): events and processes during an eruption beneath a thin glacier and implications for volcanic hazards. In *Volcano-Ice Interaction on Earth and Mars*, ed. J. L. Smellie and M. G. Chapman. Geological Society of London Special Publication, **202**, pp. 59–79.

Smellie, J. L. (2006). The relative importance of supraglacial versus subglacial meltwater escape in basaltic subglacial tuya eruptions: an important unresolved conundrum. *Earth Science Reviews*, **74**, 241–268.

Smellie, J. L. (2007). Quaternary vulcanism: subglacial landforms. In *Encyclopedia of Quaternary science*, ed. S. A. Elias. Amsterdam: Elsevier, pp. 784–798.

Smellie, J. L. (2008). Basaltic subglacial sheet-like sequences: evidence for two types with different implications for the inferred thickness of associated ice. *Earth Science Reviews*, **88**, 60–88.

Smellie, J. L. (2009). Terrestrial sub-ice volcanism: landform morphology, sequence characteristics and environmental influences, and implications for candidate Mars examples. In *Preservation of Random Mega-scale Events on Mars and Earth: Influence on Geologic History*, eds. M. G. Chapman and L. Keszthelyi. Geological Society of America Special Paper, **453**, pp. 55–76.

Smellie, J. L. and Hole, M. J. (1997). Products and processes in Pliocene–Recent, subaqueous to emergent volcanism in the Antarctic Peninsula: examples of englacial Surtseyan volcanic construction. *Bulletin of Volcanology*, **58**, 628–646.

Smellie, J. and Skilling, I. P. (1994). Products of subglacial volcanic eruptions under different ice thickness: Two examples from Antarctica. *Sedimentary Geology*, **91**, 115–129.

Smellie, J. L., Hole, M. J. and Nell, P. A. R. (1993). Late Miocene valley-confined subglacial volcanism in northern Alexander Isand, Antarctic Peninsula. *Bulletin of Volcanology*, **55**, 273–288.

Stevenson, J. A., McGarvie, D. W., Smellie, J. L. and Gilbert, J. S. (2006). Subglacial and ice-contact volcanism at Vatnafjall, Öraefajökull, Iceland. *Bulletin of Volcanology*, **68**, 737–752.

Thouret, J.-C. (1990). Effects of the November 13, 1985 eruption on the snow pack and icecap of Nevado del Ruiz volcano, Colombia. *Journal of Volcanology and Geothermal Research*, **41**, 177–201.

Tuffen, H. (2007). Models of ice melting and edifice growth at the onset of subglacial basaltic eruptions. *Journal of Geophysical Research*, **112**, B03203, doi:10.1029/2006JB004523.

Tuffen, H. and Castro, J. (2009). The emplacement of an obsidian dyke though thin ice: Hraftntinnuhryggur, Krafla, Iceland. *Journal of Volcanology and Geothermal Research*, **85**, 352–366.

Tuffen, H., McGarvie, D., Pinkerton, H., Gilbert, J. S. and Brooker, R. A. (2008). An explosive-intrusive subglacial rhyolite eruption at Dalakvísl, Torfajökull, Iceland. *Bulletin of Volcanology*, **70**, 841–860.

Tuffen, H., Owen, J. and Denton, J. S. (2010). Magma degassing during subglacial eruptions and its use to reconstruct palaeo-ice thicknesses. *Earth-Science Reviews*, **99**, 1–18.

Vinogradov, V. N. and Murav'ev, Ya. D. (1988). Lava-ice interaction during the 1983 Kluchevskoi eruption. *Volcanology and Seismology*, **7**, 39–61.

Walder, J. S. (2000a). Pyroclast/snow interactions and thermally driven slurry formation. Part 1: Theory for monodisperse grain beds. *Bulletin of Volcanology*, **62**, 105–118.

Walder, J. S. (2000b). Pyroclast/snow interactions and thermally driven slurry formation. Part 2: Experiments and theoretical extension to polydisperse tephra. *Bulletin of Volcanology*, **62**, 119–129.

Werner, R. and Schmincke, H.-U. (1999). Englacial vs. lacustrine origin of volcanic table mountains: evidence from Iceland. *Bulletin of Volcanology*, **60**, 335–354.

White, J. D. L. and Houghton, B. F. (2006). Primary volcaniclastic rocks. *Geology*, **34**, 677–680.

Wilson, L. (1980). Relationships between pressure, volatile content and ejecta velocity in three types of volcanic explosion. *Journal of Volcanology and Geothermal Research*, **8**, 297–313.

Wilson, L. and Head, J. W. (1981). Ascent and eruption of basaltic magma on the Earth and Moon. *Journal of Geophysical Research*, **86**, 2971–3001.

Wilson, L. and Head, J. W. (2002). Heat transfer and melting in subglacial basaltic volcanic eruptions: implications for volcanic deposit morphology and meltwater volumes. In *Volcano-Ice Interaction on Earth and Mars*, ed. J. L. Smellie and M. G. Chapman.

Geological Society of London Special Publication, **202**, pp. 5–26.

Wilson, L. and Head, J. W. (2007). Heat transfer in volcano-ice interactions: synthesis and applications to processes and landforms on Earth. *Annals of Glaciology*, **45**, 83–86.

Wilson, L. and Head, J. W. (2009). Tephra deposition on glaciers and ice sheets on Mars: Influence on ice survival, debris content and flow behavior. *Journal of Volcanology and Geothermal Research*, **185**, 290–297.

Wilson, L., Fagents, S. A., Robshaw, L. E. and Scott, E. D. (2007). Vent geometry and eruption conditions of the mixed rhyolite-basalt Námshraun lava flow, Iceland. *Journal of Volcanology and Geothermal Research*, **164**, 127–141.

Wohletz, K. H. (1986). Explosive magma-water interactions: thermodynamics, explosive mechanisms, and field studies. *Bulletin of Volcanology*, **48**, 245–264.

Wohletz, K. H. and McQueen, R. G. (1984). Experimental studies of hydromagmatic volcanism. In *Explosive Volcanism: Inception, Evolution, and Hazards*, Studies in Geophysics. Washington DC: National Academy Press, pp. 158–169.

Wright, A. C. (1978). A reconnaissance study of the McMurdo Volcanics north-west of Koettlitz Glacier. *New Zealand Antarctic Record*, **1**, 10–15.

Zimanowski, B., Frohlich, G. and Lorenz, V. (1991). Quantitative experiments on phreatomagmatic explosions. *Journal of Volcanology and Geothermal Research*, **48**, 341–358.

Eyjafjallajökull Links

http://en.wikipedia.org/wiki/Air_travel_disruption_after_the_2010_Eyjafjallaj%C3%B6kull_eruption
http://en.wikipedia.org/wiki/Consequences_of_the_April_2010_Eyjafjallaj%C3%B6kull_eruption
http://en.wikinews.org/wiki/Ash-triggered_flight_disruptions_cost_airlines_$1.7_billion

Exercises

13.1 Why might mafic eruptions under ice be more likely to have an explosive component than subglacial rhyolitic eruptions?

13.2 What is the minimum pressure required at the point where a sill is injected under an ice sheet and why does this suggest that dikes may potentially penetrate a significant distance into ice sheets?

13.3 When incandescent lava flows over ice or snow there is no immediate flood of meltwater from beneath the lava. Why is this?

Online resources available at www.cambridge.org/fagents

- Links to websites of interest
- Answers to exercises

Chapter 14

Modeling lahar behavior and hazards

Vernon Manville, Jon J. Major, and Sarah A. Fagents

Overview

Lahars are highly mobile mixtures of water and sediment of volcanic origin that are capable of traveling tens to > 100 km at speeds exceeding tens of km hr^{-1}. Such flows are among the most serious ground-based hazards at many volcanoes because of their sudden onset, rapid advance rates, long runout distances, high energy, ability to transport large volumes of material, and tendency to flow along existing river channels where populations and infrastructure are commonly concentrated. They can grow in volume and peak discharge through erosion and incorporation of external sediment and/or water, inundate broad areas, and leave deposits many meters thick. Furthermore, lahars can recur for many years to decades after an initial volcanic eruption, as fresh pyroclastic material is eroded and redeposited during rainfall events, resulting in a spatially and temporally evolving hazard. Improving understanding of the behavior of these complex, gravitationally driven, multiphase flows is key to mitigating the threat to communities at lahar-prone volcanoes. However, their complexity and evolving nature pose significant challenges to developing the models of flow behavior required for delineating their hazards and hazard zones.

14.1 | Introduction

The Indonesian word "lahar"' refers to a highly mobile mixture of water and sediment, other than normal stream flow, originating from a volcano (e.g., Fig. 14.1; Smith and Fritz, 1989). The term has genetic connotations rather than implying any particular flow behavior, which can range from dilute hyperconcentrated flows, in which particle concentrations greater than those of normal streamflow conditions are transported chiefly as suspended and bedload sediment, to debris flows in which a high-concentration particulate phase transports sediment en masse with fluid in its interstices (Vallance, 2000). Lahars vary greatly in volume ($\sim$10^2–10^9 m^3), peak discharge (< 10–10^7 m^3 s^{-1}), advance rate ($\sim$2–80 m s^{-1}) and runout (a few to > 100 km; Pierson, 1998). They are common on many intermediate to felsic volcanoes having relatively steep flanks and abundant volcaniclastic material, and are among the most dangerous volcanic hazards during episodes of volcanic unrest, eruptive activity, and post-eruption periods (Neall, 1996; Vallance, 2000). Their capacity to rapidly inundate large areas is a great threat to property, infrastructure (Fig. 14.2) and life, causing > 30 000 casualties in the twentieth century alone (Witham, 2005). Population growth

Modeling Volcanic Processes: The Physics and Mathematics of Volcanism, eds. Sarah A. Fagents, Tracy K. P. Gregg, and Rosaly M. C. Lopes. Published by Cambridge University Press. © Cambridge University Press 2013.

Figure 14.1 Oblique aerial photograph of snowmelt-triggered debris-flow lahar at Mount St. Helens, March 1982. The flow began as a water flood from the crater, entrained sediment along the steep north flank, and became a lahar that moved west (right) down the North Fork Toutle River valley. Photograph by T. J. Casadevall, US Geological Survey, 21 March 1982.

and economic development, particularly in the developing world, are increasing societal vulnerability to lahars as terraces bordering river channels that drain active volcanoes become increasingly colonized and urbanized. As a result, better knowledge and understanding of lahar magnitude, probability of occurrence, and dynamics are needed in order to develop the improved hazards assessments vital to hazards mitigation. Predictions of parameters such as runout distance, maximum flow depth, inundation area, flow velocity (and hence travel time to a downstream point), and energy distribution along a flow path provide information useful for mitigation purposes. The physics underlying the propagation of lahars has many commonalities with that of pyroclastic density currents (Chapter 10), but with water replacing gas as the fluid phase. This chapter reviews approaches that have been adopted for modeling lahar behavior, which are variously informed by theoretical analyses, laboratory experiments, and field observations of sediment–water mixtures, and assesses their strengths and weaknesses.

Figure 14.2 The bridge on Boyong River, close to the village of Boyong on the flanks of Merapi volcano, Java, Indonesia. (a) View in December 1994 – the bridge stands 7 m above the river valley (people for scale), which is cultivated for crops. (b) Same view, 20 February 1995 – the river valley has been filled with lahar sediment. Reprinted from Lavigne *et al.* (2000a), with permission of Elsevier.

14.2 | Lahar initiation

Initiation mechanisms can greatly influence lahar volumes, compositions, hydrographs, flow behaviors, and hazards. Lahars can be broadly subdivided into two classes based on their initiation: primary lahars that are caused directly by eruptive activity, and secondary lahars that occur during post-eruptive or quiescent periods. Both classes of lahars can occur during a single eruptive sequence (e.g., Cronin et al., 1997), or can be separated by hundreds to thousands of years (Zanchetta et al., 2004). Lahar initiation generally requires certain preconditions, including: (1) an adequate water supply; (2) abundant unconsolidated sediment; (3) gravitational potential; and (4) a triggering mechanism. Common triggers include the mobilization of pyroclastic sediment by flowing water, due chiefly to rainfall runoff (Rodolfo and Arguden, 1991), eruptive expulsion of a crater lake (Suryo and Clarke, 1985; Kilgour et al., 2010), release of an impounded water body (Björnsson, 1992; Umbal and Rodolfo, 1996; Manville and Cronin, 2007), melting of snow and ice by interaction with hot eruptive products (Major and Newhall, 1989), and liquefaction of debris avalanches and debris-avalanche deposits (Scott, 1988; Stoopes and Sheridan, 1992; Capra et al., 2002; Scott et al., 2005). Initiation mechanism therefore influences the source boundary conditions (e.g., source hydrograph) used in lahar models.

14.3 | The sediment-concentration continuum

Lahars exhibit complex flow behavior governed by fluid and particle properties, and fluid–particle and particle–particle interactions. These interactions vary according to the amount (volume concentration), type, and size distribution of the particles (Fig. 14.3; Pierson and Costa, 1987; Iverson, 2009). At intermediate solids concentrations (~20–50% by volume; bulk densities of 1300–1800 kg m^{-3}), hyperconcentrated flows (Beverage and Culbertson, 1964; Pierson, 2005) have larger sediment concentrations than are found in normal water floods. Hyperconcentrated flows typically possess a small but measurable static yield strength, an apparent viscosity that is strain-rate dependent, commonly transport large quantities of sand-sized material in full

suspension and some gravel as bedload, and have distinctive smooth, oily-appearing surfaces that may display polygonal convection cells (Fig. 14.4(a); Pierson, 2005; also see online supplementary materials listed at end of chapter). They exhibit density-stratified, turbulent, two-phase behavior in which the solid particles are supported by turbulence and buoyancy, but as solids concentrations increase, turbulence becomes damped and particle–particle interactions become more important. The volumetric concentration at which this transition occurs strongly depends on the grain-size distribution and clay-mineral content of the sediment load. Measurable static yield strengths can be achieved at concentrations as low as 3 vol.% of smectite clays (Hampton, 1975; Fig. 14.3(b)) or as high as 50 vol.% of coarse, neutrally buoyant particles (Bagnold, 1954; Pierson, 2005). Hyperconcentrated-flow deposits tend to be more poorly sorted than normal fluvial deposits, but better sorted than debris-flow deposits; they are typically weakly stratified to massively textured, and are clast-supported (Fig. 14.4(b)), reflecting particle settling through the flow and step-wise accretion at the base (Smith, 1986).

At solids concentrations greater than ~50% by volume, lahars transition from hyperconcentrated flows to debris flows. Debris flows have volumetric solids contents of ~50–80%, bulk densities of ~1800–2300 kg m^{-3}, and are comprised of particles that range in size from clay to boulders (Costa, 1988). Such mixtures are 10^4–10^5 times more viscous than water and typically possess a yield strength that must be exceeded before flow is possible. Once moving, they can achieve velocities double those of water floods of comparable depth and gradient due to greater bulk density, suppression of energy-dissipating internal turbulence, and modification of channel geometry through erosion and deposition in order to efficiently convey flow. Interstitial pore fluid (water or a water-borne slurry of fines) facilitates flow by bearing some or all of the solids load through buoyant support and increased fluid pressure, thereby reducing the effective stress acting on the solids and hence the intergranular friction resisting flow. Debris-flow mixtures commonly exhibit both dilatancy

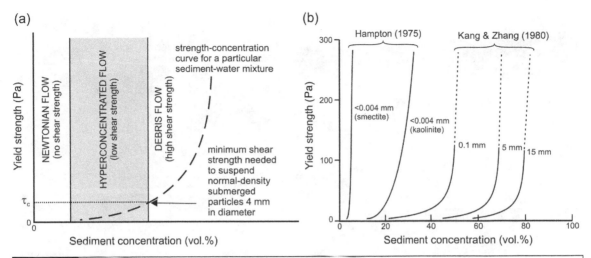

(a)

Figure 14.3 Yield strength of sediment–water mixtures as a function of suspended sediment concentration. (a) Definitions of flow type based on an idealized yield-strength–concentration curve for a poorly sorted sediment–water mixture. (b) Measured yield-strength–concentration curves for a range of sediment–water mixtures illustrating the effect of grain-size distributions (curves are marked with the median particle diameter) and compositions. Reprinted from Pierson (2005), with kind permission of Springer Science+Business Media.

and contraction (where shear causes expansion or collapse of pore space), intergranular friction (due to grain–grain contacts), fluidization (due to pore-fluid pressure exceeding hydrostatic), particle segregation, and minimal to moderate cohesion when stationary. Such flows commonly develop a relatively coarse and dry flow front, which behaves as a moving boulder dam (Fig. 14.4(c); Sharp and Nobles, 1953; Major and Iverson, 1999; Lavigne and Suwa, 2004) that can increase maximum stage height, and a body that usually appears to behave as a coherent but liquefied single-phase mass, or as a hyperconcentrated flow. Debris flows commonly develop steep lobate flow fronts, produce lateral levées, transport cobble- to boulder-sized clasts within the flow as well as along the channel bed, and form massive, very poorly sorted and ungraded matrix- to clast-supported deposits (Fig. 14.4(d)). Although debris flows can selectively deposit their coarsest clasts, they can also accrete sediment incrementally (Major, 1997) or deposit it en masse as shear stresses decline and intergranular friction, chiefly at the flow front, locks up the flow (Major and Iverson, 1999). The presence of clay-sized material can strongly influence flow behavior. *Cohesive* flows (defined as containing > 5% clay) are typically more mobile than

non-cohesive flows (< 5% clay), and they can maintain their high-concentration integrity for great distances because they are resistant to dilution and transformation. Non-cohesive flows typically contain a narrower and coarser distribution of grain sizes, entrain water and deposit sediment more easily, and commonly transform distally to hyperconcentrated flow (Figs. 14.5, 14.6; Pierson and Scott, 1985; Scott, 1988).

14.4 | Lahar characteristics

Real-time field measurements of the dynamic properties of lahars are rare, owing largely to their size, the challenges of predicting the timing of their occurrence, and financial constraints in many of the countries where lahars frequently occur. Nevertheless, a few field monitoring stations around the world have gathered data on individual events (Manville and Cronin, 2007) or on a recurrent series of flows (Suwa and Okuda, 1985; Pierson, 1986; Suwa, 1989; Ohsumi Works Office, 1995; Lavigne et al., 2000b), variously recording discharge hydrographs, compositions, depths, volumes, flow-front and surface velocities, surface-velocity distributions, basal stresses

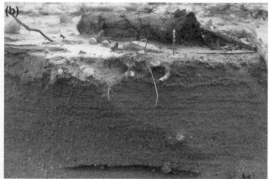

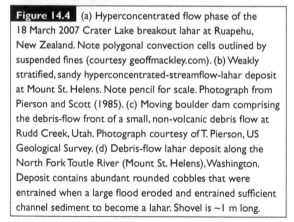

Figure 14.4 (a) Hyperconcentrated flow phase of the 18 March 2007 Crater Lake breakout lahar at Ruapehu, New Zealand. Note polygonal convection cells outlined by suspended fines (courtesy geoffmackley.com). (b) Weakly stratified, sandy hyperconcentrated-streamflow-lahar deposit at Mount St. Helens. Note pencil for scale. Photograph from Pierson and Scott (1985). (c) Moving boulder dam comprising the debris-flow front of a small, non-volcanic debris flow at Rudd Creek, Utah. Photograph courtesy of T. Pierson, US Geological Survey. (d) Debris-flow lahar deposit along the North Fork Toutle River (Mount St. Helens), Washington. Deposit contains abundant rounded cobbles that were entrained when a large flood eroded and entrained sufficient channel sediment to become a lahar. Shovel is ~1 m long.

and fluid pressures, and sediment erosion. Additional information comes from photographic records (still images and video recordings) of flow character and boulder motion, from seismic (acoustic) signals generated during flow passage (e.g., Cole *et al.*, 2009; Kumagai *et al.*, 2009), and from detailed examination of lahar deposits cross-correlated with flow observations (Pierson and Scott, 1985; Cronin *et al.*,

2000). Valuable data germane to the behavior of lahars have also come from similar monitoring of non-volcanic debris flows ranging in volume from 10^2 to 10^5 m^3 (e.g., Marchi *et al.*, 2002; Cui *et al.*, 2005; McArdell *et al.*, 2007; McCoy *et al.*, 2010; C. Berger *et al.*, 2011) and from various scale laboratory experiments.

14.4.1 Insights from small-scale laboratory experiments

A variety of laboratory studies have investigated the rheological characteristics of concentrated, small-volume (a few cm^3 to ~1 m^3) sediment–water suspensions containing natural or artificial particles ranging in size from < 0.063 mm (silt and clay) up to ~120 mm (e.g., O'Brien and Julien, 1988; Phillips and Davies, 1991; Major and Pierson, 1992; Coussot *et al.*, 1998; Contreras and Davies, 2000; Armanini

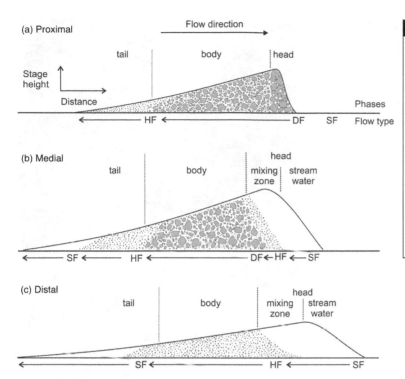

(a) Proximal

Flow direction →

tail | body | head

Stage height

Distance

Phases

← HF ← | → DF | SF | Flow type

(b) Medial

head

tail | body | mixing zone | stream water

← SF ← | HF ← | → DF ← HF ← SF

(c) Distal

head

tail | body | mixing zone | stream water

← SF ← | HF ← | SF

Figure 14.5 Schematic lahar profiles showing idealized downstream evolution of a flow as a non-cohesive debris flow at (a) proximal, (b) medial, and (c) distal locations. Diagram depicts longitudinal segregation, interaction with ambient water in the channel, and reversible transformation processes. An increasing time lag can develop between the peak water discharge and the peak sediment discharge due to differences in their respective waveform celerities. SF, HF, and DF denote, respectively, streamflow, hyperconcentrated flow, and debris flow.

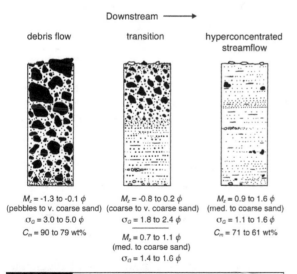

Downstream ⟶

debris flow | transition | hyperconcentrated streamflow

M_z = -1.3 to -0.1 ϕ (pebbles to v. coarse sand)
σ_G = 3.0 to 5.0 ϕ
C_m = 90 to 79 wt%

M_z = -0.8 to 0.2 ϕ (coarse to v. coarse sand)
σ_G = 1.8 to 2.4 ϕ

M_z = 0.7 to 1.1 ϕ (med. to coarse sand)
σ_G = 1.4 to 1.6 ϕ

M_z = 0.9 to 1.6 ϕ (med. to coarse sand)
σ_G = 1.1 to 1.6 ϕ
C_m = 71 to 61 wt%

Figure 14.6 Schematic representation of deposit facies of the March 1982 snowmelt-triggered lahar at Mount St. Helens showing distal transformation from debris-flow lahar to hyperconcentrated-streamflow lahar. Ranges in mean grain size (M_z), sorting (σ_G), and sediment mass concentration (C_m) are indicated. Units are given in phi scale. Redrawn from Pierson and Scott (1985). See Figure 14.5 for associated parts of the lahar profile that produce these deposits.

et al., 2005; Kaitna *et al.*, 2007; Forterre and Pouliquen, 2008). Results of these experiments are diverse, with the rheological behavior varying, commonly nonlinearly, with composition, sediment concentration, particle-size distribution, shear rate, and time. Mixture properties are commonly extremely sensitive to sediment concentration; changes of as little as 2–4 vol.% sediment can produce order-of-magnitude changes in rheological parameters. Coarse particles clearly affect slurry behavior, but the precise effects of particle-size distribution on granular interactions and the rheological behavior of natural sediment slurries are ambiguous. Although small-scale experiments on relatively fine-grained slurries indicate that granular interactions are important to debris-flow behavior, such experiments are unlikely to adequately capture the behavior of fundamentally multiphase, coarse-grained natural flows. The range of behaviors seen in the laboratory suggests that it is unlikely that debris flows can be characterized by a simple, single-phase rheological model of homogeneous material (Major and Pierson, 1992; Iverson, 2003). Scaling arguments further indicate a need for larger-scale experimental analyses of debris flows (Iverson *et al.*, 2004).

14.4.2 Insights from large-scale experimental debris flows

Limitations on the applicability of small-scale experiments to natural phenomena, together with the difficulty of making real-time observations on natural debris flows, prompted development of a series of large-scale experiments in a 95-m-long concrete flume (Fig. 14.7; Iverson *et al.*, 2010; see also online resources listed at end of chapter). Approximately 10 m³ volumes of water-saturated mixtures of mud, sand, and gravel (up to 3 cm in diameter) were released onto a fixed 31° slope that flattens distally to a 3° runout surface. Measured basal normal stresses and fluid pressures correlated with, and responded rapidly to, variations in flow depth (Iverson, 1997a; Iverson *et al.*, 2010). Although total basal normal stress increased proportionally with flow depth, basal fluid pressure was close to zero near the flow front and increased to a nearly lithostatic value in the body behind the flow front. In the runout area of the flume, measured basal normal stress and pore-fluid pressure reflected flow deceleration and deposit accumulation. Pore-fluid pressure within the body of a debris flow dissipated significantly only after sediment deposition, suggesting that debris-flow deposition was controlled chiefly by frictional resistance along the low-fluid-pressure flow margin (Major and Iverson, 1999). Measurements of basal stress and fluid pressure in natural debris flows have confirmed that fluid pressure exceeding hydrostatic is long-lived and contributes to debris-flow dynamics and mobility (McArdell *et al.*, 2007; McCoy *et al.*, 2010).

14.4.3 Summary of key lahar characteristics

Although flow volumes span nine orders of magnitude between the laboratory and field, and individual flow characteristics vary significantly, key observations highlight the fundamental physical characteristics of lahars in the hyperconcentrated-flow to debris-flow continuum:

Figure 14.7 US Geological Survey experimental debris-flow flume, H. J. Andrews Experimental Forest, Oregon. Flume is 95 m long, 2 m wide, and sits on a fixed 31° slope. Note people for scale; runout surface has 1 m grid spacing. Photograph courtesy of R. M. Iverson, US Geological Survey.

- Lahars are fundamentally unsteady flows whose characteristics vary spatially and temporally, depending on solids concentration, dilatancy, particle-size distribution and segregation, pore-fluid pressure, and solid–fluid interactions.
- Lahars can undergo multiple and reversible spatial and temporal transformations in flow behavior (e.g., from hyperconcentrated flow to debris flow, or vice versa) caused by changes in sediment concentration and type (Fig. 14.8). Sediment concentration can increase by incorporation of external sediment (commonly called *bulking*) or, albeit less importantly, through loss of water by *infiltration* into the channel bed. Sediment concentration can

decrease through *dilution* by entrained water or by selective sediment deposition (sometimes called *debulking*) (Pierson and Scott, 1985; Pierson, 1997).

- Lahar volumes and peak discharges can increase by a factor of three or more relative to initial values (Manville, 2004) as a result of sediment entrainment during flow.
- Variability in flow-path topography can strongly influence flow velocity, sediment erosion and deposition, and hence flow-character evolution. The tendency for flows to entrain sediment on steep slopes and deposit on shallow slopes (e.g., Pierson, 1995; Iverson *et al.*, 2011) is modulated by variations in channel geometry (e.g., transitions from narrow, confined flow to shallow, unconfined flow) and substrate characteristics.
- Particle segregation can occur by size, shape, and density during flow, causing the development of vertical and longitudinal stratification in the lahar.
- Similar to those of unsteady water flows, lahar discharge hydrographs (which can be very steep at the source) typically flatten and lengthen as flows propagate downstream, and may be preceded by a rising stage of normal streamflow that may represent displaced river water (e.g., Cronin *et al.*, 1999).
- Debris-flow lahars typically travel along a channel as a series of steep-fronted surges (e.g., Pierson, 1986; Iverson, 1997a; McArdell *et al.*, 2007). Particle segregation can produce a concentration of the coarsest particles dominated by solid forces at a surge front, followed by a more fluid but nevertheless high-sediment-concentration body that typically transitions to a dilute, hyperconcentrated or low-concentration-streamflow tail (e.g., Figs. 14.4(c), 14.5; Lavigne and Suwa, 2004).
- Pore-fluid pressure in a debris-flow lahar can be spatially and temporally heterogeneous and particle–particle contacts can be both frictional and collisional.
- Debris-flow lahars can deposit sediment en masse (Major and Iverson, 1999) or through step-wise vertical accretion (Major, 1997), whereas hyperconcentrated-flow lahars deposit sediment progressively by accretion

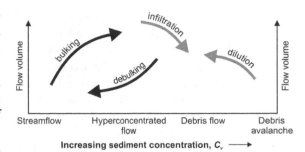

Figure 14.8 Addition and loss of sediment (black arrows) and water (gray arrows) during lahar flow causes volume and sediment concentration to vary, leading to transformations among flow regimes.

of bedload and settling from suspension (e.g., Pierson, 2005). Deposits of debris-flow lahars are typically massively textured and very poorly sorted, whereas those of hyperconcentrated-flow lahars can be moderately stratified to massively textured, and are usually moderately well to moderately poorly sorted (e.g., Fig. 14.4; Pierson and Scott, 1985; Cronin *et al.*, 2000; Major *et al.*, 2005; Pierson, 2005). Transitional deposits can also occur.

14.5 | Lahar modeling

The complexity and variability of lahars make them challenging both to characterize and to model. The rheology of such multiphase flows is composition-, scale-, time-, and possibly shear-rate dependent, with a primarily nonlinear and potentially hysteretic (i.e., behavior under increasing strain rate differing from that under decreasing strain rate) relationship between shear stress and strain rate. Existing lahar models thus draw on many disciplines, including surface water hydrology, engineering, sedimentology, geomorphology, physics and applied mathematics, and from studies of non-volcanic debris flows, landslides, and avalanches (e.g., Coussot and Meunier, 1996; Hutter *et al.*, 1996; Iverson, 1997a, 2003; Ancey, 2007; Hürlimann *et al.*, 2008; Berzi *et al.*, 2010; Christen *et al.*, 2010).

The complexity of the dynamic behavior of lahars has given rise to a diversity of modeling approaches, ranging from simple empirical models lacking explicit flow physics to sophisticated physics-based models founded on conservation laws. Thus, lahar models fall into a number of principal, partially overlapping classes. These classes include: (1) empirical models based on observed correlations among lahar parameters such as volume, flow velocity, and cross-section or inundation area, but which lack treatment of flow physics (e.g., Iverson *et al.*, 1998; Pierson, 1998); (2) simple rheological models that assume a constant stress–strain-rate relationship and composition-independent flow behavior (e.g., Fink *et al.*, 1981; Manville *et al.*, 1998); (3) hydrologic models that assume Newtonian behavior but are calibrated to lahars through modification of the flow resistance term (e.g., Costa, 1997; Manville, 2004; Carrivick *et al.*, 2009); (4) sophisticated theoretical formulations that seek to describe the constitutive behavior of multiphase mixtures (Chen, 1988a,b; Iverson, 1997a; Iverson and Denlinger, 2001; Takahashi, 2001; Pitman and Le, 2005; Berzi *et al.*, 2010; George and Iverson, 2011); and (5) mass-flow models that seek to combine elements of approaches (3) and (4) with the inclusion of processes such as sediment entrainment and deposition (e.g., Fagents and Baloga, 2006; Carrivick *et al.*, 2010) or development of vertical grain-size stratification (Takahashi *et al.*, 1992; Zanuttigh and Ghilardi, 2010).

Except for the simple empirical models, lahar models generally have four principal components: (1) a set of terms that describe conservation of mass and (sometimes) momentum of the bulk flow or its constituent components; (2) a description of channel geometry; (3) a means of quantifying flow resistance; and (4) a means of solving the resulting suite of partial differential equations numerically. The greatest differences among the more physics-based models lie in their descriptions and treatments of flow resistance and faithfulness to the underlying physics, which commonly represent a compromise between mathematical accuracy and computational tractability.

14.5.1 Empirical lahar models

Empirical models are typically founded on statistical analyses of field estimates of various parameters from past lahar events and deposits. They predict the average behavior of future lahars without regard for the underlying flow physics. Outputs from such models typically focus on parameters such as flow-front velocity (i.e., arrival time), maximum discharge, stage, and inundation area.

LAHARZ

One of the most widely used empirical models relates the area inundated by a lahar to its volume. On the basis of data compiled from 27 lahars worldwide, Iverson *et al.* (1998) identified the following relationships between lahar volume V (m³), channel cross-sectional area A (m²) filled by a flow, and planimetric inundation area B (m²; notation is summarized in Section 14.9; Fig. 14.9(a)):

$$A = 0.05\, V^{2/3}, \tag{14.1}$$

$$B = 200\, V^{2/3}. \tag{14.2}$$

Solutions of Eqs. (14.1) and (14.2) were automated in the computer algorithm LAHARZ (Iverson *et al.*, 1998; Schilling, 1998), a routine run within the ARCINFO™ Geographical Information System (GIS). The technique was designed for rapid, objective, and reproducible construction of hazard maps. The model requires user-input flow volumes and a digital elevation model (DEM) from which hydrologic grids are derived for stream network identification. Lahar deposition begins either at the intersection of a proximal hazard-zone boundary, defined as the distal extent of the distance over which all potential sediment entrainment occurs, and a stream centerline cell (Fig. 14.9(a)), or at a user-specified stream cell. The algorithm fills the cross-sectional topography at the depositional starting point according to Eq. (14.1), then marches downstream on a cell-by-cell basis repeating the channel-filling process and summing the planimetric area until Eq. (14.2) is satisfied. Output consists of maps delineating nested zones of inundation potential based on the range of user-supplied flow volumes (Fig. 14.9(b)). The model has been applied widely because of its simplicity and low data requirements

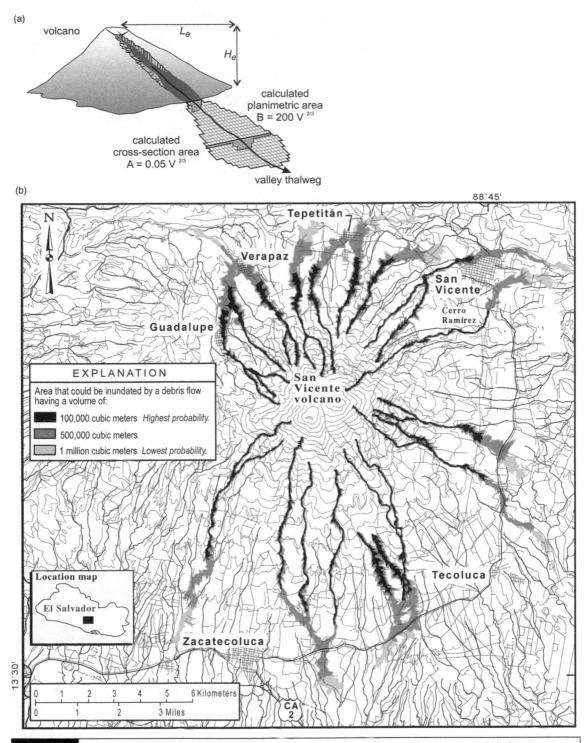

(a)
volcano
L_e
H_e
calculated planimetric area
$B = 200\ V^{2/3}$
calculated cross-section area
$A = 0.05\ V^{2/3}$
valley thalweg

(b)
88°45'

N

Tepetitán
Verapaz
San Vicente
Cerro Ramírez
Guadalupe
San Vicente volcano

EXPLANATION
Area that could be inundated by a debris flow having a volume of:
■ 100,000 cubic meters. *Highest probability.*
■ 500,000 cubic meters.
□ 1 million cubic meters. *Lowest probability.*

Tecoluca

Location map
El Salvador

Zacatecoluca

13°30'

0 1 2 3 4 5 6 Kilometers
0 1 2 3 Miles

CA 2

Figure 14.9 (a) Schematic illustration of the LAHARZ model methodology, modified from Iverson *et al.* (1998). H_e represents the drop height, and L_e the horizontal distance over which potential sediment entrainment is assumed to occur before deposition begins. (b) Example of nested lahar-inundation hazard zones predicted by LAHARZ for lahars of various volumes originating on the flanks of San Vicente volcano, El Salvador. In this analysis, deposition begins at a user-specified location. Reprinted from Major *et al.* (2003), with permission of IOS Press.

(e.g., Canuti et al., 2002; Major et al., 2004; Hubbard et al., 2007; Muñoz-Salinas et al., 2009; Worni et al., 2012). It is, however, sensitive to DEM accuracy and resolution (Iverson et al., 1998; Stevens et al., 2002; Huggel et al., 2008).

Expressions similar to Eqs. (14.1) and (14.2) have been developed for rock avalanches and non-volcanic debris flows (e.g., Hungr, 1990; Griswold and Iverson, 2008) as well as for lahars at individual volcanoes. These studies show that the proportionality coefficients can be refined for particular flow types or settings to yield more accurate results.

Travel time, discharge, volume and runout relationships

Pierson (1998) developed an empirical model to assess lahar travel times as functions of distance and flow discharge. He summarized kinematic, volumetric and hydraulic characteristics of a number of historical flows at different volcanoes and derived a series of polynomial relationships for predicting arrival times $T(x)$ of flows of different near-source peak discharges at distances x downstream from source (Fig. 14.10), and a power-law relationship for predicting near-source peak discharge Q_p from lahar volume V:

$$T(x) = a_2 x^2 + a_1 x + a_0, \qquad (14.3)$$

$$Q_p = c_1 V^a. \qquad (14.4)$$

These two equations (see also Tables 14.1 and 14.2) can be used to forecast proximal peak discharge and consequent travel times to specific distances from source if the initial volume of a future flow can be estimated. For the case of a lahar triggered by a release of water, Pierson (1998) suggested multiplying the initial flow volume of water by a factor of three to estimate the volume of a fully bulked debris-flow lahar.

For a range of volcanic and non-volcanic debris flows exhibiting well-defined distal termini, and having volumes between 6 m³ and 8 × 10⁶ m³, Rickenmann (1999) found that the runout distance L (m), flow volume V (m³), and elevation difference H (m) between the crown of the source and the deposit terminus can be linked empirically by the expression

Table 14.1 | Coefficient c_1 and exponent a in Eq. (14.4) for different types of mass flow (from Rickenmann, 1999).

Flow type	c_1	a
Granular debris flow	0.135	0.78
Muddy debris flow	0.0188	0.78
Merapi lahars	0.00558	0.831
Sakurajima lahars	0.00135	0.870
Landslide dambreak	0.293	0.56
Glacial dambreak	0.0163	0.64

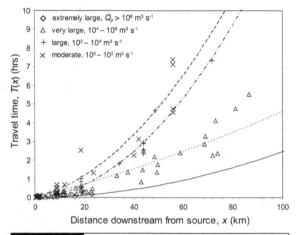

Figure 14.10 Travel time versus downstream distance for four size classes of lahar, fitted with polynomial curves. Extremely large, peak discharge > 10⁶ m³ s⁻¹; very large, 10⁴–10⁶ m³ s⁻¹; large, 10³–10⁴ m³ s⁻¹; moderate, 10²–10³ m³ s⁻¹. Replotted using data from Pierson (1998).

$$L = 1.9 V^{0.16} H^{0.83}. \qquad (14.5)$$

Such a relation is not useful, however, for dilute (hyperconcentrated flow) lahars that typically do not have coherent deposit termini, but rather lose sediment load with distance and transform gradually to normal streamflow.

14.5.2 Single-phase rheological models

A suite of single-phase rheological models exists to describe lahar behavior, particularly for the more sediment-laden, debris-flow end of the lahar continuum. These models build on observations of debris-flow behavior and deposit geometry (Yano and Daido, 1965; Johnson, 1970; Johnson and Rodine, 1984), supplemented

Peak discharge, Q_p (m³ s⁻¹)	a_0	a_1	a_2
Extremely large, $>10^6$ m³ s⁻¹	0.00909467	0.000090487	0.000282666
Very large, 10^4–10^6 m³ s⁻¹	−0.271086	0.0378719	0.000110375
Large, 10^3–10^4 m³ s⁻¹	0.087511	−0.00889418	0.0015254
Moderate, 10^2–10^3 m³ s⁻¹	0.300674	−0.0179581	0.00209817

Table 14.2 Coefficients for lahar travel-time equation (14.3).

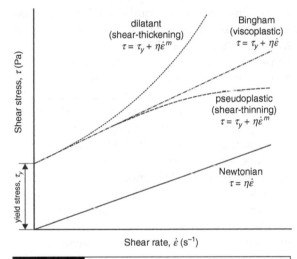

Figure 14.11 Relationships between shear stress and shear-strain rate for common formulations of the generalized Herschel–Bulkley rheological model. For dilatant fluids, $m > 1$; for pseudoplastic fluids, $m < 1$.

by rheological measurements on sub-samples of deposits reconstituted into slurries (e.g., O'Brien and Julien, 1988; Phillips and Davies, 1991; Major and Pierson, 1992; Arattano *et al.*, 2006). The models all assume that a constant stress–strain-rate relationship, suitable for an idealized, homogeneous, single-phase material, adequately describes debris-flow behavior. This relationship is described using the generalized Herschel–Bulkley equation (Fig. 14.11):

$$\tau = \tau_y + \eta \dot{\varepsilon}^m, \tag{14.6}$$

where τ is the applied shear stress, τ_y is the yield strength, η is the dynamic viscosity (i.e., the gradient of the stress–strain-rate relationship), and $\dot{\varepsilon}$ is the shear strain rate ($= \partial u / \partial y$). The simplest non-Newtonian rheological model describes a Bingham viscoplastic material ($\tau_y > 0$ and $m = 1$

in Eq. (14.6); Fig. 14.11), with a viscosity η that is independent of shear rate. Such a material behaves as an elastic solid at shear stresses lower than the yield strength, and as a linearly viscous fluid at stresses exceeding the yield strength.

Partitioning the yield strength of the Bingham model explicitly into components that combine Coulomb's friction rule with Terzaghi's effective-stress principle, related to material cohesion c, normal stress σ, intergranular friction φ, and pore-fluid pressure p_f (which modifies the applied normal stress; see online supplement), produces the Coulomb-viscous model proposed by Johnson (1970) and Johnson and Rodine (1984):

$$\tau = \left(c + (\sigma - p_f)\tan\varphi\right) + \eta\dot{\varepsilon}. \tag{14.7}$$

Viscoplastic models, however, fail to account for obvious particle–particle interactions inherent in debris flows.

To account for particle collisions in debris flows, Takahashi (1980) developed a grain-flow theory that is founded on the basis of experiments (Bagnold, 1954) relating normal and shear stresses in particulate suspensions. The central tenet of Takahashi's theory involves a dispersive stress τ_d that arises in shearing granular flows due to collisions and momentum exchanges between adjacent particles. Two regimes were identified on the basis of the ratio of inertial to viscous shear stresses in a suspension: a macroviscous regime (Eq. (14.8)) dominant at low shear rates; and a grain-inertia (collisional) regime (Eq. (14.9)) that prevails at higher rates. These regimes are expressed as:

$$\tau_d = \lambda^{3/2}\eta\dot{\varepsilon} \quad \text{for } \dot{\varepsilon} < 40\eta \, / \, (\lambda^{1/2}\rho_s d^2), \tag{14.8}$$

$$\tau_d = \rho_s(\lambda d)^2 \dot{\varepsilon}^2 \quad \text{for } \dot{\varepsilon} > 450\eta \, / \, (\lambda^{1/2}\rho_s d^2). \tag{14.9}$$

In these equations, ρ_s is the particle density, d is a characteristic particle diameter, and λ is a linear grain concentration, defined as:

$$\lambda = \frac{1}{\left(C_v^*/C_v\right)^{1/3} - 1},$$ (14.10)

in which C_v^* is the maximum possible static-grain volume concentration. Equation (14.9) shows that dispersive stress related to particle collisions increases as a function of particle concentration, grain size, and shear rate. Although Bagnold's results provide a tantalizing foundation upon which to build a debris-flow theory that incorporates effects of particle interactions, including particle collisions that support grain-size sorting, Takahashi's (1980) development of the model assumes a vertically uniform particle-concentration profile of monodisperse particles. These assumptions cannot explain grain-size sorting, and they place unrealistic restrictions on development of stresses and the velocity profile in a flow, and on the slope angle over which material can flow (Iverson and Denlinger, 1987).

Extensions of the rheometric approach to modeling lahars led to hybrid theories that combined equations in various forms. Chen (1988a,b) developed a hybrid approach, but one which required multiple adjustable coefficients. O'Brien et al. (1993) and O'Brien (2007) summed the viscoplastic and grain-inertial resistance terms in an energy slope form:

$$S_f = S_y + S_v + S_{td},$$ (14.11)

where S_f is the total friction slope, S_y is the yield slope, S_v is the viscous slope, and S_{td} is the turbulent-dispersive slope component. These energy-slope components are written in more explicit dimensionless form as

$$S_f = \frac{\tau_y}{\gamma_m h} + \frac{K \eta u}{8 \gamma_m h^2} + \frac{n^2 u^2}{h^{4/3}},$$ (14.12)

where γ_m is the specific weight of the mixture, h is flow depth, K is an empirical resistance parameter, u is depth-averaged flow velocity, n is Manning's roughness coefficient, τ_y is yield

strength, and η is the viscoplastic viscosity. Of these parameters, h and u are calculated by the model, whereas K, n, τ_y, and η are specified by the user. The specific weight γ_m of a mixture is obtained from its sediment concentration C_v and the densities of the solid and liquid phases. Values of yield strength and viscosity are determined from empirical field and laboratory observations relating C_v to τ_y and η (O'Brien, 2007). Combination of these flow-resistance terms with a two-dimensional hydraulic model (discussed below) enables a program (FLO-2D) that routes a hydrograph across a DEM. Expressions for mass conservation of both sediment and water, and a diffusive expression for momentum conservation (Section 14.5.4) are solved numerically. FLO-2D thus assumes lahars can be described as simple, single-phase rheological materials, is data intensive, and requires detailed fieldwork and laboratory analyses to estimate necessary input parameters. The model is restricted to fixed-bed (non-erodible) topography, and does not simulate rapidly varying flow, flow surges, or other flow discontinuities (such as hydraulic jumps or steep flow fronts), but instead broadly smoothes the flow profile. Therefore, it may not be suitable for high-concentration debris flows.

Although the rheometric approach to characterizing and modeling debris flows (and debris-flow lahars) has been widely adopted, such an approach has significant flaws and limitations (e.g., Iverson and Denlinger, 1987; Major and Iverson, 1999; Hunt et al., 2002; Iverson, 2003). In particular, the approach assumes that debris flows are rheologically simple and characterized by constant-valued strength and viscosity, and that particle interactions are subject to Bagnold's dispersive-stress relations. Field and experimental measurements have shown that properties of debris flows such as sediment concentration, normal stress, shear stress, and pore-fluid pressure, all of which affect flow resistance, are spatially and temporally variable and that they evolve during flow. Hence, debris flows are not rheometrically simple materials, as properties of adjacent regions of a flow can be quite different. Furthermore, laboratory-measured shear resistance owing to intrinsic strength and viscosity of liquefied slurries

$(\tau_y \sim 10\text{–}400$ Pa and $\eta \sim 10\text{–}50$ Pa s) is typically tens to hundreds of times smaller than frictional shear resistance of granular materials relevant to field-scale debris flows (Iverson, 2003). An alternative approach (Section 14.5.5) that describes evolving disparities in rate-independent shear resistance between friction-dominated and liquefied regions of debris flows, rather than rate-dependent shear resistance linked to constant-valued intrinsic rheometric properties, has been proposed for characterizing and modeling debris flows.

14.5.3 Hydrologic flow models

Another approach to modeling lahars is the use of conventional surface-water flood-routing models that treat the flow as a Newtonian fluid. Two main techniques are used: hydrologic and hydraulic routing. Hydrologic (storage) routing is based on mass conservation on a reach-by-reach basis, where the difference between volume inflow I into and outflow O from a reach is a function of storage s within the reach and time t:

$$I - O = \frac{ds}{dt}. \qquad (14.13)$$

One hydrologic routing model, the Muskingum Method, which combines Eq. (14.13) with an assumed relationship between stage (flow depth h) and discharge Q, was used to model hypothetical lahars in several catchments at Mount St. Helens (USA) following its 18 May 1980 eruption (Dunne and Leopold, 1981; Dunne and Fairchild, 1983; Fairchild, 1986). Although this model is potentially useful for roughly approximating the broad-scale evolution of a lahar hydrograph and for estimating reaches of sediment storage, the technique assumes that a lahar can be adequately characterized as a Newtonian fluid, and it is unsuitable for rapidly rising hydrographs, channels subject to backwater effects, or dense flows.

14.5.4 Hydraulic flow models

Hydraulic flow-routing models simultaneously solve equations for continuity (conservation of mass or volume: Eq. (14.14); Fig. 14.12) and momentum (Eq. (14.15)) to calculate a hydrograph

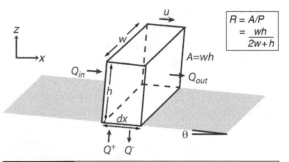

Figure 14.12 Schematic of representative control volume used to establish conservation equations. The time rate of change of mass in the control volume is balanced by the fluxes into Q_{in} and out Q_{out} of its vertical faces ($A = wh$). The hydraulic radius R is defined as the ratio of the flow cross-sectional area, A, to its wetted perimeter, P. In the absence of additional sources or sinks of material, Q^+ and Q^- are zero (as in the right-hand side of Eq. (14.15)), but they take on finite values in the case of incorporation of external sediment/water or deposition of sediment load. After Fagents and Baloga (2006).

that propagates downstream. One-dimensional, depth-averaged (i.e., shallow water), partial differential equations were first derived by Saint-Venant (1871) from the Navier–Stokes equations that describe the motion of an incompressible, linearly viscous fluid of constant viscosity under a pressure gradient (e.g., Middleton and Wilcock, 1994). In one dimension, the Saint-Venant equations can be written as:

$$\frac{\partial A}{\partial t} + \frac{\partial Q}{\partial x} = 0, \qquad (14.14)$$

$$S_f = S_0 - \frac{\partial h}{\partial x} - \frac{u}{g}\frac{\partial u}{\partial x} - \frac{1}{g}\frac{\partial u}{\partial t}, \qquad (14.15)$$

i.e., $S_f = S_0$ – pressure differential – convective acceleration – local acceleration,

where A is flow cross-sectional area, h is depth, u is depth-averaged velocity, Q is discharge (= Au), x is distance downstream, S_f is friction slope, a term taking into account energy dissipation, S_0 is channel bed slope, g is acceleration due to gravity, and t is time. Key assumptions underlying these equations are that vertical velocity gradients are negligible (velocity is depth-averaged), vertical pressure gradients are hydrostatic, and flow depth is small compared to flow length. Hydraulic models that solve the full Saint-Venant equations are referred to as fully dynamic models.

Flow resistance is usually described by a term that encompasses viscous and turbulent dissipation, frictional losses along the channel margins, and changes in channel geometry. The most common resistance terms are the empirical Manning coefficient n and Chézy coefficient C:

$$n = \frac{1}{u} R^{2/3} S_f^{1/2}, \tag{14.16}$$

$$C = \frac{u}{\sqrt{RS_f}}. \tag{14.17}$$

In these terms, the hydraulic radius ($R = A/P$, where P is wetted perimeter) is equivalent to flow depth for wide, shallow channels. When combined with a description of channel geometry, an upstream boundary condition (typically an input hydrograph, $Q(t)$), an energy-loss expression, and a downstream boundary condition such as a rating curve that relates h and Q, the Saint-Venant equations can be solved numerically to calculate the downstream evolution of a flow hydrograph. Outputs typically include stage, discharge, and velocity. This modeling approach assumes that a lahar can be adequately characterized as a viscous Newtonian fluid and that flow mass and density remain constant during flow. This is very unlikely to be the case for a highly concentrated debris-flow lahar, but the approach may provide a reasonable approximation of the behavior of a dilute, hyperconcentrated-flow lahar undergoing little sediment entrainment or deposition.

Chézy- and Manning-type one-dimensional models

Although flow resistance parameters are well characterized empirically for normal Newtonian stream flows across a range of channel morphologies (e.g., Arcement and Schneider, 1989), they are less understood for high-sediment-concentration flows approximated as Newtonian fluids. Nevertheless, lahars have been simulated using one-dimensional, unsteady streamflow models through manipulation of the resistance factor, based on observations of flow behavior and deposit geometry (Swift and Kresh, 1983; Laenen and Hansen, 1988; Costa, 1997; Manville, 2004). Resistance factors can be derived empirically from simple hydraulic equations (e.g., Eq. (14.16)) if local flow velocity or discharge estimates can be obtained and a unique relationship between stage and discharge is assumed. Discharge can be reconstructed from estimates of flow cross-sectional area and velocity. Cross-sectional area can be reconstructed from channel geometry, although possible scour and deposition during flow must be carefully assessed, and mean velocity can be estimated in various ways, including: (1) analysis of super-elevation (tilting) of flow through a channel bend, such that

$$u = \sqrt{\frac{k_e g \cos \theta \Delta e \, R_c}{w}}, \tag{14.18}$$

where θ is the channel gradient, Δe the difference in flow elevation across the channel profile, R_c the radius of curvature of the bend, w the flow width, and k_e is an empirical coefficient (Apmann, 1973); (2) run-up against obstacles; (3) expressions of the form $u = b_1 h^{b_2} S_0^{b_3}$ (Table 14.3), where h is flow depth and S_0 is channel bed slope; (4) from travel-time data; or (5) from independent estimates of local surface velocity from video-frame analysis or tracking of floating markers. Interpretation of super-elevation features can be complicated by the non-Newtonian behavior of sediment-laden flows and assumptions about the profile of the tilted flow surface; therefore, cited values of k_e in Eq. (14.18) vary between 0.1 and 1 (Costa, 1984; Chen, 1987; Bulmer et al., 2002) depending on sediment concentration and channel slope. Furthermore, estimates of mean velocity from run-up against obstacles can be influenced by the momentum of trailing flow. Discharge can also be estimated from stage–discharge rating curves.

A few studies suggest that values of Manning's n used to characterize flow resistance for lahars may vary with channel slope and geometry. Costa (1997) used the National Weather Service

Table 14.3 Coefficient b_1 and exponents b_2 and b_3 in velocity expression for different types of mass flow (from Rickenmann, 1999).

Flow type	b_1	b_2	b_3
Newtonian laminar flow	$\rho g/3\eta$	2	1
Dilatant grain shearing	$2\xi/3$	3/2	1
Newtonian turbulent flow:			
Manning–Strickler equation	$1/n$	2/3	1/2
Chézy equation	C	1/2	1/2
Empirical equation	c_2	0.3	1/2

model DAMBRK (Wetmore and Fread, 1984; Fread, 1996) to model lahars at a number of Cascade Range (USA) volcanoes, and obtained a reasonable fit to field-documented flow stages and inundation areas using a Manning's n of 0.15 (cf. $n \sim 0.025$–0.075 for water flowing in natural stream channels). However, he found that n increased with hydraulic radius. This trend, opposite to that for water floods, highlights the very different energy-dissipation processes operative in lahars. Manville (2004) varied n according to channel gradient to simulate a historical lahar at Mt. Ruapehu. He used $n = 0.15$ for the steepest, proximal reaches of channel and $n = 0.04$ for more distal, lower gradient parts of the flow path in order to obtain the best fit between observed and modeled flow behavior.

Kinematic wave models

If local and convective acceleration and the pressure-differential terms are ignored in Eq. (14.15) so that $S_0 = S_f$, then combination of that expression with the continuity equation (14.14) yields the kinematic-wave approximation of flow, in which gravitational accelerations are balanced by frictional losses. Flow propagation is then described by conservation of mass

$$\frac{\partial h}{\partial t} + \frac{\partial q}{\partial x} = 0, \qquad (14.19)$$

and a flux law, which relates discharge per unit channel width q to depth h, that performs a role similar to that of the Manning or Chézy functions in hydraulic flows:

$$q = \alpha(x)h^k. \qquad (14.20)$$

The coefficient $\alpha(x)$ accounts for the effects of channel slope, changes in channel width, and energy dissipation, and k is a constant. The technique is attractive for the description of flow propagation in channels because of its simplicity, with only one dependent variable appearing in the continuity equation (either stage or discharge). In addition, the initial rising limb and monotonically declining falling limb of the kinematic waveform emulate typical lahar hydrographs. Kinematic-wave theory breaks down at the flow front because the flow depth increases rapidly to a maximum value over a short distance. This problem may be overcome by introducing a kinematic shock (Weir, 1982) to the simulation.

The kinematic-wave method has been used to model lahars at Mount Ruapehu (Weir, 1982; Vignaux and Weir, 1990), Nevado del Ruiz (Pierson et al., 1990), and Mount St. Helens (Arattano and Savage, 1994). Weir (1982) suggested that the exponent k, which relates discharge to flow depth, is a function of sediment concentration and varies between ~ 1.2–1.5 for dilute flows and is ~ 3 for debris flows. Limitations of the model include: (1) an unchanging wave profile as the simulated flow propagates downstream; (2) an inability to model the wave front or surges; (3) assumption of a constant channel slope; and (4) a need for a means of calibrating the time parameter (usually empirically, based on previous flows in the same catchment).

Diffusion models

If only the local and convective acceleration terms are neglected in Eq. (14.15), the momentum equation simplifies to:

$$S_0 - S_f = \frac{dh}{dx}. \qquad (14.21)$$

Combining this form of the momentum equation with the continuity equation (Eq. (14.14)) produces a diffusion (or zero-inertia) equation, which has been used as the basis for at least one debris-flow model (FLO-2D, O'Brien et al., 1993; O'Brien, 2007). Diffusion-wave equations can treat changes in the shape of the waveform with distance (e.g., attenuation), but are deficient for flows having rapidly rising hydrographs or surges. They can adequately characterize slowly varying unsteady Newtonian flow in long reaches of channel, but are poor at modeling flow at channel transitions such as hydraulic jumps.

Fully dynamic flow models

Advances in computer processing power allow fully dynamic (and multi-dimensional) flow-routing models (i.e., models retaining all terms in Eqs. (14.14) and (14.15)) to be applied to lahar problems. Early work (Macedonio and Pareschi, 1992) demonstrated the use of this approach, but also highlighted the importance of channel geometry and topographic variability, and the need to adequately characterize the boundary conditions (e.g., source hydrographs). More recently, Hancox *et al.* (2001) used proprietary (Mike11) software to forecast the maximum discharge, stage, and arrival time at key points along the flow path of the March 2007 lake-breakout lahar from the summit Crater Lake of Mount Ruapehu. Reconstructions of that event were also made using the commercial Delft3D program (Fig. 14.13; Carrivick *et al.*, 2009). That program solves the shallow-water equations for an incompressible Newtonian fluid flowing over fixed channel boundaries using the Boussinesq approximation (an assumption that momentum transfer caused by turbulent eddies can be modeled with a user-specified "eddy viscosity"). Reconstruction of the Ruapehu lahar using this model reasonably matched the peak stage and inundation area of the flow, but poorly matched flow-front velocity and time to peak stage (Manville and Cronin, 2007). This poor replication may be due to various factors, including the inability of the model to treat stage-dependent roughness effects or to accommodate adequately hydraulic jumps produced by significant topographic variability, the assumption of Newtonian flow behavior for what was likely a non-Newtonian flow, neglect of sediment entrainment and transport, and computational difficulties associated with routing a large, unsteady flow over a fine grid (Carrivick *et al.*, 2009).

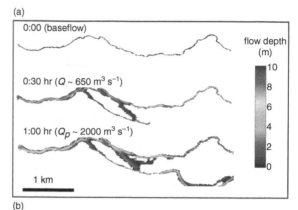

(a)

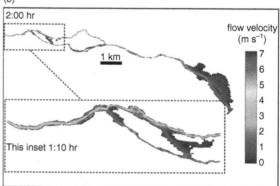

(b)

Figure 14.13 Sample outputs from Delft3D™ simulations of the March 2007 Crater Lake break-out lahar at Mt. Ruapehu, New Zealand. (a) Snapshots of flow depth at different time intervals showing downstream propagation of lahar and flooding of channel bifurcations. (b) Depth-averaged flow velocity illustrating higher flow speeds in deeper parts of the channel. Courtesy of J. Carrivick, unpublished data. See color plates section.

14.5.5 Coulomb mixture theory

Owing to the shortcomings of modeling debris flows using fixed rheological relationships between stress and strain rate, an alternative approach based on mixture theory for granular media has been developed. This approach uses separate equations for the conservation of mass and momentum of the solid and fluid components of a flow of granular material, and coupling terms that link the momentum equations of the separate phases (Hutter *et al.*, 1996; Iverson, 1997a, 2009; Iverson and Denlinger, 2001). This approach builds on descriptions of dry, cohesionless avalanches as depth-averaged (i.e., thin-layer) granular continua governed by

Coulomb-type frictional interactions (Savage, 1984; Savage and Hutter, 1989, 1991). That concept was subsequently generalized to wet debris flows by Iverson (1997a,b) using a simple mixture theory to describe the one-dimensional flow dynamics of a two-phase debris flow. Since its introduction, the Coulomb mixture theory for debris flows has been refined to include Darcy's law, which allows for laminar relative motion between the fluid and solid phases, and to allow coupled evolution of dilatancy, solid and fluid volume fractions, pore-fluid pressure, flow depth, and velocity (Savage and Iverson, 2003; George and Iverson, 2011). Initial versions of the model were expanded from one to three dimensions and solved using refined numerical methods (Iverson and Denlinger, 2001; Denlinger and Iverson, 2001).

Key assumptions of the Coulomb mixture model are: (1) the solids behave as a Coulomb frictional material such that the shear stress driving flow is proportional to effective normal stress ($\sigma' = \sigma - p_f$) and is independent of shear rate; (2) the intergranular fluid (including fines in suspension) behaves as a Newtonian viscous fluid; (3) coupling between solid and fluid components obeys Terzaghi's effective stress principle (Terzaghi, 1943; Lambe and Whitman, 1969; see online supplement) and Darcy's Law for drag owing to laminar relative motion of the solid and fluid phases (Iverson, 2003); (4) pore-fluid pressure p_f, which reduces the intergranular effective stresses, is a "state" variable that evolves through a forced advection–diffusion relation (Savage and Iverson, 2003; George and Iverson, 2011), and the timescales for diffusion of excess pore fluid pressure are long relative to the timescales of flow (Iverson, 1997a; Iverson and Denlinger, 2001; George and Iverson, 2011); (5) frictional intergranular contacts dominate shear resistance in nonliquefied regions of the flow (Iverson, 1997a); and (6) frictional shear resistance is at least an order of magnitude greater than viscous resistance even in fully liquefied flows (Iverson, 1997a).

Under the specified assumptions, the Coulomb mixture model describes: (1) the initial transition of a debris flow from a static to a flowing state; (2) the time- and space-dependent

interactions of the solid and fluid constituents in heterogeneous debris flows having high-friction, coarse-grained snouts and low-friction, liquefied interiors (Iverson, 2003); (3) the evolution of flow depth, velocity, solids volume fraction, dilatancy, and pore-fluid pressure throughout the flow event (George and Iverson, 2011); and (4) post-depositional sediment consolidation and dissipation of non-hydrostatic pore pressure (Major and Iverson, 1999; Savage and Iverson, 2003; George and Iverson, 2011). In the limiting case of dry flow, the mixture reduces to a simple granular Coulomb material, whereas when the flow is saturated, excess pore-fluid pressure can balance the solids load, cause complete liquefaction, and the mixture can behave as a purely viscous fluid.

The initial Coulomb-mixture model for debris flows (Iverson and Denlinger, 2001; Iverson, 2009) consists of depth-averaged, frame-invariant equations that form a system of conservation laws for mass (Eq. (14.22)) and momentum (Eq. (14.23)), here referred to as the debris flow equations (DFE):

$$\frac{\partial h}{\partial t} + \frac{\partial \left(h\bar{u}_x \right)}{\partial x} + \frac{\partial \left(h\bar{u}_y \right)}{\partial y} = 0, \qquad (14.22)$$

$$\rho\left[\frac{\partial \left(h\bar{u}_x \right)}{\partial t} + \frac{\partial \left(h\bar{u}_x^2 \right)}{\partial x} + \frac{\partial \left(h\bar{u}_x\bar{u}_y \right)}{\partial y} \right] =$$

$$-\operatorname{sgn}\left(\bar{u}_x\right)\left(\rho g_z h - p_{bed}\right)\left(1 + \frac{\bar{u}_x^2}{r_x g_z}\right)\tan\varphi_{bed} - 3v_f\eta_f\frac{\bar{u}_x}{h}$$

(basal shear stresses)

$$-hK_{act/pass}\frac{\partial}{\partial x}\left(\rho g_z h - p_{bed}\right) - h\frac{\partial p_{bed}}{\partial x} + v_f\eta_f h\frac{\partial^2\bar{u}_x}{\partial x^2}$$

(longitudinal normal stresses)

$$-\operatorname{sgn}\left(\frac{\partial\bar{u}_x}{\partial y}\right)hK_{act/pass}\frac{\partial}{\partial y}\left(\rho g_z h - p_{bed}\right)\sin\varphi + v_f\eta_f h\frac{\partial^2\bar{u}_x}{\partial y^2}$$

(transverse shear stresses)

$$+\rho g_x h \qquad \text{(gravitational body force stresses)}$$

$$(14.23)$$

Equation (14.23) is the x-direction (downslope) momentum equation; the y-direction (transverse) momentum equation is obtained by interchanging x and y. The direction normal to the flow bed is noted by subscript z. Overbars indicate depth-averaged values. Pore-fluid pressure is assumed to increase linearly with depth to a maximum p_{bed} at the bed, and $K_{act/pass}$ is an earth-pressure coefficient that relates depth-averaged lateral stresses to vertical stress and has a value that varies with extending or compressing flow. Other terms include the fluid volume fraction v_f, the pore-fluid viscosity η_f, local radius of bed curvature r_x, internal φ and basal φ_{bed} friction angles, downslope and normal components of gravitational acceleration g_x, g_z, and the function sgn, which changes the sign of the argument so that shear stresses always oppose shear strain in the x–y plane.

More recent advancements using the mixture-theory model (George and Iverson, 2011) have focused on feedbacks among and evolution of mixture dilatancy, solid and fluid volume fractions, and pore-fluid pressure. Feedbacks among these physical parameters affect whether the initial motion of a mass of granular material from a static state evolves into a rapidly moving debris flow or a slowly creeping landslide, the manner and rate at which the mass moves, and the spatial and temporal variations in flow depth, flow resistance, and propensity for liquefaction.

Numerical solutions of the mixture-theory equations have developed in parallel with theoretical advances. Solutions to early versions of the DFE (Iverson and Denlinger, 2001) were initially formulated using Riemann integration (Denlinger and Iverson, 2001), but were time-intensive. Numerical solution of the most recent versions of the mixture-theory equations using the simulation code DIGCLAW (George and Iverson, 2011) adopts the latest adaptive refinements to solve Riemann problems (e.g., M. Berger et al., 2011). Incorporation of an adaptive grid enables use of high-resolution DEMs and increases the sensitivity of the model at the flow front. Input parameters for the mixture-theory model include initial flow volume, starting location, bed friction angle, initial solids volume fraction and grain density, pore-fluid density and viscosity, mixture hydraulic permeability and compressibility, coefficients that describe mixture dilatancy, and topography in the form of a DEM. Outputs from the model include flow

runout, inundation area, velocity, flow depth, and evolution of the solid volume fraction and pore-fluid pressure. Predicted relaxation of pore-fluid pressure and flow speed in the George and Iverson (2011) model are highly sensitive to the hydraulic permeability of the mixture, a parameter that can have a value far different during flow motion than is measured in static debris.

The Coulomb mixture-theory approach developed by Iverson and Denlinger (2001) was adapted to produce the simulation code TITAN2D (Pitman et al., 2003; Sheridan et al., 2005). For simplicity, however, fluid terms were dropped to develop a single-phase, dry-limit model suitable for rock and dry debris avalanches, and adaptive mesh refinement methods were adopted for computational efficiency. A two-phase, depth-averaged formulation of TITAN2D was developed using engineering concepts of mixture theory (Pitman and Le, 2005). Mass and momentum conservation are calculated for each phase independently using a Coulomb frictional model for the solids and an assumption that the only fluid stress is pressure. If fluid motion relative to the solid particles is not large and if pressure is the only fluid stress, Pitman and Le (2005) assume that fluid accelerations can be ignored and a Darcy-type (fluid drag) approximation can be used as a replacement for the fluid momentum equation. Input parameters for TITAN2D include initial flow volume, internal and basal friction values, the ratio of solids to pore fluid, and topography. Model output includes the evolution of flow depth with time, from which various parameters such as runout distance, velocity, inundation area, and discharge can be derived. Figure 14.14 shows an example of output from TITAN2D.

14.5.6 Models of lahar bulking and debulking

Key features of lahar behavior that most models fail to address are the changes in volume, sediment loading, and downstream evolution that commonly occur as a result of sediment entrainment or deposition during transit. Changes owing to variations in sediment volume affect flow dynamics and potential areas of inundation

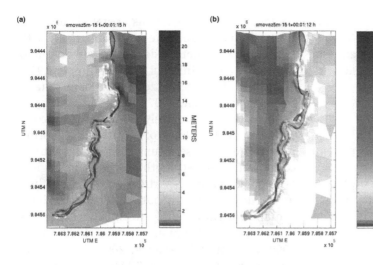

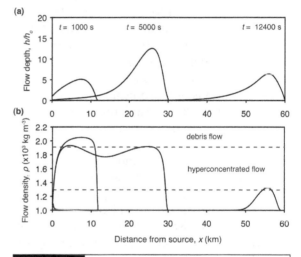

Figure 14.14 Simulations of the 2005 Vazcún Valley lahar, Ecuador, generated using the two-phase version of TITAN2D. Flow depth distributions at final time-steps for simulations of a 70 000 m^3 volume for (a) 40 vol.% solids, and (b) 60 vol.% solids. Reprinted from Williams *et al.* (2008) with permission from Elsevier. See color plates section.

and thus the consequent hazards to downstream communities. Field and experimental observations suggest that erosion and entrainment can be significant (a factor of three or more; Manville, 2004; Scott *et al.*, 2005; C. Berger *et al.*, 2011) and that assimilation of external wet sediment can cause flow momentum, mass and velocity to increase (Iverson *et al.*, 2011).

To model the influence of sediment entrainment and loss on lahar behavior, Fagents and Baloga (2006) developed volume conservation formulations similar to the shallow-water equations (e.g., Eqs. (14.14) and (14.15)), but permitted density to vary. Terms representing the addition and loss of volume (mass) were added to the right-hand side of the conservation equations to represent sediment entrainment and sediment deposition by a flow. The model was formulated for dilute, lake-breakout flows that rapidly entrain sediment on proximal slopes of a volcano, and it generates downstream profiles of flow depth, discharge, and density that evolve due to sediment entrainment and subsequent sediment deposition. Model results were qualitatively consistent with observed lake-breakout events in 1953 and 2007 at Ruapehu volcano, New Zealand (Manville, 2004; Manville and Cronin, 2007). Importantly, the model predicts erosion depths and deposit thicknesses, amounts of sediment entrained, transformations between streamflow, hyperconcentrated flow, and debris flow (Fig. 14.15), and highlights

Figure 14.15 (a) Dimensionless flow depth (normalized to initial flow depth at source, h_0) and (b) flow bulk density (a proxy for solids concentration) at three different times as a function of distance from source, applying the model of Fagents and Baloga (2006) to a lake-breakout lahar. The simulated flow starts out as a water flood but rapidly entrains sufficient sediment to become a lahar. As the flow slows on encountering shallower slopes with distance downstream, deposition of sediment causes the flow to lose sediment and decrease both its depth and bulk density.

the influence of source conditions (i.e., the form of time-dependence of the source hydrograph) and topography on flow evolution.

Disadvantages of the approach used by Fagents and Baloga (2006) include: (1) the lack of a coupled momentum equation (flow velocity is specified as a function of slope); (2) the one-dimensional nature of the governing equations; and (3) inclusion of sediment entrainment and loss parameters that need to be calibrated.

14.6 | Discussion

Models depicting the behavior of lahars span the spectrum from simple, empirical models that predict the average behavior of future lahars on the basis of relationships among characteristics and behavior of past lahars, to sophisticated mathematical and numerical models that describe and simulate fundamental physical interactions among the constituent components of flows. All models, regardless of their complexity, involve compromises between physical fidelity and computational tractability and efficiency. Key questions to consider when selecting a model concern the reasons for using the model, the data available, and the needs and expectations of those using the model output. Emergency planners and response management agencies are likely to have very different needs and model requirements than scientists. During a volcanic crisis, models that emphasize simplicity, ease of use, and practical outputs such as maximum flow depth, inundation area, and downstream arrival times will take preference over physically faithful models that require extended computational timeframes, careful calibration, and an extensive set of initial and boundary conditions. In some cases, improvements in computing power, which can favor use of more complex models, have been negated by increases in model sophistication and data requirements, particularly in the form of high-resolution DEMs.

The simplest models are heuristic, and based on rule-of-thumb or experience-based empirical relationships among various lahar parameters. However, caution should be exercised when using an empirical model that has been calibrated for a limited number of catchments or events as it may not be directly applicable elsewhere.

Rheological models that assume a constant relationship between stress and strain rate in an idealized homogeneous material are relatively simple to calibrate and implement, with key input parameters derived from field and laboratory data (Fink et al., 1981; Whipple, 1997; Coussot et al., 1998; Glaze et al., 2002; Arattano et al., 2006; O'Brien, 2007). However, such models have been extensively criticized for incompatibility with flow observations, ignoring physical reality, use of unrealistic constant viscosity and yield strength values, and neglect of dynamic particle–particle and particle–fluid interactions that contribute to vertical and horizontal segregation and stratification in flows (Iverson, 1997a; Major and Iverson, 1999; Major, 2000; Iverson and Vallance, 2001; Iverson and Denlinger, 2001; Iverson, 2003). Similarly, constitutive rheological models that seek to combine macroviscous and collisional flow regimes (e.g., Chen, 1988a,b; Takahashi, 1991) have also been criticized for imposing unrealistic assumptions and restrictions on flow properties, and for attempting to unify physically incompatible behaviors (Iverson and Denlinger, 1987; Iverson, 2003). Despite those criticisms, models invoking these approaches can replicate some features of liquefied, high-concentration mass flows, such as plug flow, unsteady flow behavior and surging, flow depth and velocity, and deposit characteristics. Consequently, it can be argued that single-phase rheological models might be useful for modeling lahars reaching debris-flow concentrations if they consist of liquefied sediment mixtures having a relatively uniform and fine grain size, particularly if they have a significant clay content; in such cases, viscous flow resistance will dominate frictional and collisional flow resistance (Iverson, 2003). However, such flows are likely to represent only a small fraction of natural lahar events.

Shallow-water flow models draw on hydrologic or hydraulic flow-routing techniques common in open-channel hydraulics, and utilize empirical resistance terms such as Manning's coefficient or Chézy's friction factor. Comparison of predicted behavior with measured behavior of dilute hyperconcentrated-flow lahars suggests that shallow-water models may acceptably simulate such flows (Manville, 2004; Carrivick et al., 2009), although these types of models are highly sensitive to the value of the resistance coefficient selected and to flow-path topography. In contrast to the resistance coefficients typical of normal Newtonian river floods, the Manning's *n* for dilute lahars appears to increase with stage height and sediment concentration; therefore, calibration is an issue.

A more rigorous, physically faithful model based on Coulomb mixture theory (Iverson, 1997a, 2009; Iverson and Denlinger, 2001; George and Iverson, 2011) describes the conservation of mass and momentum for the fluid and solid phases separately, and specifies the interactions and feedbacks among mixture dilatancy, solid volume fraction, and pore-fluid pressure which govern internal flow resistance, momentum transfer, and energy dissipation. The model accounts for much of the observed heterogeneity and behavior of real debris-flow lahars, in which inertial and frictional effects are significant and non-hydrostatic pore-fluid pressure modulates flow behavior. However, it requires a number of detailed input parameters, is computationally intensive, and its latest version (George and Iverson, 2011) is presently restricted to one-dimensional numerical solution. Another numerical model that utilizes the mixture-theory approach, TITAN2D, has been used to hindcast lahar behavior (Sheridan *et al.*, 2005; Williams *et al.*, 2008; Procter *et al.*, 2010). Simulations from that model have highlighted its sensitivity to user-specified input parameters and topographic uncertainties. Furthermore, it requires simplifying assumptions such as neglecting the liquid phase (Pitman *et al.*, 2003; Patra *et al.*, 2005) or imposing a Darcy-type fluid flow that may be inappropriate for dynamic flows or flows having very high permeability (Pitman and Le, 2005).

Regardless of the model used to simulate lahars, output results are sensitive to uncertainties in input parameters, to the quality and resolution of topographic models, and to propagation of uncertainties and errors through a model to the output. For lahars and other geophysical mass flows, uncertainties in initial volume and starting location are of first-order importance. Although initial volumes may be reasonably well constrained for lahars

triggered by lake breakouts, the forms of the source hydrographs may not be. Furthermore, the volumes and source hydrographs of lahars triggered by debris avalanches, rainfall runoff, or snow-and-ice melt are more problematic to predict. Consequently, simulations of lahars for purposes of assessing hazards and producing hazard maps (as opposed to reconstructing events) should consider a range of initial input volumes (or hydrographs) that are guided by analyses of past events. Dalbey *et al.* (2008) discuss methods for sampling uncertain input parameters, propagating those uncertainties through a mass-flow model founded on conservation laws, and quantifying the effects of uncertainty on model output. Those methods allow for construction of a hazard map assessing the spatial probability of some parameter exceeding a threshold value given a range of input parameters. Error in DEMs can also affect model output. Initial efforts at assessing the effects of uncertainties and errors in DEMs on geophysical mass-flow output are discussed by Stefanescu *et al.* (2010).

Depending on the needs of the user, a tiered approach for assessing lahar hazards can be adopted. For example, a rapidly produced one-dimensional model, constrained by adequate channel-geometry data, might be used to assess lahar impacts over a broad area. The output from that model can then be coupled to a two- or three-dimensional model simulating lahar impact on focused sub-reaches of critical interest.

14.7 | Future research directions

Significant progress has been made developing models of lahars. For example, our understanding of the physics of debris flows has advanced considerably, from conceptualizations of flows as single-phase materials having a constant stress–strain-rate relationship to more sophisticated treatments of flows as granular, multiphase materials in which the mass, momentum, and interactions of the constituent phases strongly influence flow dynamics. Recognition that boundary conditions are extremely important for understanding flow behavior and evolution has expanded recent attention from analyses of internal flow dynamics to interactions of flows with their boundaries. In addition to initial efforts at modeling sediment entrainment and loss by lahars (Fagents and Baloga, 2006; Suzuki et al., 2009), recent experiments have examined sediment entrainment and its influence on debris-flow mobility (e.g., Iverson et al., 2011), and real-time field measurements have documented sediment entrainment by debris flows (McCoy et al., 2010; C. Berger et al., 2011). Better understanding of the conditions that lead to sediment entrainment and loss (e.g., the influence of topography, channel geometry, and substrate composition and saturation), and the dynamic feedbacks that lead to particle segregation and levée formation during flow, will improve the physical fidelity of future lahar models. An ability to predict, or at least better constrain, initial lahar volumes, hydrographs, and volumes of entrainment and loss as functions of time, space, and flow behavior will greatly improve implementation of lahar models and lead to more accurate assessments of downstream hazards.

An increase in model sophistication and the use of high-resolution topography have tended to nullify simulation efficiencies gained by improved computational power. Therefore, models adopting improved numerical techniques over high-resolution, three-dimensional terrain are needed. Numerical finite-volume methods that allow the computational grid to dynamically adapt to rapidly changing conditions (e.g., M. Berger et al., 2011; George, 2010; George and Iverson, 2011) are providing steps toward improved numerical efficiency.

As models become more sophisticated at addressing lahar physics, there is an urgent need to obtain well-constrained measurements of parameters relevant for testing those models. Large-scale experiments have helped derive several types of measurements needed for understanding debris-flow physics (e.g., Iverson et al., 2010, 2011), and real-time field measurements of flow properties are providing important data for comparing natural events with experiments and models (e.g., McArdell et al., 2007; Manville

and Cronin, 2007; McCoy et al., 2010; C. Berger et al., 2011). However, significant challenges remain. These include capturing, in real time, the processes of vertical, lateral and longitudinal particle segregation by grain size or density. Furthermore, a broader base of real-time measurements over greater flow-path distances is needed to better resolve dynamical parameters and flow–boundary interactions as lahars evolve during transit and substrate conditions vary. Increased use of airborne and terrestrial laser scanning has the potential to provide finer resolution of lahar–channel interactions, and the roles of sediment entrainment and deposition in mass conservation, than can be obtained from discrete section measurements. Continued innovation in experimental and field techniques will refine the measurements needed to fully understand the interactions among the constituent components of a flow, between a flow, its source characteristics and its boundary, and the mechanisms by which a flow evolves as it moves downstream. Coupling of remote-sensing analyses of geodetic deformation of volcanoes, of volumes of crater lakes and potential release rates, or analyses of soil-thickness distributions and landslide-stability models (such as TRIGRS; Baum et al., 2002), with lahar models may provide one way forward toward predicting initial material volumes in some cases, but predicting initial lahar volumes remains an elusive goal.

14.8 | Summary

- Lahars are complex multiphase phenomena whose behavior varies as a function of sediment concentration, grain-size distribution, and feedbacks among the solid and fluid components of the mixture.
- Low-sediment-concentration lahars may be adequately modeled using standard shallow-flow techniques developed for Newtonian water flows.
- High-sediment-concentration lahars are best simulated using formulations based on mixture theories that account for behaviors of,

and interactions between, the solid and fluid constituents of a flow.

- Ranges of uncertain input parameters, especially initial lahar volume, should be used when assessing downstream hazards. Methods for sampling a range of uncertain input parameters to lahar models and for propagating those uncertainties through a model can be employed to develop probabilistic hazard maps.

14.9 | Notation

a	exponent in regression equation
a_0, a_1, a_2	coefficients in travel-time regression equation
A	cross-sectional area of flow or deposit (m^2)
b_1, b_2, b_3	coefficient and exponents in velocity expressions
B	planimetric area of inundation (m^2)
c_1, c_2	coefficients in regression equations
c	cohesive strength (Pa)
C	Chézy resistance term ($m^{1/2}$ s^{-1})
C_m	sediment mass concentration (%)
C_v	volumetric sediment concentration (% or fraction)
C_v^*	maximum static sediment volume concentration
d	particle diameter (m)
Δe	elevation difference across flow (m)
g	gravitational acceleration (m s^{-2})
g_x, g_z	downslope and normal components of gravitational acceleration (m s^{-2})
h	flow depth (m)
H	elevation difference between flow source and deposit toe (m)
H_e	elevation range on a volcano over which sediment entrainment is assumed in LAHARZ model (m)
I	inflow (m^3 s^{-1})
k	exponent in flux equation
k_e	coefficient in superelevation equation
K	empirical resistance parameter for viscous flow
$K_{act/pass}$	Rankine earth pressure coefficient

L	runout distance (m)
L_e	horizontal distance from volcano summit over which sediment entrainment is assumed in LAHARZ model (m)
m	exponent in Herschel–Bulkley equation
M_z	mean grain diameter (phi scale)
n	Manning's resistance coefficient
O	outflow (m^3 s^{-1})
p_{bed}	maximum pore-fluid pressure at flow base (Pa)
p_f	pore-fluid pressure (Pa)
p_h	hydrostatic fluid pressure (Pa)
p_t	total fluid pressure (Pa)
p^*	excess fluid pressure (Pa)
P	wetted perimeter (m)
q	discharge per unit width (m^2 s^{-1})
Q	discharge (m^3 s^{-1})
Q^+, Q^-	volumetric source and sink terms (m^3 s^{-1})
Q_p	peak discharge (m^3 s^{-1})
r_x	local radius of bed curvature (m)
R	hydraulic radius (m)
R_c	centerline radius of curvature (m)
s	storage (m^3)
S_f	friction slope
S_0	channel slope
S_{td}	turbulent-dispersive slope
S_v	viscous slope
S_y	yield slope
t	time (s)
$T(x)$	travel time to a point x km downstream (hours)
u	mean, depth-averaged flow velocity (m s^{-1})
u_x, u_y	x- and y-components of depth-averaged velocity (m s^{-1})
v_f	fluid volume fraction
V	lahar volume (m^3)
w	channel width (m)
x, y, z	longitudinal, lateral, and vertical spatial coordinates (m)
$\alpha(x)$	coefficient in flux equation
γ_m	specific weight of particle–fluid mixture (N m^{-3})
$\dot{\varepsilon}$	strain rate (s^{-1})
η	dynamic viscosity (Pa s)
η_f	viscosity of fluid phase (Pa s)

θ	channel gradient (°)
λ	linear grain concentration
ξ	lumped resistance coefficient in dilatant grain shearing ($m^{3/2}$ s^{-1})
ρ	flow density (kg m^{-3})
ρ_f	fluid density (kg m^{-3})
ρ_s	sediment/particle density (kg m^{-3})
ρ_w	density of water (kg m^{-3})
σ	normal stress (Pa)
σ'	effective normal stress (Pa)
σ_G	sorting parameter (phi scale)
τ	shear stress (Pa)
τ_d	dispersive stress (Pa)
τ_y	yield strength (Pa)
φ	internal friction angle (°)
φ_{bed}	bed friction angle (°)
ϕ	porosity

References

Ancey, C. (2007). Plasticity and geophysical flows. *Journal of Non-Newtonian Fluid Mechanics*, **142**, 4–35.

Apmann, R. P. (1973). Estimating discharge from superelevation in bends. *Journal of the Hydraulics Division, American Society of Civil Engineers*, **99**(HY1), 65–79.

Arattano, M. and Savage, W. Z. (1994). Modelling debris flows as kinematic waves. *Bulletin of the International Association of Engineering Geologists*, **49**, 3–13.

Arattano, M., Franzi, L. and Marchi, L. (2006). Influence of rheology on debris-flow simulation. *Natural Hazards and Earth System Sciences*, **6**, 519–528.

Arcement, G. J. J. and Schneider, V. R. (1989). *Guide for Selecting Manning's Roughness Coefficients for Natural Channels And Flood Plains*. United States Geological Survey Water-Supply Paper, 2339.

Armanini, A., Capart, H., Fraccarollo, L. and Larcher, M. (2005). Rheological stratification in experimental free-surface flows of granular-fluid mixtures. *Journal of Fluid Mechanics*, **532**, 269–319.

Bagnold, R. A. (1954). Experiments on a gravity-free dispersion of large solid spheres in a Newtonian fluid under shear. *Proceedings of the Royal Society of London A*, **225**, 49–63.

Baum, R. L., Savage, W. Z. and Godt, J. W. (2002). TRIGRS – a Fortran program for transient rainfall infiltration and grid-based regional slope-stability analysis. U.S. Geological Survey Open-file Report 02-0424.

Berger, C., McArdell, B. W. and Schlunegger, F. (2011). Direct measurement of channel erosion by debris flows, Ilgraben, Switzerland. *Journal of Geophysical Research*, **116**, F01002, doi:10.1029/2010JF001722.

Berger, M. J., George, D. L., LeVeque, R. J. and Mandli, K. T. (2011). The GEOCLAW software for depth-averaged flows with adaptive refinement. *Advances in Water Resources*, **34**, 1195–1206, doi:10.1016/j.advwatres.2011.02.016.

Berzi, D., Jenkins, J. T. and Larcher, M. (2010). Debris flows: Recent advances in experiments and modeling. *Advances in Geophysics*, **52**, 103–138.

Beverage, J. P. and Culbertson, J. K. (1964). Hyperconcentrations of suspended sediment. *Journal of the Hydraulics Division, American Society of Civil Engineers*, **90**(HY6), 117–128.

Björnsson, H. (1992). Jökulhlaups in Iceland: prediction, characteristics and simulation. *Annals of Glaciology*, **16**, 95–106.

Bulmer, M. H., Barnouin-Jha, O., Peitersen, M. N. and Bourke, M. (2002). An empirical approach to studying debris flows: Implications for planetary modeling studies. *Journal of Geophysical Research*, **107**(B5), 9.1–9.16.

Canuti, P., Casagli, N., Catani, F. and Falorni, G. (2002). Modeling of the Guagua Pichincha volcano (Ecuador) lahars. *Physics and Chemistry of the Earth*, **27**, 1587–1599.

Capra, L., Macías, J. L., Scott, K. M., Abrams, M. and Garduño-Monroy, V. H. (2002). Debris avalanches and debris flows transformed from collapses in the Trans-Mexican Volcanic Belt, Mexico – behaviour, and implications for hazard assessment. *Journal of Volcanology and Geothermal Research*, **113**, 81–110.

Carrivick, J. L., Manville, V. and Cronin, S. J. (2009). A fluid dynamics approach to modeling the 18th March 2007 lahar at Mt. Ruapehu, New Zealand. *Bulletin of Volcanology*, **71**, 153–169.

Carrivick, J. L., Manville, V., Graettinger, A. H. and Cronin, S. J. (2010). Coupled fluid dynamics-sediment transport modelling of a Crater Lake break-out lahar: Mt. Ruapehu, New Zealand. *Journal of Hydrology*, **388**, 399–413.

Chen, C. L. (1987). Comprehensive review of debris flow modelling concepts in Japan. In *Debris Flows and Avalanches: Process, Recognition, and Mitigation*, ed. J. E. Costa and G. F. Wieczorek. Geological Society of America, pp. 13–29.

Chen, C. L. (1988a). Generalized visco-plastic modelling of debris flow. *Journal of Hydraulic Engineering*, **114**, 237–258.

Chen, C. L. (1988b). General solutions for viscoplastic debris flow. *Journal of Hydraulic Engineering*, **114**, 259–282.

Christen, M., Kowalski, J. and Bartelt, P. (2010). RAMMS: Numerical simulation of dense snow avalanches in three-dimensional terrain. *Cold Regions Science and Technology*, **63**, 1–14, doi: 10.1016/j.coldregions.2010.04.005

Cole, S. E., Cronin, S. J., Sherburn, S. and Manville, V. (2009). Seismic signals of snow-slurry lahars in motion: 25 September 2007, Mt. Ruapehu, New Zealand. *Geophysical Research Letters*, **36**, L09405, doi:10.1029/2009GL038030.

Contreras, S. M. and Davies, T. R. H. (2000). Coarse-grained debris-flows: Hysteresis and time-dependent rheology. *Journal of Hydraulic Engineering*, **126**, 938–941.

Costa, J. E. (1984). Physical geomorphology of debris flows. In *Development and Applications in Geomorphology*, ed. J. E. Costa and P. J. Fleischer. Berlin: Springer, pp. 263–317.

Costa, J. E. (1988). Rheologic, geomorphic, and sedimentologic differentiation of water floods, hyperconcentrated flows and debris flows. In *Flood Geomorphology*, ed. V. R. Baker, R. C. Kochel and P. C. Patton. New York: John Wiley and Sons, pp. 113–122.

Costa, J. E. (1997). Hydraulic modeling for lahar hazards at Cascades volcanoes. *Environmental and Engineering Geoscience*, **3**, 21–30.

Coussot, P. and Meunier, M. (1996). Recognition, classification and mechanical description of debris flows. *Earth Science Reviews*, **40**, 209–227.

Coussot, P., Laigle, D., Arattano, M., Deganutti, A. and Marchi, L. (1998). Direct determination of rheological characteristics of debris flow. *Journal of Hydraulic Engineering*, **124**, 865–868.

Cronin, S. J., Neall, V. E., Lecointre, J. A. and Palmer, A. S. (1997). Changes in Whangaehu River lahar characteristics during the 1995 eruption sequence, Ruapehu volcano, New Zealand. *Journal of Volcanology and Geothermal Research*, **76**, 47–61.

Cronin, S. J., Neall, V. E., Lecointre, J. A. and Palmer, A. S. (1999). Dynamic interactions between lahars and stream flow: A case study from Ruapehu volcano, New Zealand. *Geological Society of America Bulletin*, **111**, 28–38.

Cronin, S. J., Lecointre, J. A., Palmer, A. S. and Neall, V. E. (2000). Transformation, internal stratification,

and depositional processes within a channelised, multi-peaked lahar flow. *New Zealand Journal of Geology and Geophysics*, **43**, 117–128.

Cui, P., Chen, X., Wang, Y., Hu, K. and Li, Y. (2005). Jiangjia Ravine debris flows in southwestern China. In *Debris-Flow Hazards and Related Phenomena*, ed. M. Jakob and O. Hungr. Chichester, UK: Springer-Praxis, pp. 565–594.

Dalbey, K., Patra, A. K., Pitman, E. B., Bursik, M. I. and Sheridan, M. F. (2008). Input uncertainty propagation methods and hazard mapping of geophysical mass flows. *Journal of Geophysical Research*, **113**, B05203, doi:10.1029/2006JB004471.

Denlinger, R. P. and Iverson, R. M. (2001). Flow of variably fluidized granular masses across three-dimensional terrain 2: Numerical predictions and experimental tests. *Journal of Geophysical Research*, **106**, 553–566.

Dunne, T. and Fairchild, L. H. (1983). Estimation of flood and sedimentation hazards around Mt. St. Helens. *Shin Sabo, Journal of the Japan Society of Erosion Control Engineering*, **36**, 12–22.

Dunne, T. and Leopold, L. B. (1981). Flood and sedimentation hazards in the Toutle and Cowlitz River system as a result of the Mt. St. Helens eruption. Federal Emergency Management Agency Report, Region X.

Fagents, S. A. and Baloga, S. M. (2006). Toward a model for the bulking and debulking of lahars. *Journal of Geophysical Research*, **111**, B10201, doi:10.1029/2005JB003986.

Fairchild, L. H. (1986). Quantitative analysis of lahar hazard. In *Mount St. Helens: 5 years later*, ed. S. A. C. Keller. Washington D.C.: Washington University Press, pp. 61–67.

Fink, J. H., Malin, M. C., D'alli, R. E. and Greeley, R. (1981). Rheological properties of mudflows associated with the spring 1980 eruptions of Mount St. Helens volcano, Washington. *Geophysical Research Letters*, **8**, 43–46.

Forterre, Y. and Pouliquen, O. (2008). Flows of dense granular media. *Annual Reviews of Fluid Mechanics*, **40**, 1–24.

Fread, D. L. (1996). Dam-breach floods. In *Hydrology of Disasters*, ed. V. P. Singh. Dordrecht: Kluwer, pp. 85–126.

George, D. L. (2010). Adaptive finite volume methods with well-balanced Riemann solvers for modeling floods in rugged terrain – application to the Malpasset dam-break flood (France, 1959). *International Journal of Numerical Methods in Fluids*, **66**, 1000–1018, doi:10.1020/fld.2298.

George, D. L. and Iverson, R. M. (2011). A two-phase debris-flow model that includes coupled evolution of volume fractions, granular dilatancy, and pore-fluid pressure. In *Proceedings of the 5th International Conference on Debris Flow Hazards Mitigation, 14–17 June, 2011, Padova, Italy,* ed. R. Genevois, D. L. Hamilton and A. Prestininzi, *Italian Journal of Engineering Geology and Environment,* 415–424, doi:10.4908/IJEGE.2011-03.b-047.

Glaze, L. S., Baloga, S. M. and Barnouin-Jha, O. S. (2002). Rheologic inferences from high-water marks of debris flows. *Geophysical Research Letters,* **29**(8), doi: 10.1029/2002GL014757.

Griswold, J. P. and Iverson, R. M. (2008). *Mobility Statistics and Automated Hazard Mapping for Debris Flows and Rock Avalanches.* United States Geological Survey Scientific Investigations Report 2007–5276.

Hampton, M. A. (1975). Competence of fine-grained debris flows. *Journal of Sedimentary Petrology,* **45**, 834–844.

Hancox, G. T., Keys, H. and Webby, M. G. (2001). Assessment and mitigation of dam-break lahar hazards from Mt. Ruapehu Crater Lake following the 1995–96 eruptions. In *Engineering and Development in Hazardous Terrains,* ed. K. J. McManus. Christchurch: New Zealand Institute of Professional Engineers.

Hubbard, B. E., Sheridan, M. F., Carrasco-Núñez, G., Díaz-Castellón, R. and Rodriguez, S. R. (2007). Comparative lahar hazard mapping at Volcan Citlaltépetl, Mexico using SRTM, ASTER and DTED-1 digital topographic data. *Journal of Volcanology and Geothermal Research,* **160**, 99–124.

Huggel, C., Schneider, D., Julio Miranda, P., Delgado, H. and Kääb, A. (2008). Evaluation of ASTER and SRTM DEM for lahar modeling: A case study on lahars from Popocatépetl Volcano, Mexico. *Journal of Volcanology and Geothermal Research,* **170**, 99–110.

Hungr, O. (1990). Mobility of rock avalanches. *Report of the National Research Institute for Earth Science and Disaster Prevention, Japan,* **46**, 11–20.

Hunt, M. L., Zenit, R., Campbell, C. S. and Brennen, C. E. (2002). Revisiting the 1954 suspension experiments of R. A. Bagnold. *Journal of Fluid Mechanics,* **452**, 1–24.

Hürlimann, M., Rickenmann, D., Medina, V. and Bateman, A. (2008). Evaluation of approaches to calculate debris-flow parameters for hazard assessment. *Engineering Geology,* **102**, 152–163.

Hutter, K., Svendsen, B. and Rickenmann, D. (1996). Debris flow modeling: A review. *Continuum Mechanics and Thermodynamics,* **8**, 1–35.

Iverson, R. M. (1997a). The physics of debris flows. *Reviews of Geophysics,* **35**, 245–296.

Iverson, R. M. (1997b). Hydraulic modelling of unsteady debris-flow surges with solid-fluid interactions. In *Debris-Flow Hazards Mitigation: Mechanics, Prediction, and Assessment,* ed. C.-L. Chen. San Francisco: American Society of Civil Engineers, pp. 550–560

Iverson, R. M. (2003). The debris-flow rheology myth. In *Debris-Flow Hazards Mitigation: Mechanics, Prediction, and Assessment,* ed. D. Rickenmann and C.-L. Chen. Rotterdam: Millpress, pp. 303–314.

Iverson, R. M. (2009). Elements of an improved model of debris-flow motion. In *Powders and Grains 2009,* ed. M. Nakagawa and S. Luding. Melville, NY: American Institute of Physics, pp. 9–16.

Iverson, R. M. and Denlinger, R. P. (1987). The physics of debris flows – a conceptual assessment. In *Erosion and Sedimentation in the Pacific Rim,* ed. R. L. Beschta, T. Blinn, G. E. Grant, G. G. Ice, and F. J. Swanson. International Association of Hydrological Sciences Publication 165, pp. 155–165.

Iverson, R. M. and Denlinger, R. P. (2001). Flow of variably fluidized granular masses across three-dimensional terrain 1: Coulomb mixture theory. *Journal of Geophysical Research,* **106**(B1), 537–552.

Iverson, R. M. and Vallance, J. W. (2001). New views of granular mass flows. *Geology,* **29**, 115–118.

Iverson, R. M., Schilling, S. P. and Vallance, J. W. (1998). Objective delineation of lahar-inundation hazard zones. *Geological Society of America Bulletin,* **110**, 972–984.

Iverson, R. M., Logan, M. and Denlinger, R. P. (2004). Granular avalanches across irregular three-dimensional terrain: 2. Experimental tests. *Journal of Geophysical Research,* **109**, F01015, doi:10.1029/2003JF000084.

Iverson, R. M., Logan, M., Lahusen, R. G. and Berti, M. (2010). The perfect debris flow? Aggregated results from 28 large-scale experiments. *Journal of Geophysical Research,* **115**, F03005, doi:10.1029/2009JF001514.

Iverson, R. M., Reid, M. E., Logan, M. *et al.* (2011). Positive feedback and momentum growth during debris-flow entrainment of wet bed sediment. *Nature Geoscience,* **4**, 116–121.

Johnson, A. M. (1970). *Physical Processes in Geology.* San Francisco: Freeman Cooper and Company.

Johnson, A. M. and Rodine, J. R. (1984). Debris flow. In *Slope Instability,* ed. D. Brunsden and D. B. Prior. Wiley, pp. 257–361.

Kang, Z. and Zhang, S. (1980). A preliminary analysis of the characteristics of debris flow. In *Proceedings of the International Symposium of River Sedimentation*. Beijing: Chinese Society for Hydraulic Engineering, pp. 225–226.

Kaitna, R., Rickenmann, D. and Schatzmann, M. (2007). Experimental study on rheologic behaviour of debris flow material. *Acta Geotechnica*, **2**, 71–85.

Kilgour, G., Manville, V., Della Pasqua, F. *et al*. (2010). The 25 September 2007 eruption of Mt. Ruapehu, New Zealand: Directed ballistics, Surtseyan jets, and ice-slurry lahars. *Journal of Volcanology and Geothermal Research*, **191**, 1–14.

Kumagai, H., Palacios, P., Maeda, T., Barba Castillo, D. and Nakano, M. (2009). Seismic tracking of lahars using tremor signals. *Journal of Volcanology and Geothermal Research*, **183**, 112–121.

Laenen, A. and Hansen, R. P. (1988). *Simulation of Three Lahars in the Mount St. Helens Area, Washington, Using a One-Dimensional, Unsteady-State Streamflow Model*. United States Geological Survey Water-Resources Investigation Report, 88-4004.

Lambe, T. W. and Whitman, R. V. (1969). *Soil Mechanics*. New York: Wiley.

Lavigne, F. and Suwa, H. (2004). Contrasts between debris flows, hyperconcentrated flows and stream flows at a channel of Mount Semeru, East Java, Indonesia. *Geomorphology*, **61**, 41–58.

Lavigne, F., Thouret, J.-C., Voight, B., Suwa, H. and Sumaryono, A. (2000a). Lahars at Merapi volcano: an overview. *Journal of Volcanology and Geothermal Research*, **100**, 423–456.

Lavigne, F., Thouret, J.-C., Voight, B. *et al*. (2000b). Instrumental lahar monitoring at Merapi Volcano, Central Java, Indonesia. *Journal of Volcanology and Geothermal Research*, **100**, 457–478.

Macedonio, G. and Pareschi, M. T. (1992). Numerical simulation of some lahars from Mount St. Helens. *Journal of Volcanology and Geothermal Research*, **54**, 65–80.

Major, J. J. (1997). Depositional processes in large-scale debris-flow experiments. *Journal of Geology*, **105**, 345–366.

Major, J. J. (2000). Gravity-driven consolidation of granular slurries – Implications for debris-flow deposition and deposit characteristics. *Journal of Sedimentary Research*, **70**, 64–83.

Major, J. J. and Iverson, R. M. (1999). Debris-flow deposition: effects of pore-fluid pressure and friction concentrated at flow margins. *Geological Society of America Bulletin*, **111**, 1424–1434.

Major, J. J. and Newhall, C. G. (1989). Snow and ice perturbation during historical volcanic eruptions and the formation of lahars and floods – a global review. *Bulletin of Volcanology*, **52**, 1–27.

Major, J. J. and Pierson, T. C. (1992). Debris flow rheology: experimental analysis of fine-grained slurries. *Water Resources Research*, **28**, 841–857.

Major, J. J., Schilling, S. P. and Pullinger, C. R. (2003). Volcanic debris flows in developing countries: The extreme need for public education and awareness of debris-flow hazards. In *Debris-flow Hazards Mitigation: Mechanics, Prediction, and Assessment*, ed. D. Rickenmann and C. -L. Chen. Rotterdam: Millpress, pp. 1185–1196.

Major, J. J., Schilling, S. P., Pullinger, C. R. and Escobar, C. D. (2004). Debris-flow hazards at San Salvador, San Vicente, and San Miguel Volcanoes, El Salvador. In *Natural Hazards in El Salvador*, ed. W. I. Rose, J. J. Bommer, D. L. Lopez, M. J. Carr and J. J. Major. *Geological Society of America Special Paper 375*, pp. 89–108.

Major, J. J., Pierson, T. C. and Scott, K. M. (2005). Debris flows at Mount St. Helens, Washington, USA. In *Debris-flow Hazards and Related Phenomena*, ed. M. Jakob and O. Hungr. Chichester, UK: Springer-Praxis, pp. 685–731.

Manville, V. (2004). Palaeohydraulic analysis of the 1953 Tangiwai lahar: New Zealand's worst volcanic disaster. *Acta Vulcanologica*, **XVI**(1/2), 137–152.

Manville, V. and Cronin, S. J. (2007). Break-out lahar from New Zealand's Crater Lake. *Eos, Transactions of the American Geophysical Union*, **88**, 441–442.

Manville, V., Hodgson, K. A. and White, J. D. L. (1998). Rheological properties of a remobilised-tephra lahar associated with the 1995 eruption of Ruapehu Volcano, New Zealand. *New Zealand Journal of Geology and Geophysics*, **41**, 157–164.

Marchi, L., Arattano, M. and Deganutti, A. M. (2002). Ten years of debris-flow monitoring in the Moscardo Torrent (Italian Alps). *Geomorphology*, **46**, 1–17.

McArdell, B. W., Bartelt, P. and Kowalski, J. (2007). Field observations of basal forces and fluid pore pressure in a debris flow. *Geophysical Research Letters*, **34**, L07406, doi:10.1029/2006GL029183.

McCoy, S. W., Kean, J. W., Coe, J. A. *et al*. (2010). Evolution of a natural debris flow: In situ measurements of flow dynamics, video imagery, and terrestrial laser scanning. *Geology*, **38**, 735–738.

Middleton, G. V. and Wilcock, P. R. (1994). *Mechanics in the Earth and Environmental Sciences*. Cambridge: Cambridge University Press.

Muñoz-Salinas, E., Castillo-Rodriguez, M., Manea, V. C., Manea, M. and Palacios, D. (2009). Lahar flow simulations using LAHARZ program: Application for the Popocatépetl volcano, Mexico. *Journal of Volcanology and Geothermal Research*, **182**, 13–22.

Neall, V. E. (1996). Hydrological disasters associated with volcanoes. In *Hydrology of Disasters*, ed. V. P. Singh. Kluwer, pp. 395–425.

O'Brien, J. S. (2007). FLO-2D User's Manual, Version 2007.06 (http://www.flo-2d.com).

O'Brien, J. S. and Julien, P. Y. (1988). Laboratory analysis of mudflow properties. *Journal of Hydraulic Engineering*, **114**, 877–887.

O'Brien, J. S., Julien, P. Y. and Fullerton, W. T. (1993). Two-dimensional water flood and mudflow simulation. *Journal of Hydraulic Engineering*, **119**, 246–261.

Ohsumi Works Office (1995). *Debris Flow at Sakurajima, Japan*. Kyushu Regional Construction Bureau.

Patra, A. K., Bauer, A. C., Nichita, C. C. *et al.* (2005). Parallel adaptive numerical simulation of dry avalanches over natural terrain. *Journal of Volcanology and Geothermal Research*, **139**, 1–21.

Phillips, C. J. and Davies, T. R. H. (1991). Determining rheological parameters of debris flow material. *Geomorphology*, **4**, 101–110.

Pierson, T. C. (1986). Flow behavior of channelized debris flows, Mount St. Helens, Washington. In *Hillslope Processes*, ed. A. D. Abrahams. Boston: Allen & Unwin, pp. 269–296.

Pierson, T. C. (1995). Flow characteristics of large eruption-triggered debris flows at snow-clad volcanoes: constraints for debris-flow models. *Journal of Volcanology and Geothermal Research*, **66**, 283–294.

Pierson, T. C. (1997). Transformation of water flood to debris flow following the eruption-triggered transient-lake breakout from the crater on 19 March 1982. In *Hydrologic Consequences of Hot-Rock/Snowpack Interactions at Mount St. Helens Volcano, Washington, 1982–84*, ed. T. C. Pierson. United States Geological Survey Professional Paper 1586, pp. 19–36.

Pierson, T. C. (1998). An empirical method for estimating travel times for wet volcanic mass flows. *Bulletin of Volcanology*, **60**, 98–109.

Pierson, T. C. (2005). Hyperconcentrated flow – transitional process between water flow and debris flow. In *Debris-flow Hazards and Related Phenomena*, ed. M. Jakob and O. Hungr. Chichester, UK: Springer-Praxis, pp. 159–202.

Pierson, T. C. and Costa, J. E. (1987). A rheologic classification of subaerial sediment-water flows. In *Debris Flows/Avalanches: Process, Recognition, and Mitigation*, ed. J. E. Costa and G. F. Wieczorek. Geological Society of America, pp. 1–12.

Pierson, T. C. and Scott, K. M. (1985). Downstream dilution of a lahar: transition from debris to hyperconcentrated streamflow. *Water Resources Research*, **21**, 1511–1524.

Pierson, T. C., Janda, R. J., Thouret, J.-C. and Borrero, C. A. (1990). Perturbation and melting of snow and ice by the 13 November 1985 eruption of Nevado del Ruiz, Columbia, and consequent mobilization, flow and deposition of lahars. *Journal of Volcanology and Geothermal Research*, **41**, 17–66.

Pitman, E. B. and Le, L. (2005). A two-fluid model for avalanche and debris flows. *Philosophical Transactions of the Royal Society A*, **363**, 1573–1601.

Pitman, E. B., Nichita, C. C., Patra, A. *et al.* (2003). Computing granular avalanches and landslides. *Physics of Fluids*, **15**, 3638–3646.

Procter, J. N., Cronin, S. J., Fuller, I. C. *et al.* (2010). Lahar hazard assessment using TITAN2D for an alluvial fan with rapidly changing geomorphology: Whangaehu River, Mt. Ruapehu. *Geomorphology*, **116**, 162–174.

Rickenmann, D. (1999). Empirical relationships for debris flows. *Natural Hazards*, **19**, 47–77.

Rodolfo, K. S. and Arguden, A. T. (1991). Rain-lahar generation and sediment-delivery systems at Mayon Volcano, Philippines. In *Sedimentation in Volcanic Settings*, ed. R. V. Fisher and G. A. Smith. Society for Sedimentary Geology Special Publication, **45**, 71–87.

de Saint-Venant, A. J. C. B. (1871). Theory of the nonpermanent movement of waters with the application to the floods of rivers and to the introduction of the tides within their beds. *Comptes Rendus de l'Academie des Sciences*, **73**, 147–154, 237–240.

Savage, S. B. (1984). The mechanics of rapid granular flows. *Advances in Applied Mechanics*, **24**, 289–366.

Savage, S. B. and Hutter, K. (1989). The motion of a finite mass of granular material down a rough incline. *Journal of Fluid Mechanics*, **199**, 177–215.

Savage, S. B. and Hutter, K. (1991). The dynamics of avalanches of granular materials from initiation to runout. Part I: Analysis. *Acta Mechanica*, **86**, 201–223.

Savage, S. B. and Iverson, R. M. (2003). Surge dynamics coupled to pore-pressure evolution in debris flows. In *Debris-flow Hazards Mitigation:*

Mechanics, Prediction, and Assessment, ed. D. Rickenmann and C.-L. Chen. Rotterdam: Millpress, pp. 503–514.

Schilling, S. P. (1998). *LAHARZ: GIS Programs for Automated Mapping of Lahar-Inundation Hazard Zones.* United States Geological Survey Open-File Report 98-638.

Scott, K. M. (1988). Origins, behavior, and sedimentology of lahars and lahar-runout flows in the Toutle-Cowlitz River system. *United States Geological Survey Professional Paper*, **1447-A**, 1–74.

Scott, K. M., Vallance, J. W., Kerle, N. *et al.* (2005). Catastrophic precipitation-triggered lahar at Casita volcano, Nicaragua: occurrence, bulking and transformation. *Earth Surface Processes and Landforms*, **30**, 59–79.

Sharp, R. P. and Nobles, L. H. (1953). Mudflow of 1941 at Wrightwood, southern California. *Geological Society of America Bulletin*, **64**, 547–590.

Sheridan, M. F., Stinton, A. J., Patra, A. *et al.* (2005). Evaluating Titan2D mass-flow model using the 1963 Little Tahoma Peak avalanches, Mount Rainier, Washington. *Journal of Volcanology and Geothermal Research*, **139**, 89–102.

Smith, G. A. (1986). Coarse-grained nonmarine volcaniclastic sediment: Terminology and depositional processes. *Geological Society of America Bulletin*, **97**, 1–10.

Smith, G. A. and Fritz, W. J. (1989). Volcanic influences on terrestrial sedimentation. *Geology*, **17**, 375–376.

Stefanescu, E. R., Bursik, M., Dalbey, K. *et al.* (2010). DEM uncertainty and hazard analysis using a geophysical flow model. In *International Congress on Environmental Modelling and Software, Ottawa, Canada*, ed. D.A. Swayne, W. Yang, A. A. Voinov, A. Rizzoli, T. Filatova, S.03.01, www.iemss.org/iemss2010/index.php?n=Main.Proceedings.

Stevens, N. F., Manville, V. and Heron, D. W. (2002). The sensitivity of a volcanic flow model to digital elevation model accuracy: experiments with digitised map contours and interferometric SAR at Ruapehu and Taranaki volcanoes, New Zealand. *Journal of Volcanology and Geothermal Research*, **119**, 89–105.

Stoopes, G. R. and Sheridan, M. R. (1992). Giant debris avalanches from the Colima Volcanic Complex, Mexico: Implications for long runout landslides (> 100 km) and hazard assessment. *Geology*, **20**, 299–302.

Suryo, I. and Clarke, M. C. G. (1985). The occurrence and mitigation of volcanic hazards in Indonesia as exemplified at the Mount Merapi, Mount Kelut and Mount Galunggung volcanoes. *Quarterly Journal of Engineering Geology*, **18**, 79–98.

Suwa, H. (1989). Field observations of debris flow. *Proceedings of the Japan-China (Taipei) Joint Seminar on Natural Hazard Mitigation*, Kyoto, Japan, 343–352.

Suwa, H. and Okuda, S. (1985). Measurement of debris flows in Japan. *Proceedings of the Fourth International Conference and Field Workshop on Landslides*, Tokyo, Japan, 391–400.

Suzuki, T., Hotta, N. and Miyamoto, K. (2009). Numerical simulation method of debris flow introducing the non-entrainment erosion rate equation at the transition point of the riverbed gradient or the channel width and in the area of Sabo dam. *Shin Sabo, Journal of the Japan Society of Erosion Control Engineering*, **62**,14–22.

Swift, C. H. I. and Kresh, D. L. (1983). *Mudflow Hazards Along the Toutle and Cowlitz Rivers from a Hypothetical Failure of Spirit Lake Blockage.* United States Geological Survey Water-Resources Investigation Report 82-4125.

Takahashi, T. (1980). Debris flow on prismatic open channel. *Journal of the Hydraulics Division, American Society of Civil Engineers*, **106**(HY3), 381–398.

Takahashi, T. (1991). *Debris Flow.* IAHR Monograph Series. Rotterdam: A. A. Balkema.

Takahashi, T. (2001). Mechanics and simulation of snow avalanches, pyroclastic flows and debris flows. In *Particulate Gravity Currents*, ed. W. D. McCaffrey, B. C. Kneller and J. Peakall. Oxford: Blackwell, pp. 11–44.

Takahashi, T., Nakagawa, H., Harada, T. and Yamashiki, Y. (1992). Routing debris flows with particle segregation. *Journal of Hydraulic Engineering*, **118**, 1490–1507.

Terzaghi, K. (1943). *Theoretical Soil Mechanics.* New York: Wiley.

Umbal, J. V. and Rodolfo, K. S. (1996). The 1991 lahars of southwestern Mount Pinatubo and evolution of the lahar-dammed Mapanuepe Lake. In *Fire and Mud: Eruptions and Lahars of Mount Pinatubo, Philippines*, ed. C. G. Newhall and R. S. Punongbayan. Seattle: University of Washington Press, pp. 951–970.

Vallance, J. W. (2000). Lahars. In *Encyclopedia of Volcanoes*, ed. H. Sigurdsson, B. Houghton, S. McNutt, H. Rymer and J. Stix. New York: Academic Press, pp. 601–616.

Vignaux, M. and Weir, G. J. (1990). A general model for Mt. Ruapehu lahars. *Bulletin of Volcanology*, **52**, 381–390.

Weir, G. J. (1982). Kinematic wave theory for Ruapehu lahars. *New Zealand Journal of Science*, **25**, 197–203.

Wetmore, J. N. and Fread, D. L. (1984). *The NWS Simplified Dam Break Flood Forecasting Model for Desk-top and Hand-held Microcomputers*. Federal Emergency Management Agency.

Whipple, K. X. (1997). Open-channel flow of Bingham fluids: Applications in debris-flow research. *Journal of Geology*, **105**, 243–262.

Williams, R., Stinton, A. J. and Sheridan, M. F. (2008). Evaluation of the Titan2D two-phase flow model using an actual event: Case study of the 2005 Vazcún Valley lahar. *Journal of Volcanology and Geothermal Research*, **177**, 760–766.

Witham, C. S. (2005). Volcanic disasters and incidents: A new database. *Journal of Volcanology and Geothermal Research*, **148**, 191–233.

Worni, R., Huggel, C., Stoffel, M. and Pulgarin, B. (2012). Challenges of modeling current very large lahars at Nevado del Huila, Colombia. *Bulletin of Volcanology*, **74**, 309–324.

Yano, K. and Daido, A. (1965). Fundamental study on mud-flow. *Bulletin of the Disaster Prevention Research Institute*, **14**, 69–83.

Zanchetta, G., Sulpizio, R., Pareschi, M. T., Leoni, F. M. and Santacroce, R. (2004). Characteristics of May 5–6, 1998 volcaniclastic debris flows in the Sarno area (Campania, southern Italy): relationships to structural damage and hazard zonation. *Journal of Volcanology and Geothermal Research*, **133**, 377–393.

Zanuttigh, B. and Ghilardi, P. (2010). Segregation process of water-granular mixtures released down a steep chute. *Journal of Hydrology*, **391**, 175–187.

Exercises

14.1 The pressure, or normal stress, acting on any plane in a mass of saturated sediment depends on the fluid that fills the pore space and the solid grains that make up the sediment skeleton. Thus, we must distinguish the fluid pressure (exerted only by the fluid), the skeleton pressure or intergranular stress (exerted only by the solids), and the total stress (which reflects a combination of the fluid and skeleton pressure).

(a) Fluid pressure and total stress within saturated sediment vary with depth. Consider a one-dimensional, water-saturated lahar deposit in which the vertical coordinate direction, y, is defined positive upward. If water statically fills the pore space, derive an expression for the hydrostatic pressure (p_h) of a column of water extending from the deposit surface, h, to a depth $h - y$.

(b) Next derive an expression for the total stress σ_z of a column of water-saturated sediment extending from the deposit surface to the same depth. Follow the sign convention that total stress is negative in compression whereas fluid pressure is defined positive.

(c) Derive an expression for total mass density of the deposit in terms of water density, ρ_w, sediment density, ρ_s, and porosity, ϕ.

(d) Using the expressions derived for total stress and total mass density, show that

$$\sigma_z = -\left[\rho_w + (\rho_s - \rho_w)(1 - \phi)\right]g(h - y)$$

This expression shows that total stress in a column of water-saturated debris depends on the weight of the column of overlying water plus the buoyant weight of the overlying solids.

(e) If the fluid pressure in a lahar exactly balances the weight of the overlying solids, the lahar is liquefied. For that case, show that the total fluid pressure, p_t can be expressed as $p_t = \rho_w g(h - y) + (\rho_s - \rho_w)(1 - \phi)g(h - y)$, or $p_t = p_h + p^*$ where p^* is the nonequilibrium, or excess, fluid pressure.

Online resources available at www.cambridge.org/fagents

- Additional exercises
- Answers to exercises
- Additional reading
- Links to websites of interest
- Video footage of lahar at Semeru volcano, Indonesia
- Illustration of Terzaghi's stress principle

Chapter 15

Introduction to quantitative volcano seismology: fluid-driven sources

Bernard Chouet

Overview

Recent technological developments and increases in the seismological instrumentation of volcanoes now allow the surface effects of subterranean volcanic processes to be imaged in unprecedented detail. The wealth and accuracy of resulting seismic data have allowed the identification of oscillatory behaviors that are intimately related to magma transport dynamics. A critically important area of research in volcano seismology today is aimed at the quantification of the source properties of these oscillatory signals, which typically include Long-Period (LP) events and tremor with periods in the range 0.2–2 s, and Very-Long-Period (VLP) events with periods in the range 2–100 s (see Glossary at end of chapter). The two types of events provide information about the seismic source geometry, fluid properties at the source, and source excitation process, all of which represent critical elements in the assessment of volcanic behavior and associated hazards. To fully exploit the potential of seismic observations, however, we must learn how to translate seismic source mechanisms into quantitative information about fluid dynamics, and we also must determine the underlying physics that governs vesiculation, fragmentation, and the collapse of bubble-rich suspensions to form

separate melt and vapor. Refined understanding of such processes requires multidisciplinary research involving detailed field measurements, coupled with laboratory experiments and numerical modeling.

15.1 | Introduction

Magma transport is fundamentally episodic in character as a result of the inherent instability of magmatic systems at all timescales. This episodicity is reflected in seismic activity, which originates in dynamic interactions between gas, liquid, and solid along geometrically complex magma transport paths. The geometrical complexity plays a central role in controlling flow disturbances and also providing specific sites where pressure and momentum changes in the fluid are effectively coupled to the Earth. In concert with this activity originating in the fluid are processes occurring in the solid rock, which manifest themselves mainly in the form of earthquakes associated with shear failures in the volcanic edifice. Whereas events originating in the fluid represent volumetric modes of deformation involving a localized conduit response to flow processes, the shear failures act as gauges that map stress concentrations

Modeling Volcanic Processes: The Physics and Mathematics of Volcanism, eds. Sarah A. Fagents, Tracy K. P. Gregg, and Rosaly M. C. Lopes. Published by Cambridge University Press. © Cambridge University Press 2013.

distributed over large volumes surrounding magma conduits and reservoirs. These are called Volcano-Tectonic (VT) earthquakes to differentiate them from pure tectonic earthquakes, although they are indistinguishable from the latter in their broadband spectral characteristics and failure mechanisms.

Seismic signals originating from the dynamics of magmatic and hydrothermal fluids typically include Long-Period (LP) events and tremor (Chouet, 1996a). This terminology stems from the appearance of these signals on the short-period seismometers that have traditionally been used in volcano monitoring. LP events resemble small tectonic earthquakes in duration but differ in their characteristic frequency range and harmonic signature (see Section 15.4). Tremor is characterized by a signal of sustained amplitude lasting from minutes to days, and sometimes for months or even longer. In many instances, LP events and tremor are found to have essentially the same temporal and spectral components (Latter, 1979; Fehler, 1983), suggesting that a common source process, differing only in duration, underlies these two types of events. Accordingly, LP events and tremor are often grouped under the common appellation LP seismicity. The periods at which LP seismicity is observed typically range from 0.2 to 2 seconds (Chouet, 1996a), and the characteristic oscillations of LP signals are commonly viewed as a result of acoustic resonance in a fluid-filled cavity or crack (see Chapter 16).

It is fairly straightforward to understand why resonance is such a pervasive phenomenon in volcanoes. The presence of bubbles in magma and hydrothermal fluids lowers the sound speed of these fluids, inducing a sharp contrast in velocity between the fluid and encasing solid, which favors the entrapment of acoustic energy in the fluid volume source region. For short-lived excitation, energy losses due to elastic radiation and dissipation processes at the source are the main factors affecting the duration of resonance, hence longer-duration signals are naturally enhanced in low-viscosity bubbly liquids. Other types of gaseous fluid mixtures may be even more efficient at sustaining source resonance. For example, gases laden

with solid particles, or gases mixed with liquid droplets, may produce velocity contrasts that are similar to, or stronger than, those associated with bubbly liquids. In particular, dusty gases made of micron-sized particles, or misty gases made of micron-sized droplets, can sustain resonance at the source over durations that far exceed those achieved with bubbly fluids (Kumagai and Chouet, 2000).

With the increased use of portable broadband seismometers in the 1990s, slower processes associated with unsteady mass transport began to be observed at many volcanoes. These types of signals, with periods extending over the range from 2 to 100 seconds, fall under the appellation of Very-Long-Period (VLP) seismicity (see Section 15.4). They are typically attributed to fluid–rock interactions, as with LP events, and may involve resonance at much longer periods than commonly observed in LP events (Kumagai, 2006), or may result from longer-term inertial volume changes in fluid-filled conduits.

Forces associated with very large eruptions may also produce signals with periods extending beyond 100 s. For example, mantle Rayleigh waves with periods near 230 s observed during the Mount Pinatubo eruption of 15 June 1991 were linked by Kanamori and Mori (1992) to an oscillatory vertical single force applied at the surface of Mount Pinatubo, which was attributed to the acoustic coupling of atmospheric oscillations induced by the sustained energy flux from the volcano. Kawakatsu et al. (2000) observed signals with periods > 100 s associated with minor phreatic activity at Aso, which they attributed to a slow increase in fluid pressure in a source located 1–1.5 km below the west side of Naka-dake first crater on the central cone of Aso. Signals falling in this category were classified as Ultra-Long-Period (ULP) signals in the terminology of Chouet (1996b). Beyond the ultra-long periods are processes associated with mass transport over timescales of minutes, hours, and days that are more effectively observed with geodetic techniques. An example is the 30-min scale of the injection process recorded by Linde et al. (1993) during the 1991 Hekla eruption.

It is clear that oscillatory processes are ubiquitous to magma flow, and feature a large variety of signals over a wide range of periods, which require very-wide-band measurements for their study. The present chapter offers a brief review of the state of the art in volcano seismology and addresses basic issues in the quantitative interpretation of processes operative in active volcanic systems. Our focus deals specifically with the quantitative analysis of signals originating in the movement of magma and/or hydrothermal fluids. Other areas in the very broad discipline of volcano seismology, such as the tomographic method used in the elaboration of the three-dimensional velocity structures of volcanoes, array-processing methods used in tracking tremor sources, or eruption monitoring involving detailed investigations of seismicity, are not addressed here. Perspectives on these and further aspects of the field may be found in the reviews by Chouet (1996b, 2003), McNutt (2002, 2005), Konstantinou and Schlindwein (2002), and Kawakatsu and Yamamoto (2007).

15.2 | Description of seismic sources in volcanoes

A general kinematic description of seismic sources in volcanoes is commonly based on a moment tensor and single force representation of the source (Aki and Richards, 2002). The seismic moment tensor allows a description of any generally oriented discontinuity in the Earth (such as slip across a fracture plane, or the opening of a crack) in terms of equivalent body forces. As mass-advection processes can also generate forces on the Earth, a complete description of volcanic sources further requires the consideration of single forces in addition to the geometrical source components described by the moment tensor. The standard approach to estimate the source-time histories of the moment and force components at the source is based on the Green's function, which describes the signal propagating through a known velocity structure that would be observed at a receiver if the source-time function were a perfect impulse.

When the wavelengths of observed seismic waves are much longer than the spatial extent of the source, the source may be approximated by a point source and the force system represented by the moment-tensor and single-force components is localized at this point. The displacement field generated by the point source may be written as (e.g., Chouet, 1996b)

$$u_n(\mathbf{x}, t) = M_{pq}(t) * G_{np,q}(\mathbf{x}, \boldsymbol{\xi}, t) + F_p(t) * G_{np}(\mathbf{x}, \boldsymbol{\xi}, t),$$
(15.1)

where $u_n(\mathbf{x}, t)$ is the n-component of seismic displacement observed at a point $\mathbf{x}$ at time t, $M_{pq}(t)$ is the time history of the pq-component of the moment tensor at position $\boldsymbol{\xi}$ of the source, $F_p(t)$ is the time history of the force applied in the p-direction at $\boldsymbol{\xi}$, and $G_{np}(\mathbf{x}, \boldsymbol{\xi}, t)$ is the Green tensor which relates the x_n-component of displacement at $\mathbf{x}$ with the x_p-component of impulsive force applied at $\boldsymbol{\xi}$. The notation $,q$ indicates spatial differentiation with respect to the ξ_q-coordinate at the source and the symbol $*$ denotes convolution. The summation convention is assumed throughout for repeated subscripts. All notation is given in Section 15.7.

In Eq. (15.1), the derivative of G_{np} with respect to the source coordinate ξ_q represents the contribution from a single couple with arm in the ξ_q-direction at $\boldsymbol{\xi}$. The strength of the $p - q$ couple is given by M_{pq}, which has dimensions of moment (N m). The sum over q implied by the repeated indices states that each displacement component at $\mathbf{x}$ is made of a sum of couples located at $\boldsymbol{\xi}$. With three components of force and three possible arm directions, nine force couples can be defined at $\boldsymbol{\xi}$, with each corresponding to one set of opposing forces (dipoles or shear couples) (Fig. 15.1). The combination of the nine couples M_{pq} characterizes all the temporal information about a fault or conduit that can be extracted from the observation of waves whose wavelengths are much larger than the spatial extent of the source. The components of the moment tensor describing a fault or conduit embedded in an isotropic medium are given by the expression (Aki and Richards, 2002)

$$M_{pq} = \iint_{\Sigma} [u_i] v_j [\lambda \delta_{ij} \delta_{pq} + \mu(\delta_{ip} \delta_{jq} + \delta_{iq} \delta_{jp})] d\Sigma. \quad (15.2)$$

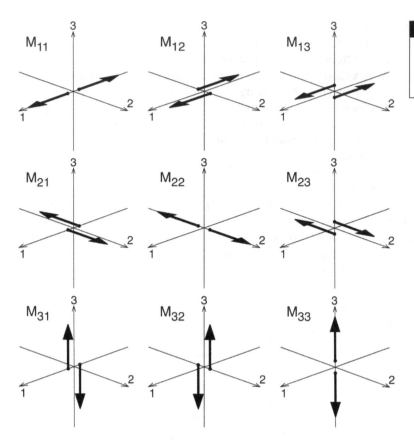

Figure 15.1 The nine possible couples corresponding to the moment-tensor components describing the equivalent force system for a seismic source in the Earth.

In this equation, Σ represents the surface of the buried fault or conduit across which the displacements are discontinuous, v_j is the j-th component of the unit vector normal to the surface element $d\Sigma$, $[u_i]$ denotes the i-th component of the displacement discontinuity between the Σ^+ side of Σ and the Σ^- side of Σ, with square brackets referring to the difference $\mathbf{u}|_{\Sigma^+} - \mathbf{u}|_{\Sigma^-}$, λ and μ are the Lamé parameters (elastic moduli) of the rock matrix, and δ_{ij} is the Kronecker symbol ($\delta_{ij} = 0$ for $i \neq j$, and $\delta_{ij} = 1$ for $i = j$). The seismic moment tensor is a symmetric second-order tensor ($M_{pq} = M_{qp}$) and thus has six independent components. Examples of moment tensors for fault slip and volumetric sources are discussed below.

15.2.1 Seismic moment tensor for representative sources

To illustrate fault slip associated with the shear failure of rock, let us consider a Cartesian coordinate system (ξ_1, ξ_2, ξ_3) at the source and assume a

slip $[u_1]$ parallel to Σ lying in the plane $\xi_3 = 0$ (Fig. 15.2 (a)). In this case, $\mathbf{v}$ has components $(0,0,1)$, $[\mathbf{u}]$ has components $([u_1],0,0)$, and the moment tensor is obtained as

$$\mathbf{M} = M(t) \begin{pmatrix} 0 & 0 & 1 \\ 0 & 0 & 0 \\ 1 & 0 & 0 \end{pmatrix}, \qquad (15.3)$$

where

$$M(t) = \mu \iint_{\Sigma} [u_1(\xi, t)] d\Sigma \qquad (15.4)$$

represents the moment source-time function. Equation (15.3) is the familiar double couple representing a slip source (Fig. 15.2 (b)). The static seismic moment for the effective point source of slip is given by

$$M_0 = \mu \overline{[u_1]} S, \qquad (15.5)$$

where $\overline{[u_1]}$ is final value of slip averaged over the fault surface ($\iint_{\Sigma} [u_1(\xi, t \to \infty)] d\Sigma / S$) and S is the fault area. The product area × slip characterizes

(a)

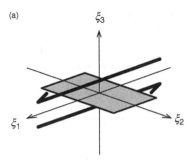

(b)

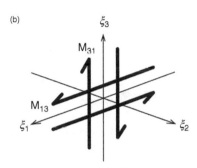

(c)

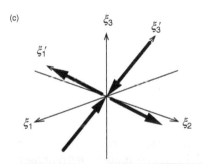

Figure 15.2 Slip on a fault (a) can be described by a superposition of two force couples, in which each force couple is represented by a pair of forces offset in the direction normal to the force (b). Sources involving slip on a fault thus have an equivalent force system in the form of a double couple composed of four forces. (c) A double couple is equivalent to a pair of vector dipoles of equal magnitudes and opposite in sign.

the strength of the source and is called "potency" (Ben-Menahem and Singh, 1981). The potency and scalar seismic moment M_0 ($\mu \times$ *potency*) represent two alternate ways used to quantify earthquake sources. Diagonalization of the matrix in Eq. (15.3) yields the eigenvalues

$$\mathbf{M'} = M(t)\begin{pmatrix} 1 & 0 & 0 \\ 0 & 0 & 0 \\ 0 & 0 & -1 \end{pmatrix}. \tag{15.6}$$

These eigenvalues, and their associated ortho-normal eigenvectors, show that a double couple is equivalent to two vector dipoles equal in magnitude and opposite in sign (Fig. 15.2 (c)). The isotropic component of the diagonalized moment tensor,

$$I = \frac{1}{3}\mathrm{tr}(\mathbf{M'}), \tag{15.7}$$

is zero in a double-couple source. Thus, a pure source of shear has no volumetric component.

Other sources of seismicity relevant to volcanic environments are volumetric sources associated with mass transfer. To illustrate, let us consider a sudden volume increase of ΔV due to an injection of fluid or a thermal expansion in a reservoir. End-member geometries for a reservoir are a crack, pipe, or sphere. For a thin crack opening in the direction $\mathbf{v}$ represented by the angles θ and ϕ (Fig. 15.3 (a)), we have $\mathbf{v} = (v_1, v_2, v_3) = (\sin\theta\cos\phi, \sin\theta\sin\phi, \cos\theta)$, $[\mathbf{u}] = ([u_1], [u_2], [u_3]) = ([u]\sin\theta\cos\phi, [u]\sin\theta\sin\phi, [u]\cos\theta)$, and the corresponding moment tensor is given by (Chouet, 1996b)

$$\mathbf{M} = \Delta V(t)\begin{pmatrix} \lambda + 2\mu\sin^2\theta\cos^2\phi & 2\mu\sin^2\theta\sin\phi\cos\phi & 2\mu\sin\theta\cos\theta\cos\phi \\ 2\mu\sin^2\theta\sin\phi\cos\phi & \lambda + 2\mu\sin^2\theta\sin^2\phi & 2\mu\sin\theta\cos\theta\sin\phi \\ 2\mu\sin\theta\cos\theta\cos\phi & 2\mu\sin\theta\cos\theta\sin\phi & \lambda + 2\mu\cos^2\theta \end{pmatrix}, \tag{15.8}$$

where $\Delta V(t) = \overline{[u]}(t) \cdot S$ represents the volume change of the crack with time, S is the area of the crack, and $\overline{[u]}(t)$ is the opening discontinuity averaged over S. Diagonalization of Eq. (15.8) yields the components of $\mathbf{M}$ referred to the principal axes of the moment tensor as coordinate axes. The eigenvalues of the tensile crack consist of three dipoles with magnitudes $\lambda\Delta V$, $\lambda\Delta V$, and $(\lambda + 2\mu)\Delta V$, where the dominant dipole is oriented normal to the crack plane (Fig. 15.3(b)). If we assume the rock matrix behaves as a Poisson solid ($\lambda = \mu$), the dipole magnitude ratios become [1 : 1 : 3]. A moment tensor of a crack embedded in a rock characterized by $\lambda = 2\mu$, which may be appropriate for volcanic rock at or near liquidus

(a)

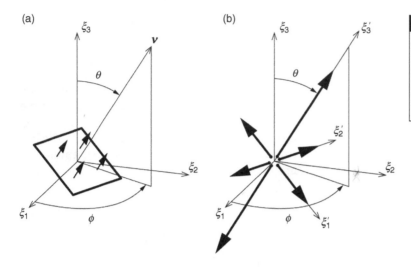

(b)

Figure 15.3 (a) Tensile crack opening in the direction *v* defined by the angles θ and ϕ. (b) Representation of the crack as three vector dipoles, in which each dipole consists of a pair of forces offset in the direction of the force.

(a)

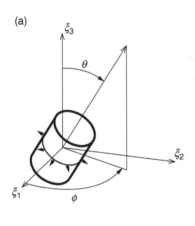

(b)

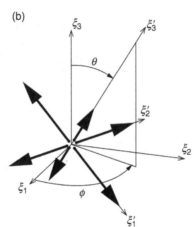

Figure 15.4 (a) Radial expansion of a pipe with axis orientation given by θ and ϕ. (b) Representation of the expanding pipe by three vector dipoles.

temperatures (Chouet *et al.*, 2003, 2005; Waite *et al.*, 2008), would have ratios of [1 : 1 : 2].

The moment tensor corresponding to the radial expansion of a cylinder whose axis orientation is given by θ and ϕ (Fig. 15.4 (a)) is obtained in a similar manner as (Chouet, 1996b)

$$\mathbf{M} = \Delta V(t) \times$$
$$\begin{pmatrix} \lambda + \mu(\cos^2\theta\cos^2\phi + \sin^2\phi) & -\mu\sin^2\theta\sin\phi\cos\phi & -\mu\sin\theta\cos\theta\cos\phi \\ -\mu\sin^2\theta\sin\phi\cos\phi & \lambda + \mu(\cos^2\theta\sin^2\phi + \cos^2\phi) & -\mu\sin\theta\cos\theta\sin\phi \\ -\mu\sin\theta\cos\theta\cos\phi & -\mu\sin\theta\cos\theta\sin\phi & \lambda + \mu\sin^2\theta \end{pmatrix},$$

$$(15.9)$$

where $\Delta V(t) = L\Delta S(t)$, in which L is the pipe length and $\Delta S(t)$ is the increase in the pipe cross-sectional area. Referred to the principal

axes as coordinate axes, the components of $\mathbf{M}$ are represented by three dipoles with magnitudes $\lambda\Delta V$, $(\lambda + \mu)\Delta V$, and $(\lambda + \mu)\Delta V$, where the smallest dipole is oriented along the pipe axis (Fig. 15.4(b)). For $\lambda = \mu$ (Poisson solid), the dipole magnitude ratios become [1 : 2 : 2], while for $\lambda = 2\mu$ (rock near liquidus temperatures) these reduce to [1 : 3/2 : 3/2].

The moment tensor for a spherical source is given by (Aki and Richards, 2002)

$$\mathbf{M} = \Delta V(t) \begin{pmatrix} \lambda + \frac{2}{3}\mu & 0 & 0 \\ 0 & \lambda + \frac{2}{3}\mu & 0 \\ 0 & 0 & \lambda + \frac{2}{3}\mu \end{pmatrix}, \quad (15.10)$$

where $\Delta V(t) = (4/3)\pi R^3 \Delta\Theta(t)$, in which R is the radius and $\Delta\Theta(t)$ is the fractional change in volume of the sphere. The spherical source has an equivalent force system made of three equivalent vector dipoles.

When estimating the magnitude of ΔV, a clear distinction must be made between a source that is unconstrained and free to expand and a source confined by the pressure of the surrounding medium. The volume change ΔV obtained from the moment tensor represents the stress-free volumetric strain introduced by Eshelby (1957) (see Aki and Richards, 2002, pp. 53–54 for a discussion) and is strictly applicable to a source that is free to expand. Under confining pressure from the surrounding medium, the actual volume increase may be smaller than ΔV. That is, rather than expanding by ΔV and being subjected to zero pressure, the source may expand by δV subjected to a pressure increase δp, where the ratio $\delta V/\Delta V$ depends on the geometry of the reservoir. In a spherical reservoir, the actual volumetric strain is related to the stress-free volume increase by (Aki and Richards, 2002)

$$\delta V = \frac{\lambda + \frac{2}{3}\mu}{\lambda + 2\mu} \Delta V , \qquad (15.11)$$

and the corresponding pressure increase is given by (Aki and Richards, 2002)

$$\delta p = \frac{4}{3}\mu \frac{\delta V}{V} . \qquad (15.12)$$

The volume change in this case may range from approximately half the size of the stress-free volume change when $\lambda = \mu$, to two-thirds of the stress-free volume change when $\lambda = 2\mu$. For a cylindrical source, one obtains (Kawakatsu and Yamamoto, 2007)

$$\delta V = \frac{\lambda + \mu}{\lambda + 2\mu} \Delta V , \qquad (15.13)$$

and

$$\delta p = \mu \frac{\delta V}{V} . \qquad (15.14)$$

For this geometry, δV may range from two-thirds of the stress-free volume change when $\lambda = \mu$, to three-quarters of the stress-free volume when $\lambda = 2\mu$.

In the case of a thin crack, the two volume changes are the same ($\delta V = \Delta V$). In a penny-shaped crack with radius R, the pressure and volume changes are related through the expression (Sneddon and Lowengrub, 1969)

$$\delta p = \frac{3}{4} \frac{\mu(\lambda + \mu)}{\lambda + 2\mu} \frac{\delta V}{R^3} . \qquad (15.15)$$

Kawakatsu and Yamamoto (2007) noted that δV represents the volume change used in geodetic analyses (e.g., Mogi, 1958) and proposed to call this volume the "Mogi volume" to distinguish it from the stress-free volume change.

In general, magma movement between adjacent segments of conduit or reservoir can be represented through a combination of volumetric sources such as described above. For example, let us consider the transfer of magma from a spherical chamber into a vertical dike. Assuming a crack opening in the direction $\theta = \pi/2$, $\phi = 0$ (see Fig. 15.3(a)), the moment tensor given by Eq. (15.8) is purely diagonal with components $(\lambda + 2\mu)\Delta V$, $\lambda\Delta V$, and $\lambda\Delta V$. Using Eq. (15.10), with shorthand notation $(1,1,1)$ to represent the diagonal matrix in this equation, and applying mass conservation, we obtain the moment components in the spherical chamber as $-(\lambda + (2/3)\mu)\Delta V(1,1,1)$. If the two volume changes share the same source-time function, the resulting source should be observed as the sum of the two sources with tensor components $(2/3)\mu\Delta V(2,-1,-1)$, which shows expansion in the ξ_1 direction and contraction in the ξ_2 and ξ_3 directions (see source coordinates in Fig. 15.3(a)). The isotropic component of the moment tensor is zero and this source is known as a compensated linear vector dipole (CLVD) (Knopoff and Randall, 1970). Other geometrical configurations of sources may produce a similar CLVD mechanism (Chouet, 1996b). In practice, however, observations are limited to those portions of conduit that provide specific sites where pressure and momentum changes in the fluid are effectively coupled to the Earth, and the isotropic component of the associated diagonalized moment tensor is non-zero.

15.2.2 Single force

The single force in Eq. (15.1) represents the exchange of linear momentum between the

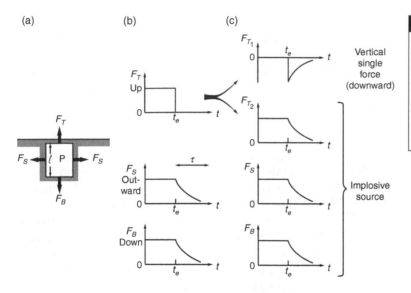

Figure 15.5 Force system equivalent to a volcanic eruption (after Kanamori et al., 1984). (a) Pressurized cavity model for a volcanic eruption. (b) Time histories of forces acting on the top, side, and bottom walls of the cavity. (c) Decomposition of the force system into a downward single force and implosive source.

source and the Earth (Takei and Kumazawa, 1994). Whenever some mass gains momentum, the counter force due to this accelerating mass is felt by the Earth. When the mass eventually decelerates and comes to rest, it induces another force in the Earth in the same direction. Momentum conservation requires that the net change of linear momentum in the overall source system must cancel out over the total duration of an event (Takei and Kumazawa, 1994), so that the two pulse-like single forces associated with the acceleration and deceleration phases must counterbalance each other.

Landslides are examples of external single-force sources (Kanamori and Given, 1982). On a timescale of minutes, a landslide may be viewed as a box sliding down a slope. As the sliding mass gains momentum, a counter force is felt by the Earth in the up-dip direction. When the sliding mass eventually loses momentum, a frictional force is applied to the underlying ground in the down-dip sliding direction. Kawakatsu (1989) developed a centroid single force (CSF) inversion method and applied it to analyses of landslide and slump events recorded by the global seismic network. The CSF inversion method was applied by Ekström et al. (2003) to quantify the stick-slip, downhill sliding of a glacial ice mass.

Another example of external force is the recoil force in the equivalent force system representing a volcanic eruption in the model of Kanamori et al. (1984). In this model, Kanamori et al. (1984) consider a shallow vertically oriented cylindrical cavity initially sealed at the top by a lid (Fig. 15.5(a)). A pressurized inviscid fluid in the cavity exerts an upward vertical force F_T on the lid, a horizontal outward force F_S on the sidewall, and a vertical downward force F_B on the bottom of the cylinder. The eruption is simulated by the sudden removal of the lid at time $t = t_e$, at which point the force F_T vanishes instantaneously and the fluid pressure in the cylinder starts to decrease with a characteristic time constant τ fixed by the mass flux of the eruption, i.e., $\tau \sim \ell/v$, where ℓ is the length of the cylinder and v is the mean fluid velocity inside the cylinder (assuming purely vertical flow and no turbulence). Since the forces F_S and F_B are both proportional to pressure, they decrease with the same time constant τ. The time histories of F_T, F_S, and F_B are sketched in Figure 15.5(b). The force F_T can be decomposed into a vertical downward component F_{T1} and vertical upward component F_{T2} in such a way that F_{T2} has the same time history as F_S and F_B (Fig. 15.5(c)). As a result, the three forces F_{T2}, F_S, and F_B form an implosive source so that the eruption mechanism is represented by the superposition of a downward vertical force (the reaction force of the volcanic jet) with this volumetric implosion. In the model considered here, the implosive source has a moment tensor

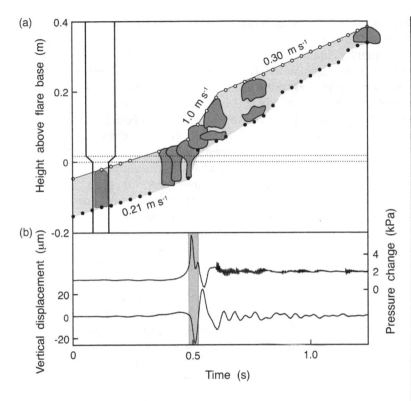

(a)

(b)

Figure 15.6 Slug ascent from small to large diameter tube in the laboratory experiments of James et al. (2006). (a) Sketches of bubble at different positions in the tube shown at the left (diameter drawn to scale, but only short vertical segment of tube illustrated). The bubble is shown by the outline filled with dark gray. Circles and solid dots, respectively, mark the positions of the nose and tail of the bubble obtained from video data. Dotted lines mark the inlet and outlet of the flare. The slug is disrupted by turbulence shortly after entering the larger tube, with resulting daughter bubbles rapidly coalescing. (b) Pressure change (upper trace) measured at the base of the apparatus, and vertical displacement (lower trace) of the apparatus. The gray stripe represents the interval between video frames during which the slug tail passed through the flare. Note the increase of pressure synchronous with the downward displacement (downward acceleration) of the tube. The higher frequency signal starting near 0.6 s in the pressure trace probably results from the turbulence responsible for the disruption and break up of the bubble. The fluid used is a 0.1 Pa s sugar solution. Slug outlines and positions modified from James et al. (2006, Fig. 5b); pressure and displacement data supplied by Mike James (see James et al., 2006, for details).

given by Eq. (15.9), but other geometries are possible, as given by Eqs. (15.8) and (15.10). The above model was used by Uhira and Takeo (1994) to quantify the source parameters of explosions at Sakurajima, and a simplified version of this model, which approximates a volcanic eruption by a single counter force, was used by Nishimura and Hamaguchi (1993) to investigate the scaling law of volcanic explosions. A single-force model was also used by Nishimura (1995) to evaluate the source parameters of explosions at Mount Tokachi, and by Johnson et al. (2008) to estimate the reaction force induced by the rapid uplift of a crater-filling lava dome coincident with explosive degassing bursts at Santaguito.

Internally, a single force may originate in the process of mass advection at the source. An example is the transient movement of liquid accompanying the disruption of a large slug of gas ascending through a region of changing conduit cross section. As demonstrated in analog experiments with a vertical, liquid-filled tube featuring a flare (James et al., 2006), gas slugs undergo an abrupt flow pattern change upon rising through the flare (Fig. 15.6(a)). Rapid

expansion of the slug into the wider cross section of the tube is accompanied by the downward and inward motion of a liquid piston formed by the thickening film of liquid falling past the slug. As it impinges on the narrower inlet to the lower segment of tube the liquid piston rapidly decelerates, generating a pressure pulse in the liquid and a downward force on the apparatus holding the tube (Fig. 15.6(b)).

Internal forces can also result from liquid movements accompanying the rapid near-surface expansion and burst of a gas slug in a conduit (James et al., 2008; see also Chapter 6, Section 6.3). Using laboratory experiments and flow simulations, James et al. (2008) modeled the final ascent of gas slugs in low-viscosity magmas. Their model, featuring a slug ascending a vertical fluid-filled tube sealed at the base, shows that as the slug ascends and expands, viscous shear along the tube wall provides support for an increasing mass of liquid. Consequently, the static pressure below the slug decreases, reducing the downward force exerted by pressure on the upward-facing base of the tube and effectively imparting an upward force on the tube (see Fig. 6.6(b)). During most of the ascent, this increasing upward force on the tube due to decreasing basal pressure is compensated by the increasing downward shear force exerted on the tube wall by the descending film of liquid surrounding the slug body. Downward shear gradually increases as the slug expands and lengthens, until rapid expansion of the slug just prior to reaching the surface induces a component of upward-directed shear in the region around and above the slug nose (see Fig. 6.6(a)), which opposes most of the downward shear due to the descending liquid film around the slug body. Upon slug burst, expansion no longer drives upward shear. Upward shear is then provided solely by liquid inertia and declines rapidly, leaving only the downward shear force, which subsequently decays slowly back to zero as the draining film thins out. The slumping of the liquid film surrounding the slug back to the top of the liquid column after the slug has burst induces a simultaneous rise in static pressure and increase in basal pressure. The slug expansion and burst thus result in a net upward force transient with both shear and pressure components being exerted on the tube. This model was used by Chouet et al. (2010) to interpret the seismic source mechanism of degassing bursts at Kilauea.

15.3 | Waveform inversion

With knowledge of the Earth's structure and the associated Green's functions, the source-time histories of the time-dependent moment tensor and force components can be retrieved through least squares inversion of observed waveforms. The frequency-domain version of Eq. (15.1) is

$$u_n(\mathbf{x}, \omega) = M_{pq}(\omega) \cdot G_{np,q}(\mathbf{x}, \xi, \omega) + F_p(\omega) \cdot G_{np}(\mathbf{x}, \xi, \omega), \tag{15.16}$$

which can be recast in matrix form as

$$\mathbf{U}(\omega) = \mathbf{G}(\omega)\mathbf{S}(\omega), \tag{15.17}$$

where $\mathbf{U}$ is a $N_t \times 1$ vector of Fourier-transformed ground displacement components, $\mathbf{G}$ is the $N_t \times 9$ matrix of Fourier transforms of the Green's functions, $\mathbf{S}$ is the 9×1 vector of Fourier-transformed force and moment-tensor components (three force components and six independent moment components), and N_t is the number of observed seismic traces. Inverting for each frequency component separately by minimizing the least squares residuals between data and synthetics, one obtains estimates for the source:

$$\mathbf{S}^{est}(\omega) = [\mathbf{G}^H(\omega)\,\mathbf{G}(\omega)]^{-1}\,\mathbf{G}^H(\omega)\,\mathbf{U}(\omega), \tag{15.18}$$

where the symbol H indicates the conjugate transpose (Hermitian) and $\mathbf{S}^{est}$ is the vector of calculated Fourier-transformed source components. After solving for all frequencies of interest, one obtains the time domain estimates of the force and moment-tensor components of the source with the inverse Fourier transforms of the relevant components of the vectors $\mathbf{S}^{est}$. Synthetic seismograms for all the traces are computed in the frequency domain using Eq. (15.16) followed by application of the inverse Fourier transform.

Working in the frequency domain reduces the computational load by allowing the inversion of many small matrices, which is more efficient than inverting a single very large matrix as required in the time-domain approach (Auger et al., 2006). It also reduces the number of samples required to model the signal. For example, a trace of 200 s sampled at 50 samples per second yields 10 000 time samples. In the frequency domain, the same trace represents 200 samples per Hz. For a VLP data inversion, in which we may typically consider a limited frequency band from 0.01 to 0.5 Hz, the frequency-domain inversion requires just 100 spectral components.

For an accurate determination of the source location and associated source mechanism, one needs to compare observed seismic data to synthetic data calculated for a realistic model of the volcanic edifice. A standard approach is to use a discretized representation of the edifice based on a digital elevation model. Using this discretized model, synthetics may then be calculated by the three-dimensional finite-difference method (Ohminato and Chouet, 1997). The procedure involves calculation of Green's functions for individual moment and single force components applied at a preset position representing the anticipated location of the point source in the edifice. As the actual position of the source is unknown a priori, the calculations are repeated for different source positions and the best-fitting point source is determined by minimizing the residual error between calculated and observed seismograms. Calculations are usually carried out for point sources distributed over a uniform mesh encompassing the anticipated source region and the number of point sources considered may be quite large. To reduce the number of calculations required to derive the Green's functions, use is made of the reciprocal relation between source and receiver (Aki and Richards, 2002),

$$G_{mn}(\mathbf{x}_1, \mathbf{x}_2) = G_{nm}(\mathbf{x}_2, \mathbf{x}_1).$$ (15.19)

This relation states that the m-component of displacement at $\mathbf{x}_1$ due to a unit impulse applied in the n-direction at $\mathbf{x}_2$, shown as $G_{mn}(\mathbf{x}_1,\mathbf{x}_2)$ in Eq. (15.19), is the same as the n-component of displacement at $\mathbf{x}_2$ due to a unit impulse applied in the m-direction at $\mathbf{x}_1$ ($G_{nm}(\mathbf{x}_2,\mathbf{x}_1)$ in Eq. (15.19)). Using reciprocity, one calculates the three components of displacement at each source node generated by impulsive forces applied in the x, y, and z directions at each receiver location in the network. The Green's functions of the moment components are then derived by spatial differentiation of the results obtained for the forces. For a network of N three-component receivers, only $3N$ finite-difference runs are required to generate Green's functions for all the point sources, resulting in a dramatic reduction in computational load.

The selection of an optimum solution is based on the residual error, the relevance of the free parameters used in the model, and the physical significance of the resulting source mechanism. The following measure of squared error is commonly used to evaluate the waveform fits in the time domain (Ohminato et al., 1998):

$$E = \frac{1}{N_r} \sum_{n=1}^{N_r} \left[\frac{\sum_{1}^{3}\sum_{p=1}^{N_s}(u_n^0(p\Delta t) - u_n^s(p\Delta t))^2}{\sum_{1}^{3}\sum_{p=1}^{N_s}(u_n^0(p\Delta t))^2} \right],$$ (15.20)

where $u_n^0(p\Delta t)$ is the p-th sample of the n-th data trace, $u_n^s(p\Delta t)$ is the p-th sample of the n-th synthetic trace, N_s is the number of samples in each trace, and N_r is the number of three-component receivers. In this expression, the squared error is normalized receiver by receiver, so that stations with lower-amplitude signals contribute equally to the squared error as stations with large-amplitude signals. This definition of error accounts for stations from all distances equally.

As the actual number of parameters in the source mechanism is unknown, a typical investigation generally consists of waveform inversions carried out for: (1) a point source including six moment and three force components to fully account for the possible expansion or contraction of a conduit/reservoir and single force due to mass advection, as well as potential shearing sources; (2) a point source including six moment tensor components but no single force components; and (3) a point source consisting of three single force components. Multiple point sources may be considered as well if warranted by adequate data.

The significance of the number of free parameters is evaluated by calculating Akaike's Information Criterion (AIC) (Akaike, 1974), which is defined as

$$\text{AIC} = N_{obs}\ln E + 2N_{par},$$ (15.21)

where $N_{obs} = N_t N_s$ is the number of independent observations, E is the squared error defined by Eq. (15.20), and N_{par} is the number of free parameters used to fit the model (N_{par} is the number of source mechanisms considered times the number of spectral components used in the frequency-domain inversion). Additional free parameters in the source mechanism are considered to be physically relevant when both the residual error and AIC are minimized.

A final consideration is the physical relevance of the solution. As formulated, the source inversion (Eq. (15.18)) does not use the constraint of a shared time history for the moment tensor. Rather, the time histories of individual moment-tensor components are obtained independently of each other (e.g., Figs. 15.8(a) and 15.8(b)). For a realistic interpretation of the source mechanism, consistent waveform shapes among individual moment tensor components are required. Differences among the time histories of individual moment tensor components may arise due to the presence of noise in the data, inadequate receiver coverage, or an inadequate starting assumption concerning the source mechanism. In the latter case, the introduction of single force components may help minimize distortion of the source-time functions of moment components (e.g., Chouet et al., 2003). For a robust estimation of the source mechanism, appropriate network coverage of the entire volcanic edifice is required. Ideally, ten or more three-component receivers ringing the edifice within a range of 5 km from the source should be considered. At the very least, five three-component receivers surrounding the source at close range are required to gain a rough idea of the source mechanism, provided noise in the data is not an issue (Dawson et al., 2011). The moment tensor estimated from waveform inversion may be interpreted through a comparison with the theoretical moment tensors given by Eqs. (15.8), (15.9), and (15.10), or geometries made of composites of such sources (see example in Section 15.3.1 below).

Inversions of VLP waveforms have imaged crack geometries at the source in the form of dikes or sills (Ohminato and Chouet, 1997; Chouet et al., 2003; Kumagai et al., 2003), as well as more complicated geometrical configurations involving a composite of a dike intersecting a sill (Chouet et al., 2005), composites of intersecting dikes (Chouet et al., 2008, 2010), or two chambers linked to each other by a narrow channel (Nishimura et al., 2000). Contrasting with these findings are results obtained by Ohminato et al. (2006) from waveform inversions of VLP signals produced by vulcanian explosions at Asama. At Asama, the contribution from a vertical single force with magnitude $10^{10} - 10^{11}$ N was found to dominate the observed waveforms, and no obvious volumetric component was identified at the source. The depth of the source is ~200 m below the summit crater, and the source-time history features two downward force components separated by an upward force component. The initial downward force component was attributed by Ohminato et al. (2006) to the sudden removal of the lid capping the pressurized conduit (this force is analogous to the force F_{T1} in Fig. 15.5(c)), and the subsequent upward force was interpreted by these authors as a drag force induced by viscous magma moving up the conduit. Ohminato et al. (2006) attributed the final downward force component to an explosive fragmentation of the magma, whose effect is to effectively cancel viscous drag so that the downward force due to jet recoil again dominates the signal. An application of the waveform inversion method to VLP data recorded at Stromboli Volcano, Italy, is presented below.

15.3.1 Shallow conduit dynamics at Stromboli imaged from VLP data

Detailed broadband measurements were carried out at Stromboli in September 1997 by Chouet et al. (2003) using a network of 21 three-component broadband (0.02–60 s) seismometers (Fig. 15.7). Eruptive activity at that time was limited to two vents located at the northern and southern perimeters of the crater, and two characteristic types of waveforms representative of eruptions from these vents were observed. The signals associated with eruptions from the northern vent were subsequently named Type-1 events, and those related to eruptions from the southern vent were named Type-2 events. Using these data, Chouet et al. (2003) carried out systematic inversions of eruption signals band-pass filtered in the 2–20 s band for a Type-1 event, or 2–30 s band for a Type-2 event. Two best-fit models, based on residual error E (Eq. (15.20)) and using six moment-tensor components plus three single-force components, were imaged for two point sources located at elevations of 520 m (Type-1) and 480 m (Type-2) above sea level, approximately 160 m northwest of the vents (Chouet et al., 2003).

Figures 15.8(a) and 15.8(b) show the source-time functions of moment and force components obtained by Chouet et al. (2003) for Type-1 and Type-2 events, respectively. The consistency

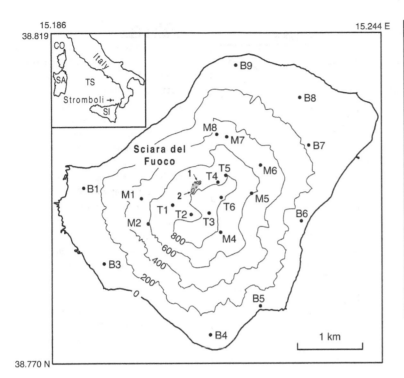

Figure 15.7 Map of Stromboli Volcano showing locations of three-component broadband stations deployed during the seismic experiment in 1997 (solid dots). Stations prefixed by "T" denote those of the "T" ring of sensors, by "M" those of the "M" ring, and by "B" those of the "B" ring, located at the crater level, mid-level, and base of the volcano, respectively. The crater is marked by the shaded area, which encompasses distinct vents. The arrows point to two eruptive vents that were active at the time. Contour lines represent 200-m contour intervals. The inset shows the location of Stromboli in the Tyrrhenian Sea (TS) in relation to Italy, Sicily (SI), Sardegna (SA), and Corsica (CO).

of waveform shapes seen among individual moment tensor components points to a robust underlying mechanism. The principal axes of the moment tensor obtained by eigenvalue decomposition of the source-time functions of the moment components are illustrated in Figures 15.8(c) and 15.8(d). The three eigenvectors identified by the bold black arrows in these figures are obtained from measurements of the maximum peak-to-trough amplitudes in the individual tensor components.

The force system in Figures 15.8(c) and 15.8(d) consists of three dipoles with amplitude ratios [1 : 0.8 : 2] and [1.1 : 1 : 2] in the Type-1 and Type-2 events, respectively. These ratios closely match the amplitude ratios [1 : 1 : $(\lambda + 2\mu)/\lambda$] for a crack, in which $\lambda = 2\mu$ is assumed – a value appropriate for volcanic rock at or near liquidus temperatures (Murase and McBirney, 1973). A simple crack model, illustrated by the gray-shaded planes in Figures 15.8(c) and 15.8(d), therefore constitutes an adequate first-order representation of the source mechanism producing the moment components shown in Figures 15.8(a) and 15.8(b).

Accompanying the volumetric source components in Figures 15.8(a) and 15.8(b) is a dominantly vertical single force. The force is initially down, then up in both event types. In the Type-1 event, the downward force is synchronous with the initial inflation of the source volume, whereas the following upward force coincides with a deflation of the source volume. Although less clear, a similar synchronicity is manifest in the Type-2 event. Taken together, this force and volume change may be viewed as the result of a piston-like action of the liquid in response to the disruption of a gas slug transiting through this particular location in the conduit (James et al., 2006).

As noted above, the amplitude ratios of the principal axes of the moment tensor are not exactly [1 : 1 : 2], implying additional complexity beyond the simple first-order mechanism illustrated in Figures 15.8(a) and 15.8(b). To investigate this complexity, Chouet et al. (2008) considered a point source composed of two intersecting cracks and performed a search for the best-fitting model by systematically varying the azimuth ϕ and polar angle θ defining

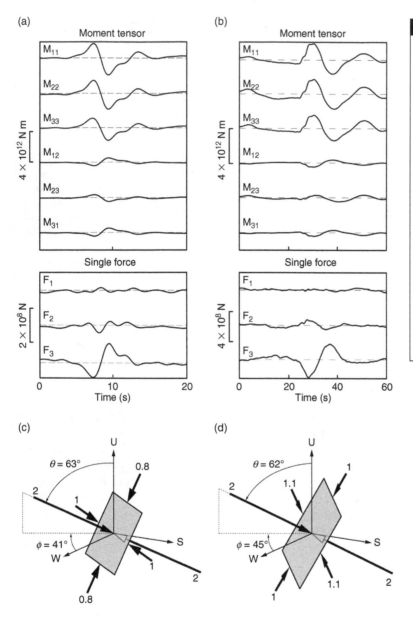

Figure 15.8 Source mechanisms of the two types of explosions occurring at Stromboli in September 1997. (a) Source-time functions obtained for a Type-1 event, and (b) source-time functions derived for a Type-2 event, in which six moment-tensor components and three single-force components are assumed for the source mechanism. (c) Source mechanism imaged for the Type-1 event, and (d) source mechanism imaged for the Type-2 event. The reference coordinates for the eigenvectors are W (west), S (south), and U (up). The eigenvectors, marked by bold arrows, have been normalized to a maximum length of 2 and represent the main deflation phase of the source seen during the interval 7–10 s in the Type-1 event in (a), or interval 30–40 s in the Type-2 event in (b).

the orientation of the normal vector to each crack plane (see Eq. (15.8); note that this source inversion solves for a single moment tensor with shared time history in each crack, plus three single forces with variable time histories). Figure 15.9 shows the result of their reconstruction of the source mechanism for the Type-1 event. This solution points to a dominant crack whose orientation is within a few degrees of the crack imaged in Figure 15.8(a), and subdominant crack with similar dip but distinct azimuth.

The moment tensor and single force components for the two-crack mechanism very closely match the moment and single force components illustrated in Figure 15.8(a), and the waveform fits obtained by these two approaches are virtually indistinguishable (Chouet *et al.*, 2008). A similar dual crack mechanism was derived for the Type-2 event.

Although the point sources imaged in the above studies provide a very good match for the VLP waveforms recorded on receivers located in

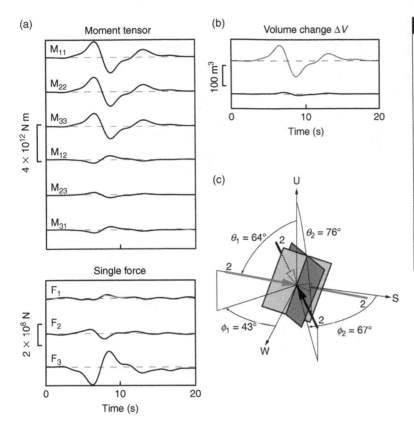

(a) Moment tensor

M_{11}
M_{22}
M_{33}
M_{12}
M_{23}
M_{31}

4×10^{12} N m

Single force

F_1
F_2
F_3

2×10^8 N

0   10   20
Time (s)

(b) Volume change ΔV

100 m³

0   10   20
Time (s)

(c)   U

$\theta_1 = 64°$ 2 $\theta_2 = 76°$

2

$\phi_1 = 43°$ 2 $\phi_2 = 67°$

W   S

Figure 15.9 Reconstruction of the source mechanism of Type-1 event based on the assumption that the source is a composite of two cracks. The moment tensor in (a) represents the sum of the moment tensors representing each crack, where each moment tensor shares a single time history according to Eq. (15.8) (i.e., each crack features its own distinct time history). The black and gray curves representing volume changes in the two cracks, shown in (b), correspond to the black and gray arrows in (c), which also shows the dominant crack (largest volume change) in dark gray.

the upper part of the edifice, noticeable misfits were observed to remain on receivers surrounding the edifice near sea level. These misfits were interpreted by Chouet et al. (2008) as evidence of contributions from deeper source components, and to quantify these components they conducted a search for another composite mechanism at a second point source. In this search, the positions of the two cracks in the original source remained fixed, while the second point source was again assumed to be a composite of two intersecting cracks combined with three single force components. Once the position and mechanism of the second point source were identified, a fine adjustment of the coupled mechanisms of the two sources was carried out to find the absolute minimum of residual error between fitted synthetics and data.

Figure 15.10 shows the two conduit structures compatible with the two point sources (marked by white circles) reconstructed from inversion of waveform data for Type-1 and Type-2 events according to this procedure. The two conduit structures are obtained simply by extending the two crack planes imaged at the upper point source up to the surface, extending the plane of the dominant crack imaged at the lower point source up to the upper point source, and viewing the subsidiary crack at the lower point source as the downward extension of the conduit. The main branch of conduit activated during eruptions at the northern vent is composed of a dike dipping steeply to the northwest and extending essentially straight from 80 m below sea level to the crater floor, 760 m above sea level (Fig. 15.10(a)). At a depth of 80 m below sea level the conduit features a sharp corner leading into a dike segment dipping to the southeast. The main dike segment above, and deep segment below the abrupt corner, both strike northeast-southwest along a direction parallel to the elongation of the volcanic edifice. A subsidiary dike segment branches off the main crack-like conduit at elevations near 440 m (dark

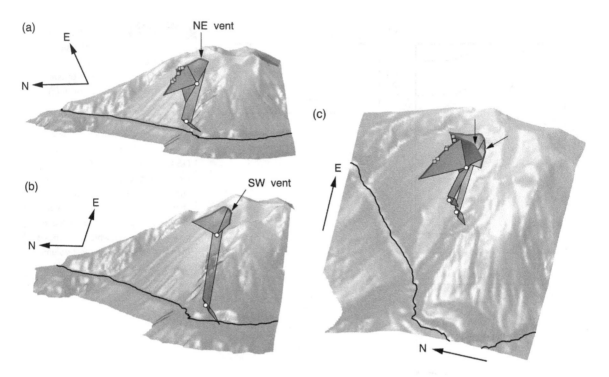

Figure 15.10 (a) Geometry of the upper 1 km of conduit underlying the northern vent area of Stromboli. A semi-transparent perspective cut-away view of the northwest quadrant of the edifice including the sector graben of the Sciara Del Fuoco, provides the reference for the location and geometry of the conduit, which is derived from the seismic source mechanisms obtained from inversions of VLP signals associated with explosions. Surface illumination from an external light source provides color contrasts, emphasizing topographical features. A black line indicates sea level. The summit of the volcano is 924 m above sea level (no vertical exaggeration). The two flow disruption sites that are sources of VLP elastic radiation are indicated by small white circles. The irregular red and blue lines, respectively, mark the surface traces of the dominant and subdominant dike segments constituting the shallowest portions of the conduit system. The eruptive vent is marked by a vertical arrow, and temporary vents active during the flank eruption in 2002–2003 are indicated by green squares. The lateral extents of individual dike segments are unknown and are shown for illustrative purpose only. (b) Same as (a) for conduit underlying the southern vent area. A slanted arrow points to the southern vent. (c) Plunging view of the Sciara Del Fuoco showing the two dike systems underlying the summit crater. The conduit structure underlying the northern vent (marked by a vertical arrow) is shaded green and that underlying the southern vent (marked by a slanted arrow) is colored blue (see (a) for details). See color plates section.

green segment in Fig. 15.10(a); see color plates section). The latter segment strikes approximately northwest and dips to the west. The surface trace of the main dike segment trends through the northern vent area (red trace in Fig. 15.10(a)), while the surface trace of the subsidiary segment extends northwest-southeast in rough alignment with several vents active in the northwest quadrant of Stromboli in 2002–2003 (blue trace in Fig. 15.10(a)). Below the southern vent, the uppermost conduit geometry features a northeast-striking, northwest-dipping dike (red trace in Fig. 15.10(b)), intersecting a west-striking dike dipping to the north (blue trace in Fig. 15.10(b)). At 520 m elevation the two dikes merge into a sub-vertical dike striking northeast, and at a depth of 280 m below sea level the conduit features a second, more abrupt corner leading into a dike dipping to the south (see Chouet *et al.* (2008) for a discussion of uniqueness and angular resolution of these solutions).

As noted above, the temporal features of the upper point sources imaged by Chouet *et al.* (2008) reflect a sequence of pressurization, depressurization, and repressurization, which may be viewed as the result of a funnel-like flow disturbance induced by the transit of a gas slug through the shallow bifurcations in the two conduits (Fig. 15.10(c)). In contrast, the temporal features of the lower point sources associated with the conduit corners below sea level were interpreted by Chouet *et al.* (2008) as a passive response of the conduit to the action of the upper source. Thus, each discontinuity in the conduit provides a site where pressure and momentum changes resulting from flow processes associated with the passage of a gas slug through the discontinuity are coupled to the Earth, or where the elastic response of the conduit can couple back into pressure and momentum changes in the fluid. Support for this view is provided by laboratory simulations (James *et al.*, 2006) investigating the ascent of a slug of gas in a vertical liquid-filled tube featuring a sharp flare (see Fig. 15.6). The pressure pulse in the liquid and downward force on the apparatus holding the tube observed as the slug clears the flare are both consistent with the mechanisms imaged for the upper sources at Stromboli (compare volume changes and vertical force F_3 in Fig. 15.9 with pressure and vertical tube displacement in Fig. 15.6(b)). The force observed under laboratory conditions also satisfactorily scales to the magnitude of the force seen in the volcanic environment, lending further support for this interpretation.

15.4 | Sources of long-period seismicity

The gaseous fluid mixtures (gas, liquid and/or solid particles, and associated mixtures) composing volcanic fluids often lead to strong velocity contrasts between the fluid and encasing solid. This in turn favors the entrapment of acoustic energy at the source, leading to long-lasting oscillations commonly observed as LP events and tremor (Chouet, 1996a). The waveform of the LP event is characterized by simple decaying harmonic oscillations except for a brief interval at the event onset (Fig. 15.11). This signature may be viewed as the response of a fluid-filled resonator to a short-lived excitation. By the same token, tremor may be attributed to the response

of the same resonator to a sustained excitation. LP events are particularly useful in the quantification of magmatic and hydrothermal processes because the properties of the resonator system at the source of this event can be inferred from the properties of the decaying oscillations in the tail of the seismogram. The damped oscillations in the LP coda are quantified by two parameters, T, and Q, where T is the period of the dominant mode of oscillation, and Q is the quality factor of the oscillatory system representing the combined effects of intrinsic and radiation losses.

Interpretations of the oscillating characteristics of LP sources have mostly relied on a model of a fluid-driven crack (Chouet, 1986). This model, which has the most natural geometry that satisfies mass-transport conditions at depth beneath a volcano (pipe-like structures are not expected to exist under the prevailing pressure conditions at depth), is supported by results from inversions of LP waveforms recorded at several volcanoes (Kumagai *et al.*, 2002b, 2005; Waite *et al.*, 2008). A simplified two-dimensional model of a fluid-driven crack was first introduced by Aki *et al.* (1977). Although this model included both the driving excitation and geometry appropriate for transport, the fluid inside the crack was treated as a passive cushion that did not support the acoustic propagation of the pressure disturbance caused by the motion of the crack wall. An extension of this model including active fluid participation was later proposed by Chouet and Julian (1985), who considered a simultaneous solution of the elastodynamics and fluid dynamics for a two-dimensional crack. This model was further extended to three dimensions by Chouet (1986) and was extensively studied by Chouet (1988, 1992).

The three-dimensional model consists of a single isolated rectangular crack embedded in an infinite elastic body and assumes zero mass transfer into and out of the crack (Fig. 15.12). Crack resonance is excited by a pressure transient applied symmetrically on both walls over a small patch of crack wall. In this model, the crack aperture is assumed to be much smaller than the seismic wavelengths of interest and the motion of the fluid inside the crack is treated as two-dimensional in-plane motion.

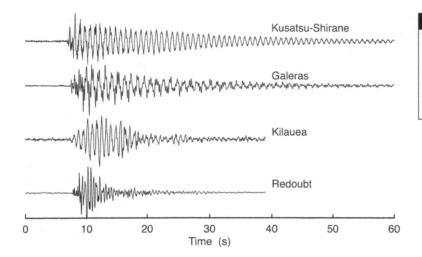

Figure 15.11 Typical signatures of long-period events observed at Kusatsu-Shirane, Galeras, Kilauea, and Redoubt Volcanoes. The signatures are all characterized by a harmonic coda following a signal onset enriched in higher frequencies.

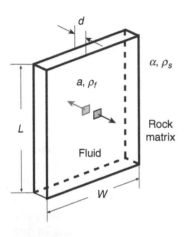

Figure 15.12 Geometry of the fluid-filled crack model of Chouet (1986). The crack has length L, width W, and aperture d, and contains a fluid with sound speed a and density ρ_f. The crack is embedded in an elastic solid with compressional wave velocity α and density ρ_s. Excitation of the crack is provided by a pressure transient applied symmetrically on both walls over the small areas indicated by the gray patches.

Assuming a Poisson solid ($\lambda = \mu$), Chouet (1986) calculated the response of the crack by solving the following dimensionless equations together with boundary conditions for stresses at the crack surface and fluid flow at the crack perimeter:

Solid:
$$\frac{\partial \bar{v}_i}{\partial \bar{t}} = \frac{1}{3}\frac{\partial \bar{\sigma}_{ik}}{\partial \bar{x}_k}, \tag{15.22}$$

$$\frac{\partial \bar{\sigma}_{ij}}{\partial \bar{t}} = \frac{\partial \bar{v}_k}{\partial \bar{x}_k}\delta_{ij} + \frac{\partial \bar{v}_i}{\partial \bar{x}_j} + \frac{\partial \bar{v}_j}{\partial \bar{x}_i}, \tag{15.23}$$

Fluid:
$$\frac{\partial \bar{V}_l}{\partial \bar{t}} = -\frac{1}{3}\frac{\rho_s}{\rho_f}\frac{\partial \bar{p}}{\partial \bar{x}_l}, \tag{15.24}$$

$$\frac{\partial \bar{p}}{\partial \bar{t}} = -\frac{b}{\mu}\frac{\partial \bar{V}_m}{\partial \bar{x}_m} - 2\frac{b}{\mu}\frac{L}{d}\bar{v}_d, \tag{15.25}$$

where the crack is set in the plane $z = 0$ in a Cartesian coordinate system x, y, z. In these equations, nondimensional variables are indicated by an overbar; $\bar{v}_i$ and $\bar{\sigma}_{ij}$ represent the particle velocity and stress components in the solid, $\bar{v}_d = \bar{v}_z(d/2)$ is the normal component of velocity of the crack wall, $\bar{V}_l$ and $\bar{p}$ are the velocity components and pressure of the fluid. The parameters ρ_s and ρ_f are the densities of the solid and fluid, μ and b are the rigidity of the solid and bulk modulus of the fluid, and L and d represent the crack length and crack aperture, respectively. The dimensionless variables are obtained through the following scaling relations:

Length $\qquad \bar{x}_i = x_i / L \qquad$ (15.26)

Time $\qquad \bar{t} = \alpha t / L \qquad$ (15.27)

Stress $\qquad \bar{\sigma}_{ij} = \sigma_{ij} / \sigma_0 \qquad$ (15.28)

Displacement $\qquad \bar{u}_i = u_i \mu / (L\sigma_0) \qquad$ (15.29)

Velocity $\qquad \bar{v}_i = v_i \mu / (\alpha\sigma_0) , \qquad$ (15.30)

where σ_0 is the effective stress, and $\alpha = \sqrt{(\lambda + 2\mu)/\rho_s}$ is the compressional wave velocity in the solid.

These studies demonstrated that the reson-ance of the crack is sustained by a slow wave termed the "crack wave." This slow wave can be understood from a consideration of Eqs. (15.24) and (15.25), which can be combined to obtain the relation:

$$\frac{\partial^2 \bar{p}}{\partial \bar{t}^2} + 2C\frac{\partial \bar{v}_d}{\partial \bar{t}} = \left(\frac{a}{\alpha}\right)^2 \left(\frac{\partial^2 \bar{p}}{\partial \bar{x}^2} + \frac{\partial^2 \bar{p}}{\partial \bar{y}^2}\right),$$ (15.31)

where $a = \sqrt{b/\rho_f}$ is the acoustic velocity of the fluid, and $C = (b/\mu)(L/d)$ is a dimensionless par-ameter called "crack stiffness" (Aki et al., 1977; Chouet, 1986). Let us consider the simple case where the normal component of displacement of the crack wall $\bar{u}_d$ is proportional to $\bar{p}$,

$$\bar{u}_d = \epsilon \bar{p},$$ (15.32)

where ϵ is a nondimensional proportionality constant. Equation (15.31) may then be recast as

$$(1 + 2C\epsilon)\frac{\partial^2 \bar{p}}{\partial \bar{t}^2} = \left(\frac{a}{\alpha}\right)^2 \left(\frac{\partial^2 \bar{p}}{\partial \bar{x}^2} + \frac{\partial^2 \bar{p}}{\partial \bar{y}^2}\right),$$ (15.33)

or equivalently

$$\frac{\partial^2 \bar{p}}{\partial \bar{t}^2} = \left(\frac{a_e}{\alpha}\right)^2 \left(\frac{\partial^2 \bar{p}}{\partial \bar{x}^2} + \frac{\partial^2 \bar{p}}{\partial \bar{y}^2}\right).$$ (15.34)

Equation (15.34) represents the dimensionless wave equation for pressure with phase velocity $a_e = \sqrt{b_e/\rho_f}$, where b_e represents an effective bulk modulus defined as

$$b_e = \frac{b}{(1 + 2C\epsilon)}.$$ (15.35)

For a perfectly rigid wall (no deformation), $\epsilon = 0$ and the phase velocity $a_e = a$. When the wall deforms in response to increasing pressure, $\epsilon > 0$ and a_e becomes smaller than a through a reduction of the effective bulk modulus. As shown in Eq. (15.35), the effective bulk modu-lus also decreases with increasing crack stiff-ness, demonstrating the critical importance of this parameter in controlling the crack dynamic behavior.

The asymptotic behavior of the crack wave was investigated analytically by Ferrazzini and Aki (1987) in a study of normal modes trapped in a liquid layer sandwiched between two elas-tic half spaces. Ferrazzini and Aki (1987) showed that the crack wave speed increases with decreas-ing wavelength and in the short-wavelength

limit reduces to the Stoneley wave propagating along the fluid–solid interface. These properties are analogous to those of tube waves propagat-ing in a fluid-filled borehole (Biot, 1952). Unlike the tube wave, however, as the wavelength increases to infinity the velocity of the crack wave approaches zero in inverse proportion to the square root of wavelength (Ferrazzini and Aki, 1987). A brief summary of the basic proper-ties of the crack wave can be found in Kawakatsu and Yamamoto (2007). Figure 15.13 illustrates the dispersion characteristics of the crack wave in the model of Chouet (1986).

Synthetic seismograms calculated with the fluid-filled crack model bear strong resem-blance to observed LP waveforms (Fig. 15.14), convincingly demonstrating the critical import-ance of active fluid participation in the source process of volcanic signals. The slow character-istics of the crack wave also lead to more realis-tic estimates of crack dimensions compared to estimates based on the sound speed of a fluid embedded in a resonator with perfectly rigid walls. Crack lengths estimated from LP data based on this model typically range from tens to several hundred meters (Saccorotti et al., 2001; Kumagai et al., 2002a, 2005).

As formulated, the fluid-filled crack model accounts for the radiation loss only. Intrinsic losses caused by dissipation mechanisms within the fluid must be treated separately. Detailed analyses of the dependence of crack resonance on fluid composition by Kumagai and Chouet (2000) show that the Q factor of the crack res-onance increases monotonically with increasing ratio a/α, and systematic investigations of LP sig-natures based on their results suggest that dusty gases or bubbly basalt are common fluids in LP events of magmatic origin (Gil Cruz and Chouet, 1997; Kumagai and Chouet, 1999), and that wet gases, steam, and bubbly water are typically rep-resentative of the source of LP events of hydro-thermal origin (Saccorotti et al., 2001; Kumagai et al., 2002a).

The model of Chouet (1986) does not address the excitation mechanism of LP events or tremor. Rather, the spatio-temporal properties of the pressure transient triggering the crack resonance are preset as kinematic conditions

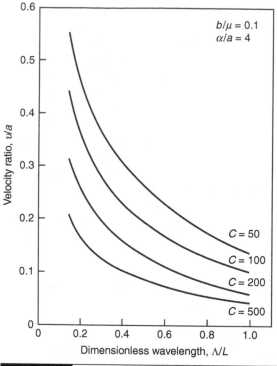

Figure 15.13 Dispersion characteristics of the crack wave derived from the model of Chouet (1986). The dispersion curves represent the ratio v/a of the phase velocity of the crack wave to acoustic velocity of the fluid plotted as a function of dimensionless wavelength Λ/L. The curves are obtained for models with ratios $b/\mu = 0.1$, $\alpha/a = 4$ for several values of the crack stiffness parameter C between 50 and 500.

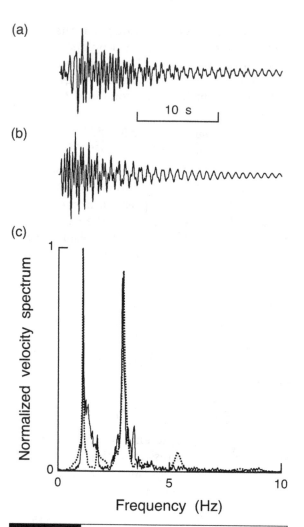

Figure 15.14 Comparison between data and synthetics for an LP event that preceded the 14 January 1993 eruption of Galeras Volcano. (a) Vertical ground velocity recorded at a distance of 1 km from the crater. (b) Vertical ground velocity calculated at the same location for the excitation of a vertical fluid-filled crack buried at a depth of 1 km beneath the vent of Galeras. (c) Spectrum for data (thin line) and synthetics (dotted line). See Chouet (1996a) for source parameters.

in the model. The usefulness of this model is thus restricted to a quantification of the crack resonance and properties of the fluids at the source of LP events. The excitation mechanism of LP seismicity may be assessed from analyses of first motions recorded by a dense network surrounding the source (Waite et al., 2008), or from a waveform inversion of the effective excitation functions of LP events (the signal components obtained after removal of the resonance characteristics in the LP signature) (Nakano et al., 2003). Further perspectives on the excitation mechanisms of LP seismicity may also be obtained from their relationship to concurrent VLP signals (Chouet, 2003; Kawakatsu and Yamamoto, 2007), and from observations of the spatio-temporal characteristics of volcanic seismicity. Observations carried out in different volcanic settings point to a wide variety of LP excitation mechanisms originating in magmatic-hydrothermal interactions, as well as magmatic instabilities.

For instance, shallow hydrothermal LP events at Kusatsu-Shirane (Kumagai et al., 2002b; Nakano et al., 2003), Kilauea (Kumagai et al., 2005), and Mount St. Helens (Waite et al., 2008) all share a common excitation mechanism involving the repeated pressurization of a steam-filled fracture, causing the venting of steam, collapse of the fracture and recharge, in response to heat transfer from an underlying magma body. At Mount St. Helens, where the LP source is located very close to the magma conduit, LP activity is also seen to trigger a passive response of the conduit itself (Waite et al., 2008). In contrast, LP seismicity observed at Redoubt was interpreted as a more energetic form of magmatic-hydrothermal interaction where an unsteady choked flow of magmatic gases provides a natural source of pressure perturbation at the origin of LP events and tremor (Chouet et al., 1994; Morrissey and Chouet, 1997). Magmatic LP events produced during bursts of ash-laden gases associated with vulcanian activity at Galeras (Gil Cruz and Chouet, 1997) and Popocatépetl (Arciniega-Ceballos et al., 2008) involve a pumping mechanism in shallow fractures similar to that inferred for hydrothermal LP events at Kusatsu-Shirane, Kilauea, and Mount St. Helens. LP events accompanying endogeneous dome growth at Soufrière Hills Volcano, Montserrat, appear to be related to a more complex process involving the production of shear fractures in highly viscous magma at the glass transition, injection of dusty gases into these fractures, and resonance of the fracture network and/or possibly resonance of a bubble-rich magma excited by the energy release from the brittle failure (Neuberg et al., 2006). LP events in the basaltic systems at Kilauea (Ohminato et al., 1998; Chouet, 2003) and Stromboli (Chouet et al., 2003, 2008) are attributed to pressure disturbances generated during the transit of large slugs of gas through conduit discontinuities.

In the model of Chouet (1986), a simultaneous solution of the equations of motion of the fluid inside the crack and the surrounding elastic solid (Eqs. (15.22)–(15.25)) is obtained numerically by using a 3D time-domain finite-difference scheme. Another promising approach to numerically simulate the dynamic response of a fluid-filled crack is to formulate this problem using the frequency-domain boundary integral method (Yamamoto and Kawakatsu, 2008). The boundary integral method relies on a formulation of governing equations in terms of integrals on boundaries,

which can be solved by using a point collocation method (Yamamoto and Kawakatsu, 2008). This approach appears particularly well-suited to address the frequency-dependent properties of materials (i.e., bubbly liquids) found in volcanic and hydrothermal processes.

Neuberg et al. (2000) and Jousset et al. (2003) investigated seismic-acoustic wave conversion and coupling in a shallow rectangular conduit embedded in a homogeneous elastic half space. They used a 2D finite-difference scheme to model major features of the LP seismic wave field and study the behavior of single LP events as well as tremor. In their model, seismic propagation in the elastic solid and acoustic propagation in the fluid-filled conduit are solved simultaneously using a single velocity-stress computational scheme. The fluid is defined by a zero shear-velocity (rigidity $\mu = 0$), and appropriate values for the density and sound speed (compressional wave velocity, α). This approach does not require explicit boundary conditions at the conduit wall to define the coupling between the fluid and solid. Acoustic-seismic conversion results from energy transmission controlled by effective material properties at the fluid–solid interface. Using a 2D finite-difference scheme and rectangular conduit model, Jousset et al. (2004) further investigated the effects of viscoelasticity and topography on the amplitudes and spectra of LP events. Their study indicates that the effects of anelastic attenuation and topography can induce significant distortion in LP spectra. Their results also suggest that the rheological properties of magmas may be constrained from detailed analyses of LP seismograms.

A nonlinear excitation mechanism of LP seismicity by fluid flow was proposed by Julian (1994). Using a simple lumped-parameter model, Julian investigated the elastic coupling of the fluid and solid as a means to produce self-excited oscillations in a viscous incompressible liquid flowing through a channel with compliant walls. In his model, an increase in flow velocity leads to a decrease in fluid pressure via the Bernoulli effect. As a result, the channel walls move inward and constrict the flow, causing an increase in fluid pressure and forcing the channel open again. The cyclic repetition of this process is the source of sustained oscillations in Julian's model. Julian demonstrated that with increasing driving pressure the model can exhibit various oscillatory behaviors resembling tremor.

Balmforth *et al.* (2005) carried out further theoretical analyses of Julian's model and explored the stability of fluid flow through a thin channel bounded by two semi-infinite elastic solids. Treating the fluid as viscous and incompressible, they demonstrated the occurrence of shock-like flow disturbances propagating in the direction of flow with phase speed similar to the background flow speed. This type of instability is analogous to what fluid dynamicists call roll waves (e.g., Balmforth and Mandre, 2004). Assuming periodic inlet–outlet flow conditions, and considering flow speeds much slower than the shear and compressional wave speeds in the solid, Balmforth *et al.* (2005) obtained the critical flow speed, $V_{crit\,roll}$, required for the generation of roll wave instability :

$$V_{crit\,roll} \approx \beta \sqrt{\frac{\rho_s}{\rho_f}}\,\varepsilon\,, \qquad (15.36)$$

where $\beta = \sqrt{\mu/\rho_s}$ is the shear wave velocity in the solid, ρ_s/ρ_f is the rock-to-fluid density ratio, and ε is the channel aspect ratio (thickness/length) with $\varepsilon \ll 1$. Equation (15.36) shows that the roll wave instability can only occur for fast flow of dense fluid through long, thin channels. For example, for flow through a channel with aspect ratio $\varepsilon = 10^{-4}$ in a solid with shear wave speed $\beta = 2$ km s^{-1} and $\rho_s = \rho_f$, Eq. (15.36) suggests that flow speeds of tens of meters per second are required for instability. Rust *et al.* (2008) discussed the conditions for roll wave instability and concluded that roll waves are unlikely to explain most volcanic tremor but could possibly occur in the flow of hot, high-pressure H$_2$O- and CO$_2$-rich fluids at sustained flow speeds on the order of 10 m s^{-1}. The wave speed of roll waves in the model of Balmforth *et al.* (2005) represents the long-wavelength limit of the dispersion relation of Ferrazzini and Aki (1987) for crack waves when the elastic and acoustic wave speeds are relatively large (Rust *et al.*, 2008). Note, however, that the crack stiffness parameter is not the proper parameter to describe these slow waves in this limit because the fluid is incompressible in the model of Balmforth *et al.* (2005).

15.5 | Future directions

Seismology alone cannot directly see into the conduit to resolve details of the fluid dynamics at the origin of the seismic source mechanisms revealed by analyses of LP or VLP signals. To develop a better understanding of fluid behavior responsible for these signals, laboratory experiments are required to explore the links between known flow processes and the resulting pressure and momentum changes. Recent laboratory studies (Lane *et al.*, 2001; James *et al.*, 2004, 2006) have elucidated self-excitation mechanisms inherent to the fluid nonlinearity that are providing new insights into the mechanisms imaged from seismic data. The results obtained by James *et al.* (2006) also demonstrate that direct links between the moment-tensor and single-force seismic-source mechanism and fluid-flow processes are possible and could potentially provide a wealth of information not available from seismic data alone.

15.6 | Summary

- Oscillatory behaviors within magmatic and hydrothermal systems are the norm rather than the exception.
- Within the bandwidths of seismometers commonly used on volcanoes today, oscillations are mainly observed as LP (0.2–2 s) and VLP (2–100 s) signals.
- At least 5, and ideally 10–15, three-component broadband receivers deployed on the volcanic edifice (preferably within a few km of the source) are necessary for a robust estimation of the seismic source mechanism for a point source.
- The sources of VLP signals represent specific sites (i.e., conduit discontinuities) where pressure and momentum changes in the fluid are

effectively coupled to the Earth, or where elastic disturbances can feed back into pressure and momentum changes in the fluid.

- LP seismicity is commonly interpreted as acoustic oscillations of fluid-filled cracks in response to pressure transients or sustained pressure fluctuations.
- LP signals have a demonstrated utility in short-term eruption forecasting because changes in the timing and vigor of these signals accompany pressurization in magmatic and hydrothermal systems prior to eruptions.

15.7 | Notation

a	acoustic velocity of fluid (m s^{-1})
a_e	effective acoustic velocity of fluid (m s^{-1})
b	bulk modulus of fluid (Pa)
b_e	effective bulk modulus of fluid (Pa)
C	crack stiffness parameter
d	crack aperture, or channel thickness (m)
E	squared error
$\mathbf{F}$	force (N)
$F_p(t)$	time history of the force applied in the p-direction at ξ
$G_{np}(\mathbf{x}, \xi, t)$	Green tensor relating the x_n-component of displacement at $\mathbf{x}$ with the x_p-component of impulsive force applied at ξ (m N^{-1})
$G_{np,q}(\mathbf{x}, \xi, t)$	spatial derivative of $G_{np}(\mathbf{x}, \xi, t)$ with respect to the ξ_q-coordinate (N^{-1})
L	crack, pipe, or channel length (m)
$\mathbf{M}$	moment tensor (N m)
$M_{pq}(t)$	time history of the pq-component of moment tensor at position ξ of the source
N_r	number of three-component receivers
N_s	number of samples in seismic trace
N_t	number of observed seismic traces
N_{obs}	number of independent observations
N_{par}	number of free parameters
$\bar{p}$	dimensionless fluid pressure
R	radius of penny-shaped crack (m)

S	fault, or crack area (m^2)
t	time (s)
$[\mathbf{u}]$	displacement discontinuity between the Σ^+ side of Σ and the Σ^- side of Σ
$[u_i]$	i-th component of the displacement discontinuity between the Σ^+ side of Σ and the Σ^- side of Σ
$\bar{u}_d$	dimensionless normal component of displacement of crack wall
$u_n(\mathbf{x}, t)$	n-component of seismic displacement observed at a point $\mathbf{x}$ at time t (m)
$\bar{V}_l$	l-component of dimensionless fluid velocity
$V_{crit\ roll}$	critical flow speed for roll wave instability (m s^{-1})
$\mathbf{x}$	coordinate vector for observation point (m)
α	compressional wave velocity in solid (m s^{-1})
β	shear wave velocity in solid (m s^{-1})
δ_{ij}	Kronecker symbol ($\delta_{ij} = 0$ for $i \neq j$, and $\delta_{ij} = 1$ for $i = j$)
δp	pressure increase (Pa)
ΔS	change in pipe cross-sectional area (m^2)
δV	volumetric change under confining pressure (m^3)
ΔV	stress-free volumetric change (m^3)
$\Delta \Theta$	fractional change in volume of sphere
ϵ	nondimensional proportionality constant
ε	channel aspect ratio (thickness/length)
θ	polar angle (°)
λ, μ	Lamé parameters (elastic moduli) of the rock matrix (Pa)
Λ	wavelength (m)
$\mathbf{v}$	unit vector normal to the surface element $d\Sigma$
v_j	j-th component of unit vector normal to the surface element $d\Sigma$
ξ	coordinate vector for general position on Σ (m)
ρ_f	density of fluid (kg m^{-3})
ρ_s	density of the solid (kg m^{-3})
$\bar{\sigma}_{ij}$	ij-component of dimensionless stress in solid

σ_0 effective stress (Pa)

Σ surface of the buried fault or conduit across which the displacements are discontinuous

Σ^+ positive side of Σ

Σ^- negative side of Σ

$d\Sigma$ surface element of buried fault or conduit (m²)

τ characteristic time constant (s)

$\bar{\upsilon}_d$ dimensionless normal component of velocity of crack wall

$\bar{\upsilon}_i$ i-th component of dimensionless particle velocity in solid

ϕ azimuth (°)

ω radial frequency (s⁻¹)

Glossary

Bubbly liquid   Liquid–gas mixture in which the gas phase is distributed as discrete bubbles dispersed in a continuous liquid phase.

Crack wave   A dispersive wave generated by fluid–solid interaction in a fluid-filled crack embedded in an elastic solid. The crack wave speed is always smaller than the acoustic speed of the fluid.

Gas slug   A gas bubble whose diameter is approximately the diameter of the pipe in which the slug is flowing. In a slug ascending a vertical pipe, the nose of the bubble has a characteristic spherical cap and the gas in the bubble is separated from the pipe wall by a falling film of liquid.

Long-period (LP) event   A seismic event originating under volcanoes with emergent onset of P waves, no distinct S waves, and a harmonic signature with periods in the range 0.2–2 s as compared to the usual tectonic earthquake of the same magnitude. LP events are attributed to the involvement of fluid such as magma and/or water in the source process (see also VLP event).

Moment tensor   A symmetric second-order tensor that completely characterizes an internal seismic point source. For an extended source, it represents a point-source approximation and can be quantified from an analysis of seismic waves whose wavelengths are much longer than the source dimensions.

Stoneley wave   A wave trapped at a plane interface between two elastic media. This wave is always possible at a fluid–solid interface with a phase velocity lower than the acoustic speed of the fluid. It can exist at a solid–solid interface under restricted conditions. See Aki and Richards (2002), pp. 156–157.

Very-long-period (VLP) event   A seismic event originating under volcanoes with typical periods in the range 2–100 s. VLP events are attributed to the involvement of fluid such as magma and/or water in the source process. Together with LP events, VLP events represent a continuum of fluid oscillations including acoustic and inertial effects resulting from perturbations in the flow of fluid through conduits.

Waveform inversion   Given an assumed model of the wave propagation medium, a procedure for determining the source mechanism and source location of a seismic event based on matching observed waveforms with synthetics calculated with the model.

References

Akaike, H. (1974). A new look at the statistical model identification. *IEEE Transactions on Automatic Control*, **AC-9**, 716–723.

Aki, K. and Richards, P. G. (2002). *Quantitative Seismology*, 2nd edn. Sausalito, CA: University Science Books.

Aki, K., Fehler, M. and Das, S. (1977). Source mechanism of volcanic tremor: Fluid-driven crack models and their application to the 1963 Kilauea eruption. *Journal of Volcanology and Geothermal Research*, **2**, 259–287.

Arciniega-Ceballos, A., Chouet, B., Dawson, P. and Asch, G. (2008). Broadband seismic measurements of degassing activity associated with lava effusion at Popocatépetl Volcano, Mexico. *Journal of Volcanology and Geothermal Research*, **170**, 12–23.

Auger, E., D'Auria, L., Martini, M. Chouet, B. and Dawson, P. (2006). Real-time monitoring and massive inversion of source parameters of very long period seismic signals: An application to

Stromboli Volcano, Italy. *Geophysical Research Letters*, **33**, L04301, doi:10.1029/2005GL024703.

Balmforth, N. J. and Mandre, S. (2004). Dynamics of roll waves. *Journal of Fluid Mechanics*, **514**, 1–33.

Balmforth, N. J., Craster, R. V. and Rust, A. C. (2005). Instability in flow through elastic conduits and volcanic tremor. *Journal of Fluid Mechanics*, **527**, 353–377.

Ben-Menahem, A. and Singh, S. J. (1981). *Seismic Waves and Sources*. New York: Springer-Verlag.

Biot, M. A. (1952). Propagation of elastic waves in a cylindrical bore containing a fluid. *Journal of Applied Physics*, **23**, 997–1005.

Chouet, B. (1986). Dynamics of a fluid-driven crack in three dimensions by the finite difference method. *Journal of Geophysical Research*, **91**, 13 967–13 992.

Chouet, B. (1988). Resonance of a fluid-driven crack: Radiation properties and implications for the source of long-period events and harmonic tremor. *Journal of Geophysical Research*, **93**, 4375–4400.

Chouet, B. (1992). A seismic source model for the source of long-period events and harmonic tremor. In *IAVCEI Proceedings in Volcanology, Volume 3*, ed. P. Gasparini, R. Scarpa and K. Aki. New York: Springer-Verlag, pp. 133–156.

Chouet, B. (1996a). Long-period volcano seismicity: its source and use in eruption forecasting. *Nature*, **380**, 309–316.

Chouet, B. (1996b). New methods and future trends in seismological volcano monitoring. In *Monitoring and Mitigation of Volcano Hazards*, ed. R. Scarpa and R. I. Tilling. New York: Springer-Verlag, pp. 23–97.

Chouet, B. (2003). Volcano seismology. *Pure and Applied Geophysics*, **160**, 739–788.

Chouet, B. and Julian, B. R. (1985). Dynamics of an expanding fluid-filled crack. *Journal of Geophysical Research*, **90**, 11 187–11 198.

Chouet, B. A., Page, R. A., Stephens, C. D., Lahr, J. C. and Power, J. A. (1994). Precursory swarms of long-period events at Redoubt Volcano (1989–1990), Alaska: Their origin and use as a forecasting tool. *Journal of Volcanology and Geothermal Research*, **62**, 95–135.

Chouet, B., Dawson, P. Ohminato, T. *et al.* (2003). Source mechanisms of explosions at Stromboli Volcano, Italy, determined from moment-tensor inversions of very-long-period data. *Journal of Geophysical Research*, **108**, 2019, doi:10.1029/2002JB001919.

Chouet, B., Dawson, P. and Arciniega-Ceballos, A. (2005). Source mechanism of Vulcanian degassing at Popocatepetl Volcano, Mexico, determined from waveform inversions of very long period signals. *Journal of Geophysical Research*, **110**, B07301, doi:10.1029/2004JB003524.

Chouet, B., Dawson, P. and Martini, M. (2008). Shallow-conduit dynamics at Stromboli Volcano, Italy, imaged from waveform inversion. In *Fluid Motions in Volcanic Conduits: A Source of Seismic and Acoustic Signals*, ed. S. J. Lane, and J. S. Gilbert, Geological Society, London, Special Publications, **307**, 57–84.

Chouet, B. A., Dawson, P. B., James, M. R. and Lane, S. J. (2010). Seismic source mechanism of degassing bursts at Kilauea Volcano, Hawaii: Results from waveform inversion in the 10–50 s band. *Journal of Geophysical Research*, **115**, B09311, doi:10.1029/2009JB006661.

Dawson, P. B., Chouet, B. A. and Power, J. (2011). Determining the seismic source mechanism and location for an explosive eruption with limited observational data: Augustine Volcano, Alaska. *Geophysical Research Letters*, **38**, L03302, doi:10.1029/2010GL045977.

Ekström, G., Nettles, M. and Abers, G. A. (2003). Glacial earthquakes. *Science*, **302**, 622–624, doi:10.1126/science.1088057.

Eshelby, J. D. (1957). The determination of the elastic field of an ellipsoidal inclusion and related problems. *Proceedings of the Royal Society*, **A241**, 376–396.

Fehler, M. (1983). Observations of volcanic tremor at Mount St. Helens volcano. *Journal of Geophysical Research*, **88**, 3476–3484.

Ferrazzini, V. and Aki, K. (1987). Slow waves trapped in a fluid-filled infinite crack: Implications for volcanic tremor. *Journal of Geophysical Research*, **92**, 9215–9233.

Gil Cruz, F., and Chouet, B. A. (1997). Long-period events, the most characteristic seismicity accompanying the emplacement and extrusion of a lava dome in Galeras Volcano, Colombia, in 1991. *Journal of Volcanolology and Geothermal Research*, **77**, 121–158.

James, M. R., Lane, S. J., Chouet, B. and Gilbert, J. S. (2004). Pressure changes associated with the ascent and bursting of gas slugs in liquid-filled vertical and inclined conduits. *Journal of Volcanolology and Geothermal Research*, **129**, 61–82.

James, M. R., Lane, S. J., and Chouet, B. A. (2006). Gas slug ascent through changes in conduit diameter: Laboratory insights into a volcano-seismic source process in low-viscosity magmas. *Journal of Geophysical Research*, **111**, B05201, doi:10.1029/2005JB003718.

James, M. R., Lane, S. J. and Corder, S. B. (2008). Modelling the rapid near-surface expansion of gas slugs in low-viscosity magmas. In *Fluid Motions in Volcanic Conduits: A Source of Seismic and Acoustic Signals*, ed S. J. Lane, and J. S. Gilbert. Geological Society, London, Special Publications, **307**, 147–167.

Johnson, J. B., Lees, J. M., Gerst, A., Sahagian, D. and Varley, N. (2008). Long-period earthquakes and co-eruptive dome inflation seen with particle image velocimetry. *Nature*, **456**, 377–381.

Jousset, P., Neuberg, J. and Sturton, S. (2003). Modelling the time-dependent frequency content of low-frequency volcanic earthquakes. *Journal of Volcanology and Geothermal Research*, **128**, 201–223.

Jousset, P., Neuberg, J. and Jolly, A. (2004). Modelling the low-frequency volcanic earthquakes in a viscoelastic medium with topography. *Geophysical Journal International*, **159**, 776–802.

Julian, B. R. (1994). Volcanic tremor, nonlinear excitation by fluid flow. *Journal of Geophysical Research*, **99**, 11 859–11 877.

Kanamori, H. and Given, J. W. (1982). Analysis of long-period waves excited by the May 18, 1980, eruption of Mount St. Helens – A terrestrial monopole? *Journal of Geophysical Research*, **87**, 5422–5432.

Kanamori. H., and Mori, J. (1992). Harmonic excitation of mantle Rayleigh waves by the 1991 eruption of Mount Pinatubo, Philippines. *Geophysical Research Letters*, **19**, 721–724.

Kanamori, H., Given, J. W. and Lay, T. (1984). Analysis of seismic body waves excited by the Mount St. Helens eruption of May 18, 1980. *Journal of Geophysical Research*, **89**, 1856–1866.

Kawakatsu, H. (1989). Centroid single force inversion of seismic waves generated by landslides. *Journal of Geophysical Research*, **94**, 12 363–12 374.

Kawakatsu, H. and Yamamoto, M. (2007). Volcano seismology. In *Treatise on Geophysics Volume 4, Earthquake Seismology*, ed. H. Kanamori and G. Schubert. New York: Elsevier, pp. 389–420.

Kawakatsu, H., Kaneshima, S., Matsubayashi, H. *et al.* (2000). Aso94: Aso seismic observation with broadband instruments. *Journal of Volcanology and Geothermal Research*, **101**, 129–154.

Knopoff, L. and Randall, M. J. (1970). The compensated linear-vector dipole: A possible mechanism for deep earthquakes. *Journal of Geophysical Research*, **75**, 4957–4963.

Konstantinou, K. I. and Schlindwein, V. (2002). Nature, wavefield properties and source mechanism of volcanic tremor: a review. *Journal of Volcanology and Geothermal Research*, **119**, 161–187.

Kumagai, H. (2006). Temporal evolution of a magmatic dike system inferred from the complex frequencies of very long period seismic signals. *Journal of Geophysical Research*, **111**, B06201, doi:10.1029/2005JB003881.

Kumagai, H. and Chouet, B. A. (1999). The complex frequencies of long-period seismic events as probes of fluid composition beneath volcanoes. *Geophysical Journal International*, **138**, F7–F12.

Kumagai, H. and Chouet, B. A. (2000). Acoustic properties of a crack containing magmatic or hydrothermal fluids. *Journal of Geophysical Research*, **105**, 25 493–25 512.

Kumagai, H., Chouet, B. A. and Nakano, M. (2002a). Temporal evolution of a hydrothermal system in Kusatsu-Shirane Volcano, Japan, inferred from the complex frequencies of long-period events. *Journal of Geophysical Research*, **107**(B10), 2236, doi:10.1029/2001JB000653.

Kumagai, H., Chouet, B. A. and Nakano, M. (2002b). Waveform inversion of oscillatory signatures in long-period events beneath volcanoes. *Journal of Geophysical Research*, **107**(B11), 2301, doi:10.1029/2001JB001704.

Kumagai, H., Miyakawa, K., Negishi, H. *et al.* (2003). Magmatic dike resonances inferred from very-long-period seismic signals. *Science*, **299**, 2058–2061.

Kumagai, H., Chouet, B. A. and Dawson, P. B. (2005). Source process of a long-period event at Kilauea Volcano, Hawaii. *Geophysical Journal International*, **161**, 243–254.

Lane, S. J., Chouet, B. A., Phillips, J. C. *et al.* (2001). Experimental observations of pressure oscillations and flow regimes in an analogue volcanic system. *Journal of Geophysical Research*, **106**, 6461–6476.

Latter, J. H. (1979). *Volcanological observations at Tongariro National Park, 2, Types and classification of volcanic earthquakes, 1976–1978*. New Zealand Department of Science and Industrial Research, Geophysical Division, Wellington, Report 150.

Linde, A. T., Agustsson, K., Sacks, I. S. and Stefansson, R. (1993). Mechanism of the 1991 eruption of Hekla from continuous borehole strain monitoring. *Nature*, **365**, 737–740.

McNutt, S. R. (2002). Volcano seismology and monitoring for eruptions. In *International Handbook of Earthquake and Engineering Seismology, Part A, Chapter 25*, ed. W. H. K. Lee, H. Kanamori, P. C. Jennings and C. Kisslinger. New York: Academic Press, pp. 383–406.

McNutt, S. R. (2005). Volcanic seismology. *Annual Review of Earth and Planetary Sciences*, **32**, 461–491.

Mogi, K. (1958). Relation between the eruptions of various volcanoes and the deformations of the ground surfaces around them. *Bulletin of the Earthquake Research Institute*, **36**, 99–134.

Morrissey, M. M. and Chouet, B. A. (1997). A numerical investigation of choked flow dynamics and its application to the triggering mechanism of long-period events at Redoubt Volcano, Alaska. *Journal of Geophysical Research*, **102**, 7965–7983.

Murase, T. and McBirney, A. R. (1973). Properties of some common igneous rocks and their melts at high temperatures. *Geological Society of America Bulletin*, **84**, 3563–3592.

Nakano, M., Kumagai, H. and Chouet, B. A. (2003). Source mechanism of long-period events at Kusatsu-Shirane Volcano, Japan, inferred from waveform inversion of the effective excitation functions. *Journal of Volcanology and Geothermal Research*, **122**, 149–164.

Neuberg, J., Luckett, R., Baptie, B. and Olsen, K. (2000). Models of tremor and low-frequency earthquake swarms on Montserrat. *Journal of Volcanology and Geothermal Research*, **101**, 83–104.

Neuberg, J. W., Tuffen, H., Collier, L. et al. (2006). The trigger mechanism of low-frequency earthquakes on Montserrat. *Journal of Volcanology and Geothermal Research*, **153**, 37–50.

Nishimura, T. (1995). Source parameters of volcanic eruption earthquakes at Mount Tokachi, Hokkaido, Japan, and magma ascending model. *Journal of Geophysical Research*, **100**, 12 465–12 473.

Nishimura, T. and Hamaguchi, H. (1993). Scaling law of volcanic explosion earthquake. *Geophysical Research Letters*, **20**, 2479–2482.

Nishimura, T., Nakamichi, H. Tanaka, S. et al. (2000). Source process of very long period seismic events associated with the 1998 activity of Iwate Volcano, northeastern Japan. *Journal of Geophysical Research*, **105**, 19 135–19 147.

Ohminato, T. and Chouet, B. A. (1997). A free-surface boundary condition for including 3D topography in the finite-difference method. *Bulletin of the Seismological Society of America*, **87**, 494–515.

Ohminato, T., Chouet, B. A., Dawson, P. B. and Kedar, S. (1998). Waveform inversion of very-long-period impulsive signals associated with magmatic injection beneath Kilauea volcano, Hawaii. *Journal of Geophysical Research*, **103**, 23 839–23 862.

Ohminato, T., Takeo, M., Kumagai, H. et al. (2006). Vulcanian eruptions with dominant single force components observed during the Asama 2004 volcanic activity in Japan. *Earth, Planets and Space*, **58**, 583–593.

Rust, A. C., Balmforth, N. J. and Mandre, S. (2008). The feasibility of generating low-frequency volcano seismicity by flow through a deformable channel. In *Fluid Motions in Volcanic Conduits: A Source of Seismic and Acoustic Signals*, ed S. J. Lane and J. S. Gilbert. Geological Society, London, Special Publications, **307**, pp. 45–56.

Saccorotti, G., Chouet, B. and Dawson, P. (2001). Wavefield properties of a shallow long-period event and tremor at Kilauea Volcano, Hawaii. *Journal of Volcanology and Geothermal Research*, **109**, 163–189.

Sneddon, I. N. and Lowengrub, M. (1969). *Crack Problems in the Classical Theory of Elasticity*. New York: John Wiley.

Takei, Y. and Kumazawa, M. (1994). Why have the single force and torque been excluded from seismic source models? *Geophysical Journal International*, **109**, 163–189.

Uhira, K. and Takeo, M. (1994). The source of explosive eruptions of Sakurajima Volcano, Japan. *Journal of Geophysical Research*, **99**, 17 775–17 789.

Waite, G. P., Chouet, B. A. and Dawson, P. B. (2008). Eruption dynamics at Mount St. Helens imaged from broadband seismic waveforms: Interaction of the shallow magmatic and hydrothermal systems. *Journal of Geophysical Research*, **113**, B02305, doi:10.1029/2007JB005259.

Yamamoto, M. and Kawakatsu, H. (2008). An efficient method to compute the dynamic response of a fluid-filled crack. *Geophysical Journal International*, **174**, 1174–1186.

Exercises

15.1 Use the expression for the moment density tensor $m_{pq} = [u_i]v_j[\lambda\delta_{ij}\delta_{pq} + \mu(\delta_{ip}\delta_{jq} + \delta_{iq}\delta_{jp})]$ in Eq. (15.2) to demonstrate that a point source representing a thin crack opening in the direction ν given by the angles θ and ϕ (see Fig. 15.3 (a)) has the moment tensor given in Eq. (15.8). Show that the eigenvalues of the tensile crack consist of three dipoles with magnitudes $\lambda\Delta V$, $\lambda\Delta V$, and $(\lambda + 2\mu)\Delta V$, where the dominant dipole is oriented normal to the crack plane and ΔV is the volume change of the crack.

15.2 Use the formula for M_{pq} in Eq. (15.2) to obtain the moment tensor in Eq. (15.9) for a point source representing the radial expansion of a cylinder with axis orientation as given in Fig. 15.4(a). Show that the eigenvalues of a pipe consist of three dipoles with magnitudes $\lambda\Delta V$, $(\lambda + \mu)\Delta V$, and $(\lambda + \mu)\Delta V$, where the smallest dipole is oriented along the pipe axis (Fig. 15.4(b)) and ΔV is the volume change of the pipe.

15.3 Assume a point source of shear on a horizontal fault plane $\xi_3 = 0$ with slip $[u_1]$ parallel to the fault as shown in Fig. 15.2(a). Use Eq. (15.2) to show that the equivalent force system for this source consists of two couples (i.e., a double couple) as illustrated in Fig. 15.2 (b), where each couple shares the same magnitude $\mu\overline{[u_1]}S$, in which S is the fault area and $\overline{[u_1]}$ is the final value of slip averaged over the fault surface. Use an eigenvalue decomposition to show that this double couple is equivalent to two vector dipoles equal in magnitude and opposite in sign and that this source has no volumetric component.

Online resources available at www.cambridge.org/fagents

• Additional reading

Chapter 16

Volcano acoustics

Milton A. Garcés, David Fee, and Robin Matoza

Overture

Acoustic waves are pressure fluctuations induced by unsteady fluid motions, and can be generated in volcanoes by diverse processes ranging from the discrete impact of a rockfall, to the harmonious reverberation of a lava tube; through the impulsive punctuations of strombolian and vulcanian explosions, and culminating with roaring jet noise from subplinian and plinian eruptions. This chapter highlights the range of volcano-acoustic source processes associated with eruptive activity ranging from hawaiian to plinian. It introduces various ways to induce magma–gas mixtures into explosive releases, aperiodic fluctuations, and harmonic oscillations. We begin by introducing some fundamental concepts in acoustics and then outline the methods used to extract useful information from complex and diverse volcanic soundscapes.

16.1 | Exposition

The science of acoustics specializes in the recording, analysis, and interpretation of sound, which in the Earth's atmosphere is caused by waves of compressed and rarefied air. The reference root mean square pressure for airborne sound is 20 μPa (20×10^{-6} Pa), which corresponds to the faint buzzing of a mosquito 3 meters away and is near the lower threshold of human hearing. In contrast, large eruptions may produce peak overpressures in excess of 100 Pa at a distance of a few kilometers. Volcanoes tend to produce sound with low frequencies (f), or long periods (period T is the reciprocal of frequency, $T = 1/f$; all notation is listed in Section 16.8). Humans can hear sound within the frequency range of 20–20 000 cycles per second or hertz (Hz). Above 20 000 Hz lies *ultrasound*, which bats famously exploit for echolocation, whereas *infrasound* is below 20 Hz. In the infrasound range we lose our perception of tonality, or pitch, so that even a pure tone would be perceived as a percussive beat with a tempo equal to the tone frequency (2 Hz = 120 beats per minute; Leventhall *et al.*, 2003).

Volcanoes generate signals across a broad sound spectrum, with large explosions and tremor at the infrasonic end and small-scale jetting and hissing from fumaroles at the audible end. However, because volcanoes have large spatial scales, the majority of the radiated acoustic power has long wavelengths. By using the fundamental relationship between wavelength λ (m), sound speed c (nominally 340 m s^{-1} in Earth's atmosphere), and frequency f (Hz),

Modeling Volcanic Processes: The Physics and Mathematics of Volcanism, eds. Sarah A. Fagents, Tracy K. P. Gregg, and Rosaly M. C. Lopes. Published by Cambridge University Press. © Cambridge University Press 2013.

$c = \lambda f$, we deduce that infrasonic frequencies correspond to spatial scales on the order of tens of meters or larger. Audible sounds from volcanoes represent the veritable tip of the iceberg compared to the majority of acoustic energy radiated at infrasonic frequencies. For example, sounds from strombolian explosions recorded close to the vent have frequencies between 1 Hz and 10 kHz (Garcés, 1995), with a range of eight orders of magnitude in acoustic intensity. Because infrasound is efficiently generated by large-scale eruption processes and may travel through the atmosphere for large distances (> 1000 km) with minimal attenuation, the infrasound frequency band is particularly well suited to the remote monitoring of eruption signals.

The field of acoustics is a branch of fluid mechanics. The equations of motion for acoustic propagation are a special case of the linearized Navier–Stokes equations for irrotational, compressible, Newtonian fluids, subject to a suite of simplifying assumptions. For further details, see Morse and Ingard (1968), Lighthill (1978), Pierce (1981), and Kinsler et al. (1982).

A propagating sound wave obeys the wave equation given by (Morse and Ingard, 1968):

$$\nabla^2 p - \frac{1}{c^2}\frac{\partial^2 p}{\partial t^2} = 0, \qquad (16.1)$$

where c is the sound speed, and the acoustic pressure p is defined as the difference between the instantaneous pressure P and the ambient (or equilibrium) pressure P_0,

$$p = P - P_0. \qquad (16.2)$$

Thus, the governing equations in acoustics are perturbation solutions, and exclude steady state or (incompressible) Bernoulli flow where $P - P_0$ would be time-invariant. *Only unsteady processes produce sound.* Of particular relevance to volcanological applications is the fundamental assumption that acoustic pressure and density fluctuations from equilibrium values are small. Earth's atmospheric equilibrium pressure (P_0) is ~0.1 MPa at sea level. In close proximity to powerful volcanic sources, acoustic pressure contributions to P can approach or even exceed the ambient value of P_0 (Eq. (16.2)). In these situations, the wave equation (Eq. (16.1)) does not hold and it is necessary to revert back to the more fundamental hydrodynamic equations. However, in practice, a distance from the source can be defined where P drops below equilibrium values and the acoustic equations are valid. Beyond this distance from the source it is possible to construct equivalent source models that capture the main features of an acoustic source process. In this chapter we concentrate on the simplest and most general relationships that may permit reasonable physical interpretations of field data and comparisons between different eruption processes.

16.1.1 Sound pressure levels

One of the most fundamental acoustic measurements is pressure amplitude (Eq. (16.2)). To date, recorded volcano acoustic signals span over 14 orders of magnitude in sound intensity, so it is useful to adopt a logarithmic scale known as the sound pressure level (*SPL*):

$$SPL = 10\log_{10}\left[\frac{p_{rms}^2}{p_{ref}^2}\right] = 20\log_{10}\left[\frac{p_{rms}}{p_{ref}}\right], \quad (16.3)$$

where

$$p_{rms}^2 = \langle p^2 \rangle = \frac{1}{T_s}\int_0^{T_s} p^2(t)\,dt, \qquad (16.4)$$

and p is the measured acoustic pressure over a specified frequency band, p_{ref} is a reference pressure, and T_s is a chosen time interval for integration. Root-mean-square (rms) pressure (p_{rms}) estimates are robust when the acoustic signal is stationary, meaning that its statistical properties do not change substantially over the period of integration. For a continuous signal with a constant period, p_{rms} can be approximated by $p_{rms}^2 = p_{max}^2/2$, where p_{max} is the peak pressure in Pa. However, for impulsive, transient signals such as explosions, the integration interval should match the signal duration to provide a sound exposure level $L_d = SPL + 10\log_{10}T_s$. The reference level of sound exposure in air is $(20\ \mu Pa)^2$ s (ANSI S12.7–1986,1986). The *SPL* or L_d are not meaningful without a specification of the frequency band used to compute Eq. (16.4), and a correction for the data acquisition response in that band (ANSI S1.13–2005, 2005).

Sound intensity, I, in W m^{-2}, is a measure of the average rate of acoustic energy flow through an area normal to the propagation direction, and is proportional to the square of the pressure far from the source. The intensity can be estimated from:

$$I \cong \frac{1}{\rho c T_s} \int_0^{T_s} p^2(t)\, dt = \frac{\langle p^2 \rangle}{\rho c}, \qquad (16.5)$$

where the sound speed c and the equilibrium density ρ of Earth's atmosphere at 20 °C and sea level pressures are 343 m s^{-1} and 1.2 kg m^{-3}, respectively. Because there is only a proportionality factor (ρc) between intensity and the square of pressure, SPL estimates can be readily converted to intensity by using a reference intensity level of 10^{-12} W m^{-2} in air. Intensity estimates are useful for comparing the acoustic response of different environments. Far from a source, the ratio of the acoustic pressure p to the associated fluid (particle) velocity u is equal to ρc, and is referred to as the characteristic acoustic impedance. We note that the relationship between acoustic pressure and particle velocity ($p = \rho c u$) shows that compressibility induces a proportionally higher pressure change than steady incompressible subsonic flow velocity $U < c$, where (from Bernoulli's equation) $P \sim \rho U^2$.

16.1.2 Source energy and power

In source modeling research, it is often useful to estimate how powerful a sound is at the volcanic source. Source levels (SL) may be estimated by removing propagation and source directivity effects, a task that in practice is fraught with complications. Although we would like to remove atmospheric and boundary (e.g., topography) effects, these effects are notoriously unstable near volcanoes and lead to additional variability in source level estimates.

Total radiated power, Π (in watts), may be estimated by integrating the intensity (Eq. (16.5)) over a surface enclosing the source:

$$\Pi_s = \frac{1}{\rho c} \int_\Omega \langle p_s^2(r,\Omega) \rangle r^2 d\Omega, \qquad (16.6)$$

where p_s is the acoustic source pressure and Ω is the solid angle of a surface. This power estimate is time-invariant for sustained, statistically stationary excitations, so it is particularly useful when comparing the energetics of processes that are stationary over the window of integration in Eq. (16.4). However, for transient events where the interval of integration T_s is restricted to the time of the main pulse, a more useful estimate is the total energy (E) radiated by the event, in joules. The energy of a transient event within the event time window can be estimated by $E = T_s \Pi$.

It is assumed that the acoustic radiation features of the source do not change over the time integration implicit in the rms estimate. Because source directivity is only a function of angle, radial and angular variables may be separated as:

$$p_s(r,\Omega) = p_{ax}(r)H(\Omega), \qquad (16.7)$$

where $p_{ax}(r)$ contains the range dependence and H contains all the angular directivity information. The simplest relationship for radial dependence is obtained when we assume that intensity decreases only as the inverse square of radial distance, a relationship known as the spherical spreading law:

$$r^2 \langle p_s^2(r) \rangle = const. \qquad (16.8)$$

In this general case, the source level (SL) at a standard $r = 1$ m from the source can be estimated from:

$$SL \equiv 10 \log_{10} \left[\frac{\langle p_{ax}^2(r=1) \rangle}{p_{ref}^2} \right] = 20 \log_{10} \left[\frac{p_{rms}}{p_{ref}} \right] \qquad (16.9)$$
$$+ 20 \log_{10} R = SPL + 20 \log_{10} R,$$

where R is the range in meters to the source and p_{rms} is the pressure at the receiver. Note that there is a predicted 6 dB drop per doubling of range to the source. Likewise, an increase by a factor of two in pressure is equivalent to a SPL increase of 6 dB. The quantity $20 \log_{10} R$ is the simplest form of *transmission loss* (TL), and provides a measure of intensity decrease due to spherical spreading. If sound is trapped in atmospheric channels defined by Earth's temperature and wind stratification (Brown and Garcés, 2008), cylindrical spreading would produce a geometrical transmission loss of $10\log_{10}R$. One may expect observed transmission losses to be somewhere in between the idealized values of cylindrical or spherical spreading, as observations are typically affected by topography, vertical temperature and wind gradients, absorption, and scattering from atmospheric turbulence, among other factors (ANSI S2.20–1983, 1983; Bowman et al., 2004). Nevertheless, Eq. (16.9) permits estimates of relative source levels referred to a calibrated microphone and range. For observations at a single volcano, Eq. (16.9) can be used to compare a variety of recorded signals with a well-characterized reference event.

16.1.3 The spectral domain

The period of an acoustic wave is the time interval between adjacent pressure maxima in a time record, or the duration of a single wavelet cycle (wavelength). For a simple harmonic wave with a single frequency f, the acoustic pressure as a function of time would be:

$$p(t) = A\sin(2\pi ft - \phi), \qquad (16.10)$$

where $f = 1/T$, T is the period, t is time in seconds, ϕ is the phase, and A is the amplitude of the sound wave.

The period of a real signal may be difficult to determine precisely because signals commonly consist of a superposition of pulses of varying periods arriving at different times. A signal $p(t)$ consisting of a sum of arbitrary individual period or frequency (harmonic) components can be divided into its individual harmonic components through Fourier decomposition. The Fourier transform F_p of the sound pressure record $p(t)$ is given by

$$F_p(f) = \int_{-\infty}^{\infty} p(t)e^{-j(2\pi f)t}dt, \qquad (16.11)$$

which has units of Pa Hz^{-1}. The power spectral density (PSD) of a digital signal (Oppenheim and Schafer, 1989), with units of Pa^2 Hz^{-1}, can be derived from the discrete version of the Fourier transform over a finite time window and provides an estimate of the mean squared acoustic pressure per frequency band in a time window. The decibel unit for PSD is dB relative to (20 μPa)^2 Hz^{-1}, which is referred to as the *spectral level*. A *spectrogram* is essentially a time-stepping PSD, and is useful for interpreting time-varying signals. To convert spectral levels to sound pressure levels, a particular frequency band has to be defined. If a spectral level is a constant S (dB) over a frequency band Δf, then sound pressure level can be estimated from $SPL = S + 10\log(\Delta f)$. If the PSD is not constant, then it is necessary to either use the mean value over the interval or integrate over the frequency band of interest to get an estimate of the mean square pressure.

PSDs and spectrograms are routinely used in acoustics for visualizing the dynamic range, quality, and character of a signal. They commonly reveal spectral features hidden in waveform displays. Figure 16.1 shows PSDs for a wide range of sustained eruptive signals; Figure 16.1(b) shows low-level activity in Kilauea and Tungurahua volcanoes, with well-defined spectral peaks. Figure 16.1(a) shows the broad spectrum of more energetic eruptions, with energy well above background levels (solid line). At a single glance it is possible to distinguish between mild and energetic eruptions, suggesting different physical processes are at play. Table 16.1 presents estimates of sustained eruption source levels derived from Figure 16.1. These estimates vary by 120 dB (10^6 Pa), yet only represent the contributions of oscillations with pressure fluctuations in the 1–2 Hz octave band and are not a measure of the total source power.

Because infrasounds are the dominant form of long-range acoustic radiation from volcanoes, the rest of this chapter discusses the techniques used to record and interpret infrasounds and presents a variety of volcanic processes that may be used to model them.

16.2 | Capture

The goal of many volcano acoustic field campaigns is to capture volcanic sound with minimal distortion in order to accurately model the source. This requires a reliable data acquisition

Table 16.1 | Estimated range, sound pressure levels (SPL), spherical spreading transmission loss (TL), and source level (SL; Eq. (16.9)) between 1 and 2 Hz for sustained volcanic eruptions, interpolated from Figure 16.1.

Source	Activity	Range (km)	SPL (dB)	TL (dB)	SL (dB)	SL (Pa rms)	Comments
Median	none	N/A	45	N/A	45	4×10^{-3}	Ambient noise model.
Pu'u 'O'o, Kilauea, USA	magma degassing	2.4	57	68	125	36	4/21/07. Underestimate, majority of acoustic energy below 1 Hz.
Halemaumau, Kilauea, USA	magma degassing	7	50	77	127	45	3/19/08. Underestimate, majority of acoustic energy below 1 Hz. Misses harmonic structure.
Tungurahua, Ecuador	strombolian	37	45	91	136	1.3×10^2	7/29/06. Background open-vent activity.
Mount St. Helens, USA	phreatic	13.4	65	83	148	5×10^2	3/9/05. Recorded for over 400 km.
Tungurahua, Ecuador	stombolian/ vulcanian	37	65	91	156	1.3×10^3	5/12/06. Misses harmonic structure and underestimates explosions.
Tungurahua, Ecuador	plinian	37	75	91	166	4×10^3	8/17/06. Underestimate, majority of acoustic energy below 1 Hz.

system. Infrasound is first transformed into an electrical analogue signal by microphones or microbarometers (Ponceau and Bosca, 2010). These signals are discretized and time stamped by a digitizer, and stored as digital waveforms. Field recordings for modeling applications require detailed knowledge of the amplitude and phase response as a function of frequency for a sensor and digitizer combination.

Ideally, the sensors and digitizers are selected so that they do not vary in the frequency band of interest, a desirable instrumental feature described as a *flat response*. If using a recording system outside its flat response, it is necessary to perform an instrument correction. It is seldom possible to accurately recover signal features at frequencies outside the flat response of the system. Other key parameters of infrasound sensors are the sensitivity, dynamic range, instrument self-noise, and mechanical sensitivity of the instrument (i.e., how sensitive the infrasound sensor is to seismic waves) (e.g., Bedard, 1971).

16.2.1 Single sensors, networks, and arrays

In practice, volcano acoustic recording systems are chosen based upon the scope of the project. For near-vent studies, single, portable, and expendable microphones are commonly selected, because the probability of these sensors being destroyed by moderate volcanic activity is high (e.g., Moran et al., 2008b). However, a single sensor cannot differentiate between wind noise or competing acoustic sources, so the source location and signal features must be unambiguous. Furthermore, all infrasound sensors have sensitivity to ground vibrations to some degree, and these are indistinguishable from atmospheric pressure oscillations on a single sensor. Co-locating a seismometer helps reduce ambiguity, and a network of single microphones can yield the propagation velocity and location of energetic impulsive signals.

A primary concern for infrasonic recordings is their high vulnerability to wind noise. Infrasonic sensors placed on top of a volcanic

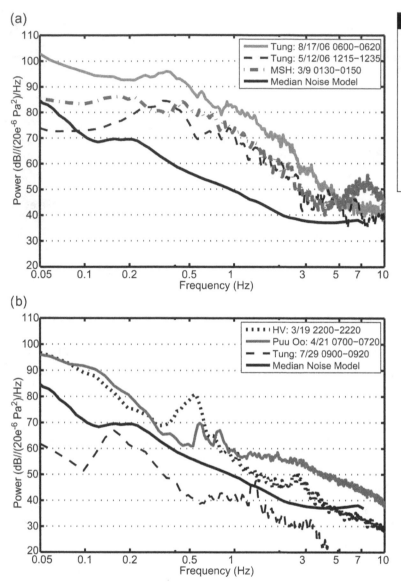

(a)

(b)

Figure 16.1 Power spectral densities for Kilauea (Halemaumau Vent–HV, Pu'u O'o), strombolian (Tungurahua 7/29/06), phreatic (Mount St. Helens–MSH), vulcanian (Tungurahua 5/12/06), and plinian (Tungurahua 8/17/06) eruption styles, not corrected for range (Table 16.1). The median ambient noise model is also shown for reference. The broad 0.2 Hz peak at HV is due to microbaroms associated with ocean processes.

edifice are exposed to strong and variable turbulent pressures induced by winds, which are recorded on the sensors as wind noise that can obscure signals of interest. Although spatial wind filters can provide some defense (Hedlin and Raspet, 2003), the most important factor in wind-noise reduction that can realistically be controlled is the site selection. Locating sensors in a wind-protected site such as a forest or the lee of a topographical barrier is usually worth the gain in wind-noise reduction in exchange for the disadvantage of being further from the source. Deployments several kilometers from exploding sources also place resources at safer distances from flying debris and facilitate maintenance, reducing data loss at critical moments.

At distances from a few kilometers to thousands of kilometers, or in situations where numerous acoustic sources are present, more sophisticated systems are advisable. Generally, the best design for remote infrasound monitoring consists of a number of calibrated sensors precisely time-stamped and arranged spatially as arrays. A single infrasonic array provides speed and direction of signal arrivals and can discriminate between dissimilar competing sources, whereas two properly sited arrays can locate and identify an arbitrary source (Olson and Szuberla, 2008; Cansi and Le Pichon, 2008). Infrasound arrays are used routinely by the international infrasound community for detection, location, identification, and monitoring of natural and man-made events (e.g., Garcés et al., 2004).

16.3 | The volcanic symphony

The captivating diversity of outgassing and fluid dynamic processes occurring at volcanoes produces a rich variety of acoustic signals. Here we present an overview of acoustic observations of volcanic eruptions. For a more comprehensive listing of volcano acoustic studies, the reader is referred to Harris and Ripepe (2007) and Johnson and Ripepe (2011).

A general linear model for a recorded volcano acoustic signal is given by:

$$p(t) = s(t) * l(t) * g(t), \tag{16.12}$$

where $p(t)$ is the observed acoustic signal, $s(t)$ is the source-time function or actual pressure time-history at the volcano (excitation mechanism), $l(t)$ denotes local resonance effects (e.g., resonance in fluid-filled cavities, cracks, conduits, etc.; Garcés, 1997), the Green's function $g(t)$ describes all propagation from the source to the recording site, and $*$ denotes convolution (Oppenheim and Schafer, 1989). In theory, $g(t)$ includes all propagation effects in the atmosphere. For shallow buried acoustic sources, $g(t)$ can also include seismoacoustic coupling through near-surface permeable material (e.g., Matoza et al., 2009a). In Sections 16.4–16.6 we take a simplified view of each of these contributions. We deal with source pressure excitation mechanisms in Section 16.4, discuss acoustic propagation away from the source in Section 16.5, and in Section 16.6 we provide an overview of resonance effects in volcanic fluid systems. In this section, we provide an introduction to the range of acoustic signal types recorded from volcanoes.

Please refer to Table 16.1 throughout this section, which lists the estimated range, sound pressure levels (SPL), spherical spreading transmission loss (TL), and source level (SL) for some of the sustained eruption signatures discussed herein.

16.3.1 Volcano acoustic nomenclature

In volcano seismology (Chapter 15), tremor is a catch-all term used to describe continuous vibration of the ground lasting from minutes to years. It is possible that different tremor-generating mechanisms exist at different volcanoes and even at the same volcano, as represented by the broad range in temporal and spectral characteristics of observed tremor. Similarly, continuous vibration of the air by volcanoes is referred to as infrasonic tremor. Descriptive terms such as harmonic (with multiple spectral peaks), spasmodic (with amplitude variations), episodic (cyclical), and broadband (covering a wide frequency range) are used to further describe infrasonic tremor. Harmonic tremor is particularly intriguing as its spectral signature is reminiscent of musical instruments, and thus suggestive of ordered spatial structures within volcanic systems. As for seismic volcanic tremor, several candidate infrasonic tremor source processes can be invoked; these are discussed in Sections 16.4 and 16.6. Coupling to the atmosphere is necessary for infrasonic tremor, or any other infrasonic signature.

Volcano seismic signals have also been classified based on their period (Chapter 15): very-long-period (VLP) events have a period of ~2–100 s (0.01–0.5 Hz); long period (LP) with a period ~0.2–2 s (0.5–5 Hz); and short period (SP) below ~0.2 s (> 5 Hz). LPs and VLPs are commonly attributed to volumetric sources associated with fluid-filled conduits and cracks, whereas SP quakes are typically attributed to brittle fracture mechanisms (volcano-tectonic earthquakes; Chapter 15). Explosions (Section 16.4) may vary substantially in their intensity and duration, and thus may have energy contributions in any of these frequency bands. Although volcano seismology nomenclature is not ideal for classifying acoustic signals, it facilitates comparison with seismic data. As discussed in Section 16.6, seismoacoustic signals in volcanoes may be coupled or decoupled in more diverse ways than previously envisioned.

16.3.2 Kilauea volcano acoustic signals

Persistent outgassing from effusive eruptions was long thought to be nearly quiescent, producing only faint audible hissing and white noise sounds at fumaroles and lava lakes (Richards, 1963). However, recent work has shown that even low-level outgassing at Kilauea Volcano, Hawaii, produces sustained infrasonic signals that can be recorded at distances of tens of kilometers (Garcés et al., 2003; Fee and Garcés, 2007). Outgassing at Pu'u 'O'o Crater, Hawaii, is dominated by infrasonic tremor lasting from minutes to years and is concentrated between ~0.4 and 10 Hz. Kilauea volcano's infrasounds appear to be strongly influenced by the efficient exsolution of the gas phase from the magma and the interaction of this separated gas with the open conduits and chambers confining it. Effusion of degassed lava produces little to no infrasound unless it encounters water or other volatiles.

Kilauea volcano has also produced other types of infrasound. Harmonic tremor from Pu'u 'O'o crater was recorded in 2007. Compressed gas intermittently excites lava tubes, and produces a higher frequency (> 10 Hz) form of tremor (Matoza et al., 2010). Fissure eruptions also produce discernable acoustic signals. During the July 2007 fissure eruption at Kilauea's East Rift Zone, fountaining along a set of four fissures that ruptured a length of ~2 km produced energetic tremor with a spectral signature similar to that of the tremor from Pu'u 'O'o. The signal onset was emergent (slowly rising), suggesting no explosive outburst of material occurred as the fissures opened (Fee et al., 2011).

The ongoing 2008–2013 activity at Halemaumau Crater produces nearly continuous harmonic infrasonic tremor in the 0.4–4 Hz frequency band. The amplitude and frequency of these spectral peaks are stable for periods of days, consistent with a roiling lava lake exciting the overlying gas-filled cavity into resonance (Fee et al., 2010a). Transient degassing bursts at Halemaumau (Fee et al., 2010a) also suggest a shallow process that induces cavity resonance.

The degassing burst oscillations were primarily in the VLP and LP bands and correlated well with video observations of ash pulses emanating from the vent.

Episodic tremor at Kilauea volcano is characterized by cyclical temporal variations in pressure amplitude (Fig. 16.2). The ~2–10 minute cyclic filling and draining of cavities at Halemaumau and Pu'u 'O'o has been observed to coincide with a "gas piston" rising and releasing an accumulated amount of gas, as well as with the rising and lowering of an exposed, degassing lava surface. The degassing portions of this episodic tremor are consistent with elevated infrasound, while the capped or quiescent lava surface correlates with little to no infrasound (Patrick et al., 2011). The spectral content of the episodic tremor at Halemaumau is the same as the aforementioned harmonic tremor, suggesting both signals are affected by the same resonating cavity.

16.3.3 Strombolian and vulcanian acoustic signals

Strombolian and vulcanian explosions are some of the most studied because of their relative abundance and availability, yet their respective waveform signatures can be difficult to differentiate using purely acoustic methods. Marchetti et al. (2009) suggest that it may be possible to better discriminate between these types of explosive signals by using a combination of thermal imaging and acoustic methods. Numerous studies at Stromboli volcano have identified two primary types of explosions: short-duration explosions (~3–5 s) with relatively large amplitudes (20–80 Pa at 350 m) and longer duration (5–15 s), more complex explosions with lower peak amplitudes (10–30 Pa at 350 m) (Ripepe and Marchetti, 2002; Harris and Ripepe, 2007). Both explosive styles occur on a periodic basis and have repeatable (stable) waveforms primarily in the LP band. Strombolian explosions are presumed to originate from overpressured gas slugs bursting at the surface (see Chapter 6). The shorter-duration explosions have sharp, compressional onsets and broadband spectra with a peak around 5 Hz, whereas the longer-duration explosions have more complex,

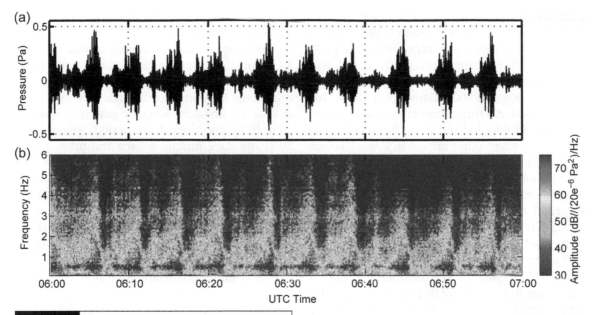

Figure 16.2 Episodic tremor (a) pressure signal and (b) PSD as a function of time (spectrogram) recorded ~7 km from the Halemaumau Vent (HV). The episodic tremor is characterized by cyclical temporal variations caused by filling and draining of the cavity at Halemaumau Vent, Kilauea Volcano. The variations are coincident with a capped surface, or "gas piston," rising and releasing an accumulated amount of gas, as well as with the rising and lowering of an exposed, degassing lava surface. The degassing portions of this episodic tremor are consistent with elevated infrasound, while the "capped" or quiescent lava surface correlates with little to no infrasound. The spectral peak around 0.5 Hz may be related to the cavity being excited into Helmholtz resonance by the release of accumulated gas. See color plates section.

peaked spectra. Waveforms of strombolian explosions from other volcanoes exhibit similar features, although explosion overpressures and durations vary (Vergniolle and Brandeis, 1996; Johnson et al., 2008).

Although infrasonic signatures from vulcanian explosions (Chapter 7) may resemble their strombolian counterparts, there are some important differences. Studies of vulcanian explosions (Garcés et al., 1998; Johnson et al., 2004; Petersen et al., 2006; Marchetti et al., 2009) show signals with energetic compressional onsets and relatively long, complex tails, referred to as codas. The source for vulcanian explosions is commonly attributed to explosive failure of a "capped" conduit (lava plug) and/or

the interaction of magma and external water (Morrissey and Mastin, 2000).

Continuous tremor is also commonly recorded at volcanoes displaying strombolian and vulcanian activity and typically follows explosions (Johnson, 2003; Fee et al., 2010b). Long-duration codas from vulcanian explosions may have signal features reminiscent of infrasonic tremor. Harmonic tremor has also been reported extensively and periodic, often audible, tremor bursts have been documented as well (Johnson and Lees, 2000). *Gliding*, or the frequency modulation of spectral peaks over time, is also observed (Garcés et al., 1998).

16.3.4 Subplinian and plinian acoustic signals

Plinian eruptions are rare, and until recently, so were high-quality acoustic recordings of these energetic events. The birth of volcano infrasound research may be traced to the cataclysmic 1883 eruption of Krakatau (Strachey, 1888). Barometric records of the Krakatau eruption observed in the US, Europe, and Russia demonstrated for the first time the ability of low-frequency sound to propagate for thousands of kilometers. Although barometers are optimally designed to capture long periods (minutes to hours), researchers were still able to make inferences on source durations and location (e.g., Goerke et al., 1965; Tahira et al., 1996). A recent pilot project with arrays at 37 and 251 km from Tungurahua Volcano recorded two subplinian eruptions and one plinian eruption between 2006 and 2008 (Garcés et al., 2008; Fee et al., 2010b). All three eruptions had extended durations (> 4 hours) and were characterized by high-amplitude tremor related to *jetting* (Section 16.4.3). The broadband spectrum of the jetting is remarkably similar among the eruptions (Fig. 16.1(b)) with the general shape resembling that recorded from man-made jet engines (Matoza et al., 2009b). A 2005 phreatic eruption signal lasting ~90 minutes at Mount St. Helens also had a similar spectral signature. The transition of the 17 August 2006 Tungurahua eruption from subplinian to plinian is marked by a distinct shift in the spectrum to lower frequencies (< 0.1 Hz; Fig. 16.1). Acoustic power was found to scale roughly with eruptive intensity and ash cloud height (Fee et al., 2010b). The 2008 plinian eruptions of Okmok and Kasatochi Volcanoes, Alaska, were also clearly identified with remote infrasound arrays (Fee et al., 2010c).

16.3.5 Other types of volcano acoustic signals

LP seismic events have been attributed to hydrothermal and magmatic activity in shallow volcanic cracks and conduits (Chouet, 1996). These events have also produced notable acoustic counterparts (Yamasato, 1998; Matoza et al., 2009a). Infrasonic LPs at Mount St. Helens have impulsive onsets, durations of 5–10 s, and relatively flat, broadband spectra. These events are highly regular in their occurrence and have stable waveform features, indicating a non-destructive, repetitive source (Matoza et al., 2009a). Acoustic records of pyroclastic density currents (PDCs) are rare or commonly masked by concurrent jetting signals. Yamasato (1997) detected PDCs acoustically at Unzen Volcano, Japan, and was able to track their progression using arrival times and Doppler shift. Ripepe et al. (2010) also tracked PDCs with a single infrasound array. The partial collapse of dacitic lava domes (Green and Neuberg, 2005; Moran et al., 2008a) and explosive blowout of gas-charged blocks impacting

the ground (Oshima and Maekawa, 2001) also generate infrasound.

Very large eruptions can also produce gravity and acoustic-gravity waves (Goerke et al., 1965; Tahira et al., 1996; Ripepe et al., 2010) that can commonly be seen as concentric cloud patterns in satellite imagery. Acoustic-gravity waves are affected by buoyancy, have periods longer than 50 s, and have unique source and propagation characteristics that are beyond the scope of this chapter. Although the possibility of sustaining VLP gravity wave oscillations in a stratified, vesiculated magma column was proposed by Garcés et al. (2000), no clear evidence for these wave types has been found to date. See Gossard and Hooke (1975) for an excellent introduction to gravity and acoustic-gravity waves.

16.4 | Excitation

In this section we discuss energetic hydrodynamic processes that may produce measurable pressure changes in volcanic fluids. The most basic type of volcanic pressure disturbance is an explosion. An explosion is defined as a rapid expansion of matter into a volume much greater than its original volume (Cooper and Kurowski, 1996). Explosions are represented as transients with a *positive* pressure onset (compression), defined by its peak pressure and rise time, and followed by a negative pressure (rarefaction) corresponding to gas overexpansion. Such bipolar explosion pulses may be described by the duration, rise time, and peak pressure of the signal (Fig. 16.3; ANSI S12.7–1986, 1986). We note that in the volcano-acoustics literature, volcanic explosion waveforms are sometimes incorrectly referred to as N-waves. True N-waves, such as those from supersonic sources (e.g., Garcés et al., 2004) and the detonation of high explosives (Gossard and Hooke, 1975), have an unambiguous N-shape. More continuous gas-release processes, such as subplinian and plinian eruptions, can last for several hours and yet are sometimes referred to as explosive. In volcanology, explosive activity refers to eruptions "where magma is torn apart and ejected from the vent as clots or blobs within a stream of gas"

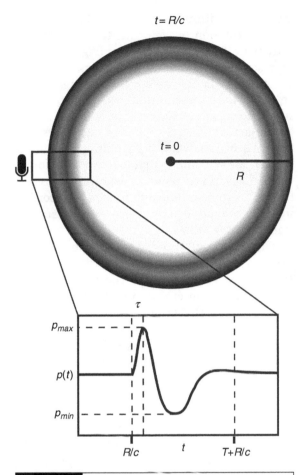

Figure 16.3 Typical explosion pulse, showing the peak pressure p_{max}, rise time τ, and duration T, where $\tau \ll T$. The pulse propagates to a station after a time R/c, where R is the range to the station and c is the sound speed.

(Parfitt and Wilson, 2008, p. 75). In acoustics, we generally reserve the word *explosion* for unambiguously impulsive acoustic signals with a very short rise time and duration, and use other terms to describe complex transients and more sustained degassing activity. In Section 16.4.3 we discuss volcanic processes that may give rise to these acoustic signatures.

16.4.1 TNT, BLEVEs, and magmatic explosions

Let us consider magma–gas mixtures (MGMs) in the context of man-made explosives, which may be roughly separated into *high explosives* and *low explosives*. The primary distinction is that high explosives *detonate* whereas low explosives

deflagrate. A detonation wave is a supersonic shock wave. In high explosives, a chemical reaction takes place at a high temperature and pressure point called the detonation shock front, which leads to rapid gas expansion and energy release. TNT is the reference compound for high explosives. Deflagration may be described as burning, and is a much slower process. Examples of low explosives are gunpowder and the fuel–air mixtures used in internal combustion engines.

Although capable of creating explosions, MGMs do not behave like man-made explosives. Magmas do not detonate or burn, and in contrast to high explosives, the primary explosive reaction in MGMs is traditionally attributed to decompression through the failure of a containment lava cap or magma skin (Parfitt and Wilson, 2008). Explosions produced by MGMs are more reminiscent of *boiling liquid expanding vapor explosions* (BLEVEs), which result from the rupture of a vessel containing a fluid that is pressurized well above its boiling point (Birk and Cunningham, 1996). Containment failure exposes the fluid to a pressure below its saturation pressure, and may trigger boiling and the explosive expansion of gas even though the fluid is non-flammable. In industrial settings, containment failures may be suppressed through pressure-release valves, or a fluid may fail to explode and instead only produce a jet release. In the case of MGM explosions, sudden decompression can trigger the rapid exsolution of gas, and may lead to a runaway process as the overlying material is removed, further decompressing the MGM. In contrast to industrial BLEVE containers, which have a limited volume, a volcano can tap into a deep MGM reservoir and thus sustain a violent eruption for hours until an equilibrium pressure is reached. As in BLEVEs, it may be possible to suppress an explosion through partial release of pressure, in which case the exsolved gas may be released only as a minor jet or an intermittent gas release.

Of note is that the sound speed of an unfragmented MGM is a strong function of the amount of bubbles, or the void fraction, of the mixture (Wood, 1964). For high void fractions, the sound speed of a MGM may be as low as tens of meters per second. Thus highly vesiculated MGMs may be vulnerable to lower fragmentation thresholds (< 70% void fraction) as relatively low flow speeds reach near-supersonic conditions and blow these foams apart, triggering an explosive eruption (Garcés, 2000).

16.4.2 Steam boilers and popping bubbles

The depressurization of a large static or pressure-compensated gas pocket that has already separated from the melt is easy to visualize. The overpressure in such a system may be estimated as the sum of the lithostatic overpressure, ΔP_l, and the rock or lava yield stress. The yield stress is the force per unit area beyond which permanent deformation will occur, and may have values on the order of 10–1000 MPa, with higher values possible for plastic materials. Lithostatic overpressure can be readily estimated from:

$$\Delta P_l = \rho_r\, g\, \Delta z \sim (2500 \text{ kg m}^{-3})(9.8 \text{ m s}^{-2})\, \Delta z$$
$$\sim \Delta z/40 \text{ MPa m}^{-1}, \qquad (16.13)$$

where Δz is depth in meters and the density ρ_r is representative of somewhat fractured and/or porous crust material. A gas pocket will remain in equilibrium with the ambient pressure or containment (skin) pressure. Because the ambient atmospheric pressure is ~0.1 MPa, it is possible to overpressurize gas with a sealed layer of rock or a magma film. Depending on the containment rupture mechanism, the pressure may be released as a jet, a small explosive burst followed by a jet, or a burst where all the gas is released simultaneously (Birk and Cunningham, 1994). If the gas is already exsolved, as in the case of a static or ascending bubble, this process will only last for as long as it takes to release the gas volume. As demonstrated by Ichihara et al. (2009), a shallow but submerged explosion triggered by sudden depressurization of an MGM would appear visually similar to a popping surface bubble, and may be similar in its source pressure signature.

16.4.3 Transients and jets

Many of the transient and continuous acoustic signals presented in Section 16.3 can be interpreted within the MGM and BLEVE framework. Explosion waveforms from magmatic activity at Tungurahua volcano (Fig. 16.4) may represent the sudden release of gas contained at extremely high pressure in the magma (i.e., an impulsive fragmentation), whereas the more sustained signals at Tungurahua associated with jetting activity (Fig. 16.5(a)) can be understood in terms of runaway fragmentation tapping a greater MGM volume that persists for hours. From the vantage point of volcano acoustics, the aim is to quantitatively relate the recorded sounds to fluid dynamic processes resulting from unstable MGMs.

Figure 16.5(b) shows the infrasonic spectrogram recorded at a range of 37 km from Tungurahua volcano during the sustained subplinian to plinian eruption of 17 August 2006 (PSD in Fig. 16.1). Figure 16.5(c) shows the estimated acoustic power above 0.1 Hz (with a peak of ~30 MW) and the associated ash cloud heights derived from satellite imagery (Fee et al., 2010b). Although there are multiple discrete impulsive explosion waveforms (Fig. 16.5(a)), the dominant signal feature is the sustained broadband, high-amplitude infrasonic tremor. Matoza et al. (2009b) proposed that these long-duration infrasonic signals were produced by turbulent jetting of ash–gas mixtures in large-scale eruption jets (> 300 m diameter at Tungurahua) and represent an infrasonic form of jet noise. Jet noise is produced at audible frequencies by man-made aircraft and rockets and results from the turbulent flow of air out of a jet engine. The acoustic frequencies at which jet noise is radiated scale with the Strouhal number (discussed further in Section 16.6.4) and are related to the length scales of the jet flow (in particular, the diameter). The jetting activity at the lower portion of eruption columns is of large diameter compared to typical man-made jets; thus, the acoustic frequencies of radiation are much lower than those observed from jet engines (Fig. 16.1(a)). Matoza et al. (2009b) used the expanded jet diameter as the characteristic length scale (Eq. (16.28), Section 16.6.4), and found the Strouhal number ranged from 0.06 to 0.4 for the volcanic jets considered in their study.

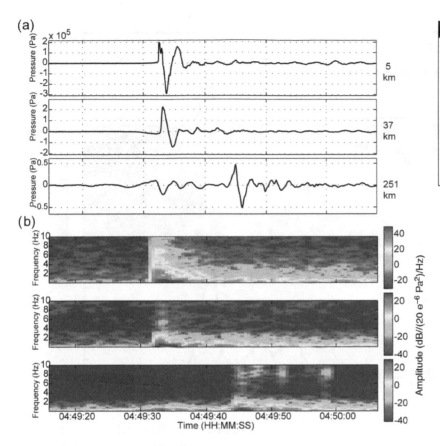

Figure 16.4 Time-corrected Tungurahua (a) explosion waveforms and (b) spectra recorded at distances of 5 km, 37 km, and 251 km. As the distance is increased, the signal loses high-frequency energy and its duration is extended. Nearest station data courtesy of H. Kumagai. See color plates section.

16.4.4 Into the sound

The candidate excitation mechanisms presented so far are nonlinear hydrodynamic processes, and can only be modeled by linear acoustics after moving some distance away from the source. The next section explains how these volcanic processes may be modeled as radiated sound.

16.5 | Acoustic radiation

The expressions presented in this section may be regarded as extremely useful fictions, applicable to ideal linearized cases where complex processes are represented as simple sources or combinations of simple sources. The primary justification for this approach is that it has performed surprisingly well in controlled acoustic environments, so we extend it to the more loosely constrained volcano acoustics problem.

16.5.1 Compact sources

As mentioned in Section 16.1, equivalent acoustic source models may be constructed even for processes that are highly nonlinear. If d_s is the length scale of the fluid region occupied by a source, the source is acoustically compact if:

$$\frac{2\pi d_s}{\lambda} \ll 1, \tag{16.14}$$

where λ is the acoustic wavelength. At a frequency of 1 Hz for air at 20 °C, this corresponds to $d_s \ll 50$ m. At steam temperatures of 700 °C that may be encountered inside magma conduits, this condition corresponds to $d_s \ll 100$ m. Although this compactness condition may not always be met, it provides a tractable starting point for discussing sources of volcanic sound.

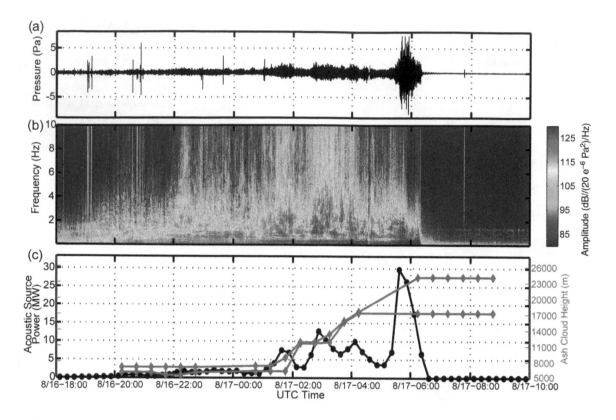

Lighthill (1978) provides an excellent discussion on the virtues of compactness. In particular, for a given mass outflow rate $m(t)$ in an infinite homogeneous volume, the acoustic pressure $p(t)$ at a distance r from the source is given by:

$$p(t) = \frac{1}{4\pi r}\frac{dm(t)}{dt}\bigg|_{t-r/c} = \frac{\rho_0}{4\pi r}\frac{d^2V(t)}{dt^2}\bigg|_{t-r/c}, \quad (16.15)$$

where V is the volume displaced by the mass of a fluid with equilibrium density ρ_0. A severe limitation on estimating mass flux from acoustic measurements is that slowly fluctuating (nearly steady-state) mass-flux contributions do not contribute to the sound field. For example,

if a fluid of density ρ_0 flows with a speed u in a section of cross-sectional area A with a time history given by:

$$u(t) = u_0 + u_s(t), \quad (16.16a)$$

where u_0 is the steady flow speed, and $u_s(t)$ is the unsteady speed as a function of time, the total rate of mass outflow $m(t)$ can be estimated from:

$$m(t) = \rho_0 A u(t) = \rho_0 A u_0 + \rho_0 A u_s(t), \quad (16.16b)$$

and hence its rate of change is

$$\frac{dm(t)}{dt} = \rho_0 A \frac{du_s(t)}{dt}. \quad (16.16c)$$

Note that Eqs. (16.15) and (16.16c) show that an increase in pressure is associated with fluid acceleration. The first term on the right-hand side of Eq. (16.16b) is the steady mass flux $\rho_0 A u_0$, which is routinely used in incompressible fluid dynamics and is generally the dominant mass contribution. Because acoustic measurements are by definition perturbations from the steady pressure field, Eq. (16.16c) shows that the time-invariant steady mass flux will not produce sound. A common assumption in compressible fluid flows is that the time-varying unsteady component $u_s(t)$ is a fraction of the steady stream velocity u_0, so that it is possible to estimate the stream velocity from the perturbation velocity. Estimates of the steady mass flux term are sensitive to the details of the models used to represent the fluid dynamics of a source process, and may not be robust because of the large number of free variables (Garcés, 1997).

Exact expressions for acoustic sources can generally be derived for only the simplest geometries and propagation conditions. Most solutions are rich in complex detail and structure near the source, but stabilize substantially far from the source. In addition, many mathematical expressions have asymptotic expansions that simplify at a distance of a few wavelengths from the source as the waves lose their curvature and start resembling plane waves. The typical criterion for the far-field is:

$$\frac{2\pi r}{\lambda} \gg 1 \Rightarrow r \gg \frac{c}{2\pi f} \sim \frac{54}{f}. \quad (16.17)$$

At 0.02, 0.1, 1, and 10 Hz, the far-field condition corresponds to $r \gg$ 3 km, 500 m, 50 m, and 5 m, respectively.

Moran et al. (2008a) exploited the compactness condition (Eq. (16.14)) to model a VLP acoustic signal resulting from a large rockfall from the 2004–08 Mount St. Helens lava extrusion. Similar modeling approaches have been applied to strombolian bubble bursts at Mount Erebus, Antarctica (Johnson et al., 2008).

16.5.2 Multipole expansions for surface and subaerial sources

An acoustic source may be represented by distributions of simple sources, an approach known as *multipole expansion* (Morse and Ingard, 1968). Multipoles permit a ready visualization of the intriguing interference patterns produced by the superposition of multiple sources. Multipole representations may be practical in industrial environments where one

may control recording conditions. Ideally, the source would be placed in an anechoic chamber and surrounded by a grid of microphones to characterize its frequency-dependent radiation pattern. Such rigor is difficult in volcanic environments.

Let us define P_m as the source pressure at 1 m radiated by a *monopole*, represented by a pulsating sphere in free space. The free-space monopole is one of the most useful approximations in the field of acoustics, as it acts as a simple source and is used as a reference for all other compact source representations. Of slightly more practical value is the expression for a source on or near Earth's surface, which is modeled as a baffled source. A baffle is an acoustic term that usually refers to the flat plate to which a speaker is mounted, but here is approximated as a solid plane bounding a half-space. The solution for the pressure radiated by a hemispherical source in a baffle is identical to that of the monopole, because both the environment and the source strength are halved. Bursting bubbles described in Section 16.4.2 are examples of hemispherical sources, as the gas overpressure would be completely released to the atmosphere – the path of least resistance – with little or no acoustic radiation into the melt or ground.

In contrast, the far-field solution for a spherical source close to an infinite baffle and radiating into a half space is given by:

$$P_{s/2} = \frac{2P_m}{r}. \quad (16.18)$$

Sound from a compact industrial source placed on a stand is commonly represented by this equation. A classic *dipole* may be represented as two monopoles pulsating in opposite phase to one another. Such a condition, which is typical of reflection of underwater sound from a shallow source near the ocean–atmosphere interface, is useful for visualizing the interference pattern caused by reflection along an infinite plane (the Lloyd's mirror effect) but is an inadequate way to model reflections in complex environments. Acoustic reflection from multiple irregular walls is better treated as a reverberation problem, to be discussed in more detail in Section 16.6.

A *quadrupole* may be constructed from two dipoles. Quadrupoles were introduced by Lighthill (1954) as possible models for acoustic radiation from turbulence, and they are infamous for being easier to visualize than to evaluate. Woulff and McGetchin (1976) used the linearized equations of motion to estimate the acoustic power radiated from monopoles, dipoles, and quadrupoles for subsonic flows. Their equations can be readily scaled in terms of the Mach number, M, defined as the ratio of the gas exit velocity to the sound speed. They estimated that monopoles, dipoles, and quadrupoles radiate as the fourth, sixth, and eighth powers of the Mach number, respectively, where $M < 1$ for subsonic flow. Their discussion illustrates how inefficient dipole and quadrupole radiation are in comparison with monopole radiation.

Although a source region may be expressed as a sum of monopole, dipole, and quadrupole representations, in the far-field the superposition of these compact sources is generally well approximated by a point monopole with an equivalent source strength equal to the sum of all the multipole contributions (Pierce, 1981). To capture a three-dimensional multipole radiation pattern in the near-field, the vertical and horizontal sensor spacing must be at least half of the spatial scale (wavelength) of interest at a given frequency. In realistic volcanic environments, in particular during large eruptions, adequately dense sampling may be exceedingly challenging. Insufficient sampling will yield spatially aliased and potentially misleading results.

16.5.3 Piston radiation for buried and surface sources

The above expressions (Eqs. (16.14)–(16.17)) for the radiated sound field are primarily pertinent to surface or near-surface sources. In the case of subsurface sources connected to the atmosphere through an open vent, the sound pressure and velocity must be compressed through the vent and the ensuing radiation may be modeled by the piston-like displacement of the fluid at the vent.

The far-field amplitude at r for a piston of radius a (vent radius) set in an infinite baffle is given by:

$$P_p = \frac{P_m}{2r}\left[\frac{2J_1(ka\sin\theta)}{ka\sin\theta}\right] = \frac{P_m}{2r}H_p(\theta), \quad (16.19)$$

where $k = 2\pi/\lambda$, θ is the source–receiver angle measured from the vertical, and J_1 is the first-order Bessel function. The expression in brackets, H_p, approximates unity as $ka\sin\theta$ approaches zero. Similar expressions may be derived for rectangular pistons or line sources in the case of fissure eruptions. In the long-wavelength and far-field approximations, these solutions converge to similarly simple expressions, lending substantial simplicity to source strength estimates.

The far-field expressions discussed thus far rely on the spherical spreading law and differ primarily in their radiation patterns. Note that it would be difficult to discriminate between dipole and piston-like radiation when $ka \approx j_{1m}$, the zeros of the Bessel function, which corresponds to atmospheric frequencies on the order of $f = j_{1,m}(340/2\pi a)$. For a vent radius on the order of 40 m, piston-source radiation patterns would be present in the 3 Hz range.

16.5.4 Useful expressions for acoustic source power

The total power Π radiated by a source, in watts, is given by integrating Eq. (16.6) over the ensonified space:

$$\Pi_s = \frac{4\pi r^2}{\rho c}\langle p^2\rangle \qquad \text{Spherical source in free space}$$
$$(16.20)$$

$$\Pi_h = \frac{4\pi r^2}{\rho c}\langle p^2\rangle \qquad \text{Hemispherical source into a half space (surface source)}$$
$$(16.21)$$

$$\Pi_{s/2} = \frac{8\pi r^2}{\rho c}\langle p^2\rangle \qquad \text{Spherical source near ground radiating into a half space}$$
$$(16.22)$$

$$\Pi_p = \frac{r^2}{\rho c}\langle p^2\rangle\int_0^{\frac{\pi}{2}}\left[\frac{2J_1(ka\sin\theta)}{ka\sin\theta}\right]^2 2\pi\sin\theta\, d\theta$$

Vented subsurface source
$$(16.23a)$$

Note that far from the source $\theta \sim \pi/2$ and $ka\sin\theta$ $\sim 2\pi a/\lambda$. If, in addition, the wavelength is much greater than the piston radius so that $\lambda \gg a$, then

$$\lim_{ka \to 0} \Pi_p = \Pi_h .\qquad (16.23b)$$

If the spatial acoustic source distribution is unknown, or if the recorded field is a superposition of subsurface, surface, and airborne sources (for example, a transient source in a deep crater accompanied by vigorous jetting) it would be judicious to use Eq. (16.20) for the free space solution, as power estimates would only be a factor of 2 (3 dB) different from Eq. (16.21) or (16.22). For explosions and other transient events, the total energy in joules for the event may be estimated by $E = T_s \Pi$, where care must be taken that the time of overpressure integration, T_s, only corresponds to the main pulse. As an example, one of the Halemaumau vent (HV) VLP degassing bursts had an estimated energy of $\sim 10^6$ J, an order of magnitude lower than the HV ensuing tremor energy radiated in one hour (~ 3 kW $= 10^7$ J hr^{-1}; Fee et al., 2010a). If the source is within the erupting vent or more than a few tens of kilometers from the recording station, other propagation effects need to be considered as well (Fee and Garcés, 2007). The effect of the atmosphere on telesonic (> 250 km) propagation is beyond the scope of this chapter, and is covered in greater detail in Le Pichon et al. (2010).

16.6 | Resonant oscillations

Resonance is an unavoidable propagation effect in most volcanic fluid systems. Magma ascends to Earth's surface through a complicated network of cracks, chambers, and conduits, and hydrothermal systems and gas accumulations can be stored in subsurface reservoirs or near-surface cavities. If pressure disturbances occur in any of these confined fluid volumes, acoustic energy is partially trapped and the volumes resonate at their natural frequencies, or eigenfrequencies. As the elastic energy leaks away from these regions, the resulting seismic and acoustic signals observed at a distance are imprinted with the resonance and attenuation characteristics of the volume. This allows for modeling of seismic and acoustic waveforms to infer the geometry and fluid composition of volcanic resonators. A critical step forward was made in volcano seismology when it was postulated that acoustic resonance effects are important in generating observed seismicity at volcanoes (e.g., Aki et al., 1977). Since then much work has been done in volcano seismology concerning solid–fluid interface waves or crack waves that occur from wave propagation in small volumes of fluid trapped in a solid (Chapter 15). Decades ago, volcano seismology identified tremor and long-period events as signals that may be indicative of resonant volcanic structures. In contrast, the first infrasonic measurements of harmonic tremor were made at Sakurajima in 1996 (Sakai et al., 1996). In the past decade, because of an increase in the quantity and quality of acoustic measurements, numerous observations of infrasonic harmonic tremor have been reported from volcanoes worldwide (Section 16.3). This section focuses on resonance effects in volcanoes. We examine several processes in a volcanic fluid system that could result in sharply defined spectral peaks in acoustic data. In particular, we discuss acoustic resonance or reverberation (standing waves), volume (Helmholtz) resonance, and flow-induced oscillations. We note that repetitive evenly spaced impulsive events (e.g., Powell and Neuberg, 2003) may also produce harmonic spectra without resonance. The combination of repetitive pulses and resonance was considered by Garcés and McNutt (1997).

16.6.1 Acoustic resonance

Acoustic resonance consists of constructively interfering echoes, and is the natural response of bounded fluids. Acoustic resonance is a familiar concept widely exploited in the design of musical instruments to produce harmonious tones (e.g., Fletcher and Rossing, 1998). Solid pipe sections of intricate geometry containing air columns (e.g., an organ pipe) produce sound with predictable spectra. By varying the effective length of a resonator, different natural sound frequencies can be produced, which are perceived as changes in pitch. The sound radiated by an instrument is a complex superposition of the different resonant oscillations of the instrument body. Unfortunately, in volcanic conduits and cracks, the fluid composition in the resonator can be highly variable and largely unknown. Consider a one-dimensional oscillator such as the classic organ pipe modes, with eigenfrequencies given by:

$$f_m = \frac{\left(m-\frac{1}{2}\right)c}{2L}, \quad \text{where } m = 1, 2, \dots \quad (16.24a)$$

or

$$f_m = \frac{mc}{2L}, \quad \text{where } m = 1, 2, 3 \dots \quad (16.24b)$$

where f_m are the resonant modes, c is the sound speed in the resonator, L is the effective length of the resonator, m is the mode number, and the form of equation depends on the whether the pipe is closed and open at one end (Eq. 16.24a) or open at both ends (Eq. 16.24b). The first mode, $m = 1$, is generally dominant and is known as the fundamental, with all higher modes called overtones. Whereas in musical instruments the resonating fluid is air and c is well known, in a volcano the fluid may be magma, gas, a dusty ash–gas mixture, juvenile or meteoric water, mud, or a multiphase combination of the above. The *in situ* acoustic properties of these volcanic fluids are critical parameters defining the resonant properties of volcanic conduits and cracks, yet these properties can only be crudely estimated from geophysical and geochemical methods. Nevertheless, a nondimensional approach enables inferences about volcanic resonators that are not tied to fixed values of key geophysical properties of the fluid.

Buckingham and Garcés (1996) developed a canonical model for the atmospheric sound field generated by a resonant magma conduit excited by an explosive compact source. To obtain a tractable analytic solution, a number of simplifying assumptions were made. A repetition of this analytic solution is beyond the scope of this chapter. However, we remark that the symmetric radial modes of a cylindrical conduit (in addition to the longitudinal pipe modes of Eqs. (16.24a) and (16.24b)) are given by:

$$f_{s,m} = \sqrt{\left(\frac{cj_{1s}}{2\pi a}\right)^2 + f_m^2}, \quad (16.25)$$

where $f_{s,m}$ is the radial resonance associated with the sth zero j_{1s} of the Bessel function J_1 and the mth longitudinal mode, c is the sound speed in the magma, a is the conduit radius, and L is the conduit length. In general, two- and three-dimensional standing wave patterns are produced inside confined volumes, leading to complex and possibly degenerate mode sequences that depend on the geometry and fluid properties of the resonator. One of the important results of Buckingham and Garcés (1996) is that the higher-order modes, which correspond to higher frequencies, radiate nearly vertically and are not efficiently propagated into the far-field. The predicted radiation pattern is reminiscent of that of a piston in a baffle, with monopole radiation dominant when the wavelength is larger than the vent aperture, and a strong vertically directed radiation pattern at high frequencies. This helps to explain why explosive sources in vents and craters rapidly lose audible frequencies with distance. Furthermore, liquid–gas interfaces are inefficient at transmitting sound, and act as good reflecting boundaries. However, sources submerged within magmatic fluids may also radiate more efficiently than anticipated because of the recently postulated anomalous transparency of infrasound at the liquid–air interface (Godin, 2006), which to date has not been incorporated into volcano-acoustic models.

The formulation of Garcés (2000) addresses the resonant properties of a variable-width tube of fluid that may be moving at high velocity relative to the sound speed of the flow. To a first-order approximation, the spacing between frequency peaks is given by:

$$\Delta f = \frac{c(1-M^2)}{2L}, \quad (16.26)$$

where c is the sound speed of the material in the conduit, $M = U/c$ is the Mach number of the flow, and L is the effective length of the conduit. Hence, the eigenfrequency variations known as *gliding* (Garcés and Le Pichon, 2009) could be explained by either a change in the effective length of the conduit L, or changes in the flow velocity U or sound speed c. Both sound speed and flow velocity vary readily with a change in the mean vesicularity of a magma–gas mixture.

16.6.2 Helmholtz resonance

For acoustic wavelengths greater than any of the linear dimensions of a volume, Helmholtz resonance can also be induced. A Helmholtz resonator is a stiff-walled cavity connected to the atmosphere through an opening with a neck. The cavity and opening would represent a vented volcanic chamber. When fluid is pushed out of the cavity, low pressure is created within the cavity, which responds by pulling the air back in, and vice versa. The system may then sustain oscillations at a frequency given by:

$$f_H = \frac{c}{2\pi}\sqrt{\frac{S_a}{L_H V}}, \qquad (16.27)$$

where S_a is the cross-sectional area of the neck, L_H is the effective length of the neck, c is the sound speed of the fluid, and V is the volume of the cavity. If there is no neck, such as in the case of a vent within a thin lava roof, the effective neck length l_H may be approximated by $1.7a$, where a is the radius of the neck opening with area $S_a = \pi a^2$ (Kinsler et al., 1982).

The dimensional requisites for Helmholtz resonance are compatible with the observed geometry of some near-surface volcanic cavities (Fee et al., 2010a). A given cavity volume can sustain a lower fundamental oscillation frequency via Helmholtz resonance than it could produce by the longitudinal modes in Eq. (16.24).

16.6.3 Echo chambers

Although a Helmholtz resonator is predicted to oscillate at a single low frequency, resonance peaks at higher frequencies are possible within a cavity in the form of three-dimensional standing acoustic waves analogous to the one-dimensional pipe modes described in Section 16.6.1. The simplest three-dimensional cavity is a rectangular volume, which would add reverberation to the sound source by acting as an echo chamber. In

rectangular coordinates, the natural oscillation frequencies of a rectangular stiff-walled chamber are given by:

$$f_{lmn} = \frac{c}{2\pi}\sqrt{\left(\frac{l\pi}{L_x}\right)^2 + \left(\frac{m\pi}{L_y}\right)^2 + \left(\frac{n\pi}{L_z}\right)^2}, \qquad (16.28)$$

where l, m, $n = 0, 1, 2, \ldots$ are the mode numbers, and L_x, L_y, and L_z are the x, y, and z longitudinal dimensions of the chamber. The eigenfrequencies in Eq. (16.28) can be considered as traveling plane waves in their respective directions (Kinsler et al., 1982).

We can rewrite Eq. (16.28) as:

$$\left(\frac{2L_z}{c}\right)f_{lmn} = \frac{f_{lmn}}{f_{1z}} = \sqrt{\left(\frac{L_z}{L_x}\right)^2 l^2 + \left(\frac{L_z}{L_y}\right)^2 m^2 + n^2}. \qquad (16.29)$$

If the vertical length L_z and the volume V of the chamber are constrained, we can specify a cross-sectional area $A = L_x L_y$ and leave only one free variable L_x

$$\frac{f_{lmn}}{f_{1z}} = \sqrt{\left(\frac{L_z}{L_x}\right)^2 l^2 + \left(\frac{L_z L_x}{A}\right)^2 m^2 + n^2}, \qquad (16.30)$$

and now it is possible to plot the normalized eigenfrequency as a function of only one dimension, L_x. Fee et al. (2010a) performed such an eigenfrequency analysis assuming Helmholtz resonance and using LIDAR measurements to constrain the volume and depth of Halemaumau's gas-filled chamber (0.5 Hz peak in Fig. 16.1(b)), and obtained a reasonably good match between theory and observation. Decades of architectural acoustic studies attest to the difficulties of deriving precise mode amplitude solutions for rooms with complex shapes. Volcanic chambers would certainly deviate from a perfect rectangular volume, and the observed mode amplitudes would depend on the source position, the walls' exact shape, texture, and sound absorption properties, and the radiation condition at the vent. However, all these effects are likely to be more aggravated as the frequency increases and the acoustic wavelength becomes comparable to the spatial scales of the irregularities in the chamber. The good match between theory and observation at low frequencies obtained by Fee *et al.* (2010a) is encouraging.

16.6.4 Flow-induced oscillations

In previous sections, we discussed how a volcanic cavity may reverberate to produce the sharply peaked tones observed in infrasonic harmonic tremor (Section 16.3). In addition, in Section 16.4 we introduced various means of exciting resonators (the $s(t)$ function in Eq. (16.12)). Thus we have treated a resonant volcanic system as separable, in which an impulsive or sustained broadband driving function can excite a cavity into producing transient resonant signals (LPs) or harmonic tremor. However, flow-induced oscillations in a volume also provide nonlinear mechanisms that can yield lower natural frequencies than predicted from purely acoustic means.

In musical instruments, the resonant properties of the pressure or force-function driving cavity resonance can be critical. For example, flute tones are a result of flow-induced *edge tones* generated by the jet of air from the player's lips onto an edge in the mouthpiece, and the amplification of these tones via the resonator body (the pipe section) (Fletcher and Rossing, 1998). When examined in more detail, the relationship between the driving mechanism and resonator response is nonlinear, and involves a complex feedback process between the acoustic modes of the resonator body and the air jet exciting the modes. This process leads to *mode-locking* in which particular tones are preferentially amplified (Fletcher, 1999). Such complex acoustic feedback processes may provide an explanation for infrasonic harmonic tremor (Matoza *et al.*, 2010).

The edge tone is just one example of a family of processes in which tones are created by the interaction of shear layers with solid boundaries (Rockwell and Naudascher, 1979). Other examples include the *hole-tone*, which is produced by flow of an axisymmetric jet from one plate impinging on a second plate with a hole in it, and the flow of air over a cavity (Rossiter, 1964). Each is believed to result from a similar feedback process. Once the flow encounters the solid object downstream, a disturbance is created that can propagate back upstream either as an acoustic or hydrodynamic disturbance.

This disturbance can interact with the shear layer upstream. Shear layers are very sensitive to minor changes in the pressure conditions around them, such that tiny acoustic pressure oscillations resulting from interaction with the solid object can result in the generation of vortices in the shear layer. These vortices are then carried back downstream in the flow where they again reach the solid object. The interaction between the upstream propagating acoustic or hydrodynamic disturbance and the downstream propagating vortices results in a closed feedback loop, which is strongest at particular frequencies. When examining flow-induced oscillations, frequencies are usually non-dimensionalized in the tonal Strouhal number, given by:

$$St = \frac{fL}{U},$$ (16.31)

where L is the length scale of the process (m), and U is the jet flow velocity (m s^{-1}). Rossiter (1964) found the Strouhal numbers of these processes agree with the empirical equation:

$$St = \frac{m - \gamma}{M + \dfrac{1}{K}},$$ (16.32)

where m is the mode number ($m = 1, 2, 3, \ldots$), $M = U/c$ is the Mach number, and γ and K are empirical constants. Rossiter (1964) then deduced that this can be expressed in terms of the physical parameters of the system as:

$$\frac{L}{U_c} + \frac{L}{c} = \frac{(m - \gamma)}{f_m},$$ (16.33)

which is known as Rossiter's equation (Howe, 1998). Here, U_c is the velocity of propagation of the vortices, c is the sound speed, f_m are the *Rossiter modes* and γ represents a phase lag. The constant K in Eq. (16.32) is the ratio of U_c to U, i.e., $U_c = KU$. It remains an empirical constant but is $\sim$0.4–0.6 for most processes (Rossiter, 1964).

The coupling of Rossiter modes of subsonic volcanic jet flows to volcanic resonator bodies such as conduits, cracks, and near-surface cavities remains an open and exciting area for future study.

16.6.5 Release

Resonance in bounded fluids produces spectral peaks whose frequencies can be scaled with ratios of the sound speed, longitudinal dimensions (normal modes), volume (Helmholtz), and flow velocity (jets). Thus five generalized parameters – c, L_x, L_y, L_z, and U – can be combined to interpret many of the tonal features that appear in acoustic signatures of volcanic origin.

16.7 | Coda

Volcanic sounds are created by manifold fluid flow instabilities, yielding often dissonant, occasionally harmonious, but invariably intriguing soundscapes. Sound arises from unsteady perturbations within the flux of volcanic fluids, and travels through conduits, chambers, vents, craters, and weather to reach our digital sentinels afield. The concepts presented in this chapter are intended as a foundation for the quantification, interpretation, and modeling of a select set of volcano acoustic signals. However, this chapter is by no means comprehensive. Each volcano has a unique voice, and one of the joys of acoustic studies is in discovering that voice and unraveling its primitive language.

16.8 | Notation

a	conduit, vent or jet radius (m)
A	cross-sectional area (m²)
c	sound speed (m s⁻¹)
d_s	fluid region length scale (m)
E	acoustic energy (J)
f	frequency (Hz)
f_H	Helmholtz resonant frequency (Hz)
f_m	resonant modes (Hz)
F_p	Fourier transform
g	Green's function
H	angular directivity function
H_p	piston directivity function
I	sound intensity (W m⁻²)
J_1	first-order Bessel function
k	wavenumber (rad m⁻¹)
K	empirical constant
l	local resonance function
L	effective length of resonator (m)
L_d	sound exposure level (Pa² s)
L_H	effective length of Helmholtz resonator neck (m)
$L_{x,y,z}$	longitudinal chamber dimensions (m)
m	mode number
$m(t)$	mass outflow rate (kg s⁻¹)
M	mach number
p	acoustic pressure (Pa)
p_{ax}	range-dependent pressure function (Pa)
p_{max}	peak pressure (Pa)
p_{ref}	reference pressure (Pa)
p_{rms}	rms pressure (Pa)
p_s	acoustic source pressure (Pa)
P	instantaneous pressure (Pa)
P_0	equilibrium pressure (Pa)
ΔP_l	lithostatic ovepressure (Pa)
P_m	monopole source pressure (Pa)
PSD	power spectral density (Pa² Hz⁻¹)
r	radial (spherical) distance (m)
R	horizontal range (m)
s	source time function
S	spectral level (dB)
S_a	area of Helmholtz resonator neck (m²)
SPL	sound pressure level (dB)
SL	source level (dB)
St	Strouhal number
t	time (s)
T	period (s)
T_s	time interval (s)
TL	transmission loss (dB)
u	particle velocity (m s⁻¹)
u_0	steady flow speed (m s⁻¹)
u_s	unsteady flow speed (m s⁻¹)
U	flow velocity (m s⁻¹)
U_c	vortex propagation velocity (m s⁻¹)
V	displaced volume, or volume of cavity (m³)
Δz	depth (m)
γ	empirical constant
θ	source-receiver angle (°)
λ	wavelength (m)
Π	acoustic power (W)
ρ	atmospheric density (kg m⁻³)
ρ_0	fluid density (kg m⁻³)
ρ_r	rock density (kg m⁻³)
ϕ	phase (rad.)
Ω	solid angle (sr)

Acknowledgments

This material is based upon work supported by National Science Foundation Grant No. 0609669, *Integrated Studies of Eruption Processes in Basaltic and Andesitic Volcanoes*, and National Oceanic and Atmospheric Administration Subcontract 09-09022, *Remote Infrasonic Monitoring of Natural Hazards*.

References

ANSI S1.13-2005 (2005) American National Standard, Measurement of Sound Pressure Levels in Air. Acoustical Society of America.

ANSI S12.7-1986 (1986) American National Standard, Methods for Measurement of Impulse Noise. Acoustical Society of America.

ANSI S2.20-1983 (1983) American National Standard for Estimating Air Blast Characteristics for Single Point Explosions in Air, with a Guide to Evaluation of Atmospheric Propagation Effects. Acoustical Society of America.

Aki, K., Fehler, M. and Das, S. (1977). Source mechanism of volcanic tremor: Fluid-driven crack models and their application to the 1963 Kilauea eruption. *Journal of Volcanology and Geothermal Research*, **2**, 259–287.

Bedard, A. (1971). Seismic response of infrasonic microphones. *Journal of Research of the National Bureau of Standards – C. Engineering and Instrumentation*, **75C**(1), 41–45.

Birk, A. M. and Cunningham, M. H. (1996). Liquid temperature stratification and its effect on BLEVEs and their hazards. *Journal of Hazardous Materials*, **48**, 219–237.

Bowman, R. J., Baker, G. E. and Bahavar, M. (2004). Infrasound station ambient noise estimates. In *Proceedings of the 26th Seismic Research Review: Trends in Nuclear Explosion Monitoring*, LA-UR-04-5801, **1**, pp. 608–617.

Brown, D. and Garcés, M. (2008). Ray tracing in an inhomogeneous atmosphere with winds. In *Handbook of Signal Processing in Acoustics*, ed. D. Havelock, S. Kuwano and M. Vorlander. New York: Springer, pp. 1437–1460.

Buckingham, M. J. and Garcés, M. A. (1996). Canonical model of volcano acoustics. *Journal of Geophysical Research*, **101**(4), 8129–8151.

Cansi Y. and Le Pichon A. (2008). Infrasound event detection using the progressive multi-channel correlation algorithm. In *Handbook of Signal Processing in Acoustics*, ed. D. Havelock, S. Kuwano and M. Vorlander. New York: Springer, pp. 1425–1435.

Chouet, B. A. (1996). New methods and future trends in seismological volcano monitoring. In *Monitoring and Mitigation of Volcano Hazards*, ed. R. Scarpa and R. I. Tilling. New York: Springer-Verlag, pp. 23–97.

Cooper, P. and Kurowski, S. (1996). *Introduction to the Technology of Explosives*. New York: Wiley.

Fee, D. and Garcés, M. (2007). Infrasonic tremor in the diffraction zone. *Geophysical Research Letters*, **34**, L16826, doi:10.1029/2007GL030616.

Fee, D., Garcés, M., Patrick, M. *et al.* (2010a). Infrasonic harmonic tremor and degassing bursts from Halema'uma'u Crater, Kilauea. *Journal of Geophysical Research*, **115**, B11316, doi:10.1029/2010JB007642.

Fee, D., Garcés, M. and Steffke, A. (2010b). Infrasound from Tungurahua Volcano 2006–2008: Strombolian to Plinian eruptive activity. *Journal of Volcanology and Geothermal Research*, **193**(1–2), 67–81.

Fee, D., Steffke, A. and Garcés, M. (2010c). Characterization of the 2008 Kasatochi and Okmok eruptions using remote infrasound arrays. *Journal of Geophysical Research*, **115**, D00L10, doi:10.1029/2009JD013621.

Fee, D., Garcés, M., Orr, T. R. and Poland, M. P. (2011). Infrasound from the 2007 fissure eruptions of Kilauea Volcano, Hawai'i. *Geophysical Research Letters*, **38**, L06309, doi:10.1029/2010GL046422.

Fletcher, N. H. (1999). The nonlinear physics of musical instruments. *Reports on Progress in Physics*, **62**, 723–764.

Fletcher, N. H. and Rossing, T. D. (1998). *The Physics of Musical Instruments*. New York: Springer.

Garcés, M. (1995). *The Acoustics of Volcanic Explosions*. PhD Dissertation, University of California, San Diego.

Garcés, M. A. (1997). On the volcanic waveguide. *Journal of Geophysical Research*, **102**, 22 547–22 564.

Garcés, M. A. (2000). Theory of acoustic propagation in a multi-phase stratified liquid flowing within an elastic-walled conduit of varying cross-sectional area. *Journal of Volcanology and Geothermal Research*, **101**, 1–17.

Garcés, M.A. and McNutt, S. R. (1997). Theory of the airborne sound field generated in a resonant magma conduit. *Journal of Volcanology and Geothermal Research*, **789**, 155–178.

Garcés, M. and Le Pichon, A. (2009). Infrasound from earthquakes, tsunamis and volcanoes. In

Encyclopedia of Complexity and Systems Science, ed. R. A. Meyers. New York: Springer.

Garcés, M. A., Hagerty, M. T. and Schwartz, S. Y. (1998). Magma acoustics and time-varying melt properties at Arenal Volcano, Costa Rica. *Geophysical Research Letters*, **25**(13), 2293–2296.

Garcés, M. A., McNutt, S. R., Hansen, R. A. and Eichelberger, J. C. (2000). Application of wave-theoretical seismoacoustic models to the interpretation of explosion and eruption tremor signals radiated by Pavlof Volcano, Alaska. *Journal of Geophysical Research*, **105**, 3039–3058.

Garcés, M., Harris, A., Hetzer, C. *et al.* (2003). Infrasonic tremor observed at Kilauea Volcano, Hawai'i. *Geophysical Research Letters*, **30**, doi:10.1029/2003GL018038.

Garcés, M., Bass, H., Drob, D. *et al.* (2004). Forensic studies of infrasound from massive hypersonic sources. *Eos, Transactions of the American Geophysical Union*, **85**(43), 443.

Garcés, M., Fee, D., Steffke, A. *et al.* (2008). Capturing the acoustic fingerprint of stratospheric ash injection. *Eos, Transactions of the American Geophysical Union*, **89**(40), 377.

Godin, O. A. (2006). Anomalous transparency of water-air interface for low-frequency sound. *Physical Review Letters*, **97**, 164301.

Goerke, V. H., Young, J. M. and Cook, R. K. (1965). Infrasonic observations of the May 16, 1963, volcanic explosion on the island of Bali. *Journal of Geophysical Research*, **70**(24), 6017–6022.

Gossard, E. E. and Hooke, W. H. (1975). *Waves in the Atmosphere: Atmospheric Infrasound and Gravity Waves – Their Generation and Propagation*. Elsevier.

Green, D. N. and Neuberg, J. (2005). Seismic and infrasonic signals associated with an unusual collapse event at the Soufriere Hills volcano, Montserrat. *Geophysical Research Letters*, **32**, L07308, doi:10.1029/2004GL022265.

Harris, A. and Ripepe, M. (2007). Synergy of multiple geophysical approaches to unravel explosive eruption conduit and source dynamics: A case study from Stromboli. *Chemie Der Erde*, **67**(1), 1–35.

Hedlin, M. A. H. and Raspet, R. (2003). Infrasonic wind-noise reduction by barriers and spatial filters. *Journal of the Acoustical Society of America*, **114**(3), 1379–1386.

Howe, M. S. (1998). *Acoustics of Fluid–Structure Interactions*. Cambridge University Press.

Ichihara, M., Ripepe, M., Goto, A. *et al.* (2009). Airwaves generated by an underwater explosion: Implications for volcanic infrasound.

Journal of Geophysical Research, **114**, B03210, doi:10.1029/2008JB005792.

Johnson, J. B. (2003). Generation and propagation of infrasonic airwaves from volcanic explosions. *Journal of Volcanology and Geothermal Research*, **121**, 1–14, doi:10.1016/ S0377-0273(02)00408-0.

Johnson, J. B. and Lees, J. M. (2000). Plugs and chugs; seismic and acoustic observations of degassing explosions at Karymsky, Russia and Sangay, Ecuador. *Journal of Volcanology and Geothermal Research*, **101**, 67–82.

Johnson, J. B. and Ripepe, M. (2011). Volcano infrasound: A review. *Journal of Volcanology and Geothermal Research*, **206**, 61–69.

Johnson, J. B., Aster, R. C. and Kyle, P. R. (2004). Volcanic eruptions observed with infrasound. *Geophysical Research Letters*, **31**, L14604, doi:10.1029/2004GL020020.

Johnson, J., Aster, R., Jones, K. R., Kyle, P. and Mcintosh, B. (2008). Acoustic source characterization of impulsive Strombolian eruptions from the Mount Erebus lava lake. *Journal of Volcanology and Geothermal Research*, **177**(3), 673–686.

Kinsler, L. E., Frey, A. R., Coppens, A. B. and Sanders, J. V. (1982). *Fundamentals of Acoustics*, 3rd edn. Wiley.

Le Pichon, A., Blanc, E. and Hauchecorne, A. (2010). *Infrasound Monitoring for Atmospheric Studies*. New York: Springer.

Leventhall, H. G., Benton, S. and Pelmear, P. (2003). *A Review of Published Research on Low Frequency Noise and its Effects*. Report for Department of Environment, Food, and Rural Affairs, UK.

Lighthill, M. J. (1954). On sound generated aerodynamically. II. Turbulence as a source of sound. *Proceeedings of the Royal Society of London*, **A 222**, 1–32.

Lighthill, M. J. (1978). *Waves in Fluids*. Cambridge University Press.

Marchetti, E., Ripepe, M., Harris, A. J. L. and Delle Donne, D. (2009). Tracing the differences between Vulcanian and Strombolian explosions using infrasonic and thermal radiation energy. *Earth and Planetary Science Letters*, **279**(3–4), 273–281.

Matoza, R. S., Garcés, M. A., Chouet, B. A. *et al.* (2009a). The source of infrasound associated with long-period events at Mount St. Helens. *Journal of Geophysical Research*, **114**, B04305, doi:10.1029/2008JB006128.

Matoza, R. S., Fee, D., Garcés, M. A. *et al.* (2009b). Infrasonic jet noise from volcanic eruptions.

Geophysical Research Letters, **36**, L08303, doi:10.1029/2008GL036486.

Matoza, R. S., Fee, D. and Garcés, M. (2010). Infrasonic tremor wavefield of the Pu'u O'o crater complex and lava tube system, Hawaii, in April 2007. *Journal of Geophysical Research*, **115**, B12312, doi:10.1029/2009JB007192.

Moran, S. C., Matoza, R. S., Garcés, M. A. *et al.* (2008a). Seismic and acoustic recordings of an unusually large rockfall at Mount St. Helens, Washington. *Geophysical Research Letters*, **35**, L19302, doi:10.1029/2008GL035176.

Moran, S. C., McChesney, P. J. and Lockhart, A. B. (2008b). Seismicity and infrasound associated with explosions at Mount St. Helens, 2004–2005. In *A Volcano Rekindled: The First Year of Renewed Eruptions at Mount St. Helens, 2004–2006*, ed. D. R. Sherrod, W. E. Scott and P. H. Stauffer. U.S. Geological Survey Professional Paper 1750, pp. 111–127.

Morrissey, M. M. and Mastin, L. G. (2000). Vulcanian eruptions. In *Encyclopedia of Volcanoes*, ed. H. Sigurdsson. San Diego, CA: Academic Press, pp. 463–475.

Morse, P. M. and Ingard, K. U. (1968). *Theoretical Acoustics*. Princeton, NJ: Princeton University Press.

Olson, J. V. and Szuberla, C. A. L. (2009). Processing infrasonic array data. In *Handbook of Signal Processing in Acoustics*, ed. D. Havelock, S. Kuwano and M. Vorlander. New York: Springer, pp. 1487–1496.

Oppenheim, A. V. and Schafer, R. W. (1989). *Discrete-Time Signal Processing*. Prentice Hall.

Oshima, H. and Maekawa, T. (2001). Excitation process of infrasonic waves associated with Merapi-type pyroclastic flow as revealed by a new recording system. *Geophysical Research Letters*, **28**(6), 1099–1102.

Parfitt, E. and Wilson, L. (2008). *Fundamentals of Physical Volcanology*. Wiley-Blackwell.

Patrick, M. R., Wilson, D., Fee, D., Orr, T. and Swanson, D. (2011). Shallow degassing events as a trigger for very-long-period seismicity at Kilauea Volcano, Hawaii. *Bulletin of Volcanology*, doi:10.1007/s00445-011-0475-y.

Petersen, T., De Angelis, S., Tytgat, G. and McNutt, S. R. (2006). Local infrasound observations of large ash explosions at Augustine Volcano, Alaska, during January 11–28, 2006. *Geophysical Research Letters*, **33**, L12303, doi:10.1029/2006GL026491.

Pierce, A. D. (1981). *Acoustics: An Introduction to its Physical Principles and Applications*. New York: McGraw-Hill.

Ponceau, D. and Bosca, L. (2010). Low-noise broadband microbarometers In *Infrasound Monitoring for Atmospheric Studies*, ed. A. Le Pichon. Netherlands: Springer, doi:10.1007/978-1-4020-9508-5_24, pp. 119–140.

Powell, T. W. and Neuberg, J. (2003). Time dependent features in tremor spectra. *Journal of Volcanology and Geothermal Research*, **128**(1–3), 177–185.

Richards, A. F. (1963). Volcanic sounds, investigation and analysis. *Journal of Geophysical Research*, **68**, 919–928.

Ripepe, M. and Marchetti, E. (2002). Array tracking of infrasonic sources at Stromboli volcano. *Geophysical Research Letters*, **29**(22), 2076, doi:10.1029/2002GL015452.

Ripepe, M., De Angelis, S., Lacanna, G. and Voight, B. (2010). Observation of infrasonic and gravity waves at Soufrière Hills Volcano, Montserrat. *Geophysical Research Letters*, **37**, L00E14, doi:10.1029/2010GL042557.

Rockwell, D. and Naudascher, E. (1979). Self-sustained oscillations of impinging free shear layers. *Annual Review of Fluid Mechanics*, **11**, 67–94.

Rossiter, J. E. (1964). Wind-tunnel experiments on the flow over rectangular cavities at subsonic and transonic speeds. *Aeronautical Research Council Reports and Memoranda 3438*. London: HMSO.

Sakai, T., Yamasato, H. and Uhira, K. (1996). Infrasound accompanying C-type tremor at Sakurajima volcano. *Bulletin of the Volcanological Society of Japan*, **41**, 181–185 (in Japanese).

Strachey, R. (1888). On the air waves and sounds caused by the eruption of Krakatoa in August, 1883. In *Krakatau 1883 (published 1983)*, ed. T. Simkin and R. S. Fiske. Smithsonian Institution Press, pp. 368–374.

Tahira, M., Nomura, M., Sawada, Y. and Kamo, K. (1996). Infrasonic and acoustic-gravity waves generated by the Mount Pinatubo eruption of June 15, 1991. In *Fire and Mud: Eruptions and Lahars of Mount Pinatubo, Philippines*, ed. C. G. Newhall and R. S. Punongbayan. Seattle, WA: University of Washington Press, pp. 601–614.

Vergniolle, S. and Brandeis, G. (1996). Strombolian explosions. 1. A large bubble breaking at the surface of a lava column as a source of sound. *Journal of Geophysical Research*, **101**(B9), 20 433–20 447.

Wood, A. B. (1964). *A Textbook of Sound*. London: Bell.

Woulff, G. and McGetchin, T. R. (1976). Acoustic noise from volcanoes: Theory and experiment.

Geophysical Journal of the Royal Astronomical Society, **45**, 601–616.

Yamasato, H. (1997). Quantitative analysis of pyroclastic flows using infrasonic and seismic data at Unzen volcano, Japan. *Journal of Physics of the Earth*, **45**(6), 397–416.

Yamasato, H. (1998). Nature of infrasonic pulse accompanying low frequency earthquake at Unzen Volcano, Japan. *Bulletin of the Volcanological Society of Japan*, **43**, 1–13.

Exercises

16.1 The mean value theorem for integrals states that, if g is continuous on $[a, b]$, there is a number c between a and b such that

$$\int_a^b g(f)df = g(c)(b-a).$$

Use the mean value theorem to derive the SPL between 1 and 2 Hz from the spectra shown in Figure 16.1, and compare to the results in Table 16.1.

16.2 Estimate the SPL for the Halemaumau Vent between 0.3 and 0.6 Hz, and the radiated source power in watts.

16.3 If a volcanic jet has an expanded diameter of 10 m and peak frequency of 1 Hz, what range of jet velocities would yield a Strouhal number range of 0.06–0.4?

Online resources available at www.cambridge.org/fagents

- Sound files and spectra for a variety of volcanic activity
- Answers to exercises

Chapter 17

Planetary volcanism

Rosaly M. C. Lopes, Sarah A. Fagents, Karl L. Mitchell, and Tracy K. P. Gregg

Overview

Volcanism is of primary importance in shaping the surfaces of many planets and satellites of the Solar System. In this chapter we show how models developed for volcanic processes on Earth can be adapted to model volcanism on other planetary bodies, including those displaying familiar silicate volcanism (such as Mars, Venus, and the Moon), as well as those with more exotic volcanic behavior (such as high-temperature volcanism on Io and "cryo-volcanism" on the icy satellites). Due to space limitations, only certain "type example" worlds are detailed here, the intent is more to give an insight into how the volcanic process varies from body to body than to discuss each. Each planet or satellite possesses a unique combination of environmental factors (gravity, atmospheric properties, surface temperature, etc.) that influence almost every aspect of magma ascent and eruption. By incorporating these parameters into models of volcanic behavior it is possible to elucidate the causes of the diversity in volcanic expression on the surfaces of other planetary bodies and hence understand the eruptive history and evolution of our Solar System neighbors.

17.1 Introduction

Volcanism has affected all solid planets and most moons in the Solar System and even some of the earliest-forming asteroids, and is therefore of key importance for the study of the evolution of planets and moons. The discovery of numerous extra-terrestrial volcanoes, including active ones, has stretched our traditional definition of "volcano" (Lopes et al., 2010a) and prompted a new understanding of how volcanism, as a process, can operate.

Prior to the Voyager 1 and 2 spacecraft observations during the late 1970s and early 1980s, the Earth was the only planet known to be volcanically active, with the Moon, Mars, Venus and possibly Mercury showing signs of past activity. Our views of volcanism were dramatically changed when Voyager 1 revealed active eruptions on Jupiter's moon Io (Morabito et al., 1979), a world about the same size as the Earth's Moon. Unlike the Earth, where volcanism is a manifestation of internal heat generated by decay of radioactive isotopes, Io's volcanism is driven by tidal dissipation due to orbital interactions with Jupiter and neighboring satellites. Moreover, no planet outside the Earth shows evidence of plate tectonics, which has consequences for the distribution of

Modeling Volcanic Processes: The Physics and Mathematics of Volcanism, eds. Sarah A. Fagents, Tracy K. P. Gregg, and Rosaly M. C. Lopes. Published by Cambridge University Press. © Cambridge University Press 2013.

volcanic landforms, and for magma chemistry (Section 17.2.4). Despite these differences, the eruption styles and products on other planets often show great similarity to Earth's. In other cases, such as *cryovolcanism* on the icy moons of outer planets, the differences are striking: instead of silicate magma, icy mixtures erupt in extremely low-temperature environments. The crusts of these moons are composed primarily of water-ice, and may host liquid water bodies in the subsurface. Therefore, ices are analogous to silicate rocks on Earth. On some of these worlds we see evidence for both active and past cryovolcanic eruptions, which are at times unexpectedly Earth-like in expression.

On the terrestrial planets (Mercury, Venus, Mars), as well as Earth's Moon and Jupiter's Io, where silicate volcanism has occurred or is currently occurring, the processes bringing material to the surface are similar to terrestrial processes. Therefore, models used to understand Earth's volcanism are generally applicable to these other bodies, although model parameters are different. Cryovolcanism requires significantly different models, in particular because of the difficulty of cryomagma ascent given the relative densities of water and ice. This chapter shows how models of volcanic processes, accounting for the different ambient conditions and magma compositions, can be used to understand volcanic behavior of planetary bodies, and uses Mars, Io, and Enceladus as examples of the diversity of volcanism in the Solar System. For consideration of environmental factors influencing volcanism on Venus, see Grosfils *et al.* (2000), Stofan and Smrekar (2005), and also the discussion of volcanism in the high-pressure sea-floor environment in Chapter 12. Recent results about volcanism on Mercury revealed by the MESSENGER spacecraft are discussed by Head *et al.* (2011) and recent lunar results from the LRO spacecraft are discussed by Joliff *et al.* (2011).

17.2 | Parametric differences among planetary bodies

Volcanic processes, from melt generation and ascent to the eruption and deposition of eruptive products, are affected by the ambient conditions on different bodies (Table 17.1). Of primary importance are the planetary gravity, atmospheric pressure and density, with ambient temperature a secondary effect. (For detailed reviews of how ambient conditions affect volcanism, see Zimbelman and Gregg, 2000.) In addition, magma compositional variations among the planets can strongly influence the expression of volcanism at the surface.

17.2.1 Gravity

In modeling terrestrial volcanic processes, it is easy to forget that gravitational acceleration is in fact a variable that depends on the mass of the individual planet (Table 17.1). In practice, this affects everything from the depth of melting in a planet's mantle to the height pyroclasts might reach in an explosive eruption. To illustrate, assuming that Earth's mantle is composed of "wet" lherzolite, this material will begin to melt at a depth of ~75 km. On Mars, a similar melting pressure is reached at a depth almost three times greater (~225 km), because of the low gravity. This has implications for the minimum ascent rate magma would require to reach the surface before solidifying, which in turn sets limits for minimum eruption rates (Wilson and Head, 1981).

Low gravity can also influence the dispersal of pyroclastic ejecta. On the Moon, for example, dark mantle deposits (DMDs) are interpreted to be pyroclasts emplaced during explosive eruptions (e.g., Wilson and Head, 1981; Weitz and Head, 1999). The most likely origin for these features is that they are pyroclasts generated from volatile-bearing magmas that would have generated hawaiian-style lava fountains (and associated scoria cones) if they had erupted on Earth. Instead, the lower lunar gravity, coupled with a lack of atmosphere, allowed the lava fragments to be dispersed over distances that precluded the building of a proximal volcanic cone.

It has long been theoretically argued that low ambient gravity should result in thicker lava flows, all other parameters being equal (e.g., Fink and Griffiths, 1990). Thus, basaltic lava flows on the Moon should be thicker than those on Mars; basaltic lava flows on Mars should be

Table 17.1 Characteristics of the main Solar System bodies on which volcanic features have been identified.

Object	Mean radius (km)	Gravity (m s^{-2})	Bulk density (kg m^{-3})	Surface temperature (K)	Atmospheric pressure (Pa)	Composition of eruptive products	Expressions of volcanism
Earth	6371	9.81	5515	288	1.0×10^5	silicate	Effusive and explosive
Moon	1737	1.6	3300	277	3×10^{-10}	silicate	Mostly effusive, some explosive
Mercury	2440	3.70	5427	100–725	5×10^{-10}	silicate	Mostly effusive, some explosive
Venus	6051	8.87	5204	733	9×10^6	silicate	Mostly effusive
Mars	3389	3.73	393	215	600	silicate	Mostly effusive, some explosive
Io	1822	1.80	3528	85 (night) to 140 (day) (110 ave.)	$< 10^{-4}$, higher at locations of plumes	silicate, some sulfur possible	Effusive and explosive
Europa	1565	1.31	3030	102	10^{-6}	H_2O, unknown other constituents	Effusive cryovolcanism, possibly some explosive
Ganymede	2634	1.43	1940	110	$< 10^{-6}$	H_2O, unknown other constituents	Effusive cryovolcanism
Enceladus	249	0.11	1120	75	trace, variable	H_2O, unknown other constituents, probably ammonia	Explosive (plumes)
Titan	2575	1.35	1881	94	1.5×10^5	H_2O, unknown other constituents, probably ammonia, methanol	Possibly both effusive and explosive cryovolcanism.
Triton	1353	0.78	2054	38	1.4–1.9	H_2O, unknown other constituents, probably nitrogen	Effusive cryovolcanism, explosive cryovolcanism as a result of solar greenhouse activity

thicker than those on Earth. However, observations do not always support these predictions, indicating that other factors (such as lava composition) play a much more important role in effusive eruptions than does gravity.

17.2.2 Atmospheric pressure and density

Atmospheric pressure varies from high (9 MPa on Venus) to essentially non-existent on planets without noticeable atmospheres (Table 17.1). For comparison, the typical atmospheric pressure at Earth's sea level is ~0.1 MPa. Energy released by decompressing volatiles during a volcanic eruption is a function of the change in pressure during decompression. High ambient pressures (such as on Venus or Earth's sea floor) will act to suppress volatile exsolution and decompression (Fig. 17.1), thereby favoring effusive rather than explosive eruptions. High ambient pressures do not preclude explosive eruptions, but do require greater volatile concentrations compared with terrestrial magmas. Therefore, a dominance of effusive over explosive eruptions on Venus may be predicted, and this prediction is supported by observations. Likewise, low ambient pressures do not guarantee explosive eruptions; only if decompression successfully drives expansion and fragmentation of a volatile-bearing magma will the explosive potential be realized.

Atmospheric density is determined by ambient pressure P_a and temperature T_a, and by both the composition and abundance of molecular compounds, such that $\rho_a = P_a/R^*T_a$, where R^* is the specific gas constant of the mixture of atmospheric gases (all notation is given in Section 17.7). Density, along with other thermophysical parameters, helps determine the rate at which a fluid will transfer heat via convection. All else being equal, a higher-density atmosphere will convect heat away from the surface of a lava flow more efficiently than a lower-density atmosphere. Thus, in the initial seconds of emplacement, the surface of a venusian flow would cool more rapidly than the surface of an identical lava erupted on Earth (Head and Wilson, 1986). In the absence of an atmosphere, no convective cooling occurs; volcanic products are cooled only by radiation and conduction and will remain hot for longer.

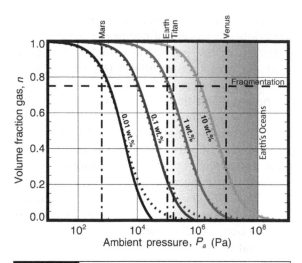

Figure 17.1 Volume fraction of exsolved gas after decompression to atmospheric pressure for magmas with different starting juvenile volatile contents, illustrating the potential influence of ambient pressure on eruption style. Note that only an ideal equilibrium system with magma and volatiles well coupled (closed system degassing), will decompress to this extent, and so the curves represent an upper limit. Fragmentation is assumed to occur at 75 vol.% gas. Approximate surface pressures are given for Mars, Earth, Titan, and Venus (vertical dashed lines), as well as a shaded range for eruptions at the bottom of Earth's oceans. Most worlds have only trace atmospheres. The magma-volatile compositional pairing is a second-order effect; to illustrate this, the solid curve in bold represents H_2O degassing from basalt and the dotted line is for CH_4 in water, based on Henry's Law approximations.

On Earth, eruption columns are driven by entrainment and heating of the surrounding air. Mars' atmosphere is less dense than Earth's and therefore there is less mass for eruption columns to entrain. The net result is that convective columns on Mars are unlikely to be as substantial as those on Earth (Section 17.3.4). Eruptions into vacuum conditions (e.g., Io) are unable to produce convecting columns at all, and instead produce umbrella-shaped plume structures controlled by particle trajectories in rarified gas conditions (17.4.4).

17.2.3 Temperature

Planetary surface temperatures vary by orders of magnitude throughout the Solar System: bodies farther from the Sun tend to be colder, and bodies without insulating atmospheres can experience hundreds of degrees in temperature variation in the course of a single revolution. Ambient temperature affects the rate of heat transfer from eruptive products to their surroundings. Heat flux due to radiation varies as $T_{hot}^4 - T_a^4$, while convective and conductive fluxes are both proportional to $T_{hot} - T_a$ (Chapter 5). In general, this implies that magmas erupted into cooler environments should cool more quickly, as might be expected; however, an exception to this rule is Venus, whose high atmospheric density implies effective convective heat transfer (Section 17.2.2), despite the high ambient temperature.

17.2.4 Magma compositions

The lack of plate tectonics on all rocky worlds other than Earth appears to have resulted in much less petrological diversity, as a result of less crustal recycling. For worlds where past or present silicate volcanism has been detected, mafic compositions (i.e., basalts) appear to dominate. Relative to more felsic (silica-rich) magmas, mafic magmas have higher densities, and lower viscosities and volatile contents. Taken together with the apparent relative paucity of water, the primary volatile in the most explosive volcanic eruptions on Earth, this suggests a tendency towards the more effusive end of the volcanic spectrum. There are, however, some exceptions.

High-temperature, low-viscosity volcanism on Io might be best explained by ultramafic compositions (Williams and Howell, 2007). Geochemical analysis of Mars shows some evidence for both ultramafic and more felsic magmas, including basaltic-andesitic, andesitic and possibly even dacitic surface chemistries (Section 17.3.2). It has been argued that weathering processes could produce high-silica surface compositions on Mars, but it seems likely, given the large inferred sizes of martian magma chambers, that chemical differentiation of magmatic bodies in the crust would produce some volume of intermediate- to high-silica magmas. On Earth, the most felsic magmas (e.g., rhyolites) are the result of plate tectonic processes and recycling of a water-rich crust. However,

high thorium concentrations of some volcanic areas on the Moon (e.g., Jolliff et al., 2011), suggest compositions that may approach rhyolitic. In this context, rhyolitic compositions suggested for Venus (Fink et al., 1993; Treiman, 2007), to explain the physical properties of both tesserae and pancake domes, may not be so far-fetched, and may have been facilitated by crustal recycling due to catastrophic resurfacing of the crust, or a past wetter environment.

There is much less known about the chemistry of cryomagmas. For most icy bodies, cryomagmas are inferred to have low-viscosity, water-rich compositions. Given the lack of atmospheres in these cases, explosive activity remains possible. An exception may be Titan, where the possibility of large abundances of liquidus-deflating ammonia in the mantle may produce rheologies similar to those of terrestrial basalts or andesites (Kargel et al., 1991). Although such properties may result in larger volatile concentrations than on other icy satellites, the presence of a thick 1.5-bar atmosphere (0.15 MPa; Table 17.1) may act to suppress explosivity.

17.3 | Volcanism on Mars

17.3.1 Observations of volcanic features on Mars

Mars has been the main focus of planetary exploration since the 1990s and, therefore, the data set available for studying its geologic processes is vast, ranging from orbital observations to in situ measurements from landers and rovers. The martian shield volcanoes are the largest in the Solar System, and among the wide variety of volcanic features are vast lava plains, many channels, shields, domes, and cones, and evidence of extensive pyroclastic deposits (Figs. 17.2, 17.3; Greeley and Spudis, 1981; Greeley et al., 2000; Carr, 2006). The presence of a frozen cryosphere, 3–5 km thick, trapping water in deep aquifers, is thought to have promoted magma–ice and magma–water interactions, and some surface features may have resulted from these interactions (Fig. 17.3(f); Fagents et al., 2002; Wilson and Mouginis-Mark, 2003a,b; Wilson and Head, 2004).

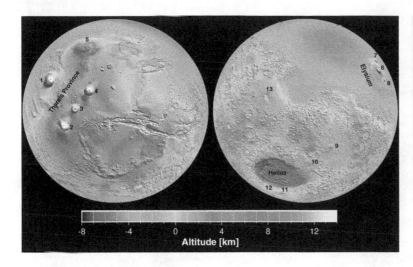

Figure 17.2 False color view of Mars topography with locations of key volcanic features labeled: 1 Olympus Mons; 2 Arsia Mons; 3 Pavonis Mons; 4 Ascraeus Mons; 5 Alba Mons; 6 Elysium Mons; 7 Hecatus Tholus; 8 Albor Tholus; 9 Tyrrhenus Mons; 10 Hadriacus Mons; 11 Amphitrites Patera; 12 Peneus Patera; 13 Nili Patera. Image: NASA/Goddard Space Flight Center. See color plates section.

The most prominent volcanoes on Mars are the giant shield volcanoes located in the Tharsis volcanic province, a vast plateau ~8000 km in diameter that covers ~25% of the planet's surface (Fig. 17.2). Within this region, Olympus Mons, the tallest volcano in the Solar System, is > 500 km wide and its summit reaches 25 km above mean planetary radius, with maximum flank slopes of 5°. It is dominated by tube- and channel-fed lava flows of likely basaltic composition (Bleacher *et al.*, 2007), with a series of nested summit calderas > 60 km across, and evidence of past glacial ice (Neukum *et al.*, 2004). Tharsis is also home to three smaller shield volcanoes – Arsia, Pavonis, and Ascraeus Montes (Fig. 17.3(a)) – and their associated rift aprons and small shield fields (Fig. 17.3(e); Crumpler and Aubele, 1978; Hauber *et al.*, 2009). The region also has vast interconnecting lava plains (Fig.17.3(d)), and several smaller (< 200 km diameter) volcanoes named *tholi* (sing.: *tholus*, Latin for "cupola" or "dome") which may be shields whose lower slopes were buried by lava (Greeley and Spudis, 1981).

A unique volcanic structure in north Tharsis is Alba Mons (Fig. 17.2), which covers an area larger than Olympus Mons, but has flank slopes of less than 1° and thus lacks the relief of shield volcanoes. The flanks are heavily fractured, indicating the influence of regional stress patterns. Alba Mons contains two discrete caldera-like features, numerous channelized, tube-fed and sheet lava flows, and deposits that have been interpreted as pyroclastic (Mouginis-Mark *et al.*, 1988). Interestingly, Alba Mons is antipodal to the Hellas impact basin, and early seismic wave propagation modeling suggested that the energy of the Hellas impact was sufficient to break the antipodal crust, creating fractures that could have acted as volcanic conduits that enabled the formation of Alba Mons at its present location (Williams and Greeley, 1994).

The second major volcanic area is the Elysium Province (Fig. 17.2), which is dominated by Elysium Mons, Hecates Tholus, and Albor Tholus, and their surrounding lava flow fields. Hecates Tholus (Fig. 17.3(b)) is ~160 × 175 km wide and extends ~6 km above mean planetary radius, and its flanks are dissected by shallow radial valleys thought to be formed fluvially as a result of melting of ice by subsurface magma bodies (Fassett and Head, 2006). The dissected nature of Hecates Tholus suggests it may be composed of ash or other easily eroded pyroclastic deposits (Mouginis-Mark *et al.*, 1982), as opposed to Elysium Mons, which appears to be composed of lava flows.

The third major volcanic area on Mars is the Circum-Hellas Volcanic Province, which is dominated by Tyrrhenus and Hadriacus Montes, and Amphitrites Patera, and their associated flow fields, surrounding the Hellas impact basin in the southern hemisphere (Fig. 17.2; Williams *et al.*, 2009). The heavily channeled and

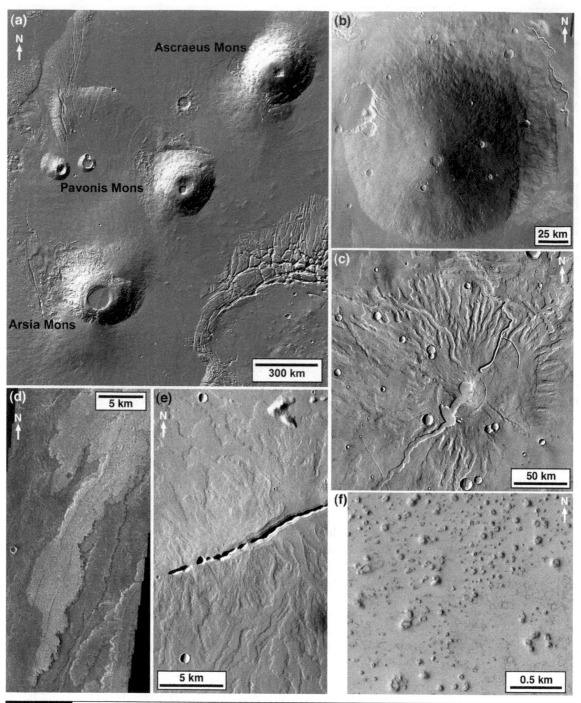

Figure 17.3 Volcanic landforms on Mars. (a) Three of the major shield volcanoes in the Tharsis volcanic Province: Arsia, Pavonis, and Ascraeus Montes. To the west of Pavonis are the smaller volcanoes Ulysses and Biblis Tholi (image credit: NASA). (b) Hecates Tholus in the Elysium Province shows significant erosional modification in the form of channeling of the volcano's flanks. Mosaic of THEMIS visible images: NASA/JPL/ASU. (c) Tyrrhenus Mons in the Circum-Hellas region is a much older, highly dissected edifice. Mosaic of THEMIS visible images: NASA/JPL/ASU. (d) Multiple large overlapping lava flows in Daedalia Planum, south of Arsia Mons. Image credit: NASA/JPL/ASU. (e) Close-up view of an eruptive fissure and numerous small flows, located in the plains northwest of Ascraeus Mons. Portion of THEMIS visible image V05484014. (f) Clusters of rootless cones formed by explosive interactions between lava and water-ice contained in the underlying substrate, western Tartarus Colles region, east of Elysium Mons. Portion of HiRISE image ESP_018668_2065_MIRB: NASA/JPL/University of Arizona.

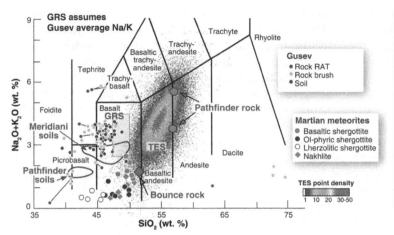

Figure 17.4 Compositions of Mars rocks derived from a variety of sources, shown on the standard total alkalis vs. silica diagram used for classification of volcanic rocks. Analyses of Gusev rocks and soils, martian meteorites, and global GRS data (calculated on a volatile-free basis) indicate a crust dominated by basalts. TES-derived data and possibly the Mars Pathfinder rock composition may reflect alteration. From McSween et al. (2009), reprinted with permission from AAAS.

dissected nature of these shields (Fig. 17.3(c)) suggests that they are composed of easily erodable, friable pyroclastic deposits rather than lava flows (Greeley and Crown, 1990; Crown and Greeley, 1993). The caldera-like depressions and surroundings of Peneus, Malea, and Pityusa Paterae suggest that they are similar to terrestrial "supervolcanoes", and produced ignimbrite deposits that have since been modified by fluvial, aeolian, and periglacial processes (Williams et al., 2009).

Age estimates based on impact crater size–frequency distributions suggest that most of the volcanic activity in the Circum-Hellas Province is significantly older than elsewhere, between ~3.5 and 4.0 Ga, with little evidence that any activity occurred more recently than 1 Ga. In contrast, effusive volcanism appears to have dominated the later part of Mars' volcanic history: crater-count ages suggest that the Tharsis shields are younger than ~1.8 Ga, whereas some flows in the volcanic plains between Tharsis and Elysium may be as young as 10 Ma (Hartmann and Berman, 2000).

17.3.2 Composition of Mars volcanic materials

There are four main bodies of evidence for the composition of the martian surface: (1) remotely sensed spectral data from orbiting spacecraft; (2) martian meteorites; (3) spectral data from Martian landers and rovers; and (4) morphologic data. Figure 17.4 summarizes compositional information derived from a variety of sources, but there remain gaps in our understanding. Nonetheless, from a physical volcanology standpoint, we can make some fairly confident assessments of the composition of Mars' surficial deposits, and hence of its magmas.

Compositional data from spacecraft
Compositional data in the form of spectra from orbiting spacecraft have been collected by Mars Global Surveyor (MGS), Mars Reconnaissance Orbiter (MRO), and Mars Express (ME). Spectral data are all consistent with a martian surface composed primarily of mafic (probably basaltic) materials that have been locally altered (Bandfield et al., 2000; Bandfield, 2002; Boynton et al., 2004, 2007; Poulet et al., 2007). Early spectral studies (Bandfield et al., 2000) suggested a component of the martian surface that is consistent with an evolved magma composition – possibly andesitic. However, an equally valid interpretation is that the component is altered basaltic glass.

Inside Nili Patera, the caldera on top of Syrtis Major, Christensen et al. (2005) found dacitic rocks, with 60–63 wt.% SiO_2. The outcrop is small (a few square kilometers). Skok et al. (2010) suggest that at least some of this enriched silica component may be caused by hydrothermal activity, and is not an outcrop of primary igneous rocks. Regardless, the total outcrop area is small enough to make this region an interesting anomaly rather than representative of martian volcanic compositions as a whole. Thus, available compositional data obtained from satellites is consistent with a surface layer being composed primarily of mafic materials – most likely basaltic.

Martian meteorites

Martian meteorites are also called the "shergottite-nakhlite-Chassigny (SNC)" group, referring to the names this class of meteorites received before it was recognized that they originated on Mars. As of 2011, 97 meteorites had been positively identified as originating on Mars. Although there are other clues to a martian origin, the "smoking gun" is the composition of gases trapped within solidified melt that is itself trapped within the meteorite: the composition of these trapped gases matches the composition of the martian atmosphere as measured by the Viking Landers (e.g., Treiman et al., 2000).

The SNC meteorites are all mafic or ultramafic igneous rocks. They range in composition from basalts (some shergottites) to dunite (Chassigny). The nahklites are ultramafic wehrlites or clinopyroxenites; some shergottites are lherzolites or harzburgites (McSween, 1994). The basalt shergottites are interpreted to come from surface lava flows, whereas the remaining martian meteorites likely came from magmatic intrusions (Nyquist et al., 2001), although the precise locations of origin are as yet unknown (Hamilton et al., 2003).

Lander and rover data

To date, data on the composition of in situ martian samples have been returned from the Viking Landers 1 and 2, Mars Pathfinder's Sojourner, the Mars Exploration Rovers Spirit and Opportunity, and Phoenix. The Viking landers' X-ray fluorescence results for Mars soil revealed a composition similar to that of iron-rich basalt. Sojourner's alpha-proton X-ray spectrometer (APXS) collected compositional data from the weathered surfaces of boulders at the landing site. McSween et al. (1999) used these data to calculate the composition of a likely end-member martian igneous rock, and determined that such a rock might be an andesite. They stress, however, that these results are model-dependent, and could be strongly

affected by sample contamination with weathering products, given that Sojourner had no means to remove the weathering rinds.

Spirit and Opportunity were each equipped with a Rock Abrasion Tool (RAT), designed to remove surficial weathering products prior to examining the fresh rock face beneath. The data collected have revealed a wealth of geologic processes affecting the surface (e.g., Ruff et al., 2006; Squyres et al., 2006). However, the measured soil and rock compositions remain most consistent with a mafic or ultramafic igneous parent.

The Phoenix lander was designed to examine the composition of martian ice rather than martian rock (Arvidson et al., 2009). However, the mechanical behavior of the soils and their appearance are consistent with those observed at the Viking 2 (Arvidson et al., 2009). These results are consistent with the soils being a weathering product from a mafic or ultramafic igneous parent rock.

Morphology

Of all the available data sets, morphology is the least reliable for determining lava composition. Observations and laboratory simulations of terrestrial volcanic behaviors indicate that lava rheology exerts a strong control on volcanic behavior and morphology (see Chapter 12, Fig. 12.6, for example). Early experiments suggested that SiO_2 content can be directly linked with lava rheology (Hulme, 1974, 1976); we now know that this is an oversimplified interpretation. Nonetheless, volcanic morphologies that are apparently generated by low-viscosity flows are most easily explained by lavas with a mafic or ultramafic composition.

Lava tubes and lava channels, for example, have been observed on Mars (Bleacher et al., 2007), as have thin, fluid flood-style flows (Jaeger et al., 2007). Although channels are observed in lava flows with a range of compositions on Earth (e.g., ultramafic, phonolitic, dacitic; Wichura et al., 2010), drained lava tubes as observed on Mars have only been seen in terrestrial basalt flows. Thus, the presence of martian lava tubes and channels supports a mafic lava composition.

In contrast, however, multiple studies have examined the morphologies of specific martian lava flows and inferred a more evolved composition, including rhyolite (e.g., Hulme, 1976; Fink, 1980; Zimbelman, 1985; Wadge and Lopes, 1991; Warner and Gregg, 2003). It is important to note, however, that the physical characteristics of lava flows (typically flow thickness and surface textures) that have led to inferred evolved lava compositions can be created by emplacement kinematics and rheology. In other words, inferring lava flow composition from flow morphology is difficult: rheology and pre-existing flow surfaces exert strong controls on lava flow morphology, and rheology is not uniquely related to lava composition. Added to these considerations are the complicating effects of Mars' environmental factors (Sections 17.2, 17.3.4).

The shield morphologies of many of the martian volcanoes are similar to those of terrestrial shield volcanoes (e.g., Mauna Loa, Hawaii) and all terrestrial shield volcanoes are basaltic. Furthermore, the morphologies of smaller volcanic features and lava flows are generally suggestive of low-viscosity magmas. Even the putative pyroclastic deposits surrounding Tyrrhenus and Hadriacus Montes could be generated by mafic magmas (Greeley and Crown, 1990; Crown and Greeley, 1993), with explosive activity driven by either magmatic volatiles or water/ice encountered by rising magma erupting into a low-pressure atmosphere.

Thus, the preponderance of evidence is that Mars has generated ultramafic to mafic volcanic products throughout its history. More evolved volcanic materials (such as the high-silica materials found within Nili Patera) are volumetrically insignificant on the Martian surface.

17.3.3 Magma ascent processes on Mars

The low pressure of the martian atmosphere ($P_a = 600$ Pa; Table 17.1), together with the low acceleration due to gravity ($g = 3.73$ m s^{-2}), exert strong controls over magma ascent and eruption. At depth, warm, buoyantly ascending diapirs may undergo pressure-release melting to produce partially molten magma reservoirs.

The buoyancy force F_b on a given volume V of melt is

$$F_b = g(\rho_r - \rho_m)V, \qquad (17.1)$$

where ρ_r and ρ_m are the country rock and magma densities, respectively. Since magma ascent rate depends on the buoyancy force (which depends upon gravity), diapirs of a given volume would ascend more slowly on Mars than on Earth. Conversely, melt bodies would have to be larger to avoid excessive cooling and stalling during buoyant ascent, implying larger volumes would be available for eruption (Wilson and Head, 1994) on Mars than on Earth.

Ascending melt may stall at a rheological barrier (e.g., the base of the lithosphere), or at a density barrier, where the country rock density becomes equal to that of the melt (i.e., the level of *neutral buoyancy*). In either case, a magma reservoir may form. The depth to the base of the lithosphere is governed by the cooling history of the planet, which is related to the size of the body. Given that Mars' diameter is roughly half that of the Earth, its lithosphere is likely to have thickened more rapidly and produced a deep rheological barrier to magma ascent. If the magma is still buoyant on encountering and stalling at such a boundary, and the stress regime (due to buoyancy pressure, reservoir excess pressure, and external stresses) allows chamber walls to rupture, magma may continue to ascend through propagating dikes. When a level of neutral buoyancy is reached at shallower levels in the crust, in the absence of excess driving pressure, magma may stall once again.

The total pressure P felt at any depth beneath the planetary surface is the sum of the lithostatic pressure, $\rho_r g z$, and the external atmospheric pressure, P_a. At depths greater than ~10 m on Earth, atmospheric pressure is a negligible component of the total pressure. On Mars, the external pressure is so low as to be insignificant (Fig. 17.5(a)). Of greater importance at depth is the low martian gravity, which produces lower pressure gradients dP/dz in the lithosphere, such that the pressure at any depth is less on Mars than on Earth (Fig. 17.5(a)). This implies that the rate of densification of country rock with depth is also lower, i.e., crustal rocks at a given depth on Mars will be less compacted than on Earth (Fig.17.5(b)). Therefore, an ascending magma will encounter country rock with a density equal to its own at greater depths on Mars than on Earth, thus implying deeper neutral buoyancy zones by a factor of ~4 (Wilson and Head, 1994).

Magma chambers forming at both deep rheological barriers and shallow neutral buoyancy zones will be correspondingly deeper on Mars than on Earth. This has a number of implications. First, higher driving pressures and wider dikes are needed for an ascending magma to reach to the surface from greater depths. The high pressure gradients driving the magma lead to higher velocities, larger mass fluxes, and correspondingly larger volume eruptions, despite the lower buoyancy forces on Mars. In addition, consideration of fracture mechanics suggests that the widths w and horizontal extents H of dikes are inversely proportional to gravity raised to some negative power that depends on the model used, e.g., $w \propto g^{-1/3}$ and $H \propto g^{-2/3}$ (Wilson and Head, 1994; see also Chapter 3). This implies that dikes on Mars would be systematically wider and longer by factors of ~1.4 and 1.9, respectively, than dikes on Earth. For laminar flow, ascent velocity is proportional to square of the dike width w and the driving pressure gradient, and the mass (or volume) flux is proportional the product of the velocity and cross-sectional area of the dike (wH). This leads to a strong inverse relationship ($\sim g^{-5/3}$) between eruption rate and gravity, indicating a systematic trend to higher eruption rates by a factor of ~5 on Mars compared to Earth (Wilson and Head, 1994). Taken together, the influence of greater dike dimensions and deeper magma chambers may account for a factor of seven times greater mass fluxes for eruptions on Mars than on Earth, all other factors being equal. This clearly has implications for the volumes of material erupted, and for the lengths of lava flows and the sizes of the edifices constructed.

As magma reaches shallower levels in the lithosphere, the pressure drops to the point at which magmatic gases (H_2O, CO_2, etc.) start to exsolve. For a given magma composition, exsolution will tend to occur at depths ~2.5–3 times greater on Mars than on Earth (Wilson and Head, 1994), primarily because of the lower lithostatic pressure gradient. However, on Mars, the low lithostatic pressure environment is likely to lead to fragmentation of magma containing relatively modest amounts of volatiles. As discussed in Chapters 4 and 7, fragmentation of low-viscosity magmas may take place due to: (1) rapid bubble expansion leading to hydrodynamic breakup; (2) expansion exceeding the structural relaxation rate of the melt, leading to: brittle fragmentation; or (3) exceeding a critical bubble volume fraction (typically taken as 65–85 vol.%). If a critical bubble volume fraction is the dominant criterion for fragmentation, the depth at which this threshold is reached (for a given volatile content and subsonic conditions), is a factor of ~3 deeper on Mars because the critical fragmentation pressure is found at greater depths (Fig. 17.1; Wilson and Head, 1994). The fragmentation depth is non-trivial to determine if the eruption is supersonic (Mitchell, 2005), but will still be greater in depth than on Earth. The rapid ascent and expansion rates expected in the low-pressure martian environment mean that the brittle and hydrodynamic breakup criteria would also be met at correspondingly greater depths on Mars. Even if martian magmas contained fewer volatiles than typical terrestrial mafic magmas, calculations using simple solubility laws suggest that CO_2 or H_2O contents exceeding ~0.01 wt.% may lead to fragmentation on Mars (cf. ~0.07 wt.% on Earth) during sustained magma ascent in which bubbles and magma are dynamically coupled (Wilson and Head, 1983).

The low lithostatic pressure gradient means that nucleation and fragmentation levels are both deeper and more widely separated on Mars than on Earth (Fig. 17.5(a); Wilson and Head, 1994). This leads to an extended period of bubble growth, the potential for numerous nucleation events and bubble populations, and ultimately a more finely fragmented magma. Once the bubbly magma has transitioned to a particle-laden gas stream, the mixture will rapidly decompress down to the low ambient pressure, leading to a

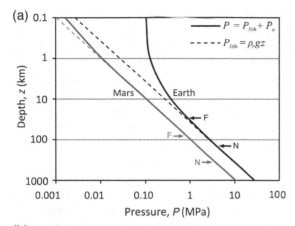

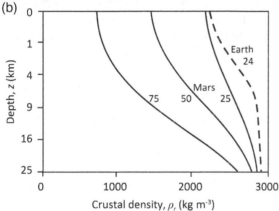

Figure 17.5 (a) Variation of pressure in the upper crust for Earth and Mars. Solid curves show the total pressure felt at any depth (the sum of lithostatic and atmospheric pressure); dashed curves show the lithostatic pressure only. Nucleation (N) and fragmentation (F) depths are indicated by arrows for a basaltic magma initially containing 0.3 wt.% dissolved H_2O. (b) Crustal density as a function of depth for Mars (solid curves) and Earth (dashed curves). Curves are marked with the porosity (%) of surface rocks.

17.3.4 Eruption styles and processes

Our understanding of the physics of magma ascent suggests that in general one might expect a greater propensity for explosive volcanism on Mars, given magmas with even modest volatile contents. The potential for ascending magma to interact with subsurface ice only acts to increase the chances for explosive volcanism.

Assuming that some Mars magmas had sufficient volatiles for fragmentation to occur, we consider the manifestation of explosive volcanism in the martian environment. Vigorous fragmentation and high-velocity eruption of finely fragmented tephra may have preferentially produced plinian eruption columns rather than the hawaiian lava fountains typical of basaltic explosive activity on Earth. Early plume models indicated that, provided the erupting mixture attained buoyancy above the vent, convective columns could rise a factor of five times higher on Mars than similar mass flux eruptions on Earth, implying rise heights of tens to > 100 km, which would serve to distribute material widely around the planet (Wilson and Head, 1994). However, the low density of the martian atmosphere would cause difficulty in achieving buoyancy because of the limited potential for heating and expansion within the column, such that collapsing columns and pyroclastic density currents (PDCs) would be common for a wider range of conditions than on Earth. A later reconsideration of assumptions underlying Morton-style convective plume models (see Chapter 8), found that some of the basic model assumptions are readily violated under present martian atmospheric conditions, such that column heights would be limited to ~10–12 km because the low-density atmosphere cannot support convection to significant altitudes (Glaze and Baloga, 2002). While PDCs could certainly have been generated from failed convective columns, one might also envision a style of particle fallout more akin to that observed on Io (Section 17.4), starting from heights above the limits of convective capability, and producing diffuse deposits (Wilson and Head, 2007). A putative, past dense martian atmosphere would have permitted convective ascent to greater

much greater energy release per unit mass of gas than would be the case on Earth, and accelerating the mixture to significantly higher velocities. Wilson and Head (1983) give a simplified expression showing that the square of the eruption velocity at the vent is proportional to the natural logarithm of the ratio of the pressure at fragmentation to atmospheric pressure. This leads to velocities in sustained eruptions being a factor of 1.6–2 times greater on Mars than on Earth for similar ascent conditions.

column heights, and more widespread deposition. It is conceivable that the ancient martian volcanoes interpreted to be composed of explosive deposits (Section 17.3.1) were produced by vigorous eruptions during periods of high atmospheric density.

Hawaiian-style lava fountains (Chapter 8), if they formed in less vigorous eruptions, are likely to have contained finer grain sizes, been erupted at greater speeds, and dispersed material over greater areas than for equivalent magma mass fluxes and volatile contents on Earth. This is likely to have led more commonly to broad tephra cones rather than steep, narrow spatter ramparts or coalescence of pyroclasts into lava flows (Wilson and Head, 1994). However, relatively few primary tephra cones have been identified. One might infer that in general Mars' lava flow fields were formed by passive lava effusion, or only very weakly fragmented, volatile-poor magmas that rapidly coalesced to produce flows.

Strombolian activity (Chapter 6) may not have been very common on Mars, based on theoretical considerations. Relatively low buoyancy forces on ascending bubbles, together with the likely more rapid magma ascent speeds, would have acted to inhibit the decoupling of the bubbles from the melt that is required for slug formation and strombolian explosions. In the case where volatiles do decouple from a stalled magma column, the great expansion potential of the released gases in the martian environment would produce wider and lower cones than would be observed for terrestrial strombolian activity.

In the case of lava effusion, eruption rates may have been considerably greater than is typical on Earth. Indeed, large shields and extensive lava plains are common on Mars (Section 17.3.1; Fig. 17.3). Once erupted on the surface of Mars, lavas would experience a combination of radiative and convective cooling from the flow surface at temperature T_{hot}, and conductive cooling to the substrate below (Section 17.2.3; Chapter 5). Given the low atmospheric density, the radiative heat flux would dominate over convective cooling. Furthermore, the total heat flux at a given surface temperature would be less than on Earth. This, together with the greater predicted thicknesses of martian flows (a combination of high effusion rates and low gravity), means that flows would in general cool more slowly and travel farther than terrestrial flows. Taken together, greater predicted dike widths (× 2) and effusion rates (× 7) for eruptions in the martian environment, imply flows five times longer than on Earth for similar magma compositions (Wilson and Head, 1994).

The discussion above deals largely in generalizations based on the expected influence of the martian gravitational and atmospheric environment; in reality there are certain to have been many variations on the styles of volcanism predicted based on theoretical treatments. However, the general picture gleaned from observations of martian volcanic features is of older explosive volcanism in the south transitioning to younger effusive volcanism in the north; this is hard to reconcile with our understanding of the physics of ascent and eruption in the martian environment. Although a denser early Mars atmosphere might have promoted widespread ash dispersal from convecting columns, it would also have acted to suppress vesiculation and fragmentation to a greater degree than the latter-day tenuous atmosphere. One possible explanation for the transition in eruptive style might lie in a general depletion of magmatic volatiles over the period that Mars was volcanically active; early volatile-rich explosive eruptions may have given way to volatile-poor effusions. Mars' small size and the lack of recycling of volatiles via plate tectonics may have promoted irreversible gas loss from the interior of the planet. Another possibility is that patterns of explosive vs. effusive volcanism are related to the availability of external volatiles such as ground ice or groundwater, and their propensity to promote explosive magma–water interactions. Spatial or temporal variations in the distribution of crustal H_2O reservoirs may therefore have had a strong influence on the manifestation of volcanism in the martian environment. Understanding this enigma remains a key objective for researchers in Mars volcanology.

17.4 | Volcanism on Io

17.4.1 Observations of volcanic features on Io

Io's colorful surface is studded with volcanic features and deposits (Fig. 17.6), and Io is so far the only body outside Earth known to exhibit active silicate volcanism. Active eruptions were discovered in 1979 when the Voyager spacecraft revealed enormous, umbrella-shaped plumes (Morabito *et al.*, 1979) and thermal anomalies (Pearl *et al.*, 1979). Results from the Galileo mission, which observed Io from 1995–2001, substantially advanced our understanding of volcanism on Io (Lopes and Williams, 2005). Distant observations of Io were also made by the Cassini spacecraft on its way to Saturn during 2000–01 (Radebaugh *et al.*, 2004) and by the New Horizons spacecraft in 2007 (Spencer *et al.*, 2007) en route to Pluto. Ground-based monitoring has also been important in the study of Io's volcanism (e.g., Marchis *et al.*, 2000; Veeder *et al.*, 1994). More than 150 active volcanic centers have been identified from these combined observations; there are likely to be many more that have not yet been observed erupting. Recent results using data from the magnetometer on Galileo (Khurana *et al.*, 2011), suggest the presence of an asthenosphere (global magma ocean) at least 50 km thick under a low-density outer crust ~30–50 km thick.

Io's volcanoes rarely build significant topographic structures. There are only a few structures, called *tholi*, scattered across Io, some of which were interpreted to be a result of shield-building basaltic volcanism (Schaber, 1980). The most common type of volcanic feature on Io is the *patera* (Fig. 17.6(a), (d), (e), (g)). Although the origin of Ionian paterae is still somewhat uncertain, they are thought to be similar to terrestrial volcanic calderas, formed by collapse over shallow magma chambers following partial removal of magma. Some paterae show angular shapes that suggest structural control, indicating that they may be structural depressions that were later used by magma to travel to the surface. More than 400 Ionian paterae have been mapped (Radebaugh

et al., 2001; Williams *et al.*, 2011). Their average diameter is ~40 km but Loki, the largest patera known in the Solar System, is > 200 km in diameter. The larger sizes of the Ionian features probably reflect large, relatively shallow magma chambers (Leone and Wilson, 2001).

Io's surface shows some remarkably large lava flow fields; the Amirani lava flow field, at ~300 km long, is the largest active flow field known in the Solar System. Repeated imaging of Amirani during the Galileo fly-bys allowed eruption rates to be estimated at 50–500 m^3 s^{-1} (Keszthelyi *et al.*, 2001). The Prometheus flow field (Fig. 17.6(e)) extended 80 km to the west between Voyager and Galileo observations; calculated resurfacing rates are an order of magnitude lower than Amirani's flow field. This range of effusion rates is surprisingly modest given the very large size of these flow fields.

Io's plumes exhibit significant variability. They range from < 100 km to > 450 km in height. Large plumes (e.g., Pele and Tvashtar; Fig. 17.6(c)) are associated with high-temperature thermal anomalies, may be quite faint (implying a vapor-rich composition), and may produce annular deposits > 1200 km in diameter, that are reddish or sometimes black in color (Fig. 17.6(b)). Smaller plumes, e.g., Prometheus (Fig. 17.6(c)), tend to be associated with relatively cool thermal anomalies, and are optically more dense than Pele-type plumes, implying a mixture of condensing gases and particles. Deposits are typically pale yellow in color. Intriguingly, during the 17-year interval between Voyager and Galileo observations, Prometheus' plume moved ~80 km to the west (see Section 17.4.4).

17.4.2 Composition of Ionian volcanic materials

The main difference between the compositions of volcanic materials on Io and Earth is that Io contains negligible amounts of water (Carlson *et al.*, 1997), and sulfur and sulfur dioxide are the main volatiles detected in both plume materials (McGrath *et al.*, 2000; Spencer *et al.*, 2000; Jessup *et al.*, 2004) and on the surface (Carlson *et al.*, 1997, 2007). Whereas Voyager

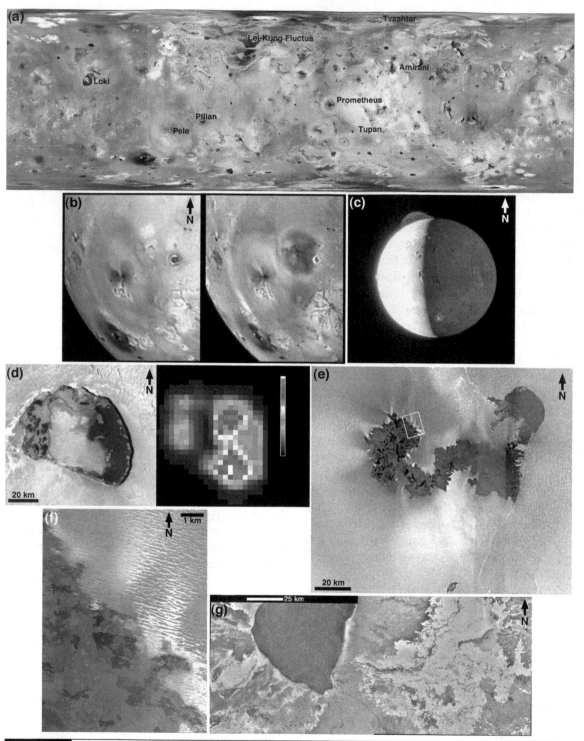

Figure 17.6 Volcanic features on Io. (a) Map of the Ionian surface showing key volcanic features. Image NASA/JPL/USGS; merged Voyager and Galileo mosaic available at http://astrogeology.usgs.gov/Projects/JupiterSatellites/io.html. (b) These two Galileo images show the appearance of a new, dark deposit 400 km in diameter that erupted from Pillan Patera in 1997, and partly covered the large red deposit from Pele Patera. Image PIA00744, NASA/JPL/University of Arizona. (c) The New Horizons Long Range Reconnaissance Imager captured this view of three of Io's plumes on 28 February 2007. In the far north is the 390-km-high Pele-type plume from Tvashtar volcano. In addition, a 60-km-high plume from Prometheus can be seen on the western limb, and the top of Masubi's plume can be seen poking into the sunlight from Io's night side, to the south of the image. Image PIA09248, NASA/Johns Hopkins Applied Physics Laboratory/Southwest Research Institute. (d) The floor of Tupan Patera

observations were unable to resolve whether sulfur or silicate was the predominant magma type, temperature measurements from the Galileo and the New Horizons spacecraft clearly showed that most active volcanoes have temperatures too high to be erupting only sulfur. A few locations show young, pale yellow or white flows (e.g., Emakong Patera; Fig. 17.6(g)) that may have been molten sulfur (Williams et al., 2001, 2004), whereas others may be silicates coated with sulfur. Greeley et al. (1984) suggested that rising silicate magma may melt near-surface sulfur-rich country rock, producing "secondary" sulfur flows, as opposed to "primary" flows that originate from molten sulfur magmas at depth.

The main question regarding the composition of Io's silicate lavas is whether they are mafic (possibly superheated) or ultramafic. With no spectroscopic measurements of lava composition (mostly due to spatial resolution limitations), temperatures detected at active volcanoes provide the best clues to magma composition. Temperatures of Io's active lavas have been calculated from observations in the near-infrared from Galileo (e.g., Lopes et al., 2001; Davies et al., 2001; Howell and Lopes, 2011), Cassini (Radebaugh et al., 2004), and New Horizons data (Spencer et al., 2007). However, temperatures determined from remote sensing data depend on the spatial resolution and wavelength range; because lava cools rapidly after exposure at the surface, measured temperatures are unlikely to represent magmatic temperatures (e.g., Lopes et al., 2001). The current consensus is that eruptions with temperatures characteristic of basalts (1400–1550 K) are common on Io, but hotter, ultramafic volcanism may be also present (Williams and Howell, 2007).

17.4.3 Magma ascent processes on Io

Initial interpretations combining optical (Solid State Imager, SSI) and infrared (Near-Infrared Mapping Spectrometer, NIMS) data of Io's eruptions showed some eruption temperatures greater than those typical of terrestrial basalts (> 1450 K), with > 1850 K for at least one eruption (McEwen et al., 1998; Davies et al., 2001), which exceeds any volcanic activity observed on Earth. Such temperatures would be consistent with an undifferentiated and almost completely molten interior, but this does not appear to be the case. Models that couple tidal heating to volcanic heat loss suggest that the interior should be no more than ~20% molten (e.g., Moore, 2001), which in turn suggests magmatic temperatures of < 1550 K, regardless of geochemistry. The discrepancy between observation and models requires explanation. Three factors may be relevant: (1) uncertainties in the thermo-spectral modeling of eruption temperatures; (2) problems with geophysical models of interior thermal evolution; and (3) unusual conduit conditions.

Reanalysis of SSI and NIMS data of two of the more anomalous eruptive sites on Io, Pele and Tvashtar, yield lower temperatures than previously reported and account for 200–300 K of the 400 K gap between observation and theory (Keszthelyi et al., 2007). The rest of the difference can be explained by a quirk of magma ascent in Io's highly compressive lithosphere. The possibility of superheating due to viscous dissipation within the magma during conduit ascent was raised by McEwen et al. (1998). This process is generally negligible for basaltic magmas on Earth (e.g., Mastin, 1995), but has been suggested as potentially important for explosive silicic eruptions (e.g., Mastin, 2005). The unique stress distribution within the

---

Figure 17.6 (cont.) (75 km across) is partly covered with dark material, presumably lava, that appears significantly warmer in Near Infrared Mapping Spectrometer data than the adjacent orange-colored "island," parts of which are cool enough for SO_2 to condense. Image PIA03601, NASA/JPL. (e) The source of the Prometheus plume lies at the distal (westernmost) end of a compound pahoehoe flow field. Image PIA02565, NASA/JPL/University of Arizona. Inset box shows the location of (f), which is a high-resolution view of the active flow front. Dark areas on the flow show recent lava breakouts, older flow surfaces brighten with time due to deposition of SO_2 from the plume. Fresh SO_2 frost is also seen on the surrounding terrain. Image PIA02557, NASA/JPL/University of Arizona. (g) Bright lobate features to the east of Emakong Patera have been interpreted as fresh sulfur flows. Image PIA02539, NASA/JPL/University of Arizona. See color plates section.

Ionian lithosphere (Jaeger *et al.*, 2003) requires an ascending magma to overcome a confining pressure of ~0.5 GPa (Keszthelyi *et al.*, 2007). For a magma ascending adiabatically against friction and gravity, g, through a pipe of radius r, the change in temperature T with decreasing depth z is given by:

$$\frac{dT}{dz} = \frac{n}{\rho_g c}\frac{dP}{dz} + \frac{u^2 f}{2rc} \qquad (17.2)$$

where P is the pressure, c is the bulk specific heat capacity, n is the mass fraction of volatiles driving the ascent, ρ_g is the density of the volatile phase, u is the velocity of ascent, and f is the wall friction factor. The first term on the right-hand side represents adiabatic cooling of any exsolved gas phases, and the second term represents viscous dissipation. Assuming turbulent flow (Wilson and Head, 2001) of an unvesiculated magma allows significant simplification:

$$\frac{dT}{dz} \sim -\frac{1}{\rho_g c}\frac{dP}{dz} - \frac{g}{c}, \qquad (17.3)$$

which is independent of velocity. The g/c term is second order, and can be ignored here. Thus, assuming 0.5 GPa overpressure (leading to ~0.34 GPa driving pressure) at 20 km depth, taking ρ_g = 2700 kg m^{-3}, g = 1.7 m s^{-2} and c = 840 J kg^{-1} K^{-1} gives a temperature increase of ~110 K.

It is important to note that this simplistic treatment is appropriate only for unvesiculated magmas. Exsolution and expansion of volatiles during ascent will inevitably produce a cooling effect that will act to counteract the viscous superheating under a wide range of circumstances. Hence, this balance between heating and cooling effects may result in considerable variation in eruptive temperatures, generally ranging between hotter effusive and cooler explosive eruptions.

17.4.4 Eruption styles and processes on Io

Io's continuous activity has allowed general classification of eruption styles (Williams and Howell, 2007), studies of plume dynamics (Geissler and Goldstein, 2007), and determination of lava effusion rates (Keszthelyi *et al.*, 2001). Three main styles of eruption have been identified: primarily explosive, primarily effusive, and those that are confined within paterae. Plumes may be associated with all three eruption styles.

Explosion-dominated eruptions, typified by Pillan volcano's 1997 eruption, are high effusion-rate, vigorous outbursts of material during short-lived (weeks-long) events. These events typically produce large (> 200 km high) plumes (known as *Pele-type plumes*; McEwen and Soderblom, 1983) likely due to release of juvenile volatiles, and result in extensive (up to ~1200 km diameter) red annular deposits of short-chain sulfur, and sometimes dark deposits, presumably silicate pyroclastics (Fig. 17.6(b)). In addition to plumes, lava fountains may also form part of the eruption sequence (e.g., at Tvashtar in 1999; Keszthelyi *et al.*, 2001). Many of the dynamical features displayed by the plumes are expected from the flow of gas out of a nozzle and into a near-vacuum (Geissler and Goldstein, 2007). Expansion of erupted gases in the low-pressure, low-gravity environment readily explains the very high plumes observed. Given the absence of an atmosphere, the erupting mixture is unable to form a convecting column, and instead erupting particles follow quasi-ballistic trajectories, influenced by particle–particle collisions and shocks in the emerging gas phase. Application of simple ballistics equations relating plume height to eruption velocity ($h = u_0^2/2g$), yields eruption velocities > 1 km s^{-1} in some cases. Theoretical treatment of plume thermodynamics (Kieffer, 1982) and numerical modeling of plumes and their entrained particulates (Zhang *et al.*, 2003) explain many of the finer details of plume characteristics.

Flow-dominated eruptions, typified by Prometheus (Fig. 17.6(e)) and Amirani volcanoes, produce extensive lava flow fields. The ability of lava to travel large distances at moderate effusion rates (< 5–500 m^3 s^{-1}), together with thermal profiles along the Prometheus and Amirani flows (Lopes *et al.*, 2001, 2004) and high spatial resolution images (Fig. 17.6(f); Keszthelyi *et al.*, 2001), suggest that these large Ionian flow fields are emplaced as insulated (tube-fed) flows of low-viscosity lava, similar to terrestrial inflated pahoehoe flow fields (Hon *et al.*, 1994). Flow-dominated eruptions are also accompanied by plumes that are generated at the active flow fronts (Fig.17.6(e)), commonly at multiple locations, and thus plume locations move as the flow advances. These plumes, generally known as *Prometheus-type plumes* (McEwen and Soderblom, 1983), are smaller (< 200 km) than those from explosive eruptions. Kieffer *et al.* (2000) and Milazzo *et al.* (2001) proposed models to explain plume formation as the explosive interaction between the hot silicate lava flow and the SO_2-frost covered substrate. In contrast to eruptions of juvenile volatiles, this plume type therefore has a rootless source in jets of vaporized SO_2 and entrained particulates emanating from the flow front; this mechanism helps to explain the smaller heights and cooler temperatures of these plumes.

The third and most common type of eruption on Io is the *intra-patera* eruption (Williams and Howell, 2007), which is effusive activity confined within a patera, commonly, but not always accompanied by plumes. These eruptions tend to last for many years without overflowing the paterae, leading to suggestions that many of the paterae contain lava lakes (Lopes *et al.*, 2004), perhaps overturning in a quasi-periodic way (Rathbun *et al.*, 2002). Alternatively, paterae may be underlain by persistent magma lenses that feed thin, temporary lava lakes, similar to the mechanism that operates at the East Pacific Rise on Earth (Gregg and Lopes, 2008). Pele's persistent plume was observed to be related to lava lake activity. An intriguing aspect of some Ionian paterae such as Loki and Tupan (Fig. 17.6(d)) is the presence of an "island" partially or completely surrounded by active lavas. How these cold islands, which appear bright because of condensed SO_2 frost, can remain unaltered for many years despite close proximity to hot lavas is still one of Io's unsolved mysteries.

17.5 | Cryovolcanism on the outer planet satellites

17.5.1 Observations of cryovolcanic features

Cryovolcanism involves the eruption of materials that would ordinarily be solid at the surface temperature of the icy bodies of the outer Solar System: cryomagmas therefore include liquid or gaseous water or aqueous solutions, possibly mixed with solid fragments. The low bulk densities of many satellites (Table 17.1) indicate that, in addition to silicates, their interiors contain some proportion of ice. Cryovolcanism may ensue if liquid subsurface reservoirs are present and able to deliver material to the surface. Active venting, in the form of erupting plumes, has been observed on only two bodies – Neptune's Triton and Saturn's Enceladus – but features interpreted to be cryovolcanic in origin are widespread on outer Solar System satellites. Many satellites have surfaces that are cratered and ancient, but other terrains are smooth and apparently young. Whereas icy tectonism, diapirism, and intrusive cryomagmatism may contribute to the diversity of features on geologically active icy bodies, cryovolcanism remains a viable mechanism for satellite resurfacing.

The Galileo spacecraft revealed that relatively recent cryovolcanic activity may have occurred on Jupiter's satellites Europa and Ganymede (Fig. 17.7(a)–(f); Fagents *et al.*, 2000; Fagents, 2003; Showman *et al.*, 2004). Both moons display geological features that can be attributed to the effects of tidal heating generated through orbital interactions with Jupiter and with their neighbor satellites. Europa has a silicate interior surrounded by a ~100-km-thick H_2O ice shell, some proportion of which is known to be in a liquid state (Kivelson *et al.*, 2000). The ice shell is pervasively fractured, and numerous pits, domes, and mottled features (Fig. 17.7(b), (d), (e)), as well as larger *chaos* terrains (Fig. 17.7(c); Pappalardo *et al.*, 1999; Greenberg *et al.*, 1999) may have been formed by cryovolcanic or intrusive cryogmagmatic processes. However, very few features exhibit strong evidence of

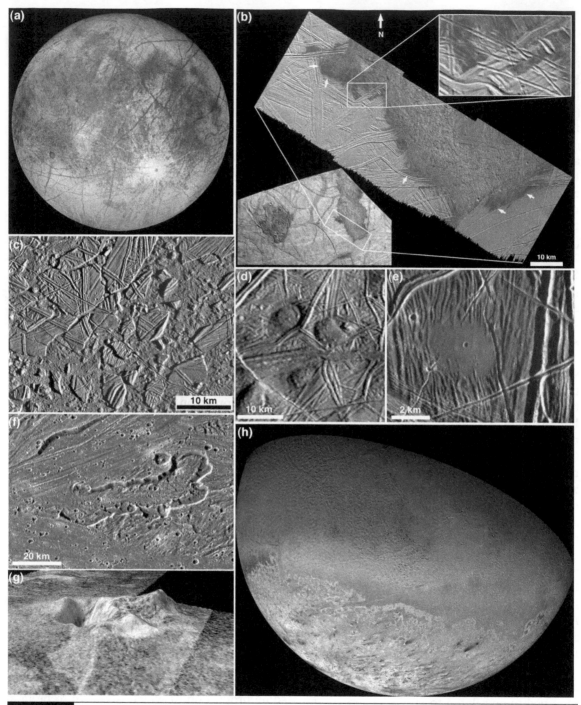

Figure 17.7 Proposed cryovolcanic features on icy satellites. (a) Galileo view of Europa's trailing hemisphere showing a surface criss-crossed with lineaments and hosting dark, mottled terrain. Image PIA00502, NASA/JPL/DLR. (b) Close-up views of Thrace Macula on Europa showing apparent releases of fluid (arrows) from the uplifted, disrupted interior of the macula. Modified from Fagents (2003). (c) Some areas of Europa's ice shell have broken apart and rafted into new positions, due to heating by warm subsurface ice or liquid water. Image PIA00591, NASA/JPL/ASU. (d) Dome-like features on the surface of Europa may represent cryovolcanic effusions or diapiric upwellings. Portion of image PIA01092, NASA/JPL/ASU. (e) Smooth, low-lying deposit on Europa may be the result of small-scale cryovolcanic flooding of the surface. Portion of image PIA00592, NASA/JPL/ASU. (f) Caldera-like feature in Sippar Sulcus on Jupiter's Ganymede. On the floor of the depression is a lobate flow-like feature with surface ridges indicating that the flow moved from the upper right to the left. Image PIA01614, NASA/JPL/Brown University. (g) Perspective view of Sotra Facula on Saturn's Titan, derived from Cassini radar and visual and infrared mapping spectrometer data. The mountains shown are >1 km high and the depressions are up to 1.5 km deep. Vertical exaggeration ×10. Portion of image PIA13695,

fluid having been delivered to the surface (Fig. 17.7(b), (e); Fagents, 2003).

Saturn's small moon Enceladus (mean radius 252 km) presents the most dramatic evidence of cryovolcanic activity observed to date (Fig. 17.8). Plumes of water vapor and icy particles are observed emanating from the highly tectonized south polar region (Fig. 17.8(b); Porco et al., 2006; Spahn et al., 2006; Spencer et al., 2006; Waite et al., 2006). This region hosts four linear, sub-parallel lineaments (known as sulci, or more informally as "tiger stripes") in the ice shell. The sulci are typically 500 m deep, 2 km wide and flanked by ridges ~100 m high (Spencer et al., 2009), and they exhibit higher surface temperatures (up to 180 K) than the surrounding regions (70–80 K; Spencer et al., 2006). Plume vent locations identified in high-resolution (~7 m/pixel) images show that these areas are mantled by smooth deposits that extend along the length of the sulci, and which thin laterally (Fig. 17.8(f)). Vent locations correlate well with the warmest locations along the sulci, and at least some of them persisted over a two-year observation period (Spitale and Porco, 2007). Cassini observations have also made it clear that Enceladus' plume activity forms Saturn's faint E-ring (Fig. 17.8(a)). Away from Enceladus' south polar region, there is a variety of terrain types (Fig. 17.8(c)), including smooth, resurfaced terrain. However, no obvious cryovolcanic landforms have been observed.

Before the Cassini-Huygens mission, geochemical and geophysical models suggested that Titan, Saturn's largest moon (mean radius of 2575 km) may be cryovolcanically active, facilitated by substantial quantities of ammonia in the interior, which would help to maintain a subsurface liquid layer. Detection of ^{40}Ar in the atmosphere (Niemann et al., 2005) supports the case for cryovolcanism, as it implies outgassing from Titan's interior. Subsequently, Cassini has observed a range of features, including lobate, flow-like morphologies, which have been interpreted as evidence of cryovolcanic activity (Sotin

et al., 2005; Barnes et al., 2006; Lopes et al., 2007; Wall et al., 2009). Furthermore, the Visible and Infrared Mapping Spectrometer (VIMS) observed periodic brightening at two locations that was attributed to active cryovolcanism (Nelson et al., 2009a, 2009b), although this interpretation has been disputed (Soderblom et al., 2009). Moore and Pappalardo (2011) argue against cryovolcanism on Titan, suggesting that the lobate features can be explained as sedimentary deposits produced by fluvial activity. A region called Sotra Facula may host the best evidence to date for cryovolcanism (Fig. 17.7(g); Lopes et al., 2010a; Kirk et al., 2010) in the form of a pair of 1-km-high mountains, possibly cryovolcanoes, with adjacent flow-like features and a pit > 1 km deep that could be a caldera or vent structure. However, the existence of cryovolcanism on Titan remains under debate – the "smoking gun" for active cryovolcanism, in the form of an enhanced thermal signature or active plumes, has not yet been detected.

Voyager observations revealed eruptive plumes as tall as 8 km on Triton (Smith et al., 1989; Soderblom et al., 1990) but these may not qualify as volcanic. Smith et al. (1989) suggested that the gas venting may be driven by solar heating and the subsequent vaporization of subsurface nitrogen. Other proposed mechanisms for gas venting include melting of or convection within the solid nitrogen polar caps (Brown and Kirk, 1994; Duxbury and Brown, 1997), but these do not explain all of the observed plume features. However, Triton also shows other evidence of cryovolcanic resurfacing, including smooth plains, quasi-circular scarp-bounded depressions similar to paterae, and various pits, domes, and channels (Fig. 17.7(h); Croft et al., 1990). The resurfaced terrains could be the manifestation of high heat production in the interior. Triton's inclined, retrograde orbit around Neptune suggests that it is a captured satellite (McKinnon, 1984). Therefore, as a result of stresses exerted during orbital circularization, Triton may have undergone significant tidal heating and interior

Figure 17.7 (cont.) NASA/JPL/USGS/University of Arizona. (h) Voyager 2 image mosaic of Neptune's Triton, showing varied terrain types. Dark, diffuse deposits in the south polar region were produced by active plumes. The "cantaloupe terrain" to the west of the image is likely the result of internal diapiric activity, while smooth surfaces in the east may be the result of cryovolcanic resurfacing. Image PIA00317, NASA/JPL/USGS.

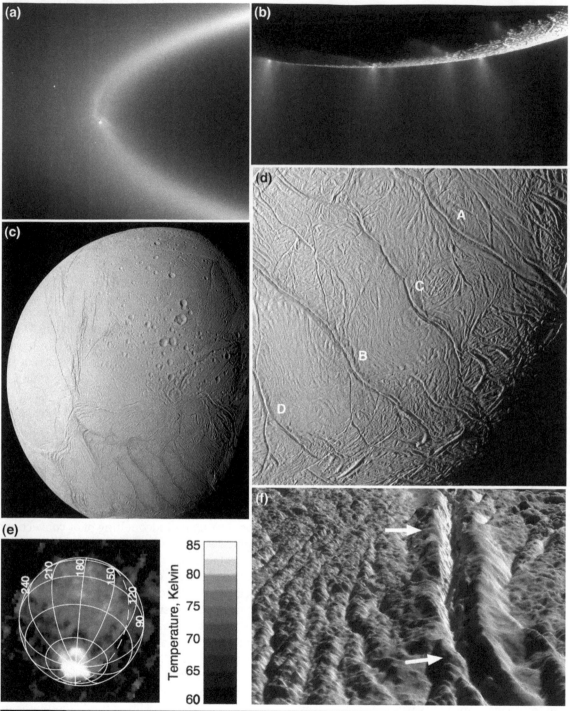

Figure 17.8 (a) Cassini view of Enceladus embedded within Saturn's E-ring. Image PIA08321, NASA/JPL/Space Science Institute. (b) Cassini ISS image showing that more than 30 jets can be identified erupting from Enceladus' south polar region. Image PIA11688, NASA/JPL/Space Science Institute. (c) ISS view of Enceladus showing ancient cratered terrain adjacent to highly tectonized regions. The active south polar region is visible towards the bottom of the image. Image PIA07800, NASA/JPL/Space Science Institute. (d) View of the south polar region showing four prominent sulci: A, Alexandria, C, Cairo, B, Baghdad, and D, Damascus. Image PIA06247. (e) Temperature map acquired by Cassini's composite infrared spectrometer in July 2005, showing enhanced thermal signature of the south polar region. Portion of image PIA09037, NASA/JPL/GSFC/Soutwest Research Institute. (f) Perspective view of Damascus Sulcus. The ridges are up to 150 m high, and the total width of the sulcus is ~5 km. Arrows indicate vent locations of active jets. Vertical exaggeration ×10. Image PIA12267, NASA/JPL/Space Science Institute/USRA Lunar and Planetary Institute.

melting to produce the diversity of geologic features observed today.

Other icy bodies, including Uranus' satellites Ariel, Miranda and Titania, also show intriguing features that may have cryovolcanic origins, perhaps facilitated by the presence of exotic cryomagma compositions. The variations in the geologic and eruptive activities among icy satellites may be attributed to differences in thermal-orbital evolution, internal composition, degree of differentiation, and ascent processes, which are presently poorly understood and may vary from world to world. Additional data from future missions are required to fully decipher the cryovolcanic histories of icy bodies. The remainder of this section focuses primarily on Enceladus, because of its spectacular and well-observed activity.

17.5.2 Compositions of cryomagmas

The compositions of cryomagmas depend on the materials that condensed from the solar nebula at a given distance from the Sun. In addition to water, at Saturn's orbit, methane and ammonia condensed, so it is likely that these exist in the moons' cryomagmas. In the further reaches of the Solar System, more volatile compounds, such as carbon monoxide, carbon dioxide, and nitrogen, may play a greater role. The viscosity of a cryomagma influences the resulting morphologies of landforms it produces. Liquid water would simply flood a surface, filling in depressions, but the inclusion of some amount of ammonia to form ammonia hydrates would produce viscosities approaching those of silicate lavas (e.g., Kargel et al., 1991). This suggests that a broad spectrum of eruption styles may be possible, and that familiar landforms, such as lobate flows, shield volcanoes, domes or cones, might grace these distant landscapes.

To date, the only direct data available on the composition of cryovolcanic products are from Enceladus' plumes. Near-infrared spectroscopy shows that Enceladus' surface is dominated by water ice, mainly in a crystalline state (Spencer et al., 2009), but the actively erupting sulci show the presence of organic compounds and carbon dioxide (Brown et al., 2006), presumably derived from the interior. Larger grain sizes (up to 100

µm) of crystalline water ice are located near the sulci (Brown et al., 2006; Porco et al., 2006), possibly due to preferential fallout of larger, slower plume particles near their sources (Jaumann et al., 2008, Hedman et al., 2009), or to sintering and growth of grains resulting from deposition of plume gases or from the enhanced temperatures near the sulci. Amorphous ice is found between sulci (Brown et al., 2006; Newmann et al., 2008), which is likely formed by condensation of plume gases at the lower (~100 K) temperatures away from the active eruptions (Spencer et al., 2006).

Direct sampling of Enceladus' plumes by the Cassini spacecraft shows that they consist predominantly (90–94 wt.%) of water vapor and water-ice grains (Waite et al., 2006; Hansen et al., 2011). Results from the Cosmic Dust Analyzer showed that particles closer to the surface tend to be large and rich in sodium salts, whereas further away from the surface the particles are small and salt-poor. Cassini's Ion Neutral Mass Spectrometer (INMS) and Ultraviolet Imaging Spectrometer (UVIS) obtained complementary information on plume composition. Apart from water, INMS detected carbon dioxide (5%), methane (0.9%) and ammonia (0.8%), as well as trace amounts of organics, H_2S and ^{40}Ar (Waite et al., 2006, 2009), and possibly molecular nitrogen or carbon monoxide. UVIS results failed to detect molecular nitrogen, however (Hansen et al., 2011).

17.5.3 Cryomagma ascent and eruption processes on Enceladus

Despite direct observations of active eruptions in the outer Solar System (Enceladus and Triton), the processes by which cryomagmas ascend on icy satellites are unclear. Many ascent models have been based on the terrestrial magmatic paradigm, and propose that cryomagmas rise buoyantly from the mantle into the lower ice shell, potentially stalling close to the surface to form a cryomagma chamber, and then erupting through conduits formed by the intrusion of dikes. Although this scenario has been used as the basis for interpretations on Titan in particular (e.g., Lopes et al., 2007), there are physical difficulties with this mechanism because the densities of all proposed unvesiculated cryo-magmas are greater than that of water-ice (specifically Ice-I, the inferred bulk composition of icy satellite crusts), and so watery cryomagmas are negatively buoyant.

There are a number of mechanisms that have been proposed to overcome ascent problems caused by negative buoyancy of cryomagma: (1) inducing positive buoyancy through exsolution of volatiles following decompression, and the subsequent ascent of fluid-filled fractures from the base of the ice shell (Crawford and Stevenson, 1988; Lorenz, 1996), or explosive eruption of sprays (Fagents et al., 2000); (2) pressurization of discrete liquid chambers (Fagents, 2003; Showman et al., 2004) or an entire ocean (Manga and Wang, 2007) during freezing and volume changes, leading to effusive eruptions; (3) partial melting of the ice shell by tidal dissipation (Mitri and Showman, 2008a; Tobie et al., 2008), facilitating near-surface reservoirs without buoyancy requirements; (4) incorporation of denser silicate material in the ice shell due to incomplete differentiation or meteoritic infall (Croft et al., 1988); and (5) solid state convection in the ice shell advecting heat and possibly chemicals upward and mobilizing near-surface pockets of salt- or ammonia-rich ices (Head and Pappalardo, 1999; Mitri et al., 2008; Choukroun et al., 2010). However, there is little direct evidence to support any of these mechanisms.

On Enceladus, the association of plumes with sulci suggests that they are erupted from elongate fractures similar to volcanic fissures on Earth, or possibly from chains of smaller sub-circular vents. Debate continues concerning the configuration and depth of the source, and whether or not a liquid reservoir is necessary to generate the plumes. The conduits could be tapping a volatile-rich liquid present as a global subsurface ocean (akin to a mantle), a local "sea" (Nimmo and Pappalardo, 2006; Collins and Goodman, 2007), or a smaller shallow reservoir (Porco et al., 2006). Tidal heating localized in a thermal plume or diapir could partially melt the ice shell and thus explain the high surface temperature of the south polar regions (Mitri and Showman, 2008a,b; Tobie et al., 2008). The presence of salt-rich grains in the plumes

is consistent with ejection from a subsurface ocean in contact with the silicate core. Such an ocean has been proposed to lie 80 km beneath the surface of the ice crust (Postberg et al., 2011). Alternatively, a cooler (< 200 K) solid source has been proposed (Kieffer et al., 2006; Gioia et al., 2007), with the eruptions driven by expansion of destabilizing methane clathrate hydrates. However, this hypothesis cannot explain the salty plume particles, and uncertainty over the detection of molecular nitrogen (which would be produced by clathrate decomposition) further weakens this argument.

Eruption velocities of Enceladus' plumes have been calculated at ~600 m s^{-1} (Hansen et al., 2011). Given the high observed mass fractions of CO_2, methane and ammonia (Waite et al., 2006, 2009), exsolution of these volatiles from a watery cryomagma reservoir is more than sufficient to drive the plumes. The initial shock of decompression as a fracture exposes the cryomagma to the ambient vacuum will cause vigorous exsolution and expansion of the volatiles and rapidly initiate explosive ascent, thus overcoming the issue of negative buoyancy, and accelerating gas-rich sprays through nozzle-like conduits to the surface (cf. Postberg et al., 2009, 2011). Activity is only likely to cease if the stress state of the crust changes considerably, or if the volatile supply is depleted.

Using very high resolution images of the plume jets (0.9 m/pixel, Porco et al., 2006), and assuming a particle size of 1 μm (the dominant particle size in the E-ring), the mass of ice in the column is estimated at 3×10^{-6} kg m^{-2}. UVIS observations indicate a column mass of vapor of 7×10^{-6} kg m^{-2} (Hansen et al., 2006), thus yielding an ice/gas ratio of ~0.4. In addition, plume brightness decreases roughly exponentially with altitude (Spencer et al., 2009). The scale height within 50 km of the surface is ~30 km, significantly less than the radius of the planet (252 km), implying that most of the particles close to the surface are falling back, consistent with the compositional results. However, the scale height is larger at higher altitudes and therefore many smaller particles are escaping to form Saturn's E-ring, facilitated by the low-gravity, vacuum conditions at Enceladus.

As the Cassini mission continues beyond 2012, it is expected that much more will be learned about the plumes which are, so far, unique manifestations of active cryovolcanism in the Solar System.

17.6 | Summary and future directions

Exploration of the Solar System has revealed the diversity of volcanic activity on different planetary bodies, which has challenged our models of volcanic processes. This has led to recognition of how significantly different planetary environments can affect volcanic ascent and eruption behavior, and how this in turn can influence the characteristics of the resulting landforms. Caution must therefore be used when interpreting eruption characteristics or magma compositions from the morphologies of volcanic features – consideration of planetary environmental factors is crucial.

An interesting consequence of the discovery of cryovolcanism in the outer Solar System, as well as Io's rampant volcanism, is the need to expand the definition of "volcano" (Lopes *et al.*, 2010b). The traditional definition of a volcano is *a place or opening from which molten rock and gas, and generally both, issue from the Earth's interior onto its surface*. But a volcano has also been defined as the hill or mountain constructed around the opening by accumulation of the material erupted (Macdonald, 1972). Lopes *et al.* (2010b) argued that this traditional definition needed to be modified to include processes such as cryovolcanism, and suggested the following definition: *A volcano is an opening on a planet or moon's surface from which magma, as defined for that body, and/or magmatic gases are erupted.*

In this broad context, we can use models of volcanic processes developed for terrestrial eruptions, but applied to the differing physical environments of other planets, to attempt to understand the nature of volcanism on other planets. In so doing we develop a greater understanding of their orbital, thermal, geologic and interior history, and therefore the evolution of the Solar System as a whole. When our models break down, we are forced to think outside of the terrestrial box, and reconsider how the interplay of gravity, atmospheric conditions, orbital dynamics, and composition can produce the observed volcanic features.

Volcanism may yet be found on currently unexplored objects in the Solar System. The New Horizons spacecraft encounter with Pluto in 2015 may reveal cryovolcanic features on its icy surface and it is possible that other trans-Neptunian objects and dwarf planets such as Quaoar could present cryovolcanic activity (Jewitt and Luu, 2004). Future spacecraft observations are expected to reveal more about the composition of extra-terrestrial magmas and the characteristics of eruptions on other planetary bodies. For example, the Mars Science Laboratory, which landed on Mars in 2012, possesses improved capabilities over previous rovers and landers to determine the composition of the martian surface. The Cassini mission is scheduled to continue making observations of the Saturn system, including Enceladus' plumes and Titan's surface, until the end of the mission in 2017. Planetary missions planned for the next decade include a Mars geophysical mission and a Ganymede orbiter that will also make observations of Io. The next decade or two should bring many new discoveries about extra-terrestrial volcanism.

17.7 | Notation

c	specific heat capacity ($J\,kg^{-1}\,K^{-1}$)
f	friction factor
F_b	buoyancy force (N)
g	acceleration due to gravity ($m\,s^{-2}$)
h	plume height (m)
H	horizontal extent of dike (m)
n	exsolved magma gas fraction
P	pressure (Pa)
P_a	atmospheric pressure (Pa)
P_{lith}	lithostatic pressure (Pa)
u	ascent velocity ($m\,s^{-1}$)
u_0	eruption velocity ($m\,s^{-1}$)
r	conduit radius (m)
R^*	specific gas constant ($J\,kg^{-1}\,K^{-1}$)

T temperature (K)
T_a atmospheric temperature (K)
T_{hot} temperature of hot material (K)
V volume (m^3)
w dike width (m)
z depth (m)
ρ_a atmospheric density (kg m^{-3})
ρ_g gas density (kg m^{-3})
ρ_m magma density (kg m^{-3})
ρ_r country rock density (kg m^{-3})

References

Arvidson, R. E., Bonitz, R. G., Robinson, M. L. *et al.* (2009). Results from the Mars Phoenix lander robotic arm experiment. *Journal of Geophysical Research*, **114**, E00E02, doi:10.1029/2009JE003408.

Bandfield, J. (2002). Global mineral distribution of Mars. *Journal of Geophysical Research* **107**(E6), 5042, doi:10.1029/2001JE001510.

Bandfield, J. L., Hamilton, V. E. and Christensen, P. R. (2000). A global view of martian surface composition from MGS-TES. *Science*, **287**, 1626–1630.

Barnes, J. W., Brown, R. H., Radebaugh, J. *et al.* (2006). Cassini observations of flow-like features in western Tui Regio, Titan. *Geophysical Research Letters*, **33**, L16204, doi:10.1029/2006GL026843.

Bleacher, J. E., Greeley, R., Williams, D. A., Cave, S. R. and Neukum, G. (2007). Trends in effusive style at the Tharsis Montes, Mars, and implications for the development of the Tharsis province. *Journal of Geophysical Research*, **112**, E09005, doi:10.1029/2006JE002873.

Boynton, W. V., Feldman, W. C., Mitrofanov, I. G. *et al.* (2004). The Mars Odyssey gamma-ray spectrometer instrument suite. *Space Science Reviews*, **110**, 37–83.

Boynton, W. V., Taylor, G. J., Evans, L. G. *et al.* (2007). Concentration of H, Si, Cl, K, Fe and Th in the low- and mid-latitude regions of Mars. *Journal of Geophysical Research*, **112**, E12S99, doi:10.1029/2007JE002887.

Brown, R. and Kirk, R. (1994). Coupling of volatile transport and internal heat flow on Triton. *Journal of Geophysical Research*, **99**, 1965–1981.

Brown, R. H., Clark, R. N., Buratti, B. J. *et al.* (2006). Composition and physical properties of Enceladus' surface. *Science*, **311**, 1425–1428.

Carlson, R., Smythe, W., Lopes-Gautier, R. *et al.* (1997). The distribution of sulfur dioxide and other infrared absorbers on the surface of Io in 1997. *Geophysical Research Letters*, **24**, 2474–2482.

Carlson, R. W., Kargel, J. S., Doute, S., Soderblom, L. A. and Dalton, J. B. (2007). Io's surface composition. In *Io After Galileo: A New View of Jupiter's Volcanic Moon*, ed. R. M. C. Lopes and J. R. Spencer. Chichester, UK: Praxis-Springer, pp. 193–229.

Carr, M. H. (2006). *The Surface of Mars*. New York, NY: Cambridge University Press.

Christensen, P. R., McSween, H. Y., Bandfield, J. L. *et al.* (2005). Evidence for magmatic evolution and diversity on Mars from infrared observations. *Nature*, **436**, 504–882.

Choukroun, M., Grasset, O., Tobie, G. and Sotin, C. (2010). Stability of methane clathrate hydrates under pressure: Influence on outgassing processes of methane on Titan. *Icarus*, **205**, 581–593

Collins, G. C. and Goodman, J. C. (2007). Enceladus' south polar sea. *Icarus*, **189**, 72–82.

Crawford, G. D. and Stevenson, D. J. (1988). Gas-driven water volcanism and the resurfacing of Europa. *Icarus*, **73**, 66–79.

Croft, S. K., Lunine, J. I. and Kargel, J. S. (1988). Equations of state of ammonia–water liquid: Derivation and planetological applications. *Icarus*, **73**, 279–293.

Croft, S. K., Kargel, J. S., Kirk, R. L. *et al.* (1990). The geology of Triton. In *Neptune and Triton*, ed. D. P. Cruikshank. Tucson: University of Arizona Press, pp. 879–947.

Crown, D. A. and Greeley, R. (1993). Volcanic geology of Hadriaca Patera and the Eastern Hellas Region of Mars. *Journal of Geophysical Research*, **98**, 3431–3451.

Crumpler, L. S. and Aubele, J. C. (1978). Structural evolution of Arsia Mons, Pavonis Mons, and Ascreus Mons: Tharsis Region of Mars. *Icarus*, **34**, 496–511.

Davies, A., Keszthelyi, L., Williams, D. *et al.* (2001). Thermal signature, eruption style and eruption evolution at Pele and Pillan on Io. *Journal of Geophysical Research*, **106**, 33 079–33 104.

Duxbury, N. S. and Brown, R. H. (1997). The role of an internal heat source for the eruptive plumes on Triton. *Icarus*, **125**, 83–93.

Fagents, S. A. (2003). Considerations for effusive cryovolcanism on Europa: The post-Galileo perspective. *Journal of Geophysical Research*, **108**, E125139, doi:10.1029/2003JE002128.

Fagents, S. A., Greeley, R., Sullivan, R. J. *et al.* (2000). Cryomagmatic mechanisms for the formation of Rhadamanthys Linea, triple band margins, and other low albedo features on Europa. *Icarus*, **144**, 54–88.

Fagents, S. A., Lanagan, P. D. and Greeley, R. (2002). Rootless cones on Mars: A consequence of lava-ground ice interaction. In *Volcano–Ice Interaction on Earth and Mars*, ed. J. L. Smellie and M. G. Chapman. Geological Society of London Special Publication, 202, pp. 295–317.

Fassett, C. I. and Head, J. W. (2006). Valleys on Hecates Tholus, Mars: origin by basal melting of summit snowpack. *Planetary and Space Science*, **54**, 370–378.

Fink, J. H. (1980). Surface folding and viscosity of rhyolite flows. *Geology*, **8**, 250–254.

Fink, J. H. and Griffiths, R. W. (1990). Radial spreading of viscous gravity currents with solidifying crust. *Journal of Fluid Mechanics*, **221**, 485–509.

Fink, J. H., Bridges, N. T. and Grimm, R. E. (1993). Shapes of Venusian "pancake" domes imply episodic emplacement and silicic composition. *Geophysical Research Letters*, **20**, 261–264.

Geissler, P. E. and Goldstein, D. B. (2007). Plumes and their deposits. In *Io After Galileo: A New View of Jupiter's Volcanic Moon*, ed. R. M. C. Lopes and J. R. Spencer. Chichester, UK: Praxis-Springer, pp. 163–192.

Gioia, G., Pinaki, C., Marshak, S. and Kieffer, S. W. (2007). Unified model of tectonics and heat transport in a frigid Enceladus. *Proceedings of the National Academy of Sciences*, **103**(34), 13 578–13 581.

Glaze, L. S. and Baloga, S. M. (2002). Volcanic plume heights on Mars: Limits of validity for convective models. *Journal of Geophysical Research*, **107**, doi:10.1029/2001JE001830.

Greeley, R. and Crown, D. A. (1990). Volcanic geology of Tyrrhena Patera, Mars. *Journal of Geophysical Research*, **95**, 7133–7149.

Greeley, R. and Spudis, P. D. (1981). Volcanism on Mars. *Reviews of Geophysics and Space Physics*, **19**, 13–41.

Greeley, R., Theilig, E. and Christensen, P. (1984). The Mauna Loa sulfur flow as an analog to secondary (?) sulfur flows on Io. *Icarus*, **60**, 189–199.

Greeley, R., Bridges, N. T., Crown, D. A. *et al.* (2000). Volcanism on the red planet: Mars. In *Environmental Effects on Volcanic Eruptions: From Deep Oceans to Deep Space*, ed. J. R. Zimbelman and T. K. P. Gregg. New York: Kluwer/Plenum, pp. 75–112.

Greenberg, R., Hoppa, G., Tufts, B. R. *et al.* (1999). Chaos on Europa. *Icarus*, **141**, 263–286.

Gregg, T. K. P. and Lopes R. M. (2008). Lava lakes on Io: New perspectives from modeling. *Icarus*, **194**, 166–172, doi:10.1016/j.icarus.2007.08.042.

Grosfils, E. B., Aubele, J., Crumpler, L., Gregg, T. K. P. and Sakimoto, S. (2000). Volcanism on Earth's seafloor and Venus. In *Environmental Effects on Volcanic Eruptions: From Deep Oceans to Deep Space*, ed. J. R. Zimbelman and T. K. P. Gregg. New York: Kluwer/Plenum, pp. 113–142.

Hamilton, V. E., Christensen, P. R., McSween, H. Y. and Bandfield, J. L. (2003). Searching for the source regions of martian meteorites using MGS TES: Integrating martian meteorites into the global distribution of igneous materials on Mars. *Meteoritics and Planetary Science*, **38**(6), 871–885.

Hansen, C. J., Esposito, L., Stewart, A. I. F. *et al.* (2006). Enceladus' water vapor plume. *Science*, **311**, 1422–1425.

Hansen, C. J., Shemansky, D. E., Esposito, L. W. *et al.* (2011). The composition and structure of the Enceladus plume. *Geophysical Research Letters*, **38**, L11202, doi:10.1029/2011GL047415.

Hartmann, W. K. and Berman, D. C. (2000). Elysium Planitia lava flows: crater count chronology and geological implications. *Journal of Geophysical Research*, **105**, 15 011–15 025.

Hauber, E., Bleacher, J., Gwinner, K., Williams, D. and Greeley, R. (2009). The topography and morphology of low shields and associated landforms of plains volcanism in the Tharsis region on Mars. *Journal of Volcanology and Geothermal Research*, **185**, 69–95.

Head, J. W. and Pappalardo, R. T. (1999). Brine mobilization during lithospheric heating on Europa: Implications for formation of chaos terrain, lenticula texture and color variations. *Journal of Geophysical Research*, **104**, 27 143–27 155.

Head, J. W. and Wilson, L. (1986). Volcanic processes and landforms on Venus: Theory, prediction and observations. *Journal of Geophysical Research*, **91**, 9407–9466.

Head, J. W., Chapman, C. R., Strom, R. G. *et al.* (2011). Flood volcanism in the northern high latitudes of Mercury Revealed by MESSENGER. *Science*, **333**, 1853–1856.

Hedman, M. M., Nicholson, P. D., Showalter, M. R. *et al.* (2009). Spectral observations of the Enceladus plume with Cassini-VIMS. *Astronomical Journal*, **693**, 1749–1762.

Hon, K., Kauahikaua, J., Denlinger, R. and Mackay, K. (1994). Emplacement and inflation of pahoehoe sheet flows: Observations and measurements of active lava flows on Kilauea Volcano, Hawaii. *Geological Society of America Bulletin*, **106**, 351–370.

Howell, R. R. and R. M. C. Lopes (2011). Morphology, temperature, and eruption dynamics at Pele. *Icarus*, **213**, 593–607, doi:10.1016/j.icarus.2011.03.008.

Hulme, G. (1974). The interpretation of lava flow morphology. *Geophysical Journal of the Royal Astronomical Society*, **39**(2), 361–383.

Hulme, G. (1976). The determination of the rheological properties and effusion rate of an Olympus Mons lava. *Icarus*, **27**(2), 207–213.

Jaeger, W. L., Turtle, E. P., Keszthelyi, L. P. *et al.* (2003). Orogenic tectonism on Io. *Journal of Geophysical Research*, **108**, doi:10.1029/2002JE001946.

Jaeger, W. L., Keszthelyi, L. P., McEwen, A. S., Dundas, C. M. and Russell, P. S. (2007). Athabasca Valles, Mars: A lava-draped channel system. *Science*, **317**, 1709–1711.

Jaumann, R., Stephan, K., Hansen, G. B. *et al.* (2008). Distribution of icy particles across Enceladus' surface as derived from Cassini-VIMS measurements. *Icarus*, **193**, 407–419.

Jessup, K. L., Spencer, J. R., Ballester, G. E. *et al.* (2004). The atmospheric signature of Io's Prometheus plume and anti-jovian hemisphere: Evidence for a sublimation atmosphere. *Icarus*, **169**, 197–215.

Jewitt, D. C. and Luu, J. (2004). Crystalline water ice in Kuiper Belt Object (50000) Quaoar. *Nature*, **432**, 731–733.

Jolliff, B. K., Wiseman, S. A., Lawrence, S. J. *et al.* (2011). Non-mare silicic volcanism on the lunar farside at Compton-Belkovich. *Nature Geoscience*, **4**, 566–571.

Kargel, J. S., Croft, S. K., Lunine, J. I. and Lewis, J. S. (1991). Rheological properties of ammonia–water liquids and crystal–liquid slurries: Planetological applications. *Icarus*, **89**, 93–112.

Keszthelyi, L., McEwen, A. S., Phillips, C. B. *et al.* (2001). Imaging of volcanic activity on Jupiter's moon Io by Galileo during GEM and GMM. *Journal of Geophysical Research*, **106**, 33 025–33 052.

Keszthelyi, L. P., Jaeger, W. and Milazzo, M. *et al.* (2007). New estimates for Io eruption temperatures: implications for the interior. *Icarus*, **192**, 491–502.

Khurana, K. K., Jia, X., Kivelson, M. G., Nimmo, F., Schubert, G. and Russell, C. T. (2011). Evidence of a global magma ocean in Io's interior. *Science*, **332**, 1186–1189.

Kieffer, S. W. (1982). Dynamics and thermodynamics of volcanic eruptions: Implications for the plumes on Io. In *Satellites of Jupiter*, ed. D. Morrison. Tuscon, AZ: University of Arizona Press, pp. 647–723.

Kieffer, S. W., Lopes-Gautier, R., McEwen, S. *et al.* (2000). Prometheus: Io's wandering plume. *Science*, **288**, 1204–1208.

Kieffer, S. W., Lu, X., Bethke, C. M. *et al.* (2006). A clathrate reservoir hypothesis for Enceladus' south polar plume. *Science*, **314**, 1764–1766, doi:10.1126/science.1133519.

Kirk, R. L., Howington-Kraus, E., Barnes, J. W. *et al.* (2010). La Sotra y las otras: Topographic evidence for (and against) cryovolcanism on Titan. Presented at the American Geophysical Union Fall Meeting, 2010, San Francisco.

Kivelson, M. G., Khurana, K. K., Russell, C. T. *et al.* (2000). Galileo magnetometer measurements: A stronger case for a subsurface ocean at Europa. *Science*, **289**, 1340–1343.

Leone, G. and Wilson, L. (2001). The density structure of Io and the migration of magma through its lithosphere. *Journal of Geophysical Research*, **106**, 32 983–32 995.

Lopes, R. and Williams, D. (2005). Io after Galileo. *Reports on Progress in Physics*, Institute of Physics Publishing, **68**, 303–340.

Lopes, R., Kamp, L. W., Douté, S. *et al.* (2001). Io in the near-infrared: NIMS results from the Galileo fly-bys in 1999 and 2000. *Journal of Geophysical Research*, **106**, 33 053–33 078.

Lopes, R., Kamp, L. W., Smythe, W. *et al.* (2004). Lava lakes on Io. Observations of Io's volcanic activity from Galileo during the 2001 fly-bys. *Icarus*, **169**, 140–174.

Lopes, R. M. C., Mitchell, K. L., Stofan, E. R. *et al.* (2007). Cryovolcanic features on Titan's surface as revealed by the Cassini Titan Radar Mapper. *Icarus*, **186**, 395–412.

Lopes, R. M. C., Stofan, E. R., Peckyno, R. *et al.* (2010a). Distribution and interplay of geologic processes on Titan from Cassini RADAR data. *Icarus*, **205**, 540–588.

Lopes, R. M. C., Mitchell, K. L., Williams, D. A., and Mitri, G. (2010b). Beyond Earth: How extra-terrestrial volcanism has changed our definition of a volcano. In *What's a volcano? New answers to an old question*, ed. E. Canon and A. Szakacs. Geological Society of America Special Paper 470, pp. 11–30.

Lorenz, R. D. (1996). Pillow lava on Titan: expectations and constraints on cryovolcanic processes. *Planetary and Space Science*, **44**, 1021–1028.

Macdonald, G. A. (1972). *Volcanoes*. New Jersey: Prentice-Hall.

Manga, M. and Wang, C. Y. (2007). Pressurized oceans and the eruption of liquid water on Europa and Enceladus. *Geophysical Research Letters*, **34**, L07202, doi:10.1029/2007GL029297.

Marchis, F., Prangé, R. and Christou, J. (2000). Adaptive optics mapping of Io's volcanism in the IR (3.8 μm). *Icarus*, **148**, 384–396.

Mastin, L. G. (1995). Thermodynamics of gas and steam-blast eruptions. *Bulletin of Volcanology*, **57**, 85–98.

Mastin, L. G. (2005). The controlling effect of viscous dissipation on magma flow in silicic conduits. *Journal of Volcanology and Geothermal Research*, **143**, 17–28.

McEwen, A. S. and Soderblom, L. (1983). Two classes of volcanic plumes on Io. *Icarus*, **58**, 197–226.

McEwen, A. S., Keszthelyi, L. P., Spencer, J. *et al.* (1998). High-temperature silicate volcanism on Jupiter's moon Io. *Science*, **281**, 87–90.

McKinnon, W. B. (1984). On the origin of Triton and Pluto. *Nature*, **311**, 355–358.

McSween, H. Y. (1994). What we have learned about Mars from SNC Meteorites. *Meteoritics*, **29**, 757–779.

McSween, H. Y., Murchie, S. L., Crisp, J. A. *et al.* (1999). Chemical, multispectral, and textural constraints on the composition and origin of rocks at the Mars Pathfinder landing site. *Journal of Geophysical Research*, **104**, 8679–8715.

McSween, H. Y., Taylor, G. J. and Wyatt, M. B. (2009). Elemental composition of the martian crust. *Science*, **324**, 736–739.

Milazzo, M., Keszthelyi, L. and McEwen, A. (2001). Observations and initial modeling of lava–SO_2 interactions at Prometheus, Io. *Journal of Geophysical Research*, **106**, 33 121–33 127.

Mitchell, K. L. (2005). Coupled conduit flow and shape in explosive volcanic eruptions. *Journal of Volcanology and Geothermal Research*, **143**, 187–203.

Mitri, G. and Showman, A. P. (2008a). A model for the temperature-dependence of tidal dissipation in convective plumes on icy satellites: implications for Europa and Enceladus. *Icarus*, **195**, 758–764.

Mitri, G. and Showman A. P. (2008b). Thermal convection in ice-I shells of Titan and Enceladus. *Icarus*, **193**, 387–396.

Mitri, G., Showman, A. P., Lunine, J. I. and Lopes, R. M. C. (2008). Resurfacing of Titan by ammonia-water cryomagma. *Icarus*, **196**, 216–224.

Moore, W. B. (2001). The thermal state of Io. *Icarus*, **154**, 548–550.

Moore, J. M. and Pappalardo, R. T. (2011). Titan: An exogenic world? *Icarus*, **212**, 790–806.

Morabito, L. A., Synnott, S. P., Kupferman, P. N. and Collins, S. A. (1979). Discovery of currently active extraterrestrial volcanism. *Science*, **204**, 972.

Mouginis-Mark, P. J., Wilson, L. and Head, J. W. (1982). Explosive volcanism on Hecates Tholus, Mars: Investigation of eruption conditions. *Journal of Geophysical Research*, **87**, 9890–9904.

Mouginis-Mark, P. J., Wilson, L. and Zimbelman, J. R. (1988). Polygenic eruptions on Alba Patera, Mars. *Bulletin of Volcanology*, **50**, 361–379.

Nelson, R. M., Kamp, L. W., Matson, D. L. *et al.* (2009a). Saturn's Titan: Surface change, ammonia, and implications for atmospheric and tectonic activity. *Icarus*, **199**, 429–441.

Nelson, R. M., Kamp, L. W., Lopes, R. M. C. *et al.* (2009b). Photometric changes on Saturn's moon Titan: Evidence for cryovolcanism. *Geophysical Research Letters*, **36**, L04202, doi:10.1029/2008GL036206.

Neukum, G., Jaumann, R., Hoffmann, H. *et al.* (2004). Recent and episodic volcanic and glacial activity on Mars revealed by the High Resolution Stereo Camera. *Nature*, **432**, 971–979.

Newman, S. F., Buratti, B. J., Brown, R. H. *et al.* (2008). Photometric and spectral analysis of the distribution of crystalline and amorphous ices on Enceladus as seen by Cassini. *Icarus*, **193**, 397–406.

Niemann, H. B., Atreya, S. K., Bauer, S. J. *et al.* (2005). The abundances of constituents of Titan's atmosphere from the GCMS instrument on the Huygens probe. *Nature*, **438**, 778–794.

Nimmo, F. and Pappalardo, R. T. (2006). Diapir-induced reorientation of Enceladus. *Nature*, **441**, 614–616.

Nyquist, L. E., Bogard, D. D., Shih, C.-Y. *et al.* (2001). Ages and geologic histories of martian meteorites. In *Chronology and Evolution of Mars*, ed. R. Kallenbach, J. Geiss and W. K. Hartmann. Boston, MA: Kluwer Academic Publishers, pp. 105–164.

Pappalardo, R. T., Belton, M. J. S., Breneman, H. H. *et al.* (1999). Does Europa have a subsurface ocean? Evaluation of the geological evidence. *Journal of Geophysical Research*, **104**, 24 015–24 055.

Pearl, J., Hanel, R., Kunde, V. *et al.* (1979). Identification of gaseous SO_2 and new upper limits for other gases on Io. *Nature*, **280**, 755–758.

Porco, C. C., Helfenstein, P., Thomas, P. C. *et al.* (2006). Cassini observes the active south pole of Enceladus. *Science*, **311**, 1393–1401.

Postberg, F., Kempf, S., Schmidt, J. *et al.* (2009). Sodium salts in E-ring ice grains from an ocean below the surface of Enceladus. *Nature*, **459**, 1098–1101, doi: 10.1038/nature08046.

Postberg, F., Schmidt, J. Hillier, J., Kempf, S. and Srama, R. (2011). A salt-water reservoir as the source of a compositionally stratified plume on Enceladus. *Nature*, **474**, 620–622, doi:10.1038/nature10175.

Poulet, F., Gomez, C., Bibring, J.-P. *et al.* (2007). Martian surface mineralogy from OMEGA/MEx: Global mineral maps. *Journal of Geophysical Research*, **112**, E08S02, doi:10.1029/2006JE002840.

Radebaugh, J., Keszthelyi, L. P., McEwen, A. S. *et al.* (2001). Paterae on Io: A new type of volcanic caldera? *Journal of Geophysical Research*, **106**, 33 005-33 020.

Radebaugh, J., McEwen, A. S., Milazzo, M. P. *et al.* (2004). Observations and temperatures of Io's Pele Patera from Cassini and Galileo spacecraft images. *Icarus*, **169**, 65-79.

Rathbun, J. A., Spencer, J. R., Davies, A. G., Howell, R. R. and Wilson, L. (2002). Loki, Io: A periodic volcano. *Geophysical Research Letters*, **29**, 1443, doi:10.1029/2002GL014747.

Ruff, S. W., Christensen, P. R., Blaney, D. L. *et al.* (2006). The rocks of Gusev crater as viewed by the Mini-TES instrument. *Journal of Geophysical Research*, **111**, E12S18, doi:10.1029/2006JE002747.

Schaber, G. G. (1980). The surface of Io: Geologic units, morphology, and tectonics. *Icarus*, **43**, 302-333.

Showman, A. P, Mosqueira, I. and Head, J. W. (2004). On the resurfacing of Ganymede by liquid-water volcanism. *Icarus*, **172**, 625-640.

Skok, J. R., Mustard, J. F., Ehlmann, B. L., Milliken, R. E. and Murchie, S. L. (2010). Silica deposits in the Nili Patera caldera on the Syrtis Major volcanic complex on Mars. *Nature Geoscience*, doi:10.1038/ngeo990.

Smith, B. A., Soderblom, L. A., Banfield, D. *et al.* (1989). Voyager 2 at Neptune: Imaging science results. *Science*, **246**, 1422-1449.

Soderblom, L., Kieffer, S. W., Becker, T. L. *et al.* (1990). Triton's geyser-like plumes: Discovery and basic characterization. *Science*, **250**, 410-415.

Soderblom, L. A., Brown, R. H., Soderblom, J. M. *et al.* (2009). The geology of Hotei Regio, Titan: Correlation of Cassini VIMS and RADAR. Icarus, **204**, 610-618.

Sotin, C., Jaumann, R., Buratti, B. J. *et al.* (2005). Release of volatiles from a possible cryovolcano from near-infrared imaging of Titan. *Nature*, **435**, 786-789.

Spahn, F., Schmidt, J., Albers, N. *et al.* (2006). Cassini dust measurements at Enceladus and implications for the origin of the E ring. *Science*, **311**, 1416-1418.

Spencer, J. R., Jessup, K. L., McGrath, M. A., Ballester, G. E. and Yelle, R. (2000). Discovery of gaseous S$_2$ in Io's Pele plume. *Science*, **288**, 1208-1210.

Spencer, J. R., Pearl, J. C., Segura, M. *et al.* (2006). Cassini encounters Enceladus: Background and the discovery of a south polar hot spot. *Science*, **311**, 1401-1405.

Spencer, J. R., Stern, S. A., Cheng, A. F. *et al.* (2007). Io volcanism seen by New Horizons: A major eruption of the Tvashtar volcano. *Science*, **318**, 240-243.

Spencer, J. R., Barr, A. C., Esposito, L. W. *et al.* (2009). Enceladus: An active cryovolcanic satellite. In *Saturn From Cassini-Huygens*, ed. M. K. Dougherty, L. W. Esposito and S. M. Krimigis. Dordrecht, Netherlands: Springer, pp. 683-724.

Spitale, J. N. and Porco, C. C. (2007). Association of the jets of Enceladus with the warmest regions on its south-polar fractures. *Nature*, **449**, 695-697.

Squyres, S. W., Arvidson, R. E., Bollen, D. *et al.* (2006). Overview of the Opportunity Mars Exploration Rover Mission to Meridiani Planum: Eagle crater to Purgatory ripple. *Journal of Geophysical Research*, **111**, E12S12, doi:10.1029/2006JE002771.

Stofan, E. R. and Smrekar, S. E. (2005). Large topographic rises, coronae, large flow fields and large volcanoes on Venus: Evidence for mantle plumes? In *Plates, Plumes, and Paradigms*, ed. G. R. Foulger, J. H. Natland, D. C. Presnall and D. L. Anderson. Geological Society of America Special Paper, 388, pp. 841-861.

Tobie, G., Čadek, O. and Sotin, C. (2008). Solid tidal friction above a liquid water reservoir as the origin of the south pole hotspot on Enceladus. *Icarus*, **196**, 642-652

Treiman, A. H. (2007). Geochemistry of Venus's surface: Current limitations as future opportunities. In *Exploring Venus as a Terrestrial Planet*. Washington, DC: American Geophysical Union Monograph 176, pp. 7-22.

Treiman, A. H., Gleason, J. D. and Bogard, D. D. (2000). The SNC meteorites are from Mars. *Planetary and Space Science*, **48**, 1213-1230.

Veeder, G. J., Matson, D. L., Johnson, T. V., Blaney, D. L. and Goguen, J. D. (1994). Io's heat flow from infrared radiometry: 1983-1993. *Journal of Geophysical Research*, **99**, 17 095-17 162.

Wadge, G. and Lopes, R. M. C. (1991). The lobes of lava flows on Earth and Olympus Mons, Mars. *Bulletin of Volcanology*, **54**, 10-24.

Wall, S. D., Lopes, R. M., Stofan, E. R. *et al.* (2009). Cassini RADAR images at Hotei Arcus and Western Xanadu, Titan: Evidence for recent cryovolcanic activity. *Geophysical Research Letters*, **36**, L04203, doi:1029/2008GL036415.

Waite, J. H., Combi, M. R., Ip, W.-H. *et al.* (2006). Cassini Ion and Neutral Mass Spectrometer: Enceladus plume composition and structure. *Science*, **311**, 1419–1422.

Waite, J. H., Lewis, W. S., Magee, B. A. *et al.* (2009). Liquid water on Enceladus from observations of ammonia and ^{40}Ar in the plume. *Nature*, **460**, 487–490.

Warner, N. and Gregg, T. K. P. (2003). Evolved lavas on Mars? Observations from southwest Arsia Mons and Sabancaya volcano, Peru. *Journal of Geophysical Research*, **108**, doi:10.1029/2002JE001969.

Weitz, C. M. and Head, J. W. (1999). Spectral properties of the Marius Hills volcanic complex and implications for the formation of lunar domes and cones. *Journal of Geophysical Research*, **104**, 18 933–18 956.

Wichura, H., Bousquet, R. and Oberhansli, R. (2010). Emplacement of the mid-Miocene Yatta lava flow, Kenya: Implications for modeling long channeled lava flows. *Journal of Volcanology and Geothermal Research*, **198**, 325–338.

Williams, D. A. and Greeley, R. (1994). Assessment of antipodal-impact terrains on Mars. *Icarus*, **110**, 196–202.

Williams, D. A. and Howell, R. R. (2007). Active volcanism: Effusive eruptions. In *Io after Galileo*, ed. R. M. C. Lopes and J. R. Spencer. Chichester, UK: Praxis, pp. 133–161.

Williams, D. A., Greeley, R. and Davies, A. G. (2001). Evaluation of sulfur flow emplacement on Io from Galileo data and numerical modeling. *Journal of Geophysical Research*, **106**, 33 161–33 174.

Williams, D. A., Turtle, E. P., Keszthelyi, L. P. *et al.* (2004). Geologic mapping of the Culann-Tohil region of Io from Galileo imaging data. *Icarus*, **169**, 80–97.

Williams, D. A., Greeley, R. Fergason, R. L. *et al.* (2009). The Circum-Hellas Volcanic Province: Overview. *Planetary and Space Science*, **57**, 895–916.

Williams, D. A., Keszthelyi, L. P., Crown, D. A. *et al.* (2011). Volcanism on Io: New insights from global geologic mapping. *Icarus*, **214**, 91–112.

Wilson, L. and Head, J. W. (1981). Ascent and eruption of basaltic magma on the Earth and Moon. *Journal of Geophysical Research*, **86**, 2971–3001.

Wilson, L. and Head, J. W. (1983). A comparison of volcanic eruption processes on Earth, Moon, Mars, Io and Venus. *Nature*, **302**, 663–669.

Wilson, L. and Head, J. W. (1994). Mars: Review and analysis of volcanic eruption theory and relationships to observed landforms. *Reviews of Geophysics*, **32**, 221–264.

Wilson, L. and Head, J. W. (2001). Lava fountains from the 1999 Tvashtar Catena fissure eruption on Io: Implications for dike emplacement mechanisms, eruption rates and crustal structure. *Journal of Geophysical Research*, **106**, 32 997–33 004.

Wilson, L. and Head, J. W. (2004). Evidence for a massive phreatomagmatic eruption in the initial stages of formation of the Mangala Valles outflow channel, Mars. *Geophysical Research Letters*, **31**, L15701, doi:10.1029/2004GL020322.

Wilson, L. and Head, J. W. (2007). Explosive volcanic eruptions on Mars: Tephra and accretionary lapilli formation, dispersal and recognition in the geologic record. *Journal of Volcanology and Geothermal Research*, **163**, 83–97.

Wilson, L. and Mouginis-Mark, P. J. (2003a). Phreatomagmatic explosive origin of Hrad Vallis, Mars. *Journal of Geophysical Research*, **108**, doi:10.10292002JE001927.

Wilson, L. and Mouginis-Mark, P. J. (2003b). Phreatomagmatic dike–cryosphere interactions as the origin of small ridges north of Olympus Mons, Mars. *Icarus*, **165**, 242–252.

Zhang, J., Goldstein, D. B., Varghese, P. L. *et al.* (2003). Simulation of gas dynamics and radiation in volcanic plumes on Io. *Icarus*, **163**, 182–197.

Zimbelman, J. R. (1985). Estimates of rheologic properties for flows on the Martian volcano Ascraeus Mons. *Proceedings of the 16th Lunar and Planetary Science Conference, Journal of Geophysical Research*, **90** (supplement), 157–162.

Zimbelman, J. R. and Gregg, T. K. P. (2000). Volcanic diversity throughout the Solar System. In *Environmental Effects on Volcanic Eruptions: From Deep Oceans to Deep Space*, ed. J. R. Zimbelman and T. K. P. Gregg. New York: Kluwer/Plenum, pp. 1–8.

Online resources available at www.cambridge.org/fagents

- Links to websites of interest
- Exercises
- Answers to exercises

Index

CPSIA information can be obtained
at www.ICGtesting.com
Printed in the USA
LVHW021152151221
706264LV00002B/15

9 781108 812658